T0406948

COMBINATORIAL METHODS
WITH COMPUTER APPLICATIONS

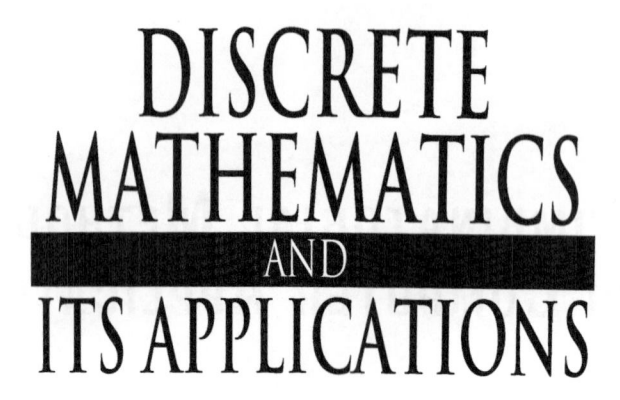

DISCRETE MATHEMATICS AND ITS APPLICATIONS

Series Editor
Kenneth H. Rosen, Ph.D.

Continued Titles

DISCRETE MATHEMATICS AND ITS APPLICATIONS

Series Editor KENNETH H. ROSEN

COMBINATORIAL METHODS WITH COMPUTER APPLICATIONS

JONATHAN L. GROSS

Columbia University
New York, U.S.A

 Chapman & Hall/CRC
Taylor & Francis Group

Boca Raton   London   New York

Chapman & Hall/CRC is an imprint of the
Taylor & Francis Group, an **informa** business

Cover photograph from the Metropolitan Museum of Art, Gift of Edgar William and Bernice Chrysler Garbisch, 1970 (1970.283.1). Photograph © 1998 the Metropolitan Museum of Art. Reprinted with permission.

Chapman & Hall/CRC
Taylor & Francis Group
6000 Broken Sound Parkway NW, Suite 300
Boca Raton, FL 33487-2742

© 2008 by Jonathan L. Gross
Chapman & Hall/CRC is an imprint of Taylor & Francis Group, an Informa business

No claim to original U.S. Government works

International Standard Book Number-13: 978-1-58488-743-0 (Hardcover)

Library of Congress Cataloging-in-Publication Data

Gross, Jonathan L.
    Combinatorial method with computer applications / Jonathan L. Gross.
        p. cm.
    Includes bibliographical references and index.
    ISBN 978-1-58488-743-0 (alk. paper)
    1. Combinatorial analysis--Data processing. I. Title.

QA164.G76 2008
511'.60285--dc22
                                                      2007032273

Visit the Taylor & Francis Web site at
http://www.taylorandfrancis.com

and the CRC Press Web site at
http://www.crcpress.com

To Susan

CONTENTS

PREFACE

Combinatorial Methods with Computer Applications provides the basis for a general sequel to a standard college course in discrete mathematics. Its objective is to enhance the ability of students to understand and to perform combinatorial computations, as they might arise in actual applications, and to use combinatorial models. It is intended for an upper-level course in a department of mathematics, computer science, or operations research, with the expectation of enrollment also of students in engineering, the sciences, and the social sciences. It is also suitable for self-study and reference by working professionals, especially in computer-related applications, and in other areas as well.

A challenge and opportunity in offering such a course is that there are, by now, dozens of different science and engineering courses that depend largely on combinatorial mathematics (sometimes blended with some calculus). Most of these courses have, of necessity, been introducing special instances of mathematical methods. In a computer science department alone, the courses on analysis of algorithms, computational complexity, computational learning, cryptography, spoken language processing, computational genomics, machine learning, and performance evaluation may all make substantial use of combinatorial methods and models beyond the level of a standard introductory discrete math course. Combinatorial methods courses have arisen and their enrollments have prospered, because of the benefit to students of prior exposure to systematic development of combinatorial methods, before encountering the specialized instances in applications.

Another challenge in offering a sequel to discrete mathematics designed for students with diverse academic backgrounds is that a lower-level discrete mathematics course is not necessarily required of students outside of computer science departments. Beyond calculus, such students have commonly taken elementary probability and linear algebra, whose prior study is quite good preparation for the study of combinatorial methods.

The elective combinatorial methods course that I teach every year in the Computer Science Department at Columbia University is taken by a mix of graduate students and upper-level undergraduates, students whose common ground is that they like mathematics. Most of them are seeking their degrees in various applied disciplines, and a few are mathematics majors. This book is written for such heterogeneous audiences.

Selection and Ordering of Contents

The selection of content for this textbook prioritizes breadth of technique and applicability. Chapters 0 through 6, which are mostly concerned with counting methods, can provide many combinatorial methods that students are most likely to need in future work within a single one-semester course, or within its self-taught equivalent. The four later chapters are a good basis for an honors-level second semester on graph theory and combinatorial designs. The entire text is woven into a unified stream of exposition, in which the chapters follow naturally upon each other. (My choice of the Hicks painting for the cover whimsically reflects my perception that the different topics presented blend well.) The most important methods appear repeatedly, underscoring their generality.

In my very fast-paced combinatorial methods course at Columbia, where I also teach a course on graph theory every year, I cover most of the content of this present book, except Chapters 7 and 8, which briefly survey most of the main topics in graph theory. Their inclusion in the book permits an instructor to craft a course that meshes well with the curricular needs of his or her department, whatever other courses it offers.

A somewhat similar selection and a roughly comparable quantity of content are offered in Liu's classic *Introduction to Combinatorial Mathematics* and also in various more recent texts on combinatorial methods for applications, all widely used, including *Applied Combinatorics* by Tucker, *Applied Combinatorics* by Roberts and Tesman, and *Introductory Combinatorics* by Brualdi. Some practical number theory is included in *Concrete Mathematics* by Graham, Knuth, and Patashnik, which offers a somewhat different eclectic combination from the others, with a distinguishing tilt toward continuous mathematics, away from algebra and graphs.

The ordering of content here also differs from that of more formal books, in the sense that several topics get a preliminary preview and other topics are developed or first presented shortly before their application, rather than strictly according to conventional mathematical taxonomy. For instance, much of the development of exponential generating functions is deferred until they are applied to the solution of the derangement recurrence. Most conspicuously, a section on the partitions of integers, a celebrated topic of number theory, appears in the midst of Chapter 9 on graph enumeration, just in time to assist in the calculation of cycle indexes for permutation groups.

How to Use This Book

Chapter 0 introduces combinatorics and the rest of the book. Beyond providing a comprehensive foundation for the systematic treatment that follows in subsequent chapters, it reviews a few topics that students may have seen already in a discrete math course and fills in some possible gaps of coverage. The pace at which it can be covered depends entirely on the background and mathematical sophistication of the students. Some of the exercises are intentionally designed for students whose background for this course is incomplete.

Past students commonly report to me after completing my course that they have come to use methods in Chapters 1 to 6 "all the time" in their professional work or in other courses. These chapters are on sequences, solving recurrences, evaluating summation expressions, binomial coefficients, partitions and permutations, and on integer methods. The techniques they present have great generality.

The level of development provided by these six chapters goes well beyond whatever prior exposure to their topics that students may have had in a discrete math course. To the extent that a student has seen some of the methods before, they may have been presented elsewhere more as a single-purpose "trick", whereas here they emerge as systematic approaches, suitable for many possible uses. Moreover, many of the methods are used not only in the chapters where introduced, but also in later chapters. The intent is to produce mathematical proficiency of great use in applications, without duplication of what is typically taught by applications courses.

Chapters 7 to 9 are designed to facilitate an optional graph theory component within a combinatorial methods course or a combination course. Thus, Chapters 7 and 8 provide a quick tour of graph theory. A student whose prior exposure to graph theory gave little attention to isomorphism and automorphism might read the first four sections of Chapter 7 before reading Chapter 9, but Chapter 9 does not otherwise depend on Chapters 7 and 8.

The last two chapters use computational methods from higher algebra, which is what I like to present at the end of my own course. Chapter 9 is concerned with using automorphism groups in algebraic counting methods, and Chapter 10 is about combinatorial designs. Since very few students except math majors have previously taken a course in abstract algebra, there is also included within the Appendix enough algebraic background to make these chapters readily accessible. To my delight, students who take my combinatorial methods course have often been inspired to later take a full course in higher algebra.

Some Features

The stylistic features of this book are similar to those that Jay Yellen and I used in *Graph Theory and Its Applications*.

- *Drawings*. There are more than 300 drawings, which serve as an aid to building intuition.

- *Exercises*. There are about 1400 exercises. The emphasis is on applying the methods taught within the body of the text, and the easiest are routine drill. Some more difficult problems require some challenging problem-solving. This book is far more concerned with using powerful methods than with deriving theorems. The proofs that are expected in the exercises are typically quite short.

- *Computational Engine*. The author's website at *www.graphtheory.com* contains a computational engine to help with calculations for some of the exercises.

- *Solutions and Hints*. Each exercise marked with a superscript[S] has a solution or hint appearing in the back of the book. Some of the solutions are detailed, and others are brief. Students may find that a detailed solution of an exercise within a grouping is of considerable help in solving other exercises in the same grouping.

- *Algorithms*. Algorithms are presented in a reader-friendly pseudocode, devoid of the details of computer implementation.

Websites

Suggestions and comments from readers are invited. They may be sent to the author's website at *www.graphtheory.com*. Thanks mostly to the efforts of my colleague Dan Sanders and my webmaster Aaron Gross, this website also maintains extensive graph theory informational resources. The general website for CRC Press is *www.crcpress.com*.

In advance, I thank my students, colleagues, and other readers for notifying me of any errors that they may find. I will post the corrections to all known errors at *www.graphtheory.com*.

Acknowledgments

Some readers of preliminary drafts of the manuscript gave me many helpful suggestions regarding the mathematical content and presentation. In particular, Mehvish Poshni, Imran Khan, and Ken Rosen are to be credited for numerous improvements throughout the manuscript. Some of the exercises appearing here were suggested by Scott Brinker when he took my combinatorial methods course at Columbia in the fall of 2003. The computational engine was implemented by Yianni Alexander. Strategic suggestions on the organization came from Jay Yellen. Special thanks to Ward Klein for his comprehensive assistance, including an extensive review of the manuscript.

Jonathan Gross

AUTHOR

Jonathan Gross is professor of computer science at Columbia University. His research in topology, graph theory, and cultural sociometry has earned him an Alfred P. Sloan Fellowship, an IBM postdoctoral fellowship, and various research grants from the Office of Naval Research, the National Science Foundation, and the Russell Sage Foundation.

His best-known mathematical invention, the *voltage graph*, is widely used in the construction of minimum graph imbeddings and of symmetric graph imbeddings. He and Thomas Tucker proved that every covering graph can be realized as a voltage graph construction. He also wrote the pioneering papers on enumerative techniques in topological graph theory, with various co-authors.

Professor Gross has created and delivered numerous software-development short courses for Bell Laboratories and for IBM. These include mathematical methods for performance evaluation at the advanced level and for developing reusable software at a basic level. He has received several awards for outstanding teaching at Columbia University, including the career Great Teacher Award from the Society of Columbia Graduates.

His previous books include *Topological Graph Theory*, co-authored with Thomas W. Tucker, *Graph Theory and Its Applications*, co-authored with Jay Yellen, and the *Handbook of Graph Theory*, co-edited with Jay Yellen. Another previous book, *Measuring Culture*, co-authored with Steve Rayner, constructs network-theoretic tools for measuring sociological phenomena.

Prior to Columbia University, Professor Gross was in the Mathematics Department at Princeton University. His undergraduate work was at the Massachusetts Institute of Technology, and he wrote his Ph.D. thesis on 3-dimensional topology at Dartmouth College.

Chapter 0

Introduction to Combinatorics

Combinatorial mathematics or, more briefly, *combinatorics*, refers to the body of mathematics developed for solving problems concerned with *discrete sets*, by which we mean finite and countably infinite sets, and with the functions to and from such sets. By way of contrast, the infinitesimal calculus (in the usual sense of differentiating and integrating) is concerned with continuous functions on the real line, which involves an uncountably infinite set of numbers. Calculus and all its generalizations are collectively called *continuous mathematics*.

Most combinatorics problems have one of three fundamental objectives: counting or calculating a sum, constructing a configuration involving two or more discrete sets (usually two) — subject to a list of constraints, or optimization, i.e., either finding the extreme values of a function or designing something with an optimal characteristic of some kind. The first section of this introductory chapter presents examples of problems of each type. The rest of the chapter surveys a few introductory methods for solving such problems and describes additional combinatorial problems. In so doing, it also provides a quick look-ahead at some concepts that are useful in subsequent chapters. Various details are deferred to those later chapters, as are most of the relevant exercises.

Some parts of mathematics, including probability, geometry, and algebra, have combinatorial aspects and continuous aspects as well. Moreover, the methods of combinatorial mathematics often have analogies in continuous mathematics.

0.1 OBJECTIVES OF COMBINATORICS

This initial section elaborates on the three fundamental objectives of combinatorial analysis: counting, constructing a configuration, and optimization. The six chapters immediately subsequent are largely concerned with counting and the final four with configurations (especially graphs). Optimization issues are sprinkled throughout. Combinatorial problems are pursued by thousands of active researchers. Enumeration, graph theory, combinatorial design, and combinatorial optimization are vast areas, each with many distinct branches. The comprehensive approach to introductory combinatorics taken in this text emphasizes topics of frequent use throughout mathematics and its applications.

The dramatic rise in the development of combinatorial mathematics in the present era largely stems from the fact that in a computer, in graphic imaging, and in many forms of data transmission and communication, information is represented by discrete bits, thereby necessitating combinatorial models. Information science and information engineering now stand side-by-side in applicability and public familiarity with physical science and physical engineering, for which continuous models are more common.

Combinatorial Enumeration

Combinatorial enumeration is concerned with the theory and methods of discrete measurement. Summing the values of a function over a finite or countable set is the prototypical discrete measurement, in which sense it is analogous to the continuous measurement of calculating the area of a region in the plane between the x-axis and a curve. The word *counting* is frequently used by combinatorialists as a minimalist synonym for combinatorial enumeration.

Most solutions to combinatorial enumeration problems depend on a relatively small number of well-understood methods for discrete summation. Applying these methods effectively requires expertise at transformation of enumeration problems into forms directly amenable to these methods. This is analogous to the kind of expertise in applying the infinitessimal calculus in which complicated-looking integrals are transformed into expressions that yield to a relatively few well-understood integration formulas.

Example 0.1.1: Our first example is concerned with evaluating the following sum:

$$1 +$$
$$2 + 1 +$$
$$3 + 2 + 1 +$$
$$\cdots$$
$$n + (n-1) + \cdots + 1$$

For $n = 12$, this sum would be the number of gifts presented by "my true love" in a well-known English holiday song,* and the value of the sum is 364. One might readily calculate that the sum of the jth row is

$$j + (j - 1) + \cdots + 1 \;=\; \frac{(j + 1)j}{2}$$

by observing that the average summand in this row is $\frac{j+1}{2}$ and that there are j summands. (This approach to summing consecutive integers is ascribed to Gauss,[†] at an early age.) Thus, the value of the original sum equals

$$\sum_{j=1}^{n} \frac{(j + 1)j}{2}$$

This latter sum rather neatly fits a standard form of what is called the *finite calculus* (see, especially, §3.4), and it can be evaluated as follows:

$$\sum_{j=1}^{n} \frac{(j + 1)j}{2} \;=\; \frac{(n + 2)(n + 1)n}{6}$$

For instance, for $n = 12$, the value is 364.

Example 0.1.1 could be generalized to summing the values of an arbitrary polynomial over a range of consecutive integers. Such summation problems arise frequently in the analysis of algorithms, in which the time to execute the body of a loop might be roughly proportional to a polynomial-valued function.

Example 0.1.2:   To evaluate the sum

$$\sum_{j=0}^{n} 4j^3 - 3j^2 + 5$$

we might use Stirling numbers (see §1.6, §5.1, and §5.2) to transform it into a sum of falling powers (see §1.5 and §3.4), for which there are simple formulas. In fact, we have additional methods for summing polynomials, such as *perturbation* (see §3.2).

In later sections of this chapter, we will see various additional kinds of counting problems.

Incidence Structures

An *incidence structure* is a combinatorial configuration that involves two or more discrete sets. Most commonly, there are exactly two sets — a set P of *points*

---

*   *The Twelve Days of Christmas,* orignally a children's rhyme, first published around 1780, according to *Wikipedia.*

[†]   For instance, see *www.mathnotes.com/aw_gauss.html.*

and a set L of *lines* — and an *incidence function* $\iota : P \times L \to \mathbb{Z}_2$. In this most common variety, the set L may optionally be represented as a family of subsets of P. Some types of combinatorial configuration have additional structure on one or both of the discrete sets.

Example 0.1.3:   An abstract model for what is called a *simple graph* is an incidence structure in which every line has exactly two points and in which no two lines have the same two points. In a spatial model, the more intuitive model for a graph, each point of the graph is called a *vertex* and identified with a point in a Euclidean space (usually the plane or 3-space), and each line of the graph is called an *edge*. An edge is represented spatially by an arc joining its two points, which are called the *endpoints* of that edge. They are said to be *adjacent vertices*. Figure 0.1.1 provides two drawings of a spatial model for the graph whose abstract model is

$$P = \{\, 1, \quad 2, \quad 3, \quad 4, \quad 5 \,\}$$
$$L = \{\, 12, \quad 14, \quad 15, \quad 23, \quad 25, \quad 45 \,\}$$

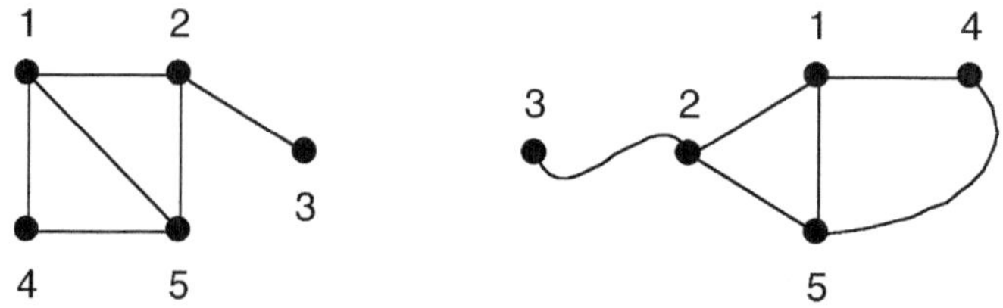

Figure 0.1.1   **Two drawings of a simple graph.**

There should be no expectation whatever that a *line* of a combinatorial configuration is represented by a straight-line of a drawing.

Practitioners of graph theory (see Chapters 7, 8, and 9) regard graphs as so interesting in themselves that there is no extrinsic need to justify their study. The same could be said for almost every area of mathematics — its practitioners are motivated more by their own intellectual curiosity than by possible applications. Nonetheless, what has made graph theory of particular importance is its many applications. Just for a start, graphs serve as models for molecules in physical chemistry and biology, for computer networks, for computer flow diagrams, for electronic networks, for networks of roads in civil engineering, for genealogy, and for social organization. Both for intrinsic interest and for their value in applications, graph theorists have solved many problems of an enumerative character.

Example 0.1.4:   While studying organic chemistry in the 19th century, Arthur Cayley encountered the problem of counting the number of different hydrocarbon isomers with the chemical formula $C_n H_{2n+2}$. The two isomers for $n = 4$, called butane and isobutane, are illustrated in Figure 0.1.2. Graph enumeration is the principal concern of Chapter 9.

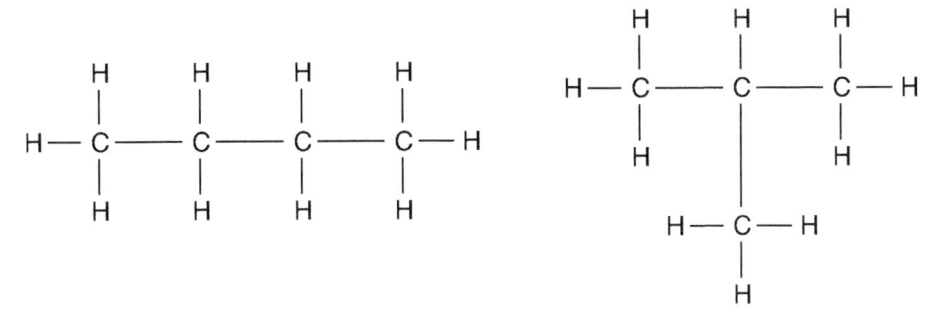

Figure 0.1.2   Butane and isobutane.

Optimization

In the present context, we mean by *combinatorial optimization* any discrete problem concerned with finding a maximum or a minimum. In some other contexts, the same phrase, combinatorial optimization, has a special meaning of finding the maximum value of a function on a region of a Euclidean space or of functions that could possibly be so represented. Even for the case in which the function is linear, there is an entire discipline and many books.

Example 0.1.5:   For instance, if one is selecting subsets of size k from a set of size n, one may wish to know the value of k for which the number of different subsets is greatest. Such a problem is clearly analogous to the problem of finding the maximum of a continuous function on a real interval, which is solved in calculus by finding the zeros of the derivative function, a technique of extensive generality. This combinatorial problem is solved in §1.5 by showing that the number of subsets first rises with increasing value of k and then falls, which follows the same pattern of reasoning as when optimizing a continuous function. However, the combinatorial technique needed to establish rising and falling of a discrete function is usually less formulaic than the maximizing and minimizing of the differential calculus, with something more of an *ad hoc* character.

Example 0.1.6:   In *extremal graph theory*, the standard type of problem is to determine the maximum number of edges that a simple graph G with n vertices may have before some property necessarily holds. For instance, how many edges may it have before there must be a set of three mutually adjacent vertices? The following solution of this problem, due to Paul Turán, appears in §8.4. (The notation $\lfloor x \rfloor$ means the largest integer less than or equal to x.)

$$|E_G| \;=\; \left\lfloor \frac{n^2}{4} \right\rfloor + 1$$

An example of the more restricted meaning of combinatorial optimization is the maximization of network flows, as described in §8.6.

0.2   ORDERING AND SELECTION

We begin with the analysis and solution of a sample counting problem involving ordering and selection, which are both fundamental ideas in combinatorics that occur throughout. The example is then generalized, and some standard artifacts of combinatorial analysis are introduced.

A Counting Problem

DEFINITION: An **ordering of a set** S of n objects is a bijection from the set

$$\{\, 1,\ 2,\ \ldots,\ n\, \}$$

to the set S. It serves as a formal model for an arrangement of the n objects into a row.

DEFINITION: An **(unordered) selection from a set** S is a subset of S.

Example 0.2.1:   In how many ways is it possible to arrange two of the letters

$$A \quad B \quad C \quad D \quad E$$

and two of the digits

$$0 \quad 1 \quad 2 \quad 3$$

into a row of four characters, such that no two digits are adjacent? For instance, the arrangement $C3A2$ meets that requirement.

It is not difficult to determine (e.g., by listing all possibilities, if no shorter method comes to mind) that there are 10 possible selections of two of the five letters and 6 possible selections of two of the four digits. Thus, there are 60 possible selections of a combination of four symbols that meets the given requirement. An arrangements of four such symbols into a row meets the requirement if it has any of the three forms

$$LDLD \quad DLDL \quad \text{and} \quad DLLD$$

where D is a digit and L is a letter. Since there are four ways that two distinct letters and two distinct digits could be placed within one of the three forms, it follows that there are 12 $(= 4 \times 3)$ ways that each of the 60 suitable selections of four symbols could be arranged so as to meet the requirement. Thus, the answer to the stated problem is 720 $(= 60 \times 12)$.

Some of the calculations in the foregoing analysis are based on a well-established counting rule, called the *Rule of Product*, to be presented in §0.3. For the time being, it is sufficient either to confirm the assertions of this section with *ad hoc* methods or to defer checking them until after reading §0.3.

Sequences and Generating Functions

A somewhat more general version of Example 0.2.1 supposes that x_n is the number of ways to form an arrangement of four symbols when there are n letters, but still only four digits. We have just calculated that $x_5 = 720$. Similar analysis yields the values

$$x_0 = 0 \quad x_1 = 0 \quad x_2 = 72 \quad x_3 = 216 \quad x_4 = 432 \quad x_5 = 720 \quad \ldots$$

The sequence over all non-negative integers n is called a *counting sequence* for this problem. Sometimes a sequence is encoded by multiplying its entries

$$g_0 \quad g_1 \quad g_2 \quad \cdots$$

by ascending powers of z (or of some other indeterminate) and summed into the form

$$g_0 + g_1 z + g_2 z^2 + \cdots$$

For this general version of Example 0.2.1, we would obtain

$$0 + 0z + 72z^2 + 216z^3 + 432z^4 + 720z^5 + \cdots$$
$$= 72z^2 + 216z^3 + 432z^4 + 720z^5 + \cdots$$

Moreover, the resulting infinite polynomial often has an equivalent closed form, called a *generating function*.

Example 0.2.2:   The closed form

$$\frac{1}{1 - 2z}$$

is equivalent to the infinite polynomial

$$1 + 2z + 4z^2 + 8z^3 + \cdots$$

Thus, it is a generating function for the sequence of powers of 2. As a generating function, such an infinite polynomial is regarded either as an encoding of its sequence of coefficients or as an algebraic expression. In this context, the issue of convergence is rarely relevant.

Generating functions are the main topic of §1.7. It is described there how they are used to solve various kinds of counting problems.

Recurrences

A sequence can be specified by giving some of its initial values and a *recurrence* that says how each later entry can be calculated from earlier entries.

Example 0.2.3:   Famously, the recurrence

$$\begin{aligned} f_0 &= 0; \quad f_1 = 1 & \text{initial values} \\ f_n &= f_{n-1} + f_{n-2} & \text{for } n \geq 2 \end{aligned}$$

gives the *Fibonacci sequence*, whose first few entries are as follows:

n	0	1	2	3	4	5	6	7	8	9 $\cdots$
f_n	0	1	1	2	3	5	8	13	21	34 $\cdots$

Generating functions are used in Chapter 2 to derive the formula

$$f_n = \frac{1}{\sqrt{5}} \cdot \left(\left(\frac{1 + \sqrt{5}}{2} \right)^n - \left(\frac{1 - \sqrt{5}}{2} \right)^n \right)$$

Such an arithmetic expression, whose evaluation can yield every value of a counting sequence, is called a **closed formula** for that sequence. A closed formula for a recurrence is called a *solution* to the recurrence, in the same sense that a differentiable function might be a solution to a differential equation.

Combination Coefficients

The number of possible selections a subset of size k from a set of size n is commonly expressed when speaking as "n-choose-k" and denoted in writing

$$\binom{n}{k}$$

which is called a *combination coefficient*. Its value is given by this equation

$$\binom{n}{k} = \frac{n!}{k!\,(n - k)!} \tag{0.2.1}$$

which is derived in §0.4. Its alternative name of *binomial coefficient* is justified in Chapter 1. For the time being, we observe that

$$\binom{n}{2} = \frac{n!}{2!\,(n - 2)!} = \frac{n(n - 1)}{2}$$

We may also perceive how combination coefficients might be used in solving still more generalized versions of Example 0.2.3.

Example 0.2.4:   The sequence of combination coefficients

$$\binom{0}{2} \quad \binom{1}{2} \quad \binom{2}{2} \quad \binom{3}{2} \quad \cdots$$

has the generating function

$$\frac{x^2}{(1 - x)^3}$$

To verify this observation, one might expand the denominator and divide it into the numerator, using the long division process on the two polynomials, which is described in more detail in §1.7.

0.3   SOME RULES FOR COUNTING

Having meaningful names for concepts, even for very simple concepts, makes it possible to state clearly and concisely what method is being used. Moreover, knowing a name for a concept makes it easier to recognize an instance of a method that it is not explicitly identified. This section introduces the names of a few principles whose applications are ubiquitous in combinatorial analysis. It also offers a glimpse at the *calculus of finite sums*, which is the discrete counterpart to the integral calculus.

NOTATION: The cardinality of a set U is denoted $|U|$. The most common binary operations on two sets U and V are denoted

$$U \cup V \quad \text{for union}$$
$$U \cap V \quad \text{for intersection}$$
$$U - V \quad \text{for difference, and}$$
$$U \times V \quad \text{for cartesian product}$$

Rules of Sum and Product

C. L. (Dave) Liu [Liu1968], then a professor of Electrical Engineering at M.I.T., gave popularity to now-standard names of two principles that relate elementary arithmetic operations to the counting of set unions and set products. They are frequently used in tandem.

DEFINITION: **Rule of Sum**: Let U and V be disjoint sets. Then

$$|U \cup V| = |U| + |V|$$

DEFINITION: **Rule of Product**: Let U and V be sets. Then

$$|U \times V| = |U| \cdot |V|$$

Example 0.3.1: The license plate numbers in a small state are five characters long. They must begin with three letters, but the other two characters may be letters or digits. According to the Rule of Product, there are 26^2 ways that the 4th and 5th characters may both be letters, $26 \cdot 10$ ways that they may be, respectively, a letter and a digit, $10 \cdot 26$ ways they may be a digit and a letter, and 10^2 ways they may both be digits. By rule of sum there are

$$26^2 + 26 \cdot 10 + 10 \cdot 26 + 10^2 = 1296$$

possibilities for the last two digits. (We notice that $1296 = (26 + 10)^2$.) Since there are, by rule of product, 26^3 possible combinations for the leading three letters, the total number of possibilities is

$$26^3 \cdot 1296$$

Sometimes the rule of product is applied in circumstances where a plausible time-sequence is imposed on the order of selection of members from the sets, without changing the resulting number of objects in the set to be counted.

Example 0.3.2:   Three six-sided dice are rolled. The dice are colored blue, red, and yellow. In how many ways can the outcome be three different numbers on the three dice? To solve this problem, we observe that whichever of the 6 possibilities occurs for the blue die, there remain 5 for the red die, and then 4 for the yellow die. Thus, the total number of possibilities is

$$6 \cdot 5 \cdot 4$$

Rule of Quotient

Another counting rule, similar in simplicity to Liu's two rules, applies to counting the number of cells in a *partition* of a set.

DEFINITION: A **partition** of a set U is a collection of mutually exclusive subsets

$$U_1, \ldots, U_p$$

called **cells of the partition**, whose union is U.

DEFINITION: **Rule of Quotient**: Let $\mathcal{P}$ be a partition of a set U into cells, each of the same cardinality k. Then the number of cells equals the quotient

$$\frac{|U|}{k}$$

Example 0.3.3:   Figure 0.3.1 shows 20 objects partitioned into cells of four each. In accordance with the Rule of Quotient, the number of cells is

$$\frac{20}{4} = 5$$

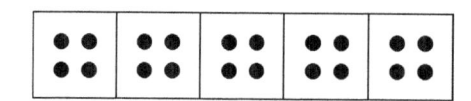

Figure 0.3.1   Partition of 20 objects into cells of four each.

Example 0.3.2, revisited: There are $6 = 3!$ ways that any given combination of three distinct numbers could occur on the three dice. If the set of all instances of three different numbers for the three dice is partitioned into cells so that each cell contains all instances of a given combination of three distinct numbers, then each cell is of cardinality 6. It follows that the total number of possible combinations of three numbers, ignoring which occurs on which die, is

$$\frac{6 \cdot 5 \cdot 4}{6} = 20$$

The Rule of Quotient cannot be applied when the cells of the partition are of different sizes.

Example 0.3.4:   Suppose that each of the squares of a 3×3 tic-tac-toe board may be filled with an "X" or an "O", or left blank, without consideration of what might arise when playing the game. Since there are three possibilities for each of the nine squares, the total number of possible configurations is 3^9. It is natural to regard two such configurations as equivalent if one could be obtained from the other by a rotation or a reflection. The equivalence classes are not all of the same size. For instance, Figure 0.3.2 illustrates an equivalence class of size four.

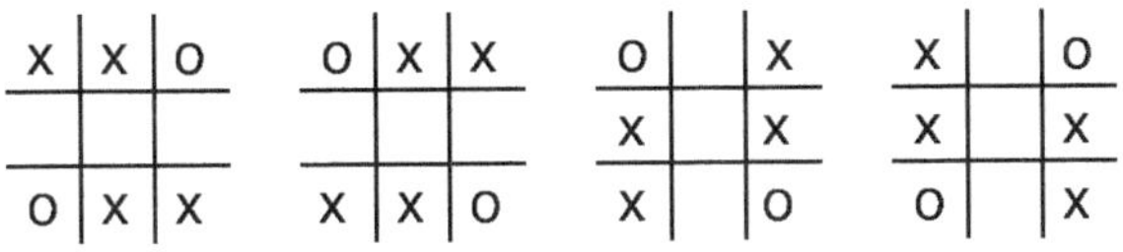

X	X	O		O	X	X		O		X		X		O
						X			X			X		X
O	X	X		X	X	O		X		O		O		X

Figure 0.3.2   Four equivalent tic-tac-toe configurations.

On the other hand, the configurations that are all blank, all "X", or all "O" are in equivalence classes of size one. There are also some equivalence classes of sizes two and eight. Thus, the Rule of Quotient cannot be applied.

Counting equivalence classes that are defined by symmetries is frequently accomplished with the aid of Burnside-Pólya counting. This method of counting is developed in Chapter 9.

When to Subtract

There are some common circumstances when calculating the cardinality of a set is achieved using a subtraction operation. One is when the set X to be counted is a subset of a larger set U and it looks easier to calculate the sizes of U and of the complement $U - X$ than the size of the set X directly.

Example 0.3.5:   To count the number of n-digit base-ten numerals that contain at least one odd numeral, we observe that there are 10^n 6-digit base-ten numerals in all, according to the Rule of Product. Of these, since there are five even digits, 5^n contain only even digits, also by the Rule of Product. Thus, the number of n-digit base-ten numerals with at least one odd digit is

$$10^n - 5^n$$

Another circumstance where subtraction is used is in calculating the size of a union of overlapping subsets. Adding the subset sizes overcounts objects that appear in more than one of the subsets, so the overcount must be subtracted.

Example 0.3.6:   To count the integers from 1 to 990 that are divisible either by 3 or 5, we first calculate that within this range, there are

$$\frac{990}{3} = 330$$

that are divisible by 3 and

$$\frac{990}{5} = 198$$

that are divisible by 5. However, the sum of these two quotient would count each of the

$$\frac{990}{15} = 66$$

integers that are divisible both by 3 and by 5 two times each. Thus, the total number of integers that are divisible either by 3 or by 5 must be

$$330 + 198 - 66 = 462$$

Reals to Integers

Three standard functions for converting a real number into a nearby integer are especially convenient when one wants to apply integer methods. Sometimes they are intrinsic to a formula, for instance in a generalization of Example 0.3.6.

DEFINITION: The **floor of a real number** x is the largest integer that is not larger than x. It is denoted $\lfloor x \rfloor$.

DEFINITION: The **ceiling of a real number** x is the smallest integer that is not smaller than x. It is denoted $\lceil x \rceil$.

DEFINITION: The **nearest integer** to a real number x is

$$round\,(x) = \begin{cases} \lfloor x \rfloor & \text{if } x - \lfloor x \rfloor < \frac{1}{2}; \\ \lfloor x \rfloor & \text{if } x - \lfloor x \rfloor = \frac{1}{2} \text{ and } \lfloor x \rfloor \text{ is even}; \\ \lceil x \rceil & \text{if } x - \lfloor x \rfloor > \frac{1}{2}; \\ \lceil x \rceil & \text{if } x - \lfloor x \rfloor = \frac{1}{2} \text{ and } \lceil x \rceil \text{ is even} \end{cases}$$

This table gives a few values of the floor function, the ceiling function, and the round function.

n	$\lfloor n \rfloor$	$\lceil n \rceil$	$round\,(n)$
4.8	4	5	5
4.5	4	5	4
3.5	3	4	4
3.2	3	4	3
-2.2	-3	-2	-2
-2.5	-3	-2	-2
-2.9	-3	-2	-3
-3.5	-4	-3	-4

Example 0.3.6, continued: In general, the number of positive integers less than or equal to n that are divisible by 3 or by 5 is

$$\frac{n}{3} + \frac{n}{5} - \frac{n}{15}$$

Pigeonhole Principle

The imagery of another elementary counting principle is that a flock of pigeons is flying in a formation that does not lend itself easily to counting the pigeons. Fortunately, however, the pigeons come to roost in a set of pigeonholes that is more easily counted, such that there is exactly one pigeon to each pigeonhole. Then the fact that the number of pigeons equals the number of pigeonholes provides a way to count the pigeons. Figure 0.3.3 illustrates this imagery.

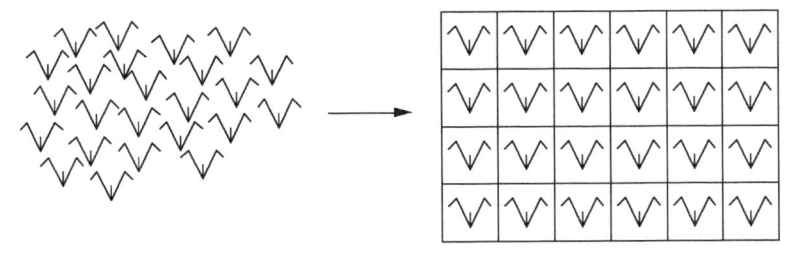

Figure 0.3.3   **Flock of pigeons neatly fills the pigeonholes.**

One widely cited informal version of the pigeonhole principle simply says that if there are more pigeons than pigeonholes, then there exists a pigeonhole with more than one pigeon. A complete formal statement of the principle is as follows. It can be proved by a straightforward induction argument. The informal version can be derived from it.

DEFINITION: **Pigeonhole Principle**: Let $f : U \rightarrow V$ be a function with finite domain and finite codomain. Let any two of the following three conditions hold:

1. f is one-to-one.
2. f is onto.
3. $|U| = |V|$.

Then the third condition also holds.

Example 0.3.7:   In any collection of 13 people, there must be two of them who were born in the same month. In this elementary example, the people are the pigeons, and the months are the pigeonholes.

Example 0.3.8:   If a baseball team scores 12 runs in a 9-inning game, then there is an inning in which they scored at least two runs.

Applications of the Pigeonhole Principle are often a bit tricky. Consider the following example.

Example 0.3.9:   Suppose there are ten pairs of socks, each pair a different color and that the socks are tossed together in a pile. It becomes necessary to pack for a business trip in total darkness, with a meeting in which it is essential to wear two socks of the same color. What is the minimum number of socks one must pack to be sure to have a matched pair?

In this case, the pigeons are socks and the pigeonholes are the 10 different colors.

Although ten pigeons might roost one per hole, this would not be possible for 11 pigeons. Thus, 11 is the minimum number of socks that guarantees the existence of a match among them.

A generalized version of the Pigeonhole Principle asserts that when there are p pigeons and h pigeonholes, there is a pigeonhole with at least

$$\left\lceil \frac{p}{h} \right\rceil$$

pigeons.

Example 0.3.10:   An equestrian asks her lawyer to write a will bequeathing her 17 horses to five beloved nieces and nephews. Then there exists a beneficiary among them who will get at least four horses.

Evaluating Sums

Complex summation expressions tend to arise quite frequently in problems concerned with counting. Some of the most useful methods for evaluating finite sums appear in Chapter 3. One such method, with a direct analogy to the calculus of integration, is called the *calculus of finite sums*.

Quite likely the most frequently used formula from the infinitessimal calculus is the formula

$$\int_{x=a}^{b} x^r = \left. \frac{x^{r+1}}{r+1} \right|_{x=a}^{b}$$

for integrating a monomial (i.e., a polynomial with only one term). The following definition is a preliminary step to expressing an analogous formula for finite summations.

DEFINITION: The rth ***falling power*** of a real number x is the product

$$x^{\underline{r}} = \overbrace{x\,(x-1)\,\cdots\,(x-r+1)}^{r \text{ factors}} \qquad \text{for } r \in \mathbb{N}$$

Remark: For nonnegative integers n and $r \leq n$,

$$n^{\underline{r}} = \frac{n!}{(n-r)!} \tag{0.3.1}$$

Example 0.3.11:   Here are some falling power evaluations.

$$x^{\underline{3}} = x\,(x-1)(x-2) = x^3 - 3x^2 + 2x$$
$$6^{\underline{3}} = 6 \cdot 5 \cdot 4 = 120$$
$$\left(\frac{2}{5}\right)^{\underline{3}} = \frac{2}{5} \cdot \frac{-3}{5} \cdot \frac{-8}{5} = \frac{48}{125}$$

The formula for summing a falling-power monomial is

$$\sum_{k=a}^{b} k^{\underline{r}} \;=\; \frac{k^{\underline{r+1}}}{r+1}\bigg|_{k=a}^{b+1} \tag{0.3.2}$$

Example 0.3.12:   We apply formula $(0.3.2)$ for exponent $r = 2$ and limits of summation $a = 3$ and $b = 5$.

$$\sum_{k=3}^{5} k^{\underline{2}} \;=\; 3^{\underline{2}} + 4^{\underline{2}} + 5^{\underline{2}}$$

$$= 6 + 12 + 20 \;=\; 38$$

$$\frac{k^{\underline{3}}}{3}\bigg|_{k=3}^{6} \;=\; \frac{6^{\underline{3}}}{3} - \frac{3^{\underline{3}}}{3}$$

$$= 40 - 2 \;=\; 38$$

Summations of ordinary powers can be achieved via a preliminary conversion to falling powers. For instance,

$$x^2 \;=\; x^{\underline{2}} + x^{\underline{1}} \quad \text{and} \quad x^3 \;=\; x^{\underline{3}} + 3x^{\underline{2}} + x^{\underline{1}}$$

The coefficients used in the conversion, which are called *Stirling numbers*, are described in §1.6 and developed in much greater detail in Chapter 5.

Empty Sums and Empty Products

In manipulating expressions with iterated sums and products, such as

$$\sum_{x_j \in S} x_j \quad \text{or} \quad \prod_{x_j \in S} x_j$$

we sometimes encounter a sum or product over the *empty set* $\emptyset$.

DEFINITION: A sum over an empty set of numbers is called an **empty sum**. Its value is taken to be 0, the additive identity of the number system.

DEFINITION: A product over an empty set of numbers is called an **empty product**. Its value is taken to be 1, the multiplicative identity of the number system.

Multisets

One of the many applications of the Rule of Quotient is to counting arrangements of *multisets*. Informally, a multiset is often described as a "set in which the same element may occur more than once".

DEFINITION: A **multiset** is a pair (S, ι) in which S is a set and $\iota : S \to \mathbb{Z}^{+}$ is a function that assigns to each element $s \in S$ a number $\iota(s)$ called its **multiplicity**. (The Greek letter iota is a mnemonic for *instances*.)

Example 0.3.13:   The letters of the word SYZYGY form a multiset in which the letter Y occurs three times, and each of the other three letters occurs once. If the six letters were all different, then the number of ways of arranging them into a row of six would be $6! = 720$. We may model this by artificially attaching distinct subscripts to each of the copies of the letter Y, so that they become Y_1, Y_2, and Y_3. We regard two arrangements of the six elements of the resulting set as equivalent if the positions of the letters G, S, and Z are the same in both arrangements. There are then $6 = 3!$ equivalent arrangements in each equivalence class. By the Rule of Quotient, the number of equivalence classes is

$$\frac{6!}{3!} \; = \; 120$$

More generally, the Rule of Quotient implies that the number of ways to arrange the elements of a finite multiset (S, ι) is

$$\frac{\left(\sum_{s \in S} \iota(s)\right)!}{\prod_{s \in S}(\iota(s)!)}$$

DEFINITION: The **cardinality of a multiset** (S, ι) is taken to be the sum

$$\sum_{s \in S} \iota(s)$$

of the multiplicities of its elements. It is denoted $|(S, \iota)|$.

A couple of additional definitions are helpful when working with multisets.

DEFINITION: A **submultiset of a multiset** (S, ι_S) is a multiset (T, ι_T) such that

$$T \subset S \qquad \text{and}$$
$$\iota_T(t) \le \iota_S(t) \quad \text{for all } t \in T$$

We shall see in §1.7 how to use generating functions to count not only the number of ways to select k elements from a given multiset, for all possible values of k, but also to count the number of strings of length k, taken from a given multiset of letters.

Example 0.3.13, continued: There are seven possible choices of three letters from the word SYZYGY. There are 34 possible strings of length 3. For the time being this can be confirmed using Rule of Sum and Rule of Product.

DEFINITION: The **restriction of a multiset** (S, ι) to a subdomain $T \subset S$ is the submultiset (T, ι_T) such that

$$\iota_T(t) \; = \; \iota(t) \quad \text{for all } t \in T$$

NOTATION: In context, the **multiplicity function of the restriction** of a multiset (S, ι) to a subdomain $T \subset S$ is simply denoted ι, since its values on the elements of T are the same as when they are regarded as elements of S.

EXERCISES for Section 0.3

0.3.1   Calculate the number of ways to arrange three 0-bits and four 1-bits into a binary string.

0.3.2   Calculate the number of functions from a set of d elements to a set of r elements.

0.3.3   Calculate the number of one-to-one functions from a set of d elements to a set of r elements.

0.3.4   How many numbers between 1 and n, inclusive, are divisible either by 2 or by 7?

0.3.5   How many numbers between 1 and n, inclusive, are divisible either by 6 or by 10?

In each of the Exercises 0.3.6 through 0.3.9, compare the two floor expressions and prove that one of them is less than or equal to the other, for all real x and y.

0.3.6[S] $\lfloor x + y \rfloor$ and $\lfloor x \rfloor + \lfloor y \rfloor$ **0.3.7** $\lfloor x - y \rfloor$ and $\lfloor x \rfloor - \lfloor y \rfloor$

0.3.8 $\lfloor x^2 \rfloor$ and $\lfloor x \rfloor^2$ **0.3.9** $\sqrt{\lfloor x^2 \rfloor}$ and $\lfloor x \rfloor$

In each of the Exercises 0.3.10 through 0.3.15, a multiset is represented by a given string of letters. Calculate the number of ways to arrange the letters of the multiset into a string.

0.3.10[S]   *BANDANA*   **0.3.11**   *FOREIGNER*

0.3.12   *HORSERADISH*   **0.3.13**   *CONSTITUTION*

0.3.14   *MISSISSIPPI*   **0.3.15**   *WOOLLOOMOOLOO*

Each of the Exercises 0.3.16 through 0.3.19 presents a possible application of the Pigeonhole Principle. Identify pigeons and pigeonholes, and calculate the answer.

0.3.16[S]   What is the minimum number of students in a class such that at least two of them were surely born on the same day of the week?

0.3.17   Suppose it is known that the maximum number of hairs on a person's head is 500,000. Show that a city with 8,000,000 people must have two persons with the same number of hairs on their heads.

0.3.18   How many times must two six-sided dice be rolled so it is certain that two of the outcomes will have the same sum?

0.3.19   What is the maximum length of a binary string such that no two of the substrings of length three are the same? (Optional: Give an example of a maximum-length sequence.)

0.3.20   List the seven possible choices of three of the letters from SYZYGY.

0.3.21   Use the Rule of Product and the Rule of Sum to verify that there are 34 possible 3-letter strings that can be formed from the word SYZYGY, if no letter may be used more often than its number of occurrences in that word.

0.4   COUNTING SELECTIONS

This section gives models for several different kinds of selection from a set S and methods for counting the number of possible selections. As defined in §0.2, an *unordered selection* from S is simply a subset of S. An *ordered selection* assigns an order to the elements of the selected subset. Some other models permit *repetition*. This discussion of selection includes the generalization to *multi-selection*, in which several disjoint subsets may be selected from the set S. In a multi-selection, sometimes the subsets are construed to be *labeled*, which serves to distinguish two subsets of the same size.

In the course of this exposition, the usefulness of constructs such as falling powers and empty products becomes evident.

Ordered Selections

DEFINITION: An **ordered selection** of k objects from a set of n objects is a function from the set

$$\{ 1, 2, \ldots, k \}$$

to the set S. It serves as a formal model for an arrangement of k objects from S into a row, or of a repetition-free list of length k of objects from S.

TERMINOLOGY NOTE: An ordered selection is sometimes elsewhere called a *permutation*. In the present context, we follow the usage of higher algebra, that a permutation is a bijection of a set to itself. Thus, here a *permutation* is the operation itself, rather than the resulting arrangement. See §0.5.

Proposition 0.4.1. *Let $P(n,k)$ be the number of possible ordered selections of k objects from a set S of n objects. Then*

$$P(n,k) \;=\; n^{\underline{k}} \tag{0.4.1}$$

Proof: By induction on k.

BASIS: For $k = 0$, the only possible ordered selection is the empty list. Thus,

$$P(n,0) \;=\; 1 \;=\; n^{\underline{0}}$$

IND HYP: Assume that $P(n,k) = n^{\underline{k}}$, for some $k \geq 0$.

IND STEP: After the first k objects have already been selected from S, the number of remaining objects from which to choose the $k + 1^{\text{st}}$ object is $n - k$. Thus,

$$
\begin{aligned}
P(n, k+1) \;&=\; P(n,k) \cdot (n-k) && \text{(Rule of Product)} \\
&=\; n^{\underline{k}} \cdot (n-k) && \text{(induction hypothesis)} \\
&=\; n(n-1) \cdots (n-k+1) \cdot (n-k) && \text{(def of falling power)} \\
&=\; n^{\underline{k+1}} &&
\end{aligned}
$$

$\Diamond$

Example 0.1.1, revisited: Each of the ways to arrange two of the letters

$$A \quad B \quad C \quad D \quad E$$

and two of the digits

$$0 \quad 1 \quad 2 \quad 3$$

into one of the forms

$$LDLD \quad DLDL \quad \text{and} \quad DLLD$$

may be regarded as a choice of one of the three forms, followed by an ordered selection of two letters from the set of five — to be placed into the two positions for letters in the chosen form, in order consistent with the order of selection — followed by an ordered selection of two digits from the set of four — to be placed into the two positions for digits in the chosen form, in order consistent with the order of selection. For instance, the arrangement $C3A2$ corresponds to the choice of the form $LDLD$, followed by the ordered selections CA and 32.

Since the number of forms is 3, the number of ordered selections of two letters is $5^{\underline{2}}$, and the number of ordered selections of digits is $4^{\underline{2}}$, it follows from the Rule of Product that the total number of arrangements is

$$3 \cdot 5^{\underline{2}} \cdot 4^{\underline{2}} = = 3 \cdot 20 \cdot 12 = 720$$

Unordered Selections

To evaluate $\binom{n}{k}$, which counts unordered selections, we regard the unordered selections as equivalence classes of ordered selections, in which two ordered selections of k objects are considered to be equivalent if they contain the exact same k objects.

Proposition 0.4.2. *The number of unordered selections of k objects from a set S of n objects is given by the rule*

$$\binom{n}{k} = \frac{n^{\underline{k}}}{k!} = \frac{n!}{k!\,(n-k)!} \tag{0.4.2}$$

Proof: By Proposition 0.4.1, the number of ordered selections of k objects from S is $n^{\underline{k}}$. Since the number of orderings of k objects is $k!$, there are $k!$ ordered selections corresponding to each unordered selection. The conclusion follows from the Rule of Quotient.   ◇

Selections with Repetitions Allowed

The number of ordered selections of k objects from a set S of n objects *with unlimited repetition allowed* is easily determined.

DEFINITION: An ***ordered selection with unlimited repetition*** of k objects from a set S of size n is a finite sequence

$$x_1, \quad x_2, \quad \ldots, \quad x_k$$

of k objects, each of which is an element of S.

Proposition 0.4.3. *The number of ordered selections of k objects from a set S of n objects is n^k.*

Proof: This is easily proved by an induction argument, involving the Rule of Product.   ◇

   Counting unordered selection with unlimited repetitions allowed seems quite difficult, if approached directly.

DEFINITION: An ***unordered selection with unlimited repetition*** of k objects from a set S of size n is a multiset (S, ι) of cardinality k, with domain S.

Example 0.4.1:   Consider counting the number of unordered selections, with unlimited repetitions allowed, of four objects from the set $\{1, 2, 3, 4\}$. There are these four selections containing only one distinct digit

$$1111 \quad 2222 \quad 3333 \quad 4444$$

these 18 with two different digits

$$
\begin{array}{cccccc}
1112 & 1113 & 1114 & 2221 & 2223 & 2224 \\
3331 & 3332 & 3334 & 4441 & 4442 & 4443 \\
1122 & 1133 & 1144 & 2233 & 2244 & 3344
\end{array}
$$

these 12 with three different digits

$$
\begin{array}{cccccc}
1123 & 1124 & 1134 & 2213 & 2214 & 2234 \\
3312 & 3314 & 3324 & 4412 & 4413 & 4423
\end{array}
$$

and only one with four different digits

$$1234$$

for a total of 35 possibilities.

The following construction greatly simplifies the task of counting unordered selections with unlimited repetitions, by representing multisets as binary strings.

DEFINITION: The ***bitcode for a multiset*** (S, ι) of cardinality k, with domain $\{1, 2, \ldots, n\}$, is defined recursively:

- If $n = 1$, then the bitcode is a string of k 0-bits.

- For $n > 1$, the bitcode for (S, ι) is the bitcode for the submultiset $(S - \{n\}, \iota)$, followed by a 1-bit, followed by a suffix of $\iota(n)$ 0-bits.

Example 0.4.1, continued: For the domain $\{1, 2, 3, 4\}$, the bitcode for the multiset

$$\{1, 1, 3, 4\}$$

is 0011010. The steps are as follows:

$$
\begin{array}{ll}
\{1, 1\} \text{ over domain } \{1\} \text{ has bitcode} & 00 \\
\{1, 1\} \text{ over domain } \{1, 2\} \text{ has bitcode} & 001 \\
\{1, 1, 3\} \text{ over domain } \{1, 2, 3\} \text{ has bitcode} & 00110 \\
\{1, 1, 3, 4\} \text{ over domain } \{1, 2, 3\} \text{ has bitcode} & 0011010
\end{array}
$$

We observe that the multiset $\{1, 1, 3, 4\}$ could be reconstructed from its bitcode 0011010. Since two 0-bits precede the first 1-bit, there must be two instances of the digit 1 in the multiset. Since there are no 0-bits between the first 1-bit and the second 1-bit of the bitcode, there must be no instances of the digit 2 in the multiset. Since there is one 0-bit between the second 1-bit and the third 1-bit of the bitcode, there must be exactly one instance of the digit 3 in the multiset. Since there is one 0-bit after the third and final 1-bit, the multiset must have exactly one instance of the digit 4.

Remark: In reconstructing a multiset of cardinality k with domain $\{1, 2, \ldots, n\}$ from its bitcode, we may regard the $k - 1$ 1-bits as separating the bitstring into k substrings of 0-bits, some of which may be nullstrings. The lengths of the k consecutive substrings of 0-bits are the multiplicities on the corresponding integers in the domain. This may be depicted as in Figure 0.4.1.

$$\underline{0\ \ 0\ \Big|\ \ \ \Big|\ 0\ \Big|\ 0\ }$$

Figure 0.4.1   A representation of the bitstring 0011010.

Proposition 0.4.4. *The correspondence between the set of multisets of cardinality k with domain $\{1, 2, \ldots, n\}$ and the set of bitstrings of length $n + k - 1$ with exactly $k - 1$ 1-bits is a bijection.*

Proof: One possible proof of this proposition is that the encoding of multisets as bitcodes is clearly invertible, which could be established by generalizing the inversion in Example 0.4.1. Another alternative is by induction. $\diamond$

Corollary 0.4.5. *The number of different multisets of cardinality k with domain $\{1, 2, \ldots, n\}$ is*

$$\binom{n + k - 1}{k - 1}$$

Proof: By Proposition 0.4.2, the number of bitstrings of length $n + k - 1$ with exactly $k - 1$ 1-bits is

$$\binom{n + k - 1}{k - 1}$$

It follows from the Pigeonhole Principle, in view of Proposition 0.4.4, that the number of different multisets of cardinality k with domain $\{1, 2, \ldots, n\}$ is the same as the number of bitstrings of length $n + k - 1$ with exactly $k - 1$ 1-bits. $\diamond$

Example 0.4.1, continued: By Corollary 0.4.5, the number of multisets of cardinality four with domain $\{1, 2, 3, 4\}$ is

$$\binom{4 + 4 - 1}{3} = \binom{7}{3} = 35$$

Thus, Corollary 0.4.5 can greatly reduce the effort needed to count multisets with repetitions.

Example 0.4.2: Consider counting the number of possible outcomes of rolling three cubic dice, with the six sides of each die marked with 1 to 6 spots. Any two outcomes with the exact same number of instances of each of the six numbers of spots are regarded as equivalent. How many different possible outcomes are there? According to Corollary 0.4.5, the answer is

$$\binom{6+3-1}{5} = \binom{8}{5} = 56$$

Distributions into Labeled Cells

Sometimes, instead of selecting a single subset of a set, a problem calls for distributing the elements of a set into disjoint cells, thereby, in effect, selecting several subsets. There are several different models.

DEFINITION: A *multicombination* from a set S of n objects is a distribution of the elements of S into k labeled cells

$$B_1 \quad B_2 \quad \ldots \quad B_k$$

(sometimes) called *boxes*. Although this does not distinguish the order of the objects with a cell, the cells are distinct.

DEFINITION: The **multicombination coefficient**

$$\binom{n}{r_1 \quad r_2 \quad \cdots \quad r_k}$$

is the number of ways to distribute a set of n objects into k labeled cells

$$B_1 \quad B_2 \quad \ldots \quad B_k$$

of respective sizes $r_1, r_2, \ldots, r_k$.

Proposition 0.4.6. *The values of the multicombination coefficients are given by the rule*

$$\binom{n}{r_1 \quad r_2 \quad \cdots \quad r_k} = \frac{n!}{r_1!\, r_2! \cdots r_k!} \qquad (0.4.3)$$

Proof: The number of ways to select r_1 for box B_1 is

$$\binom{n}{r_1}$$

The number of ways to subsequently select r_2 for box B_2 from the remaining $n - r_1$ objects is

$$\binom{n - r_1}{r_2}$$

And so on. By the Rule of Product, it follows that the number of ways to complete the distribution is

$$\binom{n}{r_1}\binom{n-r_1}{r_2}\cdots\binom{n-r_1-r_2-\cdots-r_{k-1}}{r_k}$$

$$= \frac{n!}{r_1!\,(n-r_1)!}\cdot\frac{(n-r_1)!}{r_2!\,(n-r_1-r_2)!}\cdot\ \cdots\ \cdot\frac{(n-r_1-r_2-\cdots-r_{k-1})!}{r_k!\,0!}$$

$$= \frac{n!}{r_1!\,r_2!\,\cdots\,r_k!}$$

by repeated application of the factorial formula (0.4.2) for binomial coefficients. $\diamond$

Example 0.4.3:   The ways to distribute the set $\{A,\,B,\,C,\,D\}$ into boxes of sizes $r_1 = 2$, $r_2 = 1$, and $r_3 = 1$ are given by this array

$$AB|C|D \quad AC|B|D \quad AD|B|C \quad BC|A|D \quad BD|A|C \quad CD|A|B$$
$$AB|D|C \quad AC|D|B \quad AD|C|B \quad BC|D|A \quad BD|C|A \quad CD|B|A$$

Each of of the six columns of the array shows a different possible choice for Box B_1 of 2 objects, leaving two objects, from which one object is to be chosen for box B_2, thereby leaving the remaining object for box B_3. We could calculate the total number of distributions iteratively as

$$\binom{4}{2}\binom{2}{1} \;=\; 6\cdot 2 \;=\; 12$$

or, alternatively, with a single multicombination coefficient

$$\binom{4}{2\ 1\ 1} \;=\; \frac{4!}{2!\,1!\,1!} \;=\; 12$$

TERMINOLOGY: Another name for the multicombination coefficient

$$\binom{n}{r_1\quad r_2\quad\cdots\quad r_k}$$

is the **multinomial coefficient**, since it is provably the coefficient of the term

$$x_1^{r_1}\,x_2^{r_2}\,\cdots\,x_k^{r_k}$$

in the expansion of the exponentiated multinomial

$$\left(x_1 + x_2 + \cdots + x_k\right)^n$$

Distributions into Unlabeled Cells

The difference between distributions into labeled and into unlabeled cells is best explained with concrete examples. The main idea is the cells of the same size are regarded as interchangeable.

Example 0.4.3, continued: With unlabeled boxes, each of the distributions on the top row of the array is indistinguishable from the distribution immediate below it.

Example 0.4.4:   Of four faculty in an academic department, two will be advisors to the juniors and two to the seniors. According to Proposition 0.4.6, the number of distributions meeting the requirement is

$$\frac{4!}{2!\,2!} = 6$$

If these faculty are designated A, B, C, and D, the six possible distributions are

	juniors	seniors
1.	AB	CD
2.	AC	BD
3.	AD	BC
4.	CD	AB
5.	BD	AC
6.	BC	AD

However, if we discard the labels *juniors* and *seniors* then there are only three ways that the four faculty are grouped into pairs. The distributions 1 and 4 would be indistinguishable, as would distributions 2 and 5 and distributions 3 and 6.

The following proposition gives the formula for counting distributions into unlabeled cells.

Proposition 0.4.7. *Let S be a set of n objects. Suppose that these objects are to be distributed into b_j boxes of size r_j, for $j = 1, \ldots, k$, with*

$$\sum_{j=1}^{k} b_j\, r_j = n$$

The number of ways to do this is

$$\frac{n!}{(r_1!)^{b_1}\,(r_2!)^{b_2}\,\cdots\,(r_k!)^{b_k}} \cdot \frac{1}{b_1!\,b_2!\,\cdots\,b_k!} \qquad (0.4.4)$$

Proof: This follows from Proposition 0.4.6 and the Rule of Quotient. $\diamond$

Partitions of a Set

PREVIEW OF §1.6:

- A *partition of a set* into k cells can be characterized as a distribution of that set into k unlabeled boxes with none left empty.

- The ***Stirling subset number*** $\left\{ {n \atop k} \right\}$ is the number of ways to partition a set with n objects into k cells.

Formula (0.4.4) enables us to calculate the number of partitions of a set of n objects into cells of prespecified sizes.

Example 0.4.4, continued: The number of partitions of a set of four objects into two cells, both of size two, is

$$\frac{4!}{2!\,2!} \cdot \frac{1}{2!} \; = \; 3$$

Example 0.4.5:   A set with four objects may be partitioned into two cells either with sizes 3 and 1 or with cells of sizes 2 and 2. Thus,

$$\left\{ {4 \atop 2} \right\} \; = \; \binom{4}{3\ 1} + \binom{4}{2\ 2} \cdot \frac{1}{2!} \; = \; 4 + 3 \; = \; 7$$

EXERCISES for Section 0.4

In each of the Exercises 0.4.1 through 0.4.3, calculate the number of selections with unlimited repetition for the designated problem.

0.4.1   Select eight coins from the six coins presently in circulation in the USA: 1¢, 5¢, 10¢, 25¢, 50¢, $1.

0.4.2   A bakery sells four kinds of bagels: plain, onion, garlic, and poppy seed. Select a dozen bagels.

0.4.3$^{\mathbf{S}}$ Select positive integer values for the variables x_1, x_2, and x_3 so that $x_1 + x_2 + x_3 = 11$.

0.4.4   A college schedules introductory courses in calculus, chemistry, and physics at 9:00am and requires every one of its 323 freshmen to attend one of these 9:00am courses. Calculate the number of ways to distribute the students into these three courses.

0.4.5   A wrestling team competes in a league with 14 season matches, each of which could result in a win, a loss, or a draw. Calculate the number of possible season records.

0.4.6$^{\mathbf{S}}$ Calculate the number of ways to distribute 12 indistinguishable balls into four labeled boxes.

0.4.**7**   Calculate the number of terms of the multinomial resulting from the expansion of the trinomial $(x + y + z)^4$.

In each of the Exercises 0.4.8 through 0.4.11, evaluate the given multinomial coefficient.

0.4.**8**^{S} $\begin{pmatrix} 7 \\ 3\ 2\ 2 \end{pmatrix}$ 0.4.**9** $\begin{pmatrix} 9 \\ 3\ 2\ 2\ 1\ 1 \end{pmatrix}$

0.4.**10** $\begin{pmatrix} 9 \\ 2\ 2\ 2\ 1\ 1\ 1 \end{pmatrix}$ 0.4.**11** $\begin{pmatrix} 12 \\ 3\ 3\ 2\ 2\ 1\ 1 \end{pmatrix}$

In each of the Exercises 0.4.12 through 0.4.15, calculate the number of partitions of a set of the given size into cells of the given sizes.

0.4.**12**^{S} A set of size 7 into parts of sizes 3, 2, and 2.

0.4.**13**   A set of size 9 into parts of sizes 3, 2, 2, 1, and 1.

0.4.**14**   A set of size 9 into parts of sizes 2, 2, 2, 1, 1, and 1.

0.4.**15**   A set of size 12 into parts of sizes 3, 3, 2, 2, 1, and 1.

In each of the Exercises 0.4.16 through 0.4.19, evaluate the given Stirling subset number.

0.4.**16**^{S} $\left\{ \begin{matrix} 5 \\ 3 \end{matrix} \right\}$ 0.4.**17** $\left\{ \begin{matrix} 5 \\ 2 \end{matrix} \right\}$ 0.4.**18** $\left\{ \begin{matrix} 6 \\ 3 \end{matrix} \right\}$ 0.4.**19** $\left\{ \begin{matrix} 6 \\ 4 \end{matrix} \right\}$

0.5   PERMUTATIONS

Solving problems concerned with counting configurations with symmetries, like the tic-tac-toe boards of Example 0.3.4, requires some algebra involving *permutations*, as seen in Chapter 9. It is fundamental to such algebra to know how to construct a *composition of two permutations* and how to represent a permutation in what is called *disjoint cycle form.*

DEFINITION: A **permutation** of a set S is a bijection (a one-to-one, onto function) from S to itself.

In any kind of algebra, the calculation of the effect of applying various operations depends on the representation of the objects. For instance, the rule for calculating the product of two Roman numerals is different from the rule for calculating the product of base-10 numerals. Similarly, rules for the calculation of permutation operations depend on the representation. In this section, we introduce two ways to represent permutations and the corresponding ways to calculate the *composition* of permutations.

2-Line Representation of Permutations

DEFINITION: The **2-line representation of a permutation** π of a set S is a 2-line array that lists the objects of S in its top row. Below each object x is its image $\pi(x)$ under the permutation.

Example 0.5.1:   The permutation π of the set $\{1, 2, \ldots, 9\}$ such that

$$1 \mapsto 7 \quad 2 \mapsto 4 \quad 3 \mapsto 1 \quad 4 \mapsto 8$$
$$5 \mapsto 5 \quad 6 \mapsto 2 \quad 7 \mapsto 9 \quad 8 \mapsto 6 \quad 9 \mapsto 3$$

is represented by the 2-line array

$$\pi = \begin{pmatrix} 1 & 2 & 3 & 4 & 5 & 6 & 7 & 8 & 9 \\ 7 & 4 & 1 & 8 & 5 & 2 & 9 & 6 & 3 \end{pmatrix}$$

which is illustrated by Figure 0.5.1.

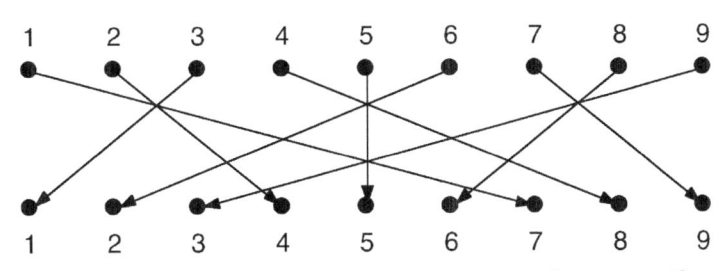

Figure 0.5.1   A permutation of the set $\{1, 2, \ldots, 9\}$.

One imagines that the nine numbers are initially in a row in ascending order. The permutation π moves whatever object is in position 1 to position 7, whatever object is in position 2 to position 4, and so on. Thus, the application of π changes their ordering to

$$3 \quad 6 \quad 9 \quad 2 \quad 5 \quad 8 \quad 1 \quad 4 \quad 7$$

DEFINITION: The **inverse of a permutation** π on a set S is the permutation π^{-1} that restores each object of S to its position before the application of π.

The 2-line representation of the inverse of a permutation can be obtained by transposing the rows, possibly sorting the columns according to the entry in the resulting first row.

Example 0.5.1, continued:

$$\pi^{-1} = \begin{pmatrix} 7 & 4 & 1 & 8 & 5 & 2 & 9 & 6 & 3 \\ 1 & 2 & 3 & 4 & 5 & 6 & 7 & 8 & 9 \end{pmatrix} \quad \text{or}$$

$$\pi^{-1} = \begin{pmatrix} 1 & 2 & 3 & 4 & 5 & 6 & 7 & 8 & 9 \\ 3 & 6 & 9 & 2 & 5 & 8 & 1 & 4 & 7 \end{pmatrix}$$

Composition of Permutations

DEFINITION: The **composition of permutations** π and τ is the permutation $\pi \circ \tau$ resulting from first applying π and then applying τ. Thus, $(\pi \circ \tau)(x) = \tau(\pi(x))$.

Obtaining the 2-line representation of the composition $\pi \circ \tau$ of two permutations is a 2-step process.

1. Rearrange the columns of the representation of τ (the permutation to be applied second) so that in each column, the top entry is the same as the bottom entry in the representation of π (the permutation to be applied first).

2. The top line of the 2-line array for the composition $\pi \circ \tau$ is the top line of the array for π. The bottom line for $\pi \circ \tau$ is the bottom line for the rearranged representation of τ.

Example 0.5.1, continued: Suppose that

$$\tau = \begin{pmatrix} 1 & 2 & 3 & 4 & 5 & 6 & 7 & 8 & 9 \\ 6 & 5 & 3 & 1 & 9 & 2 & 8 & 7 & 4 \end{pmatrix}$$

Transposing the columns of τ facilitates the computation

$$\pi = \begin{pmatrix} 1 & 2 & 3 & 4 & 5 & 6 & 7 & 8 & 9 \\ 7 & 4 & 1 & 8 & 5 & 2 & 9 & 6 & 3 \end{pmatrix}$$

$$\tau = \begin{pmatrix} 7 & 4 & 1 & 8 & 5 & 2 & 9 & 6 & 3 \\ 8 & 1 & 6 & 7 & 9 & 5 & 4 & 2 & 3 \end{pmatrix}$$

$$\pi \circ \tau = \begin{pmatrix} 1 & 2 & 3 & 4 & 5 & 6 & 7 & 8 & 9 \\ 8 & 1 & 6 & 7 & 9 & 5 & 4 & 2 & 3 \end{pmatrix}$$

For instance, since π maps whatever is in position 1 to position 7 and τ maps whatever is in position 7 to position 8, the composition $\pi \circ \tau$ maps whatever is in position 1 to position 8. This composition is illustrated in Figure 0.5.2.

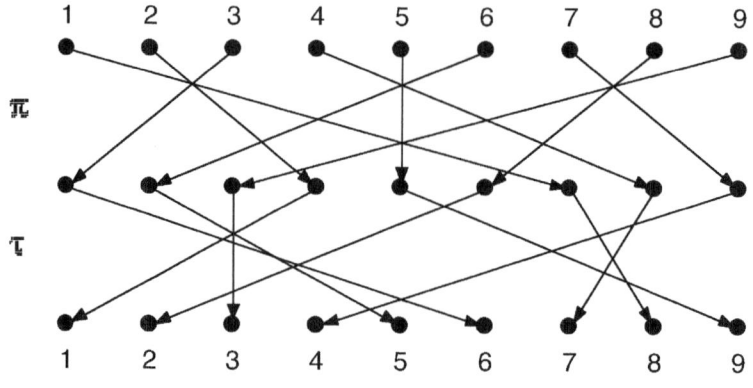

Figure 0.5.2   **A composition $\pi \circ \tau$ of permutations.**

Cyclic Permutations

A *cyclic permutation* is a permutation whose successive application would take each object of the permuted set successively through the positions of all the other objects.

DEFINITION: A permutation of the form

$$\begin{pmatrix} x & \pi(x) & \pi^2(x) & \cdots & \pi^{p-2}(x) & \pi^{p-1}(x) \\ \pi(x) & \pi^2(x) & \pi^3(3) & \cdots & \pi^{p-1}(x) & x \end{pmatrix}$$

is said to be **cyclic permutation** of **period** p.

NOTATION: A cyclic permutation is commonly represented in the **cyclic form**

$$\begin{pmatrix} x & \pi(x) & \pi^2(x) & \cdots & \pi^{p-2}(x) & \pi^{p-1}(x) \end{pmatrix}$$

Example 0.5.2:   The permutation

$$\begin{pmatrix} 1 & 2 & 3 & 4 & 5 & 6 & 7 \\ 2 & 3 & 4 & 5 & 6 & 7 & 1 \end{pmatrix} = \begin{pmatrix} 1 & 2 & 3 & 4 & 5 & 6 & 7 \end{pmatrix}$$

is cyclic of period 7. Its cyclic form is depicted by Figure 0.5.3 as a directed cycle.

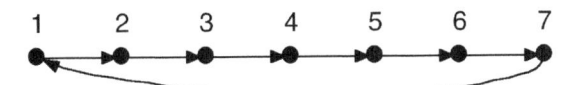

Figure 0.5.3   A cyclic permutation depicted as a directed cycle.

Example 0.5.3:   The permutation

$$\begin{pmatrix} 1 & 2 & 3 & 4 & 5 & 6 \\ 3 & 6 & 2 & 5 & 1 & 4 \end{pmatrix} = \begin{pmatrix} 1 & 3 & 2 & 6 & 4 & 5 \\ 3 & 2 & 6 & 4 & 5 & 1 \end{pmatrix} = \begin{pmatrix} 1 & 3 & 2 & 6 & 4 & 5 \end{pmatrix}$$

is cyclic of period 6. It is depicted as a directed cycle in Figure 0.5.4.

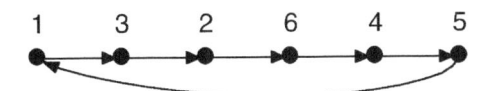

Figure 0.5.4   Another cyclic permutation.

Disjoint Cycle Representation

A fundamental way of understanding a permutation π of a finite set S is in terms of the cyclic permutations it induces on various subsets of S. Its structure is understood in terms of the lengths of these cycles of objects.

Proposition 0.5.1. *Let π be a permutation on a finite set S and let $x \in S$. Then the sequence*

$$x \quad \pi(x) \quad \pi^2(x) \quad \pi^3(x) \quad \ldots$$

eventually contains an entry $\pi^j(x)$ such that $\pi^j(x) = x$, and the sequence is periodic with period j.

Proof: Since the set S is finite, the sequence must eventually contain some entry $\pi^j(x)$ that matches a previous entry. Suppose that $\pi^i(x)$ is the previous entry such that

$$\pi^j(x) = \pi^i(x)$$

Then

$$
\begin{aligned}
\pi^{j-i}(x) &= \pi^{-i}(\pi^j(x)) \\
&= \pi^{-i}(\pi^i(x)) \\
&= \pi^0(x) = x
\end{aligned}
$$

Since $j > i \geq 0$, since $\pi^j(x)$ is the first duplicate of a previous entry, and since π^{j-i} duplicates the initial entry x, it follows that $j - i \geq j$, which implies that $i = 0$. Since $\pi^j(x) = x$, it follows that the subsequence

$$x \quad \pi(x) \quad \pi^2(x) \quad \pi^3(x) \quad \ldots \quad \pi^{j-1}(x)$$

is endlessly reiterated. $\diamond$

What now follows is a somewhat informal description of a method for representing an arbitrary permutation π on a finite set S as a composition of cyclic permutations.

Step 1: Choose an arbitrary element $x_1 \in S$. Let k_1 be the smallest integer such that $\pi^{k_1}(x_1) = x_1$. Let T_1 be the subset

$$T_1 = \left\{ x_1, \pi(x_1), \pi^2(x_1), \ldots, \pi^{k_1-1}(x_1) \right\}$$

Then the restriction $\pi|_{T_1}$ of the permutation π to the subset T_1 is the cyclic permutation

$$\pi|_{T_1} = \left(x_1 \quad \pi(x_1) \quad \pi^2(x_1) \quad \ldots \quad \pi^{k_1-1}(x_1) \right)$$

Example 0.5.1, continued: For the permutation

$$\pi = \begin{pmatrix} 1 & 2 & 3 & 4 & 5 & 6 & 7 & 8 & 9 \\ 7 & 4 & 1 & 8 & 5 & 2 & 9 & 6 & 3 \end{pmatrix}$$

consider the choice $x_1 = 1$. This leads to the subset

$$T_1 = \{1, 7, 9, 3\}$$

and to the restricted permutation

$$\pi|_{T_1} = \begin{pmatrix} 1 & 7 & 9 & 3 \end{pmatrix}$$

Step 2: In general, if $T_1 = S$, then π is cyclic on S, and $\pi = \pi|_{T_1}$. Otherwise, choose an arbitrary element

$$x_2 \in S - T_1$$

Let k_2 be the smallest integer such that $\pi^{k_2}(x_2) = x_2$. Let T_2 be the subset

$$T_2 = \{x_2,\ \pi(x_2),\ \pi^2(x_2),\ \ldots,\ \pi^{k_2-1}(x_2)\}$$

Then the restriction $\pi|_{T_2}$ of the permutation π to the subset T_2 is the cyclic permutation

$$\left(\, x_2 \quad \pi(x_2) \quad \pi^2(x_2) \quad \ldots \quad \pi^{k_2-1}(x_2)\,\right)$$

Example 0.5.1, continued: Choosing the second element $x_2 = 2$ for the permutation

$$\pi = \begin{pmatrix} 1 & 2 & 3 & 4 & 5 & 6 & 7 & 8 & 9 \\ 7 & 4 & 1 & 8 & 5 & 2 & 9 & 6 & 3 \end{pmatrix}$$

leads to the subset

$$T_2 = \{2,\ 4,\ 8,\ 6\}$$

and to the restricted permutation

$$\pi|_{T_2} = (2 \quad 4 \quad 8 \quad 6)$$

We observe that the subsets T_1 and T_2 are disjoint.

Proposition 0.5.2. *Let π be a permutation on a finite set S and let $x \in S$. Let*

$$T = \{\, \pi^i(x) \mid i \in \mathbb{N}\,\}$$

Let $y \in S - T$ and let

$$T' = \{\, \pi^j(y) \mid j \in \mathbb{N}\,\}$$

Then the subsets T and T' are disjoint.

Proof: If not, then there are nonnegative numbers i and j such that

$$\pi^i(x) = \pi^j(y) \tag{0.5.1}$$

Without loss of generality, assume that $j \leq i$. Then

$$\begin{aligned} \pi^{i-j}(x) &= \pi^{-j}(\pi^i(x)) \\ &= \pi^{-j}(\pi^j(y)) \qquad \text{by (0.5.1)} \\ &= y \end{aligned}$$

which contradicts the premise that $y \notin T$. $\diamond$

Step 3: Having selected the mutually disjoint subsets $T_1,\ T_2, \ldots,\ T_k$ in this manner, if

$$T_1 \cup T_2 \cup \cdots \cup T_k = S$$

then go to Step 4, since the decomposition of π is complete. Otherwise, choose $x_{k+1} \in S - (T_1 \cup T_2 \cup \cdots \cup T_k)$ and continue as in Step 2.

Example 0.5.1, continued: The only remaining element in the set $\{1, 2, \ldots, 9\}$, on which the permutation

$$\pi = \begin{pmatrix} 1 & 2 & 3 & 4 & 5 & 6 & 7 & 8 & 9 \\ 7 & 4 & 1 & 8 & 5 & 2 & 9 & 6 & 3 \end{pmatrix}$$

acts, is the element $x_3 = 5$, which leads to the subset

$$T_3 = \{5\}$$

and to the restricted permutation

$$\pi|_{T_3} = (5)$$

We observe that the subsets T_1, T_2, and T_3 form a partition of the set $[1:9]$.

Step 4: Arriving at this step occurs after the set S has been partitioned into subsets T_1, T_2, $\ldots$, T_k. Represent the permutation π in the form

$$\pi = \pi|_{T_1} \pi|_{T_2} \cdots \pi|_{T_k}$$

Example 0.5.1, continued: The net result of applying these steps to the permutation

$$\pi = \begin{pmatrix} 1 & 2 & 3 & 4 & 5 & 6 & 7 & 8 & 9 \\ 7 & 4 & 1 & 8 & 5 & 2 & 9 & 6 & 3 \end{pmatrix}$$

is the representation

$$\pi = (1 \ 7 \ 9 \ 3)(2 \ 4 \ 8 \ 6)(5)$$

DEFINITION: A **disjoint cycle representation** of a permutation π on a set S is as a composition of cyclic permutations on subsets of S that constitute a partition of S, one cyclic permutation for each subset in the partition.

The decomposition process described just above serves as a constructive proof of the following theorem.

Theorem 0.5.3. Let π be a permutation of a finite set S. Then π has a disjoint cycle representation. $\Diamond$

We conclude this subsection with an illustration that it is straightforward to compute the disjoint cycle representation of a composition of two permutations π and τ from the disjoint cycle representations of the factors π and τ.

Example 0.5.1, continued: The disjoint cycle forms of the permutations

$$\pi = \begin{pmatrix} 1 & 2 & 3 & 4 & 5 & 6 & 7 & 8 & 9 \\ 7 & 4 & 1 & 8 & 5 & 2 & 9 & 6 & 3 \end{pmatrix}$$

$$\tau = \begin{pmatrix} 7 & 4 & 1 & 8 & 5 & 2 & 9 & 6 & 3 \\ 8 & 1 & 6 & 7 & 9 & 5 & 4 & 2 & 3 \end{pmatrix} \quad \text{and}$$

$$\pi \circ \tau = \begin{pmatrix} 1 & 2 & 3 & 4 & 5 & 6 & 7 & 8 & 9 \\ 8 & 1 & 6 & 7 & 9 & 5 & 4 & 2 & 3 \end{pmatrix}$$

are

$$\pi = (1 \quad 7 \quad 9 \quad 3)(2 \quad 4 \quad 8 \quad 6)(5)$$
$$\tau = (1 \quad 6 \quad 2 \quad 5 \quad 9 \quad 4)(3)(7 \quad 8) \quad \text{and}$$
$$\pi \circ \tau = (1 \quad 8 \quad 2)(3 \quad 6 \quad 5 \quad 9)(4 \quad 7)$$

Starting with the disjoint cycle forms

$$\pi = (1 \quad 7 \quad 9 \quad 3)(2 \quad 4 \quad 8 \quad 6)(5)$$
$$\text{and} \quad \tau = (1 \quad 6 \quad 2 \quad 5 \quad 9 \quad 4)(3)(7 \quad 8)$$

the first cycle of $\pi \circ \tau$ is computed as follows:

$$1 \xrightarrow{\pi} 7 \xrightarrow{\tau} 8$$
$$8 \xrightarrow{\pi} 6 \xrightarrow{\tau} 2$$
$$2 \xrightarrow{\pi} 4 \xrightarrow{\tau} 1$$

That is, the first cycle of the disjoint cycle representation of $\pi \circ \tau$ may be written as

$$(1 \quad 8 \quad 2)$$

The computation then continues

$$3 \xrightarrow{\pi} 1 \xrightarrow{\tau} 6$$
$$6 \xrightarrow{\pi} 2 \xrightarrow{\tau} 5$$
$$5 \xrightarrow{\pi} 5 \xrightarrow{\tau} 9$$
$$9 \xrightarrow{\pi} 3 \xrightarrow{\tau} 3$$

which yields

$$(3 \quad 6 \quad 5 \quad 9)$$

as the second cycle of the permutation $\pi \circ \tau$. It concludes with

$$4 \xrightarrow{\pi} 8 \xrightarrow{\tau} 7$$
$$7 \xrightarrow{\pi} 9 \xrightarrow{\tau} 4$$

which yields as the third cycle

$$(4 \quad 7)$$

EXERCISES for Section 0.5

In Exercises 0.5.1 through 0.5.6, represent the indicated permutation in disjoint cycle form.

0.5.1^S $\begin{pmatrix} 1 & 2 & 3 & 4 & 5 \\ 3 & 5 & 2 & 4 & 1 \end{pmatrix}$ 0.5.2 $\begin{pmatrix} 1 & 2 & 3 & 4 & 5 & 6 & 7 \\ 4 & 7 & 2 & 5 & 6 & 1 & 3 \end{pmatrix}$

0.5.3 $\begin{pmatrix} 1 & 2 & 3 & 4 & 5 & 6 & 7 & 8 \\ 3 & 6 & 8 & 7 & 4 & 2 & 5 & 1 \end{pmatrix}$ 0.5.4 $\begin{pmatrix} 1 & 2 & 3 & 4 & 5 & 6 & 7 & 8 \\ 2 & 8 & 3 & 7 & 6 & 4 & 5 & 1 \end{pmatrix}$

0.5.5 $\begin{pmatrix} 1 & 2 & 3 & 4 & 5 & 6 & 7 & 8 \\ 3 & 8 & 6 & 7 & 4 & 5 & 2 & 1 \end{pmatrix}$ 0.5.6 $\begin{pmatrix} 1 & 2 & 3 & 4 & 5 & 6 & 7 & 8 \\ 2 & 3 & 8 & 7 & 6 & 5 & 4 & 1 \end{pmatrix}$

In Exercises 0.5.7 through 0.5.12, represent the inverse of the permutation of the designated previous exercise in 2-line form and in disjoint cycle form.

0.5.7^S Exercise 0.5.1.   0.5.10   Exercise 0.5.4.

0.5.8   Exercise 0.5.2.   0.5.11   Exercise 0.5.5.

0.5.9   Exercise 0.5.3.   0.5.12   Exercise 0.5.6.

In Exercises 0.5.13 through 0.5.18, represent the indicated composition of permutation in disjoint cycle form. In writing the disjoint cycle form of a permutation, sometimes the 1-cycles are omitted. For instance, $(1 \quad 2 \quad 5)$ means the same permutation as $(1 \quad 2 \quad 5)(3)(4)$.

0.5.13^S $(1 \quad 2 \quad 3) \circ (2 \quad 4 \quad 5)$ 0.5.14 $(1 \quad 2) \circ (2 \quad 4) \circ (3 \quad 4)$

0.5.15 $(1 \quad 2 \quad 5) \circ (1 \quad 6 \quad 3 \quad 4)$ 0.5.16 $(1 \quad 2 \quad 3 \quad 4 \quad 5) \circ (3 \quad 4 \quad 6)$

0.5.17 $(1 \quad 2 \quad 5 \quad 4) \circ (1 \quad 6 \quad 3)$ 0.5.18 $(1 \quad 3 \quad 4 \quad 5) \circ (2 \quad 3 \quad 4 \quad 6)$

0.5.19   List every permutation of $[1:5]$ that has three cycles in its disjoint cycle form.

0.5.20   List every permutation of $[1:6]$ that has four cycles in its disjoint cycle form.

DEFINITION: The **cycle structure of a permutation** π of a set of cardinality n is the monomial $t_1^{r_1} \ldots t_n^{r_n}$, such that t_j is a formal variable and r_j is the number of j-cycles in the disjoint cycle form of π. Thus, $1r_1 + 2r_2 + \cdots + nr_n = n$.

In Exercises 0.5.21 through 0.5.26, calculate the number of permutations of the given integer interval with the given cycle structure.

0.5.21^S $[1:7]$ of structure $t_1 t_2 t_4$ 0.5.22 $[1:9]$ of structure $t_2 t_3 t_4$

0.5.23 $[1:7]$ of structure $t_2^2 t_3$ 0.5.24 $[1:9]$ of structure $t_1 t_4^2$

0.5.25 $[1:7]$ of structure $t_1^3 t_2^2$ 0.5.26 $[1:9]$ of structure $t_1^3 t_2^3$

0.6   GRAPHS

One widely studied combinatorial structure is called a *graph*. Intuitively, a graph is a configuration comprising a discrete set of points in space and a discrete set of curves, each of which runs either between two points or from a point back to the same point. Formally, it is based on two abstract sets.

The beauty of various spatial models of graphs is one great attraction. Another is the capacity to serve as a practical model for applications, for instance, of network flows or of a linked database. Although this remarkably versatile structure was introduced by the Swiss mathematician Leonhard Euler (1707–1783), most of its theoretical development has occurred in relatively recent years. Chapters 7 and 8 provide a condensed survey of graph theory.

DEFINITION: A **graph** $G = (V, E)$ is a mathematical structure consisting of two finite sets V and E, called **vertices** and **edges**, respectively. Each edge has a set of one or two vertices associated to it, which are called its **endpoints**.

Example 0.6.1:   Figure 0.6.1 illustrates a graph.

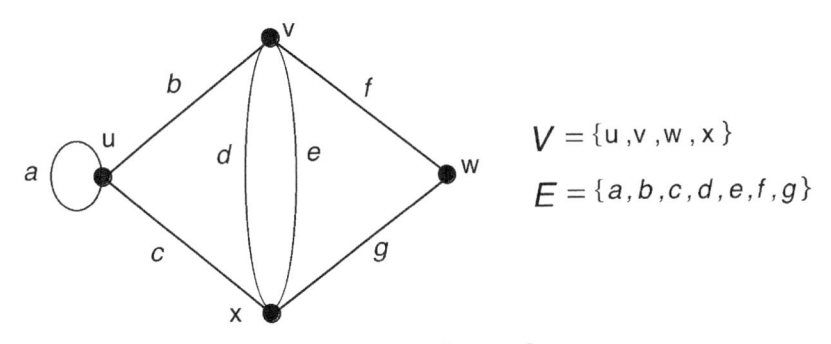

$$V = \{u, v, w, x\}$$
$$E = \{a, b, c, d, e, f, g\}$$

Figure 0.6.1   A graph.

TERMINOLOGY: An edge is said to **join** its endpoints. A vertex joined by an edge to a vertex v is said to be a **neighbor** of v. Two neighboring vertices are said to be **adjacent**.

TERMINOLOGY: In applications, the words **node** and **line** may be used for vertex and edge, respectively.

NOTATION: When G is not the only graph under consideration, the notations V_G and E_G (or $V(G)$ and $E(G)$) are used for the vertex- and edge-sets of G.

Example 0.6.1, continued: When choosing a vertex of this graph from which to send messages, vertices v and x would seem to be good choices, since from either of them, every other vertex is but an edge away, with no relay required. Numerous optimization problems arise when costs are assigned to the edges. For instance, one might want to know how to select the vertex from which the average cost of sending a message to the other vertices is the least.

TERMINOLOGY NOTE: The word *graph* is used here in an all-encompassing sense, as various attributes are tacked on. For instance, sometimes an edge is assigned a direction and/or a numerical weight. If all the edges are directed, then the graph may be called a *digraph*, and if all the edges are weighted, it may be called a *weighted graph*. However, under our philosophy of inclusivity, we may still refer to a graph with such optional attributes as a *graph*.

Simple Graphs and General Graphs

Graph theory is a source of excellent examples for combinatorial concepts. It is helpful to have some terminology in place at the outset.

DEFINITION: A **proper edge** is an edge that joins two distinct vertices. A **self-loop** is an edge that joins a single endpoint to itself.*

DEFINITION: A **multi-edge** is a collection of two or more edges having identical endpoints. The **multiplicity of a multi-edge** is the number of edges within the multi-edge.

DEFINITION: A **simple graph** is a graph with no self-loops or multi-edges. A **general graph** may have self-loops and/or multi-edges. (Thus, the graph in Figure 0.6.1 is a general graph.)

NOTATION: In a simple graph, an edge joining vertices u and v may be denoted uv, since only one such edge is possible.

Null and Trivial Graphs

DEFINITION: A **null graph** is a graph whose vertex- and edge-sets are empty.

DEFINITION: A **trivial graph** is a graph consisting of one vertex and no edges.

Degree of a Vertex

DEFINITION: The **degree** (or **valence**) of a vertex v in a graph G, denoted $deg(v)$, is the number of proper edges incident on v plus twice the number of self-loops.

TERMINOLOGY: A vertex of degree d is also called a d-**valent vertex**.

---

* We use the term "self-loop" instead of the more commonly used term "loop", because loop means something else in many applications.

Example 0.6.1, continued: The caption of Figure 0.6.2 lists the degrees of the graph from Figure 0.6.1.

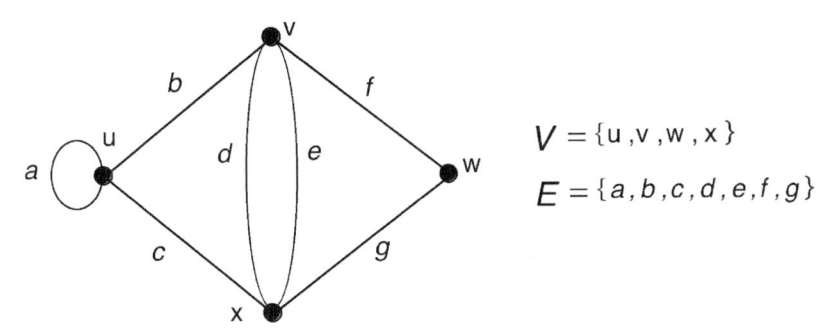

$$V = \{u, v, w, x\}$$

$$E = \{a, b, c, d, e, f, g\}$$

Figure 0.6.2   deg(u) = deg(v) = deg(x) = 4, and deg(w) = 2.

The following result of Euler establishes a fundamental relationship between the vertices and edges of a graph.

Theorem 0.6.1 [*Euler's Degree-Sum Theorem*]. *The sum of the degrees of the vertices of a graph is twice the number of edges.*

Proof: Each edge contributes two to the degree sum.   ◇

Example 0.6.1, continued: The graph of Figure 0.6.2 has 7 edges. The sum of the degrees is 14.

Corollary 0.6.2. *In a graph, the number of vertices having odd degree is even.*

Proof: Consider separately, the sum of the degrees that are odd and the sum of those that are even. The combined sum is even by Theorem 0.6.1, and since the sum of the even degrees is even, the sum of the odd degrees must also be even. Hence, there must be an even number of vertices of odd degree.   ◇

DEFINITION: The **degree sequence** of a graph is a list of the degrees of its vertices, usually given in non-increasing order.

Example 0.6.1, continued: The degree sequence of the graph of Figure 0.6.2 is

$$4 \quad 4 \quad 4 \quad 2$$

Theorem 0.6.3. *Let G be a simple n-vertex graph with $n \geq 2$. Then there are two vertices with the same degree.*

Proof: If the n vertices all had different degrees, then, by the Pigeonhole Principle, for each of the possible values $0, \ldots, n-1$, there would be a corresponding vertex. However, if some vertex has degree 0, then each other vertex could have degree at most $n-2$, precluding the existence of a vertex of degree $n-1$.   ◇

Example 0.6.2:   Suppose that on some floor of a college dormitory, each student lists the names of all the other students on that floor with whom he or she has ever shared a pizza, as represented by Figure 0.6.3. Four of the students in this sociological network — Alisa, David, Jessica, and Risa — have shared pizza with an odd number of other students in the network, in conformance with Corollary 0.6.2. We observe that Herbie and Katie have each shared pizza with four other students, which illustrates Theorem 0.6.3. There are also several other such pairs.

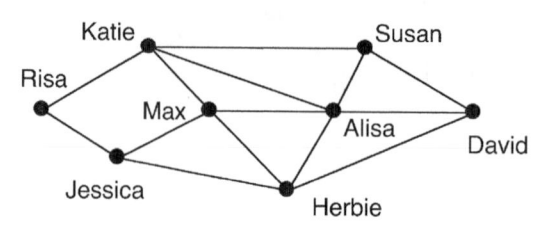

Figure 0.6.3   A sociological network.

Beyond the whimsy of Example 0.6.2, sociological networks are a matter of serious interest. For instance, so-called family trees are used in genealogy.

Complete Graphs

There are standard names for various special circumstances that arise frequently in graph-theoretic modeling. Sometimes every node is linked to every other node.

DEFINITION: A **complete graph** is a simple graph such that every pair of vertices is joined by an edge. The complete graph on n vertices is denoted K_n.

Example 0.6.3:   Complete graphs on one, two, three, four, and five vertices are shown in Figure 0.6.4.

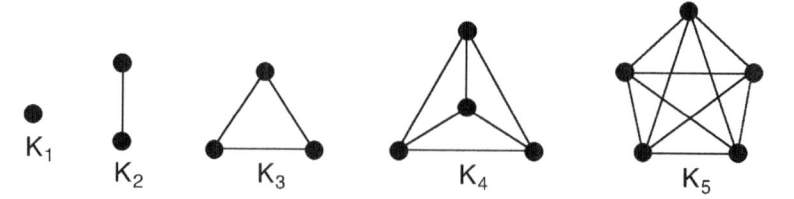

Figure 0.6.4   The first five complete graphs.

Bipartite Graphs

DEFINITION: A **bipartite graph** G is a graph whose vertex-set V can be partitioned into two subsets U and W, such that each edge of G has one endpoint in U and one endpoint in W. The pair U, W is called a **(vertex) bipartition** of G, and the sets U and W are called the **bipartition subsets** or (sometimes) the **partite sets**.

Example 0.6.4:   Two bipartite graphs are shown in Figure 0.6.5. The bipartition subsets are indicated by the solid and hollow vertices.

Figure 0.6.5   Two bipartite graphs.

Example 0.6.5:   Suppose that U is set of tasks needed for the completion of a project, that W is the set of available workers, and that there is an edge joining each worker to each task within that worker's skill set. Such a bipartite graph is quite useful in deciding how to allocate the tasks to workers.

DEFINITION: A **complete bipartite graph** is a simple bipartite graph such that every vertex in one partite set is joined to every vertex in the other partite set. Any complete bipartite graph that has m vertices in one partite set and n vertices in the other is denoted $K_{m,n}$.*

Example 0.6.6:   The complete bipartite graph $K_{3,4}$ is shown in Figure 0.6.6.

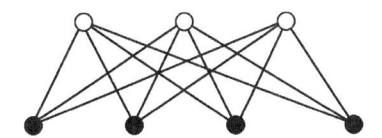

Figure 0.6.6   The complete bipartite graph $K_{3,4}$.

Representations of Graphs

It is conceptually helpful to see a small graph represented by a labeled drawing. However, for computational purposes, it is important to have a purely combinatorial specification. Various kinds of combinatorial specification have their individual merits. We briefly consider three kinds.

DEFINITION: A specification of an n-vertex, m-edge graph G by an **incidence table** has three parts:

- a list of the n vertices of G.
- a list of the m edges of G.
- a $2 \times m$ array whose columns are labeled by the edges of G, such that the endpoints of each edge appear in the column for that edge.

---

* The sense in which $K_{m,n}$ is regarded as a unique object is described in §7.4.

Example 0.6.7:   Figure 0.6.7 shows a graph G and its incidence table specification.

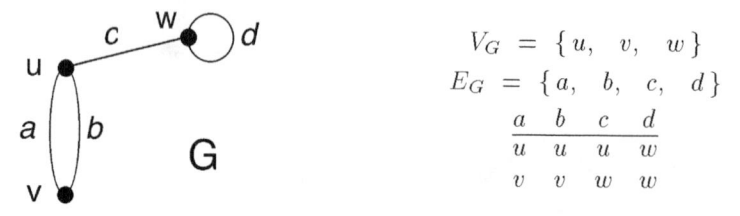

$$V_G = \{u, \ v, \ w\}$$
$$E_G = \{a, \ b, \ c, \ d\}$$

a	b	c	d
u	u	u	w
v	v	w	w

Figure 0.6.7   A graph and its incidence table specification.

COMPUTATIONAL NOTE: Specification by incidence table is an efficient representation for any kind of graph, not just for simple graphs. The space required is proportional to the number of vertices and edges. Some variations include a second table, with each row labeled by a vertex, such that the entries in that row are the edges incident on that vertex. Although this seems like redundant information, the small sacrifice of space facilitates a net improvement in algorithmic efficiency.

DEFINITION: An **incidence matrix** for an n-vertex, m-edge graph G is an $n \times m$ array I_G, whose rows and columns are labeled, respectively, by the vertices and edges of G, such that

$$I_G[v, e] = \begin{cases} 0 & \text{if } v \text{ is not an endpoint of } e \\ 1 & \text{if } v \text{ is an endpoint of } e \text{ and } e \text{ is proper} \\ 2 & \text{if } v \text{ is an endpoint of } e \text{ and } e \text{ is a self-loop} \end{cases}$$

Example 0.6.7, continued: The incidence matrix for the graph G of Figure 0.6.7 is

$$I_G = \begin{array}{c|cccc} & a. & b. & c. & d. \\ \hline u. & 1 & 1 & 1 & 0 \\ v. & 1 & 1 & 0 & 0 \\ w. & 0 & 0 & 1 & 2 \end{array}$$

We observe in this example and in general that each row-sum equals the degree of the corresponding vertex and that every column-sum is 2. Clearly, the sum of the row-sums of a matrix equals the sum of the column-sums. This provides an alternative proof of Euler's Degree-Sum Theorem (Theorem 0.6.1).

COMPUTATIONAL NOTE: The space required for an incidence matrix is proportional to the product of the numbers of vertices and edges. Moreover, the time required to retrieve the endpoints of an edge is proportional to the number of vertices, whereas it is a small constant for an incidence table specification.

DEFINITION: An **adjacency matrix** for an n-vertex simple graph G is an $n \times n$ array A_G whose rows and columns are labeled by the vertices of G, such that

$$A_G[u, v] = \begin{cases} 1 & \text{if } u \text{ and } v \text{ are adjacent} \\ 0 & \text{otherwise} \end{cases}$$

Example 0.6.8:   Figure 0.6.8 shows a simple graph G and its adjacency matrix specification.

$$A_G = \begin{array}{c|cccc} & w. & x. & y. & z. \\ \hline w. & 0 & 1 & 0 & 0 \\ x. & 1 & 0 & 1 & 1 \\ y. & 0 & 1 & 0 & 1 \\ z. & 0 & 1 & 1 & 0 \end{array}$$

Figure 0.6.8   A simple graph and its adjacency matrix.

Remark: The adjacency matrix is symmetric.

Remark: Spectral graph theory is concerned with calculation of the eigenvalues of adjacency matrices.

COMPUTATIONAL NOTE: The space required for an adjacency matrix is proportional to the square of the number of vertices.

Preview of Walks, Paths, and Distance

Graphs are commonly used to represent networks of various kinds, including networks of roads, networks of computers, and networks of people. The notion of accessibility is modeled by *walks* in graphs.

DEFINITION: A **walk in a graph** from vertex v_0 to vertex v_n is an alternating sequence

$$W = \langle v_0, e_1, v_1, e_2, \ldots, e_n, v_n \rangle$$

of vertices and edges, such that edge e_j joins vertices v_{j-1} and v_j, for $j = 1, \ldots, n$. It is a **closed walk** if it begins and ends at the same vertex and an **open walk** if it ends at a different vertex from the one at which it begins.

DEFINITION: A **path** is a walk that has no repeated vertices (or edges), except that the last vertex may possibly be the same as the first. If so, it is a **closed path**, and if not, it is an **open path**.

DEFINITION: The **length of a walk** is its number of edge-steps. (If an edge of a walk is repeated, it is counted each time it occurs. However, it follows that the length of a path is its number of edges.)

DEFINITION: The **distance between two vertices** u and v is the minimum length taken over all paths between u and v, or ∞ if there are no such paths.

Example 0.6.9:   The legendary mathematician Paul Erdős wrote about 1500 papers, and he had 509 coauthors in all. Of course, many of them had various other collaborators. The *Erdős coauthorship graph* has mathematicians as its vertices, with an edge joining two vertices if the mathematicians represented ever wrote a paper together. A mathematician's **Erdős number** is his or her distance from Paul Erdős in this graph. Erdős himself is the only person at distance 0. The 509 coauthors have an Erdős number of 1.* The concept of an Erdős number was first published in 1969 by Caspar Goffman [Goff1969].

EXERCISES for Section 0.6

In each of the Exercises 0.6.1 through 0.6.8, compare the degree sequence of the given graph with twice its number of edges.

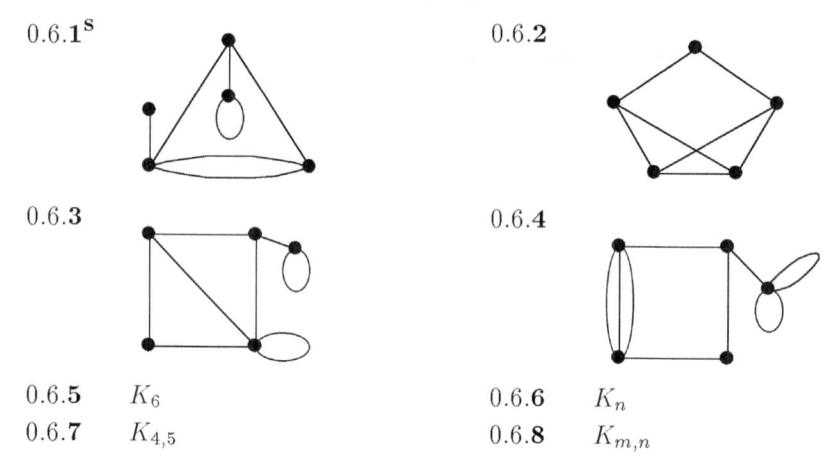

0.6.1^S

0.6.2

0.6.3

0.6.4

0.6.**5**	K_6		0.6.**6**	K_n
0.6.**7**	$K_{4,5}$		0.6.**8**	$K_{m,n}$

In each of the Exercises 0.6.9 through 0.6.12, draw a simple graph of the given degree sequence.

0.6.**9**^{S}	4 3 3 2 2	0.6.**10**	4 3 2 2 1
0.6.**11**	5 3 3 2 2 1	0.6.**12**	5 4 4 2 2 1

In each of the Exercises 0.6.13 through 0.6.16, draw two different general graphs of the given degree sequence.

0.6.**13**^{S}	4 4 3 2 1	0.6.**14**	7 4 3
0.6.**15**	6 4 3 1 0	0.6.**16**	6 5 5 4 4 1 1

Exercises 0.6.17 through 0.6.19 refer to the pizza network of Figure 0.6.3.

0.6.**17**   What is the distance between Risa and David?

0.6.**18**   What vertex or vertices have the minimum worst-case (i.e., maximum) distance to another vertex?

0.6.**19**   What vertex or vertices have the minimum average distance to the other vertices?

---

   *   The author of this textbook has an Erdős number of 2, since he wrote a paper [GrHa1980] with Frank Harary, who was a coauthor of Erdős.

PREVIEW OF Chapter 7:

- A *cycle* is a closed path of length at least 1.

- A *cycle subgraph* in a graph G is the image of a cycle in G, that is, the graph defined by the set of vertices and the set of edges that occur in that cycle.

- A *tree* is a graph that has no cycle subgraphs.

In each of the Exercises 0.6.20 through 0.6.23, draw a tree with the given degree sequence.

0.6.20^S 4 1 1 1 1   0.6.**21**   4 2 1 1 1 1

0.6.**22**   3 3 1 1 1 1   0.6.**23**   3 3 2 1 1 1 1

0.7 **NUMBER-THEORETIC OPERATIONS**

Number theory is a very large area of mathematics with connections to many other areas. It is not taxonomically classified as combinatorics. We include some number theory in this text, especially in Chapter 6, partly because of its intimate connection to the design of fast algorithms and also because we need it to help with counting and with studying graphs and other combinatorial objects. Although we presently defer nearly all details of number-theoretic methods to later chapters, we wish to make the point early that we will use whatever kind of mathematics is helpful in our problem-solving efforts.

One number-theoretic operation that occurs frequently in combinatorics is the *greatest common divisor.* Although our textbook examples are focused on small enough problems of this type to do the calculation by hand, consider trying to calculate the greatest common divisor of larger numbers, such as

$$32582657 \quad \text{and} \quad 24036583$$

DEFINITION: The **greatest common divisor** of two integers m and n, not both zero, mnemonically denoted $\gcd(m, n)$, is the largest integer that divides both m and n.

DEFINITION: The **least common multiple** of two integers m and n, mnemonically denoted $\operatorname{lcm}(m, n)$, is the smallest non-negative integer that is a non-zero multiple of both m and n.

When the prime factors are already known or easily calculated, it is quite easy to calculate a greatest common divisor by a method commonly taught in middle schools. It involves factoring the two numbers into products of primes. Although this might seem easy for small numbers, the factoring of large numbers may require considerable effort. A method called the *Euclidean algorithm*, described in Chapter 6, avoids the need to factor, and it produces the answer in time proportional to the number of digits of the larger number.

Another operation we use in trying to count or to construct all the graphs of a given kind involves listing all the ways to decompose an integer n into an iterated sum of positive integers. Such a sum is called a *partition of the integer n*.

Example 0.7.1:   The number 8 has five partitions into exactly four summands, namely

$$5 + 1 + 1 + 1 \quad 4 + 2 + 1 + 1 \quad 3 + 3 + 1 + 1 \quad 3 + 2 + 2 + 1 \quad 2 + 2 + 2 + 2$$

0.8   COMBINATORIAL DESIGNS

The final type of discrete structure presented in this book, in Chapter 10, is called a *combinatorial design*.

DEFINITION: A **combinatorial design** $\mathcal{B}$ has a non-empty domain of objects

$$X = \{ x_1, x_2, \ldots, x_v \}$$

and a non-empty collection of subsets of objects from X.

$$B = \{B_1, B_2, \ldots, B_b\}$$

For some kinds of designs, these subsets may be called *blocks*.

The art of constructing combinatorial designs is in meeting various additional requirements on the subsets B_j. In a *regular block design* the subsets B_j all have the same cardinality k, called the *blocksize*. Moreover, each object x_i occurs in the same number r of blocks, which is called the *replication number*. An example illustrates a possible application of such a design.

Example 0.8.1:   Consider how one might design a round-robin playoff[*] for 13 contestants in a competitive game for 4 players that ranks the players from 1st to 4th in each round. Such an event might plausibly have 13 rounds in which each of the players, designated as

$$0 \quad 1 \quad 2 \quad 3 \quad 4 \quad 5 \quad 6 \quad 7 \quad 8 \quad 9 \quad A \quad B \quad C$$

plays four rounds and meets each other player exactly once, as follows:

---

[*]   In a round-robin playoff, each contestant plays each other contestant.

Round	Players
1	0146
2	1257
3	2368
4	3479
5	$458A$
6	$569B$
7	$67AC$
8	$78B0$
9	$89C1$
10	$9A02$
11	$AB13$
12	$BC24$
13	$C035$

Such a playoff might be represented by the illustration of Figure 0.8.1. Twelve of the groupings of four players are represented by a curve that goes through the corresponding four points. (The thirteenth grouping is $67AC$.) Only four of these groupings are actually represented by straight lines in the drawing.

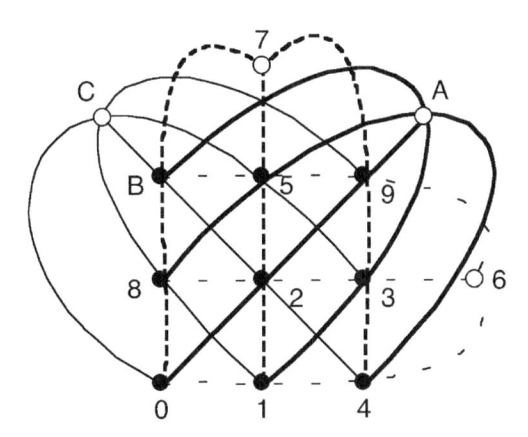

Figure 0.8.1   Geometric representation of a block design.

A *balanced block design* is a regular block design in which each pair of points occurs in the same number of lines. It is called *incomplete* if the blocksize is less than the number of points in the domain.

Example 0.8.1, continued: The playoff described here is a balanced incomplete block design, if each grouping of four players is regarded as a line.

Interestingly, the theory of balanced incomplete block designs (*BIBD's*, for short) originated largely in the design of scientific experiments in agriculture. Each block represented a different kind of treatment of the varieties of crops within it. Applying all the treatments to all the varieties would have made the experiment infeasibly large, which was the motive for constructing incomplete designs.

In a kind of combinatorial design called a *finite geometry*, the subsets of objects are called *lines*. There are numerous kinds of finite geometry. A standard general requirement is that each pair of points lies on at most one line.

Example 0.8.1, continued: As it happens, the balanced block design of Figure 0.8.1 is also a finite geometry.

GLOSSARY

adjacent vertices: two vertices joined by an edge.

balanced block design: a design in which all blocks are the same size, every element is in the same number of blocks, and every pair of elements is in the same number of blocks.

BIBD: abbreviation for *balanced incomplete block design*.

bipartite graph: a graph whose vertex-set can be partitioned into two subsets (called *partite sets*) such that every edge has one endpoint in one part and one endpoint in the other part.

block design: a combinatorial configuration with a domain X and a set B of subsets of X that are called blocks.

blocksize: the cardinality of a block of a combinatorial design.

ceiling of a real number x: the smallest integer that is not less than x; the result of "rounding up" to the next integer; denoted $\lceil x \rceil$.

calculus of finite sums: a method for evaluating finite sums, analogous to integral calculus.

cardinality of a multiset: the sum of the multiplicities of the elements in the domain.

cells of a partition of a set S: the subsets of S into which S is subdivided.

closed formula for a sequence: an algebraic formula that can produce the value of any member of the sequence.

combination: see *unordered selection*.

combinatorial design: any mathematical structure involving a primary domain and a secondary domain of designated subsets of the primary domain, or, equivalently, two domains and an *incidence function* from their cartesian product to $\mathbb{Z}_2$.

combinatorics: a collection of branches of mathematics that deal primarily with discrete sets, in contrast to continuous mathematics, the branches that deal primarily with subsets of Euclidean space.

complete bipartite graph: a simple bipartite graph such that each pair of vertices in different sides of the partition is joined by an edge.

complete graph: a simple graph such that every pair of vertices is joined by an edge.

complete block design: a block design in which every block is the entire domain.

counting: an informal reference to any kind of combinatorial enumeration.

degree of a vertex: the number of proper edges incident on that vertex plus twice the number of self-loops.

degree sequence: a list of the degrees of all the vertices in ascending order.

discrete set: a finite or countably infinite set.

edges of a graph: one of two constituent sets of the graph.

endpoints of an edge: the one or two vertices that are associated with that edge.

Euclidean algorithm: an algorithm for calculating the greatest common divisor; see Chapter 6.

Euler's degree-sum theorem: the theorem that the sum of the degrees of a graph equals twice the number of edges.

falling power $x^{\underline{n}}$ of a real number: the product $x(x-1)(x-2)\cdots(x-n+1)$.

Fibonacci sequence: the sequence $0, 1, 1, 2, 3, 5, 8, 13, \ldots$, in which each number is the sum of its two immediate predecessors.

finite calculus: a calculus of discrete differences and sums, analogous to the infinitessimal calculus for continuous real functions.

finite geometry: a *combinatorial design* in which two points of the primary domain are in at most one designated subset in the secondary domain.

floor of a real number x: the largest integer that is not greater than x; the result of "rounding down" to the next integer; denoted $\lfloor x \rfloor$.

general graph: a graph that may have self-loops and/or multi-edges.

generating function for a sequence of elements g_j: a closed form for the infinite polynomial $g_0 + g_1 z + g_2 z^2 + g_3 z^3 + \cdots$, or sometimes, that polynomial itself.

graph $G = (V, E)$: a mathematical structure consisting of two sets, V and E. The elements of V are called **vertices**, and the elements of E are called **edges**. Each edge has a set of one or two vertices associated to it, which are called its **endpoints**.

greatest common divisor $\gcd(m, n)$: for integers m and n, not both zero, the greatest common divisor is the largest positive integer that divides both of them.

incidence function: a function associated with a combinatorial structure having more than one domain; its role is to indicate for any pair of objects, one from each domain, whether each is incident on the other; for instance, a vertex of a graph and an edge of which it is an endpoint are mutually incident on each other.

least common multiple $\operatorname{lcm}(m, n)$: for integers m and n, the least common multiple is the smallest non-negative integer that is a non-zero multiple of both of them.

multi-edge: a collection (at least two) of proper edges with the same two endpoints, or of self-loops with the same endpoint.

multiset: a pair (S, m) in which S is a set and $m : S \to \mathbb{Z}^+$ is a function that assigns to each element $s \in S$ a number $m(s)$ called its multiplicity. Informally, one thinks of there being $m(s)$ copies of the element s in the multiset.

neighbor of a vertex v: any vertex that is adjacent to v.

null graph: a graph with no vertices and no edges.

ordered selection from a set S: an ordered subset of S.

partition of an integer S: a representation of a positive integer as a sum of other positive integers.

partition of a set S: a collection of mutually disjoint subsets of S whose union is S.

permutation from a set S: see *ordered selection*.

permutation of a set S: a one-to-one onto function from S to itself.

Pigeonhole Principle: a frequently applied method of counting.

proper edge: an edge with two endpoints.

regular block design: a block design in which all the blocks have the same size, and in which each element occurs in the same number of blocks.

replication number of a *BIBD*:   the number of blocks in which each element of its domain is contained.

Rule of Product: the counting rule that the size of a cartesian product of two sets is the product of the sizes of the sets.

Rule of Quotient: the counting rule that if all the cells are of the same size, then the number of cells in a partition of a set S is the quotient of the size of S by the size of the cells.

Rule of Sum: the counting rule that the size of a disjoint union of two sets is the sum of the sizes of the sets.

self-loop: an edge of a graph with only one endpoint.

simple graph: a graph with no self-loops or multi-edges.

Stirling numbers: numbers used in conversions between ordinary powers and falling powers, also for counting partitions and permutations; see Chapters 1 and 5.

trivial graph: a graph with one vertex and no edges.

unordered selection from a set S: a subset of S.

valence of a vertex of a graph: synonym for degree.

vertices: one of two constituent sets of the graph.

Chapter 1

Sequences

In combinatorial analysis, counts of selections, orderings, arrangements, or configurations for differently sized versions of a given problem are commonly given as a sequence. Alternatively, a sequence may correspond to a list of measurements of the behavior of some process over time. Even though such sequences may contain infinitely many different numerical values, there is often a finite way to represent them collectively. In particular, a *closed formula* to calculate any number in the sequence from its location in the sequence is especially convenient. A *recursion rule* for inferring later values in the sequence from earlier values is another form of finite representation. This first chapter provides acquaintance or reacquaintance with a variety of standard sequences and with these basic types of finite representations of sequences. It introduces some initial methods for manipulating such representations so that information about the properties of the sequence can be extracted efficiently.

1.1   SEQUENCES AS LISTS

In this section, we consider some common kinds of sequences and some of their attributes.

DEFINITION: A **sequence** in a set S is a list of elements of S

$$x_0 \quad x_1 \quad x_2 \quad \ldots$$

indexed by the non-negative integers, or sometimes by some other countable set. Collectively, the sequence is denoted $\langle x_n \rangle$, with angle brackets, or by variations on this basic notation.

TERMINOLOGY: A member x_j of a sequence $\langle x_n \rangle$ is also called an **entry** or a **term**. The set in which the values x_j are taken may be called the **range of the sequence**.

NOTATION: Some of the most standard sets of numbers that serve as ranges for sequences are denoted here in *blackboard bold* typeface style:

$$\mathbb{Z} = \{ \ldots, \ -2, \ -1, \ 0, \ 1, \ \ldots \} \quad \text{integers}$$
$$\mathbb{Z}^+ = \{ 1, \ 2, \ 3, \ \ldots \} \quad \text{positive integers}$$
$$\mathbb{N} = \{ 0, \ 1, \ 2, \ \ldots \} \quad \text{natural numbers}$$
$$\mathbb{R} = \text{real numbers}$$
$$\mathbb{Q} = \text{rational numbers}$$
$$\mathbb{C} = \text{complex numbers}$$

DEFINITION: An algebraic expression in the argument n for the value of the general element x_n of a sequence $\langle x_n \rangle$ is called a **closed formula** for the (elements of the) sequence.

Example 1.1.1:   The closed formula $x_n = n^3 - 5n$ specifies the sequence

$$\langle x_n \rangle \ : \ 0 \quad -4 \quad -2 \quad 12 \quad 44 \quad 100 \quad 186 \quad \ldots$$

Example 1.1.2:   The closed formula $y_n = 2^{n+2} - n^3$ specifies the sequence

$$\langle y_n \rangle \ : \ 4 \quad 7 \quad 8 \quad 5 \quad 0 \quad 3 \quad 40 \quad \ldots$$

Fast-Growing Sequences

One frequently cited attribute of a sequence is its *rate of growth*, which is understood in relation to the standard indexing sequence, i.e., the natural numbers. We refer to a sequence as a *polynomial sequence* if it is specifiable by a polynomial on the index set $\mathbb{N}$, as in Example 1.1.1, or as an *exponential sequence* if it is specifiable by an exponential. Polynomial and exponential sequences are both thought to grow rather rapidly. Precise criteria for comparing growth rates are provided later in this chapter.

Example 1.1.3:   The polynomial sequence $\langle x_n = n^2 \rangle$

n	0	1	2	3	4	5	$\cdots$
n^2	0	1	4	9	16	25	$\cdots$

grows more rapidly than the sequence of integers. Any polynomial sequence for a polynomial of degree greater than 1 grows more rapidly than the natural numbers.

Example 1.1.4:   The exponential sequence $\langle x_n = 3^n \rangle$

n	0	1	2	3	4	5	$\cdots$
3^n	1	3	9	27	81	243	$\cdots$

also grows more rapidly than the sequence of integers. Once a precise notion of comparative rate of growth is in hand in §1.4, it will be provable that any exponential sequence $\langle x_n = b^n \rangle$ with $b > 1$ grows more rapidly than any polynomial sequence. Of course, if $0 < b < 1$, then the sequence $\langle x_n = b^n \rangle$ decreases. For instance,

n	0	1	2	3	4	5	$\cdots$
$(1/3)^n$	1	$\frac{1}{3}$	$\frac{1}{9}$	$\frac{1}{27}$	$\frac{1}{81}$	$\frac{1}{243}$	$\cdots$

Example 1.1.5:   A sequence that grows even more rapidly than an exponential sequence is the factorial sequence

n	0	1	2	3	4	5	$\cdots$
$n!$	1	1	2	6	24	120	$\cdots$

This is another comparison whose meaning awaits explanation.

Slow-Growing Sequences

Various other increasing sequences grow slowly, relative to the integers. The first example here involves a fractional exponent. The second and third involve logarithms and *harmonic numbers*.

Example 1.1.6:   A sequence $\langle x_n = n^r \rangle$ grows more slowly than the natural numbers if $0 < r < 1$. Consider, for instance,

n	0	1	2	3	4	5	$\cdots$
$n^{1/2}$	0	1	$\sqrt{2}$	$\sqrt{3}$	2	$\sqrt{5}$	$\cdots$

NOTATION: The *natural logarithm* of a positive number x is denoted $\ln x$. The *logarithm to the base* 2 is denoted $\lg x$.

Example 1.1.7:   The sequence

n	1	2	3	4	5	$\cdots$
$\lg n$	0	1	$\lg 3$	2	$\lg 5$	$\cdots$

grows even more slowly than the sequence $\langle x_n = n^r \rangle$, for $r > 0$. (See Exercises.)

DEFINITION: The **harmonic number** H_n is defined as the sum

$$\sum_{k=1}^{n} \frac{1}{k} = \frac{1}{1} + \frac{1}{2} + \cdots + \frac{1}{n}$$

with $H_0 = 0$ for the empty sum.

Example 1.1.8:   The harmonic sequence

n	0	1	2	3	4	5	$\cdots$
H_n	0	1	$\frac{3}{2}$	$\frac{11}{6}$	$\frac{25}{12}$	$\frac{137}{60}$	$\cdots$

is closely related to the natural logarithm $\ln n$, as explained in §3.1.

Example 1.1.9:   The values in a sequence need not be numbers.

PREVIEW OF §8.7:   The surface S_g is the surface with g handles in the following sequence.

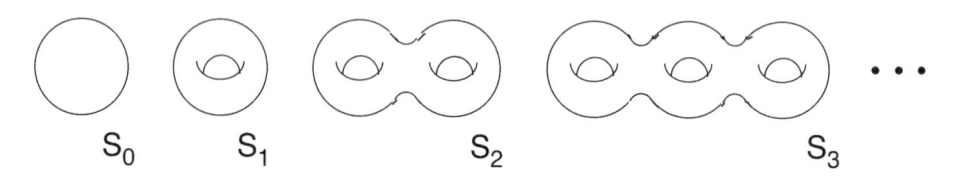

Bounded Sequences

Although all sequences considered previously in this section become arbitrarily large as the sequence continues, some sequences do not.

DEFINITION: A **bounded sequence** $\langle x_n \rangle$ is a sequence (typically of real numbers or integers) for which there is a number B (called a **bound**), such that

$$|x_n| \leq B \qquad \text{for all } n$$

Observe that the sequence is bounded in *absolute* value.

Example 1.1.10:   The real sequence

$$\left\langle x_n = 1 - \frac{1}{n+1} \right\rangle$$

is bounded. It is always non-negative, and its value never exceeds 1.

Periodic Sequences

Some sequences are repetitive. That is, the same subsequence recurs *ad infinitum*.

DEFINITION: A **periodic sequence** $\langle x_n \rangle$ is a sequence for which there is a positive integer P, such that

$$x_{j+P} \;=\; x_j \qquad \text{for all } j \in \mathbb{N}$$

The smallest such integer is called the **period of the sequence**.

Example 1.1.11:   An alternating sequence of 0's and 1's

$$0 \quad 1 \quad 0 \quad 1 \quad 0 \quad 1 \quad \cdots$$

is periodic with period 2.

DEFINITION: The **remainder function** on a pair of integers $n \in \mathbb{N}$ and $d \in \mathbb{Z}^+$ is defined as

$$n \bmod d \;=\; n - d \left\lfloor \frac{n}{d} \right\rfloor$$

It is also called the **mod function**. The arguments n and d are called the *dividend* and the *divisor*, respectively.

Example 1.1.12:   The sequence $n \bmod 3$

n	0	1	2	3	4	5	$\cdots$
$n \bmod 3$	0	1	2	0	1	2	$\cdots$

is periodic with period 3. More generally, for any fixed divisor m, the sequence

$$\langle x_n \;=\; n \bmod m \rangle$$

is periodic with period m.

Remark: Clearly, any periodic sequence is bounded, that is, by the largest number in the repeating subsequence. For instance, the sequence in Example 1.1.12 is bounded by 2.

Generalizations

At times, sequences employ sets other than the natural numbers as their subscript sets, and they sometimes have multiple subscripts. For instance, sometimes a sequence is of interest only over a finite set a, $a + 1$, $\ldots$, b of consecutive integers. At the other extreme, there may also be negative, or there may be multiple subscripts.

Example 1.1.13:   The closed formula $x(n) = 3^n$ may also be regarded as a specification of the extended sequence $\langle 3^n \mid n \in \mathbb{Z} \rangle$:

n	$\cdots$	-2	-1	0	1	2	3	$\cdots$
3^n	$\cdots$	$\frac{1}{9}$	$\frac{1}{3}$	1	3	9	27	$\cdots$

DEFINITION: An **array** of dimension d in a set S is a function from the set of d-tuples of natural numbers to the set S.

NOTATION: Array elements are commonly written in the subscripted notation

$$\begin{matrix} x_{0,0} & x_{0,1} & x_{0,2} & \cdots \\ x_{1,0} & x_{1,1} & x_{1,2} & \cdots \\ x_{2,0} & x_{2,1} & x_{2,2} & \cdots \\ \vdots & \vdots & \vdots & \ddots \end{matrix}$$

DEFINITION: The **integer interval** $[k : m]$, where $k, m \in \mathbb{Z}$, is the set

$$\{k,\ k+1,\ \ldots,\ m\}$$

The integer interval $[1 : n]$ is used as the standard set of cardinality n.

TERMINOLOGY: Extended sequences, arrays, and any of a host of other possible related mathematical structures may sometimes simply be called *sequences*.

Eventual Behavior of Sequences

Some sequences take a while before entering a permanent pattern.

DEFINITION: In general, for a property $\mathcal{P}$, we may say that a sequence $\langle x_n \rangle$ is **eventually** $\mathcal{P}$ (or related idiomatic variants of that phrasing) if there is a number N such that the subsequence $\langle x_n \mid n > N \rangle$ has that property.

Example 1.1.14:   The sequence

$$\langle x_n\ =\ n^2 - 8n + 15 \rangle$$

is eventually increasing, as illustrated in Figure 1.1.1. Its shape is an upward parabola, with its minimum at $n = 4$, after which it is strictly increasing. Thus, it is eventually increasing.

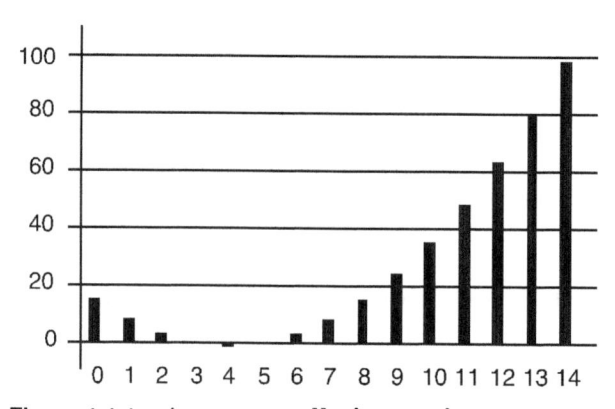

Figure 1.1.1   An eventually increasing sequence.

Example 1.1.15:   The sequence $\langle x_n = 2n^3 - 2^n \rangle$ is eventually decreasing.

Remark: Every polynomial (except a constant) is eventually increasing or eventually decreasing, depending on the sign of its term of highest degree.

Example 1.1.16:   The decimal digits of

$$\frac{4824}{8250} = 0.52412121212\ldots$$

are eventually periodic, as illustrated in Figure 1.1.2.

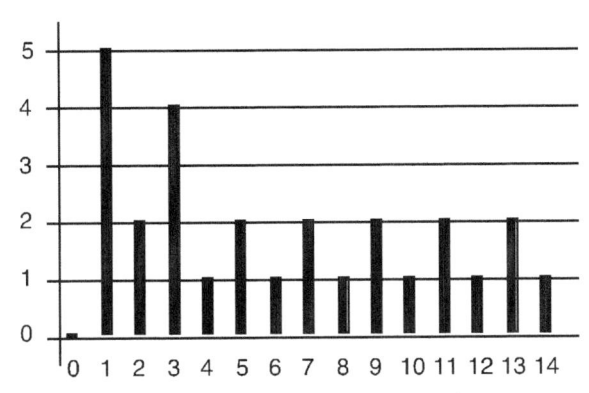

Figure 1.1.2   An eventually periodic sequence.

EXERCISES for Section 1.1

In each of the Exercises 1.1.1 through 1.1.6, write the first 12 elements of the indicated sequence, and prove that the sequence is periodic. Start at $n = 0$.

1.1.1[S] $2^n \bmod 7$		1.1.2 $3^n \bmod 7$	1.1.3 $4^n \bmod 7$
1.1.4 $n^2 \bmod 5$		1.1.5 $n^3 \bmod 4$	1.1.6 $n2^n \bmod 3$

In each of the Exercises 1.1.7 through 1.1.9, write the first 12 elements of the indicated sequence, and prove that the sequence is eventually periodic. Start at $n = 0$.

1.1.7[S] $2^n \bmod 6$ 1.1.8 $3^n \bmod 15$ 1.1.9 $2^n \bmod 12$

1.1.10   Prove that the sequence $n! \bmod 11213$ is eventually periodic.

1.1.11   Prove that the sequence $\lfloor \sqrt{n} \rfloor \bmod 3$ is not eventually periodic.

In each of the Exercises 1.1.12 through 1.1.17, find a polynomial $f(x)$ and a number P such that the sequence $a_n = f(n) \bmod P$ has the periodic pattern delimited by semi-colons. Some of these exercises may require algebraic experimentation and patience.

1.1.12   0, 1; 0, 1; ...   1.1.13[S]   0, 1, 1; 0, 1, 1; ...

1.1.**14**   0, 0, 1; 0, 0, 1; ...   1.1.**15**   0, 1, 1, 0; 0, 1, 1, 0; ...

1.1.**16**   0, 0, 1, 1; 0, 0, 1, 1; ...   1.1.**17**   1, 0, 1, 0, 1; 1, 0, 1, 0, 1; ...

1.1.**18**   Given a polynomial $f(x) = a_0 + a_1 x + a_2 x^2$ of known degree (i.e., degree 2) but unknown coefficients a_j, and given the values $f(0) = 1, f(1) = 4, f(2) = 11$, write and solve a system of linear equations in the coefficients a_j.

1.1.**19**   Given a polynomial $f(x) = a_0 + a_1 x + \cdots + a_d x^d$ of known degree d but unknown coefficients a_j, and given the values $f(0) = b_0, f(1) = b_1, \ldots, f(d) = x^d$, write a system of linear equations in the coefficients a_j.

1.1.**20**^{S} Design a polynomial $f(x)$ such that $f(x) = n^2$ for $n = 0, \ldots, 3$ and such that $f(x) > n^2$ for all $n \geq 4$.

1.1.**21**^{S} Use integral calculus to show that the sequence $\sum_{k=1}^{n} \frac{1}{n^2}$ is bounded.

1.1.**22**   Show that the sequence of decimal digits of every rational number is eventually periodic.

1.1.**23**   Show that a number whose sequence of decimal digits is eventually periodic is a rational number.

DEFINITION: A sequence $\langle u_n \rangle$ **eventually dominates** a sequence $\langle v_n \rangle$ if there is a number K such that $u_n > v_n$ for all $n \geq K$.

1.1.**24**   Show that $\ln n$ eventually grows more slowly than n^r, for any $r > 0$, in the sense of differential calculus that its derivative is eventually dominated.

1.1.**25**^{S} Show that the sequence $\langle x_n = 2^{\sqrt{n}} \rangle$ eventually dominates the sequence $\langle y_n = n^2 \rangle$.

1.2   RECURRENCES

Most of the sequences considered in §1.1 were specified by a function $j \mapsto x_j$. This section presents an alternative way that a sequence may be specified.

DEFINITION: A **standard recurrence** for a sequence prescribes a set of **initial values**

$$x_0 = b_0 \quad x_1 = b_1 \quad \ldots \quad x_k = b_k$$

and a **recursion** formula

$$x_n = \phi(x_{n-1}, x_{n-2}, \ldots, x_0) \quad \text{for } n > k$$

from which one may calculate the value of x_n, for any $n > k$, from the values of earlier entries.

Example 1.2.1:   The recurrence

$$x_0 = 0 \qquad\qquad \text{initial value}$$
$$x_n = x_{n-1} + 2n - 1 \qquad \text{recursion}$$

has as its first few values

$$x_0 = 0 \quad x_1 = 1 \quad x_2 = 4 \quad x_3 = 9 \quad x_4 = 16 \quad \ldots$$

We observe that the recursion formula here depends only on a fixed number of predecessors of x_n, specifically, only on x_{n-1}.

DEFINITION: Inferring a closed formula for a sequence from a recurrence is called *solving the recurrence*.

Example 1.2.1, continued: The first few values specified by the closed formula specification $x_n = n^2$, which are

$$0^2 = 0 \quad 1^2 = 1 \quad 2^2 = 4 \quad 3^2 = 9 \quad 4^2 = 16 \quad \ldots$$

coincide with those specified by the given recurrence

$$x_0 = 0 \quad x_1 = 1 \quad x_2 = 4 \quad x_3 = 9 \quad x_4 = 16 \quad \ldots$$

The initial value $x_0 = 0$ may be used as the basis for an induction to prove that $x_n = n^2$. Substituting $x_{n-1} = (n-1)^2$ into the recursion yields the induction step

$$x_n = (n-1)^2 + 2n - 1 = (n^2 - 2n + 1) + 2n - 1 = n^2$$

Thus, n^2 is a correct closed formula for x_n, and the recurrence is solved.

In calculating the value of x_n for a large subscript n, it is usually quicker to use a closed formula than a recurrence, since using the latter would require first calculating the values of many entries with lesser subscripts. Quite often, however, an explicit recurrence for the values of a sequence is given or readily inferrable, yet identifying the closed formula requires some analytic effort. We shall describe recurrences and closed formulas for three well-known sequences: the *Tower of Hanoi sequence*, the *Fibonacci sequence*, and the *Catalan sequence*.

A General Problem-Solving Method

Sometimes it is possible to guess the solution to a recurrence. More generally, the following approach goes a long way in mathematics, if one is good at guessing from relatively few examples.

1. Examine some small cases systematically.

2. Guess a pattern that covers all those cases.

3. Prove that the guess is correct.

Tower of Hanoi

The **Tower of Hanoi** is a puzzle invented by Edouard Lucas (1842–1891), a professor of mathematics in Paris with a keen interest in recreational mathematics. There are three pegs, a *source* peg, an *intermediate* peg, and a *target* peg. There are n drilled disks of differing diameters, initially stacked on the source peg in the order of ascending diameter, from top to bottom, as in Figure 1.2.1.

Figure 1.2.1   Tower of Hanoi puzzle.

The objective is to transfer all the disks from the source peg to the target peg, with the aid of the intermediate peg, under the following rules:

(1) Only one disk may be transfered at a time.

(2) No disk may ever lie on top of a smaller disk.

Clearly, it takes an initial value of 0 steps to transfer 0 disks. We observe that when transferring n disks from the source peg to the target peg, it is necessary first to move $n-1$ disks from the source peg to the intermediate peg (using the ultimate target peg as intermediate) — which requires h_{n-1} moves. Then the largest disk can be transferred to the ultimate target peg in a single move, after which the $n-1$ disks on the intermediate peg can be transferred in h_{n-1} moves to the target peg on top of the largest disk (using the initial peg as intermediate). Thus, the minimum number h_n of moves needed to transfer n disks satisfies the following recurrence:

RECURRENCE

$$h_0 = 0 \qquad \text{initial value}$$
$$h_n = 2h_{n-1} + 1 \qquad \text{recursion}$$

We may use the recursion to calculate the first few values of h_n and then try to guess a closed formula.

SMALL CASES

$$h_0 = 0$$
$$h_1 = 1$$
$$h_2 = 3 \qquad \qquad \text{APPARENT PATTERN}$$
$$h_3 = 7 \qquad \qquad \qquad h_n = 2^n - 1$$
$$h_4 = 15$$

Having guessed that $h_n = 2^n - 1$, we proceed to confirm the guess with a proof.

Theorem 1.2.1. *The Tower of Hanoi recurrence*

$$h_0 = 0; \quad h_n = 2h_{n-1} + 1 \ \text{ for } n \geq 1 \tag{1.2.1}$$

has the solution

$$h_n = 2^n - 1 \tag{1.2.2}$$

Proof: By induction.

BASIS: Applying the formula (1.2.2) yields the equation $h_0 = 2^0 - 1 = 1 - 1 = 0$, which agrees with the prescribed initial condition $h_0 = 0$.

IND HYP: Assume that $h_{n-1} = 2^{n-1} - 1$.

IND STEP: Starting with the recursion (1.2.1), we now complete the proof.

$$
\begin{aligned}
h_n &= 2h_{n-1} + 1 && \text{given recursion} \\
&= 2\left(2^{n-1} - 1\right) + 1 && \text{induction hypothesis} \\
&= 2^n - 2 + 1 \\
&= 2^n - 1 && \diamondsuit
\end{aligned}
$$

Fibonacci Sequence

Leonardo of Pisa (1170-1250), known as Fibonacci, championed the use of Hindu-Arabic numerals in Europe. His original contributions include the formulation and study of a well-known sequence that was mentioned in the introductory chapter.

DEFINITION: The **Fibonacci sequence** $\langle f_n \rangle$ is defined by the recurrence

$$
\begin{aligned}
f_0 &= 0; \quad f_1 = 1 && \text{initial values} \\
f_n &= f_{n-1} + f_{n-2} && \text{for } n \geq 2
\end{aligned}
\tag{1.2.3}
$$

Here are the first few entries:

n	0	1	2	3	4	5	6	7	8	9	$\cdots$
f_n	0	1	1	2	3	5	8	13	21	34	$\cdots$

DEFINITION: A **Fibonacci number** is any number that occurs in the Fibonacci sequence.

A closed formula for the Fibonacci recurrence is not easily guessed from the small cases. (However, once guessed, the solution is verifiable by a routine inductive proof.) The derivation of the following solution appears in §2.5, along with a discussion of ways that the Fibonacci sequence occurs in mathematics.

$$f_n = \frac{1}{\sqrt{5}}\left(\phi^n - \hat{\phi}^n\right) \tag{1.2.4}$$

$$\text{where } \phi = \frac{1 + \sqrt{5}}{2} \quad \text{and} \quad \hat{\phi} = \frac{1 - \sqrt{5}}{2}$$

Example 1.2.2:   For the time being, it is interesting to confirm an instance of the correctness of the formula (1.2.4) for the Fibonacci number f_n.

$$f_3 = \frac{1}{\sqrt{5}} \left(\frac{(1+\sqrt{5})^3}{8} - \frac{(1-\sqrt{5})^3}{8} \right)$$

$$= \frac{1}{\sqrt{5}} \left(\frac{1 + 3\sqrt{5} + 15 + 5\sqrt{5}}{8} \right) - \frac{1}{\sqrt{5}} \left(\frac{1 - 3\sqrt{5} + 15 - 5\sqrt{5}}{8} \right)$$

$$= \frac{1}{\sqrt{5}} \left(\frac{6\sqrt{5} + 10\sqrt{5}}{8} \right) = 2$$

Catalan Sequence

Many combinatorial objects are counted by a recurrence named for Eugène Catalan (1814–1894), a Belgian mathematician. Several examples are given in §4.4.

DEFINITION: The **Catalan sequence** $\langle c_n \rangle$ is defined by the recurrence

$$
\begin{aligned}
c_0 &= 1; & \text{initial value} \\
c_n &= c_0 c_{n-1} + c_1 c_{n-2} + \cdots + c_{n-1} c_0 & \text{for } n \geq 1
\end{aligned}
\tag{1.2.5}
$$

Here are the first few entries:

n	0	1	2	3	4	5	6	7	8	$\cdots$
c_n	1	1	2	5	14	42	132	429	1430	$\cdots$

DEFINITION: Any number that occurs in the Catalan sequence is called a **Catalan number**.

As with the Fibonacci sequence, a closed formula for the Catalan sequence is not easily guessed from the small cases. Its derivation, which is significantly more difficult than that of a closed formula for the Fibonacci numbers, appears in §4.4.

$$c_n = \frac{1}{n+1} \binom{2n}{n} \tag{1.2.6}$$

Example 1.2.3: $c_3 = \frac{1}{4} \cdot \binom{6}{3} = \frac{20}{4} = 5.$

Proving Properties of Sequences

Proof that a sequence has some given property can be derived either from a closed formula or from a recursion, with the aid of mathematical induction. As an illustration, we consider the properties of *concavity* and *convexity*.

DEFINITION: A sequence $\langle x_n \rangle$ is **concave** (on the integer interval $[a:b]$) if

$$x_n \geq \frac{x_{n-1} + x_{n+1}}{2} \qquad \text{(for } n = a+1, \ldots, b-1\text{)}$$

This means that the point (n, x_n) lies above the line segment joining the points $(n-1, x_{n-1})$ and $(n+1, x_{n+1})$ in the plane, as in Figure 1.2.2.

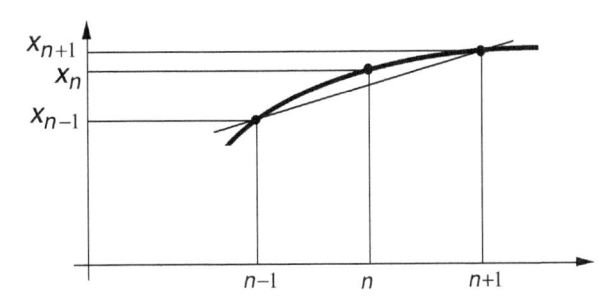

Figure 1.2.2   Concavity in a sequence.

Example 1.2.4:   Concavity of the sequence $\langle x_n = 1 - \frac{1}{n} \rangle$ follows from the observation that

$$2x_n = 2 - \frac{2}{n} = 2 - \frac{2n}{n^2} > 2 - \frac{2n}{n^2 - 1}$$

$$= \left(1 - \frac{1}{n-1}\right) + \left(1 - \frac{1}{n+1}\right) = x_{n-1} + x_{n+1}$$

Example 1.2.5:   That the Fibonacci sequence $\langle f_n \rangle$ is eventually increasing, after $n = 2$, follows easily by mathematical induction. Moreover, it is a consequence for all $n > 3$ that $f_n < 2f_{n-1}$.

DEFINITION: A sequence $\langle x_n \rangle$ is **convex** (on the integer interval $[a:b]$) if

$$x_n \leq \frac{x_{n-1} + x_{n+1}}{2} \qquad \text{for } n = a+1, \ldots, b-1$$

This means that the point (n, x_n) lies below the line segment joining the points $(n-1, x_{n-1})$ and $(n+1, x_{n+1})$ in the plane, as in Figure 1.2.3.

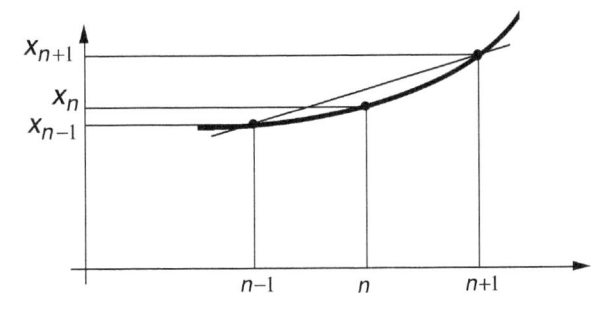

Figure 1.2.3   Convexity in a sequence.

Example 1.2.6:   The Fibonacci sequence is eventually convex, after $n = 2$. This is confirmed as follows:

$$
\begin{aligned}
f_{n+1} + f_{n-1} &= f_n + 2f_{n-1} &&\text{(by the Fibonacci recursion)} \\
&\geq 2f_n &&\text{(by Example 1.2.5)}
\end{aligned}
$$

which is equivalent to the defining condition for convexity.

EXERCISES for Section 1.2

1.2.1   Evaluate the closed formula (1.2.4) for the Fibonacci number f_4 and compare the result with the value calculated by the Fibonacci recursion.

1.2.2   Evaluate the closed formula (1.2.4) for the Fibonacci number f_5 and compare the result with the value calculated by the Fibonacci recursion.

1.2.3   Evaluate the closed formula (1.2.6) for the Catalan number c_4 and compare the result with the value calculated by the Catalan recursion.

1.2.4   Evaluate the closed formula (1.2.6) for the Catalan number c_5 and compare the result with the value calculated by the Catalan recursion.

In Exercises 1.2.5 and 1.2.6, prove that the sequence is concave.

1.2.5 H_n (harmonic number)   **1.2.6**[S] $\lg n$

In each of the Exercises 1.2.7 through 1.2.12, prove that the given sequence is convex.

1.2.7[S] n^2 **1.2.8** n^3 **1.2.9** n^{-1}

1.2.10 n^{-2} **1.2.11** 2^n **1.2.12** 2^{-n}

1.2.13[S]   Prove that the Catalan sequence is convex.

1.2.14   Can a bounded positive (infinite) sequence be convex?

1.2.15   Construct a bounded increasing sequence of positive values that is not eventually concave.

1.2.16   Prove that the sequence of values of $\sqrt{n}$ is concave.

1.2.17   Prove that the sequence of values of $\lfloor \sqrt{n} \rfloor$ is not concave.

Let p_n denote the number of regions created in the plane by n mutually intersecting straight lines, with no more than two lines interesecting at any one point. The figure at the right illustrates that $p_3 = 7$. Exercises 1.2.18 through 1.2.21 are concerned with the sequence $\langle p_n \rangle$.

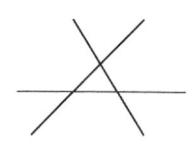

1.2.18   What are the values p_0, p_1, p_2, and p_4?

1.2.19   Give a recursion for p_n.

1.2.20   Guess a closed formula for p_n.

1.2.21   Use induction to prove that your guess is correct.

1.3 PASCAL'S RECURRENCE

A recurrence may also be used to specify an array. This section focuses on a recurrence for a doubly subscripted variable.

DEFINITION: The **combination coefficient** $\binom{n}{k}$ is the number of ways (sometimes called *combinations*) to choose a subset of cardinality k from a set of n objects.

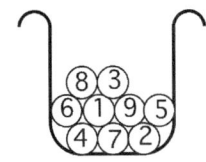

Figure 1.3.1   **There are $\binom{9}{3}$ ways to choose 3 balls from the 9 in the urn.**

Proposition 1.3.1. *The combination coefficients* $\begin{pmatrix} n \\ k \end{pmatrix}$ *satisfy the recurrence*

(I_1) $$\begin{pmatrix} n \\ 0 \end{pmatrix} = 1 \qquad \text{for all } n \geq 0$$

(I_2) $$\begin{pmatrix} 0 \\ k \end{pmatrix} = 0 \qquad \text{for all } k \geq 1$$

(R) $$\begin{pmatrix} n \\ k \end{pmatrix} = \begin{pmatrix} n-1 \\ k-1 \end{pmatrix} + \begin{pmatrix} n-1 \\ k \end{pmatrix} \quad \text{for } n \geq 1$$

Proof: Since, in any set, the empty set is the only subset of cardinality zero, the combination coefficients have initial values $\binom{n}{0} = 1,$ for all $n \geq 0$. Since there are no subsets of positive cardinality in the empty set, the combination coefficients have additional initial values $\binom{0}{k} = 0,$ for all $k \geq 1$. The first of two proofs given here for the recursion formula (R) is algebraic. The second is combinatorial.

Algebraic Proof: The algebraic proof of (R) starts with the right side of the equation and makes substitutions and arithmetic operations that result in the left side.

$$
\begin{aligned}
\begin{pmatrix} n-1 \\ k-1 \end{pmatrix} + \begin{pmatrix} n-1 \\ k \end{pmatrix} &= \frac{(n-1)!}{(k-1)!\,(n-k)!} + \frac{(n-1)!}{k!\,(n-k-1)!} \quad \text{(Formula (0.4.2))} \\[2mm]
&= \frac{k(n-1)!}{k!\,(n-k)!} + \frac{(n-k)(n-1)!}{k!\,(n-k)!} \\[2mm]
&= \frac{[k+(n-k)](n-1)!}{k!\,(n-k)!} \\[2mm]
&= \frac{n!}{k!\,(n-k)!} = \begin{pmatrix} n \\ k \end{pmatrix} \qquad\qquad \diamond
\end{aligned}
$$

Combinatorial Proof: The combinatorial proof of (R) shows that both sides of (R) count the same set of objects. The left side counts the number of ways to choose a subset of size k from the integer interval $[1:n]$. If such a subset includes object n, then it is counted by the summand $\binom{n-1}{k-1}$ on the right side, since one then chooses $k-1$ other objects from $[1:n-1]$. Alternatively, if such a subset excludes object n, then all k objects must be chosen from $[1:n-1]$, and it is counted by the summand $\binom{n-1}{k}$. Since these two cases are exclusive and exhaustive, it follows that

$$\binom{n}{k} = \binom{n-1}{k-1} + \binom{n-1}{k} \quad \text{for } n \geq 1 \qquad \qquad \Diamond$$

DEFINITION: The system $\{I_1, I_2, R\}$ is called **Pascal's recurrence**.

TERMINOLOGY: The approach used in the combinatorial proof of Proposition 1.3.1 is called the **Method of Distinguished Element**. It arises frequently in combinatorial arguments.

Remark: Offering two proofs here previews Chapter 4, where more proofs of both types are presented. Learning to understand and to create combinatorial proofs is very important preparation for Chapter 5, because there are recursions and other identities for Stirling numbers that cannot be verified by simple algebraic manipulations.

Binomial Coefficients

DEFINITION: The coefficient $b_{n,k}$ of x^k in the expansion

$$(1+x)^n = \sum_{k=0}^{n} b_{n,k}\, x^k$$

is called a **binomial coefficient**.

Example 1.3.1:   Binomial coefficients can be calculated by iteratively multiplying by $1+x$.

$$
\begin{aligned}
(1+x)^0 &= 1 \\
(1+x)^1 &= 1+x \\
(1+x)^2 &= 1+2x+x^2 \\
(1+x)^3 &= 1+3x+3x^2+x^3 \\
(1+x)^4 &= 1+4x+6x^2+4x^3+x^4
\end{aligned}
$$

Proposition 1.3.2. *The binomial coefficients $b_{n,k}$ satisfy Pascal's recurrence.*

Proof: The initial values of Pascal's recurrence are satisfied, since the values

$$
\begin{aligned}
b_{n,0} &= 1 \qquad \text{for all } n \geq 0 \\
b_{0,k} &= 0 \qquad \text{for all } k \geq 1
\end{aligned}
$$

can be verified by considering the direct expansions of $(1+x)^0$ and $(1+x)^n$, as in Example 1.3.1. To show that the recursion is satisfied, it is observed that

$$\sum_{k=0}^{n} b_{n,k}\, x^k \;=\; (1+x) \sum_{k=0}^{n-1} b_{n-1,k}\, x^k \tag{1.3.1}$$

$$= \sum_{k=0}^{n-1} b_{n-1,k}\, x^k \;+\; x \sum_{k=0}^{n-1} b_{n-1,k}\, x^k$$

$$= \sum_{k=0}^{n-1} b_{n-1,k}\, x^k \;+\; \sum_{k=0}^{n-1} b_{n-1,k}\, x^{k+1}$$

$$= \sum_{k=0}^{n} b_{n-1,k}\, x^k \;+\; \sum_{k=1}^{n} b_{n-1,k-1}\, x^k$$

$$= \sum_{k=0}^{n} \left(b_{n-1,k} \;+\; b_{n-1,k-1} \right) x^k \tag{1.3.2}$$

Thus, the coefficient $b_{n,k}$ of x^k in the sum at the left of equation (1.3.1) must equal the coefficient of x^k in the sum at the right in equation (1.3.2), i.e., it must equal the sum

$$b_{n-1,k} \;+\; b_{n-1,k-1} \qquad\qquad \diamondsuit$$

Corollary 1.3.3. *For all $n, k \geq 0$, the number $\binom{n}{k}$ of ways to choose k objects from a set of n distinct objects equals the binomial coefficient $b_{n,k}$.*

Proof: By Proposition 1.3.2, the combination coefficients $\binom{n}{k}$ and the binomial coefficients $b_{n,k}$ satisfy the exact same recurrence system. An induction argument establishes that the values must be the same. $\diamondsuit$

TERMINOLOGY NOTE: The number $\binom{n}{k}$ is commonly called a **binomial coefficient**. From here on in this book, we shall refer to it as such.

DEFINITION: If the zero values are left blank, then the array of binomial coefficients has a triangular shape and is called **Pascal's triangle**.

Table 1.3.1   **Pascal's triangle for values of $\binom{n}{r}$.**

n	$\binom{n}{0}$	$\binom{n}{1}$	$\binom{n}{2}$	$\binom{n}{3}$	$\binom{n}{4}$	$\binom{n}{5}$	$\binom{n}{6}$	Σ
0	1							1
1	1	1						2
2	1	2	1					4
3	1	3	3	1				8
4	1	4	6	4	1			16
5	1	5	10	10	5	1		32
6	1	6	15	20	15	6	1	64

In this form of Pascal's triangle, each number is the sum of the number directly above it and the number in the row above, one column to the left. Pascal's triangle also has a pyramid form:

$$
\begin{array}{ccccccccccccc}
& & & & & & 1 & & & & & & \\
& & & & & 1 & & 1 & & & & & \\
& & & & 1 & & 2 & & 1 & & & & \\
& & & 1 & & 3 & & 3 & & 1 & & & \\
& & 1 & & 4 & & 6 & & 4 & & 1 & & \\
& 1 & & 5 & & 10 & & 10 & & 5 & & 1 & \\
1 & & 6 & & 15 & & 20 & & 15 & & 6 & & 1
\end{array}
$$

In the pyramid form, each number is the sum of the two numbers just above it, one slightly to the left and the other slightly to the right.

Remark: The pyramid form of Pascal's triangle may be regarded as a *directed graph* in which there are two directed edges from each number, one to the number just below to the left, the other to the number just below to the right. It may be observed empirically that the number of directed paths from the apex of Pascal's triangle to each entry in the triangle equals the value of that entry. See the Exercises.

EXERCISES for Section 1.3

1.3.1   Calculate row $n = 7$ of Pascal's triangle from row $n = 6$.

1.3.2^S   Using only Pascal's recursion, prove the following

$$
\binom{n}{k} = \binom{2}{0}\binom{n-2}{k} + \binom{2}{1}\binom{n-2}{k-1} + \binom{2}{2}\binom{n-2}{k-2} \qquad \text{for } 0 \leq n
$$

1.3.3   By mathematical induction, generalize the equation of Exercise 1.3.2 to

$$
\binom{n}{k} = \sum_{j=0}^{r} \binom{r}{j}\binom{n-r}{k-j} \qquad \text{for } 0 \leq r \leq n
$$

1.3.4   Show that the generalized sequence $a_{n,k} = \dfrac{n!}{k!\,(n-k)!}$ satisfies Pascal's recurrence. Since the combinatorial and binomial coefficients also satisfy Pascal's recurrence, this serves as an alternative proof that

$$
\binom{n}{k} = \frac{n!}{k!\,(n-k)!} \qquad \text{for } 0 \leq k \leq n
$$

1.3.5^S   Prove that $\displaystyle\sum_{k=0}^{n} \binom{n}{k} = 2^n$, for all $n \in \mathbb{N}$.

1.3.6   Prove that $\displaystyle\sum_{k=0}^{n} \binom{n}{k} 2^k = 3^n$, for all $n \in \mathbb{N}$.

1.3.7   Prove the remark at the end of the section regarding directed paths in the pyramid form of Pascal's triangle. Hint: Prove that the number of directed paths to each entry $\binom{n}{r}$ satisfies Pascal's recursion.

1.3.8   Prove the correctness of the expansion
$$(x + y + z)^2 = x^2 + y^2 + z^2 + 2xy + 2xz + 2yz$$

In each of the Exercises 1.3.9 through 1.3.12, expand the given power of a linear multivariate polynomial, as in the statement of Exercise 1.3.8.

1.3.9 $(w + x + y)^3$ 1.3.10 $(w + x + y)^4$

1.3.11 $(w + x + y + z)^3$ 1.3.12 $(w + x + y + z)^4$

1.3.13   Recalling multinomial coefficients from §0.4, prove this generalization of Pascal's recursion:
$$\binom{n}{i_1, \ldots, i_s} =$$
$$\binom{n-1}{i_1 - 1, i_2, \ldots, i_s} + \binom{n-1}{i_1, i_2 - 1, i_3, \ldots, i_s} + \cdots + \binom{n-1}{i_1, \ldots, i_{s-1}, i_s - 1}$$

1.4   DIFFERENCES AND PARTIAL SUMS

Texts on infinitessimal calculus generally provide formulas for derivatives before deriving formulas for integrals. For similar reasons, having formulas for differences of consecutive values in a sequence provides access to formulas for partial sums of the sequence.

DEFINITION: Given a sequence $\langle a_n \rangle$, we define the **difference sequence** $\langle \triangle a_n \rangle$ by the rule
$$\triangle a_n = a_{n+1} - a_n$$

More generally, given a function $f : \mathbb{R} \to \mathbb{R}$, we define the **difference function** $\triangle f$ by the rule
$$\triangle f(x) = f(x + 1) - f(x)$$

DEFINITION: A **difference table** for a sequence $\langle a_n \rangle$ has the sequence itself in its 0th row and the difference sequence $\langle \triangle a_n \rangle$ in its 1st row. Often, the difference operation is iterated and additional rows are given. Sometimes each subsequent row is written a half-column shift to the right.

Example 1.4.1:   If $a_n = n^2$, then

$$\triangle a_n \;=\; (n+1)^2 - n^2 \;=\; 2n+1$$

and

$$\triangle^{(2)} a_n \;=\; (2(n+1)+1) - (2n+1) \;=\; 2$$

These equations yield this difference table (with a half-column rightward shift)

$a_n = \mathbf{n^2}$	0	1	4	9	16	25	36	49	$\cdots$
$\triangle a_n$		1	3	5	7	9	11	13	$\cdots$
$\triangle^{(2)} a_n$			2	2	2	2	2	2	$\cdots$

Example 1.4.2:   The sequence $\langle\, b_n \;=\; n^3 \,\rangle$ has the difference table, which was created by calculating its initial row and then iteratively taking differences.

$b_n = \mathbf{n^3}$	0	1	8	27	64	125	216	343	$\cdots$
$\triangle b_n$		1	7	19	37	61	91	127	$\cdots$
$\triangle^{(2)} b_n$			6	12	18	24	30	36	$\cdots$
$\triangle^{(3)} b_n$			6	6	6	6	6		$\cdots$

Properties of the Difference Function

In Examples 1.4.1 and 1.4.2, we observe that the second and third rows of the difference tables for the sequences $\langle n^2 \rangle$ and $\langle n^3 \rangle$, respectively, have the constant values $2 = 2!$ and $6 = 3!$. An initial aspect of our exploration is to establish that this phenomenon holds generally.

Proposition 1.4.1. *The difference operator $\triangle$ is linear. That is,*

$$\triangle\,(f(n) + cg(n)) \;=\; (\triangle f)(n) + c(\triangle g)(n)$$

Proof: The details are straightforward.

$$\begin{aligned}
\triangle\,(f(n) + cg(n)) \;&=\; (f(n+1) + cg(n+1)) \,-\, (f(n) + cg(n))\\
&=\; (f(n+1) - f(n)) \,+\, c(g(n+1) - g(n))\\
&=\; (\triangle f)(n) + c\,(\triangle g)(n) \qquad\qquad\qquad \diamondsuit
\end{aligned}$$

Proposition 1.4.2. *In the difference table for the sequence*

$$\langle n^r \mid n \in \mathbb{N} \rangle$$

the rth row has the constant value $r!$, and, accordingly, all subsequent rows are null.

Proof: By induction.

BASIS: The entries in the 0th row of the sequence $\langle n^0 \rangle$ all have the value $1 = 0!$.

IND HYP: Assume that all the entries in the $(r-1)^{\text{st}}$ row of the table for n^{r-1} have the value $(r-1)!$ and that all higher order rows are null.

IND STEP: It follows from the expansion

$$\triangle(n^r) = (n+1)^r - n^r = rn^{r-1} + b_{r-2}n^{r-2} + \cdots + b_0$$

(for appropriate coefficients b_j) and from the linearity of $\triangle$ that

$$\begin{aligned}
\triangle^{(r)}(n^r) &= \triangle^{(r-1)}(\triangle(n^r)) \\
&= \triangle^{(r-1)}(rn^{r-1} + b_{r-2}n^{r-2} + \cdots + b_0) \\
&= r\,\triangle^{(r-1)}(n^{r-1}) + \sum_{j=0}^{r-2} b_j \,\triangle^{(r-1)}(n^j)
\end{aligned}$$

By the induction hypothesis, $\triangle^{(r-1)}(n^j) = 0$, for $j \le r-2$, from which it follows that every term in the sum on the right has value 0. Thus,

$$\triangle^{(r)}(n^r) = r\,\triangle^{(r-1)}(n^{r-1})$$

It follows that the rth row of the difference table for $\langle n^r \rangle$ equals r times the $(r-1)^{\text{st}}$ row of the table for n^{r-1}, in which every entry has the value $(r-1)!$, by the induction hypothesis. $\diamond$

Summation Operator

DEFINITION: Let $\langle x_n \rangle$ be a sequence with values in an algebraic structure with an addition. Then the expression

$$\sum_{j=0}^{n} x_j$$

is called the nth **partial sum**.

DEFINITION: The **summation operator** maps a sequence $\langle x_n \mid n \in \mathbb{N} \rangle$ to the sequence of partial sums

$$\left\langle \sum_{j=0}^{n} x_j \;\middle|\; n \in \mathbb{N} \right\rangle$$

Example 1.4.3:   Under the summation operator, the integer sequence

$$\langle x_n \rangle = 1 \quad 3 \quad 5 \quad 7 \quad \cdots$$

is mapped to the integer sequence of its nth partial sums

$$\left\langle u_n = \sum_{j=0}^{n} (2j+1) \right\rangle$$

which begins with the values

$$1 \quad 4 \quad 9 \quad 16 \quad \ldots$$

It may be guessed that $u_n = (n+1)^2$, which is readily proved by induction. If one now defines

$$a_n = \sum_{j=0}^{n-1} (2j+1)$$

then the sequence $\langle a_n = n^2 \rangle$ has the values

$$0 \quad 1 \quad 4 \quad 9 \quad 16 \quad \ldots$$

which inverts Example 1.4.1. We recall from §0.3 that the empty sum is defined to be zero. This accounts for the value

$$a_0 = \sum_{j=0}^{n-1} (2j+1) = \sum_{j=0}^{-1} (2j+1) = 0$$

The next theorem establishes that the inversion is not at all a coincidence.

NOTATION: From time to time, it is convenient to use the notation x_j' as an alternative to Δx_j. This is analogous to the use of such an alternative notation in the differential calculus.

Theorem 1.4.3(a). *Let $\langle x_n \mid n \in \mathbb{N} \rangle$ be a sequence. Then*

$$\sum_{j=0}^{n-1} x_j' = x_n - x_0 \tag{1.4.1}$$

Proof: This is another straightforward calculation.

$$\sum_{j=0}^{n-1} x_j' = \sum_{j=0}^{n-1} (x_{j+1} - x_j)$$
$$= (x_1 - x_0) + (x_2 - x_1) + \cdots + (x_n - x_{n-1})$$
$$= (x_n - x_{n-1}) + (x_{n-1} - x_{n-2}) + \cdots + (x_1 - x_0)$$
$$= x_n - x_0 \qquad\qquad \diamond$$

The upper limit of the sum in equation (1.4.1) must be $n-1$, rather than n, to get the correct result. Figure 1.4.1 illustrates the proof of Theorem 1.4.3(a). The sum

$$x_0 + \sum_{j=0}^{3} x_j'$$

of the lengths along the y-axis clearly equals the height x_4 of the rightmost rectangle. Thus,

$$x_4 - x_0 = \sum_{j=0}^{3} x_j'$$

which is the total vertical distance from the top of the leftmost rectangle to the top of the rightmost rectangle.

$$(x_1 - x_0) + (x_2 - x_1) + (x_3 - x_2) + (x_4 - x_3) = x_4 - x_0$$

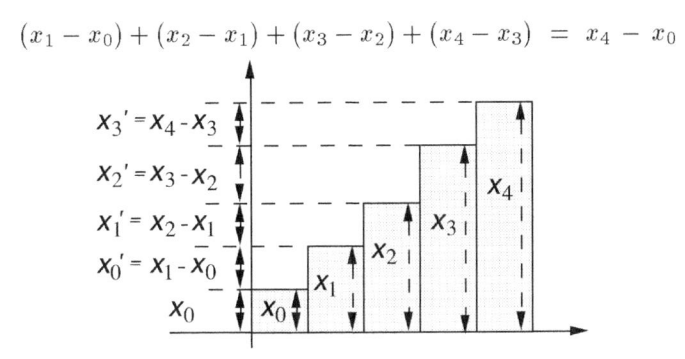

Figure 1.4.1   Accumulating consecutive differences, as in Theorem 1.4.3(a).

Theorem 1.4.3(b). *Let* $\langle x_n \mid n \in \mathbb{N} \rangle$ *be a sequence. Then*

$$\left(\sum_{j=0}^{k-1} x_j \right)_n' = x_n$$

Proof: By the definition of the difference operator,

$$\left(\sum_{j=0}^{k-1} x_j \right)_n' = \sum_{j=0}^{(n+1)-1} x_j - \sum_{j=0}^{n-1} x_j$$

$$= x_n \qquad\qquad \Diamond$$

Figure 1.4.2 illustrates the proof of Theorem 1.4.3(b). The difference of the sum $x_0 + \cdots + x_4$ of the areas of the consecutive rectangle including x_4 and the sum $x_0 + \cdots + x_3$ of the areas excluding x_4 clearly equals the area x_4 of the rightmost rectangle.

$$(x_0 + x_1 + x_2 + x_3 + x_4) \qquad - \qquad (x_0 + x_1 + x_2 + x_3) \qquad = \qquad x_4$$

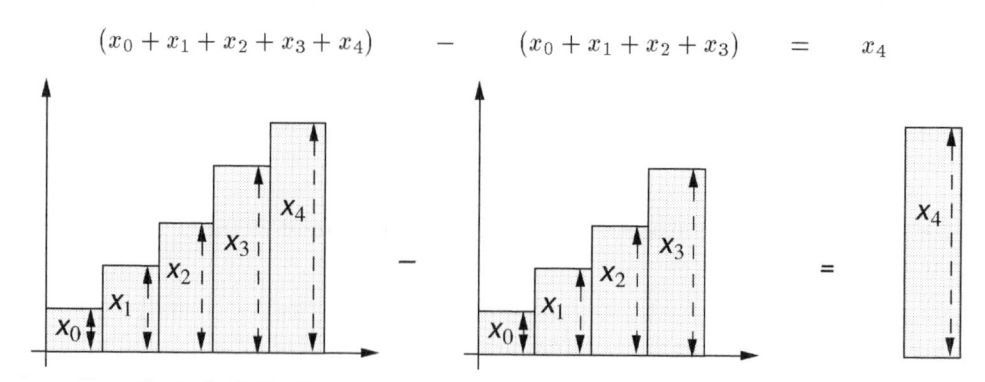

Figure 1.4.2   Subtracting consecutive sums, as in Theorem 1.4.3(b).

TERMINOLOGY: Theorem 1.4.3 is a form of what is commonly called the **Fundamental Theorem of Finite Calculus**. One sees a direct analogy to the Fundamental Theorem of Infinitessimal Calculus:

$$(a) \quad \int_0^x \tfrac{d}{dt} f(t)\, dt = f(x) - f(0)\,;$$

$$(b) \quad \tfrac{d}{dx} \int_0^x f(t)\, dt = f(x)$$

Growth Rate of Sequences

Comparison of the *growth rate* of one sequence $\langle x_n \rangle$ with that of another was mentioned informally in §1.1. The most common criterion for comparing the long term behavior of two sequences is called *asymptotic dominance*. However, by way of analogy to differential calculus, a possible measure of the growth rate of a sequence is its difference sequence.

Example 1.4.4:   To establish, in the sense of finite differences, that the sequence $\langle n^3 \rangle$ grows faster than the sequence $\langle cn^2 \rangle$, for any constant value of c, we make the following calculations.

$$\triangle n^3 \;=\; (n+1)^3 - n^3 \;=\; 3n^2 + 3n + 1$$
$$\triangle cn^2 \;=\; c(n+1)^2 - cn^2 \;=\; 2cn + c$$

For $n > c$, we have

$$3n^2 + 3n \;>\; 3cn + 3c \;>\; 2cn + c$$

Thus, $\triangle n^3$ eventually dominates $\triangle cn^2$.

Another possible measure of the growth rate of a sequence of positive values is the sequence of ratios

$$\left\langle \; \frac{x_{n+1}}{x_n} \;\middle|\; n \in \mathbb{Z}^+ \right\rangle$$

of consecutive terms.

Example 1.4.4, continued: The successive ratios of n^3 are

$$\frac{(n+1)^3}{n^3} = \frac{n^3 + 3n^2 + 3n + 1}{n^3} = 1 + \frac{3}{n} + \frac{3}{n^2} + \frac{1}{n^3}$$

and the successive ratios of cn^2 are

$$\frac{c(n+1)^2}{cn^2} = \frac{cn^2 + 2cn + c}{cn^2} = 1 + \frac{2}{n} + \frac{1}{n^2}$$

which are clearly smaller.

EXERCISES for Section 1.4

In each of the Exercises 1.4.1 through 1.4.6, construct a portion of a difference table for the given sequence, of sufficient extent to indicate the general pattern.

1.4.1[S] n^4	**1.4.2** $n(n-1)$	**1.4.3** $n(n-1)(n-2)$
1.4.4 2^n	**1.4.5** 3^n	**1.4.6** 4^n

In each of the Exercises 1.4.7 through 1.4.12, calculate the difference sequence Δa_n for the given sequence.

1.4.7[S] $a_n = c^n$	**1.4.8** $a_n = n^{-r}$ $(r > 0)$
1.4.9 $a_n = \lg n$	**1.4.10** $a_n = f_n$ (Fibonacci number)
1.4.11 $a_n = H_n$ (Harmonic number)	**1.4.12** $a_n = c_n$ (Catalan number)

In each of the Exercises 1.4.13 through 1.4.16, compare the value of the difference $a_6 - a_3$ to the value of the sum $\sum_{n=3}^{5} \Delta a_n$.

1.4.13[S] $a_n = n^2$	**1.4.14** $a_n = n(n-1)$
1.4.15 $a_n = n^3$	**1.4.16** $a_n = 2^n$

1.4.17   Prove that the difference function of a polynomial function of degree d is a polynomial function of degree $d - 1$.

1.4.18   Prove that a sequence is generated by a polynomial if and only if there is eventually a row of zeroes in the difference table.

1.4.19   Prove that a sequence is generated by an exponential c^n if and only if each row in the difference table is a multiple of other rows.

Each of the Exercises 1.4.20 through 1.4.23 gives a linear recurrence for a sequence. Write a recurrence for its difference sequence.

1.4.20 $a_0 = 1$; $a_n = 3a_{n-1} + 2$

1.4.21 $a_0 = b$; $a_n = ca_{n-1} + d$

1.4.22 $a_0 = 0$; $a_1 = 1$; $a_n = 3a_{n-1} + 2a_{n-2}$

1.4.23 $a_0 = 2$; $a_n = 4a_{n-1} + n$

1.5   FALLING POWERS

REVIEW FROM §0.2:   The nth **falling power** of a real number x is the product

$$x^{\underline{n}} \; = \; \overbrace{x\,(x-1)\,\cdots\,(x-n+1)}^{n\ \text{factors}} \qquad \text{for } n \in \mathbb{N}$$

If x is an integer, then the falling power $x^{\underline{n}}$ equals the number of ordered selections of n objects from a set of x distinct objects. Thus, for elementary combinatorial calculations, falling powers are as natural as ordinary powers.

We recall that the differential calculus has nice formulas:

$$\frac{d}{dx}\,x^2 \; = \; 2x \qquad \frac{d}{dx}\,x^3 \; = \; 3x^2 \quad \text{etc.}$$

So does the calculus of finite differences, but these are *not* examples of them:

$$\triangle(x^2) \; = \; 2x + 1$$
$$\triangle(x^3) \; = \; 3x^2 + 3x + 1$$

In the calculus of finite differences, the falling monomial $x^{\underline{n}}$ lends itself quite naturally to nice formulas that are analogous to those of the ordinary monomial x^n.

Example 1.5.1:   This illustrates what is meant by a "nice formula".

$$\begin{aligned}
\triangle(x^{\underline{3}}) \; &= \; (x+1)^{\underline{3}} - x^{\underline{3}} \\
&= \; (x+1)\,x\,(x-1) - x\,(x-1)(x-2) \\
&= \; [(x+1) - (x-2)]\,x\,(x-1) \\
&= \; 3x(x-1) \; = \; 3x^{\underline{2}}
\end{aligned}$$

The next theorem gives a difference formula for falling powers that generalizes Example 1.5.1 and is analogous to the differential calculus formula for ordinary powers.

Theorem 1.5.1. $\triangle(x^{\underline{r}}) \; = \; rx^{\underline{r-1}}$.

Proof: A straightforward approach is sufficient.

$$\begin{aligned}
\triangle\left(x^{\underline{r}}\right) \; &= \; (x+1)^{\underline{r}} - x^{\underline{r}} \\
&= \; (x+1)\,x^{\underline{r-1}} - x^{\underline{r-1}}\,(x-r+1) \\
&= \; [(x+1) - (x-r+1)]\,x^{\underline{r-1}} \; = \; rx^{\underline{r-1}} \qquad\qquad \diamond
\end{aligned}$$

Corollary 1.5.2. *For every non-negative integer r and every positive integer n,*

$$\sum_{j=0}^{n-1} j^{\underline{r}} = \frac{n^{\underline{r+1}}}{r+1}$$

Proof: By Theorem 1.5.1, we have $j^{\underline{r}} = \triangle \left(\dfrac{j^{\underline{r+1}}}{r+1} \right)$. Thus, by the Fundamental Theorem of Finite Calculus, it follows that

$$\sum_{j=0}^{n-1} j^{\underline{r}} = \left. \frac{j^{\underline{r+1}}}{r+1} \right|_{j=0}^{n} = \frac{n^{\underline{r+1}}}{r+1} \qquad\qquad \diamondsuit$$

Example 1.5.2:   Direct addition and the formula of Corollary 1.5.2 give the same result when summing $k^{\underline{2}}$.

$$\sum_{k=0}^{4} k^{\underline{2}} = 0 \cdot (-1) + 1 \cdot 0 + 2 \cdot 1 + 3 \cdot 2 + 4 \cdot 3 = 20$$

$$\frac{5^{\underline{3}}}{3} = \frac{5 \cdot 4 \cdot 3}{3} = 20$$

Unimodal Sequences

DEFINITION: A sequence $\langle x_n \rangle$ is **unimodal** if there is an index M such that

$$x_0 \le x_1 \le \cdots \le x_{M-1} \le x_M$$

and that $\langle x_n \rangle$ is non-increasing after index M. The value x_M is called the **mode** and M the **mode index**. (A tie is permitted at the mode value.)

DEFINITION: A sequence is **eventually** 0 if there is a number N such that $x_n = 0$, for all $n > N$. (Thus, there are only finitely many non-zero entries.)

Example 1.5.3:   Most of the unimodal sequences of interest in the present context are eventually 0. Figure 1.5.1 illustrates that the sequence $\binom{8}{r}$ is unimodal.

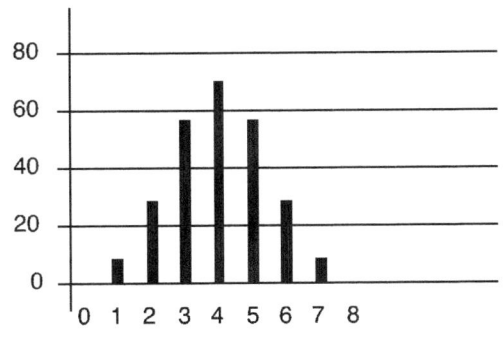

Figure 1.5.1   The unimodal sequence $\binom{8}{r}$.

Theorem 1.5.3. *For any fixed non-negative integer n, the binomial sequence*

$$\left\langle \binom{n}{r} \;\middle|\; r = 0, 1, \ldots \right\rangle$$

is unimodal with mode index $\lfloor n/2 \rfloor$ and is eventually 0.

Proof: We observe that for $r \leq \left\lfloor \dfrac{n}{2} \right\rfloor$, we have $\dfrac{n}{r} \geq 2$, and, hence, $\dfrac{n+1}{r} > 2$. Thus, $\dfrac{n-r+1}{r} > 1$. Accordingly,

$$
\begin{aligned}
\binom{n}{r-1} &= \frac{n^{\underline{r-1}}}{(r-1)!} \\[2mm]
&< \frac{n^{\underline{r-1}}}{(r-1)!} \cdot \frac{n-r+1}{r} \\[2mm]
&= \frac{n^{\underline{r}}}{r!} = \binom{n}{r}
\end{aligned}
$$

Moreover, for $r \geq \left\lfloor \dfrac{n}{2} \right\rfloor$, we have $n \leq 2r + 1$, and it follows that $n - r \leq r + 1$. Thus, $\dfrac{n-r}{r+1} \leq 1$. It follows that

$$
\binom{n}{r+1} = \frac{n^{\underline{r+1}}}{(r+1)!} = \frac{n^{\underline{r}}}{r!} \cdot \frac{n-r}{r+1} \leq \frac{n^{\underline{r}}}{r!} = \binom{n}{r}
$$

Of course, the sequence is zero for $r > n$. $\diamond$

Remark: One of the consequences of unimodality of a sequence is that it may make it possible to find the maximum by hill-climbing, for which there exist highly efficient computational strategies.

Log-Concavity and Log-Convexity

In trying to establish unimodality, curiously enough, it is often easier to prove the stronger property called *log-concavity*. For instance, this is the method used in this section to reconfirm the unimodality of the binomial coefficients in the rows of Pascal's triangle and later to prove the unimodality of some analogous sequences of Stirling numbers (see §§1.6, 5.1, 5.2).

DEFINITION: A sequence $\langle x_n \rangle$ of positive real numbers is **log-concave** (on the integer interval $[a : b]$) if, for $n = a+1, \ldots, b-1$,

$$\log x_n \;\geq\; \frac{\log x_{n-1} + \log x_{n+1}}{2} \tag{1.5.1}$$

and is **log-convex** if

$$\log x_n \;\leq\; \frac{\log x_{n-1} + \log x_{n+1}}{2} \tag{1.5.2}$$

Proposition 1.5.4. *A sequence* $\langle x_n \rangle$ *of positive real numbers is* log-concave *(on the integer interval* $[a : b]$*) if and only if, for* $n = a + 1, \ldots, b$,

$$x_n^2 \geq x_{n-1}x_{n+1} \qquad (1.5.3)$$

It is log-convex *if and only if*

$$x_n^2 \leq x_{n-1}x_{n+1} \qquad (1.5.4)$$

Proof: The defining condition (1.5.1) for log-concavity

$$\log x_n \geq \frac{\log x_{n-1} + \log x_{n+1}}{2}$$

is equivalent to the inequality

$$2 \log x_n \geq \log x_{n-1} + \log x_{n+1} \qquad (1.5.5)$$

Exponentiating both sides of the inequality (1.5.5) leads to inequality (1.5.3), i.e.,

$$x_n^2 \geq x_{n-1}x_{n+1}$$

A similar argument establishes the equivalence of inequalities (1.5.2) and (1.5.4). $\Diamond$

Theorem 1.5.5. *Let* $\langle x_n \rangle$ *be a* log-concave *sequence (over the integer interval* $[a : b]$*). Then it is* unimodal *(over that integer interval).*

Proof: It follows from Proposition 1.5.4 that the sequence of ratios

$$\frac{x_1}{x_0} \quad \frac{x_2}{x_1} \quad \frac{x_3}{x_2} \quad \ldots$$

(wherever defined) is non-increasing. That is,

$$x_n^2 \geq x_{n-1}x_{n+1} \quad \Rightarrow \quad \frac{x_n}{x_{n-1}} \geq \frac{x_{n+1}}{x_n}$$

Let M be the largest number k in the integer interval $[a : b]$ such that

$$\frac{x_k}{x_{k-1}} > 1$$

or $M = a$ if no such number k exists. Then the initial subsequence

$$x_a \quad x_{a+1} \quad \cdots \quad x_M$$

is increasing and the terminal subsequence

$$x_M \quad x_{M+1} \quad \cdots \quad x_b$$

is non-increasing, precisely the conditions for unimodality with mode index M. $\Diamond$

Theorem 1.5.6. *The binomial sequence*

$$\left\langle \binom{n}{r} \,\middle|\, r = 0, 1, \ldots \right\rangle$$

is log-concave on the integer interval $[0 : n]$.

Proof: The falling-power formula for binomial coefficients is

$$\binom{n}{r}^2 = \left(\frac{n^{\underline{r}}}{r!}\right)^2$$

Since $\dfrac{r}{r+1} < 1$ and $\dfrac{n-r}{n-r+1} < 1$, it follows that

$$\left(\frac{n^{\underline{r}}}{r!}\right)^2 > \frac{n^{\underline{r}}}{r!} \cdot \frac{n^{\underline{r}}}{r!} \cdot \frac{r}{r+1} \cdot \frac{n-r}{n-r+1}$$

and, in turn, that

$$\binom{n}{r}^2 > \frac{n^{\underline{r}}}{r!} \cdot \frac{n^{\underline{r}}}{r!} \cdot \frac{r}{r+1} \cdot \frac{n-r}{n-r+1}$$

$$= \frac{n^{\underline{r-1}}}{(r-1)!} \cdot \frac{n^{\underline{r+1}}}{(r+1)!} = \binom{n}{r-1} \cdot \binom{n}{r+1}$$

Accordingly, by Proposition 1.5.4, the binomial sequence is log-concave.   ◇

Remark: Theorems 1.5.6 and 1.5.5 can be used together to reconfirm Theorem 1.5.3, that the sequence of binomial coefficients $\binom{n}{k}$, for $k = 0, \ldots, n$, is unimodal.

EXERCISES for Section 1.5

In Exercises 1.5.1 through 1.5.4, determine which one, if either, of the two given expressions is the larger and give a proof. Assume that all the variables are integers greater than one.

1.5.1^{S} $x^{\underline{r}} y^{\underline{r}} : (xy)^{\underline{r}}$ 1.5.2 $x^{\underline{r}} x^{\underline{s}} : x^{\underline{r+s}}$

1.5.3 $x^{\underline{r^{\underline{s}}}} : (x^{\underline{r}})^{\underline{s}}$ 1.5.4 $(n^{\underline{n}})^n : (n^n)^{\underline{n}}$

In Exercises 1.5.5 through 1.5.7, evaluate the generalized binomial coefficient.

1.5.5^{S} $\displaystyle\binom{-\frac{1}{2}}{4}$ 1.5.6 $\displaystyle\binom{\frac{3}{7}}{4}$ 1.5.7 $\displaystyle\binom{\sqrt{2}}{3}$

1.5.8   From the integer interval $[1 : n]$, there are to be selected at random r numbers, without repetition. Express in falling powers the probability that such a selection includes all the numbers in the subinterval $[1 : k]$, where $k \le r$.

1.5.9   Confirm that $\begin{pmatrix} -\frac{1}{2} \\ n \end{pmatrix} = \dfrac{(-1)^n}{2^{2n}} \begin{pmatrix} 2n \\ n \end{pmatrix}$.

DEFINITION: The nth **rising power** of a real number x is the product

$$x^{\overline{n}} = \overbrace{x\,(x+1)\cdots(x+n-1)}^{n \text{ factors}}$$

1.5.10   Express $x^{\overline{n}}$ as a falling power.

1.5.11   Express $x^{\underline{n}}$ as a rising power.

1.5.12   Show that the sequence $\langle \sqrt{n} \rangle$ is concave and log-concave.

1.5.13   Show that the sequence $\langle H_n \rangle$ is concave and log-concave.

In each of the Exercises 1.5.14 through 1.5.19, decide whether the given sequence is log-convex or log-concave. Give a proof.

1.5.14[S] n^2 **1.5.15** n^3 **1.5.16** n^{-1}

1.5.17 n^{-2} **1.5.18** 2^n **1.5.19** 2^{-n}

1.5.20   Give an example to illustrate that the sum of unimodal sequences need not be unimodal.

1.6   STIRLING NUMBERS: A PREVIEW

James Stirling (1692-1770) was a Scottish mathematician. He introduced two families of numbers, now called *Stirling numbers* of the first and second kinds, for representing falling powers and ordinary powers in terms of each other. Stirling numbers are highly useful in counting partitions and permutations.

Converting Falling Powers into Ordinary Powers

The following theorem provides a recursive method for converting a falling power into ordinary powers.

Theorem 1.6.1. *Any falling power $x^{\underline{n}}$ can be expressed as a linear combination of ordinary powers, i.e., in the form*

$$x^{\underline{n}} = \sum_{k=0}^{n} s_{n,k}\, x^k \qquad \text{with } s_{n,n} = 1 \text{ and } s_{n,0} = 0 \text{ for } n \geq 1$$

Proof: By induction on the exponent n.

BASIS: For $n = 0$ and $n = 1$, we have

$$x^{\underline{0}} = 1x^0$$
$$x^{\underline{1}} = 1x^1 + 0x^0$$

Thus, we take $s_{0,0} = 1$, $s_{1,1} = 1$, and $s_{1,0} = 0$.

IND HYP: Suppose for some $n > 1$ that there exist integer coefficients

$$s_{n-1,0} \quad s_{n-1,1} \quad \cdots \quad s_{n-1,n-1}$$

for which

$$x^{\underline{n-1}} = \sum_{k=0}^{n-1} s_{n-1,k} \, x^k \qquad \text{with} \ \ s_{n-1,n-1} = 1 \ \ \text{and} \ \ s_{n-1,0} = 0$$

IND STEP: It follows that

$$
\begin{aligned}
x^{\underline{n}} &= x^{\underline{n-1}} \cdot (x - n + 1) && \text{(definition of falling power $x^{\underline{n}}$)} \\
&= (x - n + 1) \sum_{r=0}^{n-1} s_{n-1,k} \, x^k && \text{(inductive hypothesis)} \\
&= x \sum_{k=0}^{n-1} s_{n-1,k} \, x^k \ - \ (n-1) \sum_{k=0}^{n-1} s_{n-1,k} \, x^k \\
&= -(n-1)s_{n-1,0} \, x^0 \ + \ \sum_{k=1}^{n-1} (s_{n-1,k-1} - (n-1)s_{n-1,k}) \, x^k \ + \ s_{n-1,n-1}x^n \\
&= 0 \, x^0 \ + \ \sum_{k=1}^{n-1} (s_{n-1,k-1} - (n-1)s_{n-1,k}) \, x^k \ + \ 1 \, x^n
\end{aligned}
$$

Thus, we may take $s_{n,0} = 0$, $s_{n,n} = 1$, and $s_{n,k} = s_{n-1,k-1} - (n-1)s_{n-1,k}$, for $0 < k < n$. $\qquad\qquad\qquad\qquad\qquad\qquad\qquad\qquad\qquad\qquad\qquad\qquad\qquad\qquad\Diamond$

DEFINITION: The coefficients $s_{n,k}$ in the summation

$$x^{\underline{n}} = \sum_{k=0}^{n} s_{n,k} \, x^k$$

are called **Stirling numbers of the first kind**. For $k > n$ or $k < 0$, the Stirling number $s_{n,k}$ is taken to be 0, corresponding to letting the upper and lower limits of the sum go to ∞.

The Stirling numbers $s_{n,k}$ can be calculated by multiplying the factors in the expansion

$$x^{\underline{n}} = x(x-1)(x-2) \cdots (x - n + 1)$$

Example 1.6.1:

$$
\begin{aligned}
x^{\underline{2}} &= x^2 - x^1 \\
x^{\underline{3}} &= x^3 - 3x^2 + 2x^1 \\
x^{\underline{4}} &= x^4 - 6x^3 + 11x^2 - 6x^1 \\
x^{\underline{5}} &= x^5 - 10x^4 + 35x^3 - 50x^2 + 24x^1
\end{aligned}
$$

Thus, $s_{5,2} = -50$ and $s_{3,1} = 2$. We observe the alternating signs in each equation.

Converting Ordinary Powers into Falling Powers

Expressing an ordinary power as a sum of falling powers is an analogous task.

Theorem 1.6.2. *Any ordinary power x^n can be expressed as a linear combination of falling powers, i.e., in the form*

$$
x^n = \sum_{k=0}^{n} S_{n,k}\, x^{\underline{k}} \qquad \text{with } S_{n,n} = 1 \text{ and } S_{n,0} = 0 \text{ for } n \geq 1
$$

Proof: Once again, we use induction on the exponent n.

BASIS: For $n = 0$ and $n = 1$, we have

$$
\begin{aligned}
x^0 &= 1x^{\underline{0}} \\
x^1 &= 1x^{\underline{1}} + 0x^{\underline{0}}
\end{aligned}
$$

We take $S_{0,0} = 1$, $S_{1,1} = 1$, and $S_{1,0} = 0$.

IND HYP: Suppose that for some $n > 1$, the monomial x^{n-1} can be expressed as a linear combination

$$
x^{n-1} = \sum_{k=0}^{j} S_{j,k}\, x^{\underline{k}}
$$

of falling-power monomials $S_{j,k}\, x^{\underline{k}}$, each of degree less than or equal to j.

IND STEP: Then

$$x^n = x \cdot x^{n-1}$$

$$= x \cdot \sum_{k=0}^{n-1} S_{n-1,k}\, x^{\underline{k}} \qquad \text{(inductive hypothesis)}$$

$$= \sum_{k=0}^{n-1} S_{n-1,k}\, x \cdot x^{\underline{k}}$$

$$= \sum_{k=0}^{n-1} S_{n-1,k}\, (x-k) \cdot x^{\underline{k}} + \sum_{k=0}^{n-1} S_{n-1,k}\, k \cdot x^{\underline{k}}$$

$$= \sum_{k=0}^{n-1} S_{n-1,k}\, x^{\underline{k+1}} + \sum_{k=0}^{n-1} k\, S_{n-1,k}\, x^{\underline{k}}$$

$$= \sum_{k=1}^{n} S_{n-1,k-1}\, x^{\underline{k}} + \sum_{k=0}^{n-1} k\, S_{n-1,k}\, x^{\underline{k}}$$

$$= S_{n-1,n-1}\, x^{\underline{n}} + \sum_{k=1}^{n-1} (S_{n-1,k-1} + k\, S_{n-1,k})\, x^{\underline{k}} + 0\, S_{n-1,0}\, x^{\underline{0}}$$

Thus, we may take $S_{n,0} = 0$, $S_{n,n} = 1$, and $S_{n,k} = s_{n-1,k-1} + kS_{n-1,k}$, for $0 < k < n$. $\diamond$

DEFINITION: The coefficients $S_{n,k}$ in the sum

$$x^n = \sum_{k=0}^{n} S_{n,k}\, x^{\underline{k}}$$

are called **Stirling numbers of the second kind**. For $k > n$ or $k < 0$, the Stirling number $S_{n,k}$ is 0, which corresponds to letting the upper and lower limits of the sum go to ∞.

Example 1.6.2:

$$x^2 = x^{\underline{2}} + x^{\underline{1}}$$

$$x^3 = x^{\underline{3}} + 3x^{\underline{2}} + x^{\underline{1}}$$

$$x^4 = x^{\underline{4}} + 6x^{\underline{3}} + 7x^{\underline{2}} + x^{\underline{1}}$$

$$x^5 = x^{\underline{5}} + 10x^{\underline{4}} + 25x^{\underline{3}} + 15x^{\underline{2}} + x^{\underline{1}}$$

Thus, $S_{5,3} = 25$ and $S_{4,2} = 7$.

Corollary 1.5.2 provides a simple formula for the sum of the values of any falling power $n^{\underline{r}}$, over an interval of integer values of the base n. Accordingly, due to the linearity of the difference operator (Proposition 1.4.1), we could calculate the sum of the values of any ordinary power n^r, over a range of values of n, if we first express n^r as a linear combination of falling powers.

Example 1.6.3:   Notice, in particular, in Example 1.6.2, that $n^2 = n^{\underline{2}} + n^{\underline{1}}$. It follows from Theorem 1.6.2 that

$$\sum_{j=0}^{n} j^2 = \sum_{j=0}^{n} j^{\underline{2}} + \sum_{j=0}^{n} j^{\underline{1}}$$

$$= \frac{(n+1)^{\underline{3}}}{3} + \frac{(n+1)^{\underline{2}}}{2}$$

E.g., $0 + 1 + 4 + 9 + 16 + 25 + 36 = \dfrac{7^{\underline{3}}}{3} + \dfrac{7^{\underline{2}}}{2} = 70 + 21 = 91$.

In turn, this enables us to calculate the sum of the sequential values of a polynomial, since a polynomial is a linear combination of ordinary powers. This method of summing the values of polynomials will be further explored in §3.4.

Partitions

DEFINITION: A ***partition*** of a set S is a family $\mathcal{F} = \{S_1, \ldots, S_n\}$ of mutually disjoint subsets of S, called the ***cells of the partition*** $\mathcal{F}$, whose union is S.

NOTATION: Cells of partitions of a set may be indicated by the use of hyphens. If the set is small enough, then its elements can be represented by single characters, thereby avoiding potential ambiguities latent in juxtapositions of the characters.

Example 1.6.4:   The partition $\{\,\{1,3\},\, \{2,5\},\, \{4\}\,\}$ of the integer interval $[1:5]$ may be denoted

$$13 - 25 - 4$$

or also, for instance, by $4 - 52 - 13$, since the cells of a partition and the order within cells are taken to be unordered.

Stirling Subset Numbers

DEFINITION: The ***Stirling subset number***

$$\left\{ {n \atop k} \right\}$$

is the number of ways to partition the integer interval $[1 : n]$ into k non-empty non-distinct cells.[*]

In §5.1, we establish that the Stirling number $S_{n,k}$ of the second kind equals the Stirling subset number $\left\{ {n \atop k} \right\}$.

Example 1.6.2, continued: The value $S_{4,2} = 7$ is consistent with the following list of 7 partitions of $[1:4]$ into 2 cells, as an *ad hoc* calculation of $\left\{ {4 \atop 2} \right\}$.

$$1 - 234, \quad 2 - 134, \quad 3 - 124, \quad 4 - 123$$
$$12 - 34, \quad 13 - 24, \quad 14 - 23$$

---

[*] Wikipedia acknowledges D. E. Knuth for promoting usage of the user-friendly notations, $\left\{ {n \atop k} \right\}$ and $\left[{n \atop k} \right]$, of the Serbian mathematician J. Karamata (1902-1967) for Stirling numbers.

Stirling Cycle Number

DEFINITION: The **Stirling cycle number** $\left[{n \atop k}\right]$ is the number of ways to partition the integer interval $[1:n]$ into k non-empty non-distinct cycles.

In §5.2, we establish that the Stirling number $s_{n,k}$ of the first kind equals the absolute value of the Stirling cycle number $\left[{n \atop k}\right]$.

Example 1.6.1, continued: The value $s_{4,2} = 11$ of the Stirling number of the first kind is consistent with the following list of 11 partitions of the integer interval $[1:4]$ into 2 cycles, as an *ad hoc* calculation of the Stirling cycle number $\left[{4 \atop 2}\right]$.

$$(1)(2\ \ 3\ \ 4),\ \ (2)(1\ \ 3\ \ 4),\ \ (3)(1\ \ 2\ \ 4),\ \ (4)(1\ \ 2\ \ 3)$$
$$(1)(2\ \ 4\ \ 3),\ \ (2)(1\ \ 4\ \ 3),\ \ (3)(1\ \ 4\ \ 2),\ \ (4)(1\ \ 3\ \ 2)$$
$$(1\ \ 2)(3\ \ 4),\ \ (1\ \ 3)(2\ \ 4),\ \ (1\ \ 4)(2\ \ 3)$$

Remark: Since the Stirling cycle numbers

$$\left[{n \atop 1}\right]\ \ \left[{n \atop 2}\right]\ \ \cdots\ \ \left[{n \atop n}\right]$$

correspond to an inventory of all permutations of the integer interval $[1:n]$, according to the number of cycles in their disjoint cycle representation, it follows that

$$\sum_{j=1}^{n} \left[{n \atop j}\right] = n!$$

EXERCISES for Section 1.6

In Exercises 1.6.1 through 1.6.4, expand each of the falling power polynomials as a polynomial in ordinary powers.

1.6.1[S] $x^{\underline{6}}$ **1.6.2** $x^{\underline{5}} + 4x^{\underline{2}} + 3$

1.6.3 $2x^{\underline{5}} + 3x^{\underline{4}} - 2x^{\underline{1}}$ **1.6.4** $x^{\underline{7}}$

In Exercises 1.6.5 through 1.6.8, expand each of the given polynomials as a polynomial in falling powers.

1.6.5[S] x^6 **1.6.6** $x^5 + 4x^2 + 3$

1.6.7 $2x^5 + 3x^4 - 2x^1$ **1.6.8** x^7

1.6.9   List all the partitions of the integer interval $[1:5]$.

1.6.10   List all partitions of the integer interval $[1:6]$ into 3 parts.

DEFINITION: The **type of a partition** $\mathcal{F} = \{S_1, \ldots, S_n\}$ of a set of cardinality n is a string $s_1 \ldots s_n$ of positive integers (usually in ascending or descending order) such that s_j is the cardinality of the cell S_j. Thus, $s_1 + \cdots + s_n = n$.

In Exercises 1.6.11 through 1.6.16, calculate the number of partitions of the given integer interval of the given type.

1.6.11[S]   [1 : 7] of type 124   1.6.12   [1 : 9] of type 234

1.6.13   [1 : 7] of type 223   1.6.14   [1 : 9] of type 144

1.6.15   [1 : 8] of type 224   1.6.16   [1 : 10] of type 127

1.7   ORDINARY GENERATING FUNCTIONS

A sequence $\langle g_n \rangle$ can be represented by the polynomial

$$\sum_{n=0}^{\infty} g_n z^n \;=\; g_0 + g_1 z + g_2 z^2 + \cdots$$

If the sequence has infinitely many non-zero elements, then the polynomial has infinitely many terms. Generating functions have many uses, even though the motivation for introducing them is not immediately obvious. Their immediate application in this section is directly in counting. In Chapter 2, they reappear as an intermediate device in the transformation of recurrences into closed formulas for sequences. In Chapter 9, they are used in a sophisticated algebraic method for counting graphs.

DEFINITION: An *(ordinary) generating function* (abbr. **OGF**) for the sequence $\langle g_n \rangle$ is any closed form $G(z)$ such that

$$G(z) \;=\; \sum_{n=0}^{\infty} g_n z^n$$

or, sometimes, it means the polynomial itself.

Exponential Generating Functions

There is another kind of generating function, called an *exponential generating function*, that is also used directly for counting and in solving recurrences. We introduce it here and offer a brief explanation of the circumstances in which each of these two main kinds of generating function is used in counting. More extensive development of exponential generating functions appears in §5.5.

DEFINITION: An **exponential generating function** (abbr. **EGF**) for a sequence $\langle g_n \rangle$ is any closed form $\hat{G}(z)$ corresponding to the infinite polynomial

$$\sum_{n=0}^{\infty} g_n \frac{z^n}{n!}$$

or, sometimes, the polynomial itself.

Direct Counting with Ordinary Generating Functions

Ordinary generating functions are readily applicable to counting unordered selections. We now illustrate this by returning to a counting problem first raised in Example 0.3.13.

Example 1.7.1: A combination of letters from the word SYZYGY may contain at most one S. Thus, an ordinary generating function for the number of possible combinations containing no letters that are not S's is

$$1 + s$$

Similarly, ordinary generating functions for combinations containing no letters except Z's and no letters except G's are, respectively

$$1 + z \quad \text{and} \quad 1 + g$$

Since the word SYZYGY contains three Y's, the OGF for counting combinations containing no letters except Y's is

$$1 + y + y^2 + y^3$$

which signifies that there is one choice with no Y's, one choice with one Y, one with two Y's, and one with three Y's. In the product

$$(1 + s)(1 + z)(1 + g)(1 + y + y^2 + y^3)$$

of these four generating functions, the terms of degree d provide an itemization of the ways to select d letters from SYZYGY. For instance, the seven terms of degree 2 are

$$sz \quad sg \quad sy \quad zg \quad zy \quad gy \quad y^2$$

It follows that if each of the indeterminates s, z, g, and y is replaced by a single indeterminate, say x,

$$(1 + x)^3(1 + x + x^2 + x^3)$$

then the coefficient of x^d in the expansion

$$1 + 4x + 7x^2 + 8x^3 + 7x^4 + 4x^5 + x^6$$

is the number of ways to select d letters from SYZYGY. The general principle is articulated by the following proposition.

Proposition 1.7.1. Let $G(z)$ and $H(z)$ be the ordinary generating functions for counting unordered selections from two disjoint multisets S and T. Then $G(z)H(z)$ is the ordinary generating function for counting unordered selections from the union $S \cup T$.

Proof: This is a direct application of the Rule of Sum and Rule of Product. $\quad \Diamond$

Direct Counting with Exponential Generating Functions

Exponential generating functions are readily applicable to counting ordered selections. We continue the analysis of Example 1.7.1.

Example 1.7.1, continued: An ordered selection of letters from SYZYGY may contain at most one S. Thus, an exponential generating function for the number of possible combinations containing no letters that are not S's is

$$1 + s$$

Similarly, exponential generating functions for ordered selections containing no letters except Z's and no letters except G's are, respectively

$$1 + z \quad \text{and} \quad 1 + g$$

Since the word SYZYGY contains three Y's, the exponential generating function for counting ordered selections containing no letters except Y's is

$$1 + y + \frac{y^2}{2!} + \frac{y^3}{3!}$$

which signifies that there is one way with no Y's, one way with one Y, one with two Y's, and one with three Y's. In the product

$$(1+s)(1+z)(1+g)\left(1 + y + \frac{y^2}{2!} + \frac{y^3}{3!}\right)$$

of these four generating functions, the terms of degree d provide an itemization of the ways to select d letters from SYZYGY. Suppose that the multivariate indeterminate monomial of a term of degree d is given the denominator of $d!$. For instance, this would give the transformation

$$\frac{zgy^2}{2!} \quad \longrightarrow \quad \frac{4!}{2!\,1!\,1!} \cdot \frac{zgy^2}{4!} \quad = \quad \binom{4}{2\ 1\ 1} \cdot \frac{zgy^2}{4!}$$

in which the multinomial coefficient $\binom{4}{2\ 1\ 1}$ is the number of ways to order the selection ZGYY represented by the monomial zgy^2. It follows that if each of the indeterminates s, z, g, and y is replaced by a single indeterminate, say x,

$$(1+x)^3\left(1 + x + \frac{x^2}{2!} + \frac{x^3}{3!}\right)$$

then the coefficient of x^d in the expansion

$$1 + 4\frac{x}{1!} + 13\frac{x^2}{2!} + 34\frac{x^3}{3!} + 72\frac{x^4}{4!} + 120\frac{x^5}{5!} + 120\frac{x^6}{6!}$$

is the number of ordered selections of d letters from SYZYGY. The general principle is as follows.

Proposition 1.7.2. *Let $\hat{G}(z)$ and $\hat{H}(z)$ be the exponential generating functions for counting ordered selections from two disjoint multisets S and T. Then $\hat{G}(z)\hat{H}(z)$ is the exponential generating function for counting ordered selections from the union $S \cup T$.*

Proof: This is a another direct application of the Rule of Sum and Rule of Product. $\diamond$

Analyzing a Generating Function

Multiplying two or more generating functions for sequences with simple closed forms may lead to a generating function for a sequence whose closed form is not readily apparent, as seen in Example 1.7.1. Thus, to use generating functions effectively, either for direct counting or for solving recurrences, one needs to be able to analyze generating functions so as to recover a closed-form function for the list of entries. We now indicate briefly how this might be done, deferring most of the details to Chapter 2.

Example 1.7.2:   Let's consider how we might analyze the generating function

$$\frac{z}{1 - 3z + 2z^2} \tag{1.7.1}$$

to enable us to extract the coefficients. When a closed-form generating function is a quotient of polynomials, one way to extract the entries of the sequence is by *long division of polynomials*.

$$
\begin{array}{r}
z + 3z^2 + 7z^3 + 15z^4 + \cdots \\
1 - 3z + 2z^2 \overline{\smash{\big)}\ z } \\
\underline{z - 3z^2 + 2z^3 } \\
3z^2 - 2z^3 \\
\underline{3z^2 - 9z^3 + 6z^4} \\
7z^3 - 6z^4
\end{array}
$$

Long division provides a recursive procedure for generating successive entries of the sequence. This sequence corresponds to the infinite polynomial

$$z + 3z^2 + 7z^3 + 15z^4 + \cdots$$

While this is useful for the coefficients of smaller powers z^n, it is not a closed form. Factoring the denominator of the expression (1.7.1) and splitting the fraction into two parts, like this

$$
\begin{aligned}
\frac{z}{(1 - z)(1 - 2z)} &= \frac{1}{1 - 2z} - \frac{1}{1 - z} \\
&= (1 + 2z + 2^2z^2 + 2^3z^3 + \cdots) - (1 + z + z^2 + z^3 + \cdots) \\
&= \sum_{n=0}^{\infty} (2^n - 1)z^n
\end{aligned}
$$

illustrates the standard way to recover a closed-form function for arbitrary entries of a sequence with such a generating function. Techniques for splitting the fraction are developed in §2.3.

Remark: Example 1.7.2 uses the familiar algebraic identity

$$\frac{1}{1 - ay} = 1 + ay + a^2 y^2 + \cdots$$

which can be justified either by long division or by multiplying $1 - ay$ and $1 + ay + a^2 y^2 + \cdots$.

Rational Functions

Fortunately, many generating functions that arise in the course of solving direct counting problems and recursions have an essential similarity to the generating function (1.7.1).

DEFINITION: A quotient of two polynomials in z (each with finitely many terms) is called a **rational function** in z. If the degree of the numerator is less than the degree of the denominator, then it is called a **proper rational function**.

Remark: An improper rational function can be transformed by long division into the sum of a polynomial — the quotient of dividing the denominator into the numerator — and a proper rational function, whose numerator is the remainder of that division.

Long division of the denominator into the numerator transforms a generating function $G(z)$ represented as a rational function

$$G(z) = \frac{b_0 + b_1 z + \cdots + b_s z^s}{c_0 + c_1 z + \cdots + c_t z^t}$$

into its power series

$$G(z) = g_0 + g_1 z + g_2 z^2 + \cdots$$

as in Example 1.7.2. Moreover, it will be shown in Chapter 2 how to use factoring of the denominator, as in Example 1.7.2, to represent the values of the sequence by a closed function. For the time being, we consider another case of this phenomenon.

Example 1.7.3:   Here is an additional illustration of the effect of factoring the denominator and splitting the fraction into a sum of fractions with linear polynomials as denominators

$$G(z) = \frac{z - 1}{1 - 5z + 6z^2} = \frac{-2}{1 - 3z} + \frac{1}{1 - 2z}$$

$$= -2\left(1 + 3z + 3^2 z^2 + \cdots\right) + \left(1 + 2z + 2^2 z^2 + \cdots\right)$$

$$= \sum_{n=0}^{\infty} \left(2^n - 2 \cdot 3^n\right) z^n$$

Taylor Series

The fact that a rational function can be reconverted into a power series motivates the use of the terminology *generating function*, because a rational function may be regarded as *generating* its coefficients by the process of long division. Another sense in which a function $G(z)$ can generate the coefficients of a power series is by application of a **Taylor series** expansion at $z = 0$.

$$G(z) \; = \; G(0) + G'(0)\frac{z}{1!} + G''(0)\frac{z^2}{2!} + G'''(0)\frac{z^3}{3!} + \cdots$$

that assigns to the infinitely differentiable function $G(z)$ the power series

$$G(z) \; = \; g_0 + g_1 z + g_2 z^2 + \cdots$$

where

$$g_n \; = \; \frac{G^{(n)}(0)}{n!}$$

Using Taylor series permits an interpretation of a wide range of infinitely differentiable functions as generating functions.

Example 1.7.4:   For the function $G(z) = -ln(1-z)$, the value of the nth derivative at $z = 0$ is

$$G^{(n)}(0) \; = \; \left.\frac{(n-1)!}{(1-z)^n}\right|_{z=0} \; = \; (n-1)! \qquad \text{for } n \geq 1$$

and, thus,

$$G(z) \; = \; 0 + 0!\frac{z}{1!} + 1!\frac{z^2}{2!} + 2!\frac{z^3}{3!} + \cdots \; = \; \sum_{n=1}^{\infty}\frac{1}{n}z^n$$

That is, the function $-ln(1-z)$ is the OGF for the sequence $\langle x_n = \frac{1}{n}\rangle$.

Addition and Scalar Multiplication

There is a correspondence between various operations on sequences $\langle a_n \rangle$ and $\langle b_n \rangle$ and some operations on their associated generating functions

$$A(z) \; = \; \sum_{j=0}^{\infty} a_j z^j \quad \text{and} \quad B(z) \; = \; \sum_{j=0}^{\infty} b_j z^j$$

DEFINITION: The **sum of two sequences** $\langle a_n \rangle$ and $\langle b_n \rangle$ is the sequence

$$a_0 + b_0, \quad a_1 + b_1, \quad a_2 + b_2, \quad \ldots$$

This corresponds to the sum of their generating functions, i.e., to the generating function

$$(A+B)(z) \;=\; \sum_{j=0}^{\infty} (a_j + b_j) z^j$$

DEFINITION: **Multiplying the sequence $\{a_n\}$ by the scalar** c yields the sequence

$$ca_0, \quad ca_1, \quad ca_2, \quad \ldots$$

This corresponds to the generating function

$$cA(z) \;=\; \sum_{j=0}^{\infty} ca_j z^j$$

that results from multiplying the generating function $A(z)$ by that scalar.

Example 1.7.5:   Since the ordinary generating functions

$$A(z) = \frac{1}{1-5z} \quad \text{and} \quad B(z) = \frac{1}{1-7z}$$

generate the sequences $\langle a_n = 5^n \rangle$ and $\langle b_n = 7^n \rangle$, respectively, it follows that the ordinary generating function

$$2A(z) + 3B(z) \;=\; \frac{2}{1-5z} + \frac{3}{1-7z} \;=\; \frac{5-29z}{(1-5z)(1-7z)}$$

generates the sequence $\langle 2 \cdot 5^n + 3 \cdot 7^n \rangle$.

Products and Convolutions

The following two examples illustrate how one might use products of generating functions in direct computations.

Example 1.7.6:   Consider the problem of counting the number p_n of ways to make $n\cent$ postage from $3\cent$ and $5\cent$ stamps. If one had nothing but $3\cent$ stamps, the generating function would be

$$\sum_{n=0}^{\infty} a_n x^n \;=\; 1 + x^3 + x^6 + x^9 + \cdots \;=\; \frac{1}{1-x^3}$$

which signifies that there is exactly one way from $3\cent$ stamps alone to make each multiple of 3, and no way to make any other postage. Similarly, if one had nothing but $5\cent$ stamps, the generating function would be

$$\sum_{n=0}^{\infty} b_n y^n \;=\; 1 + y^5 + y^{10} + y^{15} + \cdots \;=\; \frac{1}{1-y^5}$$

In the product of these two generating functions, the number of terms of degree n would be the number of ways of making n¢ postage. For instance, the terms of degree 23 (i.e., the terms whose exponents have 23 as their sum) are

$$x^{18}y^5 \quad \text{and} \quad x^3 y^{20}$$

It follows that if z is substituted for x and y, then the coefficient of z^{23} is the number of ways. Thus, the generating function is

$$\frac{1}{1-z^3} \cdot \frac{1}{1-z^5} \;=\; \sum_{n=0}^{\infty} p_n z^n \;=\; \sum_{n=0}^{\infty} z^n \sum_{j=0}^{n} a_j b_{n-j}$$

That is, the only way to get n¢ postage is to find an $a_j = 1$ and a $b_{n-j} = 1$. The sequence $\langle p_n \rangle$ is not monotonic. For instance,

$$p_{14} = 1 \quad p_{15} = 2 \quad p_{16} = 1$$

COMPUTATIONAL NOTE: In trying to obtain actual values for such a sequence, it is useful to have the aid of a computational engine such as *Mathematica*.

DEFINITION: The **convolution of the sequences** $\langle u_n \rangle$ and $\langle v_n \rangle$ is the sequence

$$u_0 v_0, \quad u_0 v_1 + u_1 v_0, \quad u_0 v_2 + u_1 v_1 + u_2 v_0, \quad \ldots$$

Example 1.7.6, continued: Thus, the sequence $\langle p_n \rangle$ is the convolution of the sequences $\langle a_n \rangle$ and $\langle b_n \rangle$.

Example 1.7.7:   Four distinguishable six-sided dice are rolled, each marked with the numbers 1, 2, 3, 4, 5, 6. Then the generating function for the number of ways that sum of the outcomes could be n is the coefficient of z^n in the expansion of

$$(z + z^2 + z^3 + z^4 + z^5 + z^6)^4$$

Proposition 1.7.3. *The product of the generating functions*

$$U(z) = \sum_{n=0}^{\infty} u_n z^n \quad \text{and} \quad V(z) = \sum_{n=0}^{\infty} v_n z^n$$

is the generating function

$$U(z)V(z) \;=\; \sum_{n=0}^{\infty} z^n \sum_{j=0}^{n} u_j v_{n-j}$$

for the convolution of the sequences $\langle u_n \rangle$ *and* $\langle v_n \rangle$. $\diamond$

We observe that Proposition 1.7.3 provides terminology for the sum of products that occurs within the proof of Proposition 1.7.1.

Example 1.7.8:   The rational functions

$$\frac{1}{1-2z} \quad \text{and} \quad \frac{1}{1-3z}$$

generate the sequences $\langle u_n = 2^n \rangle$ and $\langle v_n = 3^n \rangle$, respectively. Their product is the generating function

$$\frac{1}{(1-2z)(1-3z)} = \frac{-2}{1-2z} + \frac{3}{1-3z}$$

$$= \sum_{n=0}^{\infty} z^n (3^{n+1} - 2^{n+1})$$

$$= 1 + 5z + 19z^2 + 69z^3 + \cdots$$

The convolution of the sequences $\langle u_n = 2^n \rangle$ and $\langle v_n = 3^n \rangle$ is the sequence whose n^{th} element (counting from the 0th element) is

$$2^0 \cdot 3^n + 2^1 \cdot 3^{n-1} + \cdots + 2^n \cdot 3^0$$

Thus, the convolution sequence begins

$$1, \quad 5, \quad 19, \quad 69, \quad \ldots$$

in affirmation of Proposition 1.7.3.

Sums and Generating Functions

Proposition 1.7.3 has a slue of useful consequences. An immediate consequence is that it provides a method for going from a counting sequence to its sequence of partial sums.

Theorem 1.7.4. *Let $B(z)$ be the ordinary generating function for a sequence $\langle b_n \rangle$. Then the ordinary generating function for the sequence*

$$\left\langle \sum_{j=0}^{n} b_j \mid n = 0, 1, \ldots \right\rangle$$

of partial sums is

$$\frac{B(z)}{1-z}$$

Proof: We observe that the total coefficient of z^n in

$$\frac{B(z)}{1-z} = \left(b_0 + b_1 z + b_2 z^2 + \cdots\right)\left(1 + z + z^2 + \cdots\right)$$

equals the sum $\sum_{j=0}^{n} b_j$, as per the following calculation:

$$b_0 + b_1 z + b_2 z^2 + b_3 z^3 + \cdots$$
$$\times \quad 1 + z + z^2 + z^3 + \cdots$$

$$b_0 + b_1 z + b_2 z^2 + b_3 z^3 + \cdots$$
$$b_0 z + b_1 z^2 + b_2 z^3 + b_3 z^4 + \cdots$$
$$b_0 z^2 + b_1 z^3 + b_2 z^4 + b_3 z^5 + \cdots$$
$$\cdots$$

$$b_0 + (b_1 + b_0)z + (b_2 + b_1 + b_0)z^2 + \cdots$$

This is just a special case of Proposition 1.7.3.   ◇

Corollary 1.7.5. $\displaystyle \frac{1}{(1-z)^r} = \sum_{n=0}^{\infty} \binom{n+r-1}{r-1} z^n$

Proof: By induction on r.

BASIS: For $r = 1$, we have

$$\frac{1}{(1-z)^1} = \sum_{n=0}^{\infty} z^n = \sum_{n=0}^{\infty} \binom{n+1-1}{1-1} z^n$$

since the value of each of the coefficients $\binom{n}{0}$ is 1.

IND HYP: Next, suppose for some $r \geq 1$ that

$$\frac{1}{(1-z)^{r-1}} = \sum_{n=0}^{\infty} \binom{n+r-2}{r-2} z^n$$

IND STEP: Then

$$\frac{1}{(1-z)^r} = \frac{1}{1-z} \cdot \frac{1}{(1-z)^{r-1}}$$

$$= \frac{1}{1-z} \sum_{n=0}^{\infty} \binom{n+r-2}{r-2} z^n \qquad \text{(inductive hypothesis)}$$

$$= \sum_{n=0}^{\infty} z^n \sum_{j=0}^{n} \binom{j+r-2}{r-2} \qquad \text{(Theorem 1.7.4)}$$

$$= \sum_{n=0}^{\infty} \frac{z^n}{(r-2)!} \sum_{j=0}^{n} (j+r-2)^{\underline{r-2}} \qquad \text{(factor inner summand)}$$

$$= \sum_{n=0}^{\infty} \frac{z^n}{(r-2)!} \frac{(n+r-1)^{\underline{r-1}} - (r-2)^{\underline{r-1}}}{r-1} \qquad \text{(Corollary 1.5.2)}$$

$$= \sum_{n=0}^{\infty} \frac{z^n}{(r-2)!} \frac{(n+r-1)^{\underline{r-1}}}{(r-1)}$$

$$= \sum_{n=0}^{\infty} \binom{n+r-1}{r-1} z^n \qquad \qquad\qquad\qquad ◇$$

Table 1.7.1 gives closed-form generating functions for some standard sequences and forms of sequences.

Table 1.7.1   Ordinary generating functions for some sequences.

sequence	closed form
$1, \ 1, \ 1, \ 1, \ \dots$	$\dfrac{1}{1-z}$
$1, \ -1, \ 1, \ -1, \ \dots$	$\dfrac{1}{1+z}$
$1, \ 0, \ 1, \ 0, \ \dots$	$\dfrac{1}{(1-z^2)}$
$1, \ 0, \ 0, \ 1, \ 0, \ 0, \ \dots$	$\dfrac{1}{(1-z^3)}$
$1, \ a, \ a^2, \ a^3, \ \dots$	$\dfrac{1}{1-az}$
$0, \ a, \ 2a^2, \ 3a^3, \ \dots$	$\dfrac{z}{1-az}$
$1, \ 2, \ 3, \ 4, \ \dots$	$\dfrac{1}{(1-z)^2}$
$1, \ \binom{m+1}{1}, \ \binom{m+2}{2}, \ \binom{m+3}{3}, \ \dots$	$\dfrac{1}{(1-z)^{m-1}}$
$\frac{1}{0!}, \ \frac{1}{1!}, \ \frac{1}{2!}, \ \frac{1}{3!} \ \dots$	e^z
$0, \ 1, \ \frac{1}{2}, \ \frac{1}{3} \ \dots$	$\ln(1-z)$

Example 1.7.9:   The rational function $\dfrac{1}{(1-z)^2}$ generates the sequence

$$\binom{n+1}{1}: \quad 1, \ 2, \ 3, \ 4, \ \dots$$

Example 1.7.10:   The rational function $\dfrac{1}{(1-z)^3}$ generates the sequence

$$\binom{n+2}{2}: \quad 1, \ 3, \ 6, \ 10, \ \dots$$

Corollary 1.7.6. $\dfrac{1}{(1-az)^r} = \displaystyle\sum_{n=0}^{\infty} \binom{n+r-1}{r-1} a^n z^n$

Proof: Substitute az for z in Corollary 1.7.5. $\diamond$

Example 1.7.11:   The rational function $\dfrac{1}{(1-2z)^2}$ generates the sequence

$$\binom{n+1}{1} 2^n : \quad 1, \quad 4, \quad 12, \quad 32, \quad \cdots$$

Example 1.7.12:   The rational function $\dfrac{1}{(1-2z)^3}$ generates the sequence

$$\binom{n+2}{2} 2^n : \quad 1, \quad 6, \quad 24, \quad 80, \quad \cdots$$

EXERCISES for Section 1.7

In each of the Exercises 1.7.1 through 1.7.6, write the OGF for the number of unordered selections of letters from the given word.

1.7.1^S	*BANDANA*	1.7.2	*FOREIGNER*
1.7.3	*HORSERADISH*	1.7.4	*CONSTITUTION*
1.7.5	*MISSISSIPPI*	1.7.6	*WOOLLOOMOOLOO*

In each of the Exercises 1.7.7 through 1.7.12, write the EGF for the number of ordered selections of letters from the given word.

1.7.7^S	*BANDANA*	1.7.8	*FOREIGNER*
1.7.9	*HORSERADISH*	1.7.10	*CONSTITUTION*
1.7.11	*MISSISSIPPI*	1.7.12	*WOOLLOOMOOLOO*

In Exercises 1.7.13 through 1.7.20, use long division on the given rational function to calculate the terms of degrees 0 through 4 of the infinite polynomial.

1.7.13^S $\dfrac{1}{1-5z+6z^2}$ 1.7.14 $\dfrac{-1+z}{1-5z+6z^2}$

1.7.15 $\dfrac{1}{1-2z+z^2}$ 1.7.16 $\dfrac{1}{1-3z+3z^2-z^3}$

1.7.17 $\dfrac{1}{1-4z+4z^2}$ 1.7.18 $\dfrac{1}{1-6z+12z^2-8z^3}$

1.7.19 $\dfrac{5-29z}{1-12z+35z^2}$ 1.7.20 $\dfrac{2-3x+5x^3}{1-4z^2+4z^4}$

Exercises 1.7.21 through 1.7.28 are concerned with a Taylor series at $z = 0$ for the given function.

   a. Calculate the first three terms of the Taylor series.

   b. Derive an expression for the nth term.

1.7.21[s] $\dfrac{1}{1-z}$ **1.7.22** $\dfrac{1}{1+z}$

1.7.23 $\dfrac{z}{1-z}$ **1.7.24** $\dfrac{1}{(1-z)^2}$

1.7.25 $\dfrac{1}{1-2z}$ **1.7.26** $\dfrac{1}{(1-2z)^2}$

1.7.27 $\dfrac{1}{1-3z+2z^2}$ **1.7.28** $\ln(1+z)$

1.7.29   Give a detailed proof of Proposition 1.7.1.

1.7.30   Give a detailed proof of Proposition 1.7.2.

1.8   SYNTHESIZING GENERATING FUNCTIONS

Synthesizing a generating function for a given sequence is a skill, like analyzing them, that is fundamental to solving counting problems with them. The approach is to recognize fundamental patterns in the sequence and to perceive how these patterns were combined.

Example 1.8.1: In the sequence

$$-4, \quad 2, \quad 5, \quad 2, \quad -6, \quad 2, \quad 7, \quad 2 \cdots \tag{1.8.1}$$

the two fundamental patterns are

$$1, \quad 2, \quad 3, \quad 4, \quad 5, \quad 6 \ \cdots \tag{1.8.2}$$

and

$$2, \quad 2, \quad 2, \quad 2, \ \cdots \tag{1.8.3}$$

It seems that sequence (1.8.2) acquired negative signs on its even elements, that the entries preceding the entry 4 were truncated, and that it was then interwoven with sequence (1.8.3) by strict alternation.

Example 1.8.1 serves as a running example for this section. Our objective is to construct its generating function.

Substitution

Proposition 1.8.1 Substitution Rule. *If $G(z)$ is a generating function for the sequence $\langle g_n \rangle$, then $G(bz)$ is a generating function for the sequence $\langle b^n g_n \rangle$.*

Proof: $\displaystyle\sum_{n=0}^{\infty} g_n (bz)^n \;=\; \sum_{n=0}^{\infty} b^n g_n z^n.$ $\hspace{4cm}$ $\diamond$

Example 1.8.1, continued: By Example 1.7.9, the generating function for the sequence (1.8.2): 1, 2, 3, 4, ... is

$$\frac{1}{(1-z)^2}$$

Substitute $(-1)z$ for z, according to Proposition 1.8.1, to obtain the OGF

$$\frac{1}{(1+z)^2}$$

for the sequence

$$1, \quad -2, \quad 3, \quad -4, \quad 5, \quad -6 \quad \cdots \tag{1.8.4}$$

Shifting Right and Left

DEFINITION: **Shifting the sequence $\langle a_n \rangle$ to the right** by k places yields the sequence

$$\overbrace{0,\; 0,\; \ldots,\; 0}^{k \text{ zeroes}},\; a_0,\; a_1,\; a_2,\; \ldots$$

The corresponding generating function is

$$z^k A(z) \;=\; \sum_{j=0}^{\infty} a_j z^{j+k}$$

DEFINITION: **Nullifying** the jth element of the sequence $\langle a_n \rangle$ means replacing a_j by 0. The corresponding generating function is

$$A(z) \;-\; a^j z^j$$

DEFINITION: **Shifting the sequence $\langle a_n \rangle$ to the left** by k places yields the sequence

$$a_k,\; a_{k+1},\; a_{k+2},\; \ldots$$

The corresponding generating function is

$$z^{-k}\left[A(z) - \sum_{j=0}^{k-1} a_j z^j \right] \;=\; \sum_{j=k}^{\infty} a_j z^{j-k}$$

The terms $a_0, a_1, \ldots, a_{k-1}$ are nullified, so that they do not end up as non-zero coefficients of negative powers of z.

Example 1.8.1, continued: Shifting sequence (1.8.4) to the left by three places yields the sequence

$$-4, \quad 5, \quad -6 \quad 7, \quad -8, \quad 9, \cdots \tag{1.8.5}$$

which corresponds to the OGF

$$z^{-3}\left(\frac{1}{(1+z)^2} - 1 + 2z - 3z^2\right) = \frac{-4 - 3z}{(1+z)^2}$$

Spacing Out

DEFINITION: **Spacing a sequence** $\langle a_n \rangle$ by k units yields the sequence

$$a_0, \overbrace{0, \ldots, 0}^{k \;\; 0's}, a_1, \overbrace{0, \ldots, 0}^{k \;\; 0's}, a_2, \overbrace{0, \ldots, 0}^{k \;\; 0's}, \ldots$$

The corresponding generating function is

$$A(z^{k+1})$$

Example 1.8.1, continued: Spacing sequence (1.8.5) by 1 place yields the sequence

$$-4, \quad 0, \quad 5, \quad 0, \quad -6, \quad 0, \quad 7, \quad 0, \quad -8, \quad 0, \quad 9, \cdots \tag{1.8.6}$$

which corresponds to the OGF

$$\left.\frac{-4 - 3z}{(1+z)^2}\right|_{z \to z^2} = \frac{-4 - 3z^2}{(1+z^2)^2}$$

Isolating a Subsequence

DEFINITION: **Isolating the subsequence** $n \equiv k \bmod m$ of the sequence $\langle a_n \rangle$ yields the sequence in which all terms are nullified, except those whose index is congruent to $k \bmod m$.

For modulus $m = 2$, the corresponding generating function is

$$\begin{cases} \dfrac{A(z) + A(-z)}{2} & \text{if } k = 0 \\[2mm] \dfrac{A(z) - A(-z)}{2} & \text{if } k = 1 \end{cases}$$

Example 1.8.1, continued: Since the rational function $\frac{1}{1-z}$ generates a sequence of 1's, the generating function for the sequence (1.8.3) is

$$\frac{2}{1-z}$$

Isolating the 1 mod 2 subsequence from sequence (1.8.3) yields the sequence

$$0, \quad 2, \quad 0, \quad 2, \quad \cdots \tag{1.8.7}$$

which corresponds to the OGF

$$\frac{1}{2}\left(\frac{2}{1-z} - \frac{2}{1+z}\right) = \frac{2z}{1-z^2}$$

which might also have been obtained by spacing sequence (1.8.3) out by 1 unit and shifting right 1 place. Sequence (1.8.1) is the sum of sequences (1.8.6) and (1.8.7). Thus, its OGF is the sum of their OGF's, i.e.,

$$\frac{-4 - 3z^2}{(1+z^2)^2} + \frac{2z}{1-z^2} = \frac{2z^5 + 3z^4 + 4z^3 + z^2 + 2z - 4}{(1-z^4)(1+z^2)}$$

Differentiation

DEFINITION: The **derivative of the generating function**

$$G(x) = \sum_{n=0}^{\infty} g_n z^n$$

is the generating function

$$G'(x) = \sum_{n=1}^{\infty} n\, g_n z^{n-1} = \sum_{n=0}^{\infty} (n+1)\, g_{n+1}\, z^n$$

Example 1.8.2:   Consider the generating function

$$G(z) = \frac{1}{1-2z} = \sum_{n=0}^{\infty} 2^n z^n$$

Then taking its derivative yields the equation

$$G'(z) = \frac{2}{(1-2z)^2} = \sum_{n=0}^{\infty} (n+1)\, 2^{n+1} z^n$$

which is consistent with Corollary 1.7.6.

EXERCISES for Section 1.8

In Exercises 1.8.1 through 1.8.8, write a closed-form generating function for the given sequence.

$1.8.1^{\text{S}}$ $1, -1, 1, -1, 1, -1, \ldots$ $1.8.2$ $1, 0, 1, 0, 1, 0, \ldots$

$1.8.3$ $1, 1, 1, 1, 1, 1, \ldots$ $1.8.4$ $1, 1, -1, -1, 1, 1, -1, -1, \ldots$

$1.8.5$ $1, 0, -1, 0, 1, 0, -1, 0, \ldots$ $1.8.6$ $1, 0, 0, 1, 0, 0, 1, 0, 0, \ldots$

$1.8.7$ $1, 1, 0, 1, 1, 0, 1, 1, 0, \ldots$ $1.8.8$ $0, 0, 0, 1, 1, 1, 1, 1, 1, \ldots$

In Exercises 1.8.9 through 1.8.18, write a closed-form generating function for the given sequence.

$1.8.9^{\text{S}}$ $1, 2, 3, 4, 5, 6, \ldots$ $1.8.10$ $1, 0, 2, 0, 3, 0, \ldots$

$1.8.11$ $1, -2, 3, -4, 5, -6, \ldots$ $1.8.12$ $1, 2, -3, -4, 5, 6, -7, -8, \ldots$

$1.8.13$ $1, 0, 3, 0, 5, 0, 7, 0, \ldots$ $1.8.14$ $1, 0, -3, 0, 5, 0, -7, 0, \ldots$

$1.8.15$ $1, 2, 0, 4, 5, 0, 7, 8, 0, \ldots$ $1.8.16$ $1, 2, 0, 3, 4, 0, 5, 6, 0, \ldots$

$1.8.17$ $1, 1, 2, 1, 3, 1, 4, 1, \ldots$ $1.8.18$ $1, 3, 6, 10, 15, 21, \ldots$

In Exercises 1.8.19 and 1.8.20, use a difference table to determine a closed formula for the nth term of the given sequence and then write the corresponding generating function.

$1.8.19^{\text{S}}$ $1, 3, 8, 17, 32, 57, 100, \ldots$ $1.8.20$ $0, 1, 4, 10, 20, 35, 56, \ldots$

$1.8.21$ The 4th roots of unity are the complex numbers $i, -1, -i,$ and 1. Given a generating function $A(z)$, show that the subsequence $n \equiv 0 \bmod 4$ of the sequence $\langle a_n \rangle$ can be isolated as

$$\frac{A(z) + A(iz) + A(-z) + A(-iz)}{4}$$

$1.8.22$ Let $1, \omega, \omega^2, \ldots, \omega^{k-1}$ be the set of kth roots of unity. Given a generating function $A(z)$, show that the subsequence $n \equiv 0 \bmod k$ of the sequence $\langle a_n \rangle$ can be isolated as

$$\frac{A(z) + A(\omega z) + A(\omega^2 z) + \cdots + A(\omega^{k-1} z)}{k}$$

1.9   ASYMPTOTIC ESTIMATES

The growth rate of a function is customarily reckoned via comparison to benchmarks. For instance, it might be said of the function nH_n that it grows faster than the function n but slower than n^2. The focus is on the long term. Computer algorithmists compare various algorithms to achieve a specific objective in their pursuit of an optimal algorithm, where optimality means using the smallest amount of computational resources as the algorithm is applied to ever larger instances of the problem at hand.

DEFINITION: Let $f(n)$ be a function such that $f(n) \neq 0$ for sufficiently large n. The sequence x_n is **asymptotic** to $f(n)$ if

$$\lim_{n \to \infty} \frac{x_n}{f(n)} = 1$$

It is often reasonably straightforward to guess or to find a well-understood function $f(n)$ such that the ratio

$$\frac{x_n}{f(n)}$$

converges. Rigorous study of *asymptotics* is concerned not only with finding a function $f(n)$ to which a given sequence $\langle x_n \rangle$ is asymptotic, but also with calculating the rate of convergence. Determining the rate of convergence tends to require a more extensive background in graduate-level continuous mathematics than can be assumed here or developed just-in-time. Thus, we focus presently on the function to which the given sequence is asymptotic.

Example 1.9.1:   How large is the Catalan number c_n? From the expansion

$$\begin{aligned}
c_n &= \frac{1}{n+1}\binom{2n}{n} = \frac{1}{n+1} \cdot \frac{(2n)^{\underline{n}}}{n!} \\
&= \frac{1}{n+1} \cdot \frac{2n}{n} \cdot \frac{2n-1}{n-1} \cdot \ \ldots \ \cdot \frac{(n+1)}{1}
\end{aligned}$$

one sees that the Catalan number c_n is a product of the value of $\frac{1}{n+1}$ and the values of n other factors, whose values form an increasing sequence from 2 to $n+1$. One surmises that

$$\frac{2^n}{n+1} < c_n < \frac{(n+1)^n}{n+1} = (n+1)^{n-1}$$

which is a very wide range of possibilities, since the lower and upper bounds are far apart. Narrowing that gap is a primary need toward improved understanding of the behavior of the Catalan sequence.

Ratio Method

Ratio Method: Considerable information about the asymptotic behavior of a sequence x_n lies in the ratio

$$\frac{x_n}{x_{n-1}}$$

of successive terms. We calculate the limit of that ratio.

Example 1.9.1, continued: The ratio of successive entries of the Catalan sequence is

$$\frac{c_n}{c_{n-1}} = \frac{1}{n+1}\binom{2n}{n} \bigg/ \frac{1}{n}\binom{2n-2}{n-1}$$

$$= \frac{1}{n+1}\cdot\frac{(2n)^{\underline{n}}}{n!} \bigg/ \frac{1}{n}\cdot\frac{(2n-2)^{\underline{n-1}}}{(n-1)!}$$

$$= \frac{1}{n+1}\cdot\frac{(2n)^{\underline{n}}}{(2n-2)^{\underline{n-1}}}$$

$$= \frac{1}{n+1}\cdot\frac{(2n)(2n-1)}{n}$$

$$\Rightarrow \quad \frac{c_n}{c_{n-1}} = \frac{4n-2}{n+1} \tag{1.9.1}$$

$$\Rightarrow \quad \lim_{n\to\infty}\frac{c_n}{c_{n-1}} = \lim_{n\to\infty}\frac{4n-2}{n+1}$$

$$= \lim_{n\to\infty}\frac{4n+4}{n+1} - \lim_{n\to\infty}\frac{6}{n+1} = 4-0$$

$$\Rightarrow \quad \lim_{n\to\infty}\frac{c_n}{c_{n-1}} = 4 \tag{1.9.2}$$

Since the ratio $\frac{c_n}{c_{n-1}}$ is everywhere less than its asymptotic upper limit of 4, and since $c_1 = 1 < 4$, it is possible to narrow the estimating range of c_n to

$$\frac{2^n}{n+1} < c_n = \frac{1}{n+1}\binom{2n}{n} < 4^n \tag{1.9.3}$$

Philosophy of Estimation: A formula in n in which the number of operations of addition, subtraction, multiplication, division, and exponentiation needed for evaluation is a constant is an easier formula to grasp than one for which that number grows with n. The number of multiplications grows in a factorial or in a falling power. Here, "grasping" includes the ability to estimate the value of the formula for a concrete value of n.

Tightening Bounds on Estimates

Concrete Substitution: Concrete early values of a sequence can often be used to improve asymptotic upper and lower bounds.

Example 1.9.1, continued: Sharpening the lower bound of (1.9.3) for the Catalan number c_n, including eliminating the denominator of $n + 1$, can begin with an observation regarding the ratio $\frac{c_n}{c_{n-1}}$ after $n = 5$.

$$4n - 2 \geq 3n + 3 \qquad \text{for } n \geq 5$$

$$\Rightarrow \quad \frac{4n - 2}{n + 1} \geq \frac{3n + 3}{n + 1} = 3$$

Recalling (1.9.1), we have

$$\frac{c_n}{c_{n-1}} \geq 3 \tag{1.9.4}$$

Using (1.9.4) and the fact that

$$c_n = c_4 \cdot \frac{c_5}{c_4} \cdot \frac{c_6}{c_5} \cdot \ \ldots \ \cdot \frac{c_n}{c_n - 1}$$

we infer that

$$c_n \geq c_4 \cdot 3^{n-4} = 14 \cdot 3^{n-4}$$

$$\Rightarrow \quad c_n = \frac{14}{81} \cdot 3^n > \frac{1}{6} \cdot 3^n \qquad \text{for } n \geq 4 \tag{1.9.5}$$

The inequality (1.9.5) also holds for c_0, c_1, c_2, and c_3. Recalling the inequality (1.9.3), it follows that

$$\frac{1}{6} \cdot 3^n < c_n < 4^n \qquad \text{for } n \geq 0 \tag{1.9.6}$$

We shall now show that the coefficient of $\frac{1}{6}$ can be removed from the lower bound of (1.9.6) for sufficiently large values of n. Since the ratio

$$\frac{c_n}{c_{n-1}}$$

is increasing monotonically to 4, it eventually exceeds $\frac{7}{2}$, say, for all $n > P$. Since $\frac{7}{6} > 1$, there is a number Q such that

$$\left(\frac{7}{2}\right)^q > \frac{3^P}{c_P} \qquad \text{for all } q \geq Q - P$$

which implies that

$$c_n = c_P \cdot \left(\frac{c_{P+1}}{c_P} \cdot \frac{c_{P+2}}{c_{P+1}} \cdot \ \ldots \ \cdot \frac{c_Q}{c_{Q-1}} \right) \cdot \left(\frac{c_{Q+1}}{c_Q} \cdot \frac{c_{Q+2}}{c_{Q+1}} \cdot \ \ldots \ \cdot \frac{c_n}{c_{n-1}} \right)$$

$$> c_P \cdot \left(\frac{7}{2}\right)^{Q-P} \cdot \left(\frac{7}{2}\right)^{n-Q}$$

$$> 3^P \cdot 3^{Q-P} \cdot 3^{n-Q}$$

$$\Rightarrow \quad c_n > 3^n \tag{1.9.7}$$

Combining (1.9.6) and (1.9.7) yields the desired result

$$3^n < c_n < 4^n \qquad \text{for } n \geq Q \tag{1.9.8}$$

Remark: In fact, this lower bound is further improvable. Since the ratio

$$\frac{c_n}{c_{n-1}}$$

is increasing monotonically to 4, it eventually surpasses $4 - \epsilon$ for any $\epsilon > 0$, say, for all $n > N(\epsilon)$. It follows, by an argument similar to that used in the derivation of (1.9.6), that

$$\frac{c_{N(\epsilon)}}{4^{N(\epsilon)}} \cdot (4 - \epsilon)^n \; < c_n \; < \; 4^n \qquad \text{for } n \geq N(\epsilon)$$

The coefficient could be removed, once again, as in the derivation of (1.9.8), to yield the asymptotic estimate

$$(4 - \epsilon)^n \; < \; c_n \; < \; 4^n$$

which is adequate for present purposes.

The following proposition formulates the method used in Example 1.9.1 as a general principle.

Proposition 1.9.1. *Let* x_n *be a sequence such that*

$$lim_{n \to \infty} \frac{x_n}{x_{n-1}} \; = \; K > 0$$

Then, for $\epsilon > 0$ *and sufficiently large values of* n,

$$(K - \epsilon)^n \; < c_n \; < \; (K + \epsilon)^n \tag{1.9.9}$$

If the ratio $\frac{x_n}{x_{n-1}}$ *is bounded above by* K, *then (1.9.9) can be sharpened to*

$$(K - \epsilon)^n \; < c_n \; < \; x_0 K^n \tag{1.9.10}$$

If bounded below by K, *then (1.9.9) can be sharpened to*

$$x_0 K^n \; < c_n \; < \; (K + \epsilon)^n \tag{1.9.11}$$

Proof: Details from Example 1.9.1 are readily transformed into a proof. This is left to the Exercises. $\diamond$

Asymptotic Dominance

DEFINITION: If there is a positive number c such that

$$f(n) \; \leq \; cg(n) \quad \text{for all } n \geq N$$

then we may write

$$f(n) \in \mathcal{O}(g(n))$$

and say "$f(n)$ is in **big-oh** of $g(n)$". The numbers c and N are called **witnesses** to the relationship.

TERMINOLOGY NOTE: Although $\mathcal{O}(g(n))$ is defined here as the class of functions that are eventually dominated by a multiple of $g(n)$, the usage "f *is* big-oh of g" (omitting the preposition "in") is quite common. The rationale is that membership in the class may be regarded as an adjectival property.

Example 1.9.2:   One way to prove that $7n^2 \in \mathcal{O}(n^3)$ is to choose the witnesses $N = 7$ and $c = 1$. Then

$$7n^2 \leq 1 \cdot n^3 \text{ for } n \geq 7$$

Another proof uses the witnesses $N = 1$ and $c = 7$. Then

$$7n^2 \leq 7 \cdot n^3 \text{ for } n \geq 1$$

In general, there tends to be a tradeoff in the size of the witnesses N and c. Choosing a larger value of witness c may enable one to choose a smaller value of witness n.

Example 1.9.3:   To prove that $n^3 \notin \mathcal{O}(7n^2)$, we observe that for any witness c, and for any number $n > 8c$,

$$n^3 > (8c)n^2 > 7c \cdot n^2$$

EXERCISES for Section 1.9

These exercises may be challenging for a reader with little prior experience at constructing proofs about limits.

1.9.1   Prove that every polynomial of degree less than d is in $\mathcal{O}(n^d)$, for $d \in \mathbb{Z}^+$.

1.9.2   Prove that every polynomial of degree d is in $\mathcal{O}(n^d)$, for $d \in \mathbb{Z}^+$.

1.9.3   Prove that if $0 < r < s$, then $n^r \in \mathcal{O}(n^s)$, but $n^s \notin \mathcal{O}(n^r)$, for $r, s \in \mathbb{R}$.

1.9.4   Prove that $\lg n \in \mathcal{O}(n^r)$, for $r > 0$.

1.9.5   Prove that $n^r \notin \mathcal{O}(\lg n)$, for $r > 0$.

1.9.6   Prove that if $0 < b < c$, then $b^n \in \mathcal{O}(c^n)$ but $c^n \notin \mathcal{O}(b^n)$, for $b, c \in \mathbb{R}$.

1.9.7   Give a proof of Proposition 1.9.1.

GLOSSARY

asymptotic to a function $f(n)$:   the property of a sequence x_n that

$$\lim_{n \to \infty} \frac{x_n}{f(n)} = 1$$

big-oh of a function $g(n)$:   the class of functions that are eventually dominated by a scalar multiple of $g(n)$.

binomial coefficient $\binom{n}{k}$:   the coefficient of x^k in the expansion of $(1 + x)^n$.

Catalan number: any number in the sequence c_n defined by the recursion

$$c_0 = 1; \quad c_n = c_0 \, c_{n-1} + c_1 \, c_{n-2} + \cdots + c_{n-1} \, c_0$$

ceiling of a real number x: the smallest integer that is not larger than x; the result of "rounding up" to the next integer; denoted $\lceil x \rceil$.

cells of a partition of a set S: the subsets into which S is subdivided.

closed formula for a sequence x_n: an algebraic expression for the value of x_n (in the argument n).

combination coefficient $\binom{n}{k}$: the number of ways to choose a subset of size k from a set of size n.

concave sequence on an integer interval $[a:b]$: a sequence x_n such that

$$x_n \geq \frac{x_{n-1} + x_{n+1}}{2} \qquad (\text{for } n = a+1,\ \ldots,\ b-1)$$

convex sequence on an integer interval $[a:b]$: a sequence x_n such that

$$x_n \leq \frac{x_{n-1} + x_{n+1}}{2} \qquad (\text{for } n = a+1,\ \ldots,\ b-1)$$

convolution of two sequences a_n and b_n: the sequence whose n^{th} entry is

$$\sum_{j=0}^{n} a_j b_{n-j}$$

difference function of a real function $f(x)$: the function $\triangle f$ given by the rule

$$\triangle f(x) = f(x+1) - f(x)$$

difference sequence of a real sequence a_n: the sequence $\triangle a_n$ given by the rule

$$\triangle a_n = a_{n+1} - a_n$$

difference table of a real sequence a_n: the table whose rows are

$$a_n,\ \triangle a_n,\ \triangle^2 a_n,\ \ldots$$

EGF: see *generating function*.

eventual dominance by a function $g(n)$: the property of a function $f(n)$ that there is a number $N \in \mathbb{Z}^+$ such that $f(n) \leq g(n)$ for all $n \geq N$.

eventually has a property: the subsequence from some index N onward has the property.

falling power $x^{\underline{n}}$: the number $x\,(x-1)\cdots(x-n+1)$.

Fibonacci number: any number in the sequence f_n defined by the recursion

$$f_0 = 0,\ f_1 = 1;\quad f_n = f_{n-1} + f_{n-2}$$

floor of a real number x: the largest integer that is not greater than x; the result of "rounding down" to the next integer; denoted $\lfloor x \rfloor$.

Fundamental Theorem of Finite Calculus: a theorem relating differencing and summation.

generating function of a sequence a_n:

___, **exponential (abbr. EGF)**: a closed form for the power series

$$a_0 + a_1 \frac{z}{1!} + a_2 \frac{z^2}{2!} + \cdots$$

or sometimes the series itself.

___, **ordinary** (abbr. **OGF**):   a closed form for the power series

$$a_0 + a_1 z + a_2 z^2 + \cdots$$

or sometimes the series itself.

growth rate of a sequence: a comparative measure of its eventual values.

harmonic number H_n: the value of the sum $\frac{1}{1} + \frac{1}{2} + \cdots + \frac{1}{n}$.

Heawood number of the surface S_g: the value of the expression

$$\left\lfloor \frac{7 + \sqrt{1 + 48g}}{2} \right\rfloor$$

integer interval $[k : m]$:   the set of integers $\{k, k+1, \ldots, m\}$.

log-concave: property of a sequence that implies unimodality.

map on a surface: a drawing of a graph on the surface, subdividing it into regions.

Method of Distinguished Element: a method used to derive combinatorial formulas.

mode of a unimodal sequence: the maximum value.

mode index of a unimodal sequence x_n: the index at which the maximum value occurs.

OGF: see *generating function*.

partial sum, n^{th}, of a sequence a_n: the sum

$$\sum_{j=0}^{n} a_j$$

partition of a set S: a family of mutually disjoint subsets whose union is S.

Pascal's recurrence: the recurrence

$$x_{n,0} = 1, \quad n_{0,k} = 0 \quad \text{for } k > 0;$$
$$x_{n,k} = x_{n-1,k-1} + x_{n-1,k} \quad \text{for } n > 0;$$

Pascal's triangle: a triangle formed by the non-zero values of the binomial coefficients.

periodic sequence: a sequence whose values are an unending reiteration of a finite initial segment.

range of a sequence: the set in which the sequence takes its values.

rational function: the quotient of two polynomials.

recursion: a formula for expressing the value of an entry of a sequence in terms of the values of earlier entries.

rising power $x^{\overline{n}}$:   the number $x(x+1) \cdots (x+n-1)$.

solving a recurrence: finding a closed formula for the entries of the sequence it specifies.

Stirling cycle number $\left[{n \atop k} \right]$:   the number of ways to partition a set of n objects into k non-empty cycles.

Stirling numbers of the first kind $s_{n,k}$:   numbers used in converting a falling power into a linear combination of ordinary powers; they are equal to the Stirling cycle numbers.

Stirling numbers of the second kind $S_{n,k}$:   numbers used in converting an ordinary power into a linear combination of falling powers; they are equal to the Stirling subset numbers.

Stirling subset number $\left\{ {n \atop k} \right\}$:   the number of ways to partition a set of n objects into k non-empty subsets.

Tower of Hanoi: a puzzle invented by Edouard Lucas, which is solved recursively.

type of a partition: a list of the sizes of its cells.

unimodal sequence: a sequence that is monotonically non-decreasing up to a maximum and monotonically non-increasing thereafter.

witnesses: two parameters that occur in establishing a big-oh relationship.

Chapter 2

Solving Recurrences

As indicated by its title, this chapter is predominantly concerned with solving recurrences. In §2.1, it identifies a basic type of recurrence, called a linear recurrence with constant coefficients, which is amenable to a fairly simple solution. The next three sections develop two approaches to solving such a recurrence. The first approach is completely general, and it applies to all kinds of recurrences, not just this special, most tractable form: one derives a generating function for the sequence specified by the recurrence, and then one analyzes that generating function so as to have a closed form for the values in the sequence. Application of the second approach is restricted to linear recurrence relations with constant coefficients: having memorized some standard patterns and their solutions, or possibly with the aid of a table of standard patterns, one sees how a given linear recurrence fits a standard pattern and adapts the solution. How to solve simultaneous recurrences is described in §2.5. Special properties of the Fibonacci numbers are featured in §2.6. The focus of §2.7 and §2.8 is on techniques for transforming a more complicated type of recurrence into a linear recurrence with constant coefficients, thus preconditioning it for solution by the well-established methods of the earlier sections.

2.1   TYPES OF RECURRENCES

REVIEW FROM §1.2:

- A **recurrence** for a sequence prescribes a set of **initial values**

$$x_0 = b_0 \quad x_1 = b_1 \quad \ldots \quad x_k = b_k$$

and a **recursion** formula

$$x_n = \phi(x_{n-1}, x_{n-2}, \ldots, x_0) \quad \text{for } n > k$$

from which one may calculate the value of x_n, for any $n > k$, from the values of earlier entries.

One top-level demarcation in the taxonomy of recursions is the distinction between *linear* and *non-linear* recursions. Another is the distinction between *homogeneous* and *non-homogeneous* recursions. This section explains these two distinctions and various other considerations that also affect the choice of a method of solution.

DEFINITION: A recursion formula of the form

$$x_n = a_{n-1}x_{n-1} + a_{n-2}x_{n-2} + \cdots + a_0x_0 + \alpha(n)$$

in which each term is linear is said to be a **linear recursion**. Each coefficient a_j may be either a **constant coefficent**, the same for all n, or a function of n, that is, a **variable coefficient**.

- It is a **recursion of degree** d if the number of coefficients a_j that are non-zero is bounded, and if the smallest subscript among the non-zero coefficients is $n - d$.

- The function $\alpha(n)$ is called the **particularity function**.

- It is a **homogeneous recursion** if the particularity function is 0.

Some Linear Recursions

It is fortunate that some of the most familiar recursions are linear, because linear recursions are usually easier to solve than non-linear recursions. It tends also to be easier to solve a recursion with constant coefficients than one with variable coefficients.

Example 2.1.1:   The *Tower of Hanoi recursion* (introduced in §1.2 and solved in §2.2)

$$h_n = 2h_{n-1} + 1 \tag{2.1.1}$$

is a non-homogeneous, linear recursion of degree 1, with a constant coefficient.

Example 2.1.2:   The *Fibonacci recursion* (introduced in §0.2 and §1.2 and solved in §2.5)

$$f_n = f_{n-1} + f_{n-2} \tag{2.1.2}$$

is a homogenous linear recursion of degree 2, with constant coefficients.

Recurrences without Fixed Degree

A recurrence of fixed degree d for a sequence $\langle x_n \rangle$ prescribes x_n as a combination of the recent past entries

$$x_{n-1} \quad x_{n-2} \quad \cdots \quad x_{n-d}$$

In the most important kind of recurrence without fixed degree, the value of x_n is a combination of entries whose indices are a fraction of n. This is called a *divide-and-conquer* recurrence. Methods for solving such recurrences appear in §2.8. This kind of recurrence arises frequently in computer science, in circumstances when completing a task on input of size n can be reduced not just to completing it for slightly smaller size input, but for input of much smaller size.

Example 2.1.3: A well-known recurrence from computer science that approximates the number of steps needed to sort a file by iterative merging does not have a fixed degree. The *merge-sort recurrence* (explained and solved in §2.8)

$$\begin{aligned} m_1 &= 1; \\ m_n &= 2m_{\lceil \frac{n}{2} \rceil} + n \end{aligned} \tag{2.1.3}$$

is a non-homogeneous linear recurrence without fixed degree, with a constant coefficient. Its recursion formula expresses that the problem of sorting a list of length n is reduced to merging two lists of size $\frac{n}{2}$.

Variable Coefficients

The three recursions (2.1.1), (2.1.2), and (2.1.3) all have constant coefficients. One of the most important linear recursions with variable coefficients arises in the study of *permutations*.

REVIEW FROM §0.5:

- A **permutation** on a set S is a one-to-one, onto function from S to itself.

- **Theorem 0.5.3.** Every permutation can be represented as the composition of disjoint cyclic permutations.

DEFINITION: A **derangement** is a permutation π with no fixed points. That is, there is no object x such that $\pi(x) = x$.

Example 2.1.4: Figure 2.1.1 illustrates a derangement.

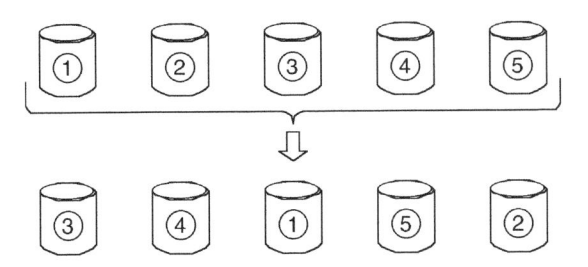

Figure 2.1.1 **The derangement** $(1\ 3)(2\ 5\ 4)$.

DEFINITION: The **derangement number** D_n is the number of derangements of the integer interval $[1 : n]$.

The derangements of the smallest integer intervals $[1 : n]$ are given in Table 2.1.1. We observe that the permutation π is a derangement if there is no 1-cycle in the disjoint cycle form of π.

Table 2.1.1   Derangements of Small Intervals $[1 : n]$.

n				D_n
1				0
2	$(1\ 2)$			1
3	$(1\ 2\ 3)$	$(1\ 3\ 2)$		2
4	$(1\ 2)(3\ 4)$	$(1\ 3)(2\ 4)$	$(1\ 4)(2\ 3)$	9
	$(1\ 2\ 3\ 4)$	$(1\ 2\ 4\ 3)$	$(1\ 3\ 2\ 4)$	
	$(1\ 3\ 4\ 2)$	$(1\ 4\ 2\ 3)$	$(1\ 4\ 3\ 2)$	

Proposition 2.1.1. *The derangement numbers* D_n *satisfy the following recursion formula.*

$$D_n \;=\; (n-1)D_{n-1} + (n-1)D_{n-2} \tag{2.1.4}$$

Proof: Every derangement of $[1 : n]$ such that n does *not* lie in a 2-cycle can be formed by inserting the number n, immediately after one of the $n - 1$ numbers in some cycle of some derangement of $[1 : n - 1]$. Every derangement of $[1 : n]$ in which n does lie in a 2-cycle can be formed from some derangement π of $[1 : n - 2]$ either by adding the 2-cycle $(n - 1\ \ n)$, or by replacing one of the $n - 2$ numbers j in some cycle of π by the number $n - 1$ and then adding the 2-cycle $(j\ \ n)$. $\diamondsuit$

Example 2.1.5:   The *derangement recurrence* (considered in more detail in §5.4)

$$D_0 \;=\; 1, \quad D_1 \;=\; 0;$$
$$D_n \;=\; (n-1)D_{n-1} + (n-1)D_{n-2}$$

is a homogenous linear recurrence of degree 2, whose coefficients are variable. The sequence it specifies is convex, since

$$\frac{D_{n+1} + D_{n-1}}{2} \;=\; \frac{(nD_n + nD_{n-1}) + D_{n-1}}{2}$$
$$\geq\; \frac{n}{2} D_n$$
$$\geq\; D_n \quad \text{for } n \geq 2$$

Even without solving the derangement recurrence, it is possible to prove inductively that most permutations have a fixed point, that is, that the ratio $\frac{D_n}{n!}$ of derangements to permutations is less than half. This is clearly true for $n = 1$ and $n = 2$.

For $n \geq 3$, we have

$$
\begin{aligned}
D_n &= (n-1)D_{n-1} + (n-1)D_{n-2} \\
&< (n-1)\frac{(n-1)!}{2} + (n-1)\frac{(n-2)!}{2} \qquad \text{(ind hyp)} \\
&= n\frac{(n-1)!}{2} - \frac{(n-1)!}{2} + (n-1)\frac{(n-2)!}{2} \\
&= \frac{n!}{2}
\end{aligned}
$$

Some Non-linear Recurrences

All of the recursions $(2.1.1)$, ..., $(2.1.4)$ are linear. Various other important recurrences are non-linear.

Example 2.1.6:   The *Catalan recurrence* (introduced in §1.2 and solved in §4.4)

$$
\begin{aligned}
c_0 &= 1; \\
c_n &= c_0 c_{n-1} + c_1 c_{n-2} + \cdots + c_{n-1} c_0
\end{aligned}
$$

is a homogenous non-linear recurrence without finite degree, with constant coefficients.

Remark: The next three sections present two basic approaches to solving a linear recurrence of fixed degree with constant coefficients. The remaining sections are concerned with reducing the solution of other kinds of recurrences to these basic approaches.

EXERCISES for Section 2.1

In each of the Exercises 2.1.1 through 2.1.4, write the first 6 values of the given recurrence, guess the closed formula, and then prove by induction that your guess is correct.

2.1.1[S] $a_n = 2a_{n-1} - 1; \quad a_0 = 3.$

2.1.2 $a_n = 4a_{n-1} - 1; \quad a_0 = 9.$

2.1.3 $a_n = a_{n-1} + 2n - 1; \quad a_0 = 0.$

2.1.4 $a_n = a_{n-1} + 2^n; \quad a_0 = 1.$

In each of the Exercises 2.1.5 through 2.1.7, write a homogeneous linear recurrence of degree 2 with constant coefficients that corresponds to the given closed formula. Hint: substitute the given solution into a general linear recurrence of degree 2.

2.1.5[S] $a_n = 3^n - 2^n.$

2.1.6 $a_n = 3^n - 2^n + 1.$

2.1.7 $a_n = n2^n - 1.$

2.1.8   List the derangements of the integer interval $[1:5]$.

2.1.9   Calculate the derangement numbers D_5 and D_6 using the derangement recurrence given in Example 2.1.5.

2.1.10   Prove that more than a third of the permutations of n objects, for $n \geq 4$, have a fixed point.

2.1.11   Write a recurrence for the number of binary strings without a pair of adjacent 0's.

2.1.12   Write a recurrence for the number of binary strings without three consecutive 0's.

2.1.13   Write a recurrence for the number of ternary strings without a pair of adjacent 0's.

2.1.14   Write a recurrence for the number of ternary strings with an even number of 1's.

2.1.15[S]   Write a recurrence for the number of ternary strings without a pair of adjacent 0's, of adjacent 1's, or of adjacent 2's.

2.1.16   Write a recurrence for the number of sequences (order matters!) of 1's and 2's whose sum is n.

2.1.17   Write a recurrence for the number of sequences (order matters!) of 1's and 2's whose sum is n, with evenly many 1's.

2.1.18[S]   Write a recurrence for the number of sequences (order matters!) of 1's and 2's whose sum is n, with evenly many 2's.

2.2   FINDING GENERATING FUNCTIONS

This section is devoted to the fundamental method for solving a recurrence of the form

$$
\begin{aligned}
g_0 &= b_0, \ \ldots, \ g_k = b_k; & \text{initial conditions} \\
g_n &= \gamma(g_{n-1}, \ldots, g_0) \ \text{for } n > k & \text{recursion}
\end{aligned}
$$

It uses three steps to determine a closed form for the corresponding generating function

$$
G(z) \ = \ \sum_{n=0}^{\infty} g_n z^n \tag{2.2.1}
$$

and then a fourth step to derive a closed formula for the coefficients g_n. We describe the four steps of this fundamental method with reference to this running example of a recurrence system.

Example 2.2.1:   This is a linear homogenous recursion of degree 2 with constant coefficients.

$$g_0 = 1, \quad g_1 = 2;$$
$$g_n = 5g_{n-1} - 6g_{n-2} \quad \text{for } n > 1$$

Step 1a. Multiply both sides of the recursion equation by z^n.

$$g_n z^n = 5g_{n-1} z^n - 6g_{n-2} z^n$$

Step 1b. Sum both sides of the resulting equation over the same range of values, with a lower bound as low as possible, and upward to ∞.

$$(1b) \qquad \sum_{n=2}^{\infty} g_n z^n = \sum_{n=2}^{\infty} 5g_{n-1} z^n - \sum_{n=2}^{\infty} 6g_{n-2} z^n$$

We start all the sums at the lower bound $n = 2$, because starting any lower would take the subscript of g_n below 0 on the left side, and this recurrence system does not specify either g_{-2} or g_{-1}.

Step 2. Recalling equation (2.2.1), we observe that

$$\sum_{n=2}^{\infty} g_n z^n = \sum_{n=0}^{\infty} g_n z^n - g_1 z - g_0 = G(z) - g_1 z - g_0$$

Thus, we can replace each infinite sum in equation (1b) by an algebraic expression involving the generating function $G(z)$.

$$(2a) \qquad \sum_{n=2}^{\infty} g_n z^n = 5z \sum_{n=2}^{\infty} g_{n-1} z^{n-1} - 6z^2 \sum_{n=2}^{\infty} g_{n-2} z^{n-2}$$

$$(2b) \qquad G(z) - g_1 z - g_0 = 5z \left(G(z) - g_0 \right) - 6z^2 G(z)$$

In $(2a)$, we factor the terms of each sum on the right, so that the power of z in the summand equals the subscript. In $(2b)$, we replace all three infinite sums.

Step 3. Solve for $G(z)$.

$$(3a) \qquad G(z) \left(1 - 5z + 6z^2 \right) = g_1 z + g_0 - 5g_0 z$$
$$= 2z + 1 - 5z$$

$$(3b) \qquad G(z) = \frac{1 - 3z}{1 - 5z + 6z^2}$$

In $(3a)$ we collect the $G(z)$ terms on the left and substitute initial values for the low-subscripted entries of the sequence. In $(3b)$, we isolate the generating function $G(z)$ on the left.

Step 4. Solve for g_n.

(4a)
$$G(z) = \frac{1 - 3z}{(1 - 2z)(1 - 3z)} = \frac{1}{1 - 2z}$$

(4b)
$$= \sum_{n=0}^{\infty} 2^n z^n \qquad \Rightarrow \qquad g_n = 2^n$$

Step (4a) converts the result of step (3b) into a more tractable form. In (4b) we extract the coefficient g_n.

Check the Answer: A better way to confirm the answer than by retracing the steps is to verify that the answer $g_n = 2^n$ satisfies the recurrence.

$$
\begin{aligned}
g_0 &= 2^0 = 1, \quad g_1 = 2^1 = 2; \qquad \text{initial conditions} \\
g_n &= 5g_{n-1} - 6g_{n-2} \qquad\qquad\qquad \text{recursion} \\
&= 5 \cdot 2^{n-1} - 6 \cdot 2^{n-2} \\
&= 5 \cdot 2^{n-1} - 3 \cdot 2^{n-1} \\
&= 2 \cdot 2^{n-1} \\
&= 2^n
\end{aligned}
$$

Step (4a) is usually not quite this simple, as illustrated by this variation on the running example.

Example 2.2.1, continued: Suppose that the initial values in the preceding problem were changed to

$$g_0 = 0, \, g_1 = 2$$

Then steps 1 and 2 would be as before. However, here is how we would finish in the modified problem.

Step 3. Solve for $G(z)$.

(3a)
$$
\begin{aligned}
G(z)(1 - 5z + 6z^2) &= g_1 z + g_0 - 5zg_0 \\
&= 2z + 0 - 0
\end{aligned}
$$

(3b)
$$G(z) = \frac{2z}{1 - 5z + 6z^2}$$

Step (3a) collects the $G(z)$ terms on the left and substitutes initial values for the low-subscripted entries of the sequence. Step (3b) isolates the generating function $G(z)$ on the left.

Step 4. Solve for g_n. Step (4a) anticipates a method called *partial fraction decomposition*, which is described in the next section. For now, we can confirm that the calculation in Step (4a) is correct, by proceeding from right to left on its top line. The next section describes how to do such a calculation from left to right.

(4a)
$$G(z) = \frac{2z}{(1 - 2z)(1 - 3z)} = \frac{-2}{1 - 2z} + \frac{2}{1 - 3z}$$

$$= \sum_{n=0}^{\infty} (-2)2^n z^n + \sum_{n=0}^{\infty} 2 \cdot 3^n z^n$$

(4b)
$$\Rightarrow \quad g_n = -2^{n+1} + 2 \cdot 3^n$$

Check the Answer: As before, we verify that the answer satisfies the recurrence. This time the answer is $g_n = -2^{n+1} + 2 \cdot 3^n$.

$$
\begin{aligned}
g_0 &= -2^1 + 2 \cdot 3^0 = -2 + 2 = 0 \qquad \text{initial conditions} \\
g_1 &= -2^2 + 2 \cdot 3^1 = -4 + 6 = 2 \\
g_n &= 5g_{n-1} - 6g_{n-2} \qquad\qquad\qquad \text{recursion} \\
&= 5\left(-2^n + 2 \cdot 3^{n-1}\right) - 6\left(-2^{n-1} + 2 \cdot 3^{n-2}\right) \\
&= (-5) \cdot 2^n + 10 \cdot 3^{n-1} + 3 \cdot 2^n - 4 \cdot 3^{n-1} \\
&= -2 \cdot 2^n + 6 \cdot 3^{n-1} \\
&= -2^{n+1} + 2 \cdot 3^n
\end{aligned}
$$

Example 2.2.2:   The method of generating functions also solves non-homogeneous recurrences. We illustrate this with a revisit to the Hanoi recurrence.

$$
h_0 = 0; \qquad h_n = 2h_{n-1} + 1 \quad \text{for } n > 0
$$

We proceed through the same four steps.

$$(1a) \qquad\qquad h_n z^n = 2h_{n-1} z^n + 1 z^n$$

$$(1b) \qquad\qquad \sum_{n=1}^{\infty} h_n z^n = \sum_{n=1}^{\infty} 2h_{n-1} z^n + \sum_{n=1}^{\infty} z^n$$

$$(2a) \qquad\qquad \sum_{n=1}^{\infty} h_n z^n = 2z \sum_{n=1}^{\infty} h_{n-1} z^{n-1} + z \sum_{n=1}^{\infty} z^{n-1}$$

$$(2b) \qquad\qquad H(z) - h_0 = 2zH(z) + \frac{z}{1-z}$$

$$(3a) \qquad\qquad H(z)(1-2z) = h_0 + \frac{z}{1-z} = 0 + \frac{z}{1-z}$$

$$(3b) \qquad\qquad H(z) = \frac{z}{(1-z)(1-2z)}$$

We explain in §2.3 how to split a rational function.

$$(4a) \qquad\qquad H(z) = \frac{1}{1-2z} - \frac{1}{1-z}$$

$$(4b) \qquad\qquad\qquad = \sum_{n=0}^{\infty} 2^n z^n - \sum_{n=0}^{\infty} z^n \qquad \Rightarrow \quad h_n = 2^n - 1$$

This solution was suggested in §1.2 by examination of small cases and then confirmed by mathematical induction.

EXERCISES for Section 2.2

In each of the Exercises 2.2.1 through 2.2.14, write a generating function for the given recurrence.

2.2.1^S $a_n = 2a_{n-1}$; $a_0 = 3$.

2.2.2 $a_n = 2a_{n-1} - 3$; $a_0 = 3$.

2.2.3 $a_n = 3a_{n-1} - 2a_{n-2}$; $a_0 = 2$, $a_1 = 1$.

2.2.4 $a_n = 3a_{n-1} - 2a_{n-2} + 2$; $a_0 = 2$, $a_1 = 1$.

2.2.5 $a_n = 3a_{n-1} - 2a_{n-2} + 2$; $a_0 = 2$, $a_1 = -1$.

2.2.6 $a_n = 3a_{n-1} - 2a_{n-2} + n$; $a_0 = 2$, $a_1 = -1$.

2.2.7 $a_n = 5a_{n-1} - 6a_{n-2} + n$; $a_0 = 1$, $a_1 = 3$.

2.2.8 $a_n = 5a_{n-1} - 6a_{n-2} + n^2$; $a_0 = 1$, $a_1 = 4$.

2.2.9^S $a_n = 7a_{n-1} + 8a_{n-2} + (-1)^n$; $a_0 = 0$, $a_1 = 1$.

2.2.10 $a_n = 4a_{n-1} - 4a_{n-2} + 2^n$; $a_0 = 3$, $a_1 = 1$.

2.2.11 $a_n = 5a_{n-1} + 6a_{n-2} + 2n + 1$; $a_0 = 2$, $a_1 = -1$.

2.2.12 $a_n = 2a_{n-2} + a_{n-3}$; $a_0 = 0$, $a_1 = 1$, $a_2 = 2$.

2.2.13 $a_n = 4a_{n-1} - a_{n-2} - 6a_{n-3}$; $a_0 = 0$, $a_1 = 1$, $a_2 = 2$.

2.2.14 $a_n = a_{n-1} + 2a_{n-2} + 3a_{n-3}$; $a_0 = 0$, $a_1 = 1$, $a_2 = 2$.

DEFINITION: For any graph G, the puzzle we will call the **Tower-of-**G has a frame that models G and a peg at every vertex. The objective is, as in the Tower of Hanoi, to move a stack of disks from a designated *source* peg s to a *target* peg t, subject to the requirements that a disk can be transferred only to a peg at an *adjacent* vertex with no smaller disks on it.

Exercises 2.2.15 through 2.2.18 all concern a Tower-of-G puzzle.

2.2.15   Find a graph H such that the Tower-of-H puzzle is equivalent to the Tower of Hanoi, and explain the equivalence.

2.2.16^S Consider the Tower-of-$K_{1,3}$ puzzle in which both the designated source peg and target peg are at vertices of degree 1. Write and solve the recurrence for the minimum number of moves.

2.2.17   Consider the Tower-of-$K_{1,3}$ puzzle in which both the designated source peg is at the vertex of degree 3 and the target peg is at a vertex of degree 1. Write and solve the recurrence for the minimum number of moves.

2.2.18   Consider the Tower-of-$K_{1,3}$ puzzle in which both the designated source peg is at a vertex of degree 1 and the target peg is at a vertex of degree 3. Write and solve the recurrence for the minimum number of moves.

2.3   PARTIAL FRACTIONS

Suppose that a linear recurrence

$$x_n = a_{n-1}x_{n-1} + a_{n-2}x_{n-2} + \cdots + a_0x_0 + \alpha(n)$$

has constant coefficients a_j and that its particularity function $\alpha(n)$ is a polynomial in n. Then the generating function constructed by Steps 1, 2, and 3 of the method of §2.2 is a proper rational function.

$$G(z) = \sum_{n=0}^{\infty} g_n z^n = \frac{b_0 + b_1 z + \cdots + b_s z^s}{c_0 + c_1 z + \cdots + c_t z^t}$$

Step 4 is to complete the solution, by deriving a closed formula for g_n. This section develops the details of Step 4. Like the previous section, this section explains the details of the method with the aid of a running example.

Example 2.3.1:   The running example now is the rational function

$$G(z) = \frac{1 - 5z}{1 - 7z + 16z^2 - 12z^3}$$

One may verify that it corresponds to the recurrence

$$g_0 = 1, \quad g_1 = 2, \quad g_2 = -2;$$
$$g_n = 7g_{n-1} - 16g_{n-2} + 12g_{n-3} \quad \text{for } n > 2$$

Step 4a-1. Factor the denominator into linear factors.

$$c_0 + c_1 z + \cdots + c_t z^t = c_0(1 - \tau_1 z)^{\varepsilon_1} \cdots (1 - \tau_k z)^{\varepsilon_k}$$

with $\varepsilon_1 + \cdots + \varepsilon_k = t$. For simplicity, we take $c_0 = 1$. For our example, we have

$$\begin{aligned} G(z) &= \frac{1 - 5z}{1 - 7z + 16z^2 - 12z^3} \\ &= \frac{1 - 5z}{(1 - 2z)^2(1 - 3z)} \end{aligned}$$

By what is called the *Fundamental Theorem of Algebra*, a polynomial with complex coefficients has a factorization into powers of linear polynomials.

Remark: There is no general method for calculating the roots of a polynomial exactly for higher degree polynomials. Nonetheless, in practice, one commonly encounters polynomials that can be factored by elementary methods.

Step 4a-2. Analyze the rational function into a sum of k rational functions, each of whose denominators is one of the factors $(1 - \tau_j)^{\varepsilon_j}$, and whose numerators are "unknown polynomials", each of the respective form

$$b_{j,0} + b_{j,1}z + \cdots + b_{j,\varepsilon_j-1}z^{\varepsilon_j-1}$$

Thus,

$$\frac{1 - 5z}{1 - 7z + 16z^2 - 12z^3} = \frac{b_{1,0} + b_{1,1}z}{(1 - 2z)^2} + \frac{b_{2,0}}{(1 - 3z)}$$

Step 4a-3. Recombine these summands, with a single denominator. For the present example,

$$\frac{1 - 5z}{1 - 7z + 16z^2 - 12z^3} = \frac{(b_{1,0} + b_{1,1}z)(1 - 3z) + b_{2,0}(1 - 2z)^2}{(1 - 2z)^2(1 - 3z)}$$

Step 4a-4. Then collect terms according to the exponent of the factor z^ℓ. For the present example,

$$= \frac{(b_{1,0} + b_{2,0}) + (-3b_{1,0} + b_{1,1} - 4b_{2,0})z + (-3b_{1,1} + 4b_{2,0})z^2}{1 - 7z + 16z^2 - 12z^3}$$

Step 4a-5. Next obtain a system of t linear equations in t unknowns $b_{j,i}$ by equating each resulting coefficient of z^ℓ in the numerator to the corresponding coefficient of z^ℓ in the numerator of the original linear function, and solve that system.

$$\left.\begin{array}{rcl} b_{1,0} \qquad\quad + b_{2,0} &=& 1 \\ -3b_{1,0} + b_{1,1} - 4b_{2,0} &=& -5 \\ -3b_{1,1} + 4b_{2,0} &=& 0 \end{array}\right\} \Rightarrow \begin{array}{rcl} b_{1,0} &=& 7 \\ b_{1,1} &=& -8 \\ b_{2,0} &=& -6 \end{array}$$

Step 4a-6. Now substitute these solutions into the right side of the equation of Step 4a-2.

$$\frac{1 - 5z}{1 - 7z + 16z^2 - 12z^3} = \frac{7 - 8z}{(1 - 2z)^2} + \frac{-6}{1 - 3z}$$

Step 4a-7. Transform each term on the right into the product of its numerator with the power series corresponding, via Corollary 1.7.4, to its denominator. Then simplify each power series.

$$= (7 - 8z) \sum_{n=0}^{\infty} \binom{n+1}{1} 2^n z^n + (-6) \sum_{n=0}^{\infty} \binom{n}{0} 3^n z^n$$

$$= (7 - 8z) \sum_{n=0}^{\infty} (n+1)2^n z^n - 6 \sum_{n=0}^{\infty} 3^n z^n$$

Step 4a-8. Finish by combining into a single power series, and then extracting a closed formula for g_n.

$$= \sum_{n=0}^{\infty} [(3n + 7) \cdot 2^n - 6 \cdot 3^n] \quad \Rightarrow \quad g_n = (3n + 7) \cdot 2^n - 6 \cdot 3^n$$

EXERCISES for Section 2.3

Each of the Exercises 2.3.1 through 2.3.14 corresponds to an exercise in §2.2, which prescribed the determination of a generating function. Analyze the corresponding generating function into partial fractions and solve the recurrence.

2.3.1[S] $a_n = 2a_{n-1};\ a_0 = 3.$

2.3.2 $a_n = 2a_{n-1} - 3;\ a_0 = 3.$

2.3.3 $a_n = 3a_{n-1} - 2a_{n-2};\ a_0 = 2,\ a_1 = 1.$

2.3.4 $a_n = 3a_{n-1} - 2a_{n-2} + 2;\ a_0 = 2,\ a_1 = 1.$

2.3.5 $a_n = 3a_{n-1} - 2a_{n-2} + 2;\ a_0 = 2,\ a_1 = -1.$

2.3.6 $a_n = 3a_{n-1} - 2a_{n-2} + n;\ a_0 = 2,\ a_1 = -1.$

2.3.7 $a_n = 5a_{n-1} - 6a_{n-2} + n;\ a_0 = 1,\ a_1 = 3.$

2.3.8 $a_n = 5a_{n-1} - 6a_{n-2} + n^2;\ a_0 = 1,\ a_1 = 4.$

2.3.9[S] $a_n = 7a_{n-1} + 8a_{n-2} + (-1)^n;\ a_0 = 0,\ a_1 = 1.$

2.3.10 $a_n = 4a_{n-1} - 4a_{n-2} + 2^n;\ a_0 = 3,\ a_1 = 1.$

2.3.11 $a_n = 5a_{n-1} + 6a_{n-2} + 2n + 1;\ a_0 = 2,\ a_1 = -1.$

2.3.12 $a_n = 2a_{n-2} + a_{n-3};\ a_0 = 0,\ a_1 = 1,\ a_2 = 2.$

2.3.13 $a_n = 4a_{n-1} - a_{n-2} - 6a_{n-3};\ a_0 = 0,\ a_1 = 1,\ a_2 = 2.$

2.3.14 $a_n = a_{n-1} + 2a_{n-2} + 3a_{n-3};\ a_0 = 0,\ a_1 = 1,\ a_2 = 2.$

2.4   CHARACTERISTIC ROOTS

To solve a homogenous linear recurrence of fixed degree d with constant coefficients, in addition to using generating functions as described in §2.2 and §2.3, there is an alternative approach called the ***method of characteristic roots***. It begins with the assumption that the recurrence has solutions of the form

$$g_n = \tau_j^n$$

We describe this alternative method with reference to the same recurrence that we used for the running example, Example 2.2.1, that illustrated the method of solution using generating functions.

Suppose that a sequence $\langle g_n \rangle$ is representable by a homogeneous linear recurrence with constant coefficients. Then its generating function $G(z)$ must be a rational function, as one might prove by analyzing the method in §2.2. Furthermore, by splitting $G(z)$ into partial fractions with denominators

$$\left(1 - \tau_j z\right)^{\varepsilon_j}$$

one can prove that the closed form for the entry g_n must be a linear combination of powers of the numbers τ_j, which are the roots of the denominator of $G(z)$.

Example 2.4.1: Applying the method of characteristic roots to this familiar recurrence provides a running example for this section.

$$g_0 = 1, \quad g_1 = 2;$$
$$g_n = 5g_{n-1} - 6g_{n-2} \quad \text{for } n > 1$$

Characteristic Equation

Step 1. Form the **characteristic equation**, as follows.

(1a) Substitute τ^n for g_n in the recurrence.

$$\tau^n = 5\tau^{n-1} - 6\tau^{n-2}$$

(1b) Factor out τ^{n-d}.

$$\tau^2 = 5\tau - 6$$

(1c) Move the non-zero terms to the left of the equals sign

$$\tau^2 - 5\tau + 6 = 0$$

thereby forming the **characteristic polynomial**.

Step 2. Factor the characteristic polynomial.
$$(\tau - 2)(\tau - 3) = 0$$

The roots of the characteristic polynomial

$$\tau_1 = 2 \quad \text{and} \quad \tau_2 = 3$$

are called the **characteristic roots**. We observe their correspondence to the linear factors of the denominator of the generating function derived in Step 4 of Example 2.2.1. We observe that

$$g_n = \tau_1^n = 2^n \quad \text{and} \quad g_n = \tau_2^n = 3^n$$

are solutions to the given recurrence.

Step 3. As a general solution to the given homogeneous recurrence, form a linear combination of the characteristic roots, using unknown coefficients. If none of the roots is repeated, the result of this step is as follows.

$$g_n = b_1 2^n + b_2 3^n$$

We shall eventually return to this step to elaborate on the case in which one or more roots is repeated.

Step 4a. Use the initial conditions to write a system of linear equations for the unknown coefficients.

$$g_0 \ = \ 1 \ = \ b_1 2^0 + b_2 3^0 \ = \ b_1 + b_2$$
$$g_1 \ = \ 2 \ = \ b_1 2^1 + b_2 3^1 \ = \ 2b_1 + 3b_2$$

Step 4b. Solve for the unknown coefficients.

$$b_1 \ = \ 1 \qquad b_2 \ = \ 0$$

Step 4c. Substitute the solutions from Step 4b into the general solution of Step 3.

$$g_n \ = \ 2^n$$

We observe that this is the same solution previously obtained for this recurrence in Example 2.2.1.

Alternative Initial Values

Suppose that we now consider, as in the continuation of Example 2.2.1, the alternative initial values

$$g_0 \ = \ 0, \quad g_1 \ = \ 2$$

Then the finish would be as follows.

Step 4a. Use the initial conditions to write a system of linear equation for the unknown coefficients.

$$g_0 \ = \ 0 \ = \ b_1 2^0 + b_2 3^0 \ = \ b_1 + b_2$$
$$g_1 \ = \ 2 \ = \ b_1 2^1 + b_2 3^1 \ = \ 2b_1 + 3b_2$$

Step 4b. Solve for the unknown coefficients.

$$b_1 \ = \ -2 \qquad b_2 \ = \ 2$$

Step 4c. Substitute the solutions from Step 4b into the general solution of Step 3.

$$g_n \ = \ -2^{n+1} + 2 \cdot 3^n$$

This is the same solution obtained previously, in Example 2.2.1, with these alternative initial values.

Repeated Roots

We now apply the method of characteristic roots to the recurrence of Example 2.3.1.

$$g_0 \ = \ 1, \quad g_1 \ = \ 2, \quad g_2 \ = \ -2;$$
$$g_n \ = \ 7g_{n-1} - 16g_{n-2} + 12g_{n-3} \quad \text{for } n > 2$$

Step 1. The characteristic equation is

$$\tau^3 - 7\tau^2 + 16\tau - 12 = 0$$

Step 2. Factor the characteristic polynomial.

$$(\tau - 2)^2 (\tau - 3) = 0$$

Step 3. If a root τ_j has multiplicity ε_j, then use

$$b_{j,0}\, \tau_j^n + b_{j,1}\, n\tau_j^n + \cdots + b_{j,\varepsilon_j-1}\, n^{\varepsilon_j-1}\tau_j^n$$

in forming the general solution with unknown coefficients. In the present example, the general solution is

$$g_n = b_{1,0}2^n + b_{1,1}n2^n + b_2 3^n$$

Step 4a. Use the initial conditions to write a system of linear equation for the unknown coefficients.

$$\begin{aligned}
g_0 = 1 &= b_{1,0}2^0 + b_{1,1}0 \cdot 2^0 + b_2 3^0 = b_{1,0} && + b_2 \\
g_1 = 2 &= b_{1,0}2^1 + b_{1,1}1 \cdot 2^1 + b_2 3^1 = 2b_{1,0} + 2b_{1,1} + 3b_2 \\
g_2 = -2 &= b_{1,0}2^2 + b_{1,1}2 \cdot 2^2 + b_2 3^2 = 4b_{1,0} + 8b_{1,1} + 9b_2
\end{aligned}$$

Step 4b. Solve for the unknown coefficients.

$$b_{1,0} = 7 \quad b_{1,1} = 3 \quad b_2 = -6$$

Step 4c. Substitute the solutions from Step 4b into the general solution of Step 3.

$$g_n = 7 \cdot 2^n + 3n \cdot 2^n - 6 \cdot 3^n$$

This is the same solution obtained in Example 2.3.1.

Remark: The proof that this method works is a matter of checking that it always yields the same solution as the method of generating functions.

Non-homogeneous Equations

To extend the method of characteristic roots to a non-homogeneous linear recurrence of degree d with constant coefficients

$$\begin{aligned}
g_0 &= b_0, \quad \ldots, \quad g_{d-1} = b_{d-1}; \\
g_n &= a_{n-1}g_{n-1} + \cdots + a_{n-d}\,g_{n-d} + \alpha(n)
\end{aligned}$$

we first isolate the **associated homogeneous recurrence**

$$\hat{g}_n = a_{n-1}\hat{g}_{n-1} + \cdots + a_{n-d}\,\hat{g}_{n-d}$$

obtained by dropping the particularity function.

We illustrate the rest of the extended method with a revisit to the Hanoi recurrence.

$$h_0 = 0;$$
$$h_n = 2h_{n-1} + 1 \quad \text{for } n > 0$$

Use Steps 1, 2, and 3 to find a general solution to the homogeneous recurrence $\hat{h}_n - 2\hat{h}_{n-1} = 0$.

Steps 1, 2. $\qquad\qquad\qquad\qquad \tau - 2 = 0$

Step 3. $\qquad\qquad\qquad\qquad\quad \hat{h}_n = b \cdot 2^n$

The result so far is called the **homogeneous part** of the general solution.

Step 3N. Find a trial function $\dot{h}_n$ that satisfies the original recurrence. Such a trial function is called the **particular solution** or the **particular part**. It usually resembles the particularity function. For instance, if the particularity function is a polynomial in n, then the trial function can be a polynomial of the same degree, with unknown coefficient. Since the particularity function for the Hanoi recurrence is a constant, the trial function can be a constant.

$$\dot{h}_n = c$$

Substitution into the original recurrence leads to a system of linear equations in the unknown coefficients.

$$\dot{h}_n = 2\dot{h}_{n-1} + 1 \qquad\qquad \text{(recurrence)}$$
$$c = 2c + 1 \qquad\qquad\quad \text{(after substitution)}$$
$$c = -1 \qquad\qquad\qquad \text{(particular solution)}$$
$$h_n = \hat{h}_n + \dot{h}_n = b \cdot 2^n - 1 \quad \text{(general solution)}$$

Step 4a. Use the initial conditions to write a system of linear equations for the unknown coefficients.

$$h_0 = 0 = b \cdot 2^0 - 1 = b - 1$$

Step 4b. Solve for the unknown coefficients.

$$b = 1$$

Step 4c. Substitute the solutions for b_1 and b_2 from Step 4b into the general solution of Step 3.

$$h_n = 2^n - 1$$

This is the same solution obtained in Example 2.2.2 by the method of generating functions.

Example 2.4.2:   We modify Example 2.4.1 by giving the recurrence a polynomial particularity function

$$g_0 = 1, \; g_1 = 2;$$
$$g_n = 5g_{n-1} - 6g_{n-2} + 4n - 3 \quad \text{for } n > 1$$

We have previously derived for the homogeneous recurrence the general solution

$$\hat{g}_n = b_1 2^n + b_2 3^n$$

Step 3N. As a particular solution we use the form

$$\dot{g}_n = c_1 n + c_0$$

and substitute it into the particularized recurrence.

$$
\begin{aligned}
4n - 3 &= \dot{g}_n - 5\dot{g}_{n-1} + 6\dot{g}_{n-2} & \text{(recurrence)} \\
&= (c_1 n + c_0) - 5(c_1(n-1) + c_0) + 6(c_1(n-2) + c_0) \\
&= n(c_1 - 5c_1 + 6c_1) + (c_0 - 5c_1 + 5c_0 - 12c_1 + 6c_0) \\
&= 2nc_1 + 2c_0 - 7c_1 & \text{(after substituting)}
\end{aligned}
$$

This leads to the linear equations and solutions

$$
\left.
\begin{aligned}
4 &= 2c_1 \\
-3 &= 2c_0 - 7c_1
\end{aligned}
\right\}
\qquad c_1 = 2 \quad c_0 = \frac{11}{2}
$$

which are combined with the general solution to the homogeneous part.

$$g_n = \hat{g}_n + \dot{g}_n = b_1 2^n + b_2 3^n + 2n + \frac{11}{2}$$

Step 4a. Use the initial conditions to write a system of linear equations for the unknowns b_1 and b_2.

$$g_0 = 1 = b_1 2^0 + b_2 3^0 + \frac{11}{2}$$
$$g_1 = 2 = b_1 2^1 + b_2 3^1 + 2 \cdot 1 + \frac{11}{2}$$

Step 4b. Solve for the unknowns b_1 and b_2.

$$
\left.
\begin{aligned}
-\frac{9}{2} &= b_1 + b_2 \\
-\frac{11}{2} &= 2b_1 + 3b_2
\end{aligned}
\right\}
\qquad
\begin{aligned}
b_1 &= -8 \\
b_2 &= \frac{7}{2}
\end{aligned}
$$

Step 4c. Substitute the solutions for b_1 and b_2 from Step 4b into the general solution from Step 3N.

$$g_n = -8 \cdot 2^n + \frac{7}{2} \cdot 3^n + 2n + \frac{11}{2}$$

Example 2.4.3: We now modify Example 2.4.1 by giving the recurrence an exponential particularity function.

$$g_0 = 1, \ g_1 = 2;$$
$$g_n = 5g_{n-1} - 6g_{n-2} + (-1)^n \quad \text{for } n > 1$$

We have previously derived for the homogeneous recurrence, as in Example 2.4.2, the general solution

$$\hat{g}_n = b_1 2^n + b_2 3^n$$

Step 3N. As a particular solution we use the form

$$\dot{g}_n = c(-1)^n$$

and substitute it into the particularized recurrence.

$$\begin{aligned}
(-1)^n &= \dot{g}_n - 5\dot{g}_{n-1} + 6\dot{g}_{n-2} && \text{(recurrence)} \\
&= c\,(-1)^n + 5c\,(-1)^n + 6c\,(-1)^n \\
&= 12c\,(-1)^n && \text{(after substituting)} \\
c &= \frac{1}{12} && \text{(solution)}
\end{aligned}$$

Combine this solution with the general solution to the homogeneous part.

$$g_n = \hat{g}_n + \dot{g}_n = b_1 2^n + b_2 3^n + \frac{1}{12}(-1)^n$$

Step 4a. Use the initial conditions to write a system of linear equations for the unknowns b_1 and b_2.

$$g_0 = 1 = b_1 2^0 + b_2 3^0 + \frac{1}{12}$$
$$g_1 = 2 = b_1 2^1 + b_2 3^1 - \frac{1}{12}$$

Step 4b. Solve for the unknowns b_1 and b_2.

$$\left.\begin{aligned}
\frac{11}{12} &= b_1 + b_2 \\
\frac{25}{12} &= 2b_1 + 3b_2
\end{aligned}\right\} \qquad
\begin{aligned}
b_1 &= \frac{8}{12} \\
b_2 &= \frac{3}{12}
\end{aligned}$$

Step 4c. Substitute the solutions for b_1 and b_2 from Step 4b into the general solution from Step 3N.

$$g_n = \frac{8}{12} \cdot 2^n + \frac{3}{12} \cdot 3^n + \frac{1}{12}(-1)^n$$

Complex Roots

A recurrence in which the initial values are real and the recursion has real coefficients has a characteristic polynomial with real coefficients. The roots of such a polynomial may be complex.

Example 2.4.4:   The recurrence

$$g_0 \; = \; 1, \; g_1 \; = \; 2;$$
$$g_n \; = \; 2g_{n-1} - 2g_{n-2} \quad \text{for } n > 1$$

has the characteristic equation

$$\tau^2 - 2\tau + 2 \; = \; 0$$

with roots

$$\tau_1 \; = \; 1 + i \quad \text{and} \quad \tau_2 \; = \; 1 - i$$

Thus, the general solution is

$$g_n \; = \; b_1(1 + i)^n \; + \; b_2(1 - i)^n$$

The initial conditions yield the complex simultaneous equations

$$g_0 \; = \; 1 \; = \; b_1(1 + i)^0 + b_2(1 - i)^0 \; = \; b_1 \; + \; b_2$$
$$g_1 \; = \; 2 \; = \; b_1(1 + i)^1 + b_2(1 - i)^1$$

with solution

$$b_1 \; = \; \frac{i + 1}{2i} \qquad b_2 \; = \; \frac{i - 1}{2i}$$

Hence, the general solution is

$$g_n \; = \; \frac{1}{2i}(1 + i)^{n+1} \; - \; \frac{1}{2i}(1 - i)^{n+1}$$

EXERCISES for Section 2.4

In each of the Exercises 2.4.1 through 2.4.14, solve the recurrence by the method of characteristic roots.

2.4.1$^\text{S}$ $a_n \; = \; 2a_{n-1}; \; a_0 \; = \; 3.$

2.4.2 $a_n \; = \; 2a_{n-1} - 3; \; a_0 \; = \; 3.$

2.4.3 $a_n \; = \; 3a_{n-1} - 2a_{n-2}; \; a_0 \; = \; 2, \; a_1 \; = \; 1.$

2.4.4 $a_n \; = \; 3a_{n-1} - 2a_{n-2} + 2; \; a_0 \; = \; 2, \; a_1 \; = \; 1.$

2.4.5 $a_n \; = \; 3a_{n-1} - 2a_{n-2} + 2; \; a_0 \; = \; 2, \; a_1 \; = \; -1.$

2.4.6 $a_n \; = \; 3a_{n-1} - 2a_{n-2} + n; \; a_0 \; = \; 2, \; a_1 \; = \; -1.$

2.4.7 $a_n = 5a_{n-1} - 6a_{n-2} + n$; $a_0 = 1,\ a_1 = 3$.

2.4.8 $a_n = 5a_{n-1} - 6a_{n-2} + n^2$; $a_0 = 1,\ a_1 = 4$.

2.4.9[S] $a_n = 7a_{n-1} + 8a_{n-2} + (-1)^n$; $a_0 = 0,\ a_1 = 1$.

2.4.10 $a_n = 4a_{n-1} - 4a_{n-2} + 2^n$; $a_0 = 3,\ a_1 = 1$.

2.4.11 $a_n = 5a_{n-1} + 6a_{n-2} + 2n + 1$; $a_0 = 2,\ a_1 = -1$.

2.4.12 $a_n = 2a_{n-2} + a_{n-3}$; $a_0 = 0,\ a_1 = 1,\ a_2 = 2$.

2.4.13 $a_n = 4a_{n-1} - a_{n-2} - 6a_{n-3}$; $a_0 = 0,\ a_1 = 1,\ a_2 = 2$.

2.4.14 $a_n = a_{n-1} + 2a_{n-2} + 3a_{n-3}$; $a_0 = 0,\ a_1 = 1,\ a_2 = 2$.

In each of the Exercises 2.4.15 through 2.4.18, solve the recurrence by the method of characteristic roots. The roots are a complex conjugate pair.

2.4.15[S] $a_n = 2a_{n-1} - 3a_{n-2}$; $a_0 = 1,\ a_1 = 2$.

2.4.16 $a_n = 2a_{n-1} - 2a_{n-2}$; $a_0 = 1,\ a_1 = 3$.

2.4.17 $a_n = 3a_{n-1} - 3a_{n-2}$; $a_0 = 1,\ a_1 = 3$.

2.4.18 $a_n = 2a_{n-1} - 4a_{n-2}$; $a_0 = 1,\ a_1 = 2$.

2.5   SIMULTANEOUS RECURSIONS

As remarked at the end of §2.1, from this point on in the chapter, we explore how to reduce other kinds of recurrences to the type for which we have good methods, that is, to linear recurrences of fixed degree with constant coefficients. Here we consider simultaneous recurrences that arise in a problem concerning growth of a rabbit population. In solving simultaneous algebraic equations, one uses a substitution from one equation in the system to reduce the number of variables in other equations. Similarly, with simultaneous recurrences, one uses a substitution from one recursion to reduce the number of different sequences occurring in other recursions. The objective is to reduce the solution of the initial system to the solution of one or more independent linear recurrences. Solving the particular system described here is reduced to solving the classical Fibonacci recurrence, which is unsurprising, because the simultaneous system presented here pertains to the rabbit population model invented by Fibonacci, depicted in Figure 2.5.1. This section solves the Fibonacci recurrence and describes how readily it pertains to other mathematical constructions and problems. Discussion of the Fibonacci sequence continues in §2.6.

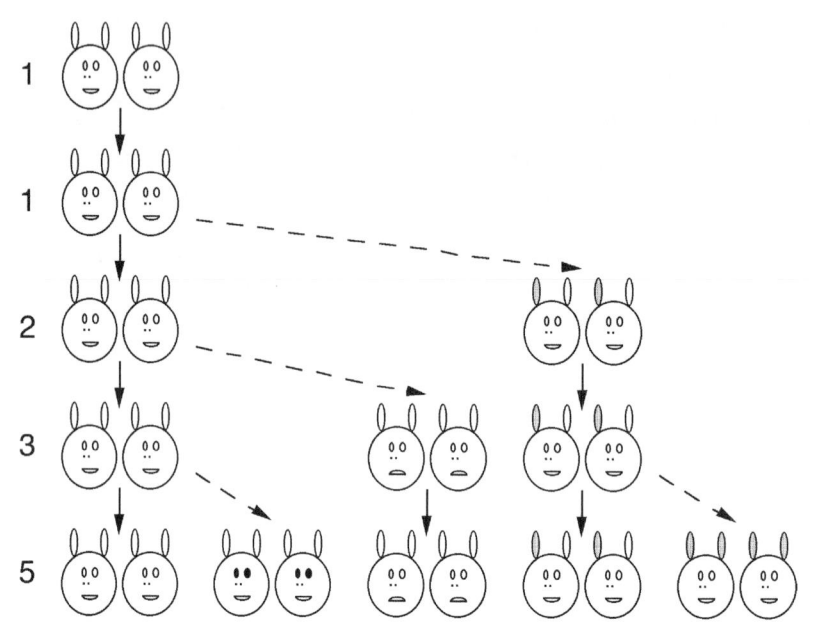

Figure 2.5.1   Fibonacci rabbit population growth.

Fibonacci Rabbits

In 1202, Fibonacci imagined a kind of rabbit that takes one month from birth to mature, with a gestation period of one month. Every mature female gives birth each month to a litter of two, with one male and one female. The population growth is described by a pair of simultaneous recurrences. Let b_n represent the number of pairs of newborn rabbits, and let a_n be the number of pairs of adult (mature) rabbits. Suppose that there are no rabbits at $n = 0$ months, and that a newborn pair initiates the system after 1 month.

We want to calculate the total number $f_n = a_n + b_n$ pairs of rabbits. This situation is modeled by a **simultaneous recursion** with initial conditions

$$a_0 = 0, \quad a_1 = 0, \quad b_0 = 0, \quad b_1 = 1;$$

and the relational equations

$$
\begin{aligned}
a_n &= a_{n-1} + b_{n-1} \\
b_n &= a_{n-1} \\
f_n &= a_n + b_n
\end{aligned}
$$

A first step in solving such a system is to use substitutions to reduce it to a recurrence with a single unknown. We see that

$$a_n = a_{n-1} + b_{n-1} = f_{n-1} \qquad (2.5.1)$$
$$b_n = a_{n-1} = f_{n-2} \qquad (2.5.2)$$
$$f_n = a_n + b_n = f_{n-1} + f_{n-2} \qquad (2.5.3)$$

and that $f_0 = a_0 + b_0 = 0$ and $f_1 = a_1 + b_1 = 0 + 1 = 1$.

The resulting single-variable recurrence

$$f_0 = 0, \ f_1 = 1$$
$$f_n = f_{n-1} + f_{n-2}$$

is recognizable as the **Fibonacci recurrence**.

Ubiquitousness of the Fibonacci Sequence

Although Fibonacci's rabbit model is Fibonacci's invention, the sequence it yields is evidently nature's invention. For instance, what follows immediately is an explanation of an occurrence of the Fibonacci sequence in the construction of a nautilus shell.

DEFINITION: A **Fibonacci rectangle** is any rectangle, subdivided into squares whose sides are of lengths that are Fibonacci numbers, in the following sequence:

- The Fibonacci rectangle r_1 is a 1×1 square.

- For each $n \geq 2$, the Fibonacci rectangle r_n is constructed by placing a square along the longer side of the rectangle r_{n-1}, as in Figure 2.5.2.

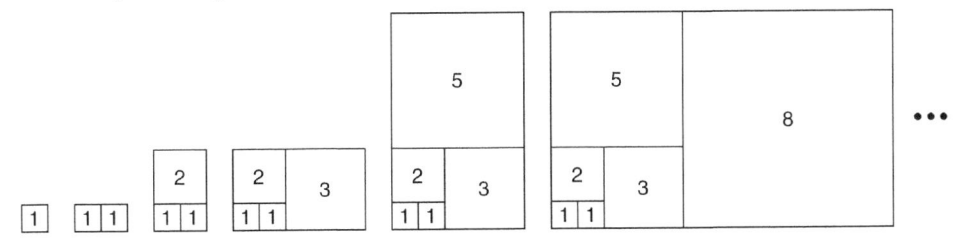

Figure 2.5.2   Fibonacci rectangles.

DEFINITION: A **spiraled Fibonacci rectangle** is a Fibonacci rectangle in which each square of size 5×5 and larger is placed so that it touches three previous squares, rather than two. Figure 2.5.3 illustrates a spiraled Fibonacci rectangle.

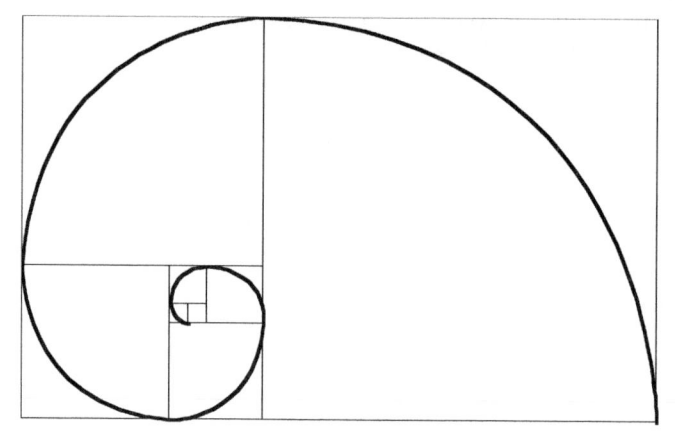

Figure 2.5.3   Fibonacci spiral.

We observe that the inscribed spiral in Figure 2.5.3 has the shape of a nautilus shell. It is called a *Fibonacci spiral*.

Solving the Fibonacci Recurrence

We now use the method of generating functions to solve the Fibonacci recurrence.

Step 1. $f_n z^n = f_{n-1} z^n + f_{n-2} z^n$.

$$\sum_{n=2}^{\infty} f_n z^n = \sum_{n=2}^{\infty} f_{n-1} z^n + \sum_{n=2}^{\infty} f_{n-2} z^n$$

Step 2. Use $F(z)$ as the generating function for f_n.

$$\sum_{n=2}^{\infty} f_n z^n = z \sum_{n=2}^{\infty} f_{n-1} z^{n-1} + z^2 \sum_{n=2}^{\infty} f_{n-2} z^{n-2}$$

$$F(z) - f_1 z - f_0 = z\left(F(z) - f_0\right) + z^2 F(z)$$

Step 3. Solve for $F(z)$.

$$F(z)(1 - z - z^2) = f_1 z + f_0 - f_0 z = 1z + 0 - 0z = z$$

$$F(z) = \frac{z}{1 - z - z^2}$$

Step 4. To solve for f_n, we use the quadratic equation

$$1 - z - z^2 = \left(1 - \frac{1 + \sqrt{5}}{2} z\right) \cdot \left(1 - \frac{1 - \sqrt{5}}{2} z\right)$$

whose roots involve the *golden mean* and its conjugate

$$\gamma \;=\; \frac{1+\sqrt{5}}{2} \quad \text{and} \quad \hat{\gamma} \;=\; \frac{1-\sqrt{5}}{2}$$

respectively. We then use partial fractions

$$F(z) \;=\; \frac{z}{1-z-z^2} \;=\; \frac{1}{\sqrt{5}} \cdot \left(\frac{1}{1-\gamma z} - \frac{1}{1-\hat{\gamma}z} \right)$$

from which we conclude

$$f_n \;=\; \frac{1}{\sqrt{5}} \cdot (\gamma^n - \hat{\gamma}^n) \tag{2.5.4}$$

which is called the **Binet formula** for the Fibonacci numbers, after Jacquet Binet, who rediscovered it in 1843, after Euler had published it in 1765. Closed forms for a_n and b_n are readily derivable from (2.5.1) and (2.5.2), respectively.

Proposition 2.5.1. *The Fibonacci number f_n is asymptotic to $\dfrac{\gamma^n}{\sqrt{5}}$.*

Proof: Since $\hat{\gamma} < 1$, it follows that $\hat{\gamma}^n$ is asymptotic to 0. Accordingly, using Eq. (2.5.1) above,

$$\lim_{n \to \infty} \frac{f_n}{\gamma^n/\sqrt{5}} \;=\; \lim_{n \to \infty} \frac{(\gamma^n + \hat{\gamma}^n)/\sqrt{5}}{\gamma^n/\sqrt{5}} \;=\; \lim_{n \to \infty} 1 + \frac{\hat{\gamma}^n}{\gamma^n} \;=\; 1 \qquad \Diamond$$

Some Tiling Problems

One of the many other contexts, besides biology, in which Fibonacci numbers arise is tiling problems. Using tiling as a model for Fibonacci numbers leads to some possibly surprising results. We visualize paving a $1 \times n$ chessboard with tiles of various lengths. A $1 \times d$ tile is called a d-tile.

Example 2.5.1: Let t_n be the number of ways to cover a $1 \times n$ chessboard with 1-tiles and 2-tiles. We have $t_0 = 1$, which represents covering a degenerate board with the empty arrangement. Figure 2.5.4 shows the possibilities for $n = 0, \ldots, 4$.

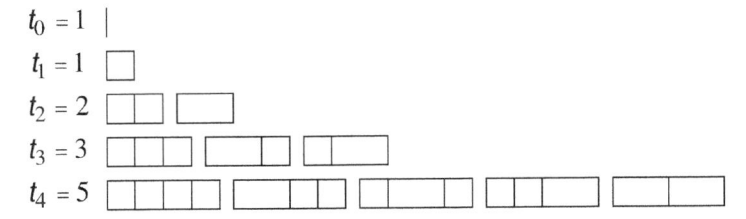

Figure 2.5.4   **Tiling a $1 \times n$ chessboard.**

The number of $1 \times n$ tilings in which the rightmost tile is a 1-tile is t_{n-1}. The number of $1 \times n$ tilings in which the rightmost tile is a 2-tile is t_{n-2}. The solution to the resulting recurrence

$$t_0 = 1, \quad t_1 = 1;$$
$$t_n = t_{n-1} + t_{n-2}$$

is clearly $t_n = f_{n+1}$, the $n + 1^{\text{st}}$ Fibonacci number.

Example 2.5.2:   Observe that any tiling in which all the tiles are of odd length can be converted to a tiling with 1-tiles and 2-tiles, whose initial tile is a 1-tile, by breaking a tile of length $2n + 1$ into a 1-tile, followed by n 2-tiles. This breakage operation can be inverted, since each maximal string of 2-tiles and the 1-tile that precedes it can be assembled into an odd-length tile. It follows that there is a one-to-one, onto correspondence between the two kinds of tiling. Since the number of tilings of a $1 \times n$ chessboard with 1-tiles and 2-tiles, and with an initial 1-tile, is the Fibonacci number $t_{n-1} = f_{n-2}$, this must also be the number of tilings with tiles of odd length.

EXERCISES for Section 2.5

2.5.1   Show that a pair of simultaneous recursions of the form

$$x_n = a x_{n-j} + b y_{n-k}$$
$$y_n = c x_{n-r} + d y_{n-s}$$

can be split into two separate linear recursions, one for the sequence $\langle x_n \rangle$, and one for the sequence $\langle y_n \rangle$.

2.5.2^{S}   Solve these simultaneous recurrences.

$$x_0 = 0, \quad y_0 = 1;$$
$$x_n = x_{n-1} + y_{n-1}$$
$$y_n = 4 x_{n-1} + y_{n-1}$$

2.5.3   Solve these simultaneous recurrences.

$$x_0 = 0, \quad y_0 = 1;$$
$$x_n = x_{n-1} + y_{n-1}$$
$$y_n = 9 x_{n-1} + y_{n-1}$$

2.5.4   Solve these simultaneous recurrences.

$$x_0 = 0, \quad x_1 = 1, \quad y_0 = 1;$$
$$x_n = x_{n-1} + y_{n-1}$$
$$y_n = 2 x_{n-2} + y_{n-1}$$

2.5.5   Solve the following recurrence.

$$x_0 = 0, \quad x_1 = 1;$$
$$x_n = 2x_{n-2} + x_{n-3}$$

2.5.6   Calculate the first five values of the following recurrence, use them to guess the solution, and then use mathematical induction to prove the correctness of your guess.

$$x_0 = 1, \quad x_1 = 2;$$
$$x_n = x_{n-1}^{-1} + 1$$

DEFINITION: The **Lucas sequence** $\langle L_n \rangle$ is defined by the recursion

$$L_0 = 2, \quad L_1 = 1;$$
$$L_n = L_{n-1} + L_{n-2}$$

Exercises 2.5.7 through 2.5.10 are concerned with the Lucas sequence, which has the same recursion formula as the Fibonacci sequence, but different initial values.

2.5.7[S]   Calculate the Lucas numbers L_0, L_1, ..., L_9.

2.5.8   Find a generating function for the Lucas sequence and a closed formula for L_n.

2.5.9[S]   Consider paving a circular $1 \times n$ track with a seam with curved 1-tiles and 2-tiles, so that two tiles meet at the seam. Let r_n be the number of ways to do this. Then $r_0 = 1$, and Figure 2.5.5 illustrates that $r_3 = 3$. Write a recurrence for the sequence r_n and solve it.

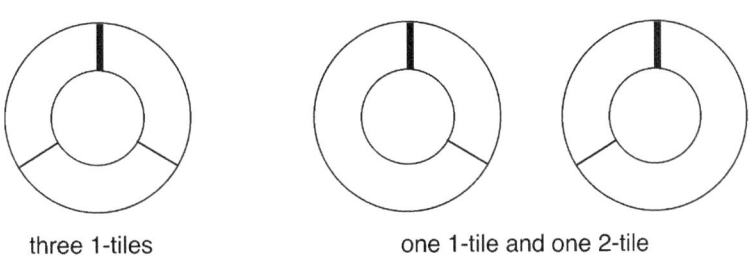

three 1-tiles   one 1-tile and one 2-tile

Figure 2.5.5   **Paving a seamed 1×3 circular track.
The seam is the dark vertical line in the track.**

2.5.10   Now consider paving such a circular $1 \times n$ track with curved 1-tiles and 2-tiles, so that the midline of a 2-tile covers the seam, for $n \geq 1$. Let s_n be the number of ways to do this. We take $s_0 = 1$, and we observe that $s_1 = 0$. Figure 2.5.6 below illustrates that $s_4 = 2$. Write a recurrence for the sequence s_n and solve it.

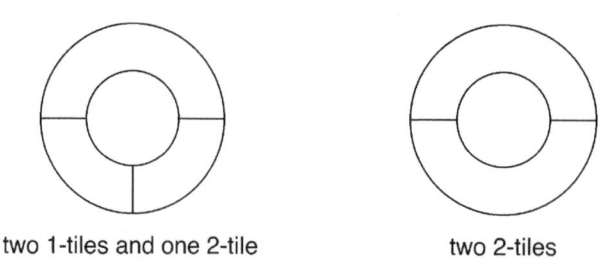

two 1-tiles and one 2-tile   two 2-tiles

Figure 2.5.6   **Covering the seam of a 1×4 circular track.**

2.5.**11**   Show that $L_n = r_n + s_n$.

Exercises 2.5.12 through 2.5.14 are concerned with the ancestry tree of a single male bee, which is called a drone. Let d_n be the number of drones and q_n the number of queens in n previous generations of bees from which that drone is descended. These numbers are given by the recurrence

$$d_0 = 1, \quad q_0 = 0;$$
$$d_n = q_{n-1};$$
$$q_n = d_{n-1} + q_{n-1}$$

which corresponds to the fact that whereas a queen has two parents — a drone and a queen — a drone has only one parent, a queen.

2.5.**12**   Draw the ancestry tree for a drone back four generations.

2.5.**13**   Draw the ancestry tree of a queen back four generations.

2.5.**14**[S]   Use substitution to split the given simultaneous recurrence for bee ancestry into two independent recurrences, one for d_n and one for q_n, and solve them.

2.6   FIBONACCI NUMBER IDENTITIES

In examining the first few entries of the Fibonacci sequence

n	0	1	2	3	4	5	6	7	8	9	10	11	12
f_n	0	1	1	2	3	5	8	13	21	34	55	89	144

we observe that for each instance of a number n and a multiple kn, the Fibonacci number f_{mn} is a multiple of f_n. For instance,

$$f_5 = 5 \quad \text{and} \quad f_{10} = 55 = 11f_5$$

Some other patterns may be immediately apparent, and some are subtle. This section is devoted to the observation and verification of such patterns.

Forward-Shift and Subscript Multipliers

As a preliminary to trying to prove that the Fibonacci number f_{mn} is a multiple of f_n, we consider a relationship between f_{k+n} and f_n. Proceeding from $k = 2$,

$$
\begin{aligned}
f_{n+2} &= f_{n+1} + f_n \\
f_{n+3} &= f_{n+2} + f_{n+1} = (f_{n+1} + f_n) + f_{n+1} = 2f_{n+1} + f_n \\
f_{n+4} &= f_{n+3} + f_{n+2} = (2f_{n+1} + f_n) + (f_{n+1} + f_n) = 3f_{n+1} + 2f_n \\
f_{n+5} &= f_{n+4} + f_{n+3} = (3f_{n+1} + 2f_n + (2f_{n+1} + f_n) = 5f_{n+1} + 3f_n \\
f_{n+6} &= f_{n+5} + f_{n+4} = (5f_{n+1} + 3f_n) + (3f_{n+1} + 2f_n) = 8f_{n+1} + 5f_n
\end{aligned}
$$

we observe that the coefficients of f_{n+1} and f_n are themselves Fibonacci numbers. The observable pattern is confirmed by the following theorem.

Theorem 2.6.1 [*Forward-Shift Identity*]. *The Fibonacci numbers satisfy the equation*

$$ f_{n+k} = f_k f_{n+1} + f_{k-1} f_n \qquad \text{for all } k \geq 1 $$

Proof: By induction on k.

BASIS: If $k = 1$, then $f_1 = 1$ and $f_0 = 0$, and, thus,

$$ f_k f_{n+1} + f_{k-1} f_n = 1 \cdot f_{n+1} + 0 \cdot f_n = f_{n+1} $$

IND HYP: Assume for all j in the interval $0 < j < k$ that

$$ f_{n+j} = f_j f_{n+1} + f_{j-1} f_n $$

IND STEP: Then

$$
\begin{aligned}
f_{n+k} &= f_{n+k-1} + f_{n+k-2} && \text{(Fibonacci recursion)} \\
&= (f_{k-1} f_{n+1} + f_{k-2} f_n) + (f_{k-2} f_{n+1} + f_{k-3} f_n) && \text{(ind hyp)} \\
&= (f_{k-1} + f_{k-2}) f_{n+1} + (f_{k-2} + f_{k-3}) f_n && \text{(regrouping)} \\
&= f_k f_{n+1} + f_{k-1} f_n && \text{(Fibonacci recursion)} \qquad \Diamond
\end{aligned}
$$

We now confirm the initial observation regarding multiples.

Corollary 2.6.2. For all $k \geq 0$, the Fibonacci number f_{kn} is a multiple of the Fibonacci number f_n.

Proof: By induction on the multiplier k.

BASIS: This is trivial for $k = 0$ and $k = 1$. That is,

$$
\begin{aligned}
f_{0n} &= 0 = 0f_n \\
f_{1n} &= f_n = 1f_n
\end{aligned}
$$

IND HYP: Assume that the Fibonacci f_{jn} is a multiple of the Fibonacci number f_n, for all j such that $0 \leq j < k$.

IND STEP: Then

$$
\begin{aligned}
f_{kn} &= f_{n+(k-1)n} \\
&= f_{(k-1)n}\, f_{n+1} + f_{(k-1)n-1}\, f_n \qquad \text{(by Theorem 2.6.1)}
\end{aligned}
$$

By the inductive hypothesis, there is a number M such that $f_{(k-1)n} = M f_n$. Thus,

$$
\begin{aligned}
f_{kn} &= M f_n f_{n+1} + f_{(k-1)n-1} f_n \\
&= \left(M f_{n+1} + f_{(k-1)n-1} \right) f_n \qquad\qquad\qquad \diamond
\end{aligned}
$$

Cassini's Identity

In returning to the early entries of the Fibonacci sequence

n	0	1	2	3	4	5	6	7	8	9	10	11	12
f_n	0	1	1	2	3	5	8	13	21	34	55	89	144

we also observe that the square of each Fibonacci number differs by 1 from the product of the Fibonacci number that follows it and the Fibonacci number that precedes it. For instance,

$$
\begin{aligned}
f_5 f_3 &= 5 \cdot 2 = 3 \cdot 3 + 1 = f_4 f_4 + 1 \\
f_6 f_4 &= 8 \cdot 3 = 5 \cdot 5 - 1 = f_5 f_5 - 1
\end{aligned}
$$

Theorem 2.6.3 [*Cassini's Identity*]. *In the Fibonacci sequence* $\langle f_n \rangle$,

$$
f_{n+1} f_{n-1} = f_n^2 + (-1)^n \qquad \text{for } n \geq 1
$$

Proof: By induction on n.

BASIS: Confirmation that the identity holds for $n = 1$ is as follows.

$$
\begin{aligned}
f_2 f_0 &= 1 \cdot 0 = 0 \\
f_1^2 + (-1)^1 &= 1 \cdot 1 - 1 = 0
\end{aligned}
$$

IND HYP: Assume that

$$
f_{k+1} f_{k-1} = f_k^2 + (-1)^k \qquad \text{for } 1 \leq k < n
$$

IND STEP: Then

$$
\begin{aligned}
f_{n+1} f_{n-1} &= (f_n + f_{n-1}) f_{n-1} \qquad \text{(Fibonacci recurrence)} \\
&= f_n f_{n-1} + f_{n-1}^2 \\
&= f_n f_{n-1} + f_n f_{n-2} - (-1)^{n-1} \qquad \text{(ind hyp)} \\
&= f_n (f_{n-1} + f_{n-2}) + (-1)^n \\
&= f_n^2 + (-1)^n \qquad\qquad \text{(Fibonacci recurrence)} \qquad\qquad \diamond
\end{aligned}
$$

Fibonacci Number System

It is clear that every non-negative integer is the sum of some Fibonacci numbers, since 1 is a Fibonacci number. The following example adds as requirements non-repetition and non-consecutiveness.

Example 2.6.1:   Each of the smallest integers that is not a Fibonacci number is the sum of two or more non-consecutive Fibonacci numbers.

$$
\begin{array}{ll}
4 = 3 + 1 & 10 = 8 + 2 \\
6 = 5 + 1 & 11 = 8 + 3 \\
7 = 5 + 2 & 12 = 8 + 3 + 1 \\
9 = 8 + 1 & 14 = 13 + 1
\end{array}
$$

Moreover, this property holds for some larger examples.

$$
100 = 89 + 8 + 3 \qquad 200 = 144 + 55 + 1
$$

Theorem 2.6.4. *Every non-negative integer n can be represented as the sum of distinct non-consecutive Fibonacci numbers.*

Proof: By induction on n.

BASIS: The number $n = 0$ is the sum of the empty set.

IND HYP: Assume for some $n > 0$ that every number less than n is representable as the sum of distinct non-consecutive Fibonacci numbers.

IND STEP: Let f_m be the largest Fibonacci number less than or equal to n. Since $f_{m+1} > n$, it follows from the Fibonacci recursion that

$$
f_{m-1} > n - f_m
$$

Thus, when the induction hypothesis is applied to $n - f_m$, the summands are non-consecutive Fibonacci numbers, each less than f_{m-1}. Accordingly, when f_m is included in the set of summands, the members of the resulting set of Fibonacci numbers remain non-consecutive, and their sum is n. $\diamond$

DEFINITION: The **Fibonacci representation of an integer** is its expression as a sum of distinct non-consecutive Fibonacci numbers.

EXERCISES for Section 2.6

In each of the Exercises 2.6.1 through 2.6.4, calculate the value of the given expression.

2.6.**1** f_{10} 2.6.**2** f_{15}

2.6.**3** $f_8 f_4 - f_7 f_3$ 2.6.**4** $f_9 f_4 - f_8 f_3$

In each of the Exercises 2.6.5 through 2.6.8, convert the given integer to its Fibonacci representation.

2.6.5[S]   202   **2.6.6**   105

2.6.7   128   **2.6.8**   243

In Exercises 2.6.9 through 2.6.20, prove the given identity.

2.6.9[S] $f_n^2 - 2f_{n-1}f_{n-2} - f_{n-1}^2 = f_{n-2}^2$ for $n \geq 2$.

2.6.10 $f_0 + f_1 + f_2 + \cdots + f_n = f_{n+2} - 1$ for $n \geq 0$.

2.6.11 $f_1 + f_3 + f_5 + \cdots + f_{2n-1} = f_{2n}$ for $n \geq 1$.

2.6.12 $f_0 + f_2 + f_4 + \cdots + f_{2n} = f_{2n+1} - 1$ for $n \geq 0$.

2.6.13 $f_0^2 + f_1^2 + f_2^2 + \cdots + f_n^2 = f_n f_{n+1}$ for $n \geq 0$.

2.6.14 $f_1 f_2 + f_2 f_3 + f_3 f_4 + \cdots + f_{2n-1}f_{2n} = f_{2n}^2$ for $n \geq 1$.

2.6.15 $\sum_{k=0}^{n} \binom{n-k}{k} = f_{n+1}$ for $n \geq 0$.

2.6.16 $f_n^2 + f_{n+1}^2 = f_{2n+1}$ for $n \geq 0$.

2.6.17 $f_{n+1}^2 - f_{n-1}^2 = f_{2n}$ for $n \geq 1$.

2.6.18 $2f_n = f_{n+1} + f_{n-2}$ for $n \geq 2$.

2.6.19 $3f_n = f_{n+2} - f_{n-2}$ for $n \geq 2$.

2.6.20 $f_1 + f_4 + f_7 + \cdots + f_{3n-2} = \frac{1}{2}f_{3n}$ for $n \geq 0$.

2.6.21   Prove that $f_n = \frac{L_{n-1}+L_{n+1}}{5}$, for $n \geq 1$, where L_n is the Lucas number.

2.6.22   Prove that the Fibonacci sequence is neither log-concave nor log-convex.

2.6.23[S]   Write the Fibonacci representation for the number $f_{2n+1} - 1$. (Suggestion: Try this first for $n = 4$.)

2.6.24   Prove that $f_{2n+1} - 1$ is the smallest integer that requires at least n summands in its Fibonacci representation.

PREVIEW OF §6.1:

* **Theorem 6.1.9.** For $n \geq 0$ and $m \geq 1$, $\gcd(f_n, f_m) = f_{\gcd(n,m)}$.

2.6.25   Confirm that Theorem 6.1.9 is correct for $\gcd(f_{12}, f_8)$.

2.7   NON-CONSTANT COEFFICIENTS

A good method for solving any recurrence that is not specified as a linear recurrence of fixed degree with constant coefficients is to transform it into such a recurrence. This is an instance of the standard mathematical strategy of reducing a given problem to a previously solved problem. Most of our attention in this section is devoted to the solution of another recurrence from computer science, called the *quicksort recurrence*.

A Reduction Strategy

Consider this general linear recursion of degree d with variable coefficients.

$$\begin{aligned}
f(n)\, x_n \;=\; & c_{n-1}\, f(n-1)\, x_{n-1} + \cdots \\
& + c_{n-d}\, f(n-d)\, x_{n-d} + p(n)
\end{aligned} \qquad (2.7.1)$$

Substituting $f(n)x_n = y_n$ yields the recursion

$$y_n \;=\; c_{n-1}\, y_{n-1} + \cdots + c_{n-d}\, y_{n-d} + p(n) \qquad (2.7.2)$$

which is linear with constant coefficients, and, therefore, is amenable to previously developed methods of solution. A solution $y_n = g(n)$ for the recursion (2.7.2) could be reverse-transformed into a solution $x_n = g(n)/f(n)$ for the recursion (2.7.1).

Example 2.7.1:   Consider the recurrence

$$\begin{aligned}
x_0 &= 0; \\
x_n &= \frac{2(n-1)}{n}\, x_{n-1} + \frac{1}{n}
\end{aligned}$$

Multiplying the recursion by n yields the recursion

$$nx_n \;=\; 2(n-1)\, x_{n-1} + 1$$

in the form of recurrence (2.7.1). The substitution $nx_n = y_n$ yields this new recurrence in the form of recurrence (2.7.2).

$$\begin{aligned}
y_0 &= 0; \\
y_n &= 2y_{n-1} + 1
\end{aligned}$$

This transformed recurrence is easily solved by the method of generating functions or by the method of characteristic roots. Indeed, if we recognize it as the Hanoi recurrence, we already have this solution for y_n:

$$y_n \;=\; 2^n - 1$$

To obtain the solution for x_n, we substitute $y_n/n = x_n$:

$$\begin{aligned}
nx_n &= 2^n - 1 \\
\Rightarrow \qquad x_n &= \frac{2^n - 1}{n}
\end{aligned}$$

Example 2.7.2:   To solve the recurrence

$$\begin{aligned}
x_0 &= 0; \\
nx_n &= \left(1 - \frac{1}{n}\right)^{n-1} x_{n-1} + (2n)^{1-n} \qquad \text{for } n \geq 1
\end{aligned}$$

we first multiply the recursion by n^{n-1}, thereby obtaining

$$n^n x_n = (n-1)^{n-1} x_{n-1} + n^{n-1}(2n)^{1-n}$$
$$\Rightarrow \quad n^n x_n = (n-1)^{n-1} x_{n-1} + 2^{1-n}$$

Substituting $n^n x_n = y_n$ yields the recurrence

$$y_0 = 0;$$
$$y_n = y_{n-1} + \left(\frac{1}{2}\right)^{n-1} \qquad \text{for } n \geq 1$$

This transformed recurrence is easily solved.

$$y_n = 1 + \frac{1}{2} + \frac{1}{4} + \cdots + \frac{1}{2^{n-1}}$$
$$= 2 - \frac{1}{2^{n-1}}$$

By reverse-substituting $y_n = n^n x_n$, we solve the given recurrence.

$$x_n = \frac{1}{n^n} \cdot \left(2 - \frac{1}{2^{n-1}}\right) = \frac{2^n - 1}{n^n 2^{n-1}}$$

Sum in a Recurrence: Quicksort

Beyond the complication of variable coefficients, the *quicksort recurrence* has no fixed degree. It involves a long sum of earlier values in the sequence. Another preliminary to applying the methods of the earlier part of this chapter is to transform it into a recurrence of fixed degree.

The quicksort recurrence arises in the analysis of the time needed to execute a well-known sorting algorithm called **quicksort**. Performing it on a sequence of numbers (which may be used as the keys to the records in a file of data) involves two signature steps, that is, steps that occur in quicksort but not in most other sorting methods. One signature step is choosing an entry of the sequence, which is called a *pivot*. The other signature step, called *tripartitioning*, is to partition the given sequence into three subsequences, as follows:

- The *front part* contains every element that is less than the pivot. This part may be empty.

- The *pivot part* contains only the pivot entry itself.

- The *back part* contains every entry not in the other two parts, all the entries that are greater than the pivot, plus any duplicates of the pivot. The back part may be empty.

If the length of a sequence is 0 or 1, then the sequence is deemed to be sorted. Otherwise, it is tripartitioned, and then its front part and its back part are quicksorted. In the implementation represented by the following algorithm, the pivot is selected at random. (This tends to produce pivots whose value is relatively near to the median of the sequence, a fortuitous event that reduces the number of subsequent iterations.) The following algorithm specifies the details of a quicksort.

Algorithm 2.7.1:   Quicksort

Input: seq $X = \langle x_j \rangle$; range limits lo, hi
Output: that same sequence in non-decreasing order

if $lo \geq hi$ **then return**
    else $pivot := random(\{lo, \ldots, hi\})$
"tripartition" $\langle x_{lo}, \ldots, x_{hi} \rangle$ into $\langle x_{pivot} \rangle$ plus
    $front := \langle x_j \mid x_j < x_{pivot} \rangle$
    $back := \langle x_j (j \neq \text{pivot}) \mid x_j \geq x_{pivot} \rangle$
$X := \text{concatenate}(\text{Qsort}(front), x_{pivot}, \text{Qsort}(back))$

Example 2.7.3:   Suppose that the given sequence is

$$(78 \quad 49 \quad 05 \quad 14 \quad 10 \quad 90 \quad 44 \quad 39 \quad 19 \quad 55)$$

and that the initial pivot is 39. Then the result of the first tripartition step is

$$(\overbrace{(05 \quad 14 \quad 10 \quad 19)}^{\text{front part}} \quad \overbrace{(39)_q}^{\text{pivot}} \quad \overbrace{(78 \quad 49 \quad 90 \quad 44 \quad 55)}^{\text{back part}})$$

The subscript q denotes a part that is fully quicksorted. Suppose that at the second stage the pivots chosen in the parts not yet fully quicksorted are 10 and 78. Then the result of the second-stage tripartitioning is

$$(((05)_q \quad (10)_q \quad (14 \quad 19)) \quad (39)_q \quad ((49 \quad 44 \quad 55) \quad (78)_q \quad (90)_q))$$

Suppose that at the third stage the pivots chosen in the parts not yet fully quicksorted are 19 and 49. Then the result of the third-stage tripartitioning is

$$(((05) \quad (10) \quad ((14) \quad (19))) \quad (39) \quad (((44) \quad (49) \quad (55)) \quad (78) \quad (90)))$$

at which point all parts are fully quicksorted. Concatenation proceeds level by level with this sequence as the final result.

$$(05 \quad 10 \quad 14 \quad 19 \quad 39 \quad 44 \quad 49 \quad 55 \quad 78 \quad 90)$$

Analysis of the Time Needed by Quicksort

Let Q_n represent the time needed to quicksort a sequence of length n. This involves the following time expenditures:

$$\begin{aligned}
1 \quad &\text{to select a pivot location} \\
n \quad &\text{to tripartition a seq of length } n \\
Q_k \quad &\text{to quicksort a front part of length } k \\
Q_{n-k-1} \quad &\text{to quicksort the back part of length } n - k - 1
\end{aligned}$$

The probability that there are exactly k items smaller than random pivot is

$$\text{pr}\,(k \text{ items } < \text{ pivot}) \;=\; \frac{1}{n}$$

This leads to the following recurrence.

$$Q_0 \;=\; 0$$

$$Q_n \;=\; 1 + n + \sum_{k=0}^{n-1} \text{pr}\,(k \text{ items} < \text{pivot}) \cdot [Q_k + Q_{n-k-1}]$$

$$Q_n \;=\; 1 + n + \sum_{k=0}^{n-1} \frac{1}{n} \cdot [Q_k + Q_{n-k-1}]$$

$$=\; 1 + n + \frac{2}{n} \sum_{k=0}^{n-1} Q_k$$

An obstacle to solving the recurrence is the unlimited number of terms in the sum. Often, such a recursion can be transformed into a recursion of fixed degree, by setting up a subtraction of sums.

$$nQ_n \;=\; n + n^2 + 2 \sum_{k=0}^{n-1} Q_k \tag{2.7.3}$$

$$(n-1)Q_{n-1} \;=\; (n-1) + (n-1)^2 + 2 \sum_{k=0}^{n-2} Q_k$$

$$=\; n^2 - n + 2 \sum_{k=0}^{n-2} Q_k \tag{2.7.4}$$

Productively, subtracting (2.7.4) from (2.7.3) yields

$$nQ_n - (n-1)Q_{n-1} \;=\; 2n + 2Q_{n-1}$$

and, thus,

$$nQ_n \;=\; (n+1)Q_{n-1} + 2n$$

which may be rewritten in the form

$$\frac{Q_n}{n+1} \;=\; \frac{Q_{n-1}}{n} + \frac{2}{n+1}$$

After making the substitution

$$\frac{Q_n}{n+1} \;=\; P_n$$

there is the following transformed recurrence

$$P_0 \;=\; 0;$$

$$P_n \;=\; P_{n-1} + \frac{2}{n+1} \qquad \text{for } n \geq 1$$

whose solution is

$$P_n \;=\; \sum_{k=1}^{n} \frac{2}{k+1} \;=\; 2\sum_{k=1}^{n}\frac{1}{k+1} \;=\; 2\sum_{j=2}^{n+1}\frac{1}{j} \;=\; 2(H_{n+1}-1)$$

which is then reverse transformed.

$$\begin{aligned}
Q_n \;&=\; (n+1)P_n \;=\; 2(n+1)(H_{n+1}-1)\\
&=\; 2(n+1)\left(H_n + \frac{1}{n+1}\right) - 2(n+1)\\
&=\; 2(n+1)H_n + 2 - 2(n+1) \;=\; 2(n+1)H_n - 2n
\end{aligned}$$

Confirming Small Cases

Direct application of the recurrence

$$Q_0 \;=\; 0;$$
$$Q_n \;=\; 1+n+\frac{2}{n}\sum_{k=0}^{n-1}Q_k$$

yields the small values

$$\begin{aligned}
Q_1 \;&=\; 1+1+\frac{2}{1}[Q_0] \;=\; 2+0 \;=\; 2\\
Q_2 \;&=\; 1+2+\frac{2}{2}[Q_0+Q_1] \;=\; 3+1\cdot 2 \;=\; 5\\
Q_3 \;&=\; 1+3+\frac{2}{3}[Q_0+Q_1+Q_2] \;=\; 4+\frac{2}{3}\cdot[2+5] \;=\; \frac{26}{3}
\end{aligned}$$

Application of the closed formula

$$Q_n \;=\; 2(n+1)H_n - 2n$$

yields the small values

$$\begin{aligned}
Q_1 \;&=\; 2\cdot(1+1)H_1 - 2\cdot 1 \;=\; 4\cdot 1 - 2 \;=\; 2\\
Q_2 \;&=\; 2\cdot(2+1)H_2 - 2\cdot 2 \;=\; 6\cdot\frac{3}{2} - 4 \;=\; 5\\
Q_3 \;&=\; 2\cdot(3+1)H_3 - 2\cdot 3 \;=\; 8\cdot\frac{11}{6} - 6 \;=\; \frac{26}{3}
\end{aligned}$$

EXERCISES for Section 2.7

In Exercises 2.7.1 through 2.7.7, transform the given recurrence into a linear recurrence with constant coefficients, and solve.

2.7.1^{S} $x_0 = 2,\ x_1 = 3;\quad x_n = \dfrac{3(n-1)x_{n-1}}{n} - \left(2 - \dfrac{4}{n}\right)x_{n-2}$ for $n \ge 2$.

2.7.2 $\quad x_0 = 0; \quad x_n = \dfrac{2(n-2)x_{n-1}}{n-1} + 1 \ $ for $n \geq 1$.

2.7.3 $\quad x_0 = 2; \quad x_n = \dfrac{2(n-3)x_{n-1}}{n-2} + n \ $ for $n \geq 1$.

2.7.4 $\quad x_0 = 1; \quad x_n = \dfrac{4nx_{n-1}}{n-1} - 1 \ $ for $n \geq 1$.

2.7.5 $\quad x_0 = 2; \quad x_n = \dfrac{2(n-1)^2 x_{n-1}}{n^2} + 2^n \ $ for $n \geq 1$.

2.7.6 $\quad x_1 = 5; \quad \lg n\, x_n = 3\lg(n-1)\,x_{n-1} + 1 \ $ for $n \geq 1$.

2.7.7 $\quad x_0 = 0, \quad x_1 = 1; \quad x_n^2 = x_{n-1}^2 + x_{n-2}^2 \ $ for $n \geq 2$.

PREVIEW OF §5.4: Some recurrences with non-constant coefficients are solved with exponential generating functions, instead of ordinary generating functions.

2.7.8 Solve the derangement recurrence by using an exponential generating function.

2.8 DIVIDE-AND-CONQUER RELATIONS

A *divide-and-conquer strategy* for solving a problem is to partition it into sub-problems, such that the total effort needed to do all the subproblems is significantly less than a direct approach to the original problem, even if includes in total effort the costs of partitioning the original problem and of recombining the solutions to the smaller problem into a solution to the original problem.

DEFINITION: A recurrence of the form

$$x_n = c\, x_{n/d} + \alpha(n)$$

is said to be a **divide-and-conquer recurrence**.

Remark: Such a recurrence represents the circumstance in which each of c sub-problems is smaller than the original by a factor of d and in which $\alpha(n)$ is the cost of partitioning and recombining.

Divide-and-conquer strategy is frequently used in the development of fast algorithms. The running time for such algorithms is often described by a divide-and-conquer recurrence. A good approach to solving a divide-and-conquer recursion is to make a substitution that transforms it into a recursion of fixed degree. This approach is applied to recursions arising from two computer science algorithms, *binary search* and *mergesort*, and to a recursion used to solve a problem of great antiquity.

Binary Search

Searching an ordered domain to find the location of a record whose key matches a given number, called the *target (of the search)*, is one of the many tasks at which a divide-and-conquer strategy yields a major reduction of work effort. A sequential search, in which one scans a list of records from one end to the other, is a naive approach. Consider the benefit of comparing the target key value to the middle key in the list.

The middle record of the search file is construed to divide the search file into the first half, which contains every record whose key precedes the key of the middle record, and the second half, which contains all the other records. The signature step of a *binary search* is that the target value is compared to the key of the middle record. If it precedes the middle record, then the target record cannot be in the second half of the file, so it is inactive for the remainder of the search. Otherwise, the first half goes inactive. This step is then applied to the active half. This continues, recursively, until there is only one active record remaining.

Example 2.8.1: Suppose we are searching for the target value $y = 74$ in the following list of length 16:

$$X \;=\; (5 \quad 18 \quad 31 \quad 34 \quad 35 \quad 39 \quad 42 \quad 47 \quad 51 \quad 53 \quad 60 \quad 74 \quad 75 \quad 80 \quad 81 \quad 96)$$

Initially, the entire list is active, with a lower limit location of $lo = 1$ and an upper limit location of $hi = 16$.

In the *first stage*, the middle location is determined to be

$$mid \;=\; \left\lceil \frac{lo + hi}{2} \right\rceil \;=\; \left\lceil \frac{1 + 16}{2} \right\rceil \;=\; 9$$

The target value $y = 74$ is compared with the middle value $x_9 = 51$. Since

$$y = 74 \;\geq\; x_9 = 51$$

and since the list is sorted, it follows that the target value $y = 74$, if present in the list, must be in the second half of the list, which becomes the only active sector. Resetting the lower limit to $lo = 9$ achieves the choice of active sector.

In the *second stage*, the middle location of the active sector $x_9, \ldots, x_{16}$ is location

$$mid \;=\; \left\lceil \frac{lo + hi}{2} \right\rceil \;=\; \left\lceil \frac{9 + 16}{2} \right\rceil \;=\; 13$$

The target value $y = 74$ is compared with the middle value $x_{13} = 75$. Since

$$y = 74 \;\leq\; x_{13} = 75$$

it follows that the target value $y = 74$, if present in the list, must be in the first half of the active sector, which becomes the new active sector. Resetting the upper limit to $hi = 12$ accomplishes this.

In the *third stage*, the middle location of the active sector x_9, ..., x_{12} is location

$$mid = \left\lceil \frac{lo + hi}{2} \right\rceil = \left\lceil \frac{9 + 12}{2} \right\rceil = 11$$

The target value $y = 74$ is compared with the value $x_{11} = 60$. Since

$$y = 74 \geq x_{11} = 60$$

it follows that the target value $y = 74$, if present in the list, must be in the second half of the active sector, which becomes the current active sector. Therefore, the lower limit is reset to $lo = 11$.

In the *fourth stage*, the middle location of the active sector x_{11}, x_{12} is location

$$mid = \left\lceil \frac{lo + hi}{2} \right\rceil = \left\lceil \frac{11 + 12}{2} \right\rceil = 12$$

The target value $y = 74$ is compared with the value $x_{12} = 74$. Since

$$y = 74 \geq x_{12} = 74$$

it follows that the target value $y = 74$, if present in the list, must be in the second half of the active sector, which becomes the final active sector, as the lower limit is reset to $lo = 12$.

The final active sector has only one item. If it were not the target item, that would imply that the target item is not in the original list. If it is the target item, as in this example, then its location is returned as the output of the search.

The following algorithm gives the general rules for a binary search.

Algorithm 2.8.1:   Recursive Binary Search (RBS)

Input: a non-decr seq $X = \langle x_j \rangle$; range limits lo, hi;
      a target value y
Output: if $y \notin \{x_{lo}, \ldots, x_{hi}\}$ then $*$ ("not found");
      else $min\{j \in \{lo, \ldots, hi\} \mid y = x_j\}$

call $RBS(X, lo, hi, y)$

$output := \begin{cases} lo & \text{if } y = x_{lo} \\ * & \text{if } y \neq x_{lo} \end{cases}$

Recursive Subroutine $RBS(X, lo, hi, y)$
if $lo = hi$ **then return**
else $mid = \lceil (hi + lo)/2 \rceil$
if $y < x_{mid}$ **then** $hi := mid - 1$ **else** $lo = mid$
call $RBS(X, lo, hi, y)$

Analysis of the Time Needed for a Binary Search

Let b_n be the number of comparisons needed to perform a binary search on an array of size n. Since at each stage, the limits of the active search space within the original sequence are reset to about half their previous range, the value of b_n is represented by the following divide-and-conquer binary-search recurrence:

$$b_1 = 2;$$
$$b_n = b_{n/2} + 2$$

The substitutions $n = 2^k$ and $b_{2^k} = c_k$ transform this to the recurrence

$$c_0 = 2;$$
$$c_k = c_{k-1} + 2$$

The solution to the transformed recurrence is evidently

$$c_k = 2k + 2$$

from which it follows (by the inverse substitutions $k = \lg n$ and $c_{\lg n} = b_n$) that the solution to the binary-search recurrence is

$$b_n = 2 \lg n + 2$$

COMPUTATIONAL NOTE: Partitioning a search space and searching the parts one at a time would not yield a net reduction of searching effort if the time to search each part were proportional to its size. Such a circumstance would be dividing-without-conquering, since there would be an added cost of subdividing the space. Nor would it be of much help if the subdivision permitted elimination only of tiny fragments of the given search space. However, in a binary search, half the given space is eliminated at each iteration, which quickly reduces the active space to one record.

Merging

Mergesort is based on repeated merging. A *merge* is conceptualized as having two input lists L_1 and L_2, both in non-decreasing order, and an output list L. It is necessary to have access to the head ends of the input lists and to the tail end of the output list. In the main step of a merge, either the lead entry of input list L_1 or the lead entry of input list L_2, whichever is lesser (either, if they are equal) is transferred to the tail of the output list L. The main step is iterated until one of the two input lists is empty, after which all remaining entries in the other input list are transferred to the tail of the output list.

Algorithm 2.8.2 prescribes a process for merging two sorted lists.

Algorithm 2.8.2:   Merge

Input: non-decreasing lists L_1 and L_2
Output: a merged non-decr list L, initially empty

while both input lists are non-empty
    move $\min(\text{head}(L_1), \text{head}(L_2))$ from its own list
       to the tail of the output list
    if that transfer makes one list empty **then** transfer
       all the remaining elements of the other list to
       the end of the output list

Example 2.8.2:   Suppose that the input lists and output list are initially

$$L_1 : \quad 2 \quad 14 \quad 30 \quad 37 \quad 55$$
$$L_2 : \quad 3 \quad 36 \quad 43 \quad 65$$
$$L :$$

After two transfers, the lists are

$$L_1 : \quad 14 \quad 30 \quad 37 \quad 55$$
$$L_2 : \quad 36 \quad 43 \quad 65$$
$$L : \quad 2 \quad 3$$

After two more transfers, the lists are

$$L_1 : \quad 37 \quad 55$$
$$L_2 : \quad 36 \quad 43 \quad 65$$
$$L : \quad 2 \quad 3 \quad 14 \quad 30$$

The final lists are

$$L_1 :$$
$$L_2 :$$
$$L : \quad 2 \quad 3 \quad 14 \quad 30 \quad 36 \quad 37 \quad 43 \quad 55 \quad 65$$

The time needed to merge the lists L_1 and L_2 is at worst proportional to the sum of their lengths.

Iterative Mergesort

A *mergesort* is a sort by iterative merging. Suppose that a file of length 2^n to be sorted is initially regarded as a list of 2^n subfiles of length 1. These subfiles are organized into a list of length 2^{n-1} of pairs of subfiles of length 1. Each pair is merged into a sorted subfile of length 2, leading to a list of 2^{n-1} sorted subfiles, each of length 2. Next, these subfiles are paired, and then the two subfiles within each pair are merged into a sorted subfile of length 4. This continues iteratively until a single sorted file of length 2^n is obtained. This method is readily modified for the case in which the length of the given initial file is not a power of 2.

Example 2.8.3: Suppose that the list to be sorted is

$$X = [82 \quad 48 \quad 03 \quad 17 \quad 11 \quad 94 \quad 41 \quad 37]$$

which has length 8. From an iterative perspective, this list is initially viewed as a list of 8 files, each of length 1.

$$X_1 = [(82) \quad (48) \quad (03) \quad (17) \quad (11) \quad (94) \quad (41) \quad (37)]$$

The files of length 1 are paired, as follows:

$$X_1' = [((82) \quad (48)) \quad ((03) \quad (17)) \quad ((11) \quad (94)) \quad ((41) \quad (37))]$$

Merging the two sublists of length 1 within each pair yields this file with 4 sorted subfiles, each of length 2.

$$X_2 = [(48 \quad 82) \quad (03 \quad 17) \quad (11 \quad 94) \quad (37 \quad 41)]$$

The sorted subfiles are paired, as follows.

$$X_2' = [((48 \quad 82) \quad (03 \quad 17)) \quad ((11 \quad 94) \quad (37 \quad 41))]$$

Merging the two sublists of length 2 within each pair yields this file with 2 sorted subfiles, each of length 4.

$$X_3 = [(03 \quad 17 \quad 48 \quad 82) \quad (11 \quad 37 \quad 41 \quad 94)]$$

These two sorted subfiles of length 4 are paired.

$$X_3' = [((03 \quad 17 \quad 48 \quad 82) \quad (11 \quad 37 \quad 41 \quad 94))]$$

Then the two subfiles of length 4 are merged, thus ultimately yielding a fully sorted list of length 8.

$$X = [03 \quad 11 \quad 17 \quad 37 \quad 41 \quad 48 \quad 82 \quad 94]$$

Recursive Mergesort

In a recursive mergesort, the order in which various pairs are merged would be slightly different from an iterative mergesort. For instance, the first two sorted sublists of length 2 would be merged into a single sublist of length 4 before the rest of the sublists of length 1 were merged into sublists of length 2. Of course, the results would be identical. Algorithm 2.8.3 represents a recursive mergesort.

Algorithm 2.8.3:   Recursive Mergesort

Input: $X = \langle x_1, x_2, \ldots, x_n \rangle$
Output: that same sequence in non-decreasing order

Recursive Subroutine $MerSo(X)$
if $n > 1$ **then**
  $m = \lceil n/2 \rceil$
  $X_1 := \langle x_1, x_2, \ldots, x_m \rangle$
  $X_2 := \langle x_{m+1}, x_2, \ldots, x_n \rangle$
  $X := Merge(X_1, X_2)$

Analysis of the Time Needed for a Mergesort

Let s_n be the number of comparisons needed to perform a mergesort on an array of size n. The value of s_n is represented by the following divide-and-conquer recurrence:

$$s_1 = 1;$$
$$s_n = 2s_{n/2} + n$$

The substitutions $n = 2^k$ and $s_{2^k} = t_k$ transform this into the recurrence

$$t_0 = 1;$$
$$t_k = 2t_{k-1} + 2^k$$

which we can solve with the method of generating functions.

$$\sum_{k=1}^{\infty} z^k t_k = 2z \sum_{k=1}^{\infty} z^{k-1} t_{k-1} + \sum_{k=1}^{\infty} z^k 2^k$$

$$T(z) - 1 = 2zT(z) + \frac{2z}{1 - 2z}$$

$$T(z) = \frac{1}{(1 - 2z)^2}$$

$$\Rightarrow \quad t_k = (k+1)2^k$$

Thus, after the inverse substitutions $k = \lg n$ and $t_{\lg n} = s_n$, the solution to the mergesort recurrence is

$$s_n = n \lg n + n$$

What enables the divide-and-conquer strategy of a mergesort to succeed at reducing the work effort, relative to naive forms of sorting, is that merging two sorted lists of equal length together takes less work than a naive sort of the union of the two lists. Naive sorts (e.g., insertion sorts and selection sorts) of n items require $\mathcal{O}(n^2)$ steps.

The Josephus Recurrence

During the Roman occupation of the Judean state, the Romans had trapped 41 Jewish rebels at a fortress called Jotapata. Rather than face likely slavery in Rome or public execution, these patriots made a suicide pact. Proceeding around a circle, every third man was to be killed, until there was only one remaining man, who would then kill himself. Joseph ben Mattiyahu ha-Cohen (who adopted the name Flavius Josephus after going over to the Romans), a survivor of several previous losses to the Romans, calculated what would be the last two positions on the circle whose occupants would remain alive, so that he and a friend could survive. This terrifying tale suggests some interesting mathematics. Walter Rouse Ball (1850-1925), a British mathematician (and also a barrister), brought attention (see [BaCo1987]) to mathematical aspects of this ancient problem.

DEFINITION: The **Josephus problem** is to calculate a closed formula for the values of the sequence $J_n^{(k)}$, the position of the last man alive, for a circle of n men in which every k^{th} man is killed.

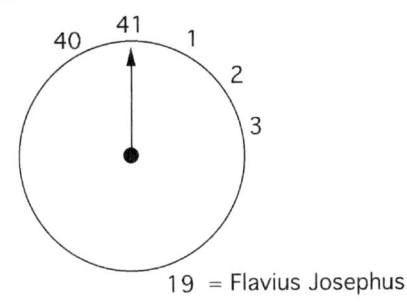

Figure 2.8.1   The Josephus problem $J_{41}^{(2)}$.

For the special case of 41 men, with every 2nd man killed (a variation from the historical event), we can readily simulate the entire process. In each cycle of this simulation, the bold numbers are those of the men who are eliminated on that cycle.

1	**2**	3	**4**	$\cdots$	39	**40**	41	0 mod 2
1	3	**5**	7	$\cdots$	**37**	39	**41**	1 mod 4
3	**7**	11	**15**	$\cdots$	**31**	35	**39**	7 mod 8
3	**11**	19	**27**	35				11 mod 16
3	19	**35**						3 mod 32

Thus, the man in position 19 is the survivor.

The survivor position $J_n^{(2)}$ for the first few values of n is given in Figure 2.8.2. Since every man in an even-numbered position is killed on the first cycle, every one of the survivor positions is an odd number.

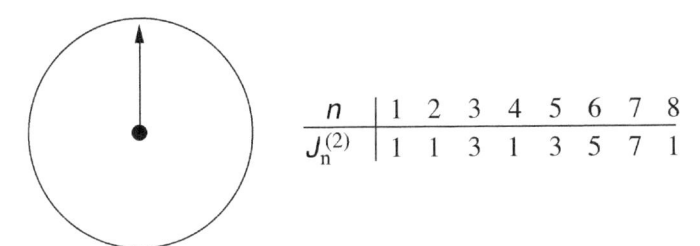

n	1	2	3	4	5	6	7	8
$J_n^{(2)}$	1	1	3	1	3	5	7	1

Figure 2.8.2   Calculating $J_n^{(2)}$ for small values of n.

After the first traversal of the elimination process around the circle, there are two possible cases, depending on whether the number of men at the outset is odd or even. If there are $2n$ men at the outset, then after eliminating the even-numbered on the first cycle, the process location immediately precedes position 1. We may regard this as location $2n - 1$, with a still-alive occupant, since the occupant of position $2n$ is gone, as shown in Figure 2.8.3. The remaining n men, all odd-numbered, are shown just outside the circle.

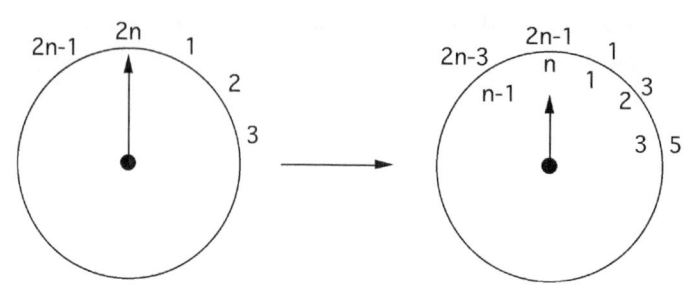

Figure 2.8.3   **After one cycle, for an even configuration.**

This is equivalent to starting with n men, whose numbers are shown just inside the circle. Each outer number is obtained by doubling the inner number and then subtracting 1. Of course, this applies to the survivor position. Thus, we have the recursion

$$J_{2n}^{(2)} \;=\; 2J_n^{(2)} - 1 \qquad \text{for } n \geq 1$$

If there are $2n+1$ men at the outset, then after eliminating the even-numbered on the first cycle, the next man to be killed is at position 1. The status of the process immediately thereafter would be as illustrated in Figure 2.8.4. Here, each outer number is obtained by doubling the inner number and adding 1, which yields the recursion

$$J_{2n+1}^{(2)} \;=\; 2J_n^{(2)} + 1 \qquad \text{for } n \geq 1$$

Thus, the recurrence problem to be solved is as follows:

$$J_1^{(2)} \;=\; 1$$
$$J_{2n}^{(2)} \;=\; 2J_n^{(2)} - 1 \qquad \text{for } n \geq 1$$
$$J_{2n+1}^{(2)} \;=\; 2J_n^{(2)} + 1 \qquad \text{for } n \geq 1$$

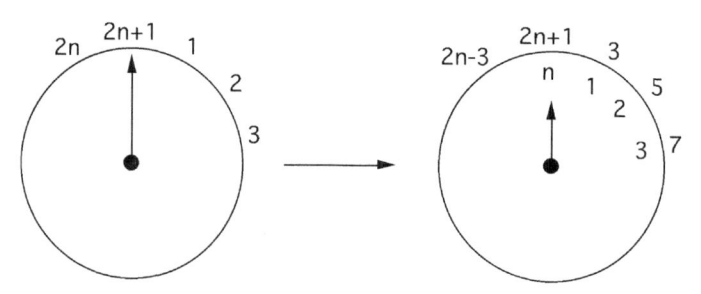

Figure 2.8.4   **After one cycle, for an odd configuration.**

Applying this divide-and-conquer recursion to $n = 41$ yields a quick solution for that case:

$$\begin{aligned}
J_{41}^{(2)} &= 2J_{20}^{(2)} + 1 \\
&= 2(2J_{10}^{(2)} - 1) + 1 \;=\; 4J_{10}^{(2)} - 1 \\
&= 4(2J_{5}^{(2)} - 1) - 1 \;=\; 8J_{5}^{(2)} - 5 \\
&= 8(2J_{2}^{(2)} + 1) - 5 \;=\; 16J_{2}^{(2)} + 3 \\
&= 16(2J_{1}^{(2)} - 1) + 3 \;=\; 32J_{1}^{(2)} - 13 \;=\; 19
\end{aligned}$$

Remark: When the Romans ultimately stormed into the fortress, all the Jews except for Josephus and his friend were dead. Upon hearing from Josephus how he and his friend had survived the suicide pact, the Romans recognized that Josephus was indeed a clever man, who could be quite valuable to them. Josephus lived out his life writing versions of history that flattered the Romans.

To solve the more general problem of calculating $J_n^{(2)}$, we extend the sample of small cases:

n	8	9	10	11	12	13	14	15	16	17	18
$J_n^{(2)}$	1	3	5	7	9	11	13	15	1	3	5

From this increased number of small cases, a pattern emerges, as indicated by the following proposition.

Proposition 2.8.1. If $n = 2^m + k$, with $0 \le k < 2^m$, then

$$J_n^{(2)} \;=\; 2k + 1 \;=\; 2\left(n \bmod 2^{\lfloor \lg n \rfloor}\right) + 1$$

Proof: By induction on n.

BASIS: The equation is clearly true for $n = 1$.

IND HYP: Assume the equation is true for all cases less than n.

IND STEP: If $n = 2^m + k$ is even, then k is even. Thus,

$$\begin{aligned}
J_n^{(2)} &= 2J_{2^{m-1} + \frac{k}{2}}^{(2)} - 1 && \text{(recursion)} \\
&= 2\left(2 \cdot \frac{k}{2} + 1\right) - 1 && \text{(induction hypothesis)} \\
&= 2k + 1
\end{aligned}$$

If $n = 2^m + k$ is odd, then $k - 1$ is even, and

$$\begin{aligned}
J_n^{(2)} &= 2J_{2^{m-1} + \frac{k-1}{2}}^{(2)} + 1 && \text{(recursion)} \\
&= 2\left(2 \cdot \frac{k-1}{2} + 1\right) + 1 && \text{(induction hypothesis)} \\
&= 2k + 1 && \diamondsuit
\end{aligned}$$

EXERCISES for Section 2.8

In Exercises 2.8.1 through 2.8.4, consider a binary search of the given list X for the given target value y. Indicate the lower and upper limits of the sequence of active sectors.

2.8.1[S] $X = (1 \quad 3 \quad 18 \quad 27 \quad 43 \quad 56 \quad 74)$ and $y = 49$.

2.8.2 $X = (12 \quad 19 \quad 43 \quad 65 \quad 78 \quad 83 \quad 91 \quad 99)$ and $y = 65$.

2.8.3 $X = (2 \quad 8 \quad 21 \quad 21 \quad 47 \quad 49 \quad 66 \quad 70 \quad 83)$ and $y = 21$.

2.8.4 $X = (16 \quad 21 \quad 32 \quad 34 \quad 36 \quad 55 \quad 67 \quad 71 \quad 79 \quad 92)$ and $y = 82$.

In Exercises 2.8.5 through 2.8.8, consider a mergesort of the given list X for the given target value y. Show the sequence of lists of sorted sublists.

2.8.5[S] $X = [92 \quad 56 \quad 83 \quad 97 \quad 72 \quad 78 \quad 15]$.

2.8.6 $X = [86 \quad 65 \quad 59 \quad 41 \quad 91 \quad 28 \quad 61 \quad 92]$.

2.8.7 $X = [22 \quad 21 \quad 85 \quad 37 \quad 29 \quad 91 \quad 25 \quad 47 \quad 96]$.

2.8.8 $X = [83 \quad 45 \quad 36 \quad 81 \quad 53 \quad 47 \quad 50 \quad 12 \quad 80 \quad 30]$.

2.8.9[S]   The number of bit-operations of the usual algorithm for multiplication of two n-bit integers is asymptotically approximate to n^2. The number of operations for divide-and-conquer multiplication (e.g., see [AhHoUl1974]) is bounded from above by the sequence $\langle t_n \rangle$, where $n = 2^k$, and

$$t_n = 3t_{n/2} + cn \text{ for } n \geq 2$$
$$t_1 = c$$

Solve for t_n.

2.8.10   The number of number multiplications of the usual algorithm for multiplication of two $n \times n$ matrices is n^3. The number of multiplications for Strassen's divide-and-conquer matrix multiplication (e.g., see [AhHoUl1974]) is bounded from above by the sequence $\langle u_n \rangle$, where $n = 2^k$, and

$$u_n = 7u_{n/2} + cn^2 \text{ for } n \geq 2$$
$$u_1 = 1$$

Solve for u_n.

In Exercises 2.8.11 through 2.8.14, suppose that a non-decreasing sequence $\langle u_n \rangle$ satisfies the recursion

$$u_n = au_{n/b} + c$$

with $a \geq 1$, with b a positive integer, and with $c > 0$. Prove the given assertion.

2.8.11   If $n = b^k$ for some positive integer k, and if $a > 1$, then the solution has the form

$$u_n = Cn^{\log_b a} + D, \text{ where}$$
$$C = u_1 + \frac{c}{a-1} \text{ and } D = \frac{-c}{a-1}$$

2.8.12   If $n = b^k$ for some positive integer k, and if $a = 1$, then the solution has the form

$$u_n = C \log_b n + D, \quad \text{where}$$
$$C = c \quad \text{and} \quad D = u_1 - c$$

2.8.13   If b divides n and $a > 1$, then $u_n \in \mathcal{O}(n \log_b a)$.

2.8.14   If b divides n and $a = 1$, then $u_n \in \mathcal{O}(\log_b n)$.

2.8.15[S]   Finding himself in position 1 among n men, Josephus gets to select the elimination parameter. Give a function of n that indiates his survival.

2.8.16   Write the terms of the Josephus sequence $J_n^{(3)}$ for $n = 1, \ldots, 10$.

2.8.17   Consider a recurrence of the form

$$x_1 = c_1, \quad x_2 = c_2;$$
$$x_n = x_{n-1}^{p} x_{n-2}^{q}$$

How would you reduce it to a linear recurrence?

GLOSSARY

binary search: a method of searching a sorted list by repeated halving.

Binet formula for the Fibonacci number f_n:

$$f_n = \frac{1}{\sqrt{5}} \cdot \left[\left(\frac{1 + \sqrt{5}}{2} \right)^n - \left(\frac{1 - \sqrt{5}}{2} \right)^n \right]$$

Cassini's Identity for the Fibonacci numbers f_n:
$$f_{n+1} f_{n-1} = f_n^{2} + (-1)^n \qquad \text{for } n \geq 1$$

Catalan recurrence: the quadratic recurrence
$$c_0 = 1;$$
$$c_n = c_0 c_{n-1} + c_1 c_{n-2} + \cdots + c_{n-1} c_0$$

characteristic polynomial: a polynomial that arises in one method for solving linear recurrences.

characteristic roots: the roots of the characteristic polynomial.

derangement: a fixed-point-free permutation.

derangement number D_n: the number of fixed-point-free permutations of a set of n objects.

divide-and-conquer recurrence: a recurrence that expresses the element x_n of a sequence in terms of some element $x_{\lceil \frac{n}{d} \rceil}$.

divide-and-conquer strategy: the strategy of reducing a problem to a set of much smaller similar problems.

Fibonacci rectangle: a rectangle partitioned into squares, such that the length of the sides of each square is a Fibonacci number.

Fibonacci recurrence:

$$f_0 = 0, \; f_1 = 1;$$
$$f_n = f_{n-1} + f_{n-2}$$

Fibonacci representation of an integer: representing that integer as the sum of an ascending sequence of Fibonacci numbers, no two of which are consecutive.

forward-shift identity for the Fibonacci numbers:

$$f_{n+k} = f_k f_{n+1} + f_{k-1} f_n \qquad \text{for all } k \geq 1$$

Fundamental Theorem of Algebra: the theorem that a polynomial of degree d has d roots over the complex numbers.

golden mean: the number $\dfrac{1 + \sqrt{5}}{2}$.

initial conditions for a recurrence: values for one or more initial elements of the specified sequence.

Josephus problem: a combinatorial problem, popularized by W. Rouse Ball, involving determination of the survivor of a sequential elimination process.

linear recursion: a recursion of the form

$$x_n = a_{n-1} x_{n-1} + a_{n-2} x_{n-2} + \cdots + a_0 x_0 + \alpha(n)$$

Lucas sequence: the sequence specified by the recurrence

$$L_0 = 2, \quad L_1 = 1;$$
$$L_n = L_{n-1} + L_{n-2}$$

mergesort: a sorting method based on repeated merging.

permutation: a bijection from a set to itself.

quicksort: a recursive method for sorting, with fast average time.

recurrence: a specification of a sequence in this form.

$$g_0 = b_0, \; \ldots, \; g_k = b_k; \qquad \text{initial conditions}$$
$$g_n = \gamma(g_{n-1}, \ldots, g_0) \text{ for } n > k \qquad \text{recursion}$$

Chapter 3

Evaluating Sums

3.1 Normalizing Summations

3.2 Perturbation

3.3 Summing with Generating Functions

3.4 Finite Calculus

3.5 Iteration and Partitioning of Sums

3.6 Inclusion-Exclusion

The concern of this chapter is a collection of methods for the evaluation of a finite sum whose summands are given as a sequence, either in a functional form $f(k)$, or in a subscripted form x_k. Analogous to the sense in which a real function may have for its *integral* over an interval an anti-derivative function evaluated at the bounds of the interval, the value of such a sum may be given by some other function of the lower and upper limits of the index k. For instance, the sum of the integers from 0 to n is given by the formula

$$\frac{n^2 + n}{2}$$

which is called a *solution* for that sum. Many summation problems of this general form can be solved by more than one method, and there is no all-encompassing way that applies to all problems, much less a best way for all problems. This chapter presents several different methods for evaluating such a sum.

There are contexts in which it is helpful to use the word *summation* to mean a formal expression

$$\sum_{k=0}^{n} f(k)$$

and *sum* to mean the value of the expression; we do not adhere to this rigidly, and we often use sum to mean either the expression or its value.

3.1   NORMALIZING SUMMATIONS

There are compelling reasons for preconditioning a given summation problem into the summation of a finite string of consecutive entries of a sequence $\langle x_n \rangle$, most especially, an initial string starting at x_0.

REVIEW FROM §1.4:

- Let $\langle x_n \rangle$ be a sequence. Then the value of the expression

$$\sum_{j=0}^{n} x_j = x_0 + x_1 + \cdots + x_n \tag{3.1.1}$$

(and sometimes the expression) is called the nth **partial sum**.

NOTATION: We sometimes use S_n to denote the nth partial sum.

Such preconditioning allows us to view evaluation of the sum (3.1.1) as solving a recurrence with initial value x_0 and recursion formula $S_n = S_{n-1} + x_n$, as declared in the following formal definition, which also gives names to various artifacts of a slightly more general form of such an expression. It also gives a precise prescription of the value of the sum.

DEFINITION: Let a and b be integers or integer-valued variables, and let $\langle x_n \rangle$ be a sequence with its values in an algebraic structure such as the integers, the reals, or the complex numbers, with an associative and commutative addition. An expression of the form

$$\sum_{k=a}^{b} x_k$$

is called a **consecutive summation**. Its value, the **sum**, is defined recursively.

$$\sum_{k=a}^{b} x_k = \begin{cases} 0 & \text{if } b < a \\ x_a & \text{if } b = a \\ \left(\sum_{k=a}^{b-1} x_k \right) + x_b & \text{if } b > a \end{cases}$$

The parameters of the expression have the following names:

- k is called the **index variable**;
- a is called the **lower limit** of the index;
- b is called the **upper limit** of the index;
- x_k is called the **summand**.

If the lower limit a and the upper limit b are both given as fixed integers, then the sum has a **definite value** within the domain of its summands. For instance, if the summands are integers, then the sum is an integer.

Example 3.1.1: $\displaystyle\sum_{k=0}^{2} k^2 \; = \; 0^2 + 1^2 + 2^2 \; = \; 5.$ $\hspace{3cm}$ (3.1.2)

Quite commonly, a summation has a lower index limit fixed at 0 and a symbolic upper limit of n, in which case summation may be regarded as an operator on a sequence

$$\langle x_n \mid n \,=\, 0,\, 1,\, \ldots \rangle$$

whose application produces a sequence of partial sums

$$\left\langle \sum_{j=0}^{n} x_j \;\middle|\; n \,=\, 0,\, 1,\, \ldots \right\rangle$$

akin to the way that integration operates on a function to produce a new function. This chapter develops methods for **evaluating the summation**, which, in this context, often means producing a closed formula for the elements of the sequence of partial sums. From a computational standpoint, viewing the preconditioning from evaluation permits us to state the methods of evaluation in concise, easy-to-apply form.

Example 3.1.1, continued: With the variable n as the upper limit, the value of the sum of the form (3.1.2) is

$$\sum_{k=0}^{n} k^2 \; = \; \frac{2n^3 + 3n^2 + n}{6}$$

This formula could be confirmed immediately by mathematical induction, or by any of several methods of summation to be introduced in subsequent sections of this chapter.

Remark: Sometimes a summation index has a variable lower limit or variables for both the lower and upper limits. The theory of such seemingly more general operators is readily reducible to sums and differences of partial sums.

Sums over Sets

In a more general expression of a summation, the *indexing set* of a given summation may be any finite set T. Given any function f with values in $\mathbb{Z}$, $\mathbb{Q}$, $\mathbb{R}$, or $\mathbb{C}$, the sum

$$\sum_{y \in T} f(t)$$

is well-defined. In a sum over an unordered indexing set, the order in which the index variable t takes its values is not specified or implied, and the value would be the same for any order of summation.

Example 3.1.2:   The sum of the weights of the edges in the graph G of Figure 3.1.1 is represented by the expression

$$\sum_{e \in E_G} w(e)$$

whose value is

$$6 + 7 + 3 + 2 + 3 + 6 + 5 + 5 + 6 + 4 + 10 + 5 \;=\; 62$$

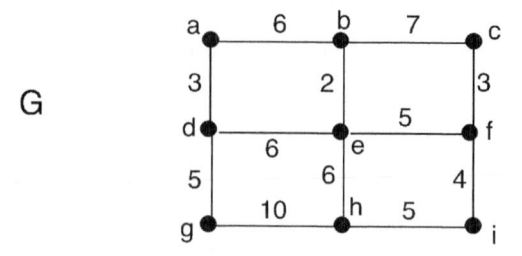

Figure 3.1.1   An edge-weighted graph.

In principle, the edges of the graph G could be indexed by integers $0, 1, \ldots, 11$, which would permit the sum of their weights to be represented by a consecutive summation. There seems to be little gained from doing so in this example. Our focus here is to do something more efficient, when possible, than successively incrementing a running total by additional summands. Such tedium is unavoidable when the summands have no discernable pattern, especially if the summands are random numbers. However, in many other cases, when the index set is a subset of the integers, a transformation may simplify the evaluation.

DEFINITION: Transformation into a consecutive summation, whose index variable ranges over consecutive integers, is called **normalizing a summation**.

Example 3.1.3:   The sum

$$\sum_{\substack{1 \le k < 20 \\ k \text{ odd}}} k$$

can be normalized to

$$\sum_{k=0}^{9} (2k + 1)$$

which is readily transformed further into

$$\sum_{k=0}^{9} (2k + 1) \;=\; \sum_{k=0}^{9} 2k + \sum_{k=0}^{9} 1$$

$$=\; 2\sum_{k=0}^{9} k \;+\; 10$$

$$=\; 2 \cdot \frac{10^{\underline{2}}}{2} \;+\; 10 \qquad \text{(by Corollary 1.5.2)}$$

$$=\; 10^{\underline{2}} + 10$$

$$=\; 100$$

Many of the methods to be introduced in this chapter are designed to work on normalized summations. Other sums one might encounter are transformed into consecutive sums to permit the application of such methods.

Iverson Truth Function

When the index variable of a summation has irregular gaps in its range, it may still be possible to normalize, by inserting into the summand an artificial multiplier that effectively cancels the summand across the gaps.

Example 3.1.4: For instance, the index variable p of the sum

$$\sum_{\substack{p \leq n \\ p \text{ prime}}} \sqrt{p} \qquad (3.1.3)$$

has gaps between consecutive primes.

DEFINITION: The **Iverson truth function** is defined by the rule

$$(\text{predicate}) = \begin{cases} 1 & \text{if the predicate is true} \\ 0 & \text{if the predicate is false} \end{cases}$$

Example 3.1.4, continued: Using the Iverson truth function facilitates the reformulation of (3.1.3) as a consecutive summation.

$$\sum_{\substack{p \leq n \\ p \text{ prime}}} \sqrt{p} = \sum_{p=1}^{n} [\,(p \text{ prime}) \cdot \sqrt{p}\,]$$

CONVENTION: The value of the product

$$(P(k)) \cdot a_k$$

is 0 whenever the value of the Iverson expression $(P(k))$ is 0, even when a_k is undefined.

Example 3.1.5: The value of the sum

$$\sum_{p=0}^{n} \frac{1}{p} \cdot (p \text{ prime})$$

is well-defined, since the "strong zero" of the Iverson expression $(p \text{ prime})$ cancels the effect of the undefined quotient $\frac{1}{p}$ when $p = 0$.

Algebraic Regrouping

Part of the art of simplifying and evaluating sums is to manipulate them so that recognizable forms emerge. The familiar algebraic properties of the number system include several principles for regrouping. These principles are applied independently and also in conjunction with the other summation methods of this chapter.

Proposition 3.1.1 [Distributive Law]. *A common factor can be distributed over all the summands.*

$$\sum_{k \in K} c a_k = c \sum_{k \in K} a_k$$

Proposition 3.1.2 [Addition Law]. *Two sums over the same index set can be combined into a single sum by adding each pair of summands with the same index.*

$$\sum_{k \in K} (a_k + b_k) = \sum_{k \in K} a_k + \sum_{k \in K} b_k$$

Proposition 3.1.3 [Permutation Law]. *The value of a sum is unchanged by permuting the order of the summands.*

$$\sum_{k \in K} a_k = \sum_{k \in K} a_{\pi(k)}$$

As a first illustration, we apply these algebraic regroupings to an *arithmetic progression*. From our present perspective, that means a sequence $\langle a_n \rangle$ given by a recurrence of the form

$$a_0 = c$$
$$a_n = a_{n-1} + b \qquad \text{for } n > 0$$

For instance, the consecutive odd numbers $3, 5, 7, 9, \ldots$ are an arithmetic progression, with initial value $c = 3$ and increment $b = 2$.

Example 3.1.6: Simplifying the sum of a finite arithmetic progression

$$S_n = \sum_{k=0}^{n} (c + bk) \tag{3.1.4}$$

can begin with application of the Permutation Law.

$$S_n = \sum_{k=0}^{n} (c + b(n - k)) \tag{3.1.5}$$

Adding equations (3.1.4) and (3.1.5) leads into the following analysis.

$$2S_n = \sum_{k=0}^{n} (c + bk) + \sum_{k=0}^{n} (c + b(n - k))$$

$$= \sum_{k=0}^{n} [(c + bk) + (c + b(n - k))] \qquad \text{(Addition Law)}$$

$$= \sum_{k=0}^{n} (2c + bn)$$

$$= (2c + bn) \sum_{k=0}^{n} 1 \qquad \text{(Distributive Law)}$$

$$= (2c + bn)(n + 1)$$

$$\Rightarrow S_n = \left(c + \frac{bn}{2} \right) \cdot (n + 1) \tag{3.1.6}$$

Example 3.1.7:   This is a special case of formula (3.1.6).

$$\sum_{k=0}^{n} k = \left(0 + \frac{1 \cdot n}{2}\right) \cdot (n+1)$$

$$= \left(\frac{n}{2}\right) \cdot (n+1)$$

$$= \binom{n+1}{2}$$

For instance,

$$0 + 1 + 2 + 3 + 4 + 5 = 15 = \binom{6}{2}$$

Harmonic Numbers

REVIEW FROM §1.2:

The sequence of **harmonic numbers** $\langle H_n \rangle$ is defined by the rule

$$H_n = \sum_{k=1}^{n} \frac{1}{k}$$

$$= \frac{1}{1} + \frac{1}{2} + \cdots + \frac{1}{n} \qquad \text{for } n \geq 0$$

The harmonic numbers are the discrete analogue of the natural logarithm

$$\ln(n) = \int_{1}^{n} \frac{1}{x} \, dx$$

Figure 3.1.2 illustrates that the harmonic number and the natural logarithm are reasonably good approximations of each other. Familiarity with upper and lower Riemann sums may add some interest here, but such familiarity is not necessary for understanding of the correctness of the approximation.

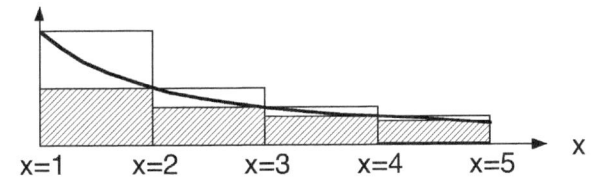

Figure 3.1.2   **Upper and lower Riemann approximations of** $\frac{1}{x}$.

Since the area under the curve $1/x$ over the interval $[1, 5]$ is $\ln(5)$, one observes that $\ln 5$ is less than the sum of the areas of the upper rectangles, i.e.,

$$\ln 5 < \frac{1}{1} + \frac{1}{2} + \frac{1}{3} + \frac{1}{4} = H_4 \qquad \text{upper sum}$$

and that $\ln 5$ is greater than the sum of the areas of the lower rectangles, i.e.,

$$H_5 - 1 = \frac{1}{2} + \frac{1}{3} + \frac{1}{4} + \frac{1}{5} < \ln 5 \qquad \text{lower sum}$$

This observation generalizes to the following:

Proposition 3.1.4. *For any positive integer* n

$$\text{(a)} \quad \ln(n+1) \; < \; H_n \; < \; \ln(n) + 1$$

$$\text{(b)} \quad H_n - 1 \; < \; \ln(n) \; < \; H_{n-1}$$

Proof: Summing the areas of the upper rectangles (i.e., taking upper Riemann sums) yields

$$\ln(n+1) \; < \; \frac{1}{1} + \cdots + \frac{1}{n} \; = \; H_n \tag{3.1.7}$$

$$\ln(n) \; < \; \frac{1}{1} + \cdots + \frac{1}{n-1} \; = \; H_{n-1} \tag{3.1.8}$$

and summing the areas of the lower rectangles (i.e., taking lower Riemann sums) yields

$$H_n - 1 \; = \; \frac{1}{2} + \cdots + \frac{1}{n-1} \; < \; \ln n \tag{3.1.9}$$

Together, (3.1.7) and (3.1.9) imply part (a). Similarly, (3.1.9) and (3.1.8) imply part (b). $\diamond$

GKP Notations

The exercises for this section use two elementary number-theoretic concepts not previously defined here and some innovative notation, introduced by [GKP1994]. Use of these notations also appears later in this book.

DEFINITION: Let n and d be integers. If there is an integer q such that $n = dq$, then we say that d **divides** n. Notation: $d \setminus n$.

Remark: The usual notation for the *divides* relation is $d \,|\, n$. [GKP1994] observes that vertical lines are already overused in mathematics, notably for absolute values, conditional probabilities, and set delimiters. Moreover, for many people, using *backslash* has mnemonic value, since it brings to mind the divides operator *slash*.

DEFINITION: Let m and n be integers whose greatest common divisor is 1. Then we say that m and n are **relatively prime**. Notation $m \perp n$.

Remark: There is no standard notation for relative primality. The *perpendicularity notation* $m \perp n$ appropriately suggests orthogonality. If two vectors are orthogonal, then the dot-product of their coordinate tuples is zero. Suppose that an integer is represented by a tuple of integers whose k^{th} coordinate is the exponent of the kth smallest prime, in its prime-power decomposition. Then two positive integers are relatively prime if and only if the dot-product of their respective representations is zero.

EXERCISES for Section 3.1

In each of the Exercises 3.1.1 through 3.1.4, rewrite the given sum as a consecutive sum without an Iverson truth function, with 0 as lower limit. The notations are defined in a remark that immediately precedes these exercises.

3.1.1^{S} $\displaystyle\sum_{7 \leq k^2 \leq 45} \frac{1}{k}$ **3.1.2** $\displaystyle\sum_{14 \leq k < 31} k \cdot (3 \backslash k)$

3.1.3 $\displaystyle\sum_{-5 \leq k \leq 29} k^{\underline{3}} \cdot (k \bmod 7 \equiv 4)$ **3.1.4** $\displaystyle\sum_{7 \leq k \leq 55} k^{\underline{2}} \cdot (k \perp 48)$

3.1.5   In the equation

$$\sum_{k=0}^{2n} 3k^2 \cdot (k \text{ even}) + \sum_{k=0}^{2n+1} 2k^2 \cdot (k \text{ odd}) \;=\; \sum_{j=0}^{n}(bj^2 + cj + d)$$

solve for b, c, and d.

3.1.6   In the equation

$$\sum_{k=0}^{n} 3j^2 + \sum_{k=0}^{2n} 2j^2 \;=\; \sum_{j=0}^{n}(bn^2 + cn + d)$$

solve for b, c, and d.

3.1.7^{S} Use Proposition 3.1.4 to prove for $n \geq 2$ that

$$1 \;<\; \frac{H_n}{\ln n} \;<\; 1 + \frac{1}{\ln n}$$

3.1.8   Use Proposition 3.1.4 to prove for $n \geq 2$ that

$$1 - \frac{1}{\ln n} \;<\; \frac{\ln n}{H_n} \;<\; 1$$

3.2 PERTURBATION

Perturbation is conceptually quite a simple method, whose correctness is transparent, without any theoretical development. Like the other methods of summation considered in this chapter, its goal is to transform a formula for the entries of a sequence $\langle x_n \rangle$ into a formula for the entries of its sequence $\langle S_n \rangle$ of partial sums. The intended result is an expression for evaluating any sum of consecutive entries within the sequence $\langle x_n \rangle$.

The initial step of a **perturbation** is to equate two expressions for S_{n+1}, the $n+1^{\text{st}}$ partial sum of the sequence $\langle x_n \rangle$.

$$S_n + x_{n+1} = x_0 + \sum_{k=1}^{n+1} x_k$$

By transparency of correctness of the method, we mean, for a start, that the sums on both sides of the equal sign are clearly equal. The summation on the right is first transformed so that its lower and upper limits are 0 and n, respectively, and then manipulated algebraically in order to recast it as a multiple of S_n plus the sum of a few other terms. The theoretical correctness of such algebraic steps was justified in §3.1. Perturbation is a practical method, and additional tricks are used as needed. What makes it interesting is not the theory behind it, but the fact that it works so effectively so often.

Example 3.2.1:   A very simple first example of applying perturbation is to evaluate the sum

$$S_n = \sum_{k=0}^{n} 2^k \tag{3.2.1}$$

Of course, the solution is easily obtainable by other methods, but the details serve as a good illustration of the technique of perturbation.

$$S_n + 2^{n+1} = 2^0 + \sum_{k=1}^{n+1} 2^k = 1 + \sum_{k=1}^{n+1} 2^k \qquad \text{(set up)}$$

$$= 1 + \sum_{k=0}^{n} 2^{k+1} \qquad \text{(change of limits)}$$

$$= 1 + 2 \sum_{k=0}^{n} 2^k$$

$$= 1 + 2 S_n$$

$$\Rightarrow \qquad S_n = 2^{n+1} - 1 \qquad \text{(solution)} \tag{3.2.2}$$

For instance, for $n = 3$, the value of the sum (3.2.1) is

$$2^0 + 2^1 + 2^2 + 2^3 = 1 + 2 + 4 + 8 = 15$$

and the value of the closed formula in (3.2.2) is

$$2^4 - 1 = 16 - 1 = 15$$

A Classical Example of Perturbation

Example 3.2.2:   A classic example to show the power of the method of perturbation is evaluating the sum

$$S_n = \sum_{k=0}^{n} k2^k \tag{3.2.3}$$

which is not so easily evaluated by the most elementary methods. The setup used here (and on Example 3.2.1) is characteristic of applications of the perturbation method.

$$S_n + (n+1)2^{n+1} = 0 \cdot 2^0 + \sum_{k=1}^{n+1} k2^k = \sum_{k=1}^{n+1} k2^k \qquad \text{(set up)}$$

$$= \sum_{k=0}^{n} (k+1)2^{k+1} \qquad \text{(change of limits)}$$

$$= \sum_{k=0}^{n} k2^{k+1} + \sum_{k=0}^{n} 2^{k+1} \qquad \text{(Addition Law)}$$

$$= 2\sum_{k=0}^{n} k2^k + 2\sum_{k=0}^{n} 2^k \qquad \text{(Distributive Law)}$$

$$= 2S_n + 2(2^{n+1} - 1) \qquad \text{(from Example 3.2.1)}$$

$$\Rightarrow \quad S_n = (n+1)2^{n+1} - 2(2^{n+1} - 1)$$

$$= (n-1)2^{n+1} + 2 \qquad \text{(solution)} \tag{3.2.4}$$

For $n = 3$, the result of the term-by-term summation (3.2.3)

$$\sum_{k=0}^{3} k2^k = 0 \cdot 2^0 + 1 \cdot 2^1 + 2 \cdot 2^2 + 3 \cdot 2^3$$

$$= 0 + 2 + 8 + 24 = 34$$

agrees with the evaluation of the formula (3.2.4) derived by perturbation.

$$(n-1)2^{n+1} = (3-1)2^4 + 2$$

$$= 2 \cdot 16 + 2 = 32 + 2 = 34$$

Indirect Perturbation

When a first attempt at perturbation misses the target, it may help to adjust what is to be perturbed and to try a second time, as indicated by the next example.

Example 3.2.3:   We evaluate the sum

$$S_n = \sum_{k=0}^{n} H_k \tag{3.2.5}$$

by perturbation, as in previous examples.

$$S_n + H_{n+1} = H_0 + \sum_{k=1}^{n+1} H_k = 0 + \sum_{k=1}^{n+1} H_k$$

$$= \sum_{k=0}^{n} H_{k+1} = \sum_{k=0}^{n} \left(H_k + \frac{1}{k+1} \right)$$

$$= \sum_{k=0}^{n} H_k + \sum_{k=0}^{n} \frac{1}{k+1}$$

$$= S_n + \sum_{k=0}^{n} \frac{1}{k+1}$$

$$\Rightarrow \quad H_{n+1} = \sum_{k=0}^{n} \frac{1}{k+1} \tag{3.2.6}$$

Formula (3.2.6) is quite correct, but it is not what was wanted, since the symbol S_n cancelled out. When this occurs, a standard maneuver is to multiply the summand by the index variable k and to perturb the result.

Example 3.2.3, continued: Multiplying the summand H_k by the index variable in this example yields the summation

$$S_n = \sum_{k=0}^{n} k H_k \tag{3.2.7}$$

which we now perturb, as follows.

$$S_n + (n+1)H_{n+1} = 0 H_0 + \sum_{k=1}^{n+1} k H_k$$

$$= 0 + \sum_{k=1}^{n+1} k H_k$$

$$= \sum_{k=0}^{n} (k+1) H_{k+1}$$

$$= \sum_{k=0}^{n} (k+1) \left(H_k + \frac{1}{k+1} \right)$$

$$= \sum_{k=0}^{n} (k+1) H_k + \sum_{k=0}^{n} \frac{k+1}{k+1}$$

$$= \sum_{k=0}^{n} k H_k + \sum_{k=0}^{n} H_k + \sum_{k=0}^{n} 1$$

$$= S_n + \sum_{k=0}^{n} H_k + n + 1$$

$$\Rightarrow \quad \sum_{k=0}^{n} H_k = (n+1)H_{n+1} - (n+1) \tag{3.2.8}$$

This time, the result is a formula (3.2.8) for the sum of consecutive harmonic numbers, the formula we actually want. For $n = 3$, directly adding the harmonic numbers, which are the summands of the sum (3.2.5)

$$\sum_{k=0}^{3} H_k = 0 + \left(\frac{1}{1}\right) + \left(\frac{1}{1} + \frac{1}{2}\right) + \left(\frac{1}{1} + \frac{1}{2} + \frac{1}{3}\right) = \frac{13}{3}$$

and applying the summation formula (3.2.8)

$$4H_4 - 4 = 4 \cdot \frac{25}{12} - 4 = \frac{25}{3} - 4 = \frac{13}{3}$$

yield the same result, thereby illustrating correctness of the formula.

As a second example of indirect perturbation, consider the problem of deriving a formula for summing k^2.

Example 3.2.4:   To evaluate the sum

$$S_n = \sum_{k=0}^{n} k^2 \tag{3.2.9}$$

we start as usual.

$$
\begin{aligned}
S_n + (n+1)^2 &= 0^2 + \sum_{k=1}^{n+1} k^2 = \sum_{k=1}^{n+1} k^2 \\
&= \sum_{k=0}^{n} (k+1)^2 = \sum_{k=0}^{n} (k^2 + 2k + 1) \\
&= \sum_{k=0}^{n} k^2 + \sum_{k=0}^{n} 2k + \sum_{k=0}^{n} 1 \qquad \text{(Addition Law)} \\
&= \sum_{k=0}^{n} k^2 + 2\sum_{k=0}^{n} k + \sum_{k=0}^{n} 1 \qquad \text{(Distributive Law)} \\
&= S_n + 2\sum_{k=0}^{n} k + (n+1) \\
\Rightarrow \quad \sum_{k=0}^{n} k &= \frac{(n+1)^2 - (n+1)}{2} = \frac{n^2 + n}{2} \tag{3.2.10}
\end{aligned}
$$

Thus, as in Example 3.2.3, direct perturbation has yielded a correct equation that is not the desired result. Seeking to remedy this situation, we once again multiply the summand by the index variable and re-perturb.

Example 3.2.4, continued: Since perturbing the sum of consecutive values of k^2 just above has yielded an evaluation for the sum of consecutive values of k, it may be less than fully surprising that perturbing the sum of values of k^3 yields a formula for the sum of values of k^2. This time, set

$$S_n = \sum_{k=0}^{n} k^3 \tag{3.2.11}$$

Then

$$S_n + (n+1)^3 = 0^3 + \sum_{k=1}^{n+1} k^3 = \sum_{k=1}^{n+1} k^3$$

$$= \sum_{k=0}^{n} (k+1)^3$$

$$= \sum_{k=0}^{n} (k^3 + 3k^2 + 3k + 1)$$

$$= \sum_{k=0}^{n} k^3 + 3 \sum_{k=0}^{n} k^2 + 3 \sum_{k=0}^{n} k + \sum_{k=0}^{n} 1$$

$$= S_n + 3 \sum_{k=0}^{n} k^2 + 3 \sum_{k=0}^{n} k + \sum_{k=0}^{n} 1$$

$$\Rightarrow \quad 3 \sum_{k=0}^{n} k^2 = (n+1)^3 - 3 \sum_{k=0}^{n} k - \sum_{k=0}^{n} 1$$

$$= (n+1)^3 - \frac{3n^2 + 3n}{2} - (n+1)$$

$$= (n+1)^3 - \frac{3n^2 + 5n + 2}{2} = \frac{2n^3 + 3n^2 + n}{2}$$

$$\Rightarrow \quad \sum_{k=0}^{n} k^2 = \frac{2n^3 + 3n^2 + n}{6} \tag{3.2.12}$$

For $n = 3$, we confirm the agreement of the value of the sum (3.2.9)

$$\sum_{k=0}^{3} k^2 = 0^2 + 1^2 + 2^2 + 3^2$$

$$= 0 + 1 + 4 + 9 = 14$$

with the value of formula (3.2.12)

$$\frac{2n^3 + 3n^2 + n}{6} = \frac{2 \cdot 3^3 + 3 \cdot 3^2 + 3}{6} = \frac{84}{6} = 14$$

Remark: As with direct perturbation, the correctness of the method of indirect perturbation is clear. Although one could plausibly memorize a list of circumstances in which the indirect form is the more helpful form, we adopt here the practical approach of trying direct perturbation first, and then indirect perturbation if it seems to be needed.

EXERCISES for Section 3.2

In each of the Exercises 3.2.1 through 3.2.14, evaluate the given sum by perturbation.

3.2.1^{S} $\displaystyle\sum_{k=0}^{n} 3^k$ **3.2.2** $\displaystyle\sum_{k=0}^{n} 4^k$

3.2.3 $\displaystyle\sum_{k=0}^{n} k3^k$ **3.2.4** $\displaystyle\sum_{k=0}^{n} k4^k$

3.2.5 $\displaystyle\sum_{k=0}^{n} k^2 3^k$ **3.2.6** $\displaystyle\sum_{k=0}^{n} k^2 4^k$

3.2.7 $\displaystyle\sum_{k=0}^{n} 3^{-k}$ **3.2.8** $\displaystyle\sum_{k=0}^{n} 4^{-k}$

3.2.9 $\displaystyle\sum_{k=0}^{n} k3^{-k}$ **3.2.10** $\displaystyle\sum_{k=0}^{n} k4^{-k}$

3.2.11 $\displaystyle\sum_{k=0}^{n} k^2 3^{-k}$ **3.2.12** $\displaystyle\sum_{k=0}^{n} k^2 4^{-k}$

3.2.13 $\displaystyle\sum_{k=0}^{n} 2^{\frac{k}{2}}$ **3.2.14** $\displaystyle\sum_{k=0}^{n} 3^{\frac{k}{2}}$

In each of the Exercises 3.2.15 through 3.2.18, evaluate the given sum by indirect perturbation.

3.2.15^{S} $\displaystyle\sum_{k=0}^{n} k^3$ **3.2.16** $\displaystyle\sum_{k=0}^{n} k^4$

3.2.17 $\displaystyle\sum_{k=0}^{n} kH_k$ **3.2.18** $\displaystyle\sum_{k=0}^{n} k^2 H_k$

3.3   SUMMING WITH GENERATING FUNCTIONS

In this section, it will be seen that most of the sums evaluated in §3.2 could easily be evaluated, alternatively, by using generating functions, as indicated by Theorem 1.7.2 and its corollaries, with the aid of partial fractions, as needed.

REVIEW FROM §1.7:

- **Theorem 1.7.2.** Let $B(z)$ be the ordinary generating function for a sequence $\langle b_n \rangle$. Then the ordinary generating function for the sequence of partial sums of $\langle b_n \rangle$ is

$$\frac{B(z)}{1-z}$$

- **Corollary 1.7.3.**　$\dfrac{1}{(1-z)^r} = \displaystyle\sum_{n=0}^{\infty} \binom{n+r-1}{r-1} z^n$

- **Corollary 1.7.4.**　$\dfrac{1}{(1-az)^r} = \displaystyle\sum_{n=0}^{\infty} \binom{n+r-1}{r-1} a^n z^n$

Revisiting Examples

Example 3.2.1, revisited: We examine how to use generating functions to re-derive the summation formula

$$\sum_{k=0}^{n} 2^k = 2^{n+1} - 1$$

for the powers of 2. As first mentioned in §1.7, the ordinary generating function for the sequence $\langle b_k = 2^k \rangle$ is

$$B(z) = \frac{1}{1 - 2z}$$

By Theorem 1.7.2, the generating function for the sequence

$$\left\langle u_n = \sum_{k=0}^{n} b_k = \sum_{k=0}^{n} 2^k \ \middle|\ n = 0, 1, \dots \right\rangle$$

is

$$U(z) = \sum_{n=0}^{\infty} u_n z^n = \frac{1}{1-z} B(z)$$

$$= \frac{1}{(1-z)(1-2z)}$$

By the method of partial fractions (described in §2.3), which here involves the solution of a pair of simultaneous linear equations, it follows that

$$\frac{1}{(1-z)(1-2z)} = \frac{-1}{(1-z)} + \frac{2}{(1-2z)}$$

$$= \sum_{n=0}^{\infty} (-1) z^n + \sum_{n=0}^{\infty} 2 \cdot 2^n z^n$$

$$= \sum_{n=0}^{\infty} (2^{n+1} - 1) z^n$$

$$\Rightarrow \quad u_n = \sum_{k=0}^{n} 2^k = 2^{n+1} - 1$$

Example 3.2.2, revisited: We now rederive the summation formula

$$\sum_{k=0}^{n} k 2^k = (n-1) 2^{n+1} + 2$$

Corollary 1.7.4 provides the formula

$$\frac{1}{(1-az)^r} = \sum_{n=0}^{\infty} \binom{n+r-1}{r-1} a^n z^n$$

into which the substitutions $a = 2$ and $r = 2$ yield

$$\frac{1}{(1-2z)^2} = \sum_{n=0}^{\infty} \binom{n+1}{1} 2^n z^n = \sum_{n=0}^{\infty} (n+1) 2^n z^n$$

from which it follows that

$$\frac{2z}{(1-2z)^2} = \sum_{n=0}^{\infty} (n+1) 2^{n+1} z^{n+1} = \sum_{n=0}^{\infty} n\, 2^n z^n$$

Thus, the ordinary generating function for the sequence $\langle b_n = n2^n \rangle$ is

$$B(z) = \frac{2z}{(1-2z)^2}$$

By Theorem 1.7.2, the generating function for its sequence

$$\left\langle v_n = \sum_{k=0}^{n} k2^k \ \middle|\ n = 0, 1, \dots \right\rangle$$

of partial sums is

$$V(z) = \sum_{n=0}^{\infty} v_n z^n = \frac{1}{1-z} B(z)$$

$$= \frac{2z}{(1-z)(1-2z)^2}$$

By the method of partial fractions, which this time requires the solution of three simultaneous linear equations, we have

$$\frac{2z}{(1-z)(1-2z)^2} = \frac{2}{(1-z)} + \frac{8z}{(1-2z)^2} - \frac{2}{(1-2z)^2}$$

Thus, by Corollaries 1.7.3 and 1.7.4, it follows that

$$v_n = \sum_{k=0}^{n} k2^k = 2 + 4n \cdot 2^n - (n+1)2^{n+1}$$

$$= (n-1)2^{n+1} + 2$$

For this example, the previous evaluation by perturbation may seem less effort than the method of generating functions, because of the linear equations and the care needed to apply Corollary 1.7.4 accurately.

Example 3.2.4, revisited: To rederive the summation formula

$$\sum_{k=0}^{n} k^2 = \frac{2n^3 + 3n^2 + n}{6}$$

the method of generating functions is easier than perturbation, since it avoids the false start encountered in perturbation, which is unlikely to be discovered until the late stages. To derive the ordinary generating function for the sequence $\langle b_k = k^2 \rangle$ we start with Corollary 1.7.3.

$$\frac{1}{(1-z)^r} = \sum_{n=0}^{\infty} \binom{n+r-1}{r-1} z^n$$

For $r = 3$, this yields

$$\frac{1}{(1-z)^3} = \sum_{n=0}^{\infty} \binom{n+2}{2} z^n$$

Therefore,

$$\frac{z^2}{(1-z)^3} = \sum_{n=0}^{\infty} \binom{n}{2} z^n = \sum_{n=0}^{\infty} \frac{n^2 - n}{2} z^n$$

and

$$\frac{z}{(1-z)^3} = \sum_{n=0}^{\infty} \binom{n+1}{2} z^n = \sum_{n=0}^{\infty} \frac{n^2 + n}{2} z^n$$

from which it follows that

$$\frac{z^2 + z}{(1-z)^3} = \sum_{n=0}^{\infty} n^2 z^n = B(z)$$

By Theorem 1.7.2, the generating function for the sequence

$$\left\langle y_n = \sum_{k=0}^{n} k^2 \,\middle|\, n = 0, 1, \ldots \right\rangle$$

is

$$Y(z) = \sum_{n=0}^{\infty} y_n z^n = \frac{B(z)}{1-z} = \frac{z^2 + z}{(1-z)^4}$$

Corollary 1.7.3 with $r = 4$ is

$$\frac{1}{(1-z)^4} = \sum_{n=0}^{\infty} \binom{n+3}{3} z^n$$

Thus,

$$Y(z) = \frac{z^2 + z}{(1-z)^4} = \sum_{n=0}^{\infty} \binom{n+3}{3} z^{n+2} + \sum_{n=0}^{\infty} \binom{n+3}{3} z^{n+1}$$

$$= \sum_{n=0}^{\infty} \binom{n+1}{3} z^n + \sum_{n=0}^{\infty} z \binom{n+2}{3} z^n$$

Therefore,

$$
\begin{aligned}
y_n &= \frac{(n+1)^{\underline{3}}}{6} + \frac{(n+2)^{\underline{3}}}{6} \\
&= \frac{n^3 - n}{6} + \frac{n^3 + 3n^2 + 2n}{6} \\
&= \frac{2n^3 + 3n^2 + n}{6}
\end{aligned}
$$

EXERCISES for Section 3.3

In each of the Exercises 3.3.1 through 3.3.20,

  a. write a generating function for the sequence of summands;

  b. write a generating function for the sequence of partial sums;

  c. split the result of part (b) by partial fractions;

  d. use part (c) to evaluate the given sum. Where appropriate, perhaps compare this result for part (d) with your solution to a corresponding exercise from §3.2.

3.3.1^{S} $\displaystyle\sum_{k=0}^{n} 3^k$ 3.3.2 $\displaystyle\sum_{k=0}^{n} 4^k$

3.3.3 $\displaystyle\sum_{k=0}^{n} k3^k$ 3.3.4 $\displaystyle\sum_{k=0}^{n} k4^k$

3.3.5 $\displaystyle\sum_{k=0}^{n} k^2 3^k$ 3.3.6 $\displaystyle\sum_{k=0}^{n} k^2 4^k$

3.3.7^{S} $\displaystyle\sum_{k=0}^{n} 3^{-k}$ 3.3.8 $\displaystyle\sum_{k=0}^{n} 4^{-k}$

3.3.9 $\displaystyle\sum_{k=0}^{n} k3^{-k}$ 3.3.10 $\displaystyle\sum_{k=0}^{n} k4^{-k}$

3.3.11 $\displaystyle\sum_{k=0}^{n} k^2 3^{-k}$ 3.3.12 $\displaystyle\sum_{k=0}^{n} k^2 4^{-k}$

3.3.13^{S} $\displaystyle\sum_{k=0}^{n} 2^{\frac{k}{2}}$ 3.3.14 $\displaystyle\sum_{k=0}^{n} 3^{\frac{k}{2}}$

3.3.15 $\displaystyle\sum_{k=0}^{n} k^3$ 3.3.16 $\displaystyle\sum_{k=0}^{n} k^4$

3.3.17^{S} $\displaystyle\sum_{k=0}^{n} \binom{k+1}{1} 3^k$ 3.3.18 $\displaystyle\sum_{k=0}^{n} k^{\underline{1}} 3^k$

3.3.19 $\displaystyle\sum_{k=0}^{n} \binom{k+2}{2} 3^k$ 3.3.20 $\displaystyle\sum_{k=0}^{n} k^{\underline{2}} 3^k$

3.4 FINITE CALCULUS

In the Fundamental Theorem of Finite Calculus (Theorem 1.4.3), Part (a) asserts how sums can be evaluated as anti-differences, analogous to way the fundamental theorem of infinitessimal calculus asserts that integrals can be evaluated as anti-derivatives. This is yet another powerful method for evaluating sums. This section develops a few of the most important formulas of the finite calculus.

REVIEW FROM §1.4:

- Given a function $f : \mathbb{R} \to \mathbb{R}$, the **difference function** $\triangle f$ is given by the rule

$$\triangle f(x) \;=\; f(x+1) - f(x) \tag{3.4.1}$$

- Given a sequence $\langle x_n \rangle$, we define the **difference sequence** $\langle \triangle x_n \rangle$ by the rule

$$\triangle x_n \;=\; x_n' \;=\; x_{n+1} - x_n$$

- **Theorem 1.4.3 [Fundamental Theorem of Finite Calculus].** Let $\langle x_n \rangle$ be any standard sequence. Then

$$(a) \; \sum_{j=0}^{n-1} x_j' \;=\; x_n - x_0; \qquad (b) \; \left(\sum_{j=0}^{k-1} x_j \right)_n' \;=\; x_n$$

Summing a Polynomial

We recall that the finite calculus formulas for differencing and summing a falling power are directly analogous to the infinitessimal calculus formula for differentiating and integrating an ordinary power.

REVIEW FROM §1.5:

- The nth **falling power** function on a real variable x, for any $n \in \mathbb{N}$, is defined by the rule

$$x^{\underline{n}} \;=\; \overbrace{x(x-1)\cdots(x-n+1)}^{n \text{ factors}}$$

- **Theorem 1.5.1.** For $r \in \mathbb{Z}^+$, we have $\triangle(x^{\underline{r}}) \;=\; r x^{\underline{r-1}}$.

- **Corollary 1.5.2.** For $r \in \mathbb{N}$, we have $\displaystyle\sum_{k=0}^{n-1} k^{\underline{r}} \;=\; \frac{n^{\underline{r+1}}}{r+1}$.

We established in §1.6 that ordinary powers can be converted into falling powers.

Review from §1.6:

- **Theorem 1.6.1**   Any ordinary power x^n can be expressed as a linear combination of falling powers, i.e., in the form

$$x^n = \sum_{r=0}^{n} S_{n,r}\, x^{\underline{r}}$$

where the coefficients $S_{n,r}$ are called *Stirling numbers of the second kind*.

We will now use the reviewed results to see how finite calculus makes many kinds of summation quite routine.

Example 3.2.4, revisited again: In this chapter, we have already derived the summation formula

$$\sum_{k=0}^{n} k^2 = \frac{2n^3 + 3n^2 + n}{6}$$

first using indirect perturbation, and then again with generating functions. Yet most practitioners of combinatorial calculations would say that using summation calculus, as we saw in Example 1.6.3, is the easiest of the three approaches. When solving this sum with finite calculus, we first express k^2 as a linear combination of falling powers. For monomials of low degree, it is easy enough to calculate the coefficients of the falling powers by *ad hoc* methods.

$$\begin{aligned}
k^2 &= S_{2,2}\, k^{\underline{2}} + S_{2,1}\, k^{\underline{1}} \\
&= k^{\underline{2}} + k^{\underline{1}}
\end{aligned} \tag{3.4.2}$$

Summing both sides of equation (3.4.2), we obtain

$$\begin{aligned}
\sum_{k=0}^{n} k^2 &= \sum_{k=0}^{n} (k^{\underline{2}} + k^{\underline{1}}) \\
&= \sum_{k=0}^{n} k^{\underline{2}} + \sum_{k=0}^{n} k^{\underline{1}}
\end{aligned}$$

Applying Corollary 1.5.2 now yields

$$\begin{aligned}
\sum_{k=0}^{n} k^2 &= \left.\frac{k^{\underline{3}}}{3}\right|_{k=0}^{n+1} + \left.\frac{k^{\underline{2}}}{2}\right|_{k=0}^{n+1} \\
&= \frac{(n+1)^{\underline{3}}}{3} + \frac{(n+1)^{\underline{2}}}{2} \\
&= \frac{n^3 - n}{3} + \frac{n^2 + n}{2} \\
&= \frac{2n^3 + 3n^2 + n}{6}
\end{aligned}$$

Formula for Summing Exponentials

The supply of useful finite summation formulas is readily extended beyond the monomial formula of Corollary 1.5.2. This begins with sums and differences of exponentiations in which the base is constant and the exponent is variable.

Theorem 3.4.1. *Let the constant value c be a real number and let x be a real or integer variable. Then*

$$\triangle c^x \;=\; (c-1)c^x$$

Proof: $\triangle c^x \;=\; c^{x+1} - c^x \;=\; (c-1)\,c^x.$ $\diamond$

Example 3.4.1:   For the case $c = 2$, Theorem 3.4.1 gives the result

$$\triangle 2^x \;=\; (2-1)\,2^x \;=\; 2^x$$

which is analogous to the differential-calculus result

$$\frac{d}{dx}\,e^x \;=\; e^x$$

This is one of numerous reasons why the number 2 is regarded as the natural base of discrete mathematics in the same sense that the real number e is the natural base for continuous mathematics. More generally, the continuous-mathematics formula

$$\frac{d}{dx}\,c^x \;=\; \ln c \cdot c^x$$

is analogous to the discrete-mathematics formula of Theorem 3.4.1.

$$\triangle c^x \;=\; (c-1)\,c^x$$

Example 3.4.2
$$\triangle 3^x \;=\; 3^{x+1} - 3^x \;=\; 2 \cdot 3^x$$
$$\triangle 4^x \;=\; 4^{x+1} - 4^x \;=\; 3 \cdot 4^x$$

This leads to a major formula of the finite-summation calculus, the formula for summing exponentials.

Corollary 3.4.2 [Exponential Formula]. *Let c be any real number except 1. Then*

$$\sum_{k=0}^{n-1} c^k \;=\; \frac{c^n - 1}{c - 1} \tag{3.4.3}$$

Proof: For $c \neq 1$, applying the Fundamental Theorem of Finite Calculus to the conclusion of Theorem 3.4.1 implies that

$$\sum_{k=0}^{n-1} c^k = \left. \frac{c^k}{c-1} \right|_{k=0}^{n}$$

$$= \frac{c^n - 1}{c - 1} \qquad\qquad\qquad \diamond$$

Remark: For the case $c = 1$, which is excluded from Corollary 3.4.2, we have the sum

$$\sum_{k=0}^{n-1} 1^k = n$$

Example 3.4.3:   We observe that when $c = 5$ and $n = 4$, the value of the sum on the left side of equation (3.4.3)

$$\sum_{k=0}^{3} 5^k = 5^0 + 5^1 + 5^2 + 5^3 = 1 + 5 + 25 + 125 = 156$$

agrees with the value of the quotient on the right side

$$\frac{5^4 - 1}{4} = \frac{625 - 1}{4} = 156$$

As easy as it was to derive the formula

$$\sum_{k=0}^{n} 2^k = 2^{n+1} - 1$$

either with perturbation or with generating functions, it is even easier with the calculus of summation, as now shown.

Example 3.2.1, revisited again: According to Theorem 3.4.1, we have

$$\triangle 2^k = 2^k$$

Summing both sides, we obtain

$$\sum_{k=0}^{n} 2^k = \sum_{k=0}^{n} \triangle 2^k$$

after which, applying the Fundamental Theorem yields the result

$$\sum_{k=0}^{n} 2^k = \left. 2^k \right|_{k=0}^{n+1}$$

$$= 2^{n+1} - 1$$

Falling Negative Powers

The extension of the list of useful differencing and summation formulas continues. We observe that non-negative falling powers can be defined recursively.

$$x^{\underline{0}} = 1$$
$$x^{\underline{r+1}} = x^{\underline{r}}(x - r) \qquad \text{for } r > 0$$

Running the recursion backward extends the utility of the falling power concept.

DEFINITION: **Non-positive falling powers** are defined as follows.

$$x^{\underline{0}} = 1$$
$$x^{\underline{r}} = \frac{x^{\underline{r+1}}}{x - r} \qquad \text{for } r < 0$$

Example 3.4.4: Here are a few evaluations of the definition of negative falling powers.

$$x^{\underline{-1}} = \frac{x^{\underline{0}}}{x - (-1)} = \frac{x^{\underline{0}}}{x + 1} = \frac{1}{x + 1}$$
$$x^{\underline{-2}} = \frac{x^{\underline{-1}}}{x - (-2)} = \frac{x^{\underline{-1}}}{x + 2} = \frac{1}{(x + 1)(x + 2)}$$
$$x^{\underline{-3}} = \frac{x^{\underline{-2}}}{x - (-3)} = \frac{x^{\underline{-2}}}{x + 3} = \frac{1}{(x + 1)(x + 2)(x + 3)}$$

Proposition 3.4.3. *For any positive number r and any real number x,*

$$x^{\underline{-r}} = \frac{1}{(x + 1) \cdots (x + r)}$$

Proof: By induction on r. $\diamond$

Although ordinary powers are additive in a product of ordinary monomials with the same base, in the sense that

$$x^r \cdot x^s = x^{r+s}$$

it is clear that falling powers are not additive in a product of falling-power monomials. For instance,

$$x^{\underline{2}} \cdot x^{\underline{3}} = x(x - 1) \cdot x(x - 1)(x - 2)$$

but

$$x^{\underline{2+3}} = x^{\underline{5}} = x(x - 1)(x - 2)(x - 3)(x - 4)$$

Thus, there is no reason to expect that $x^{\underline{-r}} = (x^{\underline{r}})^{-1}$. On the other hand, an important analogy to infinitessimal calculus is preserved.

Proposition 3.4.4. *The difference formula for negative falling powers is the same formula as for positive falling powers. That is, for every positive integer r,*

$$\triangle x^{\underline{-r}} = (-r)x^{\underline{-r-1}}$$

Proof: Start by applying the defining formula (3.4.1) for the difference operator.

$$\triangle x^{\underline{-r}} = (x+1)^{\underline{-r}} - x^{\underline{-r}}$$

$$= \frac{1}{(x+2)\cdots(x+r+1)} - \frac{1}{(x+1)\cdots(x+r)}$$

Then by routine manipulation

$$= \frac{1}{(x+2)\cdots(x+r)}\left[\frac{1}{x+r+1} - \frac{1}{x+1}\right]$$

$$= \frac{1}{(x+2)\cdots(x+r)}\left[\frac{-r}{(x+1)(x+r+1)}\right]$$

$$= \frac{-r}{(x+1)\cdots(x+r+1)}$$

we achieve the result

$$\triangle x^{\underline{-r}} = (-r)x^{\underline{-r-1}} \qquad\qquad \diamond$$

Corollary 3.4.5. *For any integer $r \neq 0$ and any real number x,*

$$\triangle x^{\underline{r}} = rx^{\underline{r-1}}$$

Proof: This combines Theorem 1.5.1 and Proposition 3.4.4. $\qquad\qquad \diamond$

Corollary 3.4.6 [Monomial Formula]. *For any integers $r \neq -1$ and n,*

$$\sum_{k=0}^{n-1} k^{\underline{r}} = \left.\frac{k^{\underline{r+1}}}{r+1}\right|_{k=0}^{n} \qquad\qquad (3.4.4)$$

Proof: This combines the Fundamental Theorem and Corollary 3.4.5. $\qquad \diamond$

Example 3.4.5:   To make a direct evaluation of the left side of Equation (3.4.4) for $r = -2, -3$ and $n = 4$, we first calculate the following partial table of the values of $k^{\underline{r}}$, i.e., of a small integer to a small falling negative power.

r	$0^{\underline{r}}$	$1^{\underline{r}}$	$2^{\underline{r}}$	$3^{\underline{r}}$	$4^{\underline{r}}$	$\cdots$
-1	$\frac{1}{1}$	$\frac{1}{2}$	$\frac{1}{3}$	$\frac{1}{4}$	$\frac{1}{5}$	$\cdots$
-2	$\frac{1}{1\cdot 2}$	$\frac{1}{2\cdot 3}$	$\frac{1}{3\cdot 4}$	$\frac{1}{4\cdot 5}$	$\frac{1}{5\cdot 6}$	$\cdots$
-3	$\frac{1}{1\cdot 2\cdot 3}$	$\frac{1}{2\cdot 3\cdot 4}$	$\frac{1}{3\cdot 4\cdot 5}$	$\frac{1}{4\cdot 5\cdot 6}$	$\frac{1}{5\cdot 6\cdot 7}$	$\cdots$

Case $r = -2$ and $n = 4$: The value on the left side is

$$0^{\underline{-2}} + 1^{\underline{-2}} + 2^{\underline{-2}} + 3^{\underline{-2}} = \frac{1}{1\cdot 2} + \frac{1}{2\cdot 3} + \frac{1}{3\cdot 4} + \frac{1}{4\cdot 5} = \frac{4}{5}$$

and the value on the right side is

$$\frac{k^{\underline{-1}}}{-1}\bigg|_{k=0}^{4} = \frac{4^{\underline{-1}}}{-1} - \frac{0^{\underline{-1}}}{-1} = \frac{1}{(-1)\cdot 5} - \frac{1}{-1} = \frac{-1}{5} - (-1) = \frac{4}{5}$$

Case r = −3 and n = 4: The value on the left side is

$$0^{\underline{-3}} + 1^{\underline{-3}} + 2^{\underline{-3}} + 3^{\underline{-3}} = \frac{1}{1\cdot 2\cdot 3} + \frac{1}{2\cdot 3\cdot 4} + \frac{1}{3\cdot 4\cdot 5} + \frac{1}{4\cdot 5\cdot 6} = \frac{7}{30}$$

and the value on the right side is

$$\frac{k^{\underline{-2}}}{-2}\bigg|_{k=0}^{4} = \frac{4^{\underline{-2}}}{-2} - \frac{0^{\underline{-2}}}{-2} = \frac{1}{(-2)\cdot 5\cdot 6} - \frac{1}{(-2)\cdot 1\cdot 2} = \frac{7}{30}$$

Harmonic Numbers

In the formula for evaluating $\sum k^{\underline{r}}$ for the special case with $r = 1$ there is another analogy between the natural logarithm $\ln n$ and the harmonic number H_n, which lies in the similarity of the derivative

$$\frac{d}{dx}\ln x = x^{-1}$$

to the difference formula

$$\begin{aligned}
\triangle H_n &= H_{n+1} - H_n \\
&= \left(\frac{1}{1} + \cdots + \frac{1}{n+1}\right) - \left(\frac{1}{1} + \cdots + \frac{1}{n}\right) \\
&= \frac{1}{n+1} = n^{\underline{-1}}
\end{aligned}$$

and, naturally enough, in the similarity of the summation formula

$$\sum_{k=0}^{n-1} k^{\underline{-1}} = H_n \tag{3.4.5}$$

to the integration formula

$$\int_{x=1}^{t} x^{-1}dx = \ln x$$

Product Formula

Analogous to the product formula for derivatives,

$$(u(x) \cdot v(x))' \; = \; u'(x) \cdot v(x) \; + \; u(x) \cdot v'(x)$$

there is a product formula for finite differences.

Proposition 3.4.7 [Product Formula]. *Let* $h(x) = g(x) \cdot f(x)$. *Then*

$$\triangle h(x) \; = \; \triangle g(x) \cdot f(x+1) \; + \; g(x) \cdot \triangle f(x) \tag{3.4.6}$$

Proof: Once again, it is sufficient to do some routine algebraic manipulation, starting from an application of the definition of the difference operator.

$$\begin{aligned}
\triangle h(x) \; &= \; h(x+1) \; - \; h(x) \\
&= \; g(x+1) \cdot f(x+1) \; - \; g(x) \cdot f(x) \\
&= \; g(x+1) \cdot f(x+1) \; - \; g(x) \cdot f(x) \\
&\qquad - \, g(x) \cdot f(x+1) \; + \; g(x) \cdot f(x+1) \\
&= \; \triangle g(x) \cdot f(x+1) \; + \; g(x) \cdot \triangle f(x) \qquad\qquad \diamond
\end{aligned}$$

Example 3.4.6:   Take $g(n) \; = \; n^2$ and $f(n) \; = \; H_n$. According to the product formula (3.4.6), we have

$$\begin{aligned}
\triangle(n^2 H_n) \; &= \; \triangle n^2 \cdot H_{n+1} \; + \; n^2 \cdot \triangle H_n \\
&= \; 2n \cdot H_{n+1} \; + \; n(n-1) \cdot \frac{1}{n+1} \\
&= \; 2n \left(H_n + \frac{1}{n+1} \right) + \frac{n^2 - n}{n+1} \\
&= \; 2n H_n \; + \; \frac{n^2 + n}{n+1} \\
&= \; 2n H_n \; + \; n
\end{aligned}$$

Unsurprisingly, evaluating the defining formula (3.4.1) for a finite difference yields the identical result.

$$\begin{aligned}
\triangle(n^2 H_n) \; &= \; (n+1)^2 H_{n+1} \; - \; n^2 H_n \\
&= \; (n^2 + n) \left(H_n + \frac{1}{n+1} \right) - (n^2 - n) H_n \\
&= \; (n^2 + n) H_n + \frac{n^2 + n}{n+1} - (n^2 - n) H_n \\
&= \; 2n H_n \; + \; n
\end{aligned}$$

Summation by Parts

From the infinitessimal calculus, we recall the following formula for integration by parts

$$\int_a^b u(x)v'(x)dx = u(x)v(x)\Big|_a^b - \int_a^b u'(x)v(x)dx$$

The finite calculus has an analogous formula, called *summation by parts*.

Proposition 3.4.8 [Summation by Parts]. *Let $g(k)$ and $f(k)$ be functions on the integers. Then*

$$\sum_{k=0}^{n-1} g(k) \triangle (f(k)) = g(k)f(k)\Big|_{k=0}^n - \sum_{k=0}^{n-1} \triangle(g(k)) f(k+1) \qquad (3.4.7)$$

Proof: This corollary to Proposition 3.4.7 follows from the Fundamental Theorem of Finite Calculus.   ◇

Example 3.2.2, revisited again: After using the substitutions

$$g(k) = k^{\underline{1}} \qquad \text{and} \qquad f(k) = 2^k$$

summing the sequence $\langle k2^k \mid k = 0, 1, 2, \ldots \rangle$ by parts takes the form

$$\sum_{k=0}^n k2^k = \sum_{k=0}^n k^{\underline{1}} 2^k$$

$$= k^{\underline{1}} 2^k \Big|_0^{n+1} - \sum_{k=0}^n k^{\underline{0}} 2^{k+1}$$

which leads to the calculations

$$= (n+1) \cdot 2^{n+1} - 2\sum_{k=0}^n 2^k$$

$$= (n+1) \cdot 2^{n+1} - 2(2^{n+1} - 1)$$

$$= (n-1) \cdot 2^{n+1} + 2$$

Example 3.2.3, revisited: Since integration by parts is helpful in evaluating the integral of $\ln x$ to $x \ln x - x$, it is unsurprising that summation by parts is helpful in summing H_n to $nH_n - n$.

$$\sum_{k=0}^{n-1} H_k = \sum_{k=0}^{n-1} k^{\underline{0}} H_k$$

$$= k^{\underline{1}} H_k \Big|_0^n - \sum_{k=0}^{n-1} (k+1)^{\underline{1}} \frac{1}{k+1} = k^{\underline{1}} H_k \Big|_0^n - \sum_{k=0}^{n-1} 1$$

$$= (nH_n - 0) - n$$

$$= nH_n - n$$

Table 3.4.1   Formulas for the calculus of differences.

function	difference function
$k^{\underline{r}}$	$r\,k^{\underline{r-1}}$
c^k	$(c-1)\,c^k$
H_n	$\dfrac{1}{n+1}$
$g(k)f(k)$	$\triangle g(x)\,f(x+1)\,+\,g(x)\,\triangle f(x)$

Table 3.4.2   Formulas for the calculus of summations.

summation	formula	reference
$\displaystyle\sum_{k=0}^{n-1} c^k, \quad c \neq 0$	$\dfrac{c^n - 1}{c - 1}$	(3.4.3)
$\displaystyle\sum_{k=0}^{n-1} k^{\underline{r}}, \quad r \neq -1$	$\dfrac{n^{\underline{r+1}}}{r + 1}$	(3.4.4)
$\displaystyle\sum_{k=0}^{n-1} k^{\underline{-1}}$	H_n	(3.4.5)
$\displaystyle\sum_{k=0}^{n-1} g(k)\,\triangle\,(f(k))$	$g(k)f(k)\Big|_{k=0}^{n} \;-\; \displaystyle\sum_{k=0}^{n-1} \triangle(g(k))\,f(k+1)$	(3.4.7)

EXERCISES for Section 3.4

In each of the Exercises 3.4.1 through 3.4.16, evaluate the given sum by finite calculus. Perhaps compare the result with your solution to the corresponding exercise from §3.2 or from §3.3.

$3.4.1^{\text{S}}$ $\displaystyle\sum_{k=0}^{n} 3^k$ $3.4.2$ $\displaystyle\sum_{k=0}^{n} 4^k$

$3.4.3$ $\displaystyle\sum_{k=0}^{n} k3^k$ $3.4.4^{\text{S}}$ $\displaystyle\sum_{k=0}^{n} k4^k$

$3.4.5$ $\displaystyle\sum_{k=0}^{n} k^2 3^k$ $3.4.6$ $\displaystyle\sum_{k=0}^{n} k^2 4^k$

$3.4.7$ $\displaystyle\sum_{k=0}^{n} 3^{-k}$ $3.4.8$ $\displaystyle\sum_{k=0}^{n} 4^{-k}$

3.4.9 $\displaystyle\sum_{k=0}^{n} k3^{-k}$ **3.4.10** $\displaystyle\sum_{k=0}^{n} k4^{-k}$

3.4.11 $\displaystyle\sum_{k=0}^{n} k^2 3^{-k}$ **3.4.12** $\displaystyle\sum_{k=0}^{n} k^2 4^{-k}$

3.4.13 $\displaystyle\sum_{k=0}^{n} 2^{\frac{k}{2}}$ **3.4.14** $\displaystyle\sum_{k=0}^{n} 3^{\frac{k}{2}}$

3.4.15 $\displaystyle\sum_{k=0}^{n} k^3$ **3.4.16** $\displaystyle\sum_{k=0}^{n} k^4$

In Exercises 3.4.17 and 3.4.18, evaluate the given sum with finite calculus. Perhaps compare each with the corresponding exercise from §3.3.

3.4.17 $\displaystyle\sum_{k=0}^{n} k^{\underline{3}}$ **3.4.18** $\displaystyle\sum_{k=0}^{n} k^{\underline{2}} 3^k$

In Exercises 3.4.19 and 3.4.20, evaluate the given sum with finite calculus. Perhaps compare with Exercises 3.2.17 and 3.2.18, respectively.

3.4.19^{S} $\displaystyle\sum_{k=0}^{n} k H_k$ **3.4.20** $\displaystyle\sum_{k=0}^{n} k^2 H_k$

In each of the Exercises 3.4.21 through 3.4.26, calculate the difference function $\triangle f(n)$ for the given function $f(n)$.

3.4.21 $k^{\underline{4}}$ **3.4.22** k^4 **3.4.23** 4^k
3.4.24^{S} $k^{\underline{-2}}$ **3.4.25** $k^3 + k^2$ **3.4.26** 4^{-k}

In each of the Exercises 3.4.27 through 3.4.32, calculate the anti-difference function for the given function $f(n)$.

3.4.27 $k^{\underline{4}}$ **3.4.28** k^4 **3.4.29** 4^k
3.4.30^{S} $k^{\underline{-2}}$ **3.4.31** $k^3 + k^2$ **3.4.32** 4^{-k}

In Exercises 3.4.33 through 3.4.37,
  a. calculate the next two terms of the given sequence.
  b. specify the function that yields the given sequence.
Hint: use difference tables, which were introduced in §1.4.

3.4.33^{S} 7 8 15 28 47 72

3.4.34 1 −2 −3 10 49 126

3.4.35 0 0 2 8 20 40

3.4.36 2 5 15 33 61 103

3.4.37 0 −1 −6 −13 −4 87 470

3.5   ITERATION AND PARTITIONING OF SUMS

This section is concerned with iterated summation. In the simplest case, the index set U of the sum

$$S = \sum_{i \in U} x_i$$

is a 2-dimensional array, such that one could first take the row sums and then add those sums to get the total. Sometimes the first summation, called the *inner summation*, is for groupings other than rows. In selecting groupings for the inner summations, the consideration is that both the inner summation and the other summation, called the *outer summation*, should be amenable to reasonably convenient methods of evaluation. Sometimes, when given a double summation to evaluate, it is helpful to swap the order of summation, as described here.

Double summation need not be twice as hard. Indeed, sometimes a single sum is recast as a double sum to make it more tractable. A possible strategy in evaluating a difficult sum

$$S = \sum_{i \in U} x_i$$

is to find a partition

$$U = \bigcup_{k \in K} U_k$$

such that each of the sub-sums

$$S_k = \sum_{i \in U_k} x_i$$

is tractable, and also such that the sum

$$\sum_{k \in K} S_k = \sum_{k \in K} \sum_{i \in U_k} x_i$$

of the sub-sums is tractable.

Independent Indices

An example from graph theory illustrates the simplest case of a double summation, in which the index of the inner sum is independent of the index of the outer sum.

Example 3.5.1:   The degree of a vertex v of a graph is the total number of edge-incidences on v. Each edge e contributes 0, 1, or 2 to that total, corresponding to the number $I(v, e)$ of times that vertex v is an endpoint of edge e. Figure 3.5.1 shows a graph, with its vertex degrees as bold numbers.

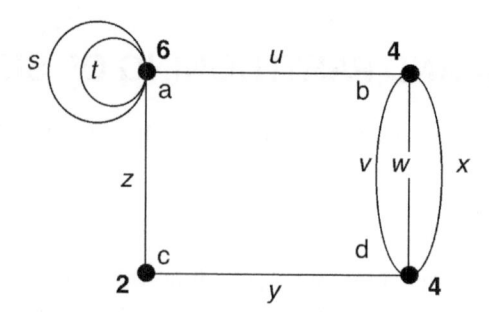

Figure 3.5.1   **Degrees of the vertices of a graph.**

Thus, the sum of all the edge-vertex incidences

$$\sum_{(v,e) \in V \times E} I(v,e)$$

is indexed by the cartesian product $V \times E$, where V is the set of vertices and E the set of edges. The obvious partition of this sum over a cartesian product of two sets is into an iterated sum

$$\sum_{(v,e) \in V \times E} I(v,e) \;=\; \sum_{v \in V} \sum_{e \in E} I(v,e)$$

over the incidence matrix, with rows labeled by vertices and columns by edges, so that the row-sums are the degrees.

	s	t	u	v	v	x	y	z	degree
a	2	2	1	0	0	0	0	1	6
b	0	0	1	1	1	1	0	0	4
c	0	0	0	0	0	0	1	1	2
d	0	0	0	1	1	1	1	0	4

Of course, the result of adding the row-sums of an array equals the result of adding the column-sums. In this case, since every column-sum is 2, adding the column-sums is equivalent to doubling the number of edges, which is faster than adding degree-sums. This observation yields an alternative proof of Euler's Degree-Sum Theorem (Theorem 0.6.1).

Theorem 3.5.1 [Euler's Degree-Sum Theorem]. *The sum of the degrees of the vertices of a graph equals twice the number of edges.*

Proof: Let $V = (V, E)$ be a graph. Then starting from row sums

$$\sum_{v \in V} \deg(v) \;=\; \sum_{v \in V} \sum_{e \in E} I(v,e) \qquad\qquad \text{sum of row sums}$$

swap the order of summation:

$$= \sum_{e \in E} \sum_{v \in V} I(v, e) \qquad \text{sum of column sums}$$

$$= \sum_{e \in E} 2 \qquad \text{every column sum is 2}$$

$$= 2\,|E| \qquad\qquad\qquad\qquad \diamondsuit$$

Interchanging the order of summation is a fundamental technique for evaluating an iterated sum over an array. It is useful when the implicit repartitioning yields inner sums and an outer sum for which the total effort of evaluation is less than that for the given iterated summation problem.

In this instance, the cost of repartitioning was trivial, because the index of the inner sum of the given iterated sum was *independent* of the index of the outer sum.

Dependent Indices

If the limits of the index of the inner sum are independent of the index of the outer sum, then the order of summation can be transposed without changing the limits of either index. However, it is quite common for the outer index to range from 1 to n, while the upper limit of the inner index equals the outer index. As illustrated in Figure 3.5.2, this amounts to summing over the rows of the lower-left triangle of an $n \times n$ array.

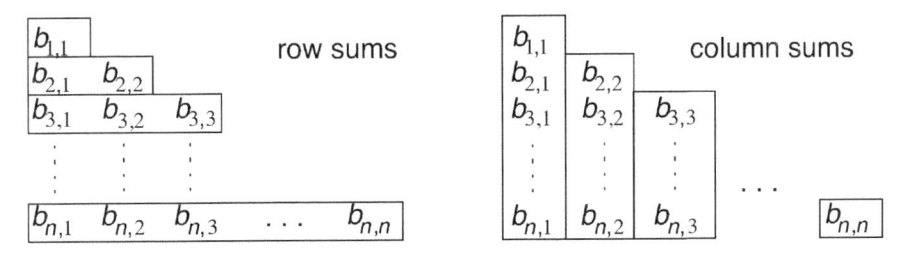

Figure 3.5.2   Row sums and column sums.

Interchanging the order of summation for this form of iterated sum turns the outer sums into column sum. The new inner index (the row index) has the outer index (the column index) as its lower limit and ranges up to n.

Example 3.2.3, revisited: The sum of the harmonic numbers has previously been evaluated by perturbation and by finite calculus. Another effective method is going to a double sum and then interchanging the order of summation.

$$\sum_{k=0}^{n-1} H_k = \sum_{k=0}^{n-1} \sum_{j=1}^{k} \frac{1}{j} \qquad \text{write as double sum}$$

$$= \sum_{j=1}^{n-1} \sum_{k=j}^{n-1} \frac{1}{j} \qquad \text{swap order of summation}$$

$$= \sum_{j=1}^{n-1} \frac{1}{j} \sum_{k=j}^{n-1} 1 \qquad\qquad\qquad \text{factor out constant}$$

$$= \sum_{j=1}^{n-1} \frac{1}{j} (n-j) \qquad\qquad\qquad \text{evaluate inner sum}$$

$$= \sum_{j=1}^{n-1} \left(\frac{n}{j} - \frac{j}{j} \right)$$

$$= \sum_{j=1}^{n} \left(\frac{n}{j} - \frac{j}{j} \right) \qquad\qquad\qquad \text{add zero}$$

$$= \sum_{j=1}^{n} \frac{n}{j} - \sum_{j=1}^{n} \frac{j}{j} \;=\; n \sum_{j=1}^{n} \frac{1}{j} - \sum_{j=1}^{n} 1$$

$$= nH_n - n$$

In circumstances when swapping rows and columns of an array does not adequately reduce the evaluation, it may help to reorganize the partitioning of summation so that the inner sum is over some tractable geometric pattern other than a row or column.

Example 3.5.2:   To envision how to repartition the double sum

$$\sum_{k=1}^{n} \sum_{j=1}^{k-1} \frac{1}{k-j}$$

as an aid in evaluation, it helps to write out the array of summands, like this.

$k \downarrow$	1	2	3	4	$j \to$
2	$\frac{1}{1}$				
3	$\frac{1}{2}$	$\frac{1}{1}$			
4	$\frac{1}{3}$	$\frac{1}{2}$	$\frac{1}{1}$		
5	$\frac{1}{4}$	$\frac{1}{3}$	$\frac{1}{2}$	$\frac{1}{1}$	

Evidently, summing rows or columns amounts to summing the harmonic numbers. However, the strategy of summing on the southeastward diagonals (in which the entries are constant) yields the following result, which is consistent with Example 3.2.3, which also sums harmonic numbers.

$$\sum_{d=1}^{n} \sum_{j=1}^{n-d} \frac{1}{d} \;=\; \sum_{d=1}^{n} \frac{n-d}{d} \;=\; \sum_{d=1}^{n} \frac{n}{d} - \sum_{d=1}^{n} \frac{d}{d}$$

$$= \; n \sum_{d=1}^{n} \frac{1}{d} - \sum_{d=1}^{n} 1 \;=\; nH_n - n$$

Linear Partitioning: Floor Sums

Sometimes a sequence of less tractable summands can be partitioned into consecutive finite subsequences with tractable sums. In particular, this may occur when the summands are the floors or ceilings of a non-decreasing sequence.

Example 3.5.3:   Seeking to evaluate a sum of floors may suggest resorting to an approximation, such as

$$
\sum_{k=0}^{n} \left\lfloor \sqrt{k} \right\rfloor \approx \sum_{k=0}^{n} \sqrt{k}
$$

$$
\approx \int_{x=0}^{n} x^{1/2} dx
$$

$$
= \left. \frac{2}{3} x^{3/2} \right|_{x=0}^{n}
$$

$$
= \frac{2}{3} n^{3/2}
$$

For $n = 9$, the value of this approximation is

$$
\left. \frac{2}{3} n^{3/2} \right|_{n=9} = \frac{2}{3} \cdot 9^{3/2} = \frac{2}{3} \cdot 27 = 18
$$

whereas the exact value is

$$
\left\lfloor \sqrt{0} \right\rfloor + \left\lfloor \sqrt{1} \right\rfloor + \left\lfloor \sqrt{2} \right\rfloor + \left\lfloor \sqrt{3} \right\rfloor
$$

$$
+ \left\lfloor \sqrt{4} \right\rfloor + \left\lfloor \sqrt{5} \right\rfloor + \left\lfloor \sqrt{6} \right\rfloor + \left\lfloor \sqrt{7} \right\rfloor + \left\lfloor \sqrt{8} \right\rfloor + \left\lfloor \sqrt{9} \right\rfloor
$$

$$
= 0 + 1 + 1 + 1 + 2 + 2 + 2 + 2 + 2 + 3 = 16
$$

Sometimes, an approximation this rough meets the purpose at hand. However, it is helpful to be in command of methods that get an exact value when it is needed. There are five steps in the derivation of an exact evaluation formula for such a sum by the method of linear partitioning.

Step 1. List the early terms of the sequence, and partition them according to the value of the summand.

Step 2. Express the size of all but the last cell.

Step 3. Express the size of the last cell, which needs individual attention, since its size might not follow the same rule as the other cells.

Step 4. Evaluate the given sum.

Step 5. Confirm for a small case.

We now demonstrate the application of this method to Example 3.5.3.

Example 3.5.3, continued: As the index k of the sum

$$\sum_{k=0}^{n} \left\lfloor \sqrt{k} \right\rfloor$$

increases, the value of the summand $\lfloor k \rfloor$ increases also, but more slowly than the index itself.

Step 1 is to partition the index values according to the value of the summand. This is represented for $k = 0, 1, \ldots 17$ as follows:

Table 3.5.1 Partitioning for the summand $\lfloor k \rfloor$.

		$\overbrace{}^{1}$	$\overbrace{}^{3}$	$\overbrace{}^{5}$	$\overbrace{}^{7}$	
k	0	1 2 3	4 5 6 7 8	9 10 11 12 13 14 15	16 17
$\lfloor \sqrt{k} \rfloor$	0	1 1 1	2 2 2 2 2	3 3 3 3 3 3 3	4 4

Step 2 is to express the sizes of all but the last cell. Each other cell in Table 3.5.1 is grouped with an overbrace, with its size written over the overbrace. The smallest number within each cell is the square m^2 of some number m. Since the number $(m+1)^2$ starts the next cell, it follows that the cell containing m^2 is

$$\left\{ m^2, \ m^2 + 1, \ \ldots, \ m^2 + 2m \right\}$$

Evidently,

$$\# \left\{ m^2, \ m^2 + 1, \ \ldots, \ m^2 + 2m \right\} \ = \ 2m + 1$$

Step 3 is to express the size of the last cell

$$\left\{ \lfloor \sqrt{n} \rfloor^2, \ \ldots, \ n \right\}$$

whose entries correspond to the uppermost summand $\lfloor \sqrt{n} \rfloor$. Its size is

$$\# \left\{ \lfloor \sqrt{n} \rfloor^2, \ \ldots, \ n \right\} \ = \ n - \lfloor \sqrt{n} \rfloor^2 + 1$$

Step 4. To evaluate the given sum, we multiply each of the realized values of the summand by the corresponding number of values of the index k and then sum the products.

$$
\begin{aligned}
\sum_{k=0}^{n} \lfloor \sqrt{k} \rfloor &= \sum_{m=0}^{\lfloor \sqrt{n} \rfloor - 1} (2m+1) \cdot m \ + \ \left(n - \lfloor \sqrt{n} \rfloor^2 + 1 \right) \cdot \lfloor \sqrt{n} \rfloor \\
&= \sum_{m=0}^{\lfloor \sqrt{n} \rfloor - 1} \left(2m^{\underline{2}} + 3m^{\underline{1}} \right) \ + \ \left(n - \lfloor \sqrt{n} \rfloor^2 + 1 \right) \cdot \lfloor \sqrt{n} \rfloor \\
&= \left(\frac{2m^{\underline{3}}}{3} + \frac{3m^{\underline{2}}}{2} \right) \Bigg|_{m=0}^{\lfloor \sqrt{n} \rfloor} \ + \ \left(n - \lfloor \sqrt{n} \rfloor^2 + 1 \right) \cdot \lfloor \sqrt{n} \rfloor \\
&= \frac{2 \lfloor \sqrt{n} \rfloor^{\underline{3}}}{3} + \frac{3 \lfloor \sqrt{n} \rfloor^{\underline{2}}}{2} \ + \ \left(n - \lfloor \sqrt{n} \rfloor^2 + 1 \right) \cdot \lfloor \sqrt{n} \rfloor
\end{aligned}
$$

Step 5. We confirm for the small case $n = 11$.
Sum values in Step 1: $0 + 1 + 1 + 1 + 2 + 2 + 2 + 2 + 2 + 3 + 3 + 3 = 22$.
Compare with the value by formula of Step 4.

$$\frac{2\lfloor\sqrt{11}\rfloor^3}{3} + \frac{3\lfloor\sqrt{11}\rfloor^2}{2} + \left(11 - \lfloor\sqrt{11}\rfloor^2 + 1\right)\cdot\lfloor\sqrt{11}\rfloor$$

$$= 4 + 9 + 3\cdot 3 = 22$$

Example 3.5.4: We now use linear partitioning to evaluate the sum

$$\sum_{k=1}^{n}\lfloor\lg k\rfloor$$

Step 1. List the early terms of the sequence, and partition them according to the value of the summand.

k	1	2 3	4 5 6 7	8 $\cdots$ 15	16 $\cdots$ 31	32 33 $\cdots$
$\lfloor\lg k\rfloor$	0	1 1	2 2 2 2	3 $\cdots$ 3	4 $\cdots$ 4	5 5 $\cdots$

Step 2. To express the size of all but the last cell, we observe that the cell corresponding to the summand m is

$$\left\{2^m,\ 2^m + 1,\ \ldots,\ 2^{m+1} - 1\right\}$$

Its size is

$$\#\left\{2^m,\ 2^m + 1,\ \ldots,\ 2^{m+1} - 1\right\} = 2^m$$

Step 3. The last cell is

$$\left\{2^{\lfloor\lg n\rfloor},\ 2^{\lfloor\lg n\rfloor} + 1,\ \ldots,\ n\right\}$$

Its size is

$$n - 2^{\lfloor\lg n\rfloor} + 1$$

Step 4. Evaluate the given sum, using the previously derived formula (e.g., see Example 3.2.2) for summing $k\cdot 2^k$.

$$\sum_{k=1}^{n}\lfloor\lg k\rfloor = \sum_{m=1}^{\lfloor\lg n\rfloor - 1} m\cdot 2^m + \lfloor\lg n\rfloor\left(n - 2^{\lfloor\lg n\rfloor} + 1\right)$$

$$= \left(\lfloor\lg n\rfloor - 2\right)\cdot 2^{\lfloor\lg n\rfloor} + 2 + \lfloor\lg n\rfloor\left(n - 2^{\lfloor\lg n\rfloor} + 1\right)$$

Step 5. Confirm for the small case $n = 9$.
Sum the values in Step 1: $0 + 1 + 1 + 2 + 2 + 2 + 2 + 3 + 3 = 16$.
Compare with the value given by the formula of Step 4.

$$\left(\lfloor\lg 9\rfloor - 2\right)\cdot 2^{\lfloor\lg 9\rfloor} + 2 + \lfloor\lg 9\rfloor\left(n - 2^{\lfloor\lg 9\rfloor} + 1\right)$$

$$= (3 - 2)\cdot 2^3 + 2 + 3(9 - 2^3 + 1)$$

$$= 1\cdot 2^3 + 2 + 3\cdot 2 = 16$$

Remark: The two evaluations just considered have an easy second step, because within each group the value of the summand is constant. If the summand were $k\lfloor\sqrt{k}\rfloor$, for instance, then an inner sum might be introduced in Step 2 for the partial sum over the interval corresponding to a group.

EXERCISES for Section 3.5

In each of the Exercises 3.5.1 through 3.5.12, evaluate the given double sum.

3.5.1 $\displaystyle\sum_{k=0}^{n}\sum_{j=0}^{n}(j+k)$ 3.5.2 $\displaystyle\sum_{k=0}^{n}\sum_{j=0}^{n}j\cdot k$

3.5.3 $\displaystyle\sum_{k=0}^{n}\sum_{j=0}^{n}(j+2k+3)$ 3.5.4 $\displaystyle\sum_{k=0}^{n}\sum_{j=0}^{n}\binom{k}{j}$

3.5.5[S] $\displaystyle\sum_{k=0}^{n}\sum_{j=0}^{k}(j+k)$ 3.5.6 $\displaystyle\sum_{k=0}^{n}\sum_{j=0}^{k}j\cdot k$

3.5.7 $\displaystyle\sum_{k=0}^{n}\sum_{j=0}^{k}(j+2k+3)$ 3.5.8 $\displaystyle\sum_{k=0}^{n}\sum_{j=0}^{k}\binom{k}{j}$

3.5.9 $\displaystyle\sum_{k=0}^{n}\sum_{j=k}^{n}(j+k)$ 3.5.10 $\displaystyle\sum_{k=0}^{n}\sum_{j=k}^{n}j\cdot k$

3.5.11 $\displaystyle\sum_{k=0}^{n}\sum_{j=k}^{n}(j+2k+3)$ 3.5.12 $\displaystyle\sum_{k=0}^{n}\sum_{j=k}^{n}\binom{k}{j}$

In each of the Exercises 3.5.13 through 3.5.20, evaluate the given sum.

3.5.13 $\displaystyle\sum_{k=0}^{n}\left\lfloor\frac{k}{2}\right\rfloor$ 3.5.14[S] $\displaystyle\sum_{k=0}^{n}(k\bmod 3)$

3.5.15 $\displaystyle\sum_{k=0}^{n}\lfloor\sqrt[3]{k}\rfloor$ 3.5.16 $\displaystyle\sum_{k=0}^{n}\lfloor\log_3 k\rfloor$

3.5.17 $\displaystyle\sum_{k=0}^{n}\lceil\sqrt{k}\rceil$ 3.5.18 $\displaystyle\sum_{k=0}^{n}\lceil\lg k\rceil$

3.5.19 $\displaystyle\sum_{k=0}^{n}\left(\lfloor\sqrt{k}\rfloor\bmod 3\right)$ 3.5.20 $\displaystyle\sum_{k=0}^{n}(k^2\bmod 3)$

3.6   INCLUSION-EXCLUSION

Sometimes, the index set for a complicated sum has subsets with tractable sums, but those subsets overlap. The strategic insight of the inclusion-exclusion method is that the partial sums over those subsets can be combined into a total sum by subtracting the overcounts.

Venn Diagrams for Two Overlapping Subsets

A *Venn diagram* provides a visual model for evaluating sums over an index set given as the union of overlapping subsets. The simplest case has two overlapping subsets, A and B, as in Figure 3.6.1. The domain from which both subsets are drawn is denoted U.

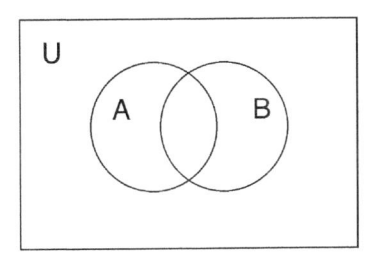

Figure 3.6.1   Venn diagram for two overlapping subsets.

Suppose that the objective is to calculate the sum $S_{A \cup B}$ over the set $A \cup B$, with partial sums S_A, S_B, and $S_{A \cap B}$ over subsets A, B, and $S_{A \cap B}$, respectively. Then

$$S_{A \cup B} = S_A + S_B - S_{A \cap B}$$

That is, to calculate $S_{A \cup B}$, we add S_A and S_B and then subtract the overlap $S_{A \cap B}$.

Example 3.6.1:   The number of integers in the range $1, \ldots, n$ that are divisible either by 2 or by 3 is expressible as a consecutive sum with indexing in the integer interval $[1 : n]$ and the summand

$$f(k) = \begin{cases} 1 & \text{if } n \text{ is divisible either by 2 or by 3} \\ 0 & \text{otherwise} \end{cases}$$

that is, by the sum

$$\sum_{k=1}^{n} (2 \backslash k \ \lor \ 3 \backslash k)$$

Every number that contributes 1 to this sum lies either in the set $\{k \in [1 : n] \mid 2 \backslash k\}$, with cardinality $\lfloor n/2 \rfloor$, or in the set $\{k \in [1 : n] \mid 3 \backslash k\}$, with cardinality $\lfloor n/3 \rfloor$. Adding these two cardinalities overcounts by $\lfloor n/6 \rfloor$, the number of integers in $[1 : n]$ that are divisible both by 2 and by 3. Thus,

$$\sum_{k=1}^{n} (2 \backslash k \ \lor \ 3 \backslash k) = \left\lfloor \frac{n}{2} \right\rfloor + \left\lfloor \frac{n}{3} \right\rfloor - \left\lfloor \frac{n}{6} \right\rfloor$$

In Figure 3.6.1,

$$A = \{k \in [1:n] \mid 2\backslash k\}$$

and

$$B = \{k \in [1:n] \mid 3\backslash k\}$$

Their intersection is

$$A \cap B = \{k \in [1:n] \mid 6\backslash k\}$$

and

$$U = [1:n]$$

In Example 3.6.1, the implicit summand is the number 1, since we counted the number of elements in a set. That is,

$$S_X = \sum_{k \in X} 1 = |X|$$

for $X = A,\ B,\ A \cap B,$ or $A \cup B$

Example 3.6.2:   A related problem is to calculate the sum of the numbers that are divisible by 2 or 3. Then, instead of having a constant value of 1, the value of the summand equals the index itself. That is,

$$S_X = \sum_{k \in X} k$$

for $X = A,\ B,\ A \cap B,$ or $A \cup B$

Thus,

$$S_A = \sum_{2\,\backslash\,k\,\leq\,n} k = \sum_{j=1}^{\lfloor n/2 \rfloor} 2j = 2 \sum_{j=1}^{\lfloor n/2 \rfloor} j = 2\,\frac{\lfloor n/2 \rfloor^2 + \lfloor n/2 \rfloor}{2}$$

Similarly,

$$S_B = \sum_{3\,\backslash\,k\,\leq\,n} k = 3\,\frac{\lfloor n/3 \rfloor^2 + \lfloor n/3 \rfloor}{2}$$

and

$$S_{A \cap B} = \sum_{6\,\backslash\,k\,\leq\,n} k = 6\,\frac{\lfloor n/6 \rfloor^2 + \lfloor n/6 \rfloor}{2}$$

Therefore,

$$S_{A \cup B} = S_a + S_B - S_{A \cap B}$$

$$= 2\,\frac{\lfloor n/2 \rfloor^2 + \lfloor n/2 \rfloor}{2} + 3\,\frac{\lfloor n/3 \rfloor^2 + \lfloor n/3 \rfloor}{2} - 6\,\frac{\lfloor n/6 \rfloor^2 + \lfloor n/6 \rfloor}{2}$$

For the small case $n = 14$, direct addition and the formula both yield $S_{A \cup B} = 68$.

Venn Diagrams for Three or More Subsets

Venn diagrams are quite commonly drawn for three overlapping subsets, and they have this general definition.

DEFINITION: A family of n simple closed curves (typically circles or ellipses) in the plane, whose interior regions represent some subsets $A_1, A_2, \ldots, A_n$ of a set A within a domain U, is called a **Venn diagram**, after the logician John Venn (1834-1923).

TERMINOLOGY: The domain U from which both the subsets A and B are drawn is commonly called the *universal set*.

Example 3.6.3:   The Eurasian Translators Company has 15 expert linguists fluent in at least two of the languages Armenian, Bulgarian, and Czech. Of these,

$$
\begin{aligned}
S_{A \cap B} &= 5 \quad \text{speak Armenian and Bulgarian} \\
S_{A \cap C} &= 7 \quad \text{speak Armenian and Czech} \\
\text{and} \quad S_{B \cap C} &= 9 \quad \text{speak Bulgarian and Czech}
\end{aligned}
$$

How many speak all three languages? Figure 3.6.2 is the relevant Venn diagram.

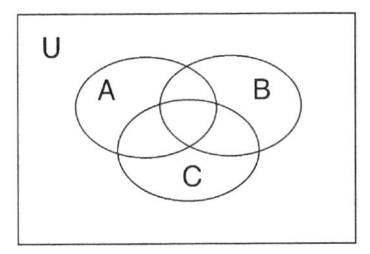

Figure 3.6.2   Venn diagram for three overlapping subsets.

Whereas 15 is the given number of linguists fluent in two or more of the three languages, the sum

$$ S_{A \cap B} + S_{A \cap C} + S_{B \cap C} = 5 + 7 + 9 = 21 $$

of the numbers corresponding to the three intersection-regions for which data are given triple-counts the contribution $S_{A \cap B \cap C}$ in the triple intersection at the center of the diagram and counts all the other translators only once. Thus, subtracting 15, thereby excluding the total number of translators who speak at least two of the languages by the calculation

$$ 2\,S_{A \cap B \cap C} = 21 - 15 = 6 $$

yields the result

$$ 2\,S_{A \cap B \cap C} = 6 $$

from which one concludes that

$$ S_{A \cap B \cap C} = 3 $$

After deriving an analytic solution to a Venn diagram problem, it is helpful to check the result by inserting numbers into the relevant regions of the diagram. In this case, the number 3 is inserted into the centermost region, representing the population of the region $A \cap B \cap C$ in Figure 3.6.3. Then it must be excluded from the populations given for composite regions $A \cap B$, $A \cap C$, and $B \cap C$, in order to obtain populations for the simple regions they contain.

NOTATION: The complement of a set X in a domain U is denoted $\overline{X}$.

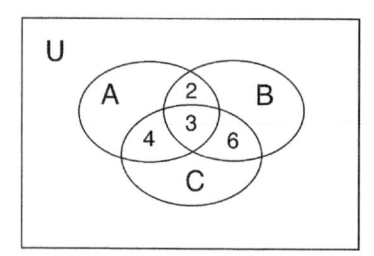

Figure 3.6.3   Inserting numbers into regions of a Venn diagram.

We observe that all of the inserted numbers are consistent with the original data as well as with the derived population of $A \cap B \cap C$.

$$
\begin{aligned}
S_{A \cap B} &= 5 = 2 + 3 = S_{A \cap B \cap \overline{C}} + S_{A \cap B \cap C} \\
S_{A \cap C} &= 7 = 4 + 3 = S_{A \cap \overline{B} \cap C} + S_{A \cap B \cap C} \\
\text{and } \ S_{B \cap C} &= 9 = 6 + 3 = S_{\overline{A} \cap B \cap C} + S_{A \cap B \cap C}
\end{aligned}
$$

Context for Inclusion-Exclusion

A more general context of *inclusion-exclusion* evaluations is a set A within a domain U, a real-valued function $f : U \to \mathbb{R}$, and a representation

$$
A = \bigcup_{k=1}^{n} A_k
$$

of set A as a union of subsets A_k, conceptualized like a Venn diagram with n mutually intersecting subsets. (Some of the regions may be empty.)

Remark: Quite often, the function $f : U \to \mathbb{R}$ is the constant function $f(x) = 1$, in which case the evaluation amounts to calculating the cardinality of a region.

NOTATION: The intersection of two sets A_i and A_j may be denoted by the juxtaposition $A_i A_j$.

DEFINITION: An intersection $A_{i_1} A_{i_2} \cdots A_{i_r}$ is called an *r-fold intersection* of the family $\{A_k\}$.

Formulas for Inclusion-Exclusion

As illustrated by Example 3.6.3, evaluating sums over combinations of regions in Venn diagrams takes some care. Fortunately, such evaluations can usually be reduced to the application of two or three general inclusion-exclusion equations for sums over single regions.

Theorem 3.6.1 [Exclude-All Equation for Set Size]. *Let $A_1, \ldots, A_n$ be subsets of a set U, with*

$$A = \bigcup_{k=1}^{n} A_k \quad \text{and} \quad S_r = \sum_{i_1, \ldots, i_r \in [1:n]} |A_{i_1} \cdots A_{i_r}|$$

so that, for $r = 1, \ldots, n$, the number S_r is the sum of the cardinalities of all r-fold intersections of the family $\{A_k\}$. Then

$$|\overline{A}| = |\overline{A_1}\,\overline{A_2}\cdots\overline{A_n}| = |U| + \sum_{k=1}^{n}(-1)^k S_k$$

Proof: First suppose that the element $x \in U$ lies in none of the sets A_j. Then x is counted once on the left side of the equation, and it is counted in the formula on the right side only by the summand $|U|$, and not by any summand S_k.

Now suppose that x lies in exactly m of the subsets A_j, with $m > 0$. Accordingly x is not counted on the left side of the equation. On the right side, it is counted $\binom{m}{k}$ times by S_k, since there are $\binom{m}{k}$ ways to choose k sets from the m sets A_j that contain x, and x is also counted once by $|U|$. Thus, its net count on the right side is

$$1 + \sum_{k=1}^{m}(-1)^k \binom{m}{k} = \sum_{k=0}^{m}(-1)^k \binom{m}{k}$$
$$= (1-z)^m\Big|_{z=1} = 0 \qquad\qquad \diamond$$

The other main inclusion-exclusion formula is derivable by the same approach, or, as shown here, as a corollary of Theorem 3.6.1.

Corollary 3.6.2 [Include-All Equation for Set Size]. *Let $A_1, \ldots, A_n$ be subsets of a set U, with*

$$A = \bigcup_{k=1}^{n} A_k \quad \text{and} \quad S_r = \sum_{i_1, \ldots, i_r \in [1:n]} |A_{i_1} \cdots A_{i_r}|$$

so that, for $r = 1, \ldots, n$, the number S_r is the sum of the cardinalities of all r-fold intersections of the family $\{A_k\}$. Then

$$|A| = \sum_{k=1}^{n}(-1)^{k-1} S_k$$

Proof: Observe that the universal set U is the disjoint union of the set A and the set $\overline{A_1}\,\overline{A_2}\,\cdots\,\overline{A_n}$. Therefore,

$$|U| \;=\; |A| + \left|\,\overline{A_1}\,\overline{A_2}\,\cdots\,\overline{A_n}\,\right|$$

and, accordingly,

$$|A| \;=\; |U| - \left|\,\overline{A_1}\,\overline{A_2}\,\cdots\,\overline{A_n}\,\right|$$

and then, by Theorem 3.6.1,

$$= \left(\left|\,\overline{A_1}\,\overline{A_2}\,\cdots\,\overline{A_n}\,\right| - \sum_{k=1}^{n}(-1)^k S_k\right) - \left|\,\overline{A_1}\,\overline{A_2}\,\cdots\,\overline{A_n}\,\right|$$

$$= \sum_{k=1}^{n}(-1)^{k-1}S_k \qquad\qquad\qquad \diamond$$

Theorem 3.6.3 provides an inclusion-exclusion formula for the sum of the values of an arbitrary function $f : U \to \mathbb{R}$ on the domain U, not simply for counting the size of a set.

Theorem 3.6.3 [General Exclude-All Equation]. *Let $A_1,\ldots,A_n$ be subsets of a set U, with*

$$A \;=\; \bigcup_{k=1}^{n} A_k$$

and let $f : U \to \mathbb{R}$ be a real-valued function. Let the sum

$$S_r \;=\; \sum_{i_1,\ldots,i_r\in[1:n]}\;\sum_{x\in A_{i_1}\cdots A_{i_r}} f(x)$$

be taken over all r-fold intersections of the family $\{A_k\}$. Then

$$\sum_{x\in\overline{A_1}\,\cdots\,\overline{A_n}} f(x) \;=\; \sum_{x\in U} f(x) + \sum_{k=1}^{n}(-1)^k S_k$$

Proof: The proof is a straightforward modification of the proof of the Exclude-All Equation for set sizes. $\diamond$

In the remainder of this section, the two main inclusion-exclusion formulas are applied to various combinatorial problems.

Stirling Subset Numbers

Although there are various similarities between Stirling numbers and binomial coefficients, there is no known closed formula for Stirling numbers of either kind, unlike the situation for binomial coefficients. However, there is a summation formula for a Stirling subset number, whose derivation by inclusion-exclusion is our immediate objective. The ideas involved are encapsulated in the following example.

Example 3.6.4:   The Stirling subset number $\left\{{5 \atop 4}\right\}$ is the number of ways to distribute a set of 5 objects into 4 cells with no cell left empty. For a problem this small, listing cases is easy enough, but it is instructive to apply inclusion-exclusion. Toward that objective, for $i = 1, 2, 3, 4$, let A_i be the set of distributions with box i left empty. Clearly,

$$
\begin{aligned}
|A_i| &= 3^5 & \text{for } i = 1, 2, 3, 4 \\
|A_i A_j| &= 2^5 & \text{for } i \neq j \\
|A_i A_j A_k| &= 1^5 & \text{for mutually distinct } i, j, k
\end{aligned}
$$

Moreover,

$$
S_k = \binom{4}{k}(4-k)^5 \tag{3.6.1}
$$

since there are $\binom{4}{k}$ ways to choose k of the subsets A_i from the collection of four such subsets, and each intersection $A_{i_1} A_{i_2} \cdots A_{i_k}$ contains $(4-k)^5$ objects. Furthermore,

$$
\left| \overline{A_1}\,\overline{A_2}\,\overline{A_3}\,\overline{A_4} \right| = \left\{{5 \atop 4}\right\} 4! \tag{3.6.2}
$$

since each distribution with none of the boxes left empty amounts to assigning the labels $1, 2, 3, 4$ to the cells of a partition. Finally, if U is the set of all ways to distribute 5 objects into 4 cells, we have

$$
|U| = 4^5 \tag{3.6.3}
$$

When we substitute into the Exclude-All Equation

$$
\left| \overline{A_1}\,\overline{A_2}\,\overline{A_3}\,\overline{A_4} \right| = |U| - S_1 + S_2 - S_3 + S_4
$$

the values from Equations (3.6.1), (3.6.2), and (3.6.3), we obtain the equation

$$
\begin{aligned}
\left\{{5 \atop 4}\right\} 4! &= 4^5 - \binom{4}{1}3^5 + \binom{4}{2}2^5 - \binom{4}{3}1^5 + \binom{4}{4}0^5 \\
&= 1024 - 972 + 192 - 4 = 240 \\
\Rightarrow \quad \left\{{5 \atop 4}\right\} &= \frac{240}{4!} = 10
\end{aligned}
$$

A confirming observation is that, since two of the elements are paired, and since the others have cells to themselves, clearly

$$
\left\{{5 \atop 4}\right\} = \binom{5}{2} = 10
$$

In a similar manner, an inclusion-exclusion analysis leads to an identity for the Stirling subset numbers

$$
\left\{{n \atop m}\right\}
$$

Proof of the following theorem simply generalizes the steps and calculations that we have just completed.

Theorem 3.6.4. *Let n and k be a pair of non-negative integers. Then*

$$\left\{ {n \atop k} \right\} k! \;=\; \sum_{j=0}^{k} (-1)^j \binom{k}{j} (k-j)^n$$

Proof: For $i = 1, 2, \ldots, k$, let A_i be the set of distributions of n distinct objects into k *distinct* boxes with box i left empty. Clearly,

$$|A_i| \;=\; (k-1)^n \qquad \text{for } i = 1, 2, \ldots, k$$

and, more generally, for any $j \in [1, k]$

$$|A_{i_1} A_{i_2} \cdots A_{i_j}| \;=\; (k-j)^n \qquad \text{for mutually distinct } i_1, i_2, \ldots, i_j$$

Since there are $\binom{k}{j}$ ways to choose the mutually distinct $i_1, i_2, \ldots, i_j$, and since S_j is the sum of the numbers of ways to leave j specific boxes empty (with others possibly empty also), it follows, by analogy to Eq. (3.6.1), that

$$S_j \;=\; \binom{k}{j} (k-j)^n \tag{3.6.1$'$}$$

Furthermore,

$$\left| \overline{A_1}\, \overline{A_2} \cdots \overline{A_k} \right| \;=\; \left\{ {n \atop k} \right\} k! \tag{3.6.2$'$}$$

since each distribution with none of the k boxes left empty amounts to assigning the labels $1, 2, \ldots, k$ to the cells of a partition. Finally, if U is the set of all ways to distribute n objects into k cells, we have

$$|U| \;=\; k^n \;=\; \binom{k}{0} (k-0)^n \tag{3.6.3$'$}$$

Substituting the values from Equations (3.6.1$'$), (3.6.2$'$), and (3.6.3$'$) just above into the Exclude-All Equation

$$\left| \overline{A_1}\, \overline{A_2} \cdots \overline{A_k} \right| \;=\; |U| - S_1 + S_2 + \cdots + (-1)^k S_k$$

we obtain the identity

$$\left\{ {n \atop k} \right\} k! \;=\; \sum_{j=0}^{k} (-1)^j \binom{k}{j} (k-j)^n \qquad\qquad \diamond$$

Derangements

Inclusion-exclusion is also helpful in analyzing the derangement recurrence.

REVIEW FROM §2.1:

- A **derangement** is a permutation π with no fixed points.
- The **derangement number** D_n is the number of derangements of the integer interval $[1 : n]$.

- The **derangement recurrence** (see also §5.4) is

$$D_0 = 1, \quad D_1 = 0; \tag{3.6.4a}$$

$$D_n = (n-1)D_{n-1} + (n-1)D_{n-2} \tag{3.6.4b}$$

Example 3.6.5: In the classical **hatcheck problem**, each of n persons leaves a hat in the cloakroom, but the hatchecks are lost, and the n hats are redistributed randomly. It asks, what is the probability that no hat goes to its rightful owner? This problem is recognizable as a homespun version of calculating the proportion of permutations of n objects that are derangements.

If U is the set of all possible hat distributions, then

$$|U| = n! \tag{3.6.5}$$

To calculate the number D_n of derangements, let A_i be the set of permutations in which hat i goes to its rightful owner. Then

$$D_n = \left| \overline{A_1}\, \overline{A_2} \, \cdots \, \overline{A_n} \right| \tag{3.6.6}$$

and

$$|A_i| = (n-1)!$$

More generally,

$$|A_{i_1} A_{i_2} \cdots A_{i_r}| = (n-r)!$$

which implies that

$$S_k = \binom{n}{k}(n-k)! \tag{3.6.7}$$

When the Exclude-All Equation

$$\left| \overline{A_1}\, \overline{A_2} \, \cdots \, \overline{A_n} \right| = |U| + \sum_{k=1}^{n}(-1)^k S_k$$

is combined with Equations (3.6.5), (3.6.6), and (3.6.7), we obtain

$$D_n = n! + \sum_{k=1}^{n}(-1)^k \binom{n}{k}(n-k)!$$

$$= \sum_{k=0}^{n}(-1)^k \frac{n!}{k!} \tag{3.6.8}$$

Substituting $n = 0$ and $n = 1$ into Eq. (3.6.8), we obtain

$$D_0 = \sum_{k=0}^{0}(-1)^0 \frac{0!}{k!}$$

$$= (-1)^0 \frac{0!}{0!} = 1$$

and

$$D_1 = \sum_{k=0}^{1}(-1)^k\frac{1!}{k!}$$

$$= (-1)^0\frac{1!}{0!} + (-1)^1\frac{1!}{0!} = 1 + (-1) = 0$$

which establishes that equation (3.6.8) satisfies the initial conditions (3.6.4a) of the derangement recurrence. Moreover, assuming that equation (3.6.8) satisfies the recurrence for D_{n-1} and D_{n-2}, we confirm by the following calculation that it also satisfies the recurrence for D_n.

$$(n-1)D_{n-1} + (n-1)D_{n-2}$$

$$= (n-1)\sum_{k=0}^{n-1}(-1)^k\frac{(n-1)!}{k!} + (n-1)\sum_{k=0}^{n-2}(-1)^k\frac{(n-2)!}{k!}$$

$$= \sum_{k=0}^{n-1}(-1)^k\frac{n!}{k!} - \sum_{k=0}^{n-1}(-1)^k\frac{(n-1)!}{k!} + \sum_{k=0}^{n-2}(-1)^k\frac{(n-1)!}{k!}$$

$$= \sum_{k=0}^{n-1}(-1)^k\frac{n!}{k!} - (-1)^{n-1}\frac{(n-1)!}{(n-1)!}$$

$$= \sum_{k=0}^{n-1}(-1)^k\frac{n!}{k!} + (-1)^n\frac{n!}{n!}$$

$$= \sum_{k=0}^{n}(-1)^k\frac{n!}{k!}$$

Remark: We observe that

$$D_n = \sum_{k=0}^{n}(-1)^k\frac{n!}{k!} \tag{3.6.8}$$

implies that

$$\frac{D_n}{n!} = \sum_{k=0}^{n}(-1)^k\frac{1}{k!} \xrightarrow[n\to\infty]{} e^{-1}$$

Thus, one might approximate the value of e^{-1}, and hence, of the number e, by generating random permutations and counting the proportion that are derangements.

Counting Bipartite Matchings

In the rest of this section at least a passing prior acquaintance with graph theory would be helpful. [GrYe2006] is recommended.

REVIEW FROM §0.6:

- A **bipartite graph** G is a graph whose vertex set can be partitioned into two subsets X and Y such that every edge has one vertex in X and the other in Y.

PREVIEW OF §8.6:

- A **matching** in a graph is a set of edges such that no two edges have an endpoint in common.

- A **perfect matching** in a graph is a matching in which every vertex is the endpoint of one of the edges.

Example 3.6.6:   In the bipartite graph G of Figure 3.6.4, there are five perfect matchings:

one with $f(1) = a$ (shown with thicker edges);

three with $f(1) = b$;

and one with $f(1) = c$.

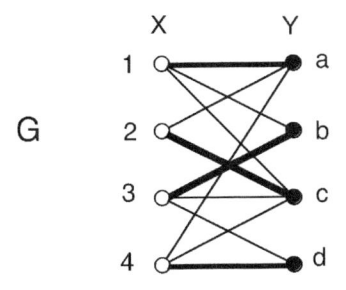

Figure 3.6.4   **Perfect matching in a bipartite graph.**

Let U be the total set of bijections $X \to Y$. Then

$$|U| \;=\; 4! \;=\; 24 \tag{3.6.9}$$

A bijection $X \to Y$ is *consistent* with the bipartite graph G if it is representable as a perfect matching in G. For $i = 1, 2, 3, 4$, let A_i be the number of bijections $f : X \to Y$ such that the assignment $i \mapsto f(i)$ is inconsistent with the graph G, that is, such that vertex $f(i)$ is not adjacent to vertex i.

For each choice of a vertex of Y that is not adjacent to vertex i, there are 3! bijections $X \to Y$, corresponding to the 3! ways to assign the other 3 vertices of X to the remaining 3 vertices of Y. Thus,

$$|A_i| = (4 - deg(i)) \cdot 3!$$

and there are similar formulas for multiple inconsistencies.

$$|A_1| \;=\; |A_3| \;=\; |A_4| \;=\; 1 \cdot 3! \text{ and } |A_2| \;=\; 2 \cdot 3!$$

$$|A_1 A_2| \;=\; |A_1 A_3| \;=\; |A_1 A_4| \;=\; |A_2 A_4| \;=\; |A_3 A_4| \;=\; 2!$$

$$\text{and } |A_2 A_3| \;=\; 2 \cdot 2!$$

$$|A_1 A_2 A_3| \;=\; |A_1 A_3 A_4| \;=\; |A_2 A_3 A_4| \;=\; 1 \text{ and } |A_1 A_2 A_4| \;=\; 0$$

$$\Rightarrow \; S_1 \;=\; 30, \; S_2 \;=\; 14, \; S_3 \;=\; 3, \; S_4 \;=\; 0 \tag{3.6.10}$$

Therefore, by using (3.6.9) and (3.6.10) with the Exclude-All Equation, the number of perfect matchings is shown to be

$$
\begin{aligned}
\left| \overline{A_1}\,\overline{A_2}\,\overline{A_3}\,\overline{A_4} \right| &= |U| - S_1 + S_2 - S_3 + S_4 \\
&= 24 - 30 + 14 - 3 + 0 \\
&= 5
\end{aligned}
$$

which agrees with our ad hoc count at the outset.

An alternative representation of this counting problem is the chessboard of Figure 3.6.5. Observe that a square is shaded if and only if it is forbidden to match the vertex corresponding to its row to the vertex corresponding to its column. Each perfect matching corresponds to a selection of one unshaded square in each row, such that there is at most one selection in each column.

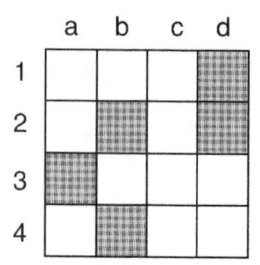

Figure 3.6.5   **Chessboard representation of a bipartite matching problem.**

Consistent with this representation, there is a well-developed theory of *rook polynomials* that uses a divide-and-conquer strategy for calculating the numbers S_i used in inclusion-exclusion.

Remark: The method given here is applicable not only to perfect matchings, but also to counting complete matchings of the vertices in one part of the bipartition to the other part when that other part has more vertices.

Chromatic Polynomials

An algebraic invariant called the *chromatic polynomial* of a graph can be calculated by inclusion-exclusion.

PREVIEW OF §8.3:

- A **vertex-coloring of a graph** G in the set $[1 : n]$, often simply called a **coloring**, is a function $f : V_G \to [1 : n]$.
- A **proper vertex-coloring of a graph** is a coloring such that no two adjacent vertices have the same color.

DEFINITION: The **chromatic polynomial** $P(G, t)$ of a graph G is the function whose value at the integer t is the number of proper colorings of G with at most t colors.

As a preliminary to a more systematic approach, we consider an *ad hoc* construction of a chromatic polynomial. It illustrates why the function $P(G, n)$ is a polynomial.

Example 3.6.7:   The graph $K_{1,2}$ requires at least two colors for a proper coloring. We observe that, given two colors, exactly two proper 2-colorings are possible, one of which is illustrated in Figure 3.6.6. We write $p_2 = 2$. Given three colors, exactly six 2-colorings (i.e., 3!) are possible, so we write $p_3 = 6$.

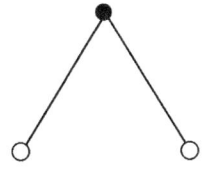

Figure 3.6.6   **A proper 2-coloring for the bipartite graph $K_{1,2}$.**

For any positive integer t, the number of ways to choose two colors is $\binom{t}{2}$ and the number of ways to choose three colors is $\binom{t}{3}$, By the Rule of Product and the Rule of Sum, it follows that the number of proper colorings with t colors is

$$
\begin{aligned}
p_2 \binom{t}{2} + p_3 \binom{t}{3} &= 2 \cdot \frac{t(t-1)}{2!} + 6 \cdot \frac{t(t-1)(t-2)}{3!} \\
&= (t^2 - t) + (t^3 - 3t^2 + 2t) \\
&= t^3 - 2t^2 + t
\end{aligned}
\tag{3.6.11}
$$

In general, it may be computationally difficult to determine the exact numbers of proper colorings

$$p_1 \quad p_2 \quad \cdots \quad p_n$$

for an n-vertex graph G, or even to decide the *chromatic number*, which is the smallest positive value. However, whatever those numbers may be, the chromatic polynomial is

$$
P(G, t) = p_1 \binom{t}{2} + p_2 \binom{t}{2} + \cdots + p_n \binom{t}{n}
$$

Example 3.6.7, continued: To recalculate the chromatic polynomial of the graph $K_{1,2}$ by inclusion-exclusion, its two edges are labeled 1 and 2, as shown in Figure 3.6.7.

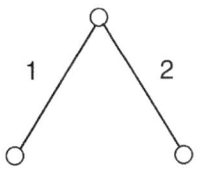

Figure 3.6.7   **An edge-labeling for the bipartite graph $K_{1,2}$.**

Let U be the set of all colorings of $K_{1,2}$ with at most t colors. Let A_1 be the set of such colorings in which the endpoints of edge 1 have the same color, and let A_2

be the set of such colorings in which the endpoints of edge 2 have the same color. Then

$$P(K_{1,2}, t) = |U| - |A_1 A_2|$$

This is a job for the Exclude-All Equation. Evidently, $|U| = t^3$.

To calculate $|A_1|$, we recognize that there are t possible choices of a color for both endpoints of edge 1, and then another t possible choices for the color of the remaining vertex. Clearly, this holds also for $|A_2|$. Thus,

$$|A_1| = |A_2| = t^2$$
$$\text{and} \quad S_1 = |A_1| + |A_2| = 2t^2$$

Any coloring in $A_1 \cap A_2$ has the same color at both endpoints of edge 1 and the same color at both ends of edge 2. Since these two edges share an endpoint, all three vertices of $K_{1,2}$ must have the same color. There are t possible choices for this color. Thus,

$$|A_1 A_2| = t$$
$$\text{and} \quad S_2 = |A_1 A_2| = t$$

We now complete the recalculation,

$$\overline{A_1} \, \overline{A_2}| = |U| - S_1 + S_2$$
$$= t^3 - 2t^2 + t \tag{3.6.12}$$

which agrees with (3.6.11).

Example 3.6.8:   To calculate the chromatic polynomial $P(K_4, t)$ of the complete graph K_4 by inclusion-exclusion, label its its six edges with numbers $1, \ldots, 6$, as shown in Figure 3.6.8.

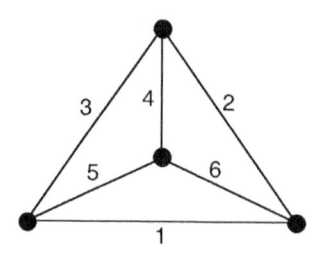

Figure 3.6.8   An edge-labeling of the complete graph K_4.

Let U be the set of all colorings of K_4 with colors in $[1 : t]$. Then

$$|U| = t^4$$

Next, let A_i be the set of colorings in $[1 : t]$ such that the endpoints of edge i have the same color. Then $|A_i| = t^3$, since there are t possibilities for the color of the endpoints of edge i and t possibilities for each of the other two vertices. Since there are 6 edges, it follows that

$$S_1 = |A_1| + |A_2| + \cdots + |A_6| = 6t^3$$

There are $\binom{6}{2} = 15$ ways to choose a pair of edges, i and j. For 3 of these pairs, edges i and j have no vertex in common, in which case a coloring in $A_i A_j$ may use t colors for the endpoints of edge i and t colors for the endpoints of edge j, yielding t^2 possibilities. For the 12 pairs of edges that have a vertex in common, there are t choices for the color of the three vertices in the union of their endpoint sets and t choices for the remaining vertex, again yielding t^2 possibilities. Thus,

$$S_2 = 15t^2$$

There are $\binom{6}{3} = 20$ ways to choose three edges. Exactly 4 of these 20 triples form a 3-cycle. There are t choices for the color of all three vertices in that 3-cycle and t choices for the remaining vertex. The other 16 edge-triples form a connected subgraph (a *spanning tree*) that contains all four vertices of K_4, so all four must get the same color, for which there are t choices. Accordingly,

$$S_3 = 4t^2 + 16t$$

A subset of four or more edges must contain all the vertices of K_4. It follows that

$$S_4 = \binom{6}{4}t = 15t$$

$$S_5 = \binom{6}{5}t = 6t$$

$$\text{and} \quad S_6 = \binom{6}{6}t = t$$

By the Exclude-All Formula,

$$\begin{aligned}
P(K_4, t) &= |U| - S_1 + S_2 - S_3 + S_4 - S_5 + S_6 \\
&= t^4 - 6t^3 + 15t^2 - (4t^2 + 16t) + 15t - 6t + t \\
&= t^4 - 6t^3 + 11t^2 - 6t
\end{aligned}$$

Remark: There are circumstances where the Exclude-All Formula is an excellent way to calculate chromatic polynomials, which are not revealed by these small examples. For instance, it can be used to prove that the chromatic polynomial of any n-vertex tree is $t(t-1)^{n-1}$.

EXERCISES for Section 3.6

PREVIEW OF §6.5:

- For a positive integer n, the **Euler phi-function** $\phi(n)$ gives the number of positive integers not exceeding n that are relatively prime to n.

In each of the Exercises 3.6.1 through 3.6.6, use inclusion-exclusion to calculate $\phi(n)$ for the given number n. (A faster method of calculation is given in §6.5.)

3.6.1^S 48   3.6.2   60   3.6.3   100

3.6.4   81   3.6.5   64   3.6.6   96

In each of the Exercises 3.6.7 through 3.6.10, use inclusion-exclusion to calculate the given Stirling subset number.

3.6.7^S $\left\{ {5 \atop 2} \right\}$ 3.6.8 $\left\{ {5 \atop 3} \right\}$ 3.6.9 $\left\{ {6 \atop 4} \right\}$ 3.6.10 $\left\{ {6 \atop 5} \right\}$

3.6.11^S Eight 6-sided dice are rolled. What is the probability that each of the numbers $1, \ldots, 6$ occurs at least once?

3.6.12   How many ways are there to select five cards from a 52-card poker deck so that each of the four suits is represented?

3.6.13   In how many ways can three 0's, three 1's, and three 2's be arranged in a row so that no three adjacent digits are the same?

3.6.14   In how many ways can the 26 letters A, B, ..., Z be arranged in a row so that none of the words YES, NO, or MAYBE occurs?

3.6.15   Calculate the number of binary sequences of length n such that no 1 is immediately adjacent to another 1.

3.6.16^S Calculate the number of permutations of the integer interval $[1:2n]$ such that no even integer is fixed.

3.6.17^S Calculate the number of permutations of n objects in which exactly k objects are fixed.

In Exercises 3.6.18 through 3.6.21, use inclusion-exclusion to calculate the number of perfect matchings in the given bipartite graph.

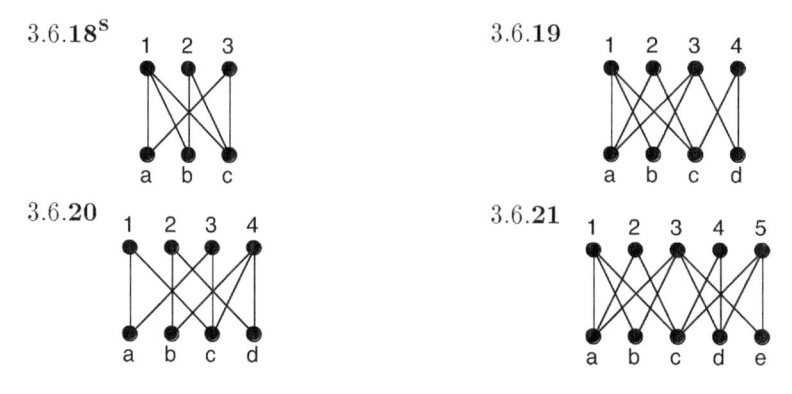

In Exercises 3.6.22 through 3.6.30, use inclusion-exclusion to calculate the chromatic polynomial of the given graph.

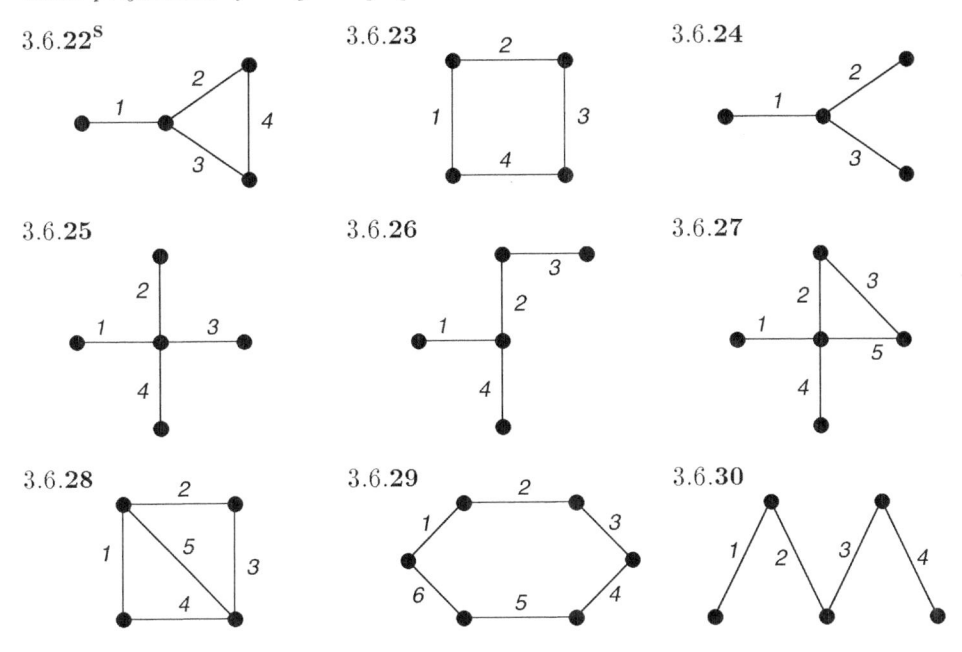

3.6.22^S 3.6.23 3.6.24

3.6.25 3.6.26 3.6.27

3.6.28 3.6.29 3.6.30

GLOSSARY

bipartite graph: a graph whose vertex-set can be partitioned into two subsets (called *partite sets*) such that every edge has one endpoint in one partite set and one endpoint in the other partite set.

chromatic polynomial of a graph G: the function $P(G, t)$ whose value at the integer t is the number of proper colorings of G with at most t colors.

coloring of a graph: usually, a *vertex-coloring*; sometimes, either a vertex-coloring or an edge-coloring.

consecutive sum: a sum $\sum_{k=a}^{b} x_k$ indexed over a consecutive set of integers.

derangement: a permutation with no fixed points.

derangement number D_n: the number of permutations of the integer interval $[1 : n]$ that are derangements.

derangement recurrence: the recurrence
$$D_0 = 1, \quad D_1 = 0;$$
$$D_n = (n - 1)D_{n-1} + (n - 1)D_{n-2}$$

difference function for a function f: the function $\triangle f$ defined by the rule
$$\triangle f(x) = f(x + 1) - f(x)$$

difference sequence for a sequence $\langle x_n \rangle$:   the sequence $\langle x_n' \rangle$ defined by the rule
$$x_n' = x_{n+1} - x_n$$

falling power function $x^{\underline{n}}$:   the polynomial
$$x^{\underline{n}} = \overbrace{x(x-1)\cdots(x-n+1)}^{n \text{ factors}}$$

harmonic number: the number
$$H_n = \frac{1}{1} + \frac{1}{2} + \cdots + \frac{1}{n}$$

inclusion-exclusion: a method for counting the number of elements in a union of sets that overlap.

Iverson truth function: the function defined by the rule
$$(\text{predicate}) = \begin{cases} 1 & \text{if the predicate is true} \\ 0 & \text{if the predicate is false} \end{cases}$$

matching in a graph: a set of edges with no endpoints in common.

normalizing a sum: manipulating it into an equivalent consecutive sum.

partial sum: usually, a sum over an initial subsequence of the index set.

perfect matching in a graph: a matching such that every vertex is an endpoint of one of the edges.

perturbation of a sum: a method for evaluating a sum that begins by equating the result of splitting off the initial summand with the result of splitting off the final summand.

product formula for finite differences: the summation formula
$$\triangle(g(x) \cdot f(x)) = \triangle g(x) \cdot f(x+1) + g(x) \cdot \triangle f(x)$$

proper vertex-coloring of a graph: a coloring such that no two adjacent vertices have the same color.

summation by parts: the summation rule
$$\sum_{k=0}^{n-1} g(k) \, \triangle(f(k)) = \left. g(k)f(k) \right|_{k=0}^{n} - \sum_{k=0}^{n-1} \triangle(g(k)) \, f(k+1)$$
that inverts the product rule.

Venn diagram: a drawing with several overlapping ovals, which represent overlapping sets.

vertex-coloring of a graph: an assignment of colors to its vertices.

Chapter 4

Binomial Coefficients

4.1 Binomial Coefficient Identities
4.2 Binomial Inversion Operation
4.3 Applications to Statistics
4.4 The Catalan Recurrence

Binomial coefficients are numbers that arise in the expansion of an exponentiated binomial, and they are among the most ubiquitous artifacts of combinatorial mathematics — indeed, they are seemingly somehow involved with almost every combinatorial construction. The fact, proved in Chapter 1, that they also express the numbers of ways to choose a subset of size k from a set of n objects, for any value of k, makes it possible to derive many of their properties using highly intuitive combinatorial counting arguments, as an alternative to computational algebraic arguments. Various identities presented in §4.1 are used in subsequent sections to simplify complicated expressions involving binomial coefficients. Some applications to mathematical statistics are examined in §4.3. It is shown in §4.4 how the Catalan numbers can be expressed in terms of binomial coefficients.

4.1   BINOMIAL COEFFICIENT IDENTITIES

The objective of this section is to derive the most important identities for binomial coefficients. These identities will be used in the later sections of this chapter and throughout the rest of the book.

REVIEW FROM §1.3 AND §1.5:   We recall the following definitions and results.

- The *combination coefficient* $\binom{n}{k}$ is the number of ways to choose a subset of k objects from a set of size n.

- **Proposition 1.3.1.**   The combination coefficients satisfy **Pascal's recurrence**:

$$\binom{n}{0} = 1 \qquad \text{for all } n \geq 0$$

$$\binom{0}{k} = 0 \qquad \text{for all } k \geq 1$$

$$\binom{n}{k} = \binom{n-1}{k-1} + \binom{n-1}{k} \quad \text{for } n \geq 1 \qquad (4.1.1)$$

- The **binomial coefficient** $b_{n,k}$ is the coefficient of x^k in the binomial expansion

$$(1+x)^n = \sum_{k=0}^{n} b_{n,k}\, x^k$$

- **Proposition 1.3.2.**   The binomial coefficients satisfy Pascal's recurrence.

- **Corollary 1.3.3.**   For all $k, n \in \mathbb{N}$,

$$\binom{n}{k} = b_{n,k}$$

- Since combination coefficients have exactly the same values as binomial coefficients, as per Corollary 1.3.3, they are commonly referred to as *binomial coefficients*.

- **Proposition 1.5.3.**   For all non-negative integers n and k,

$$\binom{n}{k} = \frac{n^{\underline{k}}}{k!} \qquad (4.1.2a)$$

- **Corollary 1.5.4.**   For all non-negative integers n and k,

$$\binom{n}{k} = \frac{n!}{k!\,(n-k)!} \qquad (4.1.2b)$$

- Table 4.1.1, below, of the binomial coefficients is called **Pascal's triangle**.

Table 4.1.1 Pascal's triangle of binomial coefficients.

n	$\binom{n}{0}$	$\binom{n}{1}$	$\binom{n}{2}$	$\binom{n}{3}$	$\binom{n}{4}$	$\binom{n}{5}$	$\binom{n}{6}$	$\binom{n}{7}$	$\binom{n}{8}$	Σ
0	1									1
1	1	1								2
2	1	2	1							4
3	1	3	3	1						8
4	1	4	6	4	1					16
5	1	5	10	10	5	1				32
6	1	6	15	20	15	6	1			64
7	1	7	21	35	35	21	7	1		128
8	1	8	28	56	70	56	28	8	1	256

Combinatorial vs. Algebraic Proofs

There are several reasons for acquiring proficiency at more than one kind of proof. In the first place, one kind of proof may be simpler for some problems and the other simpler for other problems. Futhermore (especially for people who are a bit impatient), obtaining the same result by two quite different approaches tends to be more reliable confirmation than doing the same steps a second time. In addition, the different insights of different approaches are useful in deriving analogous results and consequences.

Several binomial-coefficient identities in this section are given not just one proof, but two. The following terminology is used to distinguish these approaches.

TERMINOLOGY: An **algebraic proof of an equation** is achieved by transforming one side of the equation with the aid of substitutions and of arithmetic operations into the expression on the other side.

TERMINOLOGY: A **combinatorial proof of an equation** is achieved by showing that both sides of the equation count the same thing. Sometimes such a proof uses the pigeonhole principle.

Numerous examples of both kinds of proof follow. Sometimes we give two or more proofs of a single assertion. Various general methods, including mathematical induction, may be used with either type of proof.

Symmetry

Some identities are generalizations of properties readily noticeable in Pascal's triangle. One such property is that each row of Pascal's triangle is palindromic: it reads the same forward or backward. For instance, we observe the symmetry of row 8.

$$1 \quad 8 \quad 28 \quad 56 \quad 70 \quad 56 \quad 28 \quad 8 \quad 1$$

Proposition 4.1.1 [Row-Symmetry Property]. *For any integers n and k such that $0 \le k \le n$,*

$$\binom{n}{k} = \binom{n}{n-k} \tag{4.1.3}$$

Algebraic Proof: Using Eq. (4.1.2b) yields this easy algebraic proof.

$$\binom{n}{k} = \frac{n!}{k! \cdot (n-k)!} = \frac{n!}{(n-k)! \cdot k!} = \binom{n}{n-k} \qquad \diamond$$

Combinatorial Proof: The left side of Equation (4.1.3) is the number of ways to select k objects from the set of n, to be in the designated subset. The right side is the number of ways to select $n - k$ objects to be excluded from it. There must be the same number of ways to do either. $\diamond$

Row-Sum Property

Another property of Pascal's triangle is that the sum of the entries in each row is a power of 2. For instance, in row 8,

$$1 + 8 + 28 + 56 + 70 + 56 + 28 + 8 + 1$$
$$= 256 = 2^8$$

Proposition 4.1.2 [Row-Sum Property]. *The sum of the entries in row n of Pascal's triangle is 2^n. That is,*

$$\sum_{k=0}^{n} \binom{n}{k} = 2^n \tag{4.1.4}$$

Combinatorial Proof: According to Corollary 1.3.3, the summands on the left side are the number of ways to choose subsets of cardinality k from a set S of n objects, for respective values of k. Their total is the number of ways to select a subset from S, over all possible subset sizes, which is clearly 2^n, shown on the right side, since each of the n objects is either present or absent. $\diamond$

Algebraic Proof: Substituting $x = 1$ into both sides of the equation for a binomial expansion yields the following result.

$$(1 + x)^n = \sum_{k=0}^{n} \binom{n}{k} x^n$$

$$\Rightarrow \quad (1 + x)^n \Big|_{x=1} = \sum_{k=0}^{n} \binom{n}{k} x^n \Big|_{x=1} = \sum_{k=0}^{n} \binom{n}{k} 1^n$$

$$\Rightarrow \quad 2^n = \sum_{k=0}^{n} \binom{n}{k} \qquad \qquad \diamond$$

Algebraic Proof: Another algebraic proof is by straightforward induction, starting with row 0 of Pascal's triangle as a basis case, and then using Pascal's recursion to show that the sum in row n doubles the sum in row $n - 1$. $\diamond$

Column-Sum Property

Some other properties of Pascal's triangle emerge after further investigation. For instance, the sum of all the entries in any column, up to and including the entry in row n, can be found in the next column in row $n + 1$.

Example 4.1.1: In columns 2 and 3 of Pascal's triangle, we see the following configuration.

n	$\binom{n}{1}$	$\binom{n}{2}$	$\binom{n}{3}$
2		1	
3		3	
4		6	
5		10	
6		15	
7			$\boxed{35}$

$$1 + 3 + 6 + 10 + 15 = 35$$

Proposition 4.1.3 [*Column-Sum Property*]. *The sum of the entries in column c ($c \geq 0$) of Pascal's triangle, from row 0 down to row n, equals the entry in row $n + 1$, column $c + 1$. That is,*

$$\sum_{k=0}^{n} \binom{k}{c} = \binom{n+1}{c+1} \tag{4.1.5}$$

Proof: By induction on the row number n.

BASIS: For $n = 0$, the sum of the entries down to row 0 is 1, in column $c = 0$, and is otherwise 0; also,

$$\binom{1}{c+1} = \begin{cases} 1 & \text{if } c = 0 \\ 0 & \text{otherwise} \end{cases}$$

IND HYP: Assume for some $n \geq 1$ that

$$\sum_{k=0}^{n-1} \binom{k}{c} = \binom{n}{c+1}$$

IND STEP: Then

$$\begin{aligned}
\sum_{k=0}^{n} \binom{k}{c} &= \sum_{k=0}^{n-1} \binom{k}{c} + \binom{n}{c} \\
&= \binom{n}{c+1} + \binom{n}{c} && \text{(induction hypothesis)} \\
&= \binom{n+1}{c+1} && \text{(Pascal's recursion)} \qquad \diamondsuit
\end{aligned}$$

Diagonal-Sum Properties

DEFINITION: A diagonal from the upper left of a 2-dimensional array, toward the lower right, is called a **southeast diagonal**. A diagonal in the opposite direction is called a **northwest diagonal**.

DEFINITION: A diagonal from the lower left of a 2-dimensional array, toward the upper right, is called a **northeast diagonal**. A diagonal in the opposite direction is called a **southwest diagonal**.

We observe that the sum of the elements along a finite initial segment of the southeast diagonal in Pascal's triangle appears just below the southeasternmost entry.

Proposition 4.1.4 [Southeast-Diagonal-Sum Property]. *The sum of the first $n + 1$ entries on the southeast diagonal from row r, column 0 in Pascal's triangle equals the entry in row $r + n + 1$, column n, the entry immediately below the last entry of the diagonal. That is,*

$$\sum_{k=0}^{n} \binom{r + k}{k} = \binom{r + n + 1}{n} \tag{4.1.6a}$$

Proof: This result follows from two previously derived properties of Pascal's triangle.

$$\sum_{k=0}^{n} \binom{r + k}{k} = \sum_{k=0}^{n} \binom{r + k}{r} \qquad \text{(symmetry)}$$

$$= \binom{r + n + 1}{r + 1} \qquad \text{(column-sum property)}$$

$$= \binom{r + n + 1}{n} \qquad \text{(symmetry)} \qquad \Diamond$$

Example 4.1.2:   Here is a southeast diagonal-sum.

n	$\binom{n}{0}$	$\binom{n}{1}$	$\binom{n}{2}$	$\binom{n}{3}$	$\binom{n}{4}$
2	1				
3		3			
4			6		
5				10	
6					15
7					35

$$1 + 3 + 6 + 10 + 15 = 35$$

The following corollary simply reverses the order of summation of the elements on the diagonal.

Corollary 4.1.5 [Northwest-Diagonal-Sum Property]. *For any non-negative integer* m *such that* $0 \leq m \leq n$, *the binomial coefficients satisfy the equation*

$$\sum_{k=0}^{m} \binom{n-k}{m-k} = \binom{n+1}{m} \tag{4.1.6b}$$

Proof: Reversing the northwest diagonal sum

$$\binom{n-0}{m-0} + \binom{n-1}{m-1} + \cdots + \binom{n-m}{m-m}$$

on the left of the equation yields the southeast diagonal sum

$$\binom{n-m}{0} + \binom{n-m+1}{1} + \cdots + \binom{n}{m}$$

which starts at row $n - m$ and includes $m + 1$ entries downward, ending at row n, column m. By Proposition 4.1.4, the value of this southeast diagonal sum is the binomial coefficient

$$\binom{n+1}{m} \qquad\qquad \diamond$$

An especially fascinating pattern emerges in the sums of northeast diagonals, namely, that they are Fibonacci numbers. For instance, the sum $1 + 5 + 6 + 1$ along the northeast diagonal that starts at $\binom{6}{0}$ is the Fibonacci number $f_7 = 13$.

Example 4.1.3:   The boxed Fibonacci numbers shown here do not actually appear at the locations shown. They are simply the sums along the northeast diagonals that lead to them.

n	$\boxed{\binom{n}{0}}$	$\boxed{\binom{n}{1}}$	$\boxed{\binom{n}{2}}$	$\boxed{\binom{n}{3}}$	$\boxed{\binom{n}{4}}$		
1			$\boxed{5}$				
2			1			$1+3+1 = \boxed{5} = f_5$	
3		3		$\boxed{21}$			
4	1			4		$1+6+10+4 = \boxed{21} = f_8$	
5			10				
6		6					
7	1						

Proposition 4.1.6 [Northeast-Diagonal Fibonacci Property]. *The sum of the entries on the NE diagonal from row* n, *column 0 in Pascal's triangle equals the Fibonacci number* f_{n+1}. *That is,*

$$\sum_{k=0}^{n} \binom{n-k}{k} = f_{n+1} \tag{4.1.7}$$

Proof: BASIS: For $n = 0$ and $n = 1$ the northeast diagonal sums are $1 = f_1$ and $1 + 0 = f_2$, respectively.

IND HYP: For $n \geq 2$ assume that

$$\sum_{k=0}^{n-1} \binom{n-1-k}{k} = f_n \quad \text{and} \quad \sum_{k=0}^{n-2} \binom{n-2-k}{k} = f_{n-1}$$

IND STEP: By the Pascal recursion, we have

$$\binom{n-k}{k} = \binom{n-k-1}{k-1} + \binom{n-k-1}{k}$$

Therefore,

$$\begin{aligned}
\sum_{k=0}^{n} \binom{n-k}{k} &= \sum_{k=0}^{n} \binom{n-k-1}{k-1} + \sum_{k=0}^{n} \binom{n-k-1}{k} \\
&= \sum_{j=0}^{n-2} \binom{n-j-2}{j} + \sum_{k=0}^{n-1} \binom{n-k-1}{k} \\
&= f_{n-1} + f_n \qquad \text{(induction hypothesis)} \\
&= f_{n+1} \qquad\qquad \text{(Fibonacci recursion)} \qquad\qquad \diamondsuit
\end{aligned}$$

Products of Binomial Coefficients

Another pattern in Pascal's triangle is the relationship between each element and the element to its upper left.

Example 4.1.4:   We observe in this inset from Pascal's triangle that

$$4\binom{7}{4} = 4 \cdot 35 = 140 = 7 \cdot 20 = 7\binom{6}{3}$$

n	$\binom{n}{2}$	$\binom{n}{3}$	$\binom{n}{4}$	$\binom{n}{5}$
5				
6		20		
7			35	
8				

$$\binom{7}{4} \cdot \frac{4}{7} = 35 \cdot \frac{4}{7} = 20 = \binom{6}{3}$$

or, equivalently

$$\binom{7}{4} \cdot 4 = 140 = \binom{6}{3} \cdot 7$$

The generality of this relationship, which is called the *absorption property*, is established by the next proposition.

Proposition 4.1.7 [Absorption Property].   For $0 \leq k \leq n$,

$$\binom{n}{k} k = n\binom{n-1}{k-1} \tag{4.1.8}$$

Algebraic Proof: By algebraic manipulation, we have

$$\binom{n}{k} k \;=\; \frac{n^{\underline{k}}}{k!} k \;=\; \frac{n^{\underline{k}}}{(k-1)!} \;=\; n\frac{(n-1)^{\underline{k-1}}}{(k-1)!} \;=\; n\binom{n-1}{k-1} \qquad\qquad \diamondsuit$$

Combinatorial Proof: Alternatively, we observe that the left side

$$\binom{n}{k} k$$

is the number of ways of choosing a board of k directors from a set of n persons and then a chairperson from within that board of k. This is clearly equivalent to the number of ways to choose a chairperson from a set of n persons and then another $k-1$ persons from the remaining $n-1$ persons for the rest of the board of directors, which is the right side

$$n\binom{n-1}{k-1} \qquad\qquad \diamondsuit$$

Absorption is a special case of a relationship between an element and other elements along the northwest diagonal direction. This relationship is expressed by a highly useful combinatorial identity that generalizes the following illustration.

Example 4.1.4, continued: Observe that whereas at one position northwest of the coefficient $\binom{7}{4}$ we have

$$20 \;=\; \binom{6}{3} \;=\; \binom{7}{4}\frac{4^{\underline{1}}}{7^{\underline{1}}} \;=\; 35\cdot\frac{4}{7}$$

at three positions northwest of $\binom{7}{4}$ in Pascal's triangle, we have

$$4 \;=\; \binom{4}{1} \;=\; \binom{7}{4}\frac{4^{\underline{3}}}{7^{\underline{3}}} \;=\; 35\cdot\frac{24}{210}$$

n	$\binom{n}{1}$	$\binom{n}{2}$	$\binom{n}{3}$	$\binom{n}{4}$	$\binom{n}{5}$
4	4				
5					
6					
7			35		
8					

$$35\cdot\frac{4^{\underline{3}}}{7^{\underline{3}}} \;=\; \binom{7}{4}\cdot\frac{4^{\underline{3}}/3!}{7^{\underline{3}}/3!} \;=\; \binom{4}{1} \;=\; 4$$

or, equivalently

$$\binom{7}{4}\cdot\binom{4}{3} \;=\; \binom{7-3}{4-3}\cdot\binom{7}{3}$$

The following formulation of this property is called the *subset-of-a-subset property*.

Proposition 4.1.8 [Subset-of-a-Subset Identity]. *For $0 \leq k \leq m \leq n$,*

$$\binom{n}{m}\binom{m}{k} = \binom{n}{k}\binom{n-k}{m-k} \tag{4.1.9}$$

Algebraic Proof: By straightforward algebraic calculation, we have

$$\binom{n}{m}\binom{m}{k} = \frac{n!}{m!\,(n-m)!} \cdot \frac{m!}{k!\,(m-k)!}$$

$$= \frac{n!}{k!\,(n-m)!\,(m-k)!}$$

$$= \frac{n!}{k!\,(n-k)!} \cdot \frac{(n-k)!}{(n-m)!\,(m-k)!}$$

$$= \binom{n}{k}\binom{n-r}{m-k} \qquad \qquad \Diamond$$

Combinatorial Proof: We can also reason combinatorially that the left side

$$\binom{n}{m}\binom{m}{k}$$

is the number of ways of choosing a board of m directors from a set of n persons and then an executive committee of k persons from within that board of m. This is clearly equivalent to the number of ways to choose an executive committee of k persons from a set of n persons and then another $m-k$ persons from the remaining $n-k$ persons for the rest of the board of directors, which is the right side

$$\binom{n}{k}\binom{n-k}{m-k} \qquad \qquad \Diamond$$

Vandermonde Convolution

Theorem 4.1.9 [*Vandermonde Convolution*]. *Let n, m, and k be non-negative integers. Then*

$$\sum_{j=0}^{n} \binom{n}{j}\binom{m}{k-j} = \binom{n+m}{k} \tag{4.1.10}$$

Combinatorial Proof: A combinatorial proof supposes that there are $n+m$ objects in a set, n of them blue and m of them red, and that k objects are to be chosen, for which there are clearly $\binom{n+m}{k}$ ways in all, the number of the right side. The number of ways to select j blue objects and $k-j$ red objects is the product $\binom{n}{j}\binom{m}{k-j}$; so the sum of all these products, which is on the left side, must be the same total as the right side. $\Diamond$

Another Proof: The sum on the left of the combinatorial equation above equals the coefficient of x^k on the left side of the polynomial equation

$$(1+x)^n (1+x)^m = (1+x)^{n+m}$$

and the binomial coefficient on the right side of the combinatorial equation equals the coefficient of x^k on the right side of that polynomial equation. $\Diamond$

Summary of Binomial Coefficient Identities

Table 4.1.2   **Basic Binomial Coefficient Identities**

$$\binom{n}{k} = \binom{n-1}{k} + \binom{n-1}{k-1} \qquad \text{Pascal's Recursion} \quad (4.1.1)$$

$$\binom{n}{k} = \frac{n^{\underline{k}}}{k!} \qquad \text{Falling Power Formula} \quad (4.1.2a)$$

$$\binom{n}{k} = \frac{n!}{k!\,(n-k)!} \qquad \text{Factorial Formula} \quad (4.1.2b)$$

$$\binom{n}{k} = \binom{n}{n-k} \qquad \text{Symmetry} \quad (4.1.3)$$

$$\sum_{k=0}^{n} \binom{n}{k} = 2^n \qquad \text{Row-Sum} \quad (4.1.4)$$

$$\sum_{r=0}^{n} \binom{r}{c} = \binom{n+1}{c+1} \qquad \text{Column-Sum} \quad (4.1.5)$$

$$\sum_{k=0}^{n} \binom{r+k}{k} = \binom{r+n+1}{n} \qquad \text{Southeast Diagonal} \quad (4.1.6a)$$

$$\sum_{k=0}^{m} \binom{n-k}{m-k} = \binom{n+1}{m} \qquad \text{Northwest Diagonal} \quad (4.1.6b)$$

$$\sum_{k=0}^{n} \binom{n-k}{k} = f_{n+1} \qquad \text{Fibonacci Northeast Diagonal} \quad (4.1.7)$$

$$\binom{n}{k} k = n \binom{n-1}{k-1} \qquad \text{Absorption} \quad (4.1.8)$$

$$\binom{n}{m}\binom{m}{k} = \binom{n}{k}\binom{n-k}{m-k} \qquad \text{Subset-of-a-Subset} \quad (4.1.9)$$

$$\sum_{j=0}^{n} \binom{n}{j}\binom{m}{k-j} = \binom{n+m}{k} \qquad \text{Vandermonde Convolution} \quad (4.1.10)$$

Parity of Binomial Coefficients

Beyond the basics of binomial coefficients, there are many fascinating byways. For instance, how might one determine the parity of a given binomial coefficient, such as

$$\binom{165}{93}$$

without doing a lot of calculation? Scanning Pascal's triangle enhances the mystery. One observes that all the entries in rows 1, 3, and 7, numbers of the form $2^n - 1$, are odd. Moreover, the number of odd numbers in a row appears to be a power of 2.

Determination of the parity of a binomial coefficient was studied systematically by the British mathematician James Glaisher (1848-1928).

Theorem 4.1.10. *Let n and k be non-negative integers. Then*

$$\binom{n}{k} \equiv \begin{cases} 0 \bmod 2 & \text{if } n \text{ is even and } k \text{ is odd} \\ \binom{\lfloor n/2 \rfloor}{\lfloor k/2 \rfloor} \bmod 2 & \text{otherwise} \end{cases}$$

Proof: This proof splits naturally into four cases.

Case 1 — n even and k odd: Since n is even, it is clear that, for this case, the value of the right side of the absorption identity

$$k\binom{n}{k} = n\binom{n-1}{k-1}$$

is even. Since the product $k\binom{n}{k}$ on the left side must also be even, and since k is odd, it follows that $\binom{n}{k}$ is even.

Case 2 — n even and k even: For this case, we expand the binomial coefficient.

$$\binom{n}{k} = \frac{n^{\underline{k}}}{k!} = \frac{n(n-1)(n-2)\cdots(n-k+1)}{1\cdot 2\cdot 3\cdots k}$$

$$= \frac{(n-1)(n-3)\cdots(n-k+1)}{1\cdot 3\cdot 5\cdots(k-1)} \cdot \frac{n\,(n-2)(n-4)\cdots(n-k+2)}{2\cdot 4\cdot 6\cdots k}$$

Since the denominator has $k/2$ even factors, we obtain

$$\binom{n}{k} = \frac{(n-1)(n-3)\cdots(n-k+1)}{1\cdot 3\cdot 5\cdots(k-1)} \cdot \frac{n\,(n-2)(n-4)\cdots(n-k+2)}{2^{\frac{k}{2}}\cdot 1\cdot 2\cdot 3\cdots \frac{k}{2}}$$

And since the denominator has $k/2$ even factors, we obtain

$$\binom{n}{k} = \frac{(n-1)(n-3)\cdots(n-k+1)}{1\cdot 3\cdot 5\cdots(k-1)} \cdot \frac{2^{\frac{k}{2}}\cdot \frac{n}{2}(\frac{n}{2}-1)(\frac{n}{2}-2)\cdots(\frac{n}{2}-\frac{k}{2}+1)}{2^{\frac{k}{2}}\cdot 1\cdot 2\cdot 3\cdots \frac{k}{2}}$$

$$= \frac{(n-1)(n-3)\cdots(n-k+1)}{1\cdot 3\cdot 5\cdots(k-1)} \cdot \binom{n/2}{k/2}$$

Therefore,

$$1\cdot 3\cdot 5\cdots(k-1)\binom{n}{k} = (n-1)(n-3)\cdots(n-k+1)\binom{n/2}{k/2}$$

It follows that for n and k both even,

$$\binom{n}{k} \equiv \binom{n/2}{k/2} \equiv \binom{\lfloor n/2 \rfloor}{\lfloor k/2 \rfloor} \bmod 2 \qquad (4.1.11)$$

The first equivalence in (4.1.11) holds because each of the factors preceding the binomial coefficient in the numerator and in the denominator is odd, and multiplication of an integer by an odd number does not change its parity. The second holds because $n/2 = \lfloor n/2 \rfloor$ and $k/2 = \lfloor k/2 \rfloor$ for N and k both even.

Case 3 — n odd and k odd: As in Case 1, our starting point is the absorption identity

$$k \binom{n}{k} = n \binom{n-1}{k-1}$$

Since n and k are both odd, and once again, since multiplication of an integer by an odd number does not change the parity, it follows that

$$\binom{n}{k} \equiv \binom{n-1}{k-1} \bmod 2$$

Since $n-1$ and $k-1$ are both even, it follows from Case 2 that

$$\binom{n-1}{k-1} \equiv \binom{\lfloor n/2 \rfloor}{\lfloor k/2 \rfloor} \bmod 2$$

and, hence, that

$$\binom{n}{k} \equiv \binom{\lfloor n/2 \rfloor}{\lfloor k/2 \rfloor} \bmod 2$$

Case 4 — n odd and k even: The symmetry identity implies that

$$(n-k)\binom{n}{k} = (n-k)\binom{n}{n-k} \quad \text{and} \quad n\binom{n-1}{n-k-1} = n\binom{n-1}{k}$$

It follows from the absorption identity

$$(n-k)\binom{n}{n-k} = n\binom{n-1}{n-k-1}$$

that

$$(n-k)\binom{n}{k} = n\binom{n-1}{k}$$

Since $n-k$ and n are both odd, we have

$$\binom{n}{k} \equiv \binom{n-1}{k} \bmod 2$$

Applying Case 2 to the right side, we obtain

$$\binom{n}{k} \equiv \binom{\lfloor (n-1)/2 \rfloor}{\lfloor k/2 \rfloor} \bmod 2$$

Since n is odd, the upper index $\lfloor (n-1)/2 \rfloor$ equals $\lfloor n/2 \rfloor$.   ◇

A simple algorithm to decide the parity of a binomial coefficient is to apply Theorem 4.1.10 iteratively, either until the upper index is even and the lower index odd or until the lower index is 0.

Example 4.1.5:   Here are both possible types of termination.

$$\binom{165}{93} \equiv \binom{82}{46} \equiv \binom{41}{23} \equiv \binom{20}{11} \equiv 0 \bmod 2$$

$$\binom{75}{11} \equiv \binom{37}{5} \equiv \binom{18}{2} \equiv \binom{9}{1} \equiv \binom{4}{0} \equiv 1 \bmod 2$$

In order to see why the number of odd binary coefficients in a row of Pascal's triangle is a power of 2, we observe that, in binary numbers, the integer operation

$$n \mapsto \lfloor n/2 \rfloor$$

is achieved by erasing the rightmost bit. We observe also that Case 1 of Theorem 4.1.10, in which n is even and k is odd, is discernible by a 0-bit at the right end of the binary numeral for n and a 1-bit at the right end of the binary numeral for k. If the parity algorithm uses binary numerals, then iterative erasure of the rightmost bits is not actually necessary. It is possible, instead, to align both numerals flush right and to scan to see whether there is a 0-bit above a 1-bit.

Example 4.1.5, continued: In scanning the aligned binary numerals

$$165_{10} = 1\,0\,1\,0\,0\,1\,0\,1_2$$
$$93_{10} = 0\,1\,0\,1\,1\,1\,0\,1_2$$

leftward from the right end, the first occurrence of a 0 in the upper index occurs at the 2^1-bit. Since there is also a 0-bit immediately below it, the scan continues. The next 0 in the upper index occurs at the 2^3-bit, and there is a 1-bit below it, so the scan terminates and the decision is even parity. In scanning the aligned binary numerals

$$75_{10} = 1\,0\,0\,1\,0\,1\,1_2$$
$$11_{10} = 0\,0\,0\,1\,0\,1\,1_2$$

one observes that there is a 0-bit beneath every 0-bit in the upper index, so the decision is odd parity.

Proposition 4.1.11.   *The number of odd binary coefficients in row n of Pascal's triangle is 2^w, where w is the number of 1-bits in the binary representation of n.*

Proof: For the binomial coefficient $\binom{n}{k}$ to be odd, there must be a 0 at each bit in the binary numeral for k for which there is a 0 at the corresponding bit of the binary numeral for n. However, if there is a 1 at a bit of the binary numeral for n, there may be either a 0 or a 1 at the corresponding bit of the binary numeral for k. If there are w 1-bits for n, then there are 2^w values for k that satisfy the rule for the 0-bits.   ◇

Corollary 4.1.12.   *If the integer n is of the form $2^r - 1$, then every binomial coefficient in row n of Pascal's triangle is odd.*

Proof: There are no 0-bits in the binary representation of $2^r - 1$.

EXERCISES for Section 4.1

In each of the Exercises 4.1.1 through 4.1.4, expand the given binomial coefficient and evaluate the result.

4.1.1 $\quad \dbinom{8}{3}$

4.1.2 $\quad \dbinom{\frac{5}{3}}{4}$

4.1.3 $\quad \dbinom{-3}{4}$

4.1.4^{S} $\quad \dbinom{-\frac{1}{2}}{3}$

In each of the Exercises 4.1.5 through 4.1.8, expand the given binomial coefficient as a polynomial in n.

4.1.5 $\quad \dbinom{n}{n-2}$

4.1.6 $\quad \dbinom{n^2}{3}$

4.1.7^{S} $\quad \dbinom{n+2}{n-2}$

4.1.8 $\quad \dbinom{n^3}{2}$

In each of the Exercises 4.1.9 through 4.1.20, prove the given binomial coefficient identity, where $n \in \mathbb{N}$.

4.1.9^{S} $\quad (n-r)\dbinom{n}{r} = n\dbinom{n-1}{r}$

4.1.10 $\quad \dbinom{2n}{2} = 2\dbinom{n}{2} + n^2$

4.1.11 $\quad \displaystyle\sum_{k=0}^{n}\dbinom{n}{k}^2 = \dbinom{2n}{n}$

4.1.12 $\quad \left[\displaystyle\sum_{k=0}^{n}\dbinom{n}{k}\right]^2 = \displaystyle\sum_{k=0}^{2n}\dbinom{2n}{k}$

4.1.13 $\quad \displaystyle\sum_{k=0}^{n}k\dbinom{n}{k} = n2^{n-1}$

4.1.14 $\quad \displaystyle\sum_{k=0}^{n}k^2\dbinom{n}{k} = n(n+1)2^{n-2}$

4.1.15 $\quad \displaystyle\sum_{k=0}^{n-1}\dbinom{n}{k}\dbinom{n}{k+1} = \dbinom{2n}{n-1}$

4.1.16 $\quad \displaystyle\sum_{k=0}^{n}\dbinom{n}{k}(-1)^k = (n=0)$

4.1.17^{S} $\quad \displaystyle\sum_{k=0}^{n}(-1)^k k\dbinom{n}{k} = -(n=1)$

4.1.18 $\quad \displaystyle\sum_{k=0}^{n}\frac{1}{k+1}\dbinom{n}{k} = \frac{1}{n+1}(2^{n+1}-1)$

4.1.19 $\quad \displaystyle\sum_{k=1}^{n}(-1)^{k-1}\dbinom{n}{k} = 1$ for $n > 0$

4.1.20 $\quad \displaystyle\sum_{k=1}^{n}(-1)^{k-1}\frac{1}{k}\dbinom{n}{k} = H_n$

4.1.21 Prove that Pascal's recursion for binomial coefficients holds, even when the upper index is not an integer. That is,

$$\binom{x}{r} = \binom{x-1}{r} + \binom{x-1}{r-1}$$

4.1.22^{S} Prove that the absorption identity for binomial coefficients holds, even when the upper index is not an integer. That is,

$$\binom{x}{r}r = x\binom{x-1}{r-1}$$

In each of the Exercises 4.1.23 through 4.1.30, calculate the parity of the given binomial coefficient.

4.1.23 $\dbinom{17}{9}$ **4.1.24**[S] $\dbinom{80}{48}$ **4.1.25** $\dbinom{33}{9}$ **4.1.26** $\dbinom{100}{48}$

4.1.27 $\dbinom{1728}{323}$ **4.1.28** $\dbinom{6561}{1728}$ **4.1.29** $\dbinom{19937}{11213}$ **4.1.30** $\dbinom{5678}{1234}$

4.1.31 Derive a formula for the proportion of odd binomial coefficients in column k of Pascal's triangle.

4.1.32 Prove that the sequence of parities that occurs in any column of Pascal's triangle is periodic.

4.2 BINOMIAL INVERSION OPERATION

This section develops an incremental technique used with binomial coefficients, called *binomial inversion*. Its main application in this section is within a solution of the derangement recurrence, which was introduced in §2.1 and further explored in Example 3.6.5. In the course of developing binomial inversion and applying it, there is use of several of the binomial coefficient identities of §4.1.

DEFINITION: The **transform** of the sequence $\langle f_n \rangle$ under **binomial inversion** is the sequence $\langle g_n \rangle$ with

$$g_n = \sum_{j=0}^{n} \binom{n}{j}(-1)^j f_j \qquad (4.2.1)$$

A characteristic property of anything mathematical that is correctly called a **duality operation** is that a second application of the operation restores the original object. Theorem 4.2.1 confirms that a transformation called *binomial inversion* of sequences has this property.

Theorem 4.2.1. Let $\langle f_n \rangle$ be a sequence and $\langle g_n \rangle$ its transform under binomial inversion. Then, for all $n \geq 0$,

$$f_n = \sum_{j=0}^{n} \binom{n}{j}(-1)^j g_j \qquad (4.2.2)$$

In other words, retransformation restores the original sequence $\langle f_n \rangle$.

Proof: Start at the right side of Eq. (4.2.2) and substitute the inversion formula of Eq. (4.2.1) for g_j.

$$\sum_{j=0}^{n} \binom{n}{j}(-1)^j g_j = \sum_{j=0}^{n} \binom{n}{j}(-1)^j \sum_{i=0}^{j} \binom{j}{i}(-1)^i f_i$$

$$= \sum_{j=0}^{n}\sum_{i=0}^{j} \binom{n}{j}\binom{j}{i}(-1)^{j+i} f_i \qquad (4.2.3)$$

Exchanging the order of summation is useful here.

$$= \sum_{i=0}^{n}\sum_{j=i}^{n} \binom{n}{j}\binom{j}{i}(-1)^{j+i} f_i \qquad (4.2.4)$$

Applying the subset-of-a-subset identity (Proposition 4.1.8) reduces the number of occurrences of the summation index j.

$$= \sum_{i=0}^{n}\sum_{j=i}^{n} \binom{n}{i}\binom{n-i}{j-i}(-1)^{j+i} f_i$$

Then factor to simplify the inner summation.

$$= \sum_{i=0}^{n} \binom{n}{i} f_i \sum_{j=i}^{n} \binom{n-i}{j-i}(-1)^{j-i} \qquad \text{since } (-1)^{2i} = 1$$

Substitute $k = j - i$.

$$= \sum_{i=0}^{n} \binom{n}{i} f_i \sum_{k=0}^{n-i} \binom{n-i}{k}(-1)^{k}$$

Compress the inner summation to an exponentiated binomial.

$$= \sum_{i=0}^{n} \binom{n}{i} f_i \, (1-x)^{n-i} \Big|_{x=1}$$

Using the Iverson truth function $(i = n)$ leads to completion of the proof.

$$= \sum_{i=0}^{n} \binom{n}{i} f_i \, (i = n)$$

$$= \binom{n}{n} f_n \, (n = n) = f_n \qquad\qquad \Diamond$$

Observe at Eq. (4.2.4) above that the summation index j occurs twice in the summand within a binomial coefficient, once as an upper index, and once as a lower index. In such circumstances, as seen here, the subset-of-a-subset identity often facilitates a transformation that reduces the number of occurrences of the summation index in the summand.

Some Basic Examples of Inversions

The first three examples here of inversion are introductory, to show how inversion works.

Example 4.2.1:   The constant sequence

$$\langle f_n \rangle = 1 \quad 1 \quad 1 \quad 1 \quad \cdots$$

has the inversion

$$g_n = \sum_{j=0}^{n} \binom{n}{j} (-1)^j f_j$$

$$= \sum_{j=0}^{n} \binom{n}{j} (-1)^j$$

$$= (1-x)^n \Big|_{x=1}$$

$$= (1-1)^n = \begin{cases} 1 & \text{if } n = 0 \\ 0 & \text{if } n > 0 \end{cases}$$

$$\Rightarrow \quad \langle g_n \rangle = 1 \quad 0 \quad 0 \quad 0 \quad \cdots$$

More generally, the sequence

$$\langle f_n \rangle = c \quad c \quad c \quad c \quad \cdots$$

has the inversion

$$\langle g_n \rangle = c \quad 0 \quad 0 \quad 0 \quad \cdots$$

Example 4.2.2:   The natural number sequence

$$\langle f_n \rangle = 0 \quad 1 \quad 2 \quad 3 \quad \cdots$$

is inverted as follows.

$$g_n = \sum_{j=0}^{n} \binom{n}{j} (-1)^j f_j$$

$$= \sum_{j=0}^{n} j \binom{n}{j} (-1)^j \qquad (4.2.5)$$

Apply the absorption identity to eliminate an occurrence of the index j.

$$= \sum_{j=0}^{n} n \binom{n-1}{j-1} (-1)^j$$

$$= n \sum_{j=0}^{n} \binom{n-1}{j-1} (-1)^j$$

Substitute $j = i + 1$ to align the binomial coefficient with the summation limits.

$$= n \sum_{i=0}^{n-1} \binom{n-1}{i} (-1)^{i+1}$$

$$= -n \sum_{i=0}^{n-1} \binom{n-1}{i} (-1)^i$$

$$= \left. -n\,(1-x)^{n-1} \right|_{x=1} = \begin{cases} -1 & \text{if } n = 1 \\ 0 & \text{if } n \neq 1 \end{cases}$$

$$\Rightarrow \quad \langle g_n \rangle = 0 \quad -1 \quad 0 \quad 0 \quad \cdots$$

In Eq. (4.2.5) of this example, the summation index j occurs within a binomial coefficient and also as a multiplier. The absorption identity is the usual binomial identity by which the number of occurrences of the index variable is reduced in such a circumstance.

The sequence $0 \quad 1 \quad 2 \quad 3 \quad \cdots$ can also be represented as $\langle \binom{n}{1} \rangle$. Accordingly, it should be unsurprising if calculating the inversion of the sequence $\binom{n}{r}$ is similar to Example 4.2.2.

Example 4.2.3:   The binomial coefficient sequence

$$f_n = \binom{n}{r}$$

for a fixed non-negative number r has the inversion sequence

$$g_n = \sum_{j=0}^{n} \binom{n}{j} (-1)^j f_j$$

$$= \sum_{j=0}^{n} \binom{j}{r} \binom{n}{j} (-1)^j \qquad (4.2.6)$$

Apply the subset-of-a-subset identity and then factor.

$$= \sum_{j=0}^{n} \binom{n}{r} \binom{n-r}{j-r} (-1)^j$$

$$= \binom{n}{r} \sum_{j=0}^{n} \binom{n-r}{j-r} (-1)^j$$

Substitute $j = i + r$. Then

$$g_n = \binom{n}{r} \sum_{i=-r}^{n-r} \binom{n-r}{i} (-1)^{i+r}$$

$$= \binom{n}{r} \sum_{i=0}^{n-r} \binom{n-r}{i} (-1)^{i+r}$$

$$= (-1)^r \binom{n}{r} \sum_{i=0}^{n-r} \binom{n-r}{i} (-1)^i$$

$$= \begin{cases} (-1)^n & \text{if } n = r \\ 0 & \text{if } n \neq r \end{cases}$$

At Eq. (4.2.6), the summand has two occurrences of the summation index j. This time, both are within different binomial coefficients, with one occurrence as an upper index and the other as a lower index. The subset-of-a-subset identity is frequently used to eliminate one of the occurrences in such summands, thereby simplifying the sum.

Derangements

Of course, the point of learning how to invert sequences is not just to pose a new class of computational exercises. Binomial inversion has numerous extrinsic applications.

Example 4.2.4:   Every permutation of the integer interval $[1 : n]$ can be obtained by choosing r numbers from $[1 : n]$ and deranging them. Accordingly, if D_j is a derangement number, then

$$n! \; = \; \binom{n}{0}D_0 \, + \, \binom{n}{1}D_1 \, + \, \binom{n}{2}D_2 \, + \, \cdots \, + \, \binom{n}{n}D_n$$

It follows that the sequence

$$f_n \; = \; (-1)^n D_n$$

has the binomial inversion

$$g_n \; = \; n!$$

By the duality property of binomial inversion, we have

$$f_n \; = \; (-1)^n D_n \; = \; \sum_{j=0}^{n} \binom{n}{j}(-1)^j g_j$$

which implies that

$$D_n \; = \; (-1)^n \sum_{j=0}^{n} \binom{n}{j}(-1)^j g_j$$

$$= \; (-1)^n \sum_{j=0}^{n} \binom{n}{j}(-1)^j j!$$

$$= \; \sum_{j=0}^{n} n^{\underline{j}}(-1)^j$$

$$\Rightarrow \frac{D_n}{n!} \; = \; 1 - \frac{1}{1!} + \frac{1}{2!} - \frac{1}{3!} + \cdots + (-1)^n \frac{1}{n!} \xrightarrow[n \to \infty]{\lim} e^{-1}$$

Thus, the proportion of derangements among the permutations of a set of n objects tends to e^{-1} as n gets larger, a result that we previously derived with inclusion-exclusion in Example 3.6.5. This illustrates again our perspective that it is helpful to have a variety of mathematical tools available for the solution of a given problem.

More Examples of Inversions

The summation techniques presented in this section for transforming sequences are widely applicable. The next section of this chapter applies these methods to computations in probability and statistics. We complete the present section with two more examples that combine the method of binomial inversion with the binomial identities derived previously.

Example 4.2.5: When two factors of a summand are both binomial coefficients that contain the index of summation as a lower index, the key to simplification is to set up an application of the Vandermonde convolution, which would simplify the summand. The sequence

$$f_n = (-1)^n \binom{N}{n}$$

has as its binomial inversion the sequence

$$g_n = \sum_{j=0}^{n} \binom{n}{j}(-1)^j f_j$$

$$= \sum_{j=0}^{n} \binom{n}{j}(-1)^j (-1)^j \binom{N}{j}$$

$$= \sum_{j=0}^{n} \binom{N}{j}\binom{n}{j}$$

Apply the symmetry identity as a setup

$$= \sum_{j=0}^{n} \binom{N}{j}\binom{n}{n-j}$$

and then invoke the Vandermonde convolution.

$$= \binom{N+n}{n}$$

Example 4.2.6: Sometimes there is a quotient of two binomial coefficients both of which contain the index of summation. The sequence

$$f_n = (-1)^n \binom{N}{n}^{-1}$$

has as its transform under binomial inversion the sequence

$$g_n = \sum_{j=0}^{n} \binom{n}{j}(-1)^j f_j$$

$$= \sum_{j=0}^{n} \binom{n}{j}(-1)^j (-1)^j \binom{N}{j}^{-1}$$

$$= \sum_{j=0}^{n} \frac{\binom{n}{j}}{\binom{N}{j}}$$

Here we apply the subset-of-a-subset identity

$$\binom{N}{n}\binom{n}{j} = \binom{N}{j}\binom{N-j}{n-j}$$

thereby obtaining

$$g_n = \sum_{j=0}^{n} \frac{\binom{n}{j}}{\binom{N}{n}\binom{n}{j}/\binom{N-j}{n-j}}$$

$$= \binom{N}{n}^{-1} \sum_{j=0}^{n} \binom{N-j}{n-j}$$

which can be simplified using the diagonal-sum identity

$$= \binom{N}{n}^{-1}\binom{N+1}{n}$$

$$= \frac{N+1}{N-n+1}$$

EXERCISES for Section 4.2

In each of the Exercises 4.2.1 through 4.2.18, transform the given sequence with binomial inversion.

4.2.1	$0 \quad 1 \quad 1 \quad 1 \quad \cdots$		**4.2.2**	$0 \quad 0 \quad 1 \quad 1 \quad 1 \quad \cdots$
4.2.3	$0 \quad 1 \quad 0 \quad 0 \quad 0 \quad \cdots$		**4.2.4**	$0 \quad 0 \quad 1 \quad 0 \quad 0 \quad 0 \quad \cdots$
4.2.5[S]	$0 \quad a \quad b \quad 0 \quad 0 \quad \cdots$		**4.2.6**	$1 \quad 0 \quad 1 \quad 0 \quad 1 \quad 0 \quad \cdots$
4.2.7	$0 \quad 1 \quad 0 \quad 1 \quad 0 \quad 1 \quad \cdots$		**4.2.8**	$1 \quad -1 \quad 1 \quad -1 \quad 1 \quad -1 \quad \cdots$
4.2.9	$f_n = 2^{n-1}$		**4.2.10**	$f_n = 2^n$
4.2.11	$f_n = (-2)^n$		**4.2.12**[S]	$f_n = 3^n$
4.2.13[S]	$f_n = n^2$		**4.2.14**	$f_n = n^2$
4.2.15	$f_n = n^3$		**4.2.16**	$f_n = n^3$
4.2.17	$f_n = 4n^2 - 2n + 3$		**4.2.18**	$f_n = 4n^2 - 2n + 3$

4.3   APPLICATIONS TO STATISTICS

Binomial coefficients frequently occur within summands of sums that arise in probability and statistics. Identities derived in §4.1 are now used to simplify some of these summation expressions. We continue to seek to reduce the number of occurrences of the index of summation within the summand. There are a few additional rules of thumb to be learned here and used.

Probability and Random Variables

Some basic definitions* are now recalled from elementary statistics and probability. The pace of the exposition here presumes that the reader has some prior familiarity with these topics.

DEFINITION: A **discrete probability space** is a pair $\langle \Omega, \Pr \rangle$ as follows.

- The discrete set Ω is called a **sample space**.
- A subset of Ω is called an **event**.
- The set 2^Ω of all subsets of Ω is called the **event space**.
- The function $\Pr : 2^\Omega \to \mathbb{R}$ is called a **probability measure**, and it satisfies the following axioms.
  1. $0 \le \Pr(A) \le 1$, for every event $A \subseteq \Omega$. The number $\Pr(A)$ is called the **probability of the event** A.
  2. $\Pr(\Omega) = 1$.
  3. If the events A_s, for $s \in S$, are mutually exclusive subsets of Ω, then

$$\Pr\Big(\bigcup_{s \in S} A_s \Big) \;=\; \sum_{s \in S} \Pr(A_s)$$

DEFINITION: A **random variable** X on a sample space is a real-valued function. It is called a **discrete random variable** if the set of values it takes is finite or countably infinite.

NOTATION: Let $X : \Omega \to \mathbb{R}$ be a discrete random variable on a sample space Ω with probability measure $\Pr$. For $x \in \mathbb{R}$, the probability of the set $\{\omega \in \Omega \mid X(\omega) = x\}$ is denoted $\Pr(x)$.

Mean and Variance

The *expected value* of a random variable, also called the *mean*, is commonly described as a weighted average. This is quite distinct from various informal notions of a *typical outcome*, since the expected value might never occur. It is simply an average of what could occur. The *variance* and the *standard deviation* are measures

---

* E.g., see Devore, *Probability and Statistics*, 6th Edition, Brooks-Cole, 2004.

of dispersion from the mean. If they are large, then values relatively far from the mean are more likely to occur than in a distribution with the same mean, but in which the variance and standard deviation are small.

DEFINITION: Let $X : \Omega \to \mathbb{R}$ be a discrete random variable on a sample space Ω with probability measure Pr, and let D be the set of values that X takes. The **expected value** or **mean of the random variable** X, denoted $E(X)$ or μ_X, is the sum

$$E(X) \;=\; \mu_X \;=\; \sum_{x \in D} x \cdot \Pr(x) \tag{4.3.1}$$

DEFINITION: Let $X : \Omega \to \mathbb{R}$ be a discrete random variable on a sample space Ω with probability measure Pr, and let D be the set of values that X takes. The **variance of the random variable** X, denoted $V(X)$ or σ_X^2, is the sum

$$V(X) \;=\; \sigma_X^2 \;=\; \sum_{x \in D} (x - \mu_X)^2 \cdot \Pr(x) \;=\; E([X - \mu_X]^2) \tag{4.3.2}$$

DEFINITION: Let $X : \Omega \to \mathbb{R}$ be a discrete random variable. The **standard deviation of the random variable** X, denoted $SD(X)$ or σ_X, is the square root of the variance.

$$SD(X) \;=\; \sigma_X \;=\; \sqrt{\sigma_X^2} \tag{4.3.3}$$

NOTATION: When it is clear from context to which random variable X they pertain, the subscripts for mean and variance may be dropped, so that they are denoted μ and σ^2.

DEFINITION: In calculating the **mean of a list of numbers** or the **variance of a list of numbers**, one regards each element of the list as equally likely.

Proposition 4.3.1. Let $X : \Omega \to \mathbb{R}$ be a discrete random variable. Then

$$\sigma_X^2 \;=\; E(X^2) - \mu^2 \tag{4.3.4}$$

Proof: Let D be the set of values taken by X. We proceed straightfowardly, starting from the Equation (4.3.2).

$$\begin{aligned}
\sigma_X^2 &= \sum_{x \in D} (x - \mu_X)^2 \cdot \Pr(x) \\
&= \sum_{x \in D} (x^2 - 2x\mu_X + \mu_X^2) \Pr(x) \\
&= \sum_{x \in D} x^2 \Pr(x) - \sum_{x \in D} 2x\mu_X \Pr(x) + \sum_{x \in D} \mu_X^2 \Pr(x) \\
&= E(X^2) - 2\mu_X \cdot \sum_{x \in D} x\Pr(x) + \mu_X^2 \cdot \sum_{x \in D} \Pr(x) \\
&= E(X^2) - 2\mu_X \cdot \mu_X + \mu_X^2 \cdot 1 \\
&= E(X^2) - \mu_X^2 \qquad\qquad\qquad\qquad\qquad\qquad \diamondsuit
\end{aligned}$$

Binomial Distribution

The prototypical experiment whose outcomes have a binomial distribution is a sequence of n tosses of a coin. Taking one of the possible outcomes of an individual toss, say heads, to be a "success", what is binomially distributed is the number of heads. We now apply the binomial coefficient identities of §4.1 to the calculation of the mean and variance of the binomial distribution.

DEFINITION: Given an experiment with binary outcome (*success* or *failure*) that is to be performed n times, the **binomial random variable** X is the number of successes. Suppose that the probability of success is p, and that the n trials are independent. Then

$$Pr(X = j) \ = \ \binom{n}{j} p^j (1 - p)^{n-j} \tag{4.3.5}$$

The sample space is the sequence of outcomes of the n trials. An event is a set of possible outcomes.

Proposition 4.3.2. *The expected value of a binomial random variable X on n trials, each with probability p of success, is*

$$E(X) \ = \ np$$

Proof: Start at Eq. (4.3.1), which defines expected value.

$$E(X) \ = \ \sum_{j=0}^{n} j \cdot Pr(X = j)$$

We substitute the probability of a binomial random variable given by Eq. (4.3.5).

$$= \ \sum_{j=0}^{n} j \binom{n}{j} p^j (1 - p)^{n-j}$$

Absorption eliminates one of the four occurrences of the summation index j.

$$= \ \sum_{j=0}^{n} n \binom{n-1}{j-1} p^j (1 - p)^{n-j}$$

$$= \ np \sum_{j=0}^{n} \binom{n-1}{j-1} p^{j-1}(1 - p)^{n-j}$$

Substituting $i = j - 1$ yields the summation

$$= \ np \sum_{i=0}^{n-1} \binom{n-1}{i} p^i (1 - p)^{n-1-i}$$

that we recognize as a binomial expansion, and simplify.

$$= \ np\,[p + (1 - p)]^{n-1}$$

$$= \ np \qquad\qquad\qquad\qquad \Diamond$$

Proposition 4.3.3. *The variance of a binomial random variable X on n trials, each with probability p of success, is*

$$V(X) \;=\; np(1-p)$$

Proof: Once again, start at Eq. (4.3.1).

$$E(X^2) \;=\; \sum_{j=0}^{n} j^2 \cdot \Pr(X = j)$$

$$=\; \sum_{j=0}^{n} j^2 \binom{n}{j} p^j (1-p)^{n-j}$$

There are once again four occurrences of the index j of summation. Applying absorption reduces the exponent of j in one occurrence, a reasonable step.

$$=\; \sum_{j=0}^{n} jn \binom{n-1}{j-1} p^j (1-p)^{n-j}$$

$$=\; np \sum_{j=0}^{n} j \binom{n-1}{j-1} p^{j-1} (1-p)^{n-j}$$

Substitute $i = j - 1$ to align the indices of the binomial coefficient with the upper and lower limits of the summation, another reasonable step.

$$=\; np \sum_{i=0}^{n-1} (1+i) \binom{n-1}{i} p^i (1-p)^{n-1-i}$$

Splitting the sum like this helps here

$$=\; np \sum_{i=0}^{n-1} \binom{n-1}{i} p^i (1-p)^{n-1-i} \;+\; np \sum_{i=0}^{n-1} i \binom{n-1}{i} p^i (1-p)^{n-1-i}$$

because the summation in the first part is recognizable as a binomial expansion.

$$=\; np \;+\; np \sum_{i=0}^{n-1} i \binom{n-1}{i} p^i (1-p)^{n-1-i}$$

Applying absorption again now eliminates one occurrence of the summation index.

$$=\; np \;+\; np \sum_{i=0}^{n} (n-1) \binom{n-2}{i-1} p^i (1-p)^{n-1-i}$$

Substituting $k = i - 1$ realigns the lower index of the binomial coefficient with the lower limit of the summation.

$$=\; np \;+\; n(n-1) p^2 \sum_{k=0}^{n} \binom{n-2}{k} p^k (1-p)^{n-2-k}$$

$$=\; np \;+\; n(n-1) p^2 \;=\; np + n^2 p^2 - np^2$$

$$= n^2 p^2 + np(1-p)$$

By Propositions 4.3.1 and 4.3.2,

$$\begin{aligned} \sigma_X^2 &= E(X^2) - E(X)^2 \\ &= [n^2 p^2 + np(1-p)] - n^2 p^2 \\ &= np(1-p) \end{aligned} \qquad \Diamond$$

Unbiased Estimator of the Mean

An intuitive statistical approach to estimating the proportion of persons in a population of large size N who have a given characteristic (such as enjoying recreational mathematics) is to take a random sample and to use the proportion in that sample to estimate the proportion in the general population. We will use binomial coefficient identities in confirming the validity of this approach.

DEFINITION: An estimator $\hat{\theta}$ of a statistical characteristic θ of a population is said to be an **unbiased estimator** if the expected value $E(\hat{\theta})$ for a random sample equals θ.

Proposition 4.3.4. *The sample proportion is an unbiased estimator of the proportion of individuals in a general population that have a given characteristic.*

Proof: Suppose that in a population of size N exactly M individuals have a given trait. A sample of size n is taken. The random variables of interest are the number m of persons with that trait and the proportion

$$X = \frac{m}{n}$$

of persons with the trait. The total number of ways to choose a sample of size n is

$$\binom{N}{n}$$

The number of ways that a sample of size n could have exactly j persons with the prescribed trait is the product

$$\binom{M}{j}\binom{N-M}{n-j}$$

of the number of choices of j individuals from the population of size M with the trait and the number of choices of the remaining $n-j$ individuals from the $N-M$ persons who do not have the trait. Thus,

$$Pr\,(m=j) = \frac{\binom{M}{j}\binom{N-M}{n-j}}{\binom{N}{n}}$$

Accordingly,

$$E(X) = \sum_{j=0}^{n} \frac{j}{n} \cdot Pr\,(m = j)$$

$$= \frac{1}{n} \sum_{j=0}^{n} j \cdot \frac{\binom{M}{j}\binom{N-M}{n-j}}{\binom{N}{n}}$$

$$= \frac{1}{n} \binom{N}{n}^{-1} \sum_{j=0}^{n} j \cdot \binom{M}{j}\binom{N-M}{n-j}$$

Apply the absorption identity to eliminate one occurrence of the index of summation.

$$= \frac{1}{n} \binom{N}{n}^{-1} \sum_{j=0}^{n} M \cdot \binom{M-1}{j-1}\binom{N-M}{n-j}$$

$$= \frac{M}{n} \binom{N}{n}^{-1} \sum_{j=0}^{n} \binom{M-1}{j-1}\binom{N-M}{n-j}$$

Now use the Vandermonde convolution.

$$= \frac{M}{n} \binom{N}{n}^{-1} \binom{N-1}{n-1}$$

$$= \frac{M}{n} \cdot \frac{n!}{N^{\underline{n}}} \cdot \frac{(N-1)^{\underline{n-1}}}{(n-1)!} = \frac{M}{N}$$

Thus, the intuitive method of estimating the mean is unbiased. $\diamond$

Unbiased Estimator of the Variance

Let X be a random variable on a space Ω. The identically distributed random variables

$$X_1 \quad X_2 \quad \ldots \quad X_n$$

are the values of X on n samples from Ω, with sample mean $\overline{X}$. Statisticians use the estimator

$$\hat{\sigma}^2 = \frac{\sum(X_i - \overline{X})^2}{n-1} = \frac{\sum X_i^2 - n^{-1}\left(\sum X_i\right)^2}{n-1}$$

with $n-1$ in the denominator (rather than n), for the variance. This is explained by the next proposition.

Proposition 4.3.5. *The sample statistic*

$$\hat{\sigma}^2 = \frac{\sum (X_i - \overline{X})^2}{n-1} = \frac{\sum X_i^2 - n^{-1}\left(\sum X_i\right)^2}{n-1} \qquad (4.3.6)$$

is an unbiased estimator of the variance of the random variable X.

Proof: $E\left(\hat{\sigma}^2\right) = \dfrac{E\left(\sum X_i^2\right)}{n-1} - \dfrac{E\left[\left(\sum X_i\right)^2\right]}{n(n-1)}$

$$= \frac{1}{n-1}\sum_{i=1}^{n} E(X_i^2) - \frac{1}{n(n-1)}\sum_{i=1}^{n}\sum_{j=1}^{n} E(X_i X_j)$$

Split the double summation into two parts.

$$= \frac{1}{n-1}\sum_{i=1}^{n} E(X_i^2) - \frac{1}{n(n-1)}\sum_{i=1}^{n} E(X_i^2)$$

$$\qquad - \frac{1}{n(n-1)}\sum_{i=1}^{n}\sum_{j=1}^{n} E(X_i X_j)\cdot(j\neq i)$$

$$= \left(\frac{1}{n-1} - \frac{1}{n(n-1)}\right)\cdot\sum_{i=1}^{n} E(X^2)$$

$$\qquad - \frac{1}{n(n-1)}\sum_{i=1}^{n}\sum_{j=1}^{n} E(X)E(X)\cdot(j\neq i)$$

$$= \frac{1}{n}\cdot\sum_{i=1}^{n} E(X^2) - \frac{1}{n(n-1)}\sum_{i=1}^{n}\sum_{j=1}^{n} E(X)E(X)\cdot(j\neq i)$$

Both sums are resolvable.

$$= \frac{1}{n}\cdot n E(X^2) - \frac{n(n-1)}{n(n-1)} E(X)^2$$

$$= E(X^2) - E(X)^2$$

$$= V(X)$$

Thus, division by $n-1$ leads to an unbiased estimate. $\Diamond$

EXERCISES for Section 4.3

In Exercises 4.3.1 through 4.3.3, a fair die is to be rolled 20 times.

4.3.1   What is the probability of obtaining exactly four sixes?

4.3.2   What is the expected number of sixes?

4.3.3   What is the standard deviation of the number of sixes?

DEFINITION: The context of a **hypergeometric distribution** is a finite sample space with N objects, of which M are successes. A sample of n objects is to be chosen, without replacement. The random variable X is the number of successes in the sample.

In Exercises 4.3.4 through 4.3.7, give an algebraic formula for each of the following properties of a hypergeometric distribution.

4.3.4^S $\Pr(X = j)$ 4.3.5^S $E(X)$

4.3.6 $E(X^2)$ 4.3.7^S $V(X)$

In Exercises 4.3.8 through 4.3.10, 10 cards are drawn without replacement from a deck of 20 playing cards.

4.3.8   What is the probability of obtaining exactly 3 face cards?

4.3.9   What is the expected number of face cards?

4.3.10   What is the standard deviation of the number of face cards?

DEFINITION: The context for a **negative binomial distribution** is an infinite sequence of trials, each resulting in a success or a failure, such that success occurs on each trial with probability p. The experiment continues until r successes have been observed. The random variable X is the number of trials that precede the rth success.

In Exercises 4.3.11 through 4.3.14, give an algebraic formula for each of the following properties of a negative binomial distribution.

4.3.11^S $\Pr(X = j)$ 4.3.12 $E(X)$

4.3.13^S $E(X^2)$ 4.3.14^S $V(X)$

In Exercises 4.3.15 through 4.3.17, a fair die is to be rolled until the number six has occurred 4 times.

4.3.15   What is the probability of exactly 20 rolls before the rth six?

4.3.16   What is the expected number of rolls?

4.3.17   What is the standard deviation of the number of rolls?

DEFINITION: A random variable X has a **Poisson distribution** with parameter λ ($\lambda > 0$) if

$$\Pr(X = j) \ = \ \frac{e^{-\lambda}\lambda^j}{j!}$$

In Exercises 4.3.18 through 4.3.20, give an algebraic formula for each of the following properties of a Poisson distribution.

4.3.18^S $E(X)$ 4.3.19 $E(X^2)$ 4.3.20^S $V(X)$

4.4   THE CATALAN RECURRENCE

The Catalan numbers c_n, named for Eugène Catalan, are used to count many different kinds of combinatorial objects. Two that are prominent in computer science applications are *binary trees* and nestings of parentheses. Another is *subdiagonal paths*.

REVIEW FROM §1.2:

- The **Catalan sequence** $\{c_n\}$ is defined by the recurrence

$$c_0 = 1; \qquad\qquad\qquad \text{initial value}$$
$$c_n = c_0 c_{n-1} + c_1 c_{n-2} + \cdots + c_{n-1} c_0 \qquad \text{for } n \geq 1$$

Surprisingly, perhaps, the closed formula for c_n is a multiple of a binomial coefficient.

Binary Trees

In graph theory, the set $\mathcal{T}_\downarrow$ of **binary trees** can be defined recursively:

- The empty tree Φ is in the set $\mathcal{T}_\downarrow$.

- The tree $K_1^\bullet$ with a single vertex designated as the *root* is in the set $\mathcal{T}_\downarrow$.

- If $T \in \mathcal{T}_\downarrow$, and if v is a vertex of the tree T, then each of the following rooted trees is in the set $\mathcal{T}_\downarrow$.

  i. The tree obtained by adjoining a new vertex to v, called the *left-child* of vertex v. (A vertex has at most one left-child.)

  ii. The tree obtained by adjoining a new vertex to v, called the *right-child* of vertex v. (A vertex has at most one right-child.)

Figure 4.4.1 illustrates the binary trees with 0, 1, 2, and 3 vertices. It is easy enough to verify for these small cases that the Catalan number c_n is the number of binary trees with n vertices.

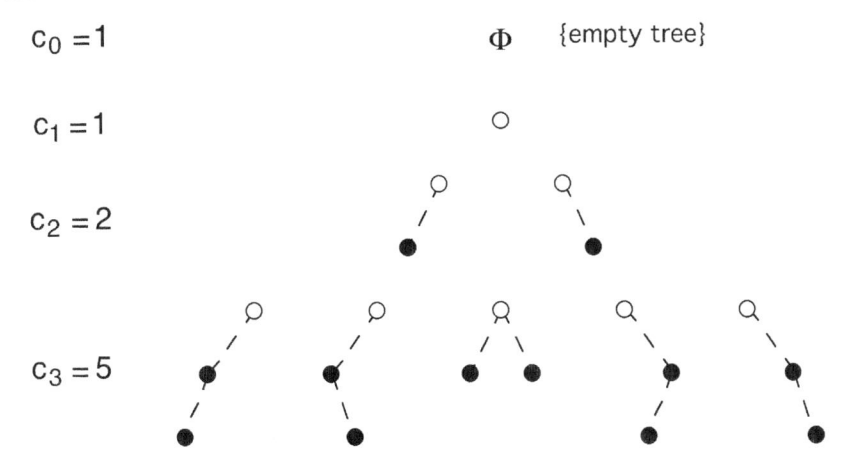

Figure 4.4.1   The smallest binary trees.

Remark: In computer science, each child of a vertex of a binary tree is designated either as a left-child or a right-child, even if there is only one child. The importance of this designation occurs in applications such as *binary search trees* and *priority trees* (see [GrYe2006]).

DEFINITION: The **left subtree** of a binary tree T is the subtree whose root is the left-child of the root of T. The **right subtree** of a binary tree T is the subtree whose root is the right-child of the root of T.

Proposition 4.4.1. For $n \geq 0$, the number of n-vertex binary trees equals the Catalan number c_n.

Proof: By induction on the number n of vertices.

BASIS: Clearly, $c_0 = 1$ and $c_1 = 1$ are the numbers of binary trees with 0 and 1 vertices, respectively.

IND HYP: Let $n > 0$. Suppose for all integers k with $0 \leq k < n$, that c_k is the number of binary trees with k vertices.

IND STEP: Suppose that a binary tree has n vertices. For $k = 0, 1, \ldots, n-1$, the number of possible left subtrees with k vertices is c_k, by the induction hypothesis. Of course, the right subtree would then have $n - k - 1$ vertices, so that there would be a total number of n vertices within the union of the two subtrees and the root, as depicted in Figure 4.4.2.

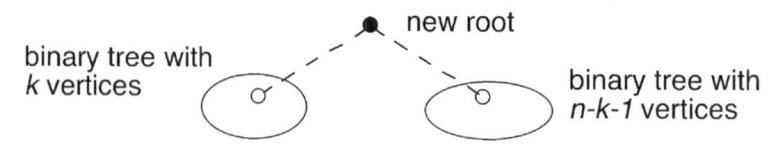

Figure 4.4.2   **Joining left and right subtrees to a root.**

The induction hypothesis also implies in this circumstance, therefore, that the number of possible right subtrees is c_{n-k-1}. Hence, by the rule of product, there are $c_k c_{n-k-1}$ n-vertex binary trees with k vertices in the left subtree. Accordingly, the total number of n-vertex binary trees is given by the sum

$$
\begin{aligned}
c_n &= \sum_{k=0}^{n-1} c_k \, c_{n-k-1} \\
&= c_0 \, c_{n-1} + c_1 \, c_{n-2} + \cdots + c_{n-1} \, c_0 \qquad\qquad \diamondsuit
\end{aligned}
$$

Nested Parentheses

The set $\mathcal{P}$ of **well-nested strings of parentheses** is defined recursively (as depicted in Figure 4.4.3 below):

- The empty string Λ is in $\mathcal{P}$.

- If $P_i, P_o \in \mathcal{P}$, then the string $(P_i)P_o$ is in $\mathcal{P}$. That is, we insert the string P_i inside a new pair and then juxtapose the string P_o at the right.

In listing the well-nested strings with 0, 1, 2, and 3 pairs of parentheses, the new pair specified by the recursion rule above is depicted by brackets.

0 pairs	Λ = empty string	$c_0 = 1$
1 pair	[ ]	$c_1 = 1$
2 pairs	[ ](), [()]	$c_2 = 2$
3 pairs	[ ]()(), [ ](()), [()](), [()()], [(())]	$c_3 = 5$

Proposition 4.4.2. *For $n \geq 0$, the number of well-nested strings of parentheses equals the Catalan number c_n.*

Proof: This proof follows the exact same lines as the proof of Proposition 4.4.1. The new pair of parentheses with well-nested substrings inside and outside in the recursive construction here corresponds to the new root with left and right binary subtrees there.   ◇

Subdiagonal Paths

DEFINITION: A **northeastward path** or **NE-path** in the array $[0 : n] \times [0 : n]$ is a path whose directed edges are each one unit in length and lead northward or eastward.

DEFINITION: A **subdiagonal path** from $(0, 0)$ to (n, n) in $[0 : n] \times [0 : n]$ is a NE-path along which each point (x, y) satisfies the inequality $x \geq y$.

The inequality in the definition means, as illustrated in Figure 4.4.3, that the path never crosses above the longest northeast diagonal.

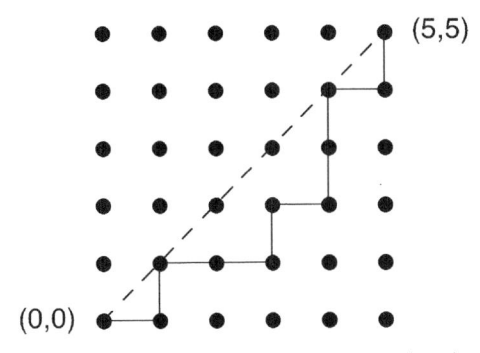

Figure 4.4.3   **A subdiagonal path from $(0, 0)$ to (n, n).**

Proposition 4.4.3. *For $n \geq 0$, the number of subdiagonal paths from $(0, 0)$ to (n, n) in the array $[0 : n] \times [0 : n]$ equals the Catalan number c_n.*

Proof: In every prefix of a well-nested string of parentheses, the number of left parentheses is greater than or equal to the number of right parentheses, and the total number of left parentheses equals the total number of right parentheses. Both these facts are provable by an induction on the length of the string. It follows that the well-nested strings of n pairs of parentheses are in bijective correspondence with the subdiagonal paths in the array $[0 : n] \times [0 : n]$.   ◇

Solving the Catalan Recurrence

Of the many methods of solving the Catalan recurrence, the one now presented, based on work of D. Andre in 1878, is probably the simplest to follow.

Theorem 4.4.4. *The Catalan recurrence*

$$c_0 = 1; \qquad\qquad\qquad \text{initial value}$$
$$c_n = c_0 c_{n-1} + c_1 c_{n-2} + \cdots + c_{n-1} c_0 \quad \text{for } n \geq 1$$

has the solution

$$c_n = \frac{1}{n+1}\binom{2n}{n} = \frac{1}{n+1} \cdot \frac{(2n)!}{n!\,n!}$$

Proof: Consider the set S_{NE} of all NE paths from $(0,0$ to (n,n) in $[0 : n] \times [0 : n]$. Suppose that each step eastward on an NE path is represented by the letter E and that each step northward is represented by the letter N. For instance, the path in Figure 4.4.3 is represented by the string

ENEENENNEN

This correspondence is evidently a bijection between the set S_{NE} of NE-paths and the set of strings in E and N of length $2n$ with n occurrences of each letter. The number of ways to choose the n locations for the N's in such a string is

$$\binom{2n}{n}$$

The bijection establishes that this is the total number of NE paths. By Proposition 4.4.3, the Catalan number c_n equals the number of subdiagonal paths in the array $[0 : n] \times [0 : n]$. Our approach is to subtract from this total the number of strings that do *not* represent subdiagonal paths.

Observe that a path is not subdiagonal if and only if, at some point, the number of northward steps has exceeded the number of eastward steps. Accordingly, the corresponding string

$$s_1 s_2 \ldots s_{2n}$$

would have a prefix in which the number of N's exceeds the number of E's. If $2j+1$ is the smallest index at which this occurs, then the number of E's in the prefix

$$s_1 s_2 \ldots s_{2j+1}$$

is j and the number of N's is $j + 1$.

It follows that in the suffix
$$s_{2j+2}s_{2j+3}\cdots s_{2n}$$
there are $n-j$ E's and $n-j-1$ N's. Suppose that in the suffix, each E is replaced by an N and each N by an E. This is called a **reflection of the subpath** or a **reflection of the substring**. The resulting string has $n-1$ E's and $n+1$ N's. It represents an NE path in $[0:n-1] \times [0:n+1]$, and there is a bijection between the set of non-subdiagonal paths from $(0,0)$ to (n,n) in $[0:n] \times [0:n]$ and the set of NE paths from $(0,0)$ to $(n-1, n+1)$ in $[0:n-1] \times [0:n+1]$, whose cardinality is

$$\binom{2n}{n-1}$$

Thus, the number of subdiagonal paths from $(0,0)$ to (n,n) in $[0:n] \times [0:n]$ is

$$\binom{2n}{n} - \binom{2n}{n-1} = \frac{(2n)^{\underline{n-1}}(n+1)}{n!} - \frac{(2n)^{\underline{n-1}}n}{n!}$$

$$= \frac{1}{n+1}\binom{2n}{n} \qquad \diamond$$

Example 4.4.1: $c_3 = \dfrac{1}{4}\dbinom{6}{3} = \dfrac{20}{4} = 5.$

Example 4.4.2: $c_4 = \dfrac{1}{5}\dbinom{8}{4} = \dfrac{70}{5} = 14.$

Generalized Binomial Theorem

An alternative proof of the solution to the Catalan recurrence uses the *generalized binomial theorem*.

NOTATION: The kth derivative of a function $f(x)$ may be denoted $f^{(k)}$.

DEFINITION: An *analytic function* is a function $f(x)$ with an nth derivative for every $n \geq 0$.

Theorem 4.4.5 (Generalized Binomial Theorem). *For any real number s, the exponentiated binomial $(1+x)^s$ has the power series*

$$(1+x)^s = \sum_{k=0}^{\infty}\binom{s}{k}x^k$$

$$= 1 + \frac{s^{\underline{1}}}{1!}\cdot x + \frac{s^{\underline{2}}}{2!}\cdot x^2 + \cdots + \frac{s^{\underline{k}}}{k!}\cdot x^k + \cdots$$

Proof: For $f(x) = (1+x)^s$, observe that

$$f(0) = (1+x)^s\Big|_{x=0} = 1^s = s^{\underline{0}}$$

$$f'(0) = s(1+x)^{s-1}\Big|_{x=0} = s\cdot 1^{s-1} = s^{\underline{1}}$$

$$f''(0) = s^{\underline{2}}(1+x)^{s-2}\Big|_{x=0} = s^{\underline{2}}\cdot 1^{s-2} = s^{\underline{2}}$$

By induction, it can be proved that

$$f^{(k)}(0) = s^{\underline{k}}$$

Recall that the Maclaurin series expansion* of an analytic function $f(x)$ is

$$f(x) = \sum_{k=0}^{\infty} \frac{f^{(k)}(0)}{k!} \cdot x^k$$

$$= \frac{f(0)}{0!} \cdot x^0 + \frac{f'(0)}{1!} \cdot x^1 + \frac{f''(0)}{2!} \cdot x^2 + \cdots$$

Thus, the substitution $f^{(n)}(0) = s^{\underline{n}}$ yields the conclusion. $\diamondsuit$

Example 4.4.3: In the solution of the Catalan recurrence below, we use this generalized binomial expansion.

$$(1 - 4z)^{1/2} = \sum_{k=0}^{\infty} \binom{\frac{1}{2}}{k} (-4z)^k$$

$$= 1 + \frac{(\frac{1}{2})^{\underline{1}}}{1!} \cdot (-4z) + \frac{(\frac{1}{2})^{\underline{2}}}{2!} \cdot (-4z)^2 + \cdots$$

Alternative Proof of the Catalan Formula

An alternative proof of the solution

$$c_n = \frac{1}{n+1} \binom{2n}{n} = \frac{1}{n+1} \cdot \frac{(2n)!}{n!n!}$$

to the Catalan recurrence provides a traditional illustration of the power of the method of generating functions in solving recurrences. We define the generating function

$$C(z) = \sum_{n=0}^{\infty} c_n z^n$$

and begin as in §2.2.

Step 1a. Multiply both sides of the Catalan recursion by z^n.

$$c_n z^n = \sum_{k=0}^{n-1} c_k c_{n-k-1} z^n \tag{4.4.1}$$

Step 1b. Sum both sides of Eq. (4.4.1) over the same range of values, as large as possible.

$$\sum_{n=1}^{\infty} c_n z^n = \sum_{n=1}^{\infty} \sum_{k=0}^{n-1} c_k c_{n-k-1} z^n \tag{4.4.2}$$

---

* This is equivalent to the Taylor series expansion at $x = 0$.

Step 2. Replace the infinite sum on the left of Eq. (4.4.2) with a finite sum involving the generating function $C(z)$.

$$C(z) - c_0 = \sum_{n=1}^{\infty} \sum_{k=0}^{n-1} c_k c_{n-k-1} z^n$$

Exchange the order of summation.

$$= \sum_{k=0}^{\infty} \sum_{n=k+1}^{\infty} c_k c_{n-k-1} z^n$$

$$= z \sum_{k=0}^{\infty} c_k z^k \sum_{n=k+1}^{\infty} c_{n-k-1} z^{n-k-1}$$

Substitute $j = n - k - 1$.

$$= z \sum_{k=0}^{\infty} c_k z^k \sum_{j=0}^{\infty} c_j z^j$$

$$= z \sum_{k=0}^{\infty} c_k z^k \, C(z)$$

$$= z \, C(z) \sum_{k=0}^{\infty} c_k z^k$$

$$= z \, C(z)^2$$

$$\Rightarrow \qquad z C(z)^2 - C(z) + 1 = 0 \qquad\qquad (4.4.3)$$

Step 3. Solve for $C(z)$ in Eq. (4.4.3) by the quadratic formula.

$$C(z) = \frac{1 \pm \sqrt{1 - 4z}}{2z} \qquad\qquad (4.4.4)$$

Step 4. To solve for the value of the general Catalan number c_n, we apply the Generalized Binomial Theorem, as in Example 4.4.3, to Eq. (4.4.4).

$$(1 - 4z)^{1/2} = \sum_{n=0}^{\infty} \binom{\frac{1}{2}}{n} (-4z)^n = 1 + \sum_{n=1}^{\infty} \frac{\left(\frac{1}{2}\right)^{\underline{n}}}{n!} (-4z)^n$$

$$= 1 + \frac{\left(\frac{1}{2}\right)^{\underline{1}}}{1!} \cdot (-4z) + \frac{\left(\frac{1}{2}\right)^{\underline{2}}}{2!} \cdot (-4z)^2 + \cdots$$

Since every term of this series except the first is signed negative, the appropriate choice is the negative root. That is,

$$C(z) = \frac{1 - \sqrt{1 - 4z}}{2z} = \frac{-1}{2z} \sum_{n=1}^{\infty} \frac{\left(\frac{1}{2}\right)^{\underline{n}}}{n!} (-4z)^n \qquad\qquad (4.4.5)$$

To simplify (4.4.5), we expand part of the summand

$$\frac{1}{n!}\left(\frac{1}{2}\right)^{\underline{n}} = \frac{1}{n!} \cdot \frac{1}{2} \cdot \frac{-1}{2} \cdot \frac{-3}{2} \cdot \ldots \cdot \frac{-(2n-3)}{2}$$

$$= \frac{1}{n!} \cdot \frac{(-1)^{n-1}}{2^n} \prod_{j=1}^{n-1}(2j-1)$$

$$= \frac{1}{n!} \cdot \frac{(-1)^{n-1}}{2^n} \prod_{j=1}^{n-1} \frac{(2j-1)(2j)}{2j}$$

$$= \frac{1}{n!} \cdot \frac{(-1)^{n-1}}{2^n} \cdot \frac{(2n-2)!}{2^{n-1}(n-1)!}$$

$$= \frac{(-1)^{n-1}(2n-2)!}{2^{2n-1}(n-1)!n!} \qquad (4.4.6)$$

and we substitute the result (4.4.6) back into Eq. (4.4.5), to obtain

$$C(z) = \frac{-1}{2z} \sum_{n=1}^{\infty} \frac{\left(\frac{1}{2}\right)^{\underline{n}}}{n!}(-4z)^n$$

$$= \frac{-1}{2z} \sum_{n=1}^{\infty} \frac{(-1)^{n-1}(2n-2)!}{2^{2n-1}(n-1)!n!}(-4z)^n$$

$$= \frac{-1}{2z} \sum_{n=1}^{\infty} \frac{(-1)^{n-1}(2n-2)!}{2^{2n-1}(n-1)!n!}(-4z)^n$$

$$= \frac{-1}{2z} \sum_{n=1}^{\infty} \frac{(-1)^{n-1}(2n-2)!}{2^{2n-1}(n-1)!n!}(-1)^n 2^{2n} z^n$$

$$= \sum_{n=1}^{\infty} \frac{(2n-2)!}{(n-1)!n!} z^{n-1} = \sum_{n=1}^{\infty} \frac{1}{n}\binom{2n-2}{n-1} z^{n-1}$$

$$= \sum_{n=0}^{\infty} \frac{1}{n+1}\binom{2n}{n} z^n$$

This yields the conclusion

$$c_n = \frac{1}{n+1}\binom{2n}{n} \qquad\qquad \diamond$$

EXERCISES for Section 4.4

The point (x, y) in the infinite northeast quadrant $\mathbb{N} \times \mathbb{N}$ represents x steps eastward and y steps northward from the origin $(0, 0)$. All of the exercises for this section are concerned with counting NE-paths to various points (x, y). Let $q(x, y)$ be the number of subdiagonal paths from $(0, 0)$ to (x, y).

4.4.1[S]   Prove that $q(n, n) = c_n$, the nth Catalan number.

4.4.2   Prove that $q(x, y)$ satisfies the following recurrence.

$$\begin{aligned} q_{x,0} &= 1 &&\text{for } x \geq 0 \\ q_{x,y} &= 0 &&\text{for } y > x \\ q_{x,y} &= q_{x,y-1} + q_{x-1,y} &&\text{for } y \geq 1 \end{aligned}$$

4.4.3   Give a formula in Catalan numbers for the value of $q(n, n-1)$.

4.4.4[S]   Give a formula in Catalan numbers for the value of $q(n, n-2)$.

4.4.5   Give a formula in Catalan numbers for the value of $q(n, n-3)$.

4.4.6   Give a formula in Catalan numbers for the value of $q(n, n-4)$.

4.4.7   Count the number of subdiagonal paths to (n, n) such that between the first E and the final N, the number of E's exceeds the number of N's.

4.4.8   Count the number of subdiagonal paths to (n, n) that touch the diagonal at least once between the origin and (n, n).

4.4.9[S]   Count the number of NE-paths to (n, n), such that in every prefix, the number of N's exceeds the number of E's by more than 1.

4.4.10   Count the number of NE-paths to (n, n), such that in every prefix, the number of N's exceeds the number of E's by at most 1.

4.4.11   Count the number of NE-paths to (n, n), such that in every prefix, the number of N's exceeds the number of E's by more than 2.

4.4.12   Count the number of NE-paths to (n, n), such that in every prefix, the number of N's exceeds the number of E's by at most 2.

4.4.13   Count the number of NE-paths to (n, n), such that in every prefix, the number of N's exceeds the number of E's by more than k.

4.4.14   Count the number of NE-paths to (n, n), such that in every prefix, the number of N's exceeds the number of E's by at most k.

4.4.15[S]   Count the number of NE-paths to (n, n), such that in every prefix, the number of N's and the number of E's differ by more than k, where $k \geq n/2$.

4.4.16   Count the number of NE-paths to (n, n), such that in every prefix, the number of N's and the number of E's differ by at most k, where $k \geq n/2$.

4.4.17   A fair coin is tossed $2n$ times. What is the probability that the number of heads equals the number of tails at the end of the sequence, and never exceeds it before then?

GLOSSARY

algebraic proof: a proof achieved by transforming expressions by substitutions and arithmetic operations.

analytic function: a real function that has an nth derivative for all $n > 0$.

binary tree: a rooted tree such that each vertex has a possible left-child, a possible right-child, and no other children.

binomial coefficient: formally, a coefficient of x^k in the expansion of $(1 + x)^n$; its value is the same as $\binom{n}{k}$, the combination coefficient.

binomial inversion: a transformation on sequences; it is its own inverse.

Catalan sequence: the sequence defined by the recurrence
$$
\begin{aligned}
c_0 &= 1; & &\text{initial value} \\
c_n &= c_0 c_{n-1} + c_1 c_{n-2} + \cdots + c_{n-1} c_0 & &\text{for } n \geq 1
\end{aligned}
$$

combination coefficient $\binom{n}{k}$:   the number of ways to choose a subset of k elements from a set of size n.

combinatorial proof: a proof achieved by exhibiting a model in which two expressions count the same thing.

derangement: a permutation that leaves no objects fixed.

derangement number D_n: the number of possible derangements of n objects.

event for a probability measure: a subset of the sample space.

event space for a probability measure: the set of all subsets of the sample space.

expected value of a discrete random variable X: a formal model for a probabilistically weighted average.

integer interval $[1 : n]$:   the set $\{1, 2, \ldots, n\}$.

left subtree of a vertex v of a binary tree: the left child of v and all its descendants.

mean of a list of numbers: the sum, divided by the length of the list.

mean of a random variable: synonym for expected value.

Pascal's recurrence: the recurrence
$$
\binom{n}{0} = 1 \quad \text{for all } n \geq 0 \qquad \binom{0}{k} = 0 \quad \text{for all } k \geq 1
$$
$$
\binom{n}{k} = \binom{n-1}{k-1} + \binom{n-1}{k} \quad \text{for } n \geq 1
$$

Pascal's triangle: a triangle formed from the non-zero binomial coefficients.

probability measure: a function on an event space that satisfies certain axioms.

random variable: a variable on a sample space with a probability measure.

right subtree of a vertex v of a binary tree: the right child of v and all its descendants.

sample space: a set with a probability measure on it.

standard deviation of a random variable: the square root of the variance.

unbiased estimator of a statistic: a statistic that when applied to a sample has an expected value equal to the true value for the entire sample space.

upper bound for a subset S of a poset: an element that dominates every element of S.

variance: a measure of the distribution of values of a random variable or of some characteristic of a sample of data.

well-nested strings of parentheses: strings with as many right parentheses as left, such that in every prefix, the number of left parentheses is at least as large as the number of right parentheses.

Chapter 5

Partitions and Permutations

5.1 Stirling Subset Numbers

5.2 Stirling Cycle Numbers

5.3 Inversions and Ascents

5.4 Derangements

5.5 Exponential Generating Functions

5.6 Posets and Lattices

The principal focus of Chapter 5 is counting the partitions and permutations of a set. An immediate connection between the partitions and the permutations of a set is that a permutation partitions the objects according to their cycles in its disjoint cycle form. This connection is of great importance within the algebraic counting methods of Chapter 9. In this chapter, we see that the two *Stirling recursions*, one used to count partitions and the other to count permutations, are both quite similar to Pascal's recursion for combination coefficients. In establishing Pascal's recursion and other identities for binomial coefficients, we have seen that it is generally possible to proceed from an algebraic expression involving factorials and/or falling powers. The virtue of the combinatorial proofs for such identities is the intuitive appeal they embody, which stems from their pertaining closely to the identification of binomial coefficients with counting selections from a set. In establishing the Stirling recursion and various other identities for counting the partitions and permutations of a set, virtue (of intuitive modeling) becomes a necessity, since the Stirling numbers have no closed algebraic formulas. A secondary topic of this chapter is partially ordered sets, known familiarly as *posets*, some of which have sufficient structure to be what are called *lattices*. Some of the most interesting lattices arise in connection with partitions and permutations, and a final section of this chapter considers various posets encountered in our explorations of counting methods from the perspective of their structure as lattices.

5.1   STIRLING SUBSET NUMBERS

Stirling subset numbers count the number of ways that a set can be partitioned. They satisfy a recurrence similar to Pascal's recurrence for binomial coefficients, and they can be placed into a triangle similar to Pascal's triangle. The Stirling subset numbers also satisfy a number of identities analogous to the identities for binomial coefficients in §4.1. We now recall some definitions and an inclusion-exclusion formula for Stirling numbers.

REVIEW FROM §1.6:

- A **partition** of a set S is a family of mutually disjoint non-empty subsets whose union is S.

- The **Stirling subset number** $\left\{ {n \atop k} \right\}$ is the number of ways to partition a set of n distinct objects into k non-empty non-distinct cells.

REVIEW FROM §3.6:

- **Theorem 3.6.4**. Let n and k be a pair of non-negative integers. Then

$$\left\{ {n \atop k} \right\} k! \;=\; \sum_{j=0}^{k} (-1)^j \binom{k}{j} (k-j)^n$$

Our immediate concern is careful attention to three properties within the definition of a partition: non-distinctness of the cells, non-emptiness of the cells, and distinctness of the objects of the set.

Non-Distinctness of Cells of a Partition

Non-distinctness of the cells means that they are regarded as a set, not as a list. If a given partition has k labeled cells, there are $k!$ ways to list them.

Example 5.1.1:   We consider an *ad hoc* calculation of the Stirling number

$$\left\{ {4 \atop 2} \right\}$$

We observe that the set $\{a, b, c, d\}$ can be partitioned into two subsets of two objects each in the following three ways.

$$1.\, \{a,b\}, \{c,d\} \qquad 2.\, \{a,c\}, \{b,d\} \qquad 3.\, \{a,d\}, \{b,c\}$$

Changing the order in which the subsets are listed in a representation of a *partition* does not change the partition. Thus, the partition

$$\{c,d\}, \{a,b\}$$

is identical to partition (1) above. By way of contrast, if the objects were to be distributed into compartments distinguished by pre-assigned names or, equivalently, by their order in the listing, each of the partitions above would correspond to two such distributions, for a total of six ways to distribute four objects into two distinct compartments with a 2-2 distribution.

In addition to partitions (1), (2), and (3), given above, of four objects into two parts of two objects each, the Stirling subset number $\left\{ {4 \atop 2} \right\}$ also counts the partitions of four objects into subsets of sizes one and three, i.e., these four partitions:

$$4.\ \{a\}, \{b, c, d\} \qquad 5.\ \{b\}, \{a, c, d\} \qquad 6.\ \{c\}, \{a, b, d\} \qquad 7.\ \{d\}, \{a, b, c\}$$

The two compartments within each of these four partitions could be ordered in two ways, if they were distinct. This would give a total of eight distributions into distinct compartments with a 1-3 (or 3-1) distribution.

It follows that, altogether, there are

$$\left\{ {4 \atop 2} \right\} \ = \ 3 + 4 \ = \ 7$$

distributions into non-distinct calls, and

$$\left\{ {4 \atop 2} \right\} \ 2! \ = \ 6 + 8 \ = \ 14$$

distributions into non-distinct calls. These two results are consistent with an application of Theorem 3.6.4.

$$\begin{aligned}
\left\{ {4 \atop 2} \right\} \ 2! \ &= \ \sum_{j=0}^{2} (-1)^j \binom{2}{j} (2 - j)^4 \\
&= \ (-1)^0 \binom{2}{0}(2 - 0)^4 \ + \ (-1)^1 \binom{2}{1}(2 - 1)^4 \ + \ (-1)^2 \binom{2}{2}(2 - 2)^4 \\
&= \ 1 \cdot 1 \cdot 2^4 \ + \ (-1) \cdot 2 \cdot 1^4 \ + \ 1 \cdot 1 \cdot 0^4 \\
&= \ 16 \ - \ 2 \ + \ 0 \ = \ 14
\end{aligned}$$

The following proposition summarizes this part of the discussion.

Proposition 5.1.1. *The number of ways to distribute n distinct objects into k distinct boxes with none left empty is*

$$k! \left\{ {n \atop k} \right\}$$

Proof: After partitioning the objects into

$$\left\{ {n \atop k} \right\}$$

non-distinct non-empty cells, we can assign k distinct labels to the k cells in $k!$ ways. $\diamond$

Every Cell of a Partition is Non-Empty

Specifying *non-emptiness of the cells* of a partition into k cells is consistent with the everyday notion of dividing a set into parts. (For instance, when Julius Caesar wrote in *The Gallic Wars* that all Gaul is divided into three parts, he meant non-empty parts.)

Example 5.1.1, continued: If one of the two cells of a distribution of the set $\{a, b, c, d\}$ could be left empty, there would be a total of 8 ways to separate the four objects into two parts, which would include the distribution

$$\{a, b, c, d\}\, \{\,\}$$

If the cells were also distinct, there would be twice as many, for a total of 16 ways. Such a distribution is achievable by assigning one of the two compartment names to each of the four objects, for which, of course, there are $2^4 = 16$ ways.

Proposition 5.1.2. *The number of ways to distribute n distinct objects into k distinct boxes with some possibly left empty is*

$$k^n \qquad\qquad\qquad \diamondsuit$$

Distinctness of Objects

Distinctness of the objects is a critical feature, since two distributions of a multiset of indistinguishable objects would differ only in the numbers of objects in the cells.

Example 5.1.1, continued: The only two possible partitions of four indistinguishable objects into two non-empty non-distinct cells have the following forms:

$$\{a\}\,\{a, a, a\} \quad \text{and} \quad \{a, a\}\,\{a, a\}$$

They are equivalent to the integer partitions

$$4 = 1 + 3 \quad \text{and} \quad 4 = 2 + 2$$

In general, partitioning n indistinguishable objects into k indistinguishable cells is equivalent to partitioning the integer n into a sum of k parts, a topic that is explored further in §9.4.

On the other hand, if the two cells are distinct, then the distribution of four non-distinct objects amounts to choosing four cells from a set of two distinct cells, with repetitions allowed. We developed a counting formula for combinations with repetitions in Chapter 0.

Proposition 5.1.3. *The number of ways to distribute n non-distinct objects into k distinct boxes with some possibly left empty is*

$$\binom{k+n-1}{n}$$

Proof: This is equivalent to choosing n objects from a set of k with repetitions allowed. The formula was derived in conjunction with Corollary 0.4.5.   ◇

The Type of a Partition

Clearly, the sum of the sizes of the cells of a partition of a set of n objects must be equal to n.

DEFINITION: An arrangement of the sizes of the cells into non-increasing order is called the ***type of a partition***.

Example 5.1.1, continued: The partitions

$$1.\ \{a,b\},\{c,d\} \qquad 2.\ \{a,c\},\{b,d\} \qquad 3.\ \{a,d\},\{b,c\}$$

are of type 22, and the partitions

$$4.\ \{a\},\{b,c,d\} \qquad 5.\ \{b\},\{a,c,d\} \qquad 6.\ \{c\},\{a,b,d\} \qquad 7.\ \{d\},\{a,b,c\}$$

are of type 31.

Stirling's Subset Number Recurrence

A recurrence similar to Pascal's recurrence provides a systematic means to calculate a Stirling subset number $\left\{{n \atop k}\right\}$, without resorting to separate counts for each partition type. Since there is no simple closed formula for a Stirling number, unlike the situation for a binomial coefficient, there is no simple algebraic proof, and we resort to a combinatorial proof.

Proposition 5.1.4 [*Stirling's recurrence for subset numbers*]. *The Stirling subset numbers satisfy the following recurrence:*

$$\left\{{0 \atop k}\right\} = (k=0) \qquad\qquad \left\{{n \atop 0}\right\} = (n=0)$$

$$\left\{{n \atop k}\right\} = \left\{{n-1 \atop k-1}\right\} + k\left\{{n-1 \atop k}\right\} \qquad \text{for } n \geq 1$$

Combinatorial Proof: The initial conditions are clear.

The recursion is verified by splitting the partitions of the integer interval $[1:n]$ into two kinds, as per the Method of Distinguished Element, which was used with Pascal's recursion in §1.3. The first kind contains every partition in which the integer

n gets a cell to itself. Since the other $n - 1$ integers must then be partitioned into $k - 1$ non-empty cells, there are

$$\left\{ \begin{matrix} n - 1 \\ k - 1 \end{matrix} \right\}$$

cases of the first kind.

In the second kind, the set $[1 : n - 1]$ is partitioned into k non-empty cells, and then one of those k cells is selected as the cell for the integer n. There are

$$k \left\{ \begin{matrix} n - 1 \\ k \end{matrix} \right\}$$

cases of the second kind. The sum of the numbers of cases in these two kinds is the total number of partitions.   ◇

Stirling's Triangle for Subset Numbers

The recurrence for the Stirling subset numbers leads to a triangle similar to Pascal's triangle, called *Stirling's triangle for subset numbers*. It appears as Table 5.1.1.

Table 5.1.1   **Stirling's triangle for values of $\left\{ {n \atop k} \right\}$.**

n	$\left\{{n \atop 0}\right\}$	$\left\{{n \atop 1}\right\}$	$\left\{{n \atop 2}\right\}$	$\left\{{n \atop 3}\right\}$	$\left\{{n \atop 4}\right\}$	$\left\{{n \atop 5}\right\}$	$\left\{{n \atop 6}\right\}$	B_n
0	1							1
1	0	1						1
2	0	1	1					2
3	0	1	3	1				5
4	0	1	7	6	1			15
5	0	1	15	25	10	1		52
6	0	1	31	90	65	15	1	203

The rest of this section is devoted to the development of formulas for Stirling's subset triangle that are analogous to the formulas of §4.1 for Pascal's triangle for binomial coefficients.

Rows Are Log-Concave

We recall from §1.5 that each row of Pascal's triangle rises to its maximum among the non-zero entries and then falls off. That is, the rows are unimodal. The rows of Stirling's triangle for subset numbers share the property of unimodality.

REVIEW FROM §1.5:

- A sequence $\langle x_n \rangle$ is a *log-concave sequence* if

$$x_{n-1}\, x_{n+1} \leq x_n^2, \quad \text{for all } n \geq 1$$

- A log-concave sequence is unimodal.

Example 5.1.2:   Figure 5.1.1 illustrates the unimodality of row 6. In fact, every row of Stirling's triangle for subset numbers (see Table 5.1.1) is unimodal.

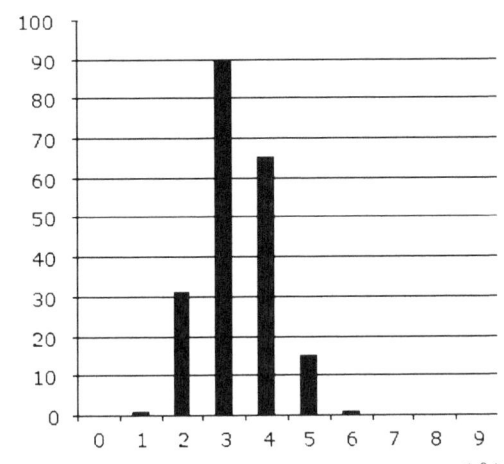

Figure 5.1.1   Graph of the values of $\left\{ {6 \atop k} \right\}$.

Lemma 5.1.5.   Let $\langle x_n \rangle$ be a *log-concave sequence*. Then

$$x_{n-2}\, x_{n+1} \leq x_{n-1}\, x_n$$

Proof: The log-concavity inequality is applied twice.

$$
\begin{aligned}
x_{n-2}\, x_{n+1} &\leq x_{n-2} \cdot \frac{x_n^{\,2}}{x_{n-1}} && \text{since } x_{n-1}\, x_{n+1} \leq x_n^2 \\
&= \frac{x_{n-2}}{x_{n-1}} \cdot x_n^2 \\
&\leq \frac{x_{n-1}}{x_n} \cdot x_n^2 && \text{since } x_{n-2}\, x_n \leq x_{n-1}^2 \\
&= x_{n-1}\, x_n && \Diamond
\end{aligned}
$$

Proposition 5.1.6.   For all $n \geq 0$, the sequence of Stirling subset numbers

$$\left\{ {n \atop 0} \right\},\ \left\{ {n \atop 1} \right\},\ \dots,\ \left\{ {n \atop n} \right\}$$

is log-concave. That is,

$$\left\{ {n \atop k-1} \right\} \left\{ {n \atop k+1} \right\} \leq \left\{ {n \atop k} \right\} \left\{ {n \atop k} \right\}$$

Proof: This is an algebraic proof by induction on n.

BASIS: Rows 0 and 1 are surely log-concave.

IND HYP: Assume that row $n - 1$ is log-concave.

IND STEP: Under Stirling's subset recurrence, the product

$$\left\{ {n \atop k-1} \right\} \left\{ {n \atop k+1} \right\}$$

has the expansion

$$\left(\left\{ {n-1 \atop k-2} \right\} + (k-1) \left\{ {n-1 \atop k-1} \right\} \right) \times \left(\left\{ {n-1 \atop k} \right\} + (k+1) \left\{ {n-1 \atop k+1} \right\} \right)$$

$$= \left\{ {n-1 \atop k-2} \right\} \left\{ {n-1 \atop k} \right\} + (k^2 - 1) \left\{ {n-1 \atop k-1} \right\} \left\{ {n-1 \atop k+1} \right\}$$

$$+ (k-1) \left\{ {n-1 \atop k-1} \right\} \left\{ {n-1 \atop k} \right\} + (k+1) \left\{ {n-1 \atop k-2} \right\} \left\{ {n-1 \atop k+1} \right\}$$

to which log-concavity and Lemma 5.1.5 are applied, under the induction hypothesis

$$\leq \left\{ {n-1 \atop k-1} \right\}^2 + (k^2 - 1) \left\{ {n-1 \atop k} \right\}^2$$

$$+ (k-1) \left\{ {n-1 \atop k-1} \right\} \left\{ {n-1 \atop k} \right\} + (k+1) \left\{ {n-1 \atop k-1} \right\} \left\{ {n-1 \atop k} \right\}$$

$$\leq \left\{ {n-1 \atop k-1} \right\}^2 + k^2 \left\{ {n-1 \atop k} \right\}^2 + 2k \left\{ {n-1 \atop k-1} \right\} \left\{ {n-1 \atop k} \right\}$$

$$= \left(\left\{ {n-1 \atop k-1} \right\} + k \left\{ {n-1 \atop k} \right\} \right)^2$$

$$= \left\{ {n \atop k} \right\}^2 \qquad\qquad\qquad \diamond$$

Bell Numbers

DEFINITION: The **Bell number** B_n is the number of partitions of a set of n distinct objects.

Thus, the nth Bell number is the sum

$$B_n = \sum_{k=0}^{n} \left\{ {n \atop k} \right\}$$

of row n of Stirling's triangle for subset numbers.

Theorem 5.1.7. *The Bell numbers satisfy the recurrence*

$$B_0 = 1$$
$$B_n = \sum_{k=0}^{n-1} \binom{n-1}{k} B_k \quad \text{for } n \geq 1$$

Proof: This proof has combinatorial steps and algebraic steps. The initial condition

$$B_0 = \left\{ {n \atop 0} \right\} = 1$$

is clearly satisfied.

For $n \geq 1$, consider the case in which there are k other objects in the cell of a partition of $[1 : n]$ that contains the number n. There are

$$\binom{n-1}{k}$$

ways to select these k numbers and then B_{n-k-1} ways to partition the remaining $n - k - 1$ numbers. Thus, the total number of partitions of n objects is

$$B_n = \sum_{k=0}^{n-1} \binom{n-1}{k} B_{n-k-1}$$

which is transformable, by symmetry of binomial coefficients, to

$$= \sum_{k=0}^{n-1} \binom{n-1}{n-k-1} B_{n-k-1}$$

Reversing the order of summation yields the conclusion

$$B_n = \sum_{k=0}^{n-1} \binom{n-1}{k} B_k \qquad\qquad \Diamond$$

Example 5.1.3:   Table 5.1.1 provides the Bell numbers

B_0	B_1	B_2	B_3	B_4	B_5	B_6
1	1	2	5	15	52	203

We observe, for instance, that

$$\binom{3}{0} B_0 + \binom{3}{1} B_1 + \binom{3}{2} B_2 + \binom{3}{3} B_3$$
$$= 1 \cdot 1 + 3 \cdot 1 + 3 \cdot 2 + 1 \cdot 5$$
$$= 15 = B_4$$

Column-Sum Formulas

There are two readily accessible summation formulas for column c of the triangle for Stirling subset numbers. They both assert that a weighted partial sum of the entries in column c can be found in column $c + 1$. In the two formulas, the weightings differ.

Proposition 5.1.8. *Let n and c be non-negative integers. Then*

$$\left\{ {n + 1 \atop c + 1} \right\} = \sum_{k=0}^{n} \binom{n}{k} \left\{ {k \atop c} \right\}$$

Proof: In partitioning the $n + 1$ numbers of the integer interval $[1 : n + 1]$ into $c + 1$ cells, there are

$$\binom{n}{k}$$

ways to select $n - k$ other numbers to be in the same cell as the number $n + 1$ and then

$$\left\{ {k \atop c} \right\}$$

ways to partition the remaining k numbers into c additional cells. $\Diamond$

Example 5.1.4:   In column $c = 1$ of the triangle for Stirling subset numbers, all the non-zero entries are 1's. Thus, Proposition 5.1.8 takes the form

$$\left\{ {n + 1 \atop 2} \right\} = \sum_{k=1}^{n} \binom{n}{k} \left\{ {k \atop 1} \right\} = \sum_{k=1}^{n} \binom{n}{k} = 2^n - 1$$

Example 5.1.5:   In column $c = 2$ of the Stirling triangle for subset numbers, there appear the consecutive entries

$$\left\{ {2 \atop 2} \right\} = 1 \quad \left\{ {3 \atop 2} \right\} = 3 \quad \left\{ {4 \atop 2} \right\} = 7$$

In row 4 of Pascal's triangle, there are the consecutive entries

$$\binom{4}{2} = 6 \quad \binom{4}{3} = 4 \quad \binom{4}{4} = 1$$

Proposition 5.1.8 asserts for this case that

$$\left\{ {5 \atop 3} \right\} = \sum_{k=2}^{4} \binom{4}{k} \left\{ {k \atop 2} \right\}$$

$$= \binom{4}{2} \left\{ {2 \atop 2} \right\} + \binom{4}{3} \left\{ {3 \atop 2} \right\} + \binom{4}{4} \left\{ {4 \atop 2} \right\}$$

$$= 6 \cdot 1 + 4 \cdot 3 + 1 \cdot 7 = 25$$

The sum in Proposition 5.1.8 can be visualized as a dot product of a row of Pascal's triangle with a column of Stirling's triangle.

$$
\begin{array}{c|ccc}
r \to & 2 & 3 & 4 \\
\hline
\dbinom{4}{r} & 6 & 4 & 1
\end{array}
\qquad
\begin{array}{c|cc}
n \downarrow & \left[\!\begin{smallmatrix} n \\ 2 \end{smallmatrix}\!\right] & \left[\!\begin{smallmatrix} n \\ 3 \end{smallmatrix}\!\right] \\
\hline
2 & 1 & \\
3 & 3 & \\
4 & 7 & \\
5 & & 25
\end{array}
$$

$$6 \cdot 1 + 4 \cdot 3 + 1 \cdot 7 = 25$$

Proposition 5.1.9. *Let n and c be non-negative integers. Then*

$$
\left\{ \begin{matrix} n+1 \\ c+1 \end{matrix} \right\} \;=\; \sum_{k=0}^{n} (c+1)^{n-k} \left\{ \begin{matrix} k \\ c \end{matrix} \right\}
$$

Proof: By induction.

BASIS: The equation is clearly true when $n = 0$.

IND HYP: Assume, for inductive purpose, that

$$
\left\{ \begin{matrix} n \\ c+1 \end{matrix} \right\} \;=\; \sum_{k=0}^{n-1} (c+1)^{n-k-1} \left\{ \begin{matrix} k \\ c \end{matrix} \right\}
$$

IND STEP: Then

$$
\begin{aligned}
\left\{ \begin{matrix} n+1 \\ c+1 \end{matrix} \right\} &= \left\{ \begin{matrix} n \\ c \end{matrix} \right\} + (c+1) \left\{ \begin{matrix} n \\ c+1 \end{matrix} \right\} & \text{(Stirling's recursion)} \\
&= \left\{ \begin{matrix} n \\ c \end{matrix} \right\} + (c+1) \sum_{k=0}^{n-1} (c+1)^{n-k-1} \left\{ \begin{matrix} k \\ c \end{matrix} \right\} & \text{(inductive hypothesis)} \\
&= \left\{ \begin{matrix} n \\ c \end{matrix} \right\} + \sum_{k=0}^{n-1} (c+1)^{n-k} \left\{ \begin{matrix} k \\ c \end{matrix} \right\} \\
&= \sum_{k=0}^{n} (c+1)^{n-k} \left\{ \begin{matrix} k \\ c \end{matrix} \right\} & \diamond
\end{aligned}
$$

Example 5.1.6:   Proposition 5.1.9 implies that

$$
\begin{aligned}
\left\{ \begin{matrix} 5 \\ 3 \end{matrix} \right\} &= \sum_{k=2}^{4} 3^{4-k} \left\{ \begin{matrix} k \\ 2 \end{matrix} \right\} \\
&= 3^{4-2} \left\{ \begin{matrix} 2 \\ 2 \end{matrix} \right\} + 3^{4-3} \left\{ \begin{matrix} 3 \\ 2 \end{matrix} \right\} + 3^{4-4} \left\{ \begin{matrix} 4 \\ 2 \end{matrix} \right\} \\
&= 3^2 \cdot 1 + 3^1 \cdot 3 + 3^0 \cdot 7 = 25
\end{aligned}
$$

Southeast Diagonal Sum

Along a southeast diagonal from column 0 to column c, multiply each entry by its column number and take the sum. This equals the number immediately below the last entry in that diagonal.

Proposition 5.1.10. *Let n and c be non-negative integers. Then*

$$\left\{ {n+c+1 \atop c} \right\} = \sum_{k=0}^{c} k \left\{ {n+k \atop k} \right\}$$

Proof: Again by induction.

BASIS: The equation is clearly true for all $n \geq 0$ when $c = 0$.

IND HYP: Assume for all $n \geq 0$ that

$$\left\{ {n+c \atop c-1} \right\} = \sum_{k=0}^{c-1} k \left\{ {n+k \atop k} \right\}$$

IND STEP: Then

$$\left\{ {n+c+1 \atop c} \right\} = \left\{ {n+c \atop c-1} \right\} + c \left\{ {n+c \atop c} \right\} \qquad \text{(Stirling's recursion)}$$

$$= \sum_{k=0}^{c-1} k \left\{ {n+k \atop k} \right\} + c \left\{ {n+c \atop c} \right\} \qquad \text{(ind hyp)}$$

$$= \sum_{k=0}^{c} k \left\{ {n+k \atop k} \right\} \qquad\qquad\qquad \diamondsuit$$

Example 5.1.7:   The sum in Proposition 5.1.10 can be visualized as a dot product of a southeast diagonal of Stirling's triangle with a vector of column numbers.

$n \downarrow$	$\left\{ {n \atop 1} \right\}$	$\left\{ {n \atop 2} \right\}$	$\left\{ {n \atop 3} \right\}$
3	1		
4		7	
5			25
6			90

$$1 \cdot 1 + 2 \cdot 7 + 3 \cdot 25 = 90$$

Stirling Numbers of the Second Kind

REVIEW FROM §1.6: The *Stirling numbers of the second kind* were defined as the coefficients $S_{n,k}$ in the sum

$$x^n = \sum_{k=0}^{n} S_{n,k} x^{\underline{k}}$$

Proposition 5.1.11. *For all non-negative integers n and k,*

$$S_{n,k} = \left\{ {n \atop k} \right\} \tag{5.1.1}$$

Proof: We use the Stirling subset-number recurrence, as verified in Proposition 5.1.4.

$$\left\{ {0 \atop k} \right\} = (k = 0) \qquad\qquad \left\{ {n \atop 0} \right\} = (n = 0)$$

$$\left\{ {n \atop k} \right\} = \left\{ {n-1 \atop k-1} \right\} + k \left\{ {n-1 \atop k} \right\} \quad \text{for } n \geq 1$$

It is sufficient to show that the Stirling numbers of the second kind satisfy the same recurrence. The initial conditions

$$S_{0,k} = (k = 0) \qquad \text{and} \qquad S_{n,0} = (n = 0)$$

hold, because

$$x^0 = 1x^{\underline{0}}$$

and because the constant term of the expansion

$$x^n = \sum_{k=0}^{n} S_{n,k} x^{\underline{k}} \tag{5.1.2}$$

is 0, unless $k = 0$.

The Stirling numbers $S_{n-1,k}$ of the second kind are defined with the specification

$$x^{n-1} = \sum_{k=0}^{n-1} S_{n-1,k} x^{\underline{k}} \tag{5.1.3}$$

Accordingly,

$$\begin{aligned}
x^n &= x \cdot x^{n-1} \\
&= x \cdot \sum_{k=0}^{n-1} S_{n-1,k}\, x^{\underline{k}} \qquad \text{(by (5.1.2))} \\
&= \sum_{k=0}^{n-1} S_{n-1,k}\, x^{\underline{k}} \cdot x \\
&= \sum_{k=0}^{n-1} S_{n-1,k}\, x^{\underline{k}} \cdot (x - k) + \sum_{k=0}^{n-1} S_{n-1,k}\, x^{\underline{k}} \cdot k \\
&= \sum_{k=0}^{n-1} S_{n-1,k}\, x^{\underline{k+1}} + \sum_{k=0}^{n-1} k S_{n-1,k}\, x^{\underline{k}} \\
&= \sum_{k=1}^{n} S_{n-1,k-1}\, x^{\underline{k}} + \sum_{k=0}^{n-1} k S_{n-1,k}\, x^{\underline{k}} \\
&= \sum_{k=0}^{n} (S_{n-1,k-1} + k S_{n-1,k})\, x^{\underline{k}} \tag{5.1.4}
\end{aligned}$$

Table 5.1.2 **Some Basic Formulas for Stirling Subset Numbers**

Stirling's recurrence:

$$\left\{ \begin{matrix} 0 \\ k \end{matrix} \right\} = (k = 0) \qquad\qquad \left\{ \begin{matrix} n \\ 0 \end{matrix} \right\} = (n = 0)$$

$$\left\{ \begin{matrix} n \\ k \end{matrix} \right\} = \left\{ \begin{matrix} n-1 \\ k-1 \end{matrix} \right\} + k \left\{ \begin{matrix} n-1 \\ k \end{matrix} \right\} \quad \text{for } n \geq 1 \tag{5.1.5}$$

Special values:

$$\left\{ \begin{matrix} n \\ 1 \end{matrix} \right\} = (n > 0) \qquad \left\{ \begin{matrix} n \\ 2 \end{matrix} \right\} = 2^{n-1} - 1 \text{ (for } n \geq 1) \qquad \left\{ \begin{matrix} n \\ n \end{matrix} \right\} = 1$$

Converting ordinary powers to falling powers:

$$x^n = \sum_{k=0}^{n} \left\{ \begin{matrix} n \\ k \end{matrix} \right\} x^{\underline{k}} \tag{5.1.6}$$

Using binomial coefficients to calculate Stirling subset numbers:

$$\left\{ \begin{matrix} n \\ k \end{matrix} \right\} k! = \sum_{j=0}^{k} (-1)^j \binom{k}{j} (m-j)^n = \sum_{j=0}^{k} (-1)^{k-j} \binom{k}{j} j^n \tag{5.1.7}$$

Bell numbers:

$$B_n = \sum_{k=0}^{n} \left\{ \begin{matrix} n \\ k \end{matrix} \right\} \tag{5.1.8}$$

$$B_0 = 1; \qquad B_n = \sum_{k=0}^{n-1} \binom{n-1}{k} B_k \quad \text{for } n \geq 1 \tag{5.1.9}$$

Column-sum formulas:

$$\left\{ \begin{matrix} n+1 \\ c+1 \end{matrix} \right\} = \sum_{k=0}^{n} \binom{n}{k} \left\{ \begin{matrix} k \\ c \end{matrix} \right\} \tag{5.1.10}$$

$$\left\{ \begin{matrix} n+1 \\ c+1 \end{matrix} \right\} = \sum_{k=0}^{n} (c+1)^{n-k} \left\{ \begin{matrix} k \\ c \end{matrix} \right\} \tag{5.1.11}$$

SE diagonal-sum formula:

$$\left\{ \begin{matrix} n+c+1 \\ c \end{matrix} \right\} = \sum_{k=0}^{c} k \left\{ \begin{matrix} n+k \\ k \end{matrix} \right\} \tag{5.1.12}$$

Since $x^{\underline{k}}$ must have the same coefficient in the two expansions (5.1.2) and (5.1.4) of x^n, it follows that

$$S_{n,k} \;=\; S_{n-1,k-1} + kS_{n-1,k}$$

Thus, the Stirling numbers of the second kind have the same recurrence as the Stirling subset numbers, which implies that they have the same values. $\diamondsuit$

Example 5.1.8:   The following recursive calculation of values of the Stirling number $S_{n,k}$ of the second kind illustrates how these numbers conform to Stirling's subset recursion.

$$
\begin{aligned}
x \;&=\; x^{\underline{1}} \\
\Rightarrow x^2 \;&=\; x \cdot x^{\underline{1}} \;=\; x^{\underline{1}}(x-1) + x^{\underline{1}} \;=\; x^{\underline{2}} + x^{\underline{1}} \\
\Rightarrow x^3 \;&=\; x \cdot x^{\underline{2}} + x \cdot x^{\underline{1}} \;=\; x^{\underline{2}}(x-2) + 2x^{\underline{2}} + x^{\underline{2}} + x^{\underline{1}} \\
\;&=\; x^{\underline{3}} + 3x^{\underline{2}} + x^{\underline{1}} \\
\Rightarrow x^4 \;&=\; x \cdot x^{\underline{3}} + 3x \cdot x^{\underline{2}} + x \cdot x^{\underline{1}} \\
\;&=\; \left[x^{\underline{3}}(x-3) + 3x^{\underline{3}}\right] + 3\left[x^{\underline{2}}(x-2) + 2x^{\underline{2}}\right] + \left[x^{\underline{1}}(x-1) + x^{\underline{1}}\right] \\
\;&=\; \left[x^{\underline{4}} + 3x^{\underline{3}}\right] + 3\left[x^{\underline{3}} + 2x^{\underline{2}}\right] + \left[x^{\underline{2}} + x^{\underline{1}}\right] \\
\;&=\; x^{\underline{4}} + \left[3x^{\underline{3}} + 3x^{\underline{3}}\right] + \left[6x^{\underline{2}} + x^{\underline{2}}\right] + x^{\underline{1}} \\
\;&=\; x^{\underline{4}} + 6x^{\underline{3}} + 7x^{\underline{2}} + x^{\underline{1}}
\end{aligned}
$$

EXERCISES for Section 5.1

Exercises 5.1.1 through 5.1.6 probe some nuances of the definition of Stirling subset numbers.

5.1.1   Give an expression for the number of distributions of n distinct objects into k non-distinct cells with exactly r cells left empty.

5.1.2$^\text{S}$ Give an expression for the number of distributions of n distinct objects into k non-distinct cells with up to r cells left empty.

5.1.3   Give an expression for the number of distributions of n distinct objects into k non-distinct cells with arbitrarily many cells left empty.

5.1.4   Give an expression for the number of distributions of n distinct objects into k distinct cells with exactly r cells left empty.

5.1.5$^\text{S}$ Give an expression for the number of distributions of n distinct objects into k distinct cells with up to r cells left empty.

5.1.6   Give an expression for the number of distributions of n distinct objects into k distinct cells with arbitrarily many cells left empty.

In each of the Exercises 5.1.7 through 5.1.14, using only Table 5.1.1 as given values, calculate the values of the following additional Stirling subset numbers.

5.1.7 $\left\{ {7 \atop 3} \right\}$ 5.1.8 $\left\{ {7 \atop 4} \right\}$ 5.1.9^{S} $\left\{ {7 \atop 5} \right\}$ 5.1.10 $\left\{ {7 \atop 6} \right\}$

5.1.11 $\left\{ {8 \atop 2} \right\}$ 5.1.12 $\left\{ {8 \atop 3} \right\}$ 5.1.13 $\left\{ {8 \atop 4} \right\}$ 5.1.14 $\left\{ {8 \atop 5} \right\}$

In each of the Exercises 5.1.15 through 5.1.22, confirm the column-sum formula of Proposition 5.1.8 for the following Stirling subset numbers.

5.1.15 $\left\{ {7 \atop 3} \right\}$ 5.1.16 $\left\{ {7 \atop 4} \right\}$ 5.1.17^{S} $\left\{ {7 \atop 5} \right\}$ 5.1.18 $\left\{ {7 \atop 6} \right\}$

5.1.19 $\left\{ {8 \atop 2} \right\}$ 5.1.20 $\left\{ {8 \atop 3} \right\}$ 5.1.21 $\left\{ {8 \atop 4} \right\}$ 5.1.22 $\left\{ {8 \atop 5} \right\}$

In each of the Exercises 5.1.23 through 5.1.30, confirm the column-sum formula of Proposition 5.1.9 for the following Stirling subset numbers.

5.1.23 $\left\{ {7 \atop 3} \right\}$ 5.1.24 $\left\{ {7 \atop 4} \right\}$ 5.1.25^{S} $\left\{ {7 \atop 5} \right\}$ 5.1.26 $\left\{ {7 \atop 6} \right\}$

5.1.27 $\left\{ {8 \atop 2} \right\}$ 5.1.28 $\left\{ {8 \atop 3} \right\}$ 5.1.29 $\left\{ {8 \atop 4} \right\}$ 5.1.30 $\left\{ {8 \atop 5} \right\}$

In each of the Exercises 5.1.31 through 5.1.38, confirm the diagonal-sum formula of Proposition 5.1.10 for the following Stirling subset numbers.

5.1.31 $\left\{ {7 \atop 3} \right\}$ 5.1.32 $\left\{ {7 \atop 4} \right\}$ 5.1.33^{S} $\left\{ {7 \atop 5} \right\}$ 5.1.34 $\left\{ {7 \atop 6} \right\}$

5.1.35 $\left\{ {8 \atop 2} \right\}$ 5.1.36 $\left\{ {8 \atop 3} \right\}$ 5.1.37 $\left\{ {8 \atop 4} \right\}$ 5.1.38 $\left\{ {8 \atop 5} \right\}$

5.1.39 Calculate the Bell number B_7.

5.1.40 Calculate the Bell number B_8.

5.2   STIRLING CYCLE NUMBERS

Stirling cycle numbers count the number of possible partitions of a set into cycles, in effect, the number of the permutations of the set. Like Stirling subset numbers, they satisfy a recurrence similar to Pascal's recurrence, and, in a further similarity, their non-zero entries form a triangle. This section derives some identities analogous to the identities for binomial coefficients and Stirling subset numbers.

REVIEW FROM §1.6:

- The **Stirling cycle number** $\left[\begin{smallmatrix} n \\ k \end{smallmatrix}\right]$ is the number of ways to partition n distinct objects into k non-empty non-distinct cycles.

- Since every permutation of a set of n objects can be represented as a composition of disjoint cycles, it follows that the **Stirling cycle number** $\left[\begin{smallmatrix} n \\ k \end{smallmatrix}\right]$ is the number of permutations with exactly k cycles.

In general,

$$\left[{n \atop k}\right] \geq \left\{{n \atop k}\right\}$$

since the number of ways to form a cycle from s objects already in a cell of a partition is $(s-1)!$. Thus, to calculate $\left[{n \atop k}\right]$, one could multiply the number of partitions of a given partition type $t_1 \, t_2 \, \cdots \, t_r$ by $(t_1 - 1)!(t_2 - 1)! \cdots (t_r - 1)!$ and sum over all such partitions.

Example 5.2.1:   The set $\{a, b, c, d\}$ can be partitioned into two cycles in 11 ways, which correspond to the 11 permutations of the set $\{1, 2, 3, 4\}$ with two cycles:

$$
\begin{array}{llll}
(a)(b\,c\,d) & (a)(b\,d\,c) & (b)(a\,c\,d) & (b)(a\,d\,c) \\
(c)(a\,b\,d) & (c)(a\,d\,b) & (d)(a\,b\,c) & (d)(a\,c\,b) \\
(a\,b)(c\,d) & (a\,c)(b\,d) & (a\,d)(b\,c) &
\end{array}
\qquad \left[{4 \atop 2}\right] = 11
$$

We observe that each of the four partitions of type 31 into cells corresponds to $(3-1)!(1-1)! = 2$ partitions into cycles. For instance, the partition

$$\{a\} \, \{b, c, d\}$$

corresponds to the two permutations

$$(a)(b\,c\,d) \qquad \text{and} \qquad (a)(b\,d\,c)$$

Each of the three partitions of type 22 into cells yields only $(2-1)!(2-1)! = 1$ partition into cycles. For instance, the partition

$$\{a\,b\}\{c\,d\}$$

yields only the permutation

$$(a\,b)(c\,d)$$

We observe, moreover, that

$$4 \cdot 2 + 3 \cdot 1 = 11$$

Non-Distinctness of the Cycles

Non-distinctness of the cycles means that changing the order in which its cycles are written does not change a permutation.

Example 5.2.2:   For instance, the disjoint cycle representations

$$(a\,b)(c\,d) \qquad \text{and} \qquad (c\,d)(a\,b)$$

are representations of the same permutation.

Remark: In the context of permutations, the notion of an empty cycle is meaningless. Moreover, although it does make sense to discuss a cyclic arrangement of the objects of a multiset, the notion of a cycle whose objects are not mutually distinct is regarded as undefined when studying permutations.

Stirling's Cycle Number Recurrence

As with Stirling subset numbers, a recurrence similar to Pascal's recurrence provides a systematic means to calculate a Stirling cycle number $\left[{n \atop k}\right]$, without resorting to separate counts for each partition type. Moreover, here too there is no simple algebraic proof, and we resort again to a combinatorial proof.

Proposition 5.2.1 [*Stirling's recurrence for cycle numbers*]. *The Stirling cycle numbers satisfy the following recurrence:*

$$\left[{0 \atop k}\right] = (k = 0) \qquad\qquad \left[{n \atop 0}\right] = (n = 0)$$

$$\left[{n \atop k}\right] = \left[{n-1 \atop k-1}\right] + (n-1)\left[{n-1 \atop k}\right] \qquad \text{for } n \geq 1$$

Combinatorial Proof: The initial conditions are clear.

As with Stirling subset numbers, the recursion is verified by splitting the permutations of the integer interval $[1:n]$ that have k cycles into two kinds. The first kind contains every permutation in which the number n gets a cycle to itself, and the other $n-1$ numbers are partitioned into $k-1$ non-empty cycles, so there are

$$\left[{n-1 \atop k-1}\right]$$

cases of the first kind. In the second kind, in which the number n does not have a cycle to itself, the other $n-1$ numbers are partitioned into k non-empty cycles, and then the number n is inserted immediately after some number j in one of those k cycles. There are

$$(n-1)\left[{n-1 \atop k}\right]$$

cases of the second kind, because there are, in total, $n-1$ other numbers after which the number n could be inserted. The sum of the numbers of cases in these two types is the total number of partitions of $[1:n]$ into k cycles.   ◇

Stirling's Triangle for Cycle Numbers

There is a triangle for the Stirling cycle numbers, like Pascal's triangle and the triangle for Stirling subset numbers. It appears as Table 5.2.1.

Table 5.2.1   **Stirling's triangle for values of $\left[{n \atop k}\right]$.**

n	$\left[{n \atop 0}\right]$	$\left[{n \atop 1}\right]$	$\left[{n \atop 2}\right]$	$\left[{n \atop 3}\right]$	$\left[{n \atop 4}\right]$	$\left[{n \atop 5}\right]$	$\left[{n \atop 6}\right]$	Σ
0	1							1
1	0	1						1
2	0	1	1					2
3	0	2	3	1				6
4	0	6	11	6	1			24
5	0	24	50	35	10	1		120
6	0	120	274	225	85	15	1	720

We observe that Column 1 of Stirling's triangle for cycle numbers is the sequence $(n-1)!$.

Proposition 5.2.2. *Let n be a positive integer. Then*

$$\left[{n \atop 1}\right] \;=\; (n-1)!$$

Proof: The number of ways to arrange n objects in a cycle with a designated starting point is $n!$. Two cycles may be regarded as equivalent if they differ only in the choice of starting point. There are n possible starting points. Thus, by the Rule of Quotient,

$$\left[{n \atop 1}\right] \;=\; (n-1)! \qquad\qquad \diamondsuit$$

It is less apparent, but not hard to prove, that Column 2 also has a tractable closed formula.

Proposition 5.2.3. *Let n be a positive integer. Then*

$$\left[{n \atop 2}\right] \;=\; (n-1)!\,H_{n-1}$$

Proof: Once again, by induction on n.

BASIS: $\left[{2 \atop 1}\right] \;=\; 1 \;=\; (2-1)!\,H_1$.

IND HYP: Assume for some $n \geq 2$ that

$$\left[{n-1 \atop 2}\right] \;=\; (n-2)!\,H_{n-2}$$

IND STEP: Then, by Stirling's recursion,

$$\begin{bmatrix} n \\ 2 \end{bmatrix} = \begin{bmatrix} n-1 \\ 1 \end{bmatrix} + (n-1) \begin{bmatrix} n-1 \\ 2 \end{bmatrix}$$

$$= (n-2)! + (n-1) \begin{bmatrix} n-1 \\ 2 \end{bmatrix} \qquad \text{(Proposition 5.2.2)}$$

$$= (n-2)! + (n-1)(n-2)!\, H_{n-2} \qquad \text{(induction hypothesis)}$$

$$= \frac{(n-1)!}{n-1} + (n-1)!\, H_{n-2}$$

$$= (n-1)! \left(\frac{1}{n-1} + H_{n-2} \right) = (n-1)!\, H_{n-1} \qquad \diamond$$

Example 5.2.3:   The following table helps to illustrate Proposition 5.2.3. In each column, the product of the entries in the row labeled H_{n-1} and the row labeled $(n-1)!$ is the entry in the row labeled $\begin{bmatrix} n \\ 2 \end{bmatrix}$.

n	1	2	3	4	5	$\cdots$
H_{n-1}	0	1	$\frac{3}{2}$	$\frac{11}{6}$	$\frac{25}{12}$	$\cdots$
$(n-1)!$	1	1	2	6	24	$\cdots$
$\begin{bmatrix} n \\ 2 \end{bmatrix}$	0	1	3	11	50	$\cdots$

Rows are Log-Concave

As with the rows of the Stirling triangle for subset numbers, the rows of the Stirling triangle for cycle numbers are log-concave and, thus, unimodal.

Proposition 5.2.4. *For all $n \geq 0$, the sequence of Stirling cycle numbers*

$$\begin{bmatrix} n \\ 0 \end{bmatrix}, \begin{bmatrix} n \\ 1 \end{bmatrix}, \ldots, \begin{bmatrix} n \\ n \end{bmatrix}$$

is log-concave. That is,

$$\begin{bmatrix} n \\ k-1 \end{bmatrix} \begin{bmatrix} n \\ k+1 \end{bmatrix} \leq \begin{bmatrix} n \\ k \end{bmatrix} \begin{bmatrix} n \\ k \end{bmatrix}$$

Proof: Rows 0 and 1 are vacuously log-concave. Assume that row $n-1$ is log-concave, and consider row n. Under Stirling's recurrence for cycle numbers, the product

$$\begin{bmatrix} n \\ k-1 \end{bmatrix} \begin{bmatrix} n \\ k+1 \end{bmatrix}$$

has the expansion

$$\left(\begin{bmatrix} n-1 \\ k-2 \end{bmatrix} + (n-1) \begin{bmatrix} n-1 \\ k-1 \end{bmatrix} \right) \times \left(\begin{bmatrix} n-1 \\ k \end{bmatrix} + (n-1) \begin{bmatrix} n-1 \\ k+1 \end{bmatrix} \right)$$

$$= \begin{bmatrix} n-1 \\ k-2 \end{bmatrix} \begin{bmatrix} n-1 \\ k \end{bmatrix} + (n-1)^2 \begin{bmatrix} n-1 \\ k-1 \end{bmatrix} \begin{bmatrix} n-1 \\ k+1 \end{bmatrix}$$

$$+ (n-1) \begin{bmatrix} n-1 \\ k-1 \end{bmatrix} \begin{bmatrix} n-1 \\ k \end{bmatrix} + (n-1) \begin{bmatrix} n-1 \\ k-2 \end{bmatrix} \begin{bmatrix} n-1 \\ k+1 \end{bmatrix}$$

to which log-concavity and Lemma 5.1.5 are applied under the induction hypothesis.

$$\leq \begin{bmatrix} n-1 \\ k-1 \end{bmatrix}^2 + (n-1)^2 \begin{bmatrix} n-1 \\ k \end{bmatrix}^2$$

$$+ (n-1) \begin{bmatrix} n-1 \\ k-1 \end{bmatrix} \begin{bmatrix} n-1 \\ k \end{bmatrix} + (n-1) \begin{bmatrix} n-1 \\ k-1 \end{bmatrix} \begin{bmatrix} n-1 \\ k \end{bmatrix}$$

$$\leq \begin{bmatrix} n-1 \\ k-1 \end{bmatrix}^2 + (n-1)^2 \begin{bmatrix} n-1 \\ k \end{bmatrix}^2 + 2(n-1) \begin{bmatrix} n-1 \\ k-1 \end{bmatrix} \begin{bmatrix} n-1 \\ k \end{bmatrix}$$

$$= \left(\begin{bmatrix} n-1 \\ k-1 \end{bmatrix} + (n-1) \begin{bmatrix} n-1 \\ k \end{bmatrix} \right)^2 = \begin{bmatrix} n \\ k \end{bmatrix}^2 \qquad \diamond$$

Figure 5.2.1 illustrates the unimodality of row 6.

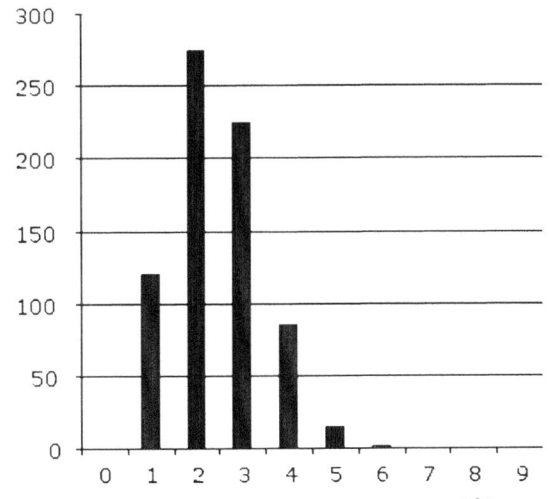

Figure 5.2.1   **Unimodality of the sequence** $\langle \begin{bmatrix} 6 \\ k \end{bmatrix} \mid k = 0, \ldots, 6 \rangle$.

Row Sums

The rows of Stirling's triangle for cycle numbers have several other interesting properties. The following property is apparent in Table 5.2.1.

Proposition 5.2.5. *Let n be a positive integer. Then*

$$\sum_{k=0}^{n} \begin{bmatrix} n \\ k \end{bmatrix} = n!$$

Proof: The simplest proof is that each row sum of Stirling's triangle for cycle numbers is the total number of permutations of a set of n objects. An alternative proof proceeds inductively on the row number, n. $\diamond$

A subtler property is how each entry in the second column is related to the row immediately above that entry, by a weighted row sum.

Proposition 5.2.6. *Let n be a positive integer. Then*

$$\sum_{j=0}^{n} j \begin{bmatrix} n \\ j \end{bmatrix} = \begin{bmatrix} n+1 \\ 2 \end{bmatrix}$$

Proof: By induction.

BASIS: If $n = 1$, then

$$0 \begin{bmatrix} 1 \\ 0 \end{bmatrix} + 1 \begin{bmatrix} 1 \\ 1 \end{bmatrix} = 0 + 1 = 1 = \begin{bmatrix} 2 \\ 2 \end{bmatrix}$$

IND HYP: For some $n \geq 2$, assume that

$$\sum_{j=0}^{n-1} j \begin{bmatrix} n-1 \\ j \end{bmatrix} = \begin{bmatrix} n \\ 2 \end{bmatrix}$$

IND STEP: Then, by Stirling's recursion,

$$\sum_{j=0}^{n} j \begin{bmatrix} n \\ j \end{bmatrix} = \sum_{j=0}^{n} j \left(\begin{bmatrix} n-1 \\ j-1 \end{bmatrix} + (n-1) \begin{bmatrix} n-1 \\ j \end{bmatrix} \right)$$

$$= \sum_{j=0}^{n} j \begin{bmatrix} n-1 \\ j-1 \end{bmatrix} + (n-1) \sum_{j=0}^{n} j \begin{bmatrix} n-1 \\ j \end{bmatrix}$$

Now split the first sum.

$$= \sum_{j=0}^{n} \begin{bmatrix} n-1 \\ j-1 \end{bmatrix} + \sum_{j=0}^{n} (j-1) \begin{bmatrix} n-1 \\ j-1 \end{bmatrix} + (n-1) \sum_{j=0}^{n} j \begin{bmatrix} n-1 \\ j \end{bmatrix}$$

Apply Proposition 5.2.5 to the first sum.

$$= (n-1)! + \sum_{j=0}^{n} (j-1) \begin{bmatrix} n-1 \\ j-1 \end{bmatrix} + (n-1) \sum_{j=0}^{n} j \begin{bmatrix} n-1 \\ j \end{bmatrix}$$

Next apply the induction hypothesis to the other two sums.

$$= (n-1)! + \begin{bmatrix} n \\ 2 \end{bmatrix} + (n-1)\begin{bmatrix} n \\ 2 \end{bmatrix}$$

$$= (n-1)! + n\begin{bmatrix} n \\ 2 \end{bmatrix}$$

Then apply Proposition 5.2.2

$$= \begin{bmatrix} n \\ 1 \end{bmatrix} + n\begin{bmatrix} n \\ 2 \end{bmatrix}$$

and conclude by applying Stirling's recursion.

$$= \begin{bmatrix} n+1 \\ 2 \end{bmatrix} \qquad\qquad\qquad \Diamond$$

Example 5.2.4:   With data from Table 5.2.1, we now illustrate Proposition 5.2.6.

$$1\begin{bmatrix} 4 \\ 1 \end{bmatrix} + 2\begin{bmatrix} 4 \\ 2 \end{bmatrix} + 3\begin{bmatrix} 4 \\ 3 \end{bmatrix} + 4\begin{bmatrix} 4 \\ 4 \end{bmatrix} = 1\cdot 6 + 2\cdot 11 + 3\cdot 6 + 4\cdot 1$$

$$= 50$$

$$= \begin{bmatrix} 5 \\ 2 \end{bmatrix}$$

Proposition 5.2.6 has Theorem 5.2.7 as a fascinating consequence.

Theorem 5.2.7. *The average number of cycles in a random permutation of n objects is H_n.*

Proof: Let the random variable X be the number of cycles in a permutation on n objects. Then

$$\Pr(k \text{ cycles}) = \frac{1}{n!}\begin{bmatrix} n \\ k \end{bmatrix}$$

Therefore, the expected number of cycles is

$$\mu = \sum_{k=0}^{n} k \cdot \frac{1}{n!}\begin{bmatrix} n \\ k \end{bmatrix}$$

$$= \frac{1}{n!}\sum_{k=0}^{n} k \cdot \begin{bmatrix} n \\ k \end{bmatrix}$$

$$= \frac{1}{n!} \cdot \begin{bmatrix} n+1 \\ 2 \end{bmatrix} \qquad\qquad \text{(Proposition 5.2.6)}$$

$$= \frac{1}{n!} \cdot n!\, H_n \qquad\qquad \text{(Proposition 5.2.3)}$$

$$= H_n \qquad\qquad\qquad\qquad\qquad \Diamond$$

Proposition 5.2.8 concerns a generalization of Proposition 5.2.6. It asserts that every entry, not just the entries in column 2, is a weighted sum of the elements of the row just above.

Proposition 5.2.8. *Let n and c be non-negative integers. Then*

$$\sum_{j=0}^{n} \binom{j}{c} \begin{bmatrix} n \\ j \end{bmatrix} = \begin{bmatrix} n+1 \\ c+1 \end{bmatrix}$$

Proof: For $c > n$, both sides of the equation are 0. Thus, in what follows, it is assumed that $c \le n$.

BASIS: If $n = 0$, then for $c = 0$,

$$\binom{0}{0} \begin{bmatrix} 0 \\ 0 \end{bmatrix} = 1 \cdot 1 = 1 = \begin{bmatrix} 1 \\ 1 \end{bmatrix}$$

IND HYP: For some $n \ge 1$, assume for all k that

$$\sum_{j=0}^{n-1} \binom{j}{k} \begin{bmatrix} n-1 \\ j \end{bmatrix} = \begin{bmatrix} n \\ c \end{bmatrix}$$

IND STEP: Then for any $c \le n$, Stirling's recursion implies

$$\sum_{j=0}^{n} \binom{j}{c} \begin{bmatrix} n \\ j \end{bmatrix} = \sum_{j=0}^{n} \binom{j}{c} \left(\begin{bmatrix} n-1 \\ j-1 \end{bmatrix} + (n-1) \begin{bmatrix} n-1 \\ j \end{bmatrix} \right)$$

which splits like this:

$$= \sum_{j=0}^{n} \binom{j}{c} \begin{bmatrix} n-1 \\ j-1 \end{bmatrix} + (n-1) \sum_{j=0}^{n} \binom{j}{c} \begin{bmatrix} n-1 \\ j \end{bmatrix}$$

which reduces, by the induction hypothesis, to

$$= \sum_{j=0}^{n} \binom{j}{c} \begin{bmatrix} n-1 \\ j-1 \end{bmatrix} + (n-1) \begin{bmatrix} n \\ c+1 \end{bmatrix}$$

Applying Pascal's recursion, we continue

$$= \sum_{j=0}^{n} \left(\binom{j-1}{c-1} + \binom{j-1}{c} \right) \begin{bmatrix} n-1 \\ j-1 \end{bmatrix} + (n-1) \begin{bmatrix} n \\ c+1 \end{bmatrix}$$

$$= \sum_{j=0}^{n} \binom{j-1}{c-1} \begin{bmatrix} n-1 \\ j-1 \end{bmatrix} + \sum_{j=0}^{n} \binom{j-1}{c} \begin{bmatrix} n-1 \\ j-1 \end{bmatrix} + (n-1) \begin{bmatrix} n \\ c+1 \end{bmatrix}$$

which reduces, by the induction hypothesis, to

$$= \begin{bmatrix} n \\ c \end{bmatrix} + \begin{bmatrix} n \\ c+1 \end{bmatrix} + (n-1) \begin{bmatrix} n \\ c+1 \end{bmatrix}$$

$$= \begin{bmatrix} n \\ c \end{bmatrix} + n \begin{bmatrix} n \\ c+1 \end{bmatrix}$$

and we finish, by applying Stirling's recursion.

$$= \begin{bmatrix} n+1 \\ c+1 \end{bmatrix} \qquad\qquad \diamond$$

Example 5.2.5:   Some data from Table 5.2.1 helps us to illustrate Proposition 5.2.8.

$$\binom{1}{2}\begin{bmatrix} 4 \\ 1 \end{bmatrix} + \binom{2}{2}\begin{bmatrix} 4 \\ 2 \end{bmatrix} + \binom{3}{2}\begin{bmatrix} 4 \\ 3 \end{bmatrix} + \binom{4}{2}\begin{bmatrix} 4 \\ 4 \end{bmatrix} = 0 \cdot 6 + 1 \cdot 11 + 3 \cdot 6 + 6 \cdot 1$$

$$= 35$$

$$= \begin{bmatrix} 5 \\ 3 \end{bmatrix}$$

Columns

Proposition 5.2.9 asserts that a weighted partial sum of the entries in column c can be found in column $c + 1$. It is analogous to Proposition 5.1.9 for Stirling subset numbers.

Proposition 5.2.9. *Let n and c be non-negative integers. Then*

$$\begin{bmatrix} n+1 \\ c+1 \end{bmatrix} = \sum_{k=0}^{n} n^{\underline{n-k}} \begin{bmatrix} k \\ c \end{bmatrix}$$

Proof: The equation is clearly true when $n = 0$. Assume, for inductive purpose, that it is true for $n - 1$. After starting with Stirling's recursion,

$$\begin{bmatrix} n+1 \\ c+1 \end{bmatrix} = \begin{bmatrix} n \\ c \end{bmatrix} + n \begin{bmatrix} n \\ c+1 \end{bmatrix}$$

we apply the inductive hypothesis.

$$= \begin{bmatrix} n \\ c \end{bmatrix} + n \sum_{k=0}^{n-1} (n-1)^{\underline{n-k-1}} \begin{bmatrix} k \\ c \end{bmatrix}$$

$$= \begin{bmatrix} n \\ c \end{bmatrix} + \sum_{k=0}^{n-1} n^{\underline{n-k}} \begin{bmatrix} k \\ c \end{bmatrix}$$

$$= \sum_{k=0}^{n} n^{\underline{n-k}} \begin{bmatrix} k \\ c \end{bmatrix} \qquad\qquad \diamond$$

The sum in Proposition 5.2.9 can be visualized as a dot product of a row of falling powers of a fixed base with a column of Stirling's triangle.

Example 5.2.6:   Consider column 2.

$k \rightarrow$	2	3	4	5
5^{5-k}	60	20	5	1

$n \downarrow$	$\begin{bmatrix} n \\ 2 \end{bmatrix}$	$\begin{bmatrix} n \\ 3 \end{bmatrix}$
2	1	
3	3	
4	11	
5	50	
6		225

$$60 \cdot 1 + 20 \cdot 3 + 5 \cdot 11 + 1 \cdot 50 = 225$$

That is,

$$\begin{bmatrix} 2 \\ 2 \end{bmatrix} \cdot 5^{\underline{3}} \; + \; \begin{bmatrix} 3 \\ 2 \end{bmatrix} \cdot 5^{\underline{2}} \; + \; \begin{bmatrix} 4 \\ 2 \end{bmatrix} \cdot 5^{\underline{1}} \; + \; \begin{bmatrix} 5 \\ 2 \end{bmatrix} \cdot 5^{\underline{0}}$$

$$= 1 \cdot 60 \; + \; 3 \cdot 20 \; + \; 11 \cdot 5 \; + \; 50 \cdot 1 \; = \; 225 \; = \; \begin{bmatrix} 6 \\ 3 \end{bmatrix}$$

Southeast Diagonal

The entries along each southeast diagonal from column 0 to column c satisfy a summation formula.

Proposition 5.2.10.   *Let n and c be non-negative integers. Then*

$$\begin{bmatrix} n + c + 1 \\ c \end{bmatrix} \; = \; \sum_{k=0}^{c} (n + k) \begin{bmatrix} n + k \\ k \end{bmatrix}$$

Proof: The equation is clearly true for all $n \geq 0$ when $c = 0$. Assume, for inductive purpose, that it is true for $c - 1$. Then, by Stirling's recursion,

$$\begin{bmatrix} n + c + 1 \\ c \end{bmatrix} \; = \; \begin{bmatrix} n + c \\ c - 1 \end{bmatrix} \; + \; (n + c) \begin{bmatrix} n + c \\ c \end{bmatrix}$$

Now apply the induction hypothesis.

$$= \; \sum_{k=0}^{c-1} (n + k) \begin{bmatrix} n + k \\ k \end{bmatrix} \; + \; (n + c) \begin{bmatrix} n + c \\ c \end{bmatrix}$$

$$= \; \sum_{k=0}^{c} (n + k) \begin{bmatrix} n + k \\ k \end{bmatrix} \qquad\qquad \diamondsuit$$

Example 5.2.7:   The sum in Proposition 5.2.10 is a dot product of a southeast diagonal of Stirling's triangle with a vector of row numbers.

$$
\begin{array}{c|ccc}
n \downarrow & \begin{bmatrix} n \\ 1 \end{bmatrix} & \begin{bmatrix} n \\ 2 \end{bmatrix} & \begin{bmatrix} n \\ 3 \end{bmatrix} \\
\hline
3 & 2 & & \\
4 & & 11 & \\
5 & & & 35 \\
6 & & & \boxed{225}
\end{array}
\qquad
3 \cdot 2 + 4 \cdot 11 + 5 \cdot 35 \;=\; 225
$$

Stirling Numbers of the First Kind

REVIEW FROM §1.6:   The *Stirling numbers of the first kind* were defined as the coefficients $s_{n,c}$ in the sum

$$
x^{\underline{n}} \;=\; \sum_{c=0}^{n} s_{n,c}\, x^{c}
$$

Proposition 5.2.11. *Let n and c be any non-negative integers. Then*

$$
s_{n,c} \;=\; (-1)^{n+c} \begin{bmatrix} n \\ c \end{bmatrix}
\tag{5.2.1}
$$

Proof: We recall the Stirling cycle recurrence of Proposition 5.2.1.

$$
\begin{bmatrix} 0 \\ k \end{bmatrix} = (k = 0) \qquad\qquad \begin{bmatrix} n \\ 0 \end{bmatrix} = (n = 0)
$$

$$
\begin{bmatrix} n \\ k \end{bmatrix} = \begin{bmatrix} n-1 \\ k-1 \end{bmatrix} + (n-1) \begin{bmatrix} n-1 \\ k \end{bmatrix} \quad \text{for } n \geq 1
$$

BASIS: The initial conditions

$$
s_{0,c} \;=\; (-1)^{0+c} \,(c = 0) \qquad \text{and} \qquad s_{n,0} \;=\; (-1)^{n+0} \,(n = 0)
$$

hold, because

$$
x^{\underline{0}} \;=\; 1 x^{0}
$$

and because the constant term of the expansion

$$
x^{\underline{n}} \;=\; \sum_{c=0}^{n} s_{n,c}\, x^{c}
\tag{5.2.2}
$$

is 0, unless $n = 0$.

IND HYP: Now assume that

$$x^{\underline{n-1}} = \sum_{c=0}^{n-1} s_{n-1,c}\, x^c$$

IND STEP: Then

$$
\begin{aligned}
x^{\underline{n}} &= (x-n+1)\cdot x^{\underline{n-1}} \\
&= x\cdot \sum_{c=0}^{n-1} s_{n-1,c}\, x^c \;-\; (n-1)\sum_{c=0}^{n-1} s_{n-1,c}\, x^c \\
&= \sum_{c=0}^{n-1} s_{n-1,c}\, x^{c+1} \;-\; (n-1)\sum_{c=0}^{n-1} s_{n-1,c}\, x^c \\
&= \sum_{c=1}^{n} s_{n-1,c-1}\, x^{c} \;-\; (n-1)\sum_{c=0}^{n-1} s_{n-1,c}\, x^c \\
&= \sum_{c=0}^{n} \big(s_{n-1,c-1} - (n-1)s_{n-1,c} \big)\, x^c \qquad\qquad (5.2.3)
\end{aligned}
$$

Since x^c must have the same coefficient in both expansions, (5.2.2) and (5.2.3), of $x^{\underline{n}}$, it follows that

$$s_{n,c} = s_{n-1,c-1} - (n-1)s_{n-1,c}$$

Thus, the absolute values of the Stirling numbers of the first kind satisfy the same recurrence as the Stirling cycle numbers. That is,

$$|s_{n,c}| = \begin{bmatrix} n \\ c \end{bmatrix}$$

This implies, by an induction, that

$$|s_{n,c}| = (-1)^{n+c} \begin{bmatrix} n \\ c \end{bmatrix}$$

$$\diamondsuit$$

Example 5.2.8: The values of $s_{n,c}$ are calculated recursively, as in the proof of Proposition 5.2.11.

$$
\begin{aligned}
x &= x^{\underline{1}} \\
\Rightarrow x^2 &= x\cdot x^{\underline{1}} \\
&= x^{\underline{1}}(x-1) + x^{\underline{1}} \\
&= x^{\underline{2}} + x^{\underline{1}} \\
\Rightarrow x^3 &= x\cdot x^{\underline{2}} + x\cdot x^{\underline{1}} \\
&= x^{\underline{2}}(x-2) + 2x^{\underline{2}} + x^{\underline{2}} + x^{\underline{1}} \\
&= x^{\underline{3}} + 3x^{\underline{2}} + x^{\underline{1}} \\
\Rightarrow x^4 &= x\cdot x^{\underline{3}} + 3x\cdot x^{\underline{2}} + x\cdot x^{\underline{1}} \\
&= [x^{\underline{4}} + 3x^{\underline{3}}] + [3x^{\underline{3}} + 6x^{\underline{2}}] + [x^{\underline{2}} + x^{\underline{1}}] \\
&= x^{\underline{4}} + 6x^{\underline{3}} + 7x^{\underline{2}} + x^{\underline{1}}
\end{aligned}
$$

Table 5.2.2   Some Basic Formulas for Stirling Cycle Numbers

Stirling's recurrence:

$$\begin{bmatrix} 0 \\ k \end{bmatrix} = (k = 0) \qquad\qquad \begin{bmatrix} n \\ 0 \end{bmatrix} = (n = 0)$$

$$\begin{bmatrix} n \\ k \end{bmatrix} = \begin{bmatrix} n-1 \\ k-1 \end{bmatrix} + (n-1)\begin{bmatrix} n-1 \\ k \end{bmatrix} \quad \text{for } n \geq 1 \qquad (5.2.4)$$

Special values for $n \geq 1$:

$$\begin{bmatrix} n \\ 1 \end{bmatrix} = (n-1)! \qquad \begin{bmatrix} n \\ 2 \end{bmatrix} = (n-1)!\, H_{n-1} \qquad \begin{bmatrix} n \\ n \end{bmatrix} = 1$$

Converting falling powers to ordinary powers:

$$x^{\underline{n}} = \sum_{k=0}^{n} \begin{bmatrix} n \\ k \end{bmatrix} (-1)^{n-k} x^k \qquad (5.2.5)$$

Row sum formulas:

$$n! = \sum_{k=0}^{n} \begin{bmatrix} n \\ k \end{bmatrix} \qquad (5.2.6)$$

$$\begin{bmatrix} n+1 \\ c+1 \end{bmatrix} = \sum_{j=0}^{n} \binom{j}{c} \begin{bmatrix} n \\ j \end{bmatrix} \qquad (5.2.7)$$

Column-sum formula:

$$\begin{bmatrix} n+1 \\ c+1 \end{bmatrix} = \sum_{k=0}^{n} n^{\underline{n-k}} \begin{bmatrix} k \\ c \end{bmatrix} \qquad (5.2.8)$$

SE diagonal-sum formula:

$$\begin{bmatrix} n+c+1 \\ c \end{bmatrix} = \sum_{k=0}^{c} (n+k) \begin{bmatrix} n+k \\ k \end{bmatrix} \qquad (5.2.9)$$

EXERCISES for Section 5.2

In each of the Exercises 5.2.1 through 5.2.8, using only Table 5.2.1 as given values, calculate the values of the following additional Stirling cycle numbers.

5.2.1	$\begin{bmatrix} 7 \\ 3 \end{bmatrix}$	5.2.2	$\begin{bmatrix} 7 \\ 4 \end{bmatrix}$	5.2.3^{s}	$\begin{bmatrix} 7 \\ 5 \end{bmatrix}$	5.2.4	$\begin{bmatrix} 7 \\ 6 \end{bmatrix}$
5.2.5	$\begin{bmatrix} 8 \\ 2 \end{bmatrix}$	5.2.6	$\begin{bmatrix} 8 \\ 3 \end{bmatrix}$	5.2.7	$\begin{bmatrix} 8 \\ 4 \end{bmatrix}$	5.2.8	$\begin{bmatrix} 8 \\ 5 \end{bmatrix}$

In each of the Exercises 5.2.9 through 5.2.16, confirm the weighted row-sum formula of Proposition 5.2.8 for the following Stirling cycle numbers.

5.2.9 $\begin{bmatrix} 6 \\ 2 \end{bmatrix}$ 　　5.2.10 $\begin{bmatrix} 6 \\ 3 \end{bmatrix}$ 　　5.2.11 $\begin{bmatrix} 6 \\ 4 \end{bmatrix}$ 　　5.2.12 $\begin{bmatrix} 6 \\ 5 \end{bmatrix}$

5.2.13 $\begin{bmatrix} 7 \\ 2 \end{bmatrix}$ 　　5.2.14 $\begin{bmatrix} 7 \\ 3 \end{bmatrix}$ 　　5.2.15 $\begin{bmatrix} 7 \\ 4 \end{bmatrix}$ 　　5.2.16^{S} $\begin{bmatrix} 7 \\ 5 \end{bmatrix}$

In each of the Exercises 5.2.17 through 5.2.24, confirm the column-sum formula of Proposition 5.2.9 for the following Stirling cycle numbers.

5.2.17 $\begin{bmatrix} 7 \\ 3 \end{bmatrix}$ 　　5.2.18 $\begin{bmatrix} 7 \\ 4 \end{bmatrix}$ 　　5.2.19^{S} $\begin{bmatrix} 7 \\ 5 \end{bmatrix}$ 　　5.2.20 $\begin{bmatrix} 7 \\ 6 \end{bmatrix}$

5.2.21 $\begin{bmatrix} 8 \\ 3 \end{bmatrix}$ 　　5.2.22 $\begin{bmatrix} 8 \\ 4 \end{bmatrix}$ 　　5.2.23 $\begin{bmatrix} 8 \\ 5 \end{bmatrix}$ 　　5.2.24 $\begin{bmatrix} 8 \\ 6 \end{bmatrix}$

In each of the Exercises 5.2.25 through 5.2.32, confirm the diagonal-sum formula of Proposition 5.2.10 for the following Stirling cycle numbers.

5.2.25 $\begin{bmatrix} 7 \\ 3 \end{bmatrix}$ 　　5.2.26 $\begin{bmatrix} 7 \\ 4 \end{bmatrix}$ 　　5.2.27^{S} $\begin{bmatrix} 7 \\ 5 \end{bmatrix}$ 　　5.2.28 $\begin{bmatrix} 7 \\ 6 \end{bmatrix}$

5.2.29 $\begin{bmatrix} 8 \\ 4 \end{bmatrix}$ 　　5.2.30 $\begin{bmatrix} 8 \\ 5 \end{bmatrix}$ 　　5.2.31 $\begin{bmatrix} 8 \\ 6 \end{bmatrix}$ 　　5.2.32 $\begin{bmatrix} 8 \\ 7 \end{bmatrix}$

5.3   INVERSIONS AND ASCENTS

Stirling cycle numbers provide an inventory for the partition of the set of all $n!$ permutations of the integer interval $[1:n]$, according to the number of cycles. In particular, the Stirling cycle number

$$\begin{bmatrix} n \\ k \end{bmatrix}$$

is the number of partitions with k cycles. This section is concerned with two other ways of partitioning those $n!$ permutations, one according to their number of *inversions* and the other according to their number of *ascents*.

NOTATION: Specifying a permutation

$$\pi = \begin{pmatrix} 1 & 2 & \cdots & n \\ a_1 & a_2 & \cdots & a_n \end{pmatrix}$$

of the integer interval $[1:n]$ by its lower line

$$a_1 \, a_2 \cdots a_n$$

is called the **one-line representation** of π.

Inversions

DEFINITION: In a permutation π of the integer interval $[1 : n]$, an **inversion** is a pair of integers $i < j$ with $\pi(j) < \pi(i)$.

In any permutation π of the integer interval $[1 : n]$, each instance of an inversion corresponds to some larger integer preceding an integer j in the one-line representation of π, so they would appear to be inverted in that line. There are $\binom{n}{2}$ pairs of integers in $[1 : n]$, each of which could possibly be inverted. At the low end, the identity permutation of $[1 : n]$ has no inversions. At the high end, the permutation that reverses the order of $[1 : n]$ has $\binom{n}{2}$ inversions.

DEFINITION: The **inversion vector** of a permutation π is the vector

$$b_1 \, b_2 \, \cdots \, b_n$$

such that b_j equals the number of larger integers preceding j in the one-line representation of π.

Example 5.3.1:   The permutation

$$\pi = 3\,5\,1\,6\,2\,4$$

has the inversion vector

$$2\,3\,0\,2\,0\,0$$

We observe that the coordinate b_j of the inversion vector $b_1 \, b_2 \, \cdots \, b_n$ is an integer in the range $[0 : n - j]$. Moreover, the total number of inversions of a permutation is the sum of the coordinates of its inversion vector.

Example 5.3.1, continued: The permutation $\pi = 3\,5\,1\,6\,2\,4$ has a total of 7 inversions, the sum of the coordinates of its inversion vector $2\,3\,0\,2\,0\,0$.

DEFINITION: The **inversion coefficient** $I_n(k)$ is the number of permutations of the integer interval $[1 : n]$ with exactly k inversions.

Table 5.3.1 gives the values of some inversion coefficients.

Table 5.3.1   Inversion coefficients.

n	$I_n(0)$	$I_n(1)$	$I_n(2)$	$I_n(3)$	$I_n(4)$	$I_n(5)$	$I_n(6)$	$I_n(7)$	$I_n(8)$	$I_n(9)$	$I_n(10)$
0	1										
1	1										
2	1	1									
3	1	2	2	1							
4	1	3	5	6	5	3	1				
5	1	4	9	15	20	22	20	15	9	4	1
6	1	5	14	29	49	71	90	101	101	90	71 $\cdots$

The table of inversion coefficients can be constructed using the following proposition. We take $I_n(c)$ to be 0 if $c < 0$.

Proposition 5.3.1. *The inversion coefficients satisfy the following recurrence.*

$$I_0(0) = 1$$

$$I_n(c) = \sum_{j=0}^{n-1} I_{n-1}(c-j) \quad \text{for } n \geq 1$$

Proof: The initial condition is true, since the null permutation on the empty set has no inversions.

To affirm the recursion inductively, assume that the recursion holds for the permutations of $[1 : n - 1]$. Now consider the one-line representation of a permutation π on $[1 : n]$ with c inversions

$$\pi : \quad \pi_1 \pi_2 \cdots \pi_n$$

Then the number of inversions contributed by the placement of the integer n within this line equals the number j of integers that follow n on that line. Thus, if n is erased from that line, then the number of inversions in the permutation corresponding to the resulting line equals $c - j$. There are exactly $I_{n-1}(c-j)$ such permutations of $[1 : n - 1]$. Thus, $I_n(c)$ is the sum of the numbers $I_{n-1}(c - j)$ over the possible values of j.   ◇

Example 5.3.2:   We observe in Table 5.3.1 that

$$
\begin{aligned}
I_4(3) &= I_3(3) + I_3(2) + I_3(1) + I_3(0) \\
&= 1 + 2 + 2 + 1 = 6 \\
I_4(4) &= I_3(4) + I_3(3) + I_3(2) + I_3(1) \\
&= 0 + 1 + 2 + 2 = 5 \\
I_4(5) &= I_3(5) + I_3(4) + I_3(3) + I_3(2) \\
&= 0 + 0 + 1 + 2 = 3
\end{aligned}
$$

Donald Knuth (see [Knut1973], p.12) regards the following observation of Marshall Hall as the most important single fact about inversions.

Theorem 5.3.2 [Hall1956]. *A permutation π on the integer interval $[1 : n]$ is reconstructible from its inversion vector*

$$b_1 \, b_2 \, \cdots \, b_n$$

Proof: To reconstruct a one-line representation of the permutation π, begin by writing the number n. After the integers

$$k, \ldots, n$$

have been written as directed here, insert the integer $k - 1$ so that it immediately follows the first b_{k-1} integers.   ◇

Corollary 5.3.3. *There is a bijective correspondence between permutations on* $[1:n]$ *and inversion vectors* $b_1 b_2 \cdots b_n$ *with* $b_j \in [0:n-j]$ *for* $j = 1, \ldots, n$.

Proof: The number of permutations of $[1:n]$ and the number of such inversion vectors are both equal to $n!$. By Theorem 5.3.2, the correspondence of permutations to inversion vectors is one-to-one. It follows by the pigeonhole principle that it is onto.   ◇

Example 5.3.3:   The one-line representation of the permutation of the integer interval $[1:7]$ corresponding to the inversion vector

$$4\,5\,1\,2\,0\,1\,0$$

is reconstructed as follows:

$$
\begin{array}{ccccccc}
7 \\
7 & 6 \\
5 & 7 & 6 \\
5 & 7 & 4 & 6 \\
5 & 3 & 7 & 4 & 6 \\
5 & 3 & 7 & 4 & 6 & 2 \\
5 & 3 & 7 & 4 & 1 & 6 & 2 \\
\end{array}
$$

Ascents

DEFINITION: An index j of a permutation

$$\pi = a_1 a_2 \cdots a_n$$

is an **ascent** if $a_j < a_{j+1}$ and a **descent** if $a_j > a_{j+1}$.

Remark: Thus, an ascent is a special kind of non-inversion.

Example 5.3.4:   The ascents of the permutation

$$\pi = 3\,5\,1\,6\,2\,4$$

are as follows:

$$
\begin{array}{ll}
1: & 3 < 5 \\
3: & 1 < 6 \\
5: & 2 < 4 \\
\end{array}
$$

Example 5.3.5:   The partition of the permutations of $[1:4]$ according to number of ascents is as follows:

$$
\begin{array}{ll}
3: & 1234 \\
2: & 1243 \ \ 1423 \ \ 1324 \ \ 1342 \ \ 2134 \ \ 2314 \ \ 2341 \ \ 2413 \\
   & 3124 \ \ 3412 \ \ 4123 \\
1: & 3421 \ \ 3241 \ \ 4231 \ \ 2431 \ \ 4312 \ \ 4132 \ \ 1432 \ \ 3142 \\
   & 4213 \ \ 2143 \ \ 3214 \\
0: & 4321 \\
\end{array}
$$

Eulerian Numbers

DEFINITION: The **Eulerian number**

$$\left\langle {n \atop k} \right\rangle$$

is the number of permutations of $[1 : n]$ with exactly k ascents.

Proposition 5.3.4. *The Eulerian numbers satisfy the recurrence*

$$\left\langle {0 \atop k} \right\rangle = \begin{cases} 1 & \text{if } k = 0 \\ 0 & \text{if } k > 0 \end{cases}$$

$$\left\langle {n \atop k} \right\rangle = (k+1) \left\langle {n-1 \atop k} \right\rangle + (n-k) \left\langle {n-1 \atop k-1} \right\rangle \quad \text{for } n > 0$$

Combinatorial Proof: The basis for the recurrence is clear. The first summand in the right side of the recursion follows from the fact that a permutation of $[1 : n]$ with k ascents is obtained from a permutation of $[1 : n-1]$ with k ascents by prepending the integer n at the start of the one-line representation or inserting it between the integers of an ascending pair. The second summand corresponds to the $n-k$ ways to increase the number of ascents by 1 in a permutation of $[1 : n-1]$ with $k-1$ ascents either by interposing n between any of the $n-k-1$ descending pairs or by appending n at the end of the line.   ◇

As with Pascal's recursion and the Stirling recursions, the Euler recursion leads to a triangular table.

Table 5.3.2 **Euler's triangle for values of** $\left\langle {n \atop r} \right\rangle$.

n	$\left\langle {n \atop 0} \right\rangle$	$\left\langle {n \atop 1} \right\rangle$	$\left\langle {n \atop 2} \right\rangle$	$\left\langle {n \atop 3} \right\rangle$	$\left\langle {n \atop 4} \right\rangle$	$\left\langle {n \atop 5} \right\rangle$	$\left\langle {n \atop 6} \right\rangle$	B_n
0	1							1
1	1	0						1
2	1	1	0					2
3	1	4	1	0				6
4	1	11	11	1	0			24
5	1	26	66	26	1	0		120
6	1	57	302	302	57	1	0	720

We observe that each row of Euler's triangle is symmetric. This observation is confirmed for all n as follows.

Proposition 5.3.5 *Symmetry for Eulerian Numbers.*

$$\left\langle {n \atop k} \right\rangle = \left\langle {n \atop n-1-k} \right\rangle$$

Proof: A permutation π of $[1 : n]$ with k ascents has $n-1-k$ descents. Accordingly, the permutation whose one-line representation is the reverse of the representation for π has $n-1-k$ ascents.   ◇

EXERCISES for Section 5.3

5.3.1[S]   Write all the permutations of $[1:4]$ with exactly 2 inversions.

5.3.2   Write all the permutations of $[1:4]$ with exactly 3 inversions.

5.3.3   Write all the permutations of $[1:4]$ with exactly 4 inversions.

5.3.4   Write all the permutations of $[1:4]$ with exactly 5 inversions.

5.3.5   Write all the permutations of $[1:5]$ with exactly 1 inversion.

5.3.6   Write all the permutations of $[1:5]$ with exactly 2 inversions.

5.3.7   Write all the permutations of $[1:5]$ with exactly 3 inversions.

5.3.8   Write all the permutations of $[1:6]$ with exactly 2 inversions.

In each of the Exercises 5.3.9 through 5.3.16, calculate the value of the given inversion coefficient.

5.3.9	$I_6(11)$	**5.3.10**	$I_6(12)$	**5.3.11**	$I_7(6)$	**5.3.12**	$I_7(7)$
5.3.13[S]	$I_7(8)$	**5.3.14**	$I_7(9)$	**5.3.15**	$I_7(10)$	**5.3.16**	$I_7(15)$

In each of the Exercises 5.3.17 through 5.3.20, reconstruct the permutation from the given inversion vector.

5.3.17[S]	3 2 1 2 0 0	**5.3.18**	5 1 3 0 1 0
5.3.19	1 2 3 2 1 1 0	**5.3.20**	4 1 2 3 2 0 0

In each of the Exercises 5.3.21 through 5.3.28, using only Table 5.3.2 as given values, calculate the values of the following additional Eulerian numbers.

5.3.21	$\left\langle {7 \atop 1} \right\rangle$	**5.3.22**[S]	$\left\langle {7 \atop 2} \right\rangle$	**5.3.23**	$\left\langle {7 \atop 3} \right\rangle$	**5.3.24**	$\left\langle {7 \atop 4} \right\rangle$
5.3.25	$\left\langle {8 \atop 1} \right\rangle$	**5.3.26**	$\left\langle {8 \atop 2} \right\rangle$	**5.3.27**	$\left\langle {8 \atop 3} \right\rangle$	**5.3.28**	$\left\langle {8 \atop 4} \right\rangle$

5.4 DERANGEMENTS

We recall that a *derangement* is a permutation in which none of the objects is fixed. The **derangement recurrence** (from §2.1)

$$D_0 = 1; \quad D_1 = 0;$$
$$D_n = (n-1)D_{n-1} + (n-1)D_{n-2} \quad \text{for } n \geq 2 \tag{5.4.1}$$

is second-degree linear with variable coefficients. From it, a first degree recurrence can be derived.

Proposition 5.4.1. *The derangement sequence satisfies the recurrence*

$$D_0 = 1;$$
$$D_n = nD_{n-1} + (-1)^n \quad \text{for } n \geq 1 \tag{5.4.2}$$

Proof: Recursion (5.4.1) above implies that

$$D_n - nD_{n-1} = -[D_{n-1} - (n-1)D_{n-2}] \quad \text{for } n \geq 2 \tag{5.4.3}$$

We now apply recursion (5.4.3) recursively.

$$\begin{aligned}
D_n - nD_{n-1} &= (-1)[D_{n-1} - (n-1)D_{n-2}] \\
&= (-1)^2[D_{n-2} - (n-2)D_{n-3}] \\
&= \cdots \\
&= (-1)^{n-1}[D_1 - D_0] = (-1)^{n-1}[0-1] \\
&= (-1)^n \\
\Rightarrow D_n &= nD_{n-1} + (-1)^n \qquad\qquad\qquad\qquad\qquad \Diamond
\end{aligned}$$

Using either derangement recurrence, (5.4.1) or (5.4.2), we can calculate the *derangement number* D_n. The ratio $D_n/n!$ is the proportion of permutations that are derangements. Some values for the ratios $D_n/n!$ and $n!/D_n$ appear in Table 5.4.1.

Table 5.4.1   Ratios of derangements to permutations.

n	$n!$	D_n	$D_n/n!$	$n!/D_n$
0	1	1	1	
1	1	0	0	
2	2	1	0.5	2.0
3	6	2	0.333333	3.0
4	24	9	0.375	2.666667
5	120	44	0.366667	2.727273
6	720	265	0.368055	2.716981
7	5040	1854	0.367857	2.718447
8	40320	14833	0.367881	2.718263
9	362880	133496	0.367879	2.718284

Seemingly, the ratios $D_n/n!$ and $n!/D_n$ converge rapidly to e^{-1} and e, respectively. The following proposition and its corollary confirm this reasonable suspicion. This is an application of the familiar technique of guessing the solution to a recurrence and proving the correctness by induction.

Theorem 5.4.2. *For every non-negative integer* n,

$$D_n = n!\left[\frac{1}{0!} - \frac{1}{1!} + \frac{1}{2!} - \frac{1}{3!} + \cdots + (-1)^n\frac{1}{n!}\right] \tag{5.4.4}$$

Proof: For $n = 0$, both sides of equation (5.4.4) have the value 1. We asssume inductively that equation (5.4.4) holds for $n - 1$. Then

$$
\begin{aligned}
D_n &= nD_{n-1} + (-1)^n &\text{(by (5.4.2))}\\
&= n(n-1)!\left[\frac{1}{0!} - \frac{1}{1!} + \frac{1}{2!} + \cdots + (-1)^{n-1}\frac{1}{(n-1)!}\right] + (-1)^n\\
&= n!\left[\frac{1}{0!} - \frac{1}{1!} + \frac{1}{2!} + \cdots + (-1)^{n-1}\frac{1}{(n-1)!}\right] + \frac{n!(-1)^n}{n!}\\
&= n!\left[\frac{1}{0!} - \frac{1}{1!} + \frac{1}{2!} + \cdots + (-1)^n\frac{1}{n!}\right] \qquad\qquad \diamond
\end{aligned}
$$

In §3.6, the derangement numbers were calculated by inclusion-exclusion. In the proof of Theorem 5.4.2, we verified the solution as a "guessed solution" to a recursion. In the next section, the derangement recurrence is solved by generating functions, without resort to guessing.

Corollary 5.4.3. $\displaystyle\lim_{n\to\infty}\frac{D_n}{n!} = e^{-1}.$ $\qquad\qquad\qquad\qquad\qquad\qquad\diamond$

Remark: By running a Monte Carlo experiment on a computer, we could use Corollary 5.4.3 to approximate the value of e.

Every permutation of n objects may be regarded as a choice of j objects to fix and a derangement of the other $n - j$ objects. This leads immediately to the following assertion, which was previously noted with Example 4.2.4.

Proposition 5.4.4. *Let n be a non-negative integer. Then*

$$
n! = \sum_{j=0}^{n}\binom{n}{j}D_{n-j} \qquad\qquad\qquad\qquad\diamond
$$

Example 5.4.1: For $n = 4$, Proposition 5.4.4 corresponds to the equation

$$
\begin{aligned}
24 &= \binom{4}{0}D_4 + \binom{4}{1}D_3 + \binom{4}{2}D_2 + \binom{4}{3}D_1 + \binom{4}{4}D_0\\
&= 1\cdot D_4 + 4\cdot D_3 + 6\cdot D_2 + 4\cdot D_1 + 1\cdot D_0\\
&= 1\cdot 9 + 4\cdot 2 + 6\cdot 1 + 4\cdot 0 + 1\cdot 1
\end{aligned}
$$

5.5   EXPONENTIAL GENERATING FUNCTIONS

Ordinary generating functions are well-adapted to problems about counting unordered selections. This section develops the other main variety of generating function, called an *exponential generating function*, which is especially useful in counting ordered selections. We will see also how exponential generating functions can be used in solving certain recurrences with variable coefficients.

REVIEW FROM §1.7:

- The **ordinary generating function** (abbr. **OGF**) for a sequence $\langle g_n \rangle$ is any closed form $G(z)$ corresponding to the infinite polynomial

$$\sum_{n=0}^{\infty} g_n z^n$$

  or sometimes, the polynomial itself.

- The **exponential generating function** (abbr. **EGF**) for a sequence $\langle g_n \rangle$ is any closed form $\hat{G}(z)$ corresponding to the infinite polynomial

$$\sum_{n=0}^{\infty} g_n \frac{z^n}{n!}$$

  or sometimes, the polynomial itself.

- **Proposition 1.7.1**. Let $G(z)$ and $H(z)$ be the ordinary generating functions for counting unordered selections from two disjoint multisets S and T. Then $G(z)H(z)$ is the ordinary generating function for counting unordered selections from the union $S \cup T$.

- The **convolution of the sequences** $\langle a_n \rangle$ and $\langle b_n \rangle$ is the sequence

$$a_0 b_0, \quad a_0 b_1 + a_1 b_0, \quad a_0 b_2 + a_1 b_1 + a_2 b_0, \quad \ldots$$

- **Proposition 1.7.3**. The product of the generating functions

$$A(z) = \sum_{n=0}^{\infty} a_n z^n \quad \text{and} \quad B(z) = \sum_{n=0}^{\infty} b_n z^n$$

  is the generating function

$$A(z)B(z) = \sum_{n=0}^{\infty} \left(\sum_{j=0}^{n} a_j b_{n-j} \right) z^n$$

  for the convolution of the sequences $\langle a_n \rangle$ and $\langle b_n \rangle$.

The following example reviews how Proposition 1.7.1 can be used to count unordered selections with ordinary generating functions.

Example 5.5.1:   Let a_n and b_n be the numbers of ways to select n letters from the multi-sets represented by the strings

$$\text{“ADD”}\quad\text{and}\quad\text{“SPICE”}$$

respectively. Thus, the ordinary generating functions for the sequences $\langle a_n \rangle$ and $\langle b_n \rangle$ are

$$A(z) \;=\; \sum_{n=0}^{\infty} a_n z^n \;=\; 1 + 2z + 2z^2 + z^3$$

$$B(z) \;=\; \sum_{n=0}^{\infty} b_n z^n \;=\; 1 + 5z + 10z^2 + 10z^3 + 5z^4 + z^5$$

The set of possibilities counted by the sequence $\langle a_i \rangle$ is completely disjoint from the set counted by the sequence $\langle b_n \rangle$, because the set of letters of “ADD” is disjoint from the set of letters of “SPICE”. It follows that the number c_n of ways to choose n letters from the multi-set represented by the string

$$\text{“ADDSPICE”}$$

is the sum

$$a_0 b_n \,+\, a_1 b_{n-1} \,+\, \ldots \,+\, a_n b_0$$

More generally, it follows that the sequence $\langle c_n \rangle$ is the convolution of the sequences $\langle a_n \rangle$ and $\langle b_n \rangle$. Therefore, according to Proposition 1.7.3, the generating function for the sequence $\langle c_n \rangle$ is the product

$$\begin{aligned} A(z)B(z) \;=\;\; & 1 + 7z + 22z^2 + 41z^3 + 50z^4 + 41z^5 \\ & + 22z^6 + 7z^7 + z^8 \end{aligned}$$

For instance, there are 21 ways to choose two letters from the seven different letters and 1 way to choose the same two letters, for a total of 22, the coefficient of z^2.

Counting Ordered Selections

To count ordered selections from a disjoint union of multisets, we use Proposition 1.7.2.

REVIEW FROM §1.7:

- **Proposition 1.7.2.** Let $\hat{G}(z)$ and $\hat{H}(z)$ be the exponential generating functions for counting ordered selections from two disjoint multisets S and T. Then $\hat{G}(z)\hat{H}(z)$ is the exponential generating function for counting ordered selections from the union $S \cup T$.

Example 5.5.2:   Let r_n and s_n be the numbers of ways to select a sequence of n letters (without repetition) from the multi-sets represented by the strings

$$\text{“ADD”}\quad\text{and}\quad\text{“SPICE”}$$

respectively. Thus, the exponential generating functions for the sequences $\langle r_n \rangle$ and $\langle s_n \rangle$ are

$$\hat{R}(z) = \sum_{n=0}^{\infty} r_n \frac{z^n}{n!} = 1 + 2\frac{z}{1!} + 3\frac{z^2}{2!} + 3\frac{z^3}{3!}$$

$$\hat{S}(z) = \sum_{n=0}^{\infty} s_n \frac{z^n}{n!} = 1 + 5\frac{z}{1!} + 20\frac{z^2}{2!} + 60\frac{z^3}{3!} + 120\frac{z^4}{4!} + 120\frac{z^5}{5!}$$

The coefficient of z^2 in the product $\hat{R}(z)\hat{S}(z)$ is

$$\frac{1 \cdot 20}{0! \, 2!} + \frac{2 \cdot 5}{1! \, 1!} + \frac{3 \cdot 1}{2! \, 0!}$$

$$= \frac{1}{2!}\left[\binom{2}{0} 1 \cdot 20 + \binom{2}{1} 2 \cdot 5 + \binom{2}{2} 3 \cdot 1 \right]$$

$$= \frac{43}{2!}$$

from which it follows that the coefficient of $\frac{z^2}{2}$ in $\hat{R}(z)\hat{S}(z)$ is

$$43$$

This corresponds to $7\underline{2} = 42$ possible ordered selections of two different letters from the seven in the string "ADDSPICE", plus 1 way to choose the same two letters, for a total of 43.

Giving a name to the construction appearing within Example 5.5.2 facilitates the use of a generalization of that method, via Proposition 5.5.1, which is analogous to Proposition 1.7.3.

DEFINITION: The **binomial convolution** of two sequences $\langle r_n \rangle$ and $\langle s_n \rangle$ is the sequence $\langle t_n \rangle$ whose n^{th} entry is

$$t_n = \sum_{j=0}^{n} \binom{n}{j} r_j s_{n-j}$$

Proposition 5.5.1. *The product of the exponential generating functions for the sequences $\langle r_n \rangle$ and $\langle s_n \rangle$ is the exponential generating function for their binomial convolution.*

Proof: The coefficient of z^n in the product of the exponential generating functions

$$\hat{R}(z) = \sum_{n=0}^{\infty} r_n \frac{z^n}{n!}$$

and

$$\hat{S}(z) \;=\; \sum_{n=0}^{\infty} s_n \frac{z^n}{n!}$$

is

$$
\begin{aligned}
\frac{r_0 s_n}{0!\,n!} &+ \frac{r_1 s_{n-1}}{1!\,(n-1)!} + \cdots + \frac{r_n s_0}{n!\,0!} \\
&= \frac{1}{n!} \left[\frac{n!}{0!\,n!}\, r_0 s_n + \frac{n!}{1!\,(n-1)!}\, r_1 s_{n-1} + \cdots + \frac{n!}{n!\,0!}\, r_n s_0 \right] \\
&= \frac{1}{n!} \left[\binom{n}{0} r_0 s_n + \binom{n}{1} r_1 s_{n-1} + \cdots + \binom{n}{n} r_n s_0 \right] \\
&= \frac{1}{n!} \sum_{j=0}^{n} \binom{n}{j} r_j s_{n-j}
\end{aligned}
$$

Thus, the coefficient of $\frac{z^n}{n!}$ in the product $\hat{R}(z)\hat{S}(z)$ is

$$\sum_{j=0}^{n} \binom{n}{j} r_j s_{n-j} \qquad\qquad \Diamond$$

We complete this section by considering several applications in which using EGF's is a highly convenient way to count.

Counting Strings with Various Symbol Requirements

If a set of symbols has cardinality k, then, of course, there are k^n strings of length n. The examples in the sequence to follow impose various rules on the strings and count the strings that satisfy those rules. The first examples are easy enough, as an intended warmup, that solution without EGF's is well within grasp, and as the complications increase, the usefulness of EGF's becomes ever more clear.

Example 5.5.3:   We count the number b_n of binary strings of length n with at least one 1. Of the 2^n binary strings of length n, only one has no 1's. Thus,

$$b_n \;=\; 2^n - 1$$

Alternatively, we could observe that the EGF for the number of all-0 strings of length n is e^z. Accordingly, the EGF for the number of all-1 strings of length n with at least one 1 is $e^z - 1$. Thus, by Proposition 1.7.2, the EGF for b_n is

$$\hat{B}(z) \;=\; \sum_{n=0}^{\infty} b_n \frac{z^n}{n!} \;=\; e^z \cdot (e^z - 1) \;=\; e^{2z} - e^z$$

The coefficient of z^n in $e^{2z} - e^z$ is

$$\frac{2^n}{n!} - \frac{1}{n!}$$

Thus, the coefficient of $\frac{z^n}{n!}$ is

$$b_n = 2^n - 1$$

Example 5.5.4:   Next we count the number t_n of ternary strings (i.e., base-3) of length n in which the digits 1 and 2 must each occur at least once. Of the 3^n ternary strings of length n, there are 2^n strings with no 1's and 2^n strings with no 2's and exactly 1 string with no 1's or 2's. Thus, by Inclusion-Exclusion,

$$t_n = 3^n - 2 \cdot 2^n + 1$$

Alternatively, we could write the EGF for t_n, which is

$$\hat{T}(z) = \sum_{n=0}^{\infty} t_n \frac{z^n}{n!} = e^z \cdot (e^z - 1)^2 = e^{3z} - 2e^{2z} + e^z$$

Thus, the coefficient of $\frac{z^n}{n!}$ is

$$t_n = 3^n - 2 \cdot 2^n + 1$$

For $n = 3$, for instance, the formula $t_n = 3^n - 2 \cdot 2^n + 1$ yields

$$\begin{aligned}
t_3 &= 3^3 - 2 \cdot 2^3 + 1 \\
&= 27 - 16 + 1 \\
&= 12
\end{aligned}$$

This corresponds to $3! = 6$ arrangements of the digits within the string 012, plus 3 arrangements of the digits within the string 112, plus 3 arrangements of the digits within the string 122.

Using exponential generating functions on such simple problems seems not to expedite the calculation. However, for more complicated restrictions on the occurrences of some of the symbols in a string, EGF's are of considerable assistance.

Example 5.5.5:   Let u_n be the number of ternary strings with at least one 1 and at least two 2's. Then the EGF for strings of 2's with at least two 2's is

$$e^z - z - 1$$

It follows that

$$\hat{U}(z) = \sum_{n=0}^{\infty} u_n \frac{z^n}{n!} = e^z(e^z - 1)(e^z - z - 1)$$

$$= e^{3z} - 2e^{2z} - ze^{2z} + ze^z + e^z$$

Therefore,

$$u_n = 3^n - 2^{n+1} - n2^{n-1} + n + 1$$

For instance,

$$u_3 = 3^3 - 2^4 - 3 \cdot 2^2 + 3 + 1$$
$$= 27 - 16 - 12 + 3 + 1$$
$$= 3$$

This corresponds to the three possible arrangements of the digits within the string 122.

Example 5.5.6:   Let v_n be the number of ternary strings with evenly many 2's and at least one 1. Then the EGF for strings of evenly many 2's is

$$1 + \frac{z^2}{2!} + \frac{z^4}{4!} + \frac{z^6}{6!} + \cdots$$
$$= \frac{e^z + e^{-z}}{2}$$

Accordingly,

$$\hat{V}(z) = \sum_{n=0}^{\infty} v_n \frac{z^n}{n!} = e^z(e^z - 1) \cdot \frac{e^z + e^{-z}}{2}$$
$$= \frac{1}{2} \cdot \left(e^{3z} - e^{2z} + e^z - 1\right)$$

Therefore,

$$v_n = \begin{cases} 0 & \text{if } n = 0 \\ \frac{1}{2}(3^n - 2^n + 1) & \text{if } n \geq 1 \end{cases}$$

For instance, this formula yields

$$v_3 = \frac{27 - 8 + 1}{2} = 10$$

which corresponds to the 7 binary strings with at least one 1, plus the 3 strings

$$022 \quad 202 \quad 220$$

An Application To Stirling Subset Numbers

Continuing as in the immediately previous examples, the EGF for the number of ternary strings with at least one 0, at least one 1, and at least one 2 is

$$(e^z - 1)^3 = e^{3z} - 3e^{2z} + 3e^z - 1$$
$$= \sum_{n=1}^{\infty}(3^n - 3 \cdot 2^n + 3)\frac{z^n}{n!}$$

If we identify the distinct positions $1, \ldots, n$ in the sequence with n distinct objects, then this is also the generating function for partitioning n distinct objects into three distinct boxes, with no box left empty. This is 3! times as many as if the boxes were indistinguishable, so that we were counting partitions into three subsets. Thus,

$$\frac{(e^z - 1)^3}{3!} = \sum_{n=0}^{\infty} \left\{ {n \atop 3} \right\} \frac{z^n}{n!}$$

is an EGF for column 3 of Stirling's subset triangle. This calculation has an immediate generalization with a corollary that is equivalent to Theorem 3.6.4.

Proposition 5.5.2. *Let n and k be non-negative integers. Then*

$$\frac{(e^z - 1)^k}{k!} = \sum_{n=0}^{\infty} \left\{ {n \atop k} \right\} \frac{z^n}{n!} \qquad \diamond$$

Corollary 5.5.3. *Let n and k be non-negative integers. Then*

$$\left\{ {n \atop k} \right\} = \frac{1}{k!} \sum_{j=0}^{k} \binom{k}{j} j^n (-1)^{k-j}$$

Proof: The sides of the equation are the coefficients of

$$\frac{z^n}{n!}$$

in Proposition 5.5.2. $\diamond$

Example 5.5.7:   Applying the formula of Corollary 5.5.3 yields the evaluation

$$
\begin{aligned}
\left\{ {4 \atop 2} \right\} &= \frac{1}{2!} \sum_{j=0}^{2} \binom{2}{j} j^4 (-1)^{2-j} \\
&= \frac{1}{2} \left[\binom{2}{0} 0^4 (-1)^{2-0} + \binom{2}{1} 1^4 (-1)^{2-1} \binom{2}{2} 2^4 (-1)^{2-2} \right] \\
&= \frac{1}{2} \left[1 \cdot 0 \cdot 1 + 2 \cdot 1 \cdot (-1) + 1 \cdot 16 \cdot 1 \right] \\
&= 7
\end{aligned}
$$

which agrees with our previous calculations of

$$\left\{ {4 \atop 2} \right\}$$

An EGF for Derangement Numbers

We now show an example of how, sometimes, an EGF can be used in solving a linear recurrence with a variable coefficient. We then use this technique in finding a generating function for the derangement numbers.

Example 5.5.8:   Consider the following recurrence of degree 2.

$$
\begin{aligned}
a_0 &= 0, \quad a_1 = 1; \\
a_n &= 3n\, a_{n-1} - 2n(n-1)\, a_{n-2} \quad \text{for } n \geq 2
\end{aligned}
$$

Step 1. Multiplying both sides of the recursion by $\frac{z^n}{n!}$ and then summing from $n = 2$ to ∞ leads to the equation

$$\sum_{n=2}^{\infty} a_n \frac{z^n}{n!} = \sum_{n=2}^{\infty} 3n a_{n-1} \frac{z^n}{n!} - \sum_{n=2}^{\infty} 2n(n-1) a_{n-2} \frac{z^n}{n!}$$

which simplifies to the form

$$(1) \qquad \sum_{n=2}^{\infty} a_n \frac{z^n}{n!} = 3z \sum_{n=2}^{\infty} a_{n-1} \frac{z^{n-1}}{(n-1)!} - 2z^2 \sum_{n=2}^{\infty} a_{n-2} \frac{z^{n-2}}{(n-2)!}$$

Step 2. By substituting the exponential generating function

$$\hat{A}(z) = \sum_{n=0}^{\infty} a_n \frac{z^n}{n!}$$

we obtain the equation

$$(2) \qquad \hat{A}(z) - a_1 z - a_0 = 3z \left(\hat{A}(z) - a_0 \right) - 2z^2 \hat{A}(z)$$

Step 3. We then solve for $\hat{A}(z)$.

$$\hat{A}(z) - 1z - 0 = 3z \left(\hat{A}(z) - 0 \right) - 2z^2 \hat{A}(z)$$

$$(3) \qquad \hat{A}(z) = \frac{z}{1 - 3z + 2z^2}$$

Step 4. Use partial fractions to solve for a_n.

$$\hat{A}(z) = \sum_{n=0}^{\infty} a_n \frac{z^n}{n!} = \frac{1}{1 - 2z} - \frac{1}{1 - z} = \sum_{n=0}^{\infty} (2^n - 1) z^n$$

$$(4) \qquad \Rightarrow \quad a_n = (2^n - 1) n!$$

Check the Answer: We now verify that the answer $a_n = (2^n - 1)n!$ satisfies the recurrence.

$$
\begin{aligned}
a_0 &= (2^0 - 1) \, 0! = 0, & \text{(initial condition)} \\
a_1 &= (2^1 - 1) \, 1! = 1; & \text{(initial condition)} \\
a_n &= 3n \, a_{n-1} - 2n(n-1) \, a_{n-2} & \text{(recursion)} \\
&= 3n \cdot (2^{n-1} - 1)(n-1)! - 2n(n-1) \cdot (2^{n-2} - 1)(n-2)! \\
&= 3n! \cdot (2^{n-1} - 1) - 2n! \cdot (2^{n-2} - 1) \\
&= n! \cdot (3 \cdot 2^{n-1} - 3) - n! \cdot (2^{n-1} - 2) \\
&= n! \cdot (2 \cdot 2^{n-1}) - 3n! + 2n!) \\
&= n! \, (2^n - 1)
\end{aligned}
$$

What enables the substitution of the EGF $\hat{A}(z)$ to lead to the successful conclusion of Example 5.5.8 is that in the recursion

$$a_n = 3n a_{n-1} - 2n(n-1) a_{n-2} \quad \text{for } n \geq 2$$

the variable coefficients of a_{n-1} and a_{n-2} are the falling power monomials $3n^{\underline{1}}$ and $-2n^{\underline{2}}$, of degrees 1 and 2, respectively. Fortunately, the variable coefficient of the derangement recurrence has the same property. The non-homogeneous part adds a small complication.

Theorem 5.5.4. Let $\hat{D}(z)$ be the EGF for the derangement numbers D_n. Then

$$\hat{D}(z) = \frac{e^{-z}}{1-z}$$

Proof: This proof follows the paradigm of Example 5.5.8.

$$D_n = nD_{n-1} + (-1)^n \qquad \text{(Proposition 5.4.1)}$$

$$\Rightarrow \sum_{n=1}^{\infty} D_n \frac{z^n}{n!} = \sum_{n=1}^{\infty} nD_{n-1}\frac{z^n}{n!} + \sum_{n=1}^{\infty}(-1)^n\frac{z^n}{n!}$$

$$\Rightarrow \hat{D}(z) - D_0\frac{z^0}{0!} = z\sum_{n=1}^{\infty} D_{n-1}\frac{z^{n-1}}{(n-1)!} + e^{-z} - 1$$

$$\Rightarrow \hat{D}(z) - 1 = z\hat{D}(z) + e^{-z} - 1$$

$$\Rightarrow \hat{D}(z) = \frac{e^{-z}}{1-z} \qquad\qquad\qquad\qquad\qquad\qquad \diamond$$

Corollary 5.5.5.   Let $\langle D_n \rangle$ be the derangement sequence. Then

$$D_n = n!\left[\frac{1}{0!} - \frac{1}{1!} + \frac{1}{2!} - \frac{1}{3!} + \cdots + (-1)^n\frac{1}{n!}\right]$$

Proof: To rederive Theorem 5.4.2, this time as a corollary to Theorem 5.5.5, we proceed as follows:

$$\hat{D}(z) = \sum_{n=0}^{\infty} D_n\frac{z^n}{n!} = \frac{e^{-z}}{1-z} = \frac{1}{1-z}\left[\frac{1}{0!} - \frac{z}{1!} + \frac{z^2}{2!} - \frac{z^3}{3!} + \cdots\right]$$

We recognize $(1-z)^{-1}$ as a summing operator.

$$\Rightarrow \frac{D_n}{n!} = \left[\frac{1}{0!} - \frac{1}{1!} + \frac{1}{2!} - \frac{1}{3!} + \cdots + (-1)^n\frac{1}{n!}\right]$$

$$\Rightarrow D_n = n!\left[\frac{1}{0!} - \frac{1}{1!} + \frac{1}{2!} - \frac{1}{3!} + \cdots + (-1)^n\frac{1}{n!}\right] \qquad\qquad \diamond$$

EXERCISES for Section 5.5

In each of the Exercises 5.5.1 through 5.5.4, write an exponential generating function for the given function of n.

5.5.1[S] n 5.5.2 $n^{\underline{2}}$

5.5.3 n^2 5.5.4 $n!$

In each of the Exercises 5.5.5 through 5.5.8, write an exponential generating function for the sequence whose n^{th} entry is the number of binary strings of length n with the given property.

5.5.5[S]   At least two 0's and at least two 1's.

5.5.6   An odd number of 0's and an odd number of 1's.

5.5.7   At most three 0's.

5.5.8   At most three 0's and an even number of 1's.

In each of the Exercises 5.5.9 through 5.5.12, write a closed formula for the number of binary strings of length n with the given property.

5.5.9[S]   At least two 0's and at least two 1's.

5.5.10   An odd number of 0's and an odd number of 1's.

5.5.11   At most three 0's.

5.5.12   At most three 0's and an even number of 1's.

In each of the Exercises 5.5.13 through 5.5.16, use an EGF to solve the given recurrence.

5.5.13[S] $a_0 = 0;$ $a_n = na_{n-1} + n!$ for $n \geq 1$.

5.5.14 $a_0 = 0;$ $a_n = na_{n-1} + (n+1)!$ for $n \geq 1$.

5.5.15[S] $a_0 = 1;$ $a_n = na_{n-1} + 1$ for $n \geq 1$.

5.5.16 $a_0 = 1,\ a_1 = 2;$ $a_n = 5na_{n-1} - 6n(n-1)a_{n-2}!$ for $n \geq 2$.

5.6   POSETS AND LATTICES

Within the combinatorial systems we have already explored, there have occurred in the near periphery various partially ordered sets, a.k.a. *posets*. In this section, we reexamine these systems, this time with explicit attention to their structure as posets. In so doing, we also undertake a quick survey of the basic theory of posets. A *lattice* is a highly structured kind of poset.

FROM APPENDIX A3:

- A **partial ordering** on a set P is a binary relation $\preceq$ with the following properties, for all $x, y, z \in P$:

  i. $x \preceq x$ (*reflexive*)

  ii. if $x \preceq y$ and $y \preceq x$ then $x = y$ (*antisymmetric*)

  iii. if $x \preceq y$ and $y \preceq z$ then $x \preceq z$ (*transitive*)

- Two elements x, y in a poset P such that either $x \preceq y$ or $y \preceq x$ are said to be **comparable**.

- If $x \preceq y$, we may say that y **dominates** x.

- We write $x \prec y$ if $x \preceq y$ and $x \neq y$.

- The structure $\mathcal{P} = \langle P, \preceq \rangle$ is called a **partially ordered set** or a **poset**. The set P is called the **domain of the poset**.

- Writing or saying "the poset P" (giving the *domain* of the poset, rather than the complete structure) is commonplace and convenient.

- The **order of a poset** $\mathcal{P} = \langle P, \preceq \rangle$ is the cardinality of its domain P. Informally, the word *size* is also used.

- A **subposet** of a poset $\langle P, \preceq \rangle$ is a subset $S \subseteq P$, in which $x \preceq_S y$ if and only if $x \preceq_P y$.

Products of Sets

One of the basic ways in which a poset arises in applications is when subjects are scaled in more than one attribute. Although each of the scales may be totally ordered, when considered simultaneously there is only a partial ordering. For instance, the possible College Board scores for the mathematics and verbal exams are numbers from 200 to 800. However, pairs of scores are only partially ordered, according to the rule in Example 5.6.1, unless one adds a context-dependent secondary rule for such comparisons.

Example 5.6.1: The cartesian product $[m : n] \times [r : s]$ of two integer intervals is partially ordered under the rule

$$(a, b) \leq (c, d) \text{ if } a \leq c \text{ and } b \leq d$$

This construction can also be generalized to an iterated product over arbitrarily many integer intervals or, indeed, over arbitrarily many posets.

Cover Digraph

Several digraphs and graphs are associated with a poset. The most useful is the *cover digraph*.

DEFINITION: If $x \prec t \prec y$ in a poset $\langle P, \preceq \rangle$, then t is called an **intermediate element** between x and y.

DEFINITION: If $x \prec y$ and if there is no intermediate element t, then y **covers** x.

DEFINITION: The **cover digraph** of a poset $\langle P, \preceq \rangle$ has the elements of the set P as its vertices. There is an arc from x to y if and only if x is covered by y. The **cover graph** is its underlying graph. A **Hasse diagram** for the poset is a drawing of the cover graph in which the dominant of any two comparable elements must appear above the other.

Example 5.6.1, continued: Figure 5.6.1 illustrates the cover digraph and Hasse diagram of the poset $[0:1] \times [0:2]$.

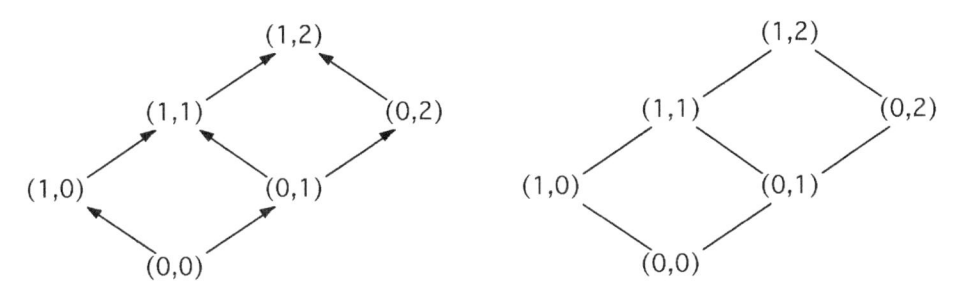

Figure 5.6.1   **Cover digraph and Hasse diagram of** $[0:1] \times [0:2]$.

For any poset $\mathcal{P} = \langle P, \preceq \rangle$, we may observe that $x \preceq y$ if and only if there is a *directed path* from x to y in the cover digraph, by which we mean a sequence of edges, aligned head to tail, proceeding from vertex x to vertex y. The digraph corresponding directly to the partial ordering itself is called the *comparability digraph*, and its underlying graph is called the *comparability graph*.

The Boolean Poset

The *boolean poset* is among the most familiar partially ordered structures.

DEFINITION: The **boolean poset**

$$\mathcal{B}_n = \langle 2^{[1:n]}, \subseteq \rangle$$

has as its domain the set of subsets of $[1:n]$. They are partially ordered by set-theoretic inclusion.

Example 5.6.2:   Figure 5.6.2 shows a cover digraph for the boolean poset $\mathcal{B}_4$.

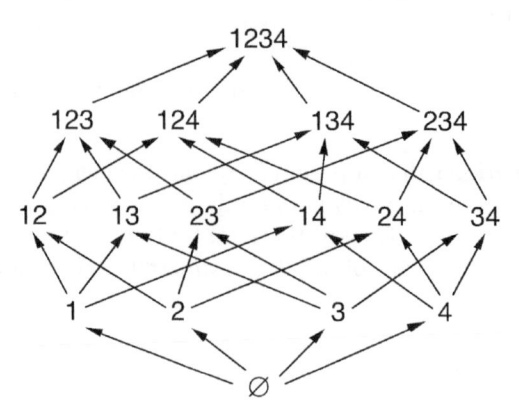

Figure 5.6.2   The boolean poset $\mathcal{B}_4$.

Various properties of the boolean poset $\mathcal{B}_n$ can be observed in Figure 5.6.2. For instance, at level k, the number of subsets is the binomial coefficient $\binom{n}{k}$. Also, the subset Y covers the subset X if $X \subseteq Y$ and if $Y - X$ is a single element of $[1 : n]$.

The Divisibility Poset

REVIEW FROM §3.1:

  • The notation $k \setminus n$ means that the integer k divides the integer n.

DEFINITION: In the **divisibility poset** $\mathcal{D}_n = \langle D_n, \, \setminus \, \rangle$, the domain is the set

$$D_n \; = \; \bigl\{ k \in [1 : n] \; \big| \; k \setminus n \bigr\}$$

and the relation is *divisor of*. The **infinite divisibility poset** $\mathcal{D} = \langle \, \mathbb{Z}^+, \, \setminus \, \rangle$ has as its domain the set of all positive integers.

Under the divisibility relation, y covers x if the quotient $\frac{y}{x}$ is prime.

Example 5.6.3:   Figure 5.6.3 illustrates the divisibility poset $\mathcal{D}_{72}$.

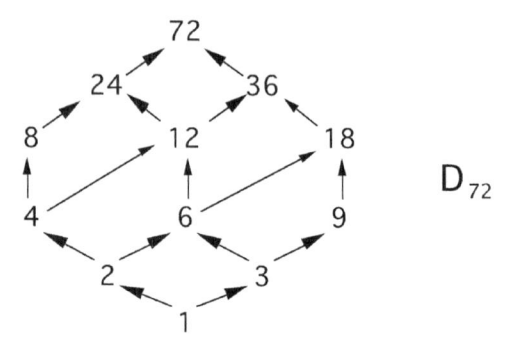

Figure 5.6.3   Cover diagram of the divisibility poset $\mathcal{D}_{72}$.

The Partition Poset

DEFINITION: A partition V of a set S is a **refinement of the partition** U if every cell of V is a subset of some cell of U. This relation is denoted $U \sqsupseteq V$.

DEFINITION: In the **partition poset** $\mathcal{P}_n = \langle P_n, \sqsupseteq \rangle$, the subsets of the integer interval $[1:n]$ are partially ordered by the refinement relation.

NOTATION: To avoid cluttering the representation of a partition, it is sometimes convenient to write the contents of the cells as strings of objects, thereby eliminating commas, and to separate the cells only by dashes, thereby eliminating the braces around each cell.

Example 5.6.4:   We now consider an *ad hoc* calculation of a Stirling subset number. The integer interval $[1:4]$ can be partitioned into 3 cells in 6 ways:

$$
\begin{array}{lll}
12\text{–}3\text{–}4 & 13\text{–}2\text{–}4 & 14\text{–}2\text{–}3 \\
23\text{–}1\text{–}4 & 24\text{–}1\text{–}3 & 34\text{–}1\text{–}2
\end{array}
\qquad
\begin{Bmatrix} 4 \\ 3 \end{Bmatrix} = 6
$$

Example 5.6.5:   A partition V covers a partition U if it splits a single cell of U into two non-empty subcells. Figure 5.6.4 illustrates a cover diagram for the partition lattice $\mathcal{P}_4$. Hyphens are used to delimit the cells.

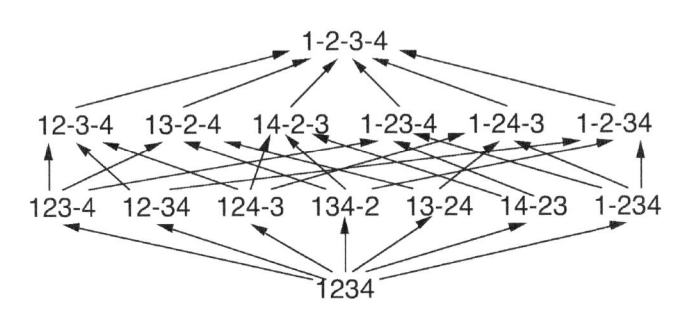

Figure 5.6.4   The partition poset $\mathcal{P}_4$.

The Inversion-Dominance Ordering on Permutations

NOTATION: The set of all permutations of the integer interval $[1:n]$ is denoted Σ_n. (Under the composition of permutations, it is a group, in the sense of Appendix A2, called the *symmetric group*.)

DEFINITION: The **inversion-dominance relation**

$$\pi \preceq \tau$$

on Σ_n means that every inversion of π is also an inversion of τ.

Example 5.6.6:   The permutation $\pi = 1342$ has two inversions, namely

$$\pi(4) < \pi(2) \quad \text{and} \quad \pi(4) < \pi(3)$$

In addition to those inversions, the permutation $\tau = 3142$ has both those inversions and the inversion

$$\tau(2) < \tau(1)$$

as well. Thus, $1342 \preceq 3142$.

DEFINITION: The **inversion poset** $\mathcal{I}_n = \langle \Sigma_n, \preceq \rangle$ is the partially ordered set whose domain is the set of partitions on $[1 : n]$, with the inversion-dominance relation $\pi \preceq \tau$ as its partial ordering.

Example 5.6.7:   A digraph representing the cover digraph of $\mathcal{I}_4$ is drawn in Figure 5.6.5 so as to embody the shape of the truncated octahedron, whose 1-skeleton is the underlying graph. Observe that the direction of the arcs is away from 1234, the least inverted permutation, and toward 4321, the most inverted.

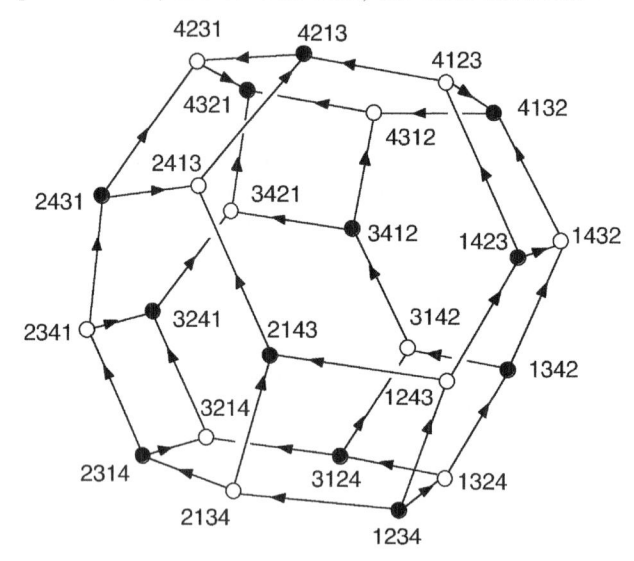

Figure 5.6.5   **Cover digraph of the inversion poset $\mathcal{I}_4$.**

The underlying graph is obtained by drawing an edge between two permutations whose one-line representations differ only by a single transposition of adjacent integers. The direction reflects increasing the number of inversions.

Minimal and Maximal Elements

DEFINITION: A **minimal element in a poset** P is an element x such that there is no element w with $w \prec x$. If $x \preceq y$ for every $y \in P$, then x is the **minimum element**.

DEFINITION: A **maximal element in a poset** P is an element y such that there is no element w with $y \prec w$. If $x \preceq y$ for every $x \in P$, then y is the **maximum element**.

Example 5.6.8:   In Figure 5.6.6, there is no minimum or maximum element. However, the elements a and b are maximal, and the elements d, j, and k are minimal.

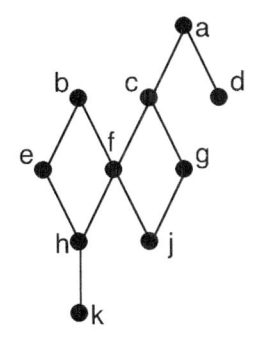

Figure 5.6.6   A poset with two maximal elements and three minimal elements.

Example 5.6.2, continued: The minimum element of the boolean poset $\mathcal{B}_n$ is the empty set $\emptyset$, and the maximum element is the entire set $\{1, 2, \ldots, n\}$.

Example 5.6.3, continued: The minimum element of the divisibility poset $\mathcal{D}_n$ is the number 1, and the maximum element is the number n. The infinite divisibility poset $\mathcal{D}$ has no maximum.

Example 5.6.5, continued: The minimum element of the partition poset $\mathcal{P}_n$ is the unpartitioned set $[1 : n]$, and the maximum element is the partition

$$1 - 2 - \cdots - n$$

into singletons.

Example 5.6.7, continued: The minimum element of the inversion poset $\mathcal{I}_n$ is the permutation $12\ldots n$, and the maximum element is the permutation $n(n-1)\ldots 1$.

Lattice Property

Informally, a *lattice* could be described as a poset with no loose ends. More rigorously, it involves the existence of upper and lower bounds for pairs of elements. It is the culmination of the following list of definitions.

DEFINITIONS:

- An **upper bound** for a subset S of a poset P is an element u such that $s \preceq u$ for all $s \in S$.

- A **lower bound** for a subset S of a poset P is an element w such that $w \preceq s$ for all $s \in S$.

- A **least upper bound** for a subset S of a poset P is an upper bound u such that if z is any other upper bound for S, then $u \preceq z$. We commonly write $lub(x, y)$ for the least upper bound of a subset of two elements, which, if it exists, must be unique, by the antisymmetry property.

- A **greatest lower bound** for a subset S of a poset P is a lower bound w for S such that if z is any other lower bound for S, then $z \preceq w$. We commonly write $glb(x, y)$ for the greatest lower bound of a subset of two elements, which, if it exists, must be unique, by the antisymmetry property.

- A **lattice** is a poset such that every pair of elements has a lub and a glb.

Example 5.6.2, continued: The boolean poset $\mathcal{B}_n$ is a lattice, in which the least upper bound of two subsets is their union and the greatest lower bound is their intersection.

Example 5.6.3, continued: The divisibility lattices $\mathcal{D}$ and $\mathcal{D}_n$ are lattices, in which the least upper bound of two numbers is their least common multiple and the greatest lower bound is their greatest common divisor.

Proving that the partition poset is a lattice involves a few details regarding the least upper and greatest lower bounds.

Example 5.6.5, continued: The partition poset $\mathcal{P}_n$ is a lattice. The constructions of the least upper bound and the greater lower bound are now given.

NOTATION: In the partition lattice $\mathcal{P}_n$, let $U \vee V$ denote the set of non-empty intersections of a cell of a partition U with a cell of another partition V.

Example 5.6.9: Let U be the partition $123 - 45 - 678$ and let V be the partition $14 - 235 - 67 - 8$. Then $U \vee V = 1 - 23 - 4 - 5 - 67 - 8$.

Proposition 5.6.1. *In the partition poset $\mathcal{P}_n$, the partition $U \vee V$ is the least upper bound of partitions U and V.*

Proof: See Exercises. $\diamond$

NOTATION: Let U and V be two partitions of the integer interval $[1 : n]$. Then

- Let $K_{U,V}$ denote the bipartite graph whose partite sets are the cells of U and the cells of V, respectively, and where a cell of U is adjacent to a cell of V if they have a vertex in common.

- Let $U \wedge V$ denote the partition of $[1 : n]$, each of whose cells is the union of the vertices in a component of $K_{U,V}$.

Example 5.6.9, continued: Let U be the partition $123 - 45 - 678$ and let V be the partition $14 - 235 - 67 - 8$. The graph $K_{U,V}$ is shown in Figure 5.6.7.

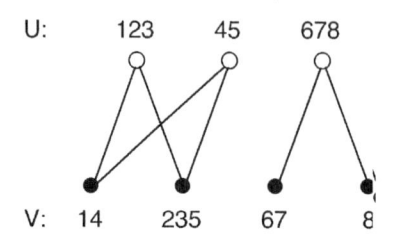

Figure 5.6.7   **The bipartite graph $K_{U,V}$ for two partitions.**

Then $U \vee V = 1 - 23 - 4 - 5 - 67 - 8$.

Proposition 5.6.2. *In the partition poset* $\mathcal{P}_n$, *the partition* $U \wedge V$ *is the greatest lower bound of partitions* U *and* V.

Proof: See Exercises. $\diamond$

Example 5.6.10: The poset whose cover diagram appears in Figure 5.6.8 is not a lattice, because although d and e are both common lower bounds for b and c, neither is a lower bound for the other.

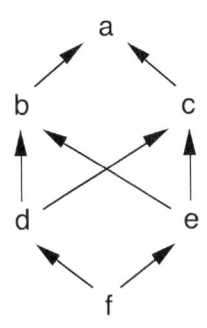

Figure 5.6.8 **A poset that is not a lattice.**

Poset Isomorphism

DEFINITION: An ***isomorphism of posets*** $\langle P, \preceq_P \rangle$ and $\langle Q, \preceq_Q \rangle$ is a bijection

$$f : P \to Q$$

such that $x \preceq_P y$ in P if and only if $f(x) \preceq_Q f(y)$ in Q.

Example 5.6.11: The divisibility poset $\mathcal{D}_{12}$ is isomorphic to the poset of integer pairs $[0:1] \times [0:2]$, under the bijection

$$1 \to (0,0) \quad 2 \to (0,1) \quad 3 \to (1,0)$$
$$4 \to (0,2) \quad 6 \to (1,1) \quad 12 \to (1,2)$$

Figure 5.6.9 shows the Hasse diagram for the poset $\mathcal{D}_{12}$.

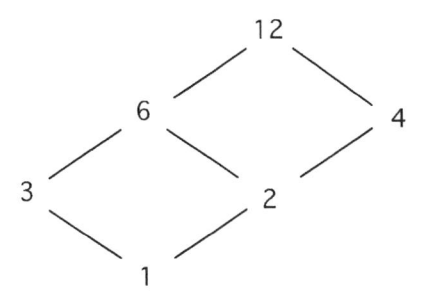

Figure 5.6.9 **Hasse diagram of the divisibility poset** $\mathcal{D}_{12}$.

Example 5.6.12: The divisibility poset $\mathcal{D}_{30}$ is isomorphic to the boolean poset $\mathcal{B}_3$ under the bijection

$$1 \rightarrow \emptyset \qquad 2 \rightarrow \{1\} \qquad 3 \rightarrow \{2\} \qquad 5 \rightarrow \{3\}$$
$$6 \rightarrow \{1,2\} \quad 10 \rightarrow \{1,3\} \quad 15 \rightarrow \{2,3\} \quad 30 \rightarrow \{1,2,3\}$$

Example 5.6.13: Figure 5.6.10 shows Hasse diagrams for the five isomorphism types of posets of size 3. The only one of them that is a lattice is at the far right.

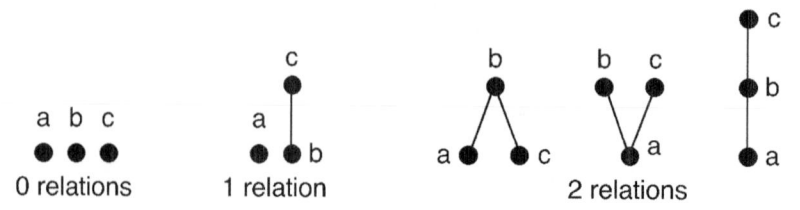

Figure 5.6.10   Hasse diagrams of the posets of size 3.

Observe that two of the posets of size 3 with 2 relations have isomorphic cover graphs (see §7.4). This complicates classifying the isomorphism types of posets of a given size. Also observe that not all simple graphs can occur as cover graphs, as indicated by Proposition 5.6.3.

Proposition 5.6.3. *The cover graph of a poset $\langle P, \preceq \rangle$ cannot contain a 3-cycle.*

Proof: Suppose that elements $u, v, w \in P$ form a 3-cycle in the cover graph. Then in each pair, one element must cover the other. By transitivity, there cannot be a cycle in the cover digraph, so one of them, say u, must cover neither of the others, and another, say w, must cover both the others. But then $u \prec v \prec w$, which implies that w does not cover u. $\diamond$

Chains and Antichains

There are two extreme forms of posets. At one extreme, in a *chain*, every pair of elements is comparable. At the other, in an *antichain*, no two elements are comparable.

DEFINITIONS: Here are a few related definitions:

- If every two elements of a poset $\langle P, \preceq \rangle$ are comparable, then $\langle P, \preceq \rangle$ is said to be **totally ordered**, **linearly ordered**, or a **chain**.

- A poset in which all elements are incomparable is called a **clutter** or an **antichain**.

- The **height of a poset** is the cardinality of a longest chain.

- The **width of a poset** is the cardinality of a maximum-size antichain.

A collection of elements of a poset forms a chain if and only if there is a directed path in the cover digraph from the vertex corresponding to one of them to the vertex corresponding to another of them, with the vertices corresponding to all the others as interior vertices along the way. A collection of elements of a poset forms an antichain if the corresponding vertices are mutually unreachable in the cover digraph.

NOTATION: It is common practice to refer to a poset, at times, by its domain, that is, writing simply P for $\langle P, \preceq \rangle$.

Posets have some general structural properties. The following two are among the most easily proved.

Proposition 5.6.4. *Let $\langle P, \preceq \rangle$ be a poset, let C be a chain in P, and let A be an antichain. Then the intersection $A \cap C$ contains at most one element.*

Proof: Let x and y be any elements of the poset $\langle P, \preceq \rangle$. If $x, y \in C$, then they are comparable. If $x, y \in A$, then they are incomparable. $\diamond$

Theorem 5.6.5. *Let $\langle P, \preceq \rangle$ be a finite poset of height h. Then P can be partitioned into h antichains, and into no fewer than h antichains.*

Proof: By Proposition 5.6.4, it follows that an antichain contains at most one element of a longest chain C. Thus, the number of antichains whose union contains C is at least h, the number of elements in chain C.

Proof that the poset $\langle P, \preceq \rangle$ can be partitioned into h antichains is by induction on the height h.

BASIS: If $h = 1$, then the poset $\langle P, \preceq \rangle$ itself is an antichain.

IND HYP: Assume that such a partition exists for $h = n - 1$.

IND STEP: Suppose that height $h = n$. Let A_1 be the antichain containing all minimal elements of the poset P. Then the longest chain in the subposet $P - A_1$ is of length $n - 1$. By the induction hypothesis, it follows that the subposet $P - A_1$ can be partitioned into $n - 1$ antichains. $\diamond$

Example 5.6.2, continued: A chain in the boolean poset $\mathcal{B}_n$ is a sequence of sets, each nested in its successor. Thus the height of the poset $\mathcal{B}_n$ is $n + 1$, corresponding to starting with the empty set and including one additional element at a time. An antichain is a collection of subsets, no two of which are nested. The collection U_k of subsets of size k is an antichain. Clearly, the boolean poset $\mathcal{B}_n$ can be partitioned into these $n + 1$ collections U_k, for $k = 0, \ldots, n$.

Example 5.6.3, continued: A chain in the divisibility poset $\mathcal{D}_n$ is a sequence of numbers, each of which is a multiple of its predecessor. It follows that the height of the divisibility poset $\mathcal{D}_n$ is 1 plus the sum of the exponents in the prime factorization of n. The subset E_k of numbers whose exponent sum is k is an antichain. Clearly, the divisibility poset $\mathcal{D}_n$ can be partitioned into these collections E_k, as illustrated in Figure 5.6.11. For instance, $12 = 2^2 3^1$, so the exponent sum is $3 = 2 + 1$, which implies that four antichains are necessary and sufficient.

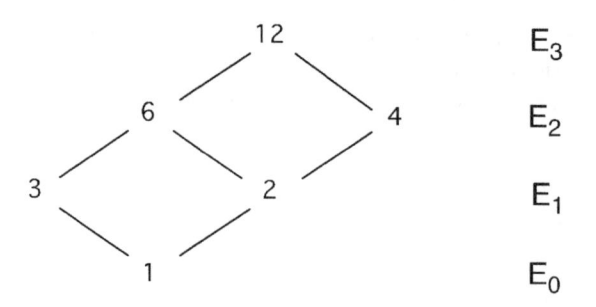

Figure 5.6.11   **Partitioning the poset $\mathcal{D}_{12}$ into four antichains.**

Ranked Posets

The various posets we have examined carefully — boolean posets, divisibility poset, partition posets, and inversion posets — have a common feature that they appear to be layered, so that any traversal of the cover graph is between adjacent ranks. The formal name used for these layers is *ranks*.

DEFINITIONS: Here is another list of related definitions:

- A **rank function** on a poset $\langle P, \preceq \rangle$ is a function $\rho : P \to \mathbb{N}$ such that if the element y covers the element x then $\rho(y) = \rho(x) + 1$.

- A **ranked poset** is a poset with a rank function.

- The kth **rank** of a ranked poset P is the antichain P_k of elements of rank k.

- The kth **Whitney number** $N_k(P)$ of a ranked poset $\langle P, \preceq \rangle$ is the cardinality of the kth rank of P.

Example 5.6.2, continued: The rank function of the boolean poset $\mathcal{B}_n$ assigns to every subset of $[1 : n]$ its number of elements. Thus, the Whitney number $N_k(\mathcal{B}_n)$ is $\binom{n}{k}$.

Example 5.6.3, continued: The rank function of the divisibility poset $\mathcal{D}_n$ assigns to every divisor of n the sum of the exponents in its prime power factorization.

Example 5.6.3, continued: The rank function of the permutation poset $\mathcal{P}_n$ is the number of cells in the partition. The Whitney number $N_r(\mathcal{P}_n)$ is the *Stirling subset number* $\left\{ {n \atop r} \right\}$. For instance, $\mathcal{P}_4$ has $\left\{ {4 \atop 2} \right\} = 7$ elements of rank 2 at the middle level of the cover diagram.

DEFINITION: A poset is **graded** if all maximal chains have the same length.

Proposition 5.6.6. *A graded poset can be ranked.*

Proof: Assign rank $\rho(x) = 0$ to every minimal element x. Then, proceeding recursively, assign rank $\rho(x) + 1$ to an element that covers x. $\diamond$

Proposition 5.6.7. *The inversion poset $\mathcal{I}_n$ is a graded poset.*

Proof: All the maximal chains extend from $12 \cdots n$ to $n(n-1) \cdots 1$ and are of length n. The rank of each permutation is the number of inversions. $\diamond$

Some posets cannot be ranked.

Example 5.6.14:   The poset of Figure 5.6.12 is unrankable. Indeed, any poset with an odd cycle in its cover graph is unrankable.

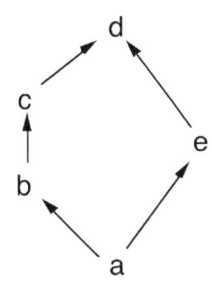

Figure 5.6.12   An unrankable poset.

Linear Extensions

Linear extension of a poset is a way to make a partially ordered set into a totally ordered set.

DEFINITION: An **extension of a poset** $\langle P, \preceq \rangle$ is a poset $\langle P, \preceq^* \rangle$ with the same domain, such that $x \preceq^* y$ whenever $x \preceq y$. Thus, an extension adds one or more relations.

DEFINITION: A **linear extension of a poset** $\langle P, \preceq \rangle$ is an extension that is totally ordered.

Example 5.6.15:   The partial orderings on a set P are partially ordered by extension. The linear extensions are the maximal orderings. The clutter is the minimum ordering.

Proposition 5.6.8.   *Every finite poset* $\langle P, \preceq \rangle$ *has a linear extension.*

Proof: Suppose that $|P| = n$.

BASIS: If $n = 1$, then $\langle P, \preceq \rangle$ is linearly ordered.

IND HYP: Assume that any poset of size $n - 1$ has a linear extension.

IND STEP: Let x be a minimal element of $\langle P, \preceq \rangle$. By the induction hypothesis, there is a linear extension of the poset $P - \{x\}$. Complete the linear extension of P by making x precede every element in the linear extension of the poset $P - \{x\}$. ◇

Example 5.6.16:   Here are three linear extensions of the boolean poset $\mathcal{B}_3$.

$$\emptyset \leq 1 \leq 2 \leq 3 \leq 12 \leq 13 \leq 23 \leq 123$$
$$\emptyset \leq 3 \leq 1 \leq 2 \leq 23 \leq 13 \leq 12 \leq 123$$
$$\emptyset \leq 1 \leq 2 \leq 12 \leq 3 \leq 23 \leq 13 \leq 123$$

DEFINITION: A **topological sort** is an algorithm whose input is a poset $\langle P, \preceq \rangle$, and whose output is a list $\langle x_j \rangle$ of the elements of the domain P of that poset that is consistent with a linear extension of the poset.

In the following algorithm for a topological sort, we take $Min(P)$ to be a function on a non-empty poset that returns a minimal element of the poset.

Algorithm 5.6.1:   Topological Sort

Input: a finite poset $\langle P, \preceq \rangle$ of size n
Output: a roster $\langle x_j \rangle$ of P such that $x_i \preceq x_j$ for $0 \leq i < j < n$

Initialize $j = 0$
while $P \neq \emptyset$
  $x_j := Min(P)$ {returns a minimal element of P}
  $P := P - x_j$
  $j := j + 1$
continue

Dilworth's Theorem

Whereas Theorem 5.6.5 concerns the decomposition of a poset into antichains, there is a complementary theorem of Robert P. Dilworth (1914-1993) that concerns a decomposition into chains. There are two preliminary lemmas.

Lemma 5.6.9. *Let $\langle P, \preceq \rangle$ be a poset, and let L be the set containing all the minimal elements of P. Then L is a maximal antichain.*

Proof: Every element of L is a minimal element in P, so no two are comparable. Thus, L is an antichain. If $y \notin L$, then since y is not a minimal element, there is an element $x \in L$ such that $x \prec y$. It follows that $L \cup \{y\}$ is not an antichain.   ◇

Lemma 5.6.10. *Let $\langle P, \preceq \rangle$ be a poset, and let U be the set that contains all the maximal elements of P. Then U is a maximal antichain.*

Proof: The proof exactly parallels the proof of Lemma 5.6.9.   ◇

Theorem 5.6.11 [Dilw1950]. *Let $\langle P, \preceq \rangle$ be a finite poset of width w. Then P can be partitioned into w chains, and into no fewer than w chains.*

Proof: By Proposition 5.6.4, each chain contains at most one element of any antichain, in particular, of a largest antichain. It follows that the width w is a lower bound on the total number of chains in a partition of P into chains.

Proof that a partition into w chains exists is by induction on the width w, with a secondary induction on the size of the poset P.

BASIS: If $w = 1$, then P itself is a chain.

IND HYP: Assume that such a partition exists, for $w = n - 1$.

IND STEP: Suppose that width $w = n$. If $|P| = n$, then each of the n elements of P serves as a chain. Assume that this is also true for all posets of width n whose size is less than the size of P.

Now let A be a maximum antichain, that is, an antichain of size n.

Case 1. Suppose the following two conditions hold:

(1.1) The antichain A is not the set of all maximal elements.

(1.2) The antichain A is not the set of all minimal elements.

We define the subposets

$$\begin{aligned} {}^{\geq}A &= \{x \in P \mid (\exists a \in A)\,[x \geq a]\} \\ {}^{\leq}A &= \{x \in P \mid (\exists a \in A)\,[x \leq a]\} \end{aligned}$$

Observe that the following two properties hold.

(i) $|{}^{\geq}A| < |P|$.

 Proof of (i). If every minimal element of P were in A, then the subset of A containing only those minimal elements of P would, by Lemma 5.6.9, already in itself be a maximal antichain. This would imply that that subset is the antichain A, which violates condition (1.2). Thus, some minimal element of P cannot be in A. Since it is a minimal element, it cannot dominate any element of A. Hence, that minimal element also cannot be in ${}^{\geq}A$.

(ii) $|{}^{\leq}A| < |P|$.

 Proof of (ii). If every maximal element of P were in A, then the subset of A containing only those maximal elements of P would, by Lemma 5.6.10, already in itself be a maximal antichain. As before, this would imply that that subset is the antichain A, in violation of condition (1.1). Thus, some maximal element of P cannot be in A. Since it is a maximal element, it cannot be dominated by any element of A. Thus, that maximal element cannot be in ${}^{\leq}A$.

Any antichain in the subposet ${}^{\geq}A$ is also an antichain in the poset P. By construction, $A \subseteq {}^{\geq}A$. Thus, the antichain A is a maximum antichain in ${}^{\geq}A$. By the induction hypothesis, it follows from (i) that the subposet ${}^{\geq}A$ can be partitioned into n chains, $B_1, \ldots, B_n$. Since every element of ${}^{\geq}A$ dominates some element of A, and since $A \subseteq {}^{\geq}A$, it follows that the minimal element of each of these chains B_j is some element $b_j \in A$. Since $\{B_1, \ldots, B_n\}$ is a partition, the elements $b_1, \ldots, b_n$ are distinct.

Similarly, it follows from (ii) that the subposet ${}^{\leq}A$ can be partitioned into n chains, $C_1, \ldots, C_n$, that the maximal element of each of these chains C_j is some element $c_j \in A$, and that the elements $c_1, \ldots, c_n$ are distinct.

Since $|A| = n$, we have $A = \{b_1, \ldots, b_n\}$ and $A = \{c_1, \ldots, c_n\}$. Hence, the minimal element b_j of each chain B_j is the maximal element $c_{\pi(j)}$ of some chain $C_{\pi(j)}$, and the union of the two chains is a chain $B_j \cup C_{\pi(j)}$ in poset P, as illustrated in Figure 5.6.13. The chains

$$B_1 \cup C_{\pi(1)}, \ldots, B_n \cup C_{\pi(n)}$$

are a partition of P.

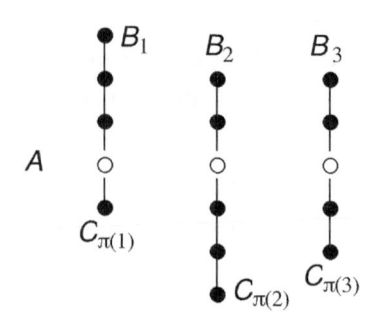

Figure 5.6.13   **Partitioning poset P into 3 chains.**

Case 2. Suppose, alternatively, that there are no antichains of maximum size n, except for either the set of all maximal elements of P or the set of all minimal elements of P (or both). In this case, let u be a minimal element and v a maximal element. Then the size of the largest antichain in the poset $P - \{u, v\}$ is $n - 1$. By the induction hypothesis, the subposet $P - \{u, v\}$ can be partitioned into $n - 1$ chains. These $n - 1$ chains, along with the chain $\{u, v\}$ give a partition of poset P into n chains. $\diamond$

EXERCISES for Section 5.6

5.6.1   Draw a Hasse diagram for the boolean poset $\mathcal{B}_3$.

5.6.2   Draw a Hasse diagram for the divisibility poset $\mathcal{D}_{450}$.

5.6.3   Draw Hasse diagrams for all 16 isomorphism types of posets on 4 elements.

In each of the Exercises 5.6.4 through 5.6.9, find a maximum chain in the given poset.

5.6.4[S]   The boolean poset $\mathcal{B}_3$.   **5.6.5**   The boolean poset $\mathcal{B}_4$.

5.6.6   The divisibility poset $\mathcal{D}_{72}$.   **5.6.7**   The divisibility poset $\mathcal{D}_{450}$.

5.6.8   The partition poset $\mathcal{P}_4$.   **5.6.9**   The partition poset $\mathcal{P}_5$.

In each of the Exercises 5.6.10 through 5.6.15, find a maximum antichain in the given poset.

5.6.10[S] The boolean poset $\mathcal{B}_3$.   **5.6.11** The boolean poset $\mathcal{B}_4$.

5.6.12   The divisibility poset $\mathcal{D}_{72}$.   **5.6.13** The divisibility poset $\mathcal{D}_{450}$.

5.6.14   The partition poset $\mathcal{P}_4$.   **5.6.15** The partition poset $\mathcal{P}_5$.

In each of the Exercises 5.6.16 through 5.6.21, partition the given poset into a minimum number of chains.

5.6.16^S The boolean poset $\mathcal{B}_3$.   5.6.17   The boolean poset $\mathcal{B}_4$.

5.6.18   The divisibility poset $\mathcal{D}_{72}$.   5.6.19   The divisibility poset $\mathcal{D}_{450}$.

5.6.20   The partition poset $\mathcal{P}_4$.   5.6.21   The partition poset $\mathcal{P}_5$.

In each of the Exercises 5.6.22 through 5.6.27, partition the given poset into a minimum number of antichains.

5.6.22^S The boolean poset $\mathcal{B}_3$.   5.6.23   The boolean poset $\mathcal{B}_4$.

5.6.24   The divisibility poset $\mathcal{D}_{72}$.   5.6.25   The divisibility poset $\mathcal{D}_{450}$.

5.6.26   The partition poset $\mathcal{P}_4$.   5.6.27   The partition poset $\mathcal{P}_5$.

In each of the Exercises 5.6.28 through 5.6.33, count the number of linear extensions of the given poset.

5.6.28^S The boolean poset $\mathcal{B}_3$.   5.6.29   The poset of Example 5.6.14.

5.6.30   The divisibility poset $\mathcal{D}_{12}$.   5.6.31   The divisibility poset $\mathcal{D}_{30}$.

5.6.32   The partition poset $\mathcal{P}_3$.   5.6.33   The poset of Example 5.6.10.

5.6.34   Prove that a lattice has a unique maximal element and a unique minimal element.

5.6.35   Prove Proposition 5.6.1, that $U \wedge V = lub(U, V)$ in the partition poset.

5.6.36   Prove Proposition 5.6.2, that $U \vee V = glb(U, V)$ in the partition poset.

5.6.37   Draw the inversion poset $\mathcal{I}_3$.

GLOSSARY

antichain in a poset: a subposet in which no two elements are comparable.

ascent in the one-line representation $a_1 a_2 \cdots a_n$ of a permutation: an index j such that $a_j < a_{j+1}$.

Bell number B_n: the number of partitions of a set of n distinct objects; thus, the sum

$$\left\{ {n \atop 0} \right\} + \left\{ {n \atop 1} \right\} + \cdots + \left\{ {n \atop n} \right\}$$

binomial convolution of two sequences $\langle r_n \rangle$ and $\langle s_n \rangle$:   the sequence $\langle t_n \rangle$ whose n^{th} entry is

$$t_n = \sum_{j=0}^{n} \binom{n}{j} r_j s_{n-j}$$

boolean poset $\mathcal{B}_n$:   the poset of subsets of the integer interval $[1 : n]$.

chain in a poset: a subposet in which any two elements are comparable.

clutter in a poset: synonym for antichain.

comparability digraph of a poset $\langle P, \preceq \rangle$:   the elements of the set P are the vertices; there is an arc from x to y if and only if $x \preceq y$.

comparability graph of a poset: the underlying graph of the comparability digraph.

comparable elements of a poset: two elements such that one dominates the other.

cover digraph of a poset $\langle P, \preceq \rangle$:   the elements of the set P are the vertices; there is an arc from x to y if and only if y covers x.

cover graph of a poset: the underlying graph of the cover digraph.

cover relation in a poset $\langle P, \preceq \rangle$: y covers x if $x \prec y$ and there is no intermediate element t such that $x \prec t \prec y$.

derangement: a permutation with no fixed points.

derangement number D_n: the number of derangements of n objects.

derangement recurrence: the recurrence
$$D_0 = 1, \quad D_1 = 0;$$
$$D_n = (n-1)D_{n-1} + (n-1)D_{n-2}$$

descent in the one-line representation $a_1 a_2 \cdots a_n$ of a permutation: an index j such that $a_j > a_{j+1}$.

divisibility poset (finite) $\mathcal{D}_n$:   the number n and its divisors, under the *divides* relation.

divisibility poset (infinite) $\mathcal{D}$:   the set of all positive integers, under the *divides* relation.

domination in a poset: y dominates x if $x \preceq y$.

dot product of the n-tuples $(x_1, x_2, \ldots, x_n)$ and $(y_1, y_2, \ldots, y_n)$:   the number $x_1 y_1 + x_2 y_2 + \cdots + x_n y_n$.

EGF: standard abbreviation for *exponential generating function*.

Eulerian number $\left\langle {n \atop k} \right\rangle$:   the number of permutations of $[1 : n]$ with exactly k *ascents*.

exponential generating function for a sequence $\langle g_n \rangle$:   any closed form $\hat{G}(z)$ corresponding to the infinite polynomial (or sometimes, the polynomial itself)
$$\sum_{n=0}^{\infty} g_n \frac{z^n}{n!}$$

extension of a poset: $\langle P, \preceq \rangle$:   another poset with the same domain in which all pairs related in P are related in the same way, but some additional pairs may be related.

graded poset: a poset in which all maximal chains have the same length.

greatest lower bound for a subset of a poset: a lower bound that dominates every other lower bound.

Hasse diagram for a poset: a plane drawing of the cover graph in which every element appears above any other element it dominates.

height of a poset: the cardinality of a longest chain.

intermediate element between two related elements x and y of a poset: an element t such that $x \prec t \prec y$.

inversion in a permutation π of the integer interval $[1:n]$: a pair of integers $i < j$ such that $\pi(j) < \pi(i)$.

inversion coefficient $I_n(k)$: the number of permutations of the integer interval $[1:n]$ with exactly k inversions.

inversion poset $\mathcal{I}_n$: the permutations on the integer interval $[1:n]$ under the relation of obtainability by adding inversions.

inversion relation $\pi \preceq \tau$ on permutations of $[1:n]$: means that every inversion of π is also an inversion of τ.

inversion vector of a permutation π of the integer interval $[1:n]$: the vector $b_1 b_2 \cdots b_n$ such that b_j equals the number of larger integers preceding j in the one-line representation of π.

isomorphism of posets: a bijection of domains that preserves all dominances and non-dominances.

lattice: a poset in which every pair of elements has a greatest lower bound and a least upper bound.

least upper bound for a subset of a poset: an upper bound that is dominated by every other upper bound.

linearly ordered poset: a poset in which every pair of elements is comparable.

log-concave sequence $\langle x_n \rangle$: a sequence such that $x_{n-1} x_{n+1} \le x_n^2$ for all $n > 1$.

lower bound for a subset S of a poset: an element that is dominated by every element of S.

maximal element in a poset: an element that is not dominated by any other.

maximum element in a poset: an element that dominates every other.

minimal element in a poset: an element that does not dominate any other.

minimum element in a poset: an element that is dominated by every other.

OGF: standard abbreviation for *ordinary generating function*.

one-line representation of a permutation π on $[1:n]$: the second line of the two-line representation.

ordinary generating function for a sequence $\langle g_n \rangle$: any closed form $G(z)$ corresponding to the infinite polynomial (or sometimes, the polynomial itself)

$$\sum_{n=0}^{\infty} g_n z^n$$

partial ordering: a relation that is reflexive, anti-symmetric, and transitive.

partially ordered set: a set with a partial ordering of its elements.

partition of a set S: a family of mutually disjoint non-empty subsets whose union is S.

partition poset $\mathcal{P}_n$:   the poset of partitions of the integer interval $[1:n]$ under the refinement relation.

Pascal's triangle: a triangle formed from the non-zero binomial coefficients.

poset: a partially ordered set.

rank function on a poset $\langle P, \preceq \rangle$:   a function $f : P \to \mathbb{N}$ such that if y covers x, then $f(y) = f(x) + 1$.

ranked poset: a poset with a rank function.

size of a poset: the cardinality of its domain.

Stirling cycle number $\left[{n \atop r} \right]$:   the number of ways to partition the integer interval $[1:n]$ into r non-empty non-distinct cycles.

Stirling numbers of the first kind: the coefficients $s_{n,k}$ in the summation

$$x^{\underline{n}} = \sum_{k=0}^{n} s_{n,k}\, x^k$$

Stirling numbers of the second kind: the coefficients $S_{n,k}$ in the summation

$$x^n = \sum_{k=0}^{n} S_{n,k}\, x^{\underline{k}}$$

Stirling subset number $\left\{ {n \atop r} \right\}$:   the number of ways to partition the integer interval $[1:n]$ into r non-empty non-distinct cells.

Stirling's triangles: triangles of the non-zero Stirling cycle and subset numbers, respectively, analogous to Pascal's triangle.

subposet of a poset $\langle P, \preceq \rangle$:   a poset whose domain is a subset of P and whose order is consistent with $\preceq$ on P.

topological sort of a poset: a process that produces a linear extension.

total ordering of a poset: same as a linear ordering.

two-line representation of a permutation π on $[1:n]$:   the $2 \times n$ array

$$\begin{pmatrix} 1 & 2 & \cdots & n \\ \pi(1) & \pi(2) & \cdots & \pi(n) \end{pmatrix}$$

Whitney number $N_k(P)$ for a ranked poset P the cardinality of the kth rank.

width of a poset: the cardinality of a maximum-sized antichain.

Chapter **6**

Integer Operators

6.1 Euclidean Algorithm

6.2 Chinese Remainder Theorem

6.3 Polynomial Divisibility

6.4 Prime and Composite Moduli

6.5 Euler Phi-Function

6.6 The Möbius Function

One reason why books on combinatorial methods attract computer science readers is that combinatorial methods are critical to the construction of fast algorithms. As a very basic example, it takes much more time to sum the values of a sequence than to apply a closed formula for that sum. As another example, using inclusion-exclusion as in §3.6 to calculate how many numbers less than n are divisible by one or more numbers in some fixed collection is much faster than examining every number from 1 to n. What motivates packaging this number theory chapter with purely combinatorial methods is largely our interest in computer applications, in which some number theory occurs in conjunction with what might elsewhere be called strictly combinatorial methods. The number theory presented here is more than just beautiful; it's useful. This chapter presents several computational problems for which integer algorithms based on number-theoretic principles are markedly faster than primitive algorithms tied more closely to the definitions.

6.1   EUCLIDEAN ALGORITHM

The *Euclidean algorithm* is a method for calculating the greatest common divisor of two integers. It is faster by far than the primitive method of successive trial divisors and methods based on factoring.

REVIEW FROM §3.1 AND APPENDIX A2:

- Let n and d be integers. If there is an integer q such that $n = dq$, then we say that d **divides** n, and we write $d \setminus n$.

  - A **prime number** is a positive integer $p > 1$ such that p has no divisors except 1 and itself.

  - Let m and n be integers whose greatest common divisor is 1. Then we say that m and n are **relatively prime**. Notation $m \perp n$.

  - The **Fundamental Theorem of Arithmetic** is that every positive integer n has a unique representation as a product of powers of ascending primes.

$$n = p_1^{e_1} p_2^{e_2} \cdots p_r^{e_r}$$

(The number 1 is representable as the empty product.)

Successive Trial-Divisors Algorithm

A primitive algorithm for calculating $\gcd(m, n)$ is to determine which of the numbers from 1 through m divides both m and n, and to save the largest common divisor, so that the trial divisors are considered in ascending order. The time required for this is proportional to $\min\{m, n\}$, regardless of the value of $\gcd(m, n)$. Algorithm 6.1.1 considers trial divisors in decreasing order, thereby terminating the first time it finds a common divisor, so it runs relatively faster than the ascending version whenever $\gcd(m, n) > 1$. Of course, the worst-case time remains proportional to m.

Algorithm 6.1.1:   Near-Primitive GCD Method

Input: non-negative integers m, n, not both 0
Output: $\gcd(m, n)$

Function $GCD1(m, n)$
if $\min\{m, n\} = 0$ **then return** $\max\{m, n\}$;
for $d := \min\{m, n\}$ **to** 1 **step** -1
    if $d \setminus m$ **and** $d \setminus n$ **then return** d;
    continue

The following minor modification of Algorithm 6.1.1 considers only the possible divisors $d = \lfloor m/k \rfloor$ for $k = 1, \ldots, \lfloor \sqrt{m} \rfloor$. This decreases the worst-case number of iterations of the loop to $\lfloor \sqrt{m} \rfloor$.

> **Algorithm 6.1.2:   Elementary GCD Method**
>
> *Input*: integers m, n, with $0 \le m \le n$ and $0 \ne n$
> *Output*: $\gcd(m, n)$
>
> Function $GCD2(m, n)$
> **if** $m = 0$ **then return** n;
> **for** $k := 1$ **to** $\lfloor \sqrt{m} \rfloor$
> $d := \lfloor m/k \rfloor$;
>   **if** $d \backslash m$ **and** $d \backslash n$ **then return** d;
>   **continue**

Prime-Decomposition Method

A different method for calculating the greatest common divisor of the numbers m and n, and their least common multiple as well, is commonly taught in an early school grade. It starts with a factorization of m and n into primes.

$$m = 2^{d_2} \cdot 3^{d_3} \cdot 5^{d_5} \cdots$$
$$n = 2^{e_2} \cdot 3^{e_3} \cdot 5^{e_5} \cdots$$

It then applies the rule

$$\gcd(m, n) = 2^{\min\{d_2,e_2\}} \cdot 3^{\min\{d_3,e_3\}} \cdot 5^{\min\{d_5,e_5\}} \cdots \quad (6.1.1)$$
$$\operatorname{lcm}(m, n) = 2^{\max\{d_2,e_2\}} \cdot 3^{\max\{d_3,e_3\}} \cdot 5^{\max\{d_5,e_5\}} \cdots \quad (6.1.2)$$

Example 6.1.1:   Here are two prime-power factorizations.

$$720 = 2^4 \cdot 3^2 \cdot 5$$
$$168 = 2^3 \cdot 3 \cdot 7$$

We now apply the elementary school method.

$$\gcd(720, 168) = 2^{\min\{4,3\}} \cdot 3^{\min\{2,1\}} \cdot 5^{\min\{1,0\}} \cdot 7^{\min\{0,1\}}$$
$$= 2^3 \cdot 3 = 24$$
$$\operatorname{lcm}(720, 168) = 2^{\max\{4,3\}} \cdot 3^{\max\{2,1\}} \cdot 5^{\max\{1,0\}} \cdot 7^{\max\{0,1\}}$$
$$= 2^4 \cdot 3^2 \cdot 5 \cdot 7 = 5040$$

This method for calculating the greatest common divisor works well if either prime factorization is known. If neither is known, it may take some effort to calculate the prime factors. When it is taught at lower school levels, the presumption is that the user *already knows the prime factorizations* of the two numbers. The following example illustrates what happens when this is not the case.

Example 6.1.2:   Hand-calculator evaluation of

$$\gcd(6469901, 11503649)$$

by prime power factorizations is daunting, because those factorizations are not immediately at hand, and they must be calculated to proceed with the easier step. This greatest common divisor is evaluated quickly by the Euclidean algorithm, as will be shown presently.

Quotient and Mod Functions

Some basic concepts from integer division are used in the Euclidean algorithm.

DEFINITION: The **integer quotient** of dividing an integer $n \geq 0$ by an integer $d > 0$ is defined recursively (in effect, by repeated subtraction)

$$\text{quotient}\,(n,\ d)\ =\ \begin{cases} 0 & \text{if } n < d \\ 1 + \text{quotient}\,(n - d,\ d) & \text{otherwise} \end{cases}$$

Remark: Equivalently, for $n \geq 0$ and $d > 0$,

$$\text{quotient}\,(n,\ d)\ =\ \left\lfloor \frac{n}{d} \right\rfloor$$

DEFINITION: The **remainder (or residue)** of dividing an integer $n \geq 0$ by an integer $d > 0$ is the number

$$n \bmod d\ =\ n - \text{quotient}\,(n,\ d) \cdot d$$

The associated binary operation is called the **mod function**, as previously noted in §1.1.

Example 6.1.1, continued: For 720 as dividend and 168 as divisor, we have

$$\text{quotient}\,(720,\ 168)\ =\ \left\lfloor \frac{720}{168} \right\rfloor\ =\ 4$$

and

$$720 \bmod 168\ =\ 720 - 4 \cdot 168$$
$$=\ 720 - 672$$
$$=\ 48$$

Proposition 6.1.1. *The integer pairs $\{m,\ n\}$ and $\{m,\ n + km\}$ have the same set of common divisors, for every integer k.*

Proof: Let d be any common divisor of m and n, say $m = rd$ and $n = sd$. Then

$$m + kn\ =\ rd + ksd\ =\ (r + ks)\,d$$

Thus, d divides $m + kn$. In the opposite direction, if $m = rd$ and $n + km = td$, then
$$n \; = \; td \, - \, krd \; = \; (t - kr)\, d \hspace{4em} \Diamond$$

Corollary 6.1.2. *For every pair of integers m and n such that $0 < m \le n$,*
$$\gcd(n, \, m) \; = \; \gcd(m, \, n \bmod m)$$

Proof: Suppose that $q \, = \,$ quotient $(n, \, m)$. Then

$$
\begin{aligned}
\gcd(n, \, m) \; &= \; \gcd(m, \, n - qm) &&\text{(by Proposition 6.1.1)} \\
&= \; \gcd(m, \, n \bmod m) &&\text{(since } n \bmod m \, = \, n - qm) \hspace{2em} \Diamond
\end{aligned}
$$

The strategy of the **Euclidean algorithm** is to apply Corollary 6.1.2 recursively. The following version captures this idea.

Algorithm 6.1.3:   Recursive Euclidean Algorithm

Input: integers $n, m \ge 0$, not both 0
Output: $\gcd(n, \, m)$

  Recursive Function gcd $(n, \, m)$
  If $n = 0$ **then return** m;
  If $m = 0$ **then return** n;
   else return gcd $(m, \, n \bmod m)$

Example 6.1.1, continued: This easy calculation illustrates the method.

$$
\begin{aligned}
\gcd(720, \, 168) \; &= \; \gcd(168, \, 48) \\
&= \; \gcd(48, \, 24) \\
&= \; \gcd(24, \, 0) \\
&= \; 24
\end{aligned}
$$

Example 6.1.2, continued: Here the calculations are mildly tedious, yet much easier than trying to factor the two numbers.

$$
\begin{aligned}
\gcd(11503649, \, 6469901) \; &= \; \gcd(6469901, \, 5033748) \\
&= \; \gcd(5033748, \, 1436153) \\
&= \; \gcd(1436153, \, 725289) \\
&= \; \gcd(725289, \, 710864) \\
&= \; \gcd(710864, \, 14425) \\
&= \; \gcd(14425, \, 4039) \\
&= \; \gcd(4039, \, 2308) \\
&= \; \gcd(2308, \, 1731) \\
&= \; \gcd(1731, \, 577) \\
&= \; \gcd(577, \, 0) \\
&= \; 577
\end{aligned}
$$

Proposition 6.1.3. *Given two numbers n and m, with $n \geq m$, let f_r be the smallest Fibonacci number that exceeds n. Then the number of recursive calls in the Euclidean algorithm is at most r.*

Proof: Suppose that there are s calls. Then let $n_0, n_1, \ldots, n_s$ be the sequence of values of the first argument in the successive calls. Thus,

$$n_0 = n \quad \text{and} \quad n_s = \gcd(n, m)$$

We observe that $n_s \geq 1 > f_0$ and that $n_{s-1} \geq 2 > f_1$. It follows by induction, in general, that

$$n_{s-k-2} > f_{k+2} = f_{k+1} + f_k$$

because

$$n_{s-k-2} \geq n_{s-k-1} + n_{s-k} > f_{k+1} + f_k$$

In particular, $n_0 > f_s$. Therefore, $s < r$. $\Diamond$

Remark: Intuitively, the number of recursive calls is at its largest, relative to the size of the numbers supplied as input, when the input supplied is two consecutive Fibonacci numbers, since then all the quotients are 1, each remainder is the next lower Fibonacci number, and the numbers passed in the recursion are reduced as little as possible at each step. Since the growth of the Fibonacci sequence is exponential, as we proved in §2.5, we conclude that in this computationally "worst case", the number of recursive calls is proportional to the logarithm of the size of the input.

Extended Euclidean Algorithm

Keeping track of the quotients and remainders at each division step of the Euclidean algorithm is useful in extending its capability. In the Euclidean computation of $\gcd(n, m)$, define

$$m_0 = m \quad \text{and} \quad n_0 = n \qquad (6.1.3)$$

and then, if after $j - 1$ steps the recursion continues, define

$$q_{j-1} = \left\lfloor \frac{n_{j-1}}{m_{j-1}} \right\rfloor$$

$$n_j = m_{j-1} \qquad (6.1.4)$$

$$m_j = n_{j-1} - q_{j-1} m_{j-1} \qquad (6.1.5)$$

Numerous applications involve the following result.

Theorem 6.1.4. *For every pair of non-negative integers m and n, not both 0, there are numbers N and M such that*

$$\gcd(n, m) = Nn + Mm$$

Proof: Suppose that the recursion of the Euclidean algorithm stops at the kth call, so that $m_k = 0$ and $n_k = \gcd(n, m)$. Then, if we define $N_k = 1$ and $M_k = 0$, we have

$$N_k n_k + M_k m_k = 1 n_k + 0 m_k = \gcd(n, m)$$

It follows from (6.1.4) and (6.1.5) that

$$\gcd(n, m) = N_k(m_{k-1}) + M_k(n_{k-1} - q_{k-1}m_{k-1})$$
$$= M_k n_{k-1} + (N_k - M_k q_{k-1})m_{k-1}$$

Whenever $k \geq j > 0$, we inductively define (with decreasing j)

$$N_{j-1} = M_j$$
$$M_{j-1} = N_j - M_j q_{j-1}$$

and, thus,

$$\gcd(n, m) = N_{j-1}n_{j-1} + M_{j-1}m_{j-1} \quad \text{for } k \geq j \geq 0$$

In particular,

$$\gcd(n, m) = N_0 n_0 + M_0 m_0$$
$$= N_0 n + M_0 m \qquad \text{by (6.1.3)} \qquad \qquad \Diamond$$

DEFINITION: The **extended Euclidean algorithm** includes the computation of N and M such that $Nn + Mm = \gcd(n, m)$, as in Theorem 6.1.4.

Example 6.1.1, continued: When preparing to apply the extension of the Euclidean algorithm, the steps of the calculation of the greatest common divisor are arranged in tabular form.

j	n_j	m_j	q_j
0	720	168	4
1	168	48	3
2	48	24	2
3	24	0	*STOP*

To continue with the extension, start by regarding the next-to-bottom row as the *current row*. Let j be its row number, in this case row 2. In that next-to-bottom row, write

$$1 \cdot n_j + 0 \cdot m_j = 1 \cdot (n_{j-1} - q_{j-1}m_{j-1})$$

with the appropriate values substituted for every subscripted variable. In this case, the substitution yields the equation

$$1 \cdot 24 + 0 \cdot 0 = 1 \cdot (168 - 3 \cdot 48)$$

which expresses the greatest common divisor as a linear combination of n_j and m_j on the left and in terms of n_{j-1} and m_{j-1} on the right, which is then simplified into a standard form of linear combination, in this case

$$1 \cdot 168 - 3 \cdot 48$$

In general, working upwards, for each row of a by-hand calculation, the substitution of $n_{j-1} - q_{j-1} m_{j-1}$ for m_j uses values from the preceding row. There is an implicit substitution of the value of m_{j-1} for the value of n_j, but since $m_{j-1} = n_j$, this does not require work. Continue upward until row 0 is reached, at which point the greatest common divisor is expressed as a linear combination of n_0 and m_0, thereby completing the objective of the extended algorithm.

j	n_j	m_j	q_j	
0	720	168	4	$(-3) \cdot 720 + 13 \cdot 168$
1	168	48	3	$1 \cdot 168 - 3 \cdot 48 = 1 \cdot 168 - 3 \cdot (720 - 4 \cdot 168)$
2	48	24	2	$1 \cdot 24 + 0 \cdot 0 = 1 \cdot (168 - 3 \cdot 48)$
3	24	0	*STOP*	

In this case, we see that

$$(-3) \cdot 720 + 13 \cdot 168 = \gcd(720,\ 168) = 24$$

Thus, $N = -3$ and $M = 13$.

Corollary 6.1.5. *For every pair of non-negative integers m and n, not both 0, if $\hat{N}$ and $\hat{M}$ are numbers such that*

$$\gcd(n,\ m) = \hat{N}n + \hat{M}m$$

then $\hat{N}n + \hat{M}m$ is the smallest positively valued combination $Nn + Mm$ with integer multipliers N and M.

Proof: By Theorem 6.1.4, $\gcd(n,\ m)$ equals some combination $Nn + Mm$. Since $\hat{N}n + \hat{M}m$ is the smallest combination of n and m, it follows that

$$\hat{N}n + \hat{M}m \leq \gcd(n, m)$$

Since $\gcd(m, n) \setminus n$ and $\gcd(m, n) \setminus m$, it follows that for every choice of integers N and M, we have

$$\gcd(m,\ n) \setminus Nn + Mm$$

In particular,

$$\gcd(m,\ n) \setminus \hat{N}n + \hat{M}m$$

It follows that $\gcd(m, n) \leq \hat{N}n + \hat{M}m$.

The GCD of Two Fibonacci Numbers

We conclude this section by combining what we know about Fibonacci numbers with what we know about greatest common divisors to produce the fascinating result that $\gcd(f_n, f_m) = f_{\gcd(n,m)}$. Some review and preliminary propositions are helpful.

REVIEW FROM §2.6:

- Theorem 2.6.1 [Forward-Shift Identity]. The Fibonacci numbers satisfy the equation
$$f_{n+k} = f_k f_{n+1} + f_{k-1} f_n \qquad \text{for all } k \geq 1$$

- Corollary 2.6.2. For all $k \geq 0$, the Fibonacci number f_{kn} is a multiple of the Fibonacci number f_n.

Proposition 6.1.6. *Let m, n, and r be integers such that $r \perp m$. Then*
$$\gcd(rn, m) = \gcd(n, m)$$

Proof: Since any common divisor of m and n is also a divisor of m and rn, it follows that $\gcd(n, m) \leq \gcd(rn, m)$. Now suppose that $Nn + Mm = \gcd(n, m)$ and that $Cr + Dm = 1$. It follows that $NCr + NDm = N$ and, thus, that
$$\begin{aligned}
\gcd(n, m) &= (NCr + NDm)n + Mm \\
&= NCrn + NDmn + Mm \\
&= (NC)rn + (NDn + M)m
\end{aligned}$$
Since $\gcd(rn, m)$ is the smallest combination of rn and m, it follows that
$$\gcd(rn, m) \leq \gcd(n, m) \qquad \Diamond$$

Proposition 6.1.7. *For $n \geq 1$, $\gcd(f_n, f_{n-1}) = 1$.*

Proof: Calculation of $\gcd(f_n, f_{n-1})$ by the Euclidean algorithm terminates with a value of 1. $\qquad \Diamond$

Corollary 6.1.8. *For $n \geq 1$ and $k \perp n$, $\gcd(f_{kn+1}, f_n) = 1$.*

Proof: By Corollary 2.6.2, f_n divides f_{kn}. Therefore,
$$\begin{aligned}
\gcd(f_n, f_{kn+1}) &= \gcd(f_{kn}, f_{kn+1}) \qquad &\text{(Prop 6.1.6)} \\
&= 1 \qquad &\text{(Prop 6.1.7)} \qquad \Diamond
\end{aligned}$$

And now for the punch line.

Theorem 6.1.9. *For $n \geq 0$ and $m \geq 1$,*
$$\gcd(f_n, f_m) = f_{\gcd(n,m)}$$

Proof: Suppose that $n = qm + r$, where $0 \leq r < m$. Then
$$\begin{aligned}
\gcd(f_n, f_m) &= \gcd(f_{qm+r}, f_m) \\
&= \gcd(f_{qm+1}f_r + f_{qm}f_{r-1}, f_m) \qquad &\text{(Thm 2.6.1)} \\
&= \gcd(f_{qm+1}f_r, f_m) \qquad &\text{(Cor 2.6.2 and Prop 6.1.1)} \\
&= \gcd(f_r, f_m) \qquad &\text{(Cor 6.1.8 and Prop 6.1.6)} \qquad \Diamond
\end{aligned}$$

EXERCISES for Section 6.1

In each of the Exercises 6.1.1 through 6.1.8: (a) use the Euclidean algorithm to calculate the greatest common divisor of the given pair of integers; (b) use prime power factorization to calculate the greatest common divisor.

6.1.1$^{\mathbf{S}}$ 89, 71   **6.1.2**   94, 85

6.1.3   210, 196   **6.1.4**   1000, 560

6.1.5   1047, 2011   **6.1.6**   11213, 19937

6.1.7   32768, 10000   **6.1.8**   6561, 1728

In each of the Exercises 6.1.9 through 6.1.16, use the extended Euclidean algorithm on the given pair of integers a and b to calculate integers M and N such that $aM + bN = \gcd(a, b)$.

6.1.9$^{\mathbf{S}}$ 89, 71   **6.1.10**   94, 85

6.1.11   210, 196   **6.1.12**   1000, 560

6.1.13   1047, 2011   **6.1.14**   11213, 19937

6.1.15   32768, 10000   **6.1.16**   6561, 1728

In each of the Exercises 6.1.17 through 6.1.21, suppose that the numbers a and b are positive integers and relatively prime.

6.1.17   Prove that the numbers M and N such that $aM + bN = 1$ may be chosen so that $1 \le M < b$ and $-a < N \le -1$.

6.1.18   Prove that the numbers M and N such that $aM + bN = 1$ may be chosen so that $-b < M < -1$ and $1 \le N < a$.

6.1.19   Prove that postage of $(a - 1)(b - 1)$ cents can be formed from a-cent and b-cent stamps. That is, there exist non-negative integers r and s such that $ra + sb = (a - 1)(b - 1)$.

6.1.20   Prove that postage of $(a - 1)(b - 1) - 1$ cents cannot be formed from a-cent and b-cent stamps.

6.1.21   Prove that postage of n cents, for any $n \ge (a - 1)(b - 1)$ cents, could be formed from a-cent and b-cent stamps. That is, there exist non-negative integers r and s such that $ra + sb = n$.

Exercises 6.1.22 through 6.1.24 anticipate the next section. Let a, b, c, d, and $m \ne 0$ be integers.

6.1.22   Prove that $a \bmod m = b \bmod m$ if and only if $m \setminus b - a$.

6.1.23   Prove that if $a \bmod m = b \bmod m$ and $c \bmod m = d \bmod m$, then $a + c \bmod m = b + d \bmod m$ and $ac \bmod m = bd \bmod m$.

6.1.24   Prove that $ab = \gcd(a, b) \operatorname{lcm}(a, b)$.

6.2   CHINESE REMAINDER THEOREM

The extended Euclidean algorithm has many applications. The application of immediate concern is in the solution of a system of *linear congruences*. The existence of solutions to certain systems is ensured by the *Chinese remainder theorem*.

Congruence Modulo m

DEFINITION: A **congruence modulo** m is a relational statement of the form

$$a \equiv b \,(\text{modulo } m)$$

It means that $m \setminus b - a$. (We sometimes omit the parentheses.)

Example 6.2.1: $17 \equiv 2 \,(\text{modulo } 5)$ and $-8 \equiv 2 \,(\text{modulo } 5)$.

The relation called congruence *modulo m* and the operator called *mod* have a similarity in their names. Their mathematical connection is as follows.

Proposition 6.2.1. *Let a and b be any integers and m a positive integer. Then*

$$a \equiv b \,(\text{modulo } m)$$

if and only if

$$a \bmod m = b \bmod m$$

Proof: Suppose that $a = qm + r$ and $b = q'm + r'$ with $0 \le r, r' < m$, so that $a \bmod m = r$ and $b \bmod m = r'$. We observe that the assertion $a \equiv b \,(\text{modulo } m)$ simply means $m \setminus b - a$, which is equivalent to the relation

$$m \setminus (q'm + r') - (qm + r)$$

which is equivalent, in turn, to the relation

$$m \setminus r' - r$$

Since $|r' - r| < m$, this holds if and only if $r' = r$, and, accordingly, if and only if $a \bmod m = b \bmod m$. $\diamond$

Linear Congruence Modulo m

Just like equations, congruences can involve indeterminates. Moreover, just as a system of linear equations may possibly have a solution, a *system of linear congruences* may possibly have a solution.

DEFINITION: For integers a, b, and $m > 0$, a **linear congruence** is a relation of the form

$$ax \equiv b \,(\text{modulo } m)$$

DEFINITION: For positive moduli $m_1, m_2, \ldots, m_k$, a **system of linear congruences** is a list

$$a_1 x \ \equiv \ b_1 \ (\text{modulo } m_1)$$
$$a_2 x \ \equiv \ b_2 \ (\text{modulo } m_2)$$
$$\vdots$$
$$a_k x \ \equiv \ b_k \ (\text{modulo } m_k)$$

A **solution to the system of congruences** is an integer x that satisfies all of them.

Example 6.2.2:   Consider the system of congruences

$$x \ \equiv \ 2 \ (\text{modulo } 3)$$
$$x \ \equiv \ 3 \ (\text{modulo } 5)$$
$$x \ \equiv \ 1 \ (\text{modulo } 7)$$

We observe that $x = 8$ is a solution.

A Lemma on Relatively Prime Numbers

The Chinese remainder theorem yields a sufficient condition for a system of linear congruences to have an essentially unique solution. Moreover, there is a systematic way to find solutions. The following proposition serves as a lemma in the proof of the Chinese remainder theorem.

Proposition 6.2.2. *Let m and n be relatively prime, and let Q be an integer such that $m \setminus Q$ and $n \setminus Q$. Then $mn \setminus Q$.*

Proof: Suppose that $Q = mr$ and $Q = ns$. Since $m \perp n$, there are integers N and M such that $Nn + Mm = 1$, by Theorem 6.1.4. Thus,

$$\begin{aligned} Q \ &= \ QNn + QMm \\ &= \ mrNn + nsMm \\ &= \ mn\,(rN + sM) \end{aligned} \qquad \diamond$$

Remark: An alternative proof of Proposition 6.2.2 requires prior proof of the uniqueness of the factorization into prime powers, which is a substantially longer proof than the proof above.

Encoding by Residues

Some aspects of number theory are quite ancient. What is now described dates back to the Chinese mathematician Sun Tsŭ in the 4th century C.E.

DEFINITION: A set of positive integers $\{m_1, \ldots, m_k\}$ is a **system of independent moduli** if $m_i \perp m_j$ whenever $i \neq j$.

DEFINITION: The **tuple of residues** of an integer n with respect to a system $\{m_1, \ldots, m_k\}$ of independent moduli is the k-tuple

$$(n \bmod m_1, \ldots, n \bmod m_k)$$

The following table shows the tuple of residues of the numbers 0 to 20 with respect to the mutually independent moduli 3, 4, and 5.

Table 6.2.1   Residues modulo 3, 4, and 5.

n	$n \bmod 3$	$n \bmod 4$	$n \bmod 5$
0	0	0	0
1	1	1	1
2	2	2	2
3	0	3	3
4	1	0	4
5	2	1	0
6	0	2	1
7	1	3	2
8	2	0	3
9	0	1	4
10	1	2	0
11	2	3	1
12	0	0	2
13	1	1	3
14	2	2	4
15	0	3	0
16	1	0	1
17	2	1	2
18	0	2	3
19	1	3	4
20	2	0	0

No two of the rows have the same list of residues, and there would be no repetition of rows until after the 60th row. This observation was generalized by Sun Tsŭ, as now indicated.

Theorem 6.2.3 [Chinese Remainder Theorem]. *Let $\{m_1, \ldots, m_k\}$ be a system of independent moduli, with $M = m_1 m_2 \cdots m_k$. Then the mapping*

$$n \;\mapsto\; (n \bmod m_1, \ldots, n \bmod m_k)$$

from the integer interval $[0 : M - 1]$ to the set of possible tuples of residues with respect to $\{m_1, \ldots, m_k\}$ is a one-to-one and onto mapping.

Proof: Since the domain $[0 : M - 1]$ and the codomain of tuples of residues with respect to $\{m_1, \ldots, m_k\}$ have the same cardinality M, it is sufficient, by the pigeonhole principle (see §0.3), to prove that no two numbers in $[0 : M - 1]$ have the same set of residues.

Suppose, to the contrary, that $0 \leq b < c < M$ and that

$$c \bmod m_j \; = \; b \bmod m_j \quad \text{for } j = 1, \ldots, k$$

Then

$$m_j \setminus (c - b) \quad \text{for } j = 1, \ldots, k$$

Accordingly, iterative application of Proposition 6.2.2 would imply that

$$m_1 m_2 \cdots m_k \setminus (c - b) \quad \text{for } j = 1, \ldots, k$$

It would follow that $M \setminus (c - b)$, since $M = m_1 m_2 \cdots m_k$. But then $c - b \geq M$, which contradicts the prior supposition that $0 \leq b < c < M$.   ◇

Arithmetic on Residue Tuples

Much of the value of encoding numbers by their residues is that arithmetic operations on the residues produce the residues of the result of the operations directly on the numbers.

DEFINITION: The **sum of two k-tuples** of residues with respect to a list of moduli $\{m_1, \ldots, m_k\}$ is the k-tuple whose j^{th} coordinate is the sum of the two j^{th} coordinates modulo m_j.

Example 6.2.3:

n	$n \bmod 3$	$n \bmod 4$	$n \bmod 5$
2	2	2	2
$+\,8$	2	0	3
$=10$	1	2	0

DEFINITION: The **product of two k-tuples** of residues with respect to a list of moduli $\{m_1, \ldots, m_k\}$ is the k-tuple whose j^{th} coordinate is the product of the two j^{th} coordinates modulo m_j.

Example 6.2.4:

n	$n \bmod 3$	$n \bmod 4$	$n \bmod 5$
2	2	2	2
$\times\,8$	2	0	3
$=16$	1	0	1

A critical feature of the encoding by residues is that it respects arithmetic. That is, the sum of the tuples for numbers r and s is the tuple of the sum $r + s$, and the product of the tuples for numbers r and s is the tuple of the product rs.

Remark: The arithmetic-preservation property enables us to add and multiply small residues instead of large numbers. If there is a large amount of arithmetic, then the cost of encoding and subsequently decoding the result of the computations may be amortized.

Residue Decoding

The following theorem provides a method by which, knowing only the residues of a number, one could recover the number itself.

Theorem 6.2.4 [Chinese Remainder Decoding]. *Let m_1 and m_2 be positive integers and let Q_1 and Q_2 be integers such that*

$$Q_1 m_1 + Q_2 m_2 = 1$$

Let n be an integer such that $0 \leq n < m_1 m_2$, and such that

$$(n \bmod m_1, n \bmod m_2) = (r_1, r_2)$$

Then

$$r_1 Q_2 m_2 + r_2 Q_1 m_1 = n$$

Proof: Since $Q_1 m_1 + Q_2 m_2 = 1$, it follows that

$$m_2 \setminus Q_1 m_1 - 1 \quad \text{and} \quad m_1 \setminus Q_2 m_2 - 1$$

and, in turn, that

$$Q_1 m_1 \bmod m_2 = 1 \quad \text{and} \quad Q_2 m_2 \bmod m_1 = 1$$

Accordingly,

$$r_1 Q_2 m_2 \bmod m_1 = r_1 \quad \text{and}$$
$$r_1 Q_2 m_2 + r_2 Q_1 m_1 \bmod m_1 = r_1 \qquad (6.2.1)$$

Similarly,

$$r_2 Q_1 m_1 \bmod m_2 = r_2 \quad \text{and}$$
$$r_1 Q_2 m_2 + r_2 Q_1 m_1 \bmod m_2 = r_2 \qquad (6.2.2)$$

By the Chinese Remainder Theorem, there is only one number in the integer interval $[0 : m_1 m_2]$ whose residues modulo m_1 and modulo m_2 are r_1 and r_2, respectively. Thus,

$$r_1 Q_2 m_2 + r_2 Q_1 m_1 = n \qquad \qquad \Diamond$$

In combination with the extended Euclidean algorithm, Theorem 6.2.4 is used to decode any tuple of moduli. It is simplest for a 2-tuple, as now illustrated.

Example 6.2.5:   Clearly, $8 \mapsto (2 \bmod 3, 3 \bmod 5)$. Either by simple observation or by an application of the extended Euclidean algorithm, we have

$$(-3) \cdot 3 + 2 \cdot 5 = 1 = Q_1 m_1 + Q_2 m_2$$

Chinese Remainder Decoding now recovers the encoded number 8.

$$\begin{aligned}
r_1 Q_2 m_2 + r_2 Q_1 m_1 &= 2 \cdot 2 \cdot 5 + 3 \cdot (-3) \cdot 3 \\
&= 20 - 27 = -7 \\
&\equiv 8 \ (\text{modulo } 15)
\end{aligned}$$

Example 6.2.6:   Decoding of the 2-tuple (4 mod 8, 2 mod 9) begins with determination of Q_1 and Q_2, easily in this case,

$$(-1) \cdot 8 + 1 \cdot 9 \; = \; 1 \; = \; Q_1 m_1 + Q_2 m_2$$

and finishes with the calculation

$$r_1 Q_2 m_2 + r_2 Q_1 m_1 \; = \; 4 \cdot 1 \cdot 9 + 2 \cdot (-1) \cdot 8$$
$$= \; 36 - 16 \; = \; 20$$

Checking that $20 \mapsto$ (4 mod 8, 2 mod 9) confirms this decoding.

Decoding 3-Tuples and Larger Tuples

Decoding a k-tuple of residues with $k \geq 3$ involves iterative application of the following principle.

Proposition 6.2.5. *Suppose that m_1, m_2, and m_3 are mutually relatively prime. Then $m_1 m_2 \perp m_3$.*

Proof: If neither of the numbers m_1 nor m_2 has a prime divisor that occurs in the prime factorization of m_3, then $m_1 m_2$ has no prime divisor that occurs in the prime factorization of m_3, since the set of prime divisors of $m_1 m_2$ is the union of the set of prime divisors of m_1 and m_2.   ◇

Example 6.2.7:   Decoding of the 3-tuple

$$(4 \bmod 8, \; 2 \bmod 9, \; 3 \bmod 5)$$

begins with the calculation of Example 6.2.6 that

$$20 \; \mapsto \; (4 \bmod 8, \; 2 \bmod 9)$$

Any number n such that $n \equiv 20 \bmod 72$ satisfies both of the conditions $n \equiv 4 \bmod 8$ and $n \equiv 3 \bmod 5$. Subsequent decoding of the 2-tuple

$$(20 \bmod 72, \; 3 \bmod 5)$$

begins with finding multipliers Q_1 and Q_2 such that

$$Q_1 \cdot 72 + Q_2 \cdot 5 \; = \; 1$$

Either by "guessing" or by the extended Euclidean algorithm, we have

$$(-2) \cdot 72 + 29 \cdot 5 \; = \; 1$$

The calculation concludes with

$$r_1 Q_2 m_2 + r_2 Q_1 m_1 \; = \; 20 \cdot 29 \cdot 5 + 3 \cdot (-2) \cdot 72$$
$$= \; 2900 - 432 \; = \; 2468$$
$$\equiv \; 308 \; (\text{modulo } 360)$$

Checking that

$$308 \; \mapsto \; (4 \bmod 8, \; 2 \bmod 9, \; 3 \bmod 5)$$

confirms this decoding.

EXERCISES for Section 6.2

6.2.1[S] Add (6 mod 8, 3 mod 9) and (7 mod 8, 5 mod 9).

6.2.2 Multiply (6 mod 8, 3 mod 9) by (7 mod 8, 5 mod 9).

6.2.3 Add (7 mod 16, 4 mod 22) and (9 mod 16, 11 mod 22).

6.2.4 Multiply (7 mod 16, 4 mod 22) by (9 mod 16, 11 mod 22).

In each of the Exercises 6.2.5 through 6.2.10, calculate the prescribed Chinese remainder encoding.

6.2.5[S]	29 mod (5, 7)	**6.2.6**	31 mod (5, 7)
6.2.7	64 mod (5, 7)	**6.2.8**	66 mod (5, 7)
6.2.9	29 mod (15, 14)	**6.2.10**	31 mod (15, 14)

In each of the Exercises 6.2.11 through 6.2.18, calculate the prescribed Chinese remainder decoding.

6.2.11[S]	(6 mod 8, 3 mod 9)	**6.2.12**	(0 mod 5, 4 mod 9)
6.2.13	(6 mod 14, 3 mod 15)	**6.2.14**	(7 mod 16, 4 mod 21)
6.2.15	(2 mod 3, 1 mod 4, 3 mod 5)	**6.2.16**	(4 mod 6, 3 mod 11, 9 mod 13)
6.2.17	(1 mod 7, 2 mod 13, 6 mod 15)	**6.2.18**	(8 mod 14, 5 mod 9, 3 mod 17)

6.2.19 Explain how to solve a system of linear congruences of the form

$$a_1 x \equiv r_1 (\text{modulo } m_1)$$
$$a_2 x \equiv r_2 (\text{modulo } m_2)$$
$$\vdots$$
$$a_n x \equiv r_n (\text{modulo } m_n)$$

when the moduli are independent and each coefficient a_j is relatively prime to the corresponding modulus m_j.

6.2.20[S] Apply the method of Exercise 6.2.19 to solving this system of linear congruences.

$$2x \equiv 1 (\text{modulo } 3)$$
$$3x \equiv 4 (\text{modulo } 5)$$
$$5x \equiv 2 (\text{modulo } 7)$$

6.2.21 Show that this system of linear congruences has the solution $x = 548$.

$$x \equiv 0 (\text{modulo } 4)$$
$$x \equiv -1 (\text{modulo } 9)$$
$$x \equiv -2 (\text{modulo } 25)$$

6.2.22 Observe that the sequence 547, 548, 549, 550 of consecutive numbers has the perfect squares 1^2, 2^2, 3^2, 5^2 as respective divisors. Generalize Exercise 6.2.21 in a proof that the sequence of integers contains arbitrarily long sequences of consecutive integers, each divisible by a perfect square.

6.3 POLYNOMIAL DIVISIBILITY

This section briefly digresses from the principal topic of integer operations to demonstrate how some of the integer operations of present interest are extendible to operations on polynomials. In particular, a pair of polynomials may have a greatest common divisor, there is a Euclidean algorithm for polynomials, and there are prime polynomials.

NOTATION: The degree of a polynomial $g(x)$ is denoted $\partial g(x)$.

DEFINITION: A **monic polynomial** is a polynomial whose coefficient on the term of largest degree is 1.

Example 6.3.1: $x^4 + 5x^3 - 4x^2 + 7x + 14$ is a monic polynomial.

The Polynomial Ring over the Integers

NOTATION: The set of polynomials of finite degree in one indeterminate x, with integer coefficients, is denoted $\mathbb{Z}[x]$.

The set $\mathbb{Z}[x]$ is closed under the operations of addition and multiplication. The null polynomial 0 serves as the additive identity, and the constant polynomial 1 serves as the multiplicative identity. Every polynomial has an additive inverse. Moreover, the multiplication of polynomials distributes over their addition. This same notation $\mathbb{Z}[x]$ is also used for the algebraic structure whose domain is $\mathbb{Z}[x]$ and whose operations are polynomial addition and polynomial multiplication.

TERMINOLOGY: In view of the algebraic properties just described, $\mathbb{Z}[x]$ is called a **polynomial ring** (see Appendix A2).

Divisibility and Mod for Polynomials

Division of polynomials is a generalization of what is commonly called *long division*, with a *quotient* and a *remainder*. The underlying mechanism procedurally is a matter of systematically subtracting multiples of the divisor from the dividend, until what is left is of lower degree than the divisor.

DEFINITION: The **quotient of dividing a polynomial** of degree r

$$g(x) \; = \; g_r x^r + g_{r-1} x^{r-1} + \cdots + g_0$$

by a polynomial of degree s

$$h(x) \; = \; h_s x^s + h_{s-1} x^{s-1} + \cdots + h_0$$

is defined recursively (using repeated subtraction) as follows:
If $r < s$ then quotient $(g(x),\, h(x)) \; = \; 0$, and, otherwise,

$$\text{quotient}\,(g(x),\, h(x)) \; = \; \frac{g_r}{h_s}\, x^{r-s} + \text{quotient}\left(g(x) - \frac{g_r}{h_s}\, x^{r-s} h(x),\; h(x) \right)$$

DEFINITION: The **remainder of division of a polynomial**

$$g(x) = g_r x^r + g_{r-1} x^{r-1} + \cdots + g_0$$

by a non-zero polynomial

$$h(x) = h_s x^s + h_{s-1} x^{s-1} + \cdots + h_0$$

is the polynomial

$$g(x) \bmod h(x) = g(x) - \text{quotient}\,(g(x),\, h(x))\, h(x)$$

DEFINITION: The non-zero polynomial $h(x)$ **divides** the polynomial $g(x)$ if there is a polynomial $f(x)$ such that

$$g(x) = h(x)f(x)$$

This relation is denoted $h(x) \setminus g(x)$.

Clearly, the polynomial $h(x)$ divides the polynomial $g(x)$ if and only if

$$g(x) \bmod h(x) = 0$$

Example 6.3.2: The polynomials $x^3 - x^2 + 1$ and $x^3 - 2$ both divide the polynomial $x^6 - x^5 - x^3 + 2x^2 - 2$, since

$$(x^3 - x^2 + 1)(x^3 - 2) = x^6 - x^5 - x^3 + 2x^2 - 2$$

Common Divisors

The notions of common divisors and greatest common divisors for polynomials are also similar to the same notions for integers.

DEFINITION: A **common divisor** of two or more polynomials is a polynomial that divides both or all of them.

The following proposition is analogous to Proposition 6.1.1.

Proposition 6.3.1. *Let $a(x)$, $b(x)$, and $c(x)$ be polynomials in the polynomial ring $\mathbb{Z}[x]$. Then the polynomial pairs $\{a(x), b(x)\}$ and $\{a(x),\ b(x) + a(x)c(x)\}$ have the exact same set of common divisors.*

Proof: Let $h(x)$ be any common divisor of $a(x)$ and $b(x)$, say $a(x) = u(x)h(x)$ and $b(x) = v(x)h(x)$. Then

$$\begin{aligned} a(x) + c(x)b(x) &= u(x)h(x) + c(x)v(x)h(x) \\ &= (u(x) + c(x)v(x))h(x) \end{aligned}$$

Conversely, if $a(x) = u(x)h(x)$ and $b(x) + a(x)c(x) = v(x)h(x)$, then

$$\begin{aligned} b(x) &= v(x)h(x) - a(x)c(x) \\ &= v(x)h(x) - u(x)h(x)c(x) \\ &= (v(x) - u(x)c(x))\, h(x) \end{aligned} \qquad \Diamond$$

DEFINITION: A *greatest common divisor of two polynomials*

$$a(x) = a_r x^r + a_{r-1} x^{r-1} + \cdots + a_0 \quad \text{and}$$
$$b(x) = b_s x^s + b_{s-1} x^{s-1} + \cdots + b_0$$

is a common divisor polynomial $g(x)$ of highest degree.

NOTATION: The notation $\gcd(g(x), h(x))$ often refers to the monic greatest common divisor of $g(x)$ and $h(x)$.

Example 6.3.3:   The polynomial $x^3 - x^2 + 1$ is a greatest common divisor of the polynomials $x^6 - x^5 - x^3 + 2x^2 - 2$ and $x^4 - x^2 + x + 1$. The polynomial $x^3 - x^2 + 1$ is monic, and we write

$$\gcd\left(x^6 - x^5 - x^3 + 2x^2 - 2, \quad x^4 - x^2 + x + 1\right)$$
$$= x^3 - x^2 + 1$$

Euclidean Algorithm for Polynomials

Theorem 6.3.2   [*Euclidean Reduction for Polynomials*]. Let $g(x)$ and $h(x)$ be polynomials such that $0 < \partial h(x) \leq \partial g(x)$. Then

$$\gcd(h(x),\, g(x)) = \gcd(h(x),\, g(x) \bmod h(x))$$

Proof: Suppose that $q(x) = quotient\,(g(x), h(x))$. Then

$$\gcd(h(x),\, g(x)) = \gcd(h(x),\, g(x) - q(x)h(x)) \qquad \text{(Proposition 6.3.1)}$$
$$= \gcd(h(x),\, g(x) \bmod h(x)) \qquad\qquad \Diamond$$

DEFINITION: The **Euclidean algorithm for polynomials** is to iterate Euclidean reduction until a residue of zero is achieved.

Example 6.3.4:   The process is directly analogous to the integer version.

$$\gcd\left(x^5 - 1,\, x^3 - 3x^2 + 3x - 1\right) = \gcd\left(x^3 - 3x^2 + 3x - 1,\, 10x^2 - 15x + 5\right)$$
$$= \gcd\left(10x^2 - 15x + 5,\, \frac{1}{4}x - \frac{1}{4}\right)$$
$$= \gcd\left(\frac{1}{4}x - \frac{1}{4},\, 0\right)$$
$$= x - 1$$

Remark: There is also an extended Euclidean algorithm for polynomials.

Prime Polynomials

DEFINITION: A monic polynomial $g(x) \neq 1$ is a **prime polynomial** if it has no monic divisors of positive degree except for itself.

Example 6.3.5:   Any linear polynomial $x + k$ is prime.

Example 6.3.6:   A quadratic polynomial $x^2 + bx + c$ is prime over the integers, unless it has two integers (perhaps both the same) as its roots. For instance, $x^2 - 2$ is prime. More generally, by the quadratic equation, it follows that for the roots to be integers, it is a necessary condition that $b^2 - 4c$ must be the square of an integer.

EXERCISES for Section 6.3

In Exercises 6.3.1 through 6.3.8, decide whether the given polynomial is prime over the integers.

6.3.1^S $x^2 - 4x + 2$ 6.3.2 $x^2 + 1$

6.3.3 $x^3 + 1$ 6.3.4 $x^3 - 1$

6.3.5 $x^3 + 2x^2 - 1$ 6.3.6 $x^3 + 2x^2 + 1$

6.3.7 $x^3 - 6x^2 + 11x - 6$ 6.3.8 $x^3 - x^2 - 4x + 4$

In Exercises 6.3.9 through 6.3.16, calculate $g(x) \bmod h(x)$ for the given pair of polynomials.

6.3.9^S $x^2 + 3x + 7 \bmod x - 2$ 6.3.10 $x^3 - 6x^2 + 11x - 6 \bmod x - 1$

6.3.11 $x^3 - 6x^2 + 11x - 6 \bmod x + 4$ 6.3.12 $x^3 - x^2 - 10 \bmod x + 3$

6.3.13 $x^3 - x^2 - 10 \bmod x^2 + 3$ 6.3.14 $x^4 - 7x^3 + 3x - 1 \bmod x^2 + 5x - 1$

6.3.15 $x^4 - 7x^3 + 3x - 1 \bmod 5x - 1$ 6.3.16 $x^5 - 8x^2 - 10 \bmod x^3 + 2$

In Exercises 6.3.17 through 6.3.22, calculate $\gcd(g(x), h(x))$ for the given pair of polynomials.

6.3.17 $\gcd(x^2 - 7x + 10, \ x - 2)$

6.3.18 $\gcd(x^2 - 7x + 10, \ x - 3)$

6.3.19 $\gcd(x^3 - 6x^2 + 11x - 6, \ x^2 - 3x + 2)$

6.3.20 $\gcd(x^3 - 6x^2 + 11x - 6, \ x^2 - 5x + 4)$

6.3.21 $\gcd(x^5 - 2x^4 + 7x^3 + 3x^2 - 6x + 21, \ x^4 - 2x^3 + 6x^2 + 2x - 7)$

6.3.22 $\gcd(x^5 - 2x^4 + 7x^3 + 3x^2 - 6x + 21, \ x^6 - x^5 + 5x^4 + 7x^3 - 2x + 7)$

6.4   PRIME AND COMPOSITE MODULI

The fast calculations toward which the methods of this section are directed mainly concern evaluating of arithmetic expressions with respect to a modulus and solving congruences. The naive approach of expanding high-valued arithmetic expressions and then dividing by the modulus can be quite tedious, relative to methods based on some understanding of number theory and algebra. Similarly, there are faster methods for solving various types of congruences than sequential trial and error.

FROM APPENDIX A2:

- The domain of the ring of **integers modulo n**, denoted $\mathbb{Z}_n$, is the set of numbers

$$\{\, 0, \quad 1, \quad \ldots, \quad n-1 \,\}$$

  We write an element of $\mathbb{Z}_n$ in the form a (modulo n) when it seems necessary or helpful to distinguish the meaning from the number $a \in \mathbb{Z}$.

- The binary operations of **addition modulo n** $(+)$ and **multiplication modulo n** $(\cdot)$ in the ring $\mathbb{Z}_n$ are given by the rules

$$b \,(\text{modulo } n) + c \,(\text{modulo } n) \;=\; b + c \,(\text{modulo } n)$$
$$b \,(\text{modulo } n) \cdot c \,(\text{modulo } n) \;=\; b \cdot c \,(\text{modulo } n)$$

  In other words, if adding or multiplying two numbers as usual for integers happens to exceed $n - 1$, then divide by n and use the remainder as the result of the operation.

- The number 0 is the additive identity of $\mathbb{Z}_n$.

- The number 1 is the multiplicative identity of $\mathbb{Z}_n$.

- Every number k has $n - k$ as its additive inverse in $\mathbb{Z}_n$.

- Some numbers have multiplicative inverses in $\mathbb{Z}_n$. For instance, 13 is the inverse of 7 in $\mathbb{Z}_{90}$, since

$$13 \cdot 7 \;=\; 91 \;\equiv\; 1 \,(\text{modulo } 90)$$

Existence of Inverses Modulo m

Whereas 1 and -1 are the only integers with multiplicative inverses (in $\mathbb{Z}$), a ring $\mathbb{Z}_n$ of integers modulo n may have more than two numbers with multiplicative inverses. However, some numbers may have no multiplicative inverse modulo n. The general context for this issue is to find all solutions to congruences of the form

$$mx \;\equiv\; 1 \,(\text{modulo } n)$$

for arbitrary positive integers m and n.

Proposition 6.4.1. *Let m and n be positive integers. Then m (modulo n) has a multiplicative inverse if and only if $m \perp n$.*

Proof: First, suppose that $m \perp n$. By the extended Euclidean algorithm, there are integers N and M such that

$$Nn + Mm = 1$$

Thus,

$$Mm \equiv 1 \; (\text{modulo } n)$$

which implies that $M \bmod n$ is a multiplicative inverse of $m \bmod n$ in $\mathbb{Z}_n$.

Conversely, if $Mm \equiv 1 \; (\text{modulo } n)$, then

$$n \setminus (Mm - 1)$$

Thus, there is an integer N such that $Nn = Mm - 1$ which implies that

$$Mm - Nn = 1$$

from which it follows that $m \perp n$. $\diamondsuit$

Corollary 6.4.2. *Let p be a prime number. Then all the numbers $1, \ldots, p-1$ have inverses in $\mathbb{Z}_p$.*

Proof: Since p is prime, all the numbers $1, \ldots, p-1$ are relatively prime to p. $\diamondsuit$

Remark: When p is prime, $\mathbb{Z}_p$ is a **field**. See Appendix A2.

The following three examples all illustrate the conclusion of Proposition 6.4.1.

Example 6.4.1: In the ring $\mathbb{Z}_6$, the numbers 1 and 5 (both relatively prime to 6) are their own inverses, but the numbers 2, 3, and 4 have no multiplicative inverses.

Example 6.4.2: In the ring $\mathbb{Z}_7$, the numbers $1, \ldots, 6$ (all of which are relatively prime to 7) all have multiplicative inverses, in accord with Corollary 6.4.2, respectively, 1, 4, 5, 2, 3, 6.

Example 6.4.3: In the ring $\mathbb{Z}_8$, the numbers 1, 3, 5, 7 (all relatively prime to 8) are their own inverses, but 2, 4, 6 (not relatively prime to 8) have no multiplicative inverses.

Calculating Inverses Modulo n

The proof of Proposition 6.4.1 provides a method for calculating the inverse modulo n of a number m such that $m \perp n$.

Step 1. Find integers N and M such that $Nn + Mm = 1$, for instance, by the extended Euclidean algorithm.

Step 2. Then take $M \bmod n$ as the multiplicative inverse of m (modulo n).

Example 6.4.4:   Since 16 and 21 are relatively prime, the number 16 must have a multiplicative inverse modulo 21. Either by inspection or by the extended Euclidean algorithm, it can be determined that

$$4 \cdot 16 \; - \; 3 \cdot 21 \; = \; 1$$

Thus, the multiplicative inverse of 16 (modulo 21) is 4.

Uniqueness of Inverse Modulo m

TERMINOLOGY: In Example 6.4.4, the number 4 is described as *the* inverse of 16 modulo 21, rather than *an* inverse. In fact, the number 25 is another multiplicative inverse of 16 modulo 21, since

$$25 \cdot 16 \; - \; 19 \cdot 21 \; = \; 1$$

However, it is proved below that a number n has at most one inverse modulo m in the range

$$1, \ldots, m - 1$$

The definite article *the* is often applied to such an inverse.

Lemma 6.4.3. *Let n be an integer and m an integer that is relatively prime to n. Then the numbers*

$$m, \quad 2m, \quad \ldots, \quad (n-1)m$$

are mutually non-congruent modulo n, i.e., a permutation of the numbers

$$1, \quad 2, \quad \ldots, \quad n - 1$$

Proof: Proposition 6.4.1 implies that m has a multiplicative inverse modulo n, that is, a number M such that

$$Mm \; = \; 1 \; + \; Nn$$

for some number N. Consider two numbers r and s such that $1 \leq r, s \leq n - 1$. Suppose that

$$rm \; \equiv \; sm \; (\text{modulo } n)$$

Then $rmM \equiv smM$ (modulo n). It follows that

$$r(1 + Nn) \; \equiv \; s(1 + Nn) \; (\text{modulo } n)$$

and, in turn, that

$$r \; \equiv \; s \; (\text{modulo } n) \hspace{4cm} \diamond$$

Corollary 6.4.4. *Let m and n be relatively prime positive integers. Then there is exactly one inverse M of m (modulo n) such that $1 \leq M < n$.* $\hspace{1cm} \diamond$

Example 6.4.5: Consider the prime $p = 7$ and the number $m = 4$. Then the sequence $\left\langle km \bmod p \mid k = 1, \ldots, p-1 \right\rangle$ is exactly the sequence

$$1 \cdot 4 = 4, \quad 2 \cdot 4 = 8, \quad 3 \cdot 4 = 12, \quad 4 \cdot 4 = 16, \quad 5 \cdot 4 = 20, \quad 6 \cdot 4 = 24$$

which reduces, modulo 7, to the sequence

$$1 \cdot 4 \equiv 4 \ (\text{modulo } n), \quad 2 \cdot 4 \equiv 1 \ (\text{modulo } n), \quad 3 \cdot 4 \equiv 5 \ (\text{modulo } n),$$
$$4 \cdot 4 \equiv 2 \ (\text{modulo } n), \quad 5 \cdot 4 \equiv 6 \ (\text{modulo } n), \quad 6 \cdot 4 \equiv 3 \ (\text{modulo } n)$$

Thus, the number 2 is the unique inverse of 4 (modulo 7) in the range $1, \ldots, 6$.

Fermat's Theorem

We now turn to the problem of *modular exponentiation*, that is, of evaluating an expression involving an exponential modulo a number, such as

$$3124^{214} \ (\text{modulo } 20)$$

This is less tedious than it at first appears, since there is no need to evaluate 3124^{214}. A first reduction is based on the following proposition.

Proposition 6.4.5. *For any integers m and $n \geq 1$,*

$$m^r \ (\text{modulo } n) \ \equiv \ (m \bmod n)^r \ (\text{modulo } n)$$

Proof: Suppose that $m = qn + (m \bmod n)$. Then

$$m^r = (qn + (m \bmod n))^r$$

In the expansion of the exponentiated binomial on the right, the only term that does not have n as a factor is $(m \bmod n)^r$. Hence,

$$m^r \ (\text{modulo } n) \ \equiv \ (m \bmod n)^r \ (\text{modulo } n) \qquad\qquad \Diamond$$

In particular,

$$3124^{214} \ (\text{modulo } 20) \ \equiv \ 4^{214} \ (\text{modulo } 20)$$

A further kind of simplification begins with the choice of a convenient power of the base number 4. For instance, choosing the exponent 3 produces the following reduction of the exponent and easy evaluation.

$$4^3 = 64 \equiv 4 \ (\text{modulo } 20)$$
$$\Rightarrow \quad 4^{214} = (4^3)^{71} \cdot 4 \equiv 4^{71} \cdot 4 \equiv 4^{72} \ (\text{modulo } 20)$$
$$\equiv (4^3)^{24} \equiv 4^{24} \equiv (4^3)^8 \equiv 4^8 \ (\text{modulo } 20)$$
$$\equiv (4^3)^2 \cdot 4^2 \equiv 4^2 \cdot 4^2 \equiv 4^4 \ (\text{modulo } 20)$$
$$\equiv 4^3 \cdot 4 \equiv 4 \cdot 4 \equiv 16 \ (\text{modulo } 20)$$

Alternatively, if we choose the exponent 5,

$$4^5 = 1024 \equiv 4 \ (\text{modulo } 20)$$
$$\Rightarrow \quad 4^{214} = (4^5)^{42} \cdot 4^4 \equiv 4^{42} \cdot 4^4 \equiv 4^{46} \equiv (4^5)^9 \cdot 4 \ (\text{modulo } 20)$$
$$\equiv 4^9 \cdot 4 \equiv 4^{10} \equiv (4^5)^2 \equiv 4^2 \equiv 16 \ (\text{modulo } 20)$$

A theorem of Fermat permits such a calculation to go even more rapidly, when the modulus is prime. Its traditional name is *Fermat's Little Theorem*.

Theorem 6.4.6 [Fermat's Little Theorem]. *Let p be a prime number and let b be any integer that is not divisible by p. Then*

$$b^{p-1} \equiv 1 \,(\text{modulo } p)$$

Proof: Lemma 6.4.3 implies that

$$\prod_{j=1}^{p-1}(jb) \equiv \prod_{j=1}^{p-1}j \equiv (p-1)! \,(\text{modulo } p) \tag{6.4.1}$$

Since multiplication modulo p retains commutativity,

$$\prod_{j=1}^{p-1}(jb) \equiv \left(\prod_{j=1}^{p-1}b\right)\prod_{j=1}^{p-1}j \,(\text{modulo } p) \tag{6.4.2}$$

Combining (6.4.1) and (6.4.2) yields

$$b^{p-1}(p-1)! \equiv (p-1)! \,(\text{modulo } p) \tag{6.4.3}$$

Applying Corollary 6.4.2 to all the factors of $(p-1)!$ in the congruence (6.4.3) implies the result

$$b^{p-1} \equiv 1 \,(\text{modulo } p) \qquad\qquad \diamond$$

Example 6.4.6: All the numbers

$$1^4 = 1, \; 2^4 = 16, \; 3^4 = 81, \; 4^4 = 256$$

are congruent to 1 modulo 5.

Example 6.4.7: Fermat's congruence cannot be used when the modulus is not prime. For instance,

$$2^{11} = 2048 \equiv 8 \,(\text{modulo } 12)$$
$$3^{11} = 177147 \equiv 3 \,(\text{modulo } 12)$$

Remark: In §6.5, there is a generalization by Euler of Fermat's Little Theorem.

Wilson's Theorem

There is still more to be harvested from Corollary 6.4.2, the principle that the numbers $1, \ldots, p-1$ all have multiplicative inverses modulo a prime p.

Proposition 6.4.7. *Let p be a prime number and let n be an integer that is not divisible by p. Then $n^2 \equiv 1$ (modulo p) if and only if $n \equiv \pm 1$ (modulo p).*

Proof: Suppose first that $n \equiv \pm 1$ (modulo p). That is, there is an integer k such that $n = kp \pm 1$. Then either

$$n^2 = (kp+1)^2 = k^2 p^2 + 2kp + 1 \equiv 1 \text{ (modulo } p)$$

or

$$n^2 = (kp-1)^2 = k^2 p^2 - 2kp + 1 \equiv 1 \text{ (modulo } p)$$

Conversely, suppose that $n^2 \equiv 1$ (modulo p). Then $p \setminus n^2 - 1$. It follows that

$$p \setminus (n-1)(n+1)$$

Thus, since p is prime, either $p \setminus n-1$ or $p \setminus n+1$. If $p \setminus n-1$, then $n \equiv 1$ (modulo p). If $p \setminus n+1$, then $n \equiv -1$ (modulo p). $\diamondsuit$

Corollary 6.4.8. *Let p be a prime number. Then $(p-2)! \equiv 1$ (modulo p).*

Proof: Let $r \in \{2, \ldots, p-2\}$. By Proposition 6.4.7, the number r cannot be its own multiplicative inverse modulo p, and that inverse must lie in that same range $\{2, \ldots, p-2\}$. It follows that the numbers $2, \ldots, p-2$ can be paired into inverses modulo p. Accordingly,

$$\prod_{j=2}^{p-2} j \equiv 1 \text{ (modulo } p)$$

Thus, $(p-2)! \equiv 1$ (modulo p). $\diamondsuit$

Theorem 6.4.9 [*Wilson's Theorem*]. *The congruence*

$$(m-1)! \equiv -1 \text{ (modulo } m)$$

holds if and only if m is prime.

Proof: If m is prime, then the congruence $(m-1)! \equiv -1$ (modulo m) follows immediately from Corollary 6.4.8.

Conversely, if m is not prime, then m has a factor r such that $r \leq \lfloor \sqrt{m} \rfloor$, say $rs = m$. If $r < s$, then

$$(m-1)! = rs \cdot \left(\prod_{j=1}^{r-1} j \right) \left(\prod_{j=r+1}^{s-1} j \right) \left(\prod_{j=s+1}^{m-1} j \right)$$

$$\equiv 0 \cdot \left(\prod_{j=1}^{r-1} j \right) \left(\prod_{j=r+1}^{s-1} j \right) \left(\prod_{j=s+1}^{m-1} j \right) \equiv 0 \not\equiv -1 \text{ (modulo } m)$$

If $r^2 = m = 4$, then

$$(m - 1)! \;=\; 3! \;=\; 6 \;\not\equiv\; -1 \;(\text{modulo } 4)$$

Otherwise, i.e., for $m \geq 6$, we have $\sqrt{m} > 2$, which implies that $2r < m$. Thus,

$$
(m - 1)! \;=\; r \cdot 2r \cdot \left(\prod_{j=1}^{r-1} j \right) \left(\prod_{j=r+1}^{2r-1} j \right) \left(\prod_{j=2r+1}^{m-1} j \right)
$$

$$
\equiv\; 2 \cdot r^2 \cdot \left(\prod_{j=1}^{r-1} j \right) \left(\prod_{j=r+1}^{2r-1} j \right) \left(\prod_{j=2r+1}^{m-1} j \right)
$$

$$
\equiv\; 2 \cdot 0 \cdot \left(\prod_{j=1}^{r-1} j \right) \left(\prod_{j=r+1}^{2r-1} j \right) \left(\prod_{j=2r+1}^{m-1} j \right) \quad (r^2 = m \equiv 0 \;(\text{modulo } m))
$$

$$
\equiv\; 0 \;(\text{modulo } m) \qquad\qquad\qquad\qquad\qquad\qquad \diamond
$$

Remark: We have proved a sharpened version of Wilson's theorem, with values for $(m - 1)!$ (modulo m) in all cases.

Quadratic Residues

DEFINITION: The integer a is a **quadratic residue** of the integer m if $a \perp m$ and if the congruence

$$x^2 \;\equiv\; a \;(\text{modulo } m)$$

has a solution. If the congruence $x^2 \equiv a \bmod m$ has no solution, then a is called a **quadratic non-residue** of m.

Remark: If c and d are congruent, then

$$c^2 \;\equiv\; d^2 \;(\text{modulo } m)$$

Thus, the set of numbers c^2 such that $1 \leq c \leq m - 1$ and $c \perp m$ is a complete set of quadratic residues of m.

Example 6.4.8:   According to the remark above, the set

$$\{\, 1 \equiv 1^2, \quad 4 \equiv 2^2, \quad 2 \equiv 3^2, \quad 2 \equiv 4^2, \quad 4 \equiv 5^2, \quad 1 \equiv 6^2 \,\}$$
$$= \{\, 1, \quad 2, \quad 4 \,\}$$

is the set of quadratic residues of 7. The numbers 3, 5, and 6 are quadratic non-residues of 7.

Example 6.4.9:   The quadratic residues of 11 are

$$1 \equiv 1^2 \equiv 10^2, \quad 4 \equiv 2^2 \equiv 9^2, \quad 9 \equiv 3^2 \equiv 8^2, \quad 5 \equiv 4^2 \equiv 7^2,$$
$$\text{and} \;\; 3 \equiv 5^2 \equiv 6^2$$

The numbers 2, 6, 7, 8, and 10 are quadratic non-residues of 11.

Example 6.4.10:   The quadratic residues of 15 are

$$1 \equiv 1^2 \equiv 4^2 \equiv 11^2 \equiv 14^2 \;\; \text{and} \;\; 4 \equiv 2^2 \equiv 7^2 \equiv 8^2 \equiv 13^2$$

Finding Solutions to a Quadratic

We now generalize some of the properties that may have been observed in these examples.

POWER OF ODD PRIME AS MODULUS

Theorem 6.4.10. *Let p be an odd prime, let n be a positive integer, and let a be an integer not divisible by p. Then the congruence*

$$x^2 \equiv a \,(\text{modulo } p^n) \tag{6.4.4}$$

has either two distinct solutions in the range $1, \ldots, p^n - 1$ or no solutions at all.

Proof: Suppose that b lies in the range $1, \ldots, p^n - 1$ and that

$$b^2 \equiv a \bmod p^n \tag{6.4.5}$$

Observe that the number $p^n - b$ lies in the range $1, \ldots, p^n - 1$, and that it is not equal to b, since p^n is odd. The calculation

$$
\begin{aligned}
(p^n - b)^2 &= p^{2n} - 2bp^n + b^2 \\
&\equiv b^2 \,(\text{modulo } p^n)
\end{aligned}
$$

establishes that $p^n - b$ is a second solution to the congruence (6.4.4).

To see that there are no more than these two solutions, consider another putative solution, i.e., a number c such that

$$c^2 \equiv a \,(\text{modulo } p^n) \tag{6.4.6}$$

Congruences (6.4.5) and (6.4.6) together imply that

$$b^2 - c^2 \equiv 0 \,(\text{modulo } p^n)$$

from which it follows that $p^n \setminus b^2 - c^2$, and, equivalently, that

$$p^n \setminus (b - c)(b + c)$$

Thus, either

$$p \setminus b - c \quad \text{or} \quad p \setminus b + c$$

If p were to divide both $b - c$ and $b + c$, then p would divide their sum $2b$. Yet, since p is an odd prime, it cannot divide 2, so it would necessarily divide b, implying that it divides a, which would contradict the choice of the number a. Accordingly, the number p does not divide both $b - c$ and $b + c$. It follows that either

$$p^n \setminus b - c \quad \text{or} \quad p^n \setminus b + c$$

If $p^n \setminus b - c$, then

$$c \equiv b \,(\text{modulo } p^n)$$

On the other hand, if $p^n \setminus b + c$, then

$$c \equiv p^n - b \,(\text{modulo } p^n)$$

We conclude that c is not an additional solution, and that either there are two solutions in the range $1, \ldots, p^n - 1$ or there are none.   ◇

Corollary 6.4.11. *Let p be an odd prime. Then the number of quadratic residues among the numbers $1, \ldots, p-1$ is*

$$\frac{p-1}{2}$$

Proof: Since none of the numbers $1, \ldots, p-1$ is divisible by p, it follows from Theorem 6.4.10 that the mapping

$$x \mapsto x^2 \bmod p$$

from $1, \ldots, p-1$ to itself is two-to-one. Thus, the image of this mapping, i.e., the set of quadratic residues, has cardinality $\frac{p-1}{2}$. $\diamond$

POWER OF 2 AS MODULUS

For modulus 2, the number 1 is the only quadratic residue, and the congruence $x^2 \equiv 1 \bmod 2$ has the unique solution $x = 1$. For modulus 4, the numbers 1 and 3 are relatively prime. The number 1 is a quadratic residue, and the number 3 is a quadratic non-residue. The congruence $x^2 \equiv 1 \bmod 4$ has the two solutions $x = 1$ and $x = 3$. For higher powers of 2, there is the following theorem.

Theorem 6.4.12. *Let n be an integer greater than 2, and let a be a quadratic residue of 2^n, whose smallest positive solution is the number b. Then in the range $1, \ldots, 2^n - 1$, the congruence*

$$x^2 \equiv a \; (\text{modulo } 2^n) \tag{6.4.7}$$

has exactly these four solutions and no others:

$$b, \quad 2^n - b, \quad 2^{n-1} - b, \quad 2^{n-1} + b \tag{6.4.8}$$

Proof: Squaring any of the three other proposed solutions implies immediately that it is a solution to the congruence (6.4.7). It is also clear that the four asserted solutions are mutually non-congruent modulo 2^n.

To see that there are no other possible solutions, consider a number c such that $c^2 \equiv a \; (\text{modulo } 2^n)$. Then, since both b and c satisfy the congruence (6.4.7), it follows that

$$2^n \setminus b^2 - c^2$$

Equivalently,

$$2^n \setminus (b - c)(b + c)$$

It may be asserted that 4 cannot divide both $b - c$ and $b + c$, since otherwise, the number 4 would divide their sum $2b$, from which it would follow that b is even, implying that a is even, contrary to the choice of a. Accordingly, either

$$2^{n-1} \setminus b - c \quad \text{or} \quad 2^{n-1} \setminus b + c \tag{6.4.9}$$

One alternative under (6.4.9) is that $2^{n-1} \setminus b - c$. Then, for some integer k, we have

$$b - c = k2^{n-1}$$
$$\Rightarrow \quad c = b - k2^{n-1}$$

If k is odd then c is one of the four solutions (6.4.8), since

$$c \equiv 2^{n-1} + b$$

and, similarly, if k is even, then

$$c \equiv 2^n + b$$

The other alternative under (6.4.9) is that $2^{n-1} \setminus b + c$. Then $c = -b + k2^{n-1}$, for some integer k. If k is odd then $c \equiv 2^{n-1} - b$, and if k is even, then $c \equiv 2^n - b$, so it is not a fifth solution.

We conclude that either there are four solutions in the range $1, \dots, p^n - 1$, as indicated, or there are none. $\diamond$

EXERCISES for Section 6.4

In Exercises 6.4.1 through 6.4.8, calculate the multiplicative inverse of the given number.

6.4.1	7 modulo 17		6.4.2	8 modulo 17
6.4.3^{S}	21 modulo 25		6.4.4	14 modulo 23
6.4.5	11 modulo 14		6.4.6	8 modulo 27
6.4.7	15 modulo 32		6.4.8	15 modulo 28

In Exercises 6.4.9 through 6.4.16, evaluate the given expression.

6.4.9	$70^{90} \bmod 17$		6.4.10	$180^{312} \bmod 17$
6.4.11^{S}	$221^{64} \bmod 25$		6.4.12	$144^{1728} \bmod 23$
6.4.13	$1111^{111} \bmod 14$		6.4.14	$88^{222} \bmod 27$
6.4.15	$155^{108} \bmod 32$		6.4.16	$515^{613} \bmod 29$

In Exercises 6.4.17 through 6.4.22, evaluate the given expression.

6.4.17^{S}	4! mod 5		6.4.18	5! mod 6
6.4.19	6! mod 7		6.4.20	7! mod 8
6.4.21	8! mod 9		6.4.22	10! mod 11

In Exercises 6.4.23 through 6.4.28, find all quadratic residues of the given number.

6.4.23	9		6.4.24	13
6.4.25^{S}	14		6.4.26	15
6.4.27	16		6.4.28	17

In Exercises 6.4.29 through 6.4.36, find all the solutions to the given quadratic congruence.

6.4.**29** $x^2 \equiv 1$ modulo 9   6.4.**30** $x^2 \equiv 4$ modulo 9

6.4.**31**[S] $x^2 \equiv 1$ modulo 24   6.4.**32** $x^2 \equiv 9$ modulo 25

6.4.**33** $x^2 \equiv 4$ modulo 15   6.4.**34** $x^2 \equiv 4$ modulo 21

6.4.**35** $x^2 \equiv 1$ modulo 30   6.4.**36** $x^2 \equiv 7$ modulo 18

6.5   EULER PHI-FUNCTION

DEFINITION: The number of positive integers not exceeding n that are relatively prime to n is given by the **Euler phi-function** $\phi(n)$.

Here are the first few values of the Euler phi-function:

n	1	2	3	4	5	6	7	8	9	$\cdots$
ϕ_n	1	1	2	2	4	2	6	4	6	$\cdots$

A preview of the Euler phi-function appears in the Exercises for §3.6, since its values could plausibly be calculated by inclusion-exclusion. However, calculating $\phi(n)$ is much simpler than that. For a start, the next proposition shows it is particularly easy to evaluate $\phi(n)$ when n is prime.

Proposition 6.5.1. *If the number p is prime, then*

$$\phi(p) \;=\; p - 1$$

Conversely, if $\phi(p) = p - 1$, then p is prime.

Proof: Suppose that p is a prime number. Then each of the numbers

$$1, \quad 2, \quad \ldots, \quad p - 1$$

is relatively prime to p, which implies that $\phi(p) = p - 1$. Conversely, if p is not a prime number, then at least one of those $p - 1$ numbers is not relatively prime to p, which implies that $\phi(p) < p - 1$.   ◇

In this section, we develop some properties of $\phi(n)$ and give a method of calculating that is much simpler than inclusion-exclusion.

Euler's Generalization of Fermat's Theorem

Euler derived the following generalization of Fermat's Theorem, as well as showing its use in evaluating exponentiation modulo a composite number.

Theorem 6.5.2 [Euler's Theorem]. *Let b and n be integers with $b \perp n$ and $n > 1$. Then*

$$b^{\phi(n)} \equiv 1 \ (\text{modulo } n)$$

Proof: We observe that if the modulus n is prime, then the conclusion reduces to Fermat's Theorem. More generally, let

$$r_1, \quad r_2, \quad \ldots, \quad r_{\phi(n)}$$

be the set of numbers less than n and relatively prime to n.

Assertion 1: Each of the numbers

$$br_1, \quad br_2, \quad \ldots, \quad br_{\phi(n)}$$

is relatively prime to the number n.

Proof of Assertion 1: Suppose that p is a prime number that divides n and also divides the product br_j. Then p would divide either b or r_j. Whichever it divides would not be relatively prime to n, a contradiction in either case. $\diamond$ Assertion 1

Assertion 2: If $i \neq j$, then $br_i \not\equiv br_j \ (\text{modulo } n)$.

Proof of Assertion 2: Suppose that $n \setminus b(r_i - r_j)$. Since $n \perp b$, none of the prime divisors of n divides b. It follows that $n \setminus r_i - r_j$. Since $|r_i - r_j| < n$, it follows that $r_i = r_j$, and thus, that $i = j$, a contradiction. $\diamond$ Assertion 2

Assertion 3: $br_1 \cdot br_2 \cdot \cdots \cdot br_{\phi(n)} \equiv r_1 r_2 \cdots \cdot r_{\phi(n)} \ (\text{modulo } n)$.

Proof of Assertion 3: It follows from Assertions 1 and 2 and the pigeonhole principle that the values

$$br_1 \bmod n, \quad \ldots, \quad br_{\phi(n)} \bmod n$$

are a permutation of the values $r_1, \ldots, r_{\phi(n)}$. $\diamond$ Assertion 3

Completion of Proof: Assertion 3 implies that

$$b^{\phi(n)} r_1 r_2 \cdots r_{\phi(n)} \equiv r_1 r_2 \cdots r_{\phi(n)} \ (\text{modulo } n)$$

and, in turn, that

$$n \setminus \left(b^{\phi(n)} - 1 \right) r_1 r_2 \cdots r_{\phi(n)}$$

Since each of the numbers r_j is relatively prime to n, it follows that

$$n \setminus \left(b^{\phi(n)} - 1 \right)$$

Thus, $b^{\phi(n)} \equiv 1 \ (\text{modulo } n)$. $\diamond$

Example 6.5.1: The numbers relatively prime to 15 are

$$1, 2, 4, 7, 8, 11, 13, 14$$

Thus, $\phi(15) = 8$. The numbers 4 and 7 are relatively prime to 15. We observe that

$$4^8 \equiv 16^4 \equiv 1^4 \equiv 1 \bmod 15$$
$$7^8 \equiv 49^4 \equiv 4^4 \equiv 16^2 \equiv 1^2 \equiv 1 \bmod 15$$

Evaluating the Phi-Function

Proposition 6.5.1 was a first step toward a general formula for $\phi(n)$. We now continue the pursuit of a formula.

Theorem 6.5.3. *Let p be a prime number and e a positive integer. Then*

$$\phi(p^e) = p^e - p^{e-1}$$

Proof: A number is not relatively prime to p^e if and only if it is divisible by p. In the integer interval $[1 : p^e]$, the numbers divisible by p are

$$p, \quad 2p, \quad \ldots, \quad p^{e-1}p$$

The cardinality of the complementary set is $p^e - p^{e-1}$. $\qquad\qquad\diamond$

Example 6.5.2: If $p = 2$, then the numbers relatively prime to 2^e are the odd numbers less than 2^e. Clearly, there are

$$\frac{2^e}{2} = 2^e - 2^{e-1}$$

such odd numbers.

DEFINITION: A function $f : \mathbb{Z}^+ \to \mathbb{Z}^+$ is a **multiplicative function** if whenever $m \perp n$

$$f(mn) = f(m)f(n)$$

Theorem 6.5.4. *The Euler phi-function is multiplicative.*

Proof: Let m and n be integers such that $m \perp n$. Then

$$
\begin{aligned}
\phi(mn) &= \sum_{b=0}^{mn-1} (b \perp mn) && \text{(definition of } \phi\text{)} \\
&= \sum_{b=0}^{mn-1} (b \perp m)(b \perp n) && \text{(Theorem A2.2)} \\
&= \sum_{b=0}^{mn-1} (b \bmod m \perp m)(b \bmod n \perp n) && \text{(Proposition 6.1.1)} \\
&= \sum_{j=0}^{m-1} \sum_{k=0}^{n-1} (j \bmod m \perp m)(k \bmod n \perp n) && \text{(Theorem 6.2.3)} \\
&= \sum_{j=0}^{m-1} (j \bmod m \perp m) \sum_{k=0}^{n-1} (k \bmod n \perp n) \\
&= \phi(m)\,\phi(n) && \diamond
\end{aligned}
$$

Example 6.5.3:   By sequential testing, we determine that the numbers relatively prime to 36 are

$$1 \quad 5 \quad 7 \quad 11 \quad 13 \quad 17 \quad 19 \quad 23 \quad 25 \quad 29 \quad 31 \quad 35$$

Thus, $\phi(36) = 12$. Either by sequential testing of the smaller positive integers or by Theorem 6.5.3, we see that $\phi(4) = 2$ and $\phi(9) = 6$, Thus

$$\phi(36) \;=\; 12 \;=\; 2 \cdot 6 \;=\; \phi(4)\phi(9)$$

Theorem 6.5.5. *Let b be a positive integer with the prime power factorization*

$$b \;=\; p_1^{e_1} \cdots p_k^{e_k}$$

Then

$$\phi(b) \;=\; \prod_{i=1}^{k} p_i^{e_i - 1}(p_i - 1)$$

Proof: This follows immediately from Theorems 6.5.3 and 6.5.4. $\diamond$

Corollary 6.5.6. *Let b be a positive integer with the prime power factorization*

$$b \;=\; p_1^{e_1} \cdots p_k^{e_k}$$

Then

$$\phi(b) \;=\; b \cdot \prod_{i=1}^{k} \left(1 - \frac{1}{p_i}\right)$$

Proof: Starting from Theorem 6.5.5,

$$\phi(b) \;=\; \prod_{i=1}^{k} p_i^{e_i - 1}(p_i - 1) \;=\; \prod_{i=1}^{k} p_i^{e_i}\left(\frac{p_i - 1}{p_i}\right)$$

$$=\; \prod_{i=1}^{k} p_i^{e_i}\left(1 - \frac{1}{p_i}\right) \;=\; b \cdot \prod_{i=1}^{k}\left(1 - \frac{1}{p_i}\right) \qquad\qquad \diamond$$

Example 6.5.4: $60 \;=\; 2^2 \cdot 3 \cdot 5$. By Corollary 6.5.6,

$$\phi(60) \;=\; 60 \cdot \left(1 - \frac{1}{2}\right)\left(1 - \frac{1}{3}\right)\left(1 - \frac{1}{5}\right)$$

$$=\; 60 \cdot \frac{1}{2} \cdot \frac{2}{3} \cdot \frac{4}{5} \;=\; 16$$

The sixteen numbers relatively prime to 60 are

$$
\begin{array}{cccccccc}
1 & 7 & 11 & 13 & 17 & 19 & 23 & 29 \\
31 & 37 & 41 & 43 & 47 & 49 & 53 & 59
\end{array}
$$

By combining Corollary 6.5.6 with Euler's theorem, we can quickly evaluate some otherwise hard-looking congruences.

Example 6.5.5:   In reducing each of the following congruences of an exponentiated expression to something more tractable, first the base is reduced by dividing by the modulus m, and then the exponent is reduced by dividing by $\phi(m)$.

$$289^{45} \bmod 15 \; = \; 4^{45} \bmod 15 \; = \; 4^5 \bmod 15 \; = \; 4$$
$$1728^{613} \bmod 35 \; = \; 13^{613} \bmod 35 \; = \; 13^{13} \bmod 35 \; = \; 13$$
$$1205^{5106} \bmod 21 \; = \; 8^{5106} \bmod 21 \; = \; 8^3 \bmod 21 \; = \; 8$$

Summing Phi over Divisors of n

We are now concerned with proving the following classical result:

$$\sum_{d \,\backslash\, n} \phi(d) \; = \; n$$

The proof is most easily understood as a generalization of an example.

Example 6.5.6:   The divisors of 12 are

$$d \; = \; 1 \quad 2 \quad 3 \quad 4 \quad 6 \quad 12$$

The sum of the values of $\phi(d)$ is

$$\sum_{d \,\backslash\, 12} \phi(d) \; = \; 1 + 1 + 2 + 2 + 2 + 4 \; = \; 12$$

This phenomenon can be explained by considering the unreduced fractions of the form $\dfrac{j}{12}$:   for $j = 1, \ldots, 12$

$$\frac{1}{12} \quad \frac{2}{12} \quad \frac{3}{12} \quad \frac{4}{12} \quad \frac{5}{12} \quad \frac{6}{12} \quad \frac{7}{12} \quad \frac{8}{12} \quad \frac{9}{12} \quad \frac{10}{12} \quad \frac{11}{12} \quad \frac{12}{12}$$

First reduce them to

$$\frac{1}{12} \quad \frac{1}{6} \quad \frac{1}{4} \quad \frac{1}{3} \quad \frac{5}{12} \quad \frac{1}{2} \quad \frac{7}{12} \quad \frac{2}{3} \quad \frac{3}{4} \quad \frac{5}{6} \quad \frac{11}{12} \quad \frac{1}{1}$$

and then regroup them according to their denominators

$$\underbrace{\frac{1}{1}}_{1=\phi(1)} \quad \underbrace{\frac{1}{2}}_{1=\phi(2)} \quad \underbrace{\frac{1}{3} \quad \frac{2}{3}}_{2=\phi(3)} \quad \underbrace{\frac{1}{4} \quad \frac{3}{4}}_{2=\phi(4)} \quad \underbrace{\frac{1}{6} \quad \frac{5}{6}}_{2=\phi(6)} \quad \underbrace{\frac{1}{12} \quad \frac{5}{12} \quad \frac{7}{12} \quad \frac{11}{12}}_{4=\phi(12)}$$

The set of numerators in each reduced subgrouping is precisely the set of numbers that are relatively prime to the common denominator of that subgrouping. Thus, the number of fractions in the subgrouping corresponding to the divisor d

of 12 equals $\phi(d)$. Since the subgroupings effectively partition the original set of unreduced fractions, it follows that

$$\sum_{d \setminus 12} \phi(d) = 12$$

Theorem 6.5.7. *Let n be any positive integer. Then*

$$\sum_{d \setminus n} \phi(d) = n$$

Proof: For each divisor d of n, the value $\phi(d)$ equals the number of unreduced fractions in the set

$$\frac{1}{n} \quad \frac{2}{n} \quad \cdots \quad \frac{n}{n}$$

whose denominator is d after reduction. Since every one of the n unreduced fractions reduces to a unique reduced fraction, the conclusion follows. $\diamondsuit$

Example 6.5.7:   The divisors of 15 are

$$d = 1 \quad 3 \quad 5 \quad 15$$

The sum of the values of $\phi(d)$ is

$$\sum_{d \setminus 15} \phi(d) = 1 + 2 + 4 + 8 = 15$$

EXERCISES for Section 6.5

In each of the Exercises 6.5.1 through 6.5.8, calculate the value of the Euler phi-function for the given argument.

6.5.1	$\phi(7)$	6.5.2	$\phi(11)$
6.5.3	$\phi(15)$	6.5.4	$\phi(24)$
6.5.5[S]	$\phi(30)$	6.5.6	$\phi(36)$
6.5.7	$\phi(48)$	6.5.8	$\phi(100)$

In Exercises 6.5.9 through 6.5.16, evaluate the given expression, using Euler's generalization of Fermat's Theorem.

6.5.9	$34^8 \bmod 15$	6.5.10	$180^{312} \bmod 15$
6.5.11[S]	$221^{64} \bmod 25$	6.5.12	$144^{1728} \bmod 25$
6.5.13	$1111^{111} \bmod 14$	6.5.14	$88^{222} \bmod 27$
6.5.15	$155^{108} \bmod 16$	6.5.16	$515^{613} \bmod 24$

In each of the Exercises 6.5.17 through 6.5.22, list all the divisors of the given number, calculate the value of the Euler phi-function on each divisor, and add those values, to confirm that their sum is the given number.

6.5.17	15	6.5.18	24

6.5.19^{S} 30   6.5.20   36

6.5.21   48   6.5.22   100

DEFINITION: The **Farey sequence** of order n, for each positive integer n, is the sequence of reduced fractions $\frac{r}{s}$, such that $0 \leq r \leq s \leq n$, in ascending order. The elements of the Farey sequence are called **Farey fractions**.

Example 6.5.8:   The Farey sequence of order 5 is

$$\frac{0}{1}, \ \frac{1}{4}, \ \frac{1}{3}, \ \frac{1}{2}, \ \frac{2}{3}, \ \frac{3}{4}, \ \frac{1}{1}$$

Exercises 6.5.23 through 6.5.26 are concerned with the Farey fractions. Let F_n denote the set of Farey fractions of order n.

6.5.23   Write the Farey sequence of order 4.

6.5.24   Write the Farey sequence of order 5.

6.5.25^{S} Prove that $|F_n| = |F_{n-1}| + \phi(n)$.

6.5.26   Prove that $|F_n| = 1 + \displaystyle\sum_{j=1}^{n} \phi(n)$

6.6   THE MÖBIUS FUNCTION

August F. Möbius (1790-1868) was a student of Gauss, later a professor of mathematics at Leipzig, whose most celebrated mathematical association is quite likely with the surface called a Möbius strip, which is one-sided when imbedded in 3-dimensional space. He was also an astronomer. This section concerns one of his contributions to classical number theory, the *Möbius function*, and its use in a summation principle called *Möbius inversion*.

DEFINITION: The **Möbius function** $\mu(n)$ is defined recursively on the positive integers as follows:

$$\mu(1) = 1$$

$$\mu(n) = -\sum_{d=1}^{n-1} (d \setminus n)\, \mu(d) \quad \text{if } n > 1$$

Example 6.6.1:   We consider the smallest cases.

$$\mu(2) = -\mu(1) = -1$$
$$\mu(3) = -\mu(1) = -1$$
$$\mu(4) = -\mu(1) - \mu(2) = -1 - (-1) = 0$$

$$\mu(5) \ = \ -\mu(1) \ = \ -1$$
$$\mu(6) \ = \ -\mu(1) - \mu(2) - \mu(3) \ = \ -1 - (-1) - (-1) \ = \ 1$$
$$\mu(7) \ = \ -\mu(1) \ = \ -1$$
$$\mu(8) \ = \ -\mu(1) - \mu(2) - \mu(4) \ = \ -1 - (-1) - 0 \ = \ 0$$
$$\mu(9) \ = \ -\mu(1) - \mu(3) \ = \ -1 - (-1) \ = \ 0$$
$$\mu(10) \ = \ -\mu(1) - \mu(2) - \mu(5) \ = \ -1 - (-1) - (-1) \ = \ 1$$
$$\mu(11) \ = \ -\mu(1) \ = \ -1$$
$$\mu(12) \ = \ -\mu(1) - \mu(2) - \mu(3) - \mu(4) - \mu(6)$$
$$= \ -1 - (-1) - (-1) - 0 - 1 \ = \ 0$$

We observe that on each of the primes 2, 3, 5, 7, and 11, the value of the Möbius function is -1. It is easy enough to prove that this is true of all primes.

Lemma 6.6.1. *Let p be a prime number. Then*

$$\mu(p) \ = \ -1$$

Proof: Since 1 is the only proper divisor of a prime number p, it follows that

$$\mu(p) \ = \ -\sum_{d=1}^{p-1} (d \setminus p) \, \mu(d)$$

$$= \ -\mu(1)$$

$$= \ -1 \qquad\qquad\qquad \Diamond$$

We observe also in Example 6.6.1 that $\mu(4) = \mu(8) = \mu(9) = 0$ and, suspecting that μ is zero-valued on every prime power, we might check a few more and then confirm our hunch.

Example 6.6.1, continued: We check the next few small cases of prime powers.

$$\mu(16) \ = \ -\mu(1) - \mu(2) - \mu(4) - \mu(8) \ = \ -1 - (-1) - 0 - 0 \ = \ 0$$
$$\mu(25) \ = \ -\mu(1) - \mu(5) \ = \ -1 - (-1) \ = \ 0$$
$$\mu(27) \ = \ -\mu(1) - \mu(3) - \mu(9) \ = \ -1 - (-1) - 0 \ = \ 0$$

Lemma 6.6.2. *Let p^k be a prime power with $k \geq 2$. Then*

$$\mu(p^k) \ = \ 0$$

Proof: Since all the divisors of p^k are of the form p^j, it follows that

$$\mu(p^k) \ = \ -\sum_{j=0}^{k-1} \mu(p^j)$$

BASIS: $k = 2$

$$\begin{aligned}
\mu(p^2) &= -\mu(1) - \mu(p) \\
&= -1 - (-1) \\
&= 0
\end{aligned}$$

IND STEP: Assume true for $j = 2, \ldots, k - 1$. Then

$$\begin{aligned}
\mu(p^k) &= -\mu(1) - \mu(p) - \mu(p^2) - \ldots - \mu(p^{k-1}) \\
&= -1 - (-1) - 0 - \ldots - 0 \\
&= 0 \qquad\qquad\qquad\qquad\qquad\qquad\qquad\qquad\qquad\qquad \Diamond
\end{aligned}$$

About Multiplicative Functions

It is proved in §6.5 that the Euler function $\phi(n)$ is multiplicative. That is, whenever $m \perp n$

$$\phi(mn) = \phi(m)\phi(n)$$

In anticipation of calculating the values of the Möbius function, we prove two general theorems about multiplicative functions, after a preparatory lemma.

Lemma 6.6.3. *Let m and n be relatively prime numbers. Then each divisor d of the product mn has a unique representation as the product $d = d_1 d_2$ of a pair of integers d_1 and d_2 such that $d_1 \setminus m$ and $d_2 \setminus n$.*

Proof: By the Fundamental Theorem of Arithmetic, the integer d has a factorization into prime powers, each of which divides either m or n, but not both, since $m \perp n$. The unique representation is

$$d_1 = \gcd(d, m) \quad \text{and} \quad d_2 = \gcd(d, n) \qquad\qquad \Diamond$$

Theorem 6.6.4. *Let $f(n)$ be a function on the positive integers, and let $F(n)$ be the function*

$$F(n) = \sum_{d \setminus n} f(d)$$

If $f(n)$ is multiplicative, then so is $F(n)$.

Proof: Let m and n be relatively prime numbers. Then

$$\begin{aligned}
F(m)\,F(n) &= \sum_{d_1 \setminus m} f(d_1) \sum_{d_2 \setminus n} f(d_2) && \text{(definition of } F) \\
&= \sum_{d_1 \setminus m} \sum_{d_2 \setminus n} f(d_1)\, f(d_2) && \text{(distribution of multiplication)} \\
&= \sum_{d_1 \setminus m} \sum_{d_2 \setminus n} f(d_1 d_2) && (f \text{ is multiplicative}) \\
&= \sum_{(d_1, d_2)\,:\, d_1 \setminus m \wedge d_2 \setminus n} f(d_1 d_2) && \\
&= \sum_{d \setminus mn} f(d) && \text{(Lemma 6.6.3)} \qquad \Diamond
\end{aligned}$$

Example 6.6.2:   To illustrate Theorem 6.6.4, let f be a multiplicative function, $m = 10$ and $n = 9$. Then

$$
\begin{aligned}
F(90) &= f(1) + f(2) + f(3) + f(5) + f(6) + f(9) \\
&\quad + f(10) + f(15) + f(18) + f(30) + f(45) + f(90) \\
&= f(1 \cdot 1) + f(2 \cdot 1) + f(1 \cdot 3) + f(5 \cdot 1) + f(2 \cdot 3) + f(1 \cdot 9) \\
&\quad + f(10 \cdot 1) + f(5 \cdot 3) + f(2 \cdot 9) + f(10 \cdot 3) + f(5 \cdot 9) + f(10 \cdot 9) \\
&= f(1)f(1) + f(2)f(1) + f(1)f(3) + f(5)f(1) + f(2)f(3) + f(1)f(9) \\
&\quad + f(10)f(1) + f(5)f(3) + f(2)f(9) + f(10)f(3) + f(5)f(9) + f(10)f(9) \\
&= [f(1) + f(2) + f(5) + f(10)] \cdot [f(1) + f(3) + f(9)] \\
&= F(10)\,F(9)
\end{aligned}
$$

The following theorem inverts the relationship of Theorem 6.6.4. It enables us to prove that the Möbius function μ is multiplicative, which is the key property in establishing a formula for the values of μ.

Theorem 6.6.5. *Let f be any function on the positive integers such that the sum*

$$
F(m) = \sum_{d \,\backslash\, m} f(d)
$$

is a multiplicative function. Then f itself is a multiplicative function.

Proof: By induction.

BASIS: Since F is multiplicative, it follows that $F(1) = 1$. Thus

$$
f(1) = \sum_{d \,\backslash\, 1} f(d) = F(1) = 1
$$

IND HYP: Assume that $f(mn) = f(m)f(n)$ for $m \perp n$ whenever $mn < s$.

IND STEP: Suppose that $m \perp n$ and that $mn = s$. Then

$$
F(mn) = \sum_{d \,\backslash\, mn} f(d) = \sum_{b \,\backslash\, m}\sum_{c \,\backslash\, n} f(bc)
$$

We infer that $b \perp c$ within the double sum, since $b \,\backslash\, m$ and $c \,\backslash\, n$, with $m \perp n$. Thus, by the induction hypothesis, we have

$$
\begin{aligned}
F(mn) &= \left(\sum_{b \,\backslash\, m}\sum_{c \,\backslash\, n} f(b)f(c) \right) - f(m)f(n) + f(mn) \\
&= \left(\sum_{b \,\backslash\, m} f(b) \sum_{c \,\backslash\, n} f(c) \right) - f(m)f(n) + f(mn) \\
&= F(m)\,F(n) - f(m)f(n) + f(mn) \qquad \text{(def of } F)
\end{aligned}
$$

It is given that F is multiplicative, which means that $F(mn) = F(m)F(n)$. It follows that

$$
f(mn) = f(m)f(n)
$$

Thus, f is multiplicative. $\Diamond$

Evaluating Mu

Theorem 6.6.6. *The Möbius function μ is multiplicative.*

Proof: Immediately from the definition of μ, the function

$$F(m) \; = \; \sum_{d \,\backslash\, m} \mu(d)$$

has the value

$$\begin{cases} 1 & \text{if } m = 1 \\ 0 & \text{otherwise} \end{cases}$$

Thus, the function $F(m)$ is multiplicative. It follows from Theorem 6.6.5 that the function μ is multiplicative. $\Diamond$

Theorem 6.6.7. *Let $p_1, \ldots, p_r$ be different primes. Then*

$$\mu(p_1^{e_1} \cdots p_r^{e_r}) \; = \; \begin{cases} (-1)^r & \text{if } e_1 = \cdots = e_r = 1 \\ 0 & \text{if } e_j \geq 2, \text{ for any } j \end{cases}$$

Proof: This follows from Lemma 6.6.1, Lemma 6.6.2, and the fact that μ is multiplicative. $\Diamond$

Example 6.6.3:   We use Theorem 6.6.7 to determine some values of $\mu(n)$.

$$\begin{aligned} \mu(1) \; &= \; 1 \\ \mu(2) \; &= \; -1 \\ \mu(4) \; &= \; \mu(2^2) \; = \; 0 \\ \mu(6) \; &= \; \mu(2 \cdot 3) \; = \; (-1)^2 \; = \; 1 \\ \mu(12) \; &= \; \mu(2^2 \cdot 3) \; = \; 0 \\ \mu(30) \; &= \; \mu(2 \cdot 3 \cdot 5) \; = \; (-1)^3 \; = \; -1 \\ \mu(210) \; &= \; \mu(2 \cdot 3 \cdot 5 \cdot 7) \; = \; (-1)^4 \; = \; 1 \end{aligned}$$

Möbius Inversion

The following identity facilitates the manipulation of a summation indexed over a lattice of divisors.

Lemma 6.6.8. *Let m and k be positive integers. Then*

$$\left\{ \frac{m}{d} \; : \; k \,\backslash\, d \,\backslash\, m \right\} \; = \; \left\{ \frac{m/k}{c} \; : \; c \,\backslash\, \frac{m}{k} \right\}$$

Proof: First suppose that $k \,\backslash\, d \,\backslash\, m$. Take $c = \dfrac{d}{k}$. Then

$$\frac{m}{d} \; = \; \frac{m}{kc} \; = \; \frac{m/k}{c} \; \text{ with } \; c \,\backslash\, \frac{m}{k}$$

Conversely, suppose that $c \setminus \dfrac{m}{k}$. Take $d = ck$. Then

$$\frac{m/k}{c} \;=\; \frac{m/k}{d/k} \;=\; \frac{m}{d} \;\; \text{with} \;\; k \setminus d \setminus m \qquad\qquad \diamond$$

Theorem 6.6.9 [Möbius Inversion Principle]. *The integer function F is related to the integer function f by the summation*

$$F(m) \;=\; \sum_{d \setminus m} f(d)$$

if and only if the function f is related to the function F by the summation

$$f(m) \;=\; \sum_{d \setminus m} \mu\left(\frac{m}{d}\right) F(d)$$

Proof: First suppose that

$$F(m) \;=\; \sum_{d \setminus m} f(d)$$

Then

$$
\begin{aligned}
\sum_{d \setminus m} \mu\left(\frac{m}{d}\right) F(d) \;&=\; \sum_{d \setminus m} \mu\left(\frac{m}{d}\right) \sum_{k \setminus d} f(k) & \text{(substitute for } F(d)) \\[2mm]
&=\; \sum_{d \setminus m}\sum_{k \setminus d} \mu\left(\frac{m}{d}\right) f(k) \\[2mm]
&=\; \sum_{k \setminus m}\sum_{k \setminus d \setminus m} \mu\left(\frac{m}{d}\right) f(k) & \text{(swap summation order)} \\[2mm]
&=\; \sum_{k \setminus m} f(k) \sum_{k \setminus d \setminus m} \mu\left(\frac{m}{d}\right) \\[2mm]
&=\; \sum_{k \setminus d} f(k) \sum_{c \setminus \frac{m}{k}} \mu\left(\frac{m/k}{c}\right) & \text{(Lemma 6.6.8)} \\[2mm]
&=\; \sum_{k \setminus d} f(k) \left(\frac{m}{k} = 1\right) & \text{(definition of } \mu) \\[2mm]
&=\; \sum_{k \setminus d} f(k)\,(k = m) \\[2mm]
&=\; f(m)
\end{aligned}
$$

This completes the "forward" direction.

Conversely, suppose that

$$f(m) \;=\; \sum_{d \setminus m} \mu\left(\frac{m}{d}\right) F(d)$$

Then

$$
\begin{aligned}
\sum_{d \,\backslash\, m} f(d) &= \sum_{d \,\backslash\, m} \sum_{k \,\backslash\, d} \mu\left(\frac{d}{k}\right) F(k) && \text{(substitute for } f(d)\text{)} \\[2mm]
&= \sum_{k \,\backslash\, d} \sum_{k \,\backslash\, d \,\backslash\, m} \mu\left(\frac{d}{k}\right) F(k) && \text{(swap summation order)} \\[2mm]
&= \sum_{k \,\backslash\, d} F(k) \sum_{k \,\backslash\, d \,\backslash\, m} \mu\left(\frac{d}{k}\right) \\[2mm]
&= \sum_{k \,\backslash\, d} F(k) \sum_{k \,\backslash\, d \,\backslash\, m} \mu\left(\frac{m}{d}\right) && \text{(rearrange summands)} \\[2mm]
&= \sum_{k \,\backslash\, d} F(k) \sum_{c \,\backslash\, \frac{m}{k}} \mu\left(\frac{m/k}{c}\right) && \text{(Lemma 6.6.8)} \\[2mm]
&= \sum_{k \,\backslash\, d} F(k) \left(\frac{m}{k} = 1\right) && \text{(definition of } \mu\text{)} \\[2mm]
&= \sum_{k \,\backslash\, d} F(k) \ (k = m) \\[2mm]
&= F(m)
\end{aligned}
$$

$\diamondsuit$

Example 6.6.4:   We recall from Theorem 6.5.7 that

$$
\sum_{d \,\backslash\, n} \phi(d) = n
$$

For $n = 6$, the sum on the left is

$$
\phi(1) + \phi(2) + \phi(3) + \phi(6) = 1 + 1 + 2 + 2 = 6 = n
$$

According to the Möbius inversion principle, one expects that

$$
\phi(6) = \sum_{d \,\backslash\, 6} \mu\left(\frac{6}{d}\right) d
$$

The value of this sum is

$$
\begin{aligned}
\mu(6) \cdot 1 + \mu(3) \cdot 2 &+ \mu(2) \cdot 3 + \mu(1) \cdot 6 \\
&= 1 \cdot 1 + (-1) \cdot 2 + (-1) \cdot 3 + 1 \cdot 6 \\
&= 1 - 2 - 3 + 6 \\
&= 2 \\
&= \phi(6)
\end{aligned}
$$

which serves as empirical confirmation.

EXERCISES for Section 6.6

In each of the Exercises 6.6.1 through 6.6.8, calculate the value of the Möbius mu-function.

6.6.1	$\mu(15)$	6.6.2	$\mu(16)$
6.6.3^S	$\mu(18)$	6.6.4	$\mu(30)$
6.6.5	$\mu(48)$	6.6.6	$\mu(210)$
6.6.7	$\mu(323)$	6.6.8	$\mu(2047)$

Exercises 6.6.9 through 6.6.14 further explore multiplicative functions.

DEFINITION: For every positive integer n, the *sum-of-all-divisors function* $\sigma(n)$ gives the sum of the divisors of n.

6.6.9   Prove that the function $\iota(n) = n$ is multiplicative.

6.6.10   Give a direct proof that the function $\sigma(n)$ is multiplicative.

6.6.11^S Use Theorem 6.6.4 to show that the function $\sigma(n)$ is multiplicative.

DEFINITION: For every positive integer n, the *number-of-divisors function* $\tau(n)$ gives the number of divisors of n.

6.6.12   Prove that the function $n \mapsto 1$ is multiplicative.

6.6.13   Give a direct proof that the function $\tau(n)$ is multiplicative.

6.6.14   Use Theorem 6.6.4 to show that the function $\tau(n)$ is multiplicative.

GLOSSARY

Chinese remainder decoding: finding a number with a given set of residues with respect to a system of independent moduli.

Chinese remainder encoding: finding the residues of a given number with respect to a system of independent moduli.

Chinese remainder theorem: the theorem that there are essentially unique solutions to certain systems of linear congruences with relatively prime moduli.

common divisor of two or more integers: a positive integer that divides each of them.

common multiple of two or more integers: a positive integer that is a multiple of each of them.

divides relation: for integers n and d, the relation $d \setminus n$ means that there is an integer q such that $dq = n$.

Euclidean algorithm: a process of iterative reductions to a greatest common divisor, using the mod operator; usually for two integers n and m, but also for polynomials.

___, **extended**:   usually, a process that represents the greatest common divisor of two integers as a linear combination of them with integer coefficients; sometimes also for polynomials.

Euler phi-function $\phi(n)$:   the number of integers from 1 to n that are relatively prime to n.

Farey sequence of order n: the sequence of reduced fractions valued from 0 to 1, with denominator at most n, in ascending order.

field: an algebraic structure with an addition and a multiplication that has multiplicative inverses as well as additive inverses.

greatest common divisor of integers n and m, not both zero: the largest common divisor.

independent moduli, system of $m_1, m_2, \cdots, m_k$:   moduli that are pairwise relatively prime.

integer quotient of a division of an integer n by an positive integer d: the unique integer q such that $0 \le n - dq < d$.

least common multiple of two or more integers: the smallest common multiple.

linear congruence: a congruence of the form $ax \equiv b(\text{modulo } m)$ with constant integers a, b, and m, and indeterminate x.

modular exponentiation: evaluating an exponential modulo some number.

monic polynomial: a polynomial whose highest-degree term has a coefficient of 1.

multiplicative function: a function f on the integers such that

$$f(mn) \;=\; f(m)f(n)$$

prime number: a positive integer larger than 1 with no proper divisors except 1.

prime polynomial: a monic polynomial with no divisors except 1 and itself.

quadratic residue of a number m: a number a, relatively prime to m, such that the congruence $x^2 \equiv a \bmod m$ has a solution.

relatively prime polynomials: polynomials whose greatest common divisor is the constant polynomial 1.

residue: the remainder of a division process.

Chapter 7

Graph Fundamentals

7.1 Regular Graphs

7.2 Walks and Distance

7.3 Trees and Acyclic Digraphs

7.4 Graph Isomorphism

7.5 Graph Automorphism

7.6 Subgraphs

7.7 Spanning Trees

7.8 Edge Weights

7.9 Graph Operations

The short preview of graph theory in Chapter 0 has enabled us to employ various graphs and graph-theoretic concepts in subsequent chapters as examples for application of some of the counting methods developed. This chapter marks a transition of focus from counting, in the preceding chapters, to configurations. Chapters 7, 8, and 9 are devoted entirely to graph theory, something of a whirlwind tour of the basics and some highlights. This chapter presents the core concepts of graph theory, with notions such as symmetry, distance, and graph isomorphism (i.e., equivalence) that are vital throughout the vast range of particular aspects and applications, plus a cluster of graph operations that are used in constructing new graphs from existing graphs. The graph model is augmented to allow for directions and edge-weights.

7.1   REGULAR GRAPHS

The late Frank Harary often referred, in his characteristically entertaining style, to "beautiful graph theory". One of the various attributes that could earn his approval of a particular graph as beautiful was symmetry. *Regularity* in a graph is a form of local symmetry.

REVIEW FROM §0.6:   The **degree** of a vertex v is the number of proper edges incident on v plus twice the number of self-loops at v.

DEFINITION: A **regular graph** is a graph in which every vertex has the same degree. If the common degree is d, then the graph may be called d-regular.

There are many interesting classes of regular graphs. Familiarity with them is very helpful, since they frequently arise as examples in discussions of graphs. Regularity also arises in practical applications. In a criss-cross grid of streets, every intersection has degree 4. Alternatively, imagine an electrical network in which each node has the same number of wire-ends incident on it, or, in larger scale, a homogeneous network of processors on a chip in which each computational node is linked to the same number of other nodes.

Geometric Regularity

The members of some classes of regular graphs are readily visualizable as geometrically regular figures. Visualization of such graphs with the geometric regularities in $\mathbb{R}^2$ or $\mathbb{R}^3$ tends to enhance intuitive conceptualization.

DEFINITION: The **cycle graph** C_n is the graph whose vertices are representable as n points spaced equally apart around the unit circle (see Figure 7.1.1) and whose edges are representable as the arcs joining adjacent vertices.

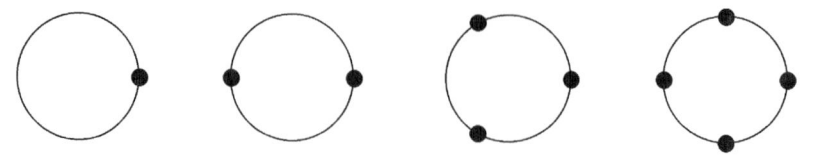

Figure 7.1.1   The cycle graphs C_1, C_2, C_3, and C_4.

Example 7.1.1:   Every cycle graph C_n is a 2-regular connected graph, and conversely, every 2-regular connected graph with n vertices is isomorphic to the cycle graph C_n. For $n \geq 3$, the cycle graph C_n can be represented as the boundary of an n-sided polygon in the plane, as in Figure 7.1.2.

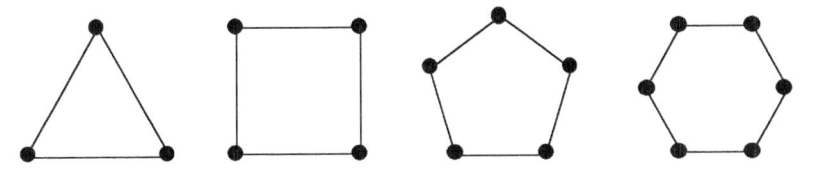

Figure 7.1.2 The cycle graphs C_3, C_4, C_5, and C_6 as polygons.

Another standard example of a regular graph is a *complete graph*.

REVIEW FROM §0.6: The **complete graph** K_n is the n-vertex simple graph such that every pair of vertices is joined by an edge.

Example 7.1.2: The complete graph K_n is $(n-1)$-regular. For $n \leq 4$, the complete graph K_n can be drawn in the plane without edge-crossings. For $n \geq 5$, drawings of K_n in the plane have edge-crossings, as, for instance, the complete graph K_6 in Figure 7.1.3.

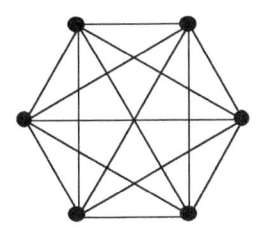

Figure 7.1.3 The complete graph K_6.

Sometimes a graph is visualized in 3-space, rather than as a plane figure. It can serve as what is intuitively the framework enclosing a *polyhedron*, which is the higher-dimensional analogue of a polygon. Just as a *regular polygon*, a 2-dimensional figure, has all its sides of the same length and perfect rotational symmetry, a *regular polyhedron* of dimension 3 has identical regular polygons as its faces and perfect rotational symmetry.

DEFINITION: For any polyhedron in 3-space, the graph representable by its corners and edges is called the **1-skeleton** (or sometimes, simply, the **skeleton**) of that polyhedron.

Example 7.1.3: The five regular polyhedra shown in Figure 7.1.4 are called the **platonic solids**. Their 1-skeletons are called **platonic graphs**. More generally, the 1-skeleton of a regular polyhedron of any dimension is a regular graph.

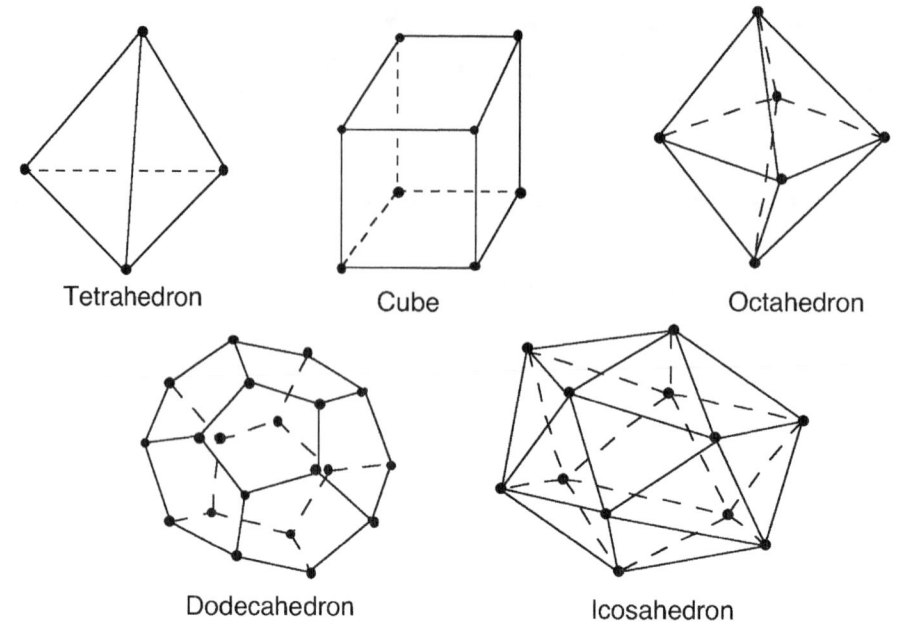

Figure 7.1.4   The five platonic graphs.

The concept of polyhedron generalizes to dimensions higher than 3. In particular, the graph K_n is also representable as the 1-skeleton of a polyhedron in $\mathbb{R}^{n-1}$ called the $(n-1)$-*simplex*.

Algebraic Regularity

Various families of regular graphs also arise from algebraic constructions. Some of the members of algebraically constructed regular classes are also geometrically realizable.

DEFINITION: To the group $\mathbb{Z}_n$ of integers modulo n and each subset S of numbers in its domain, we associate the **circulant graph** $circ\,(n:S)$, whose vertex set is

$$\{\,0,\quad 1,\quad \ldots,\quad n-1\,\}$$

and in which two vertices i and j are adjacent if and only if there is a number $s \in S$ such that $i+s=j \bmod n$ or $j+s=i \bmod n$. The elements of the set S are called **connections**.

Example 7.1.4: Figure 7.1.5 shows three circulant graphs.

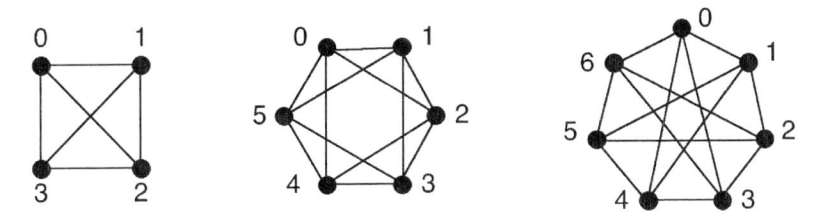

Figure 7.1.5 **The graphs** $circ\,(4:1,2)$, $circ\,(6:1,2)$, **and** $circ\,(7:1,3)$.

This extremely useful construction can be generalized.

DEFINITION: The **hypercube graph** Q_n is the n-regular graph whose vertices are the binary strings of length n, such that two vertices are adjacent if and only if the corresponding binary strings differ in a single bit. The hypercube graph Q_n is evidently n-regular, as illustrated in Figure 7.1.6.

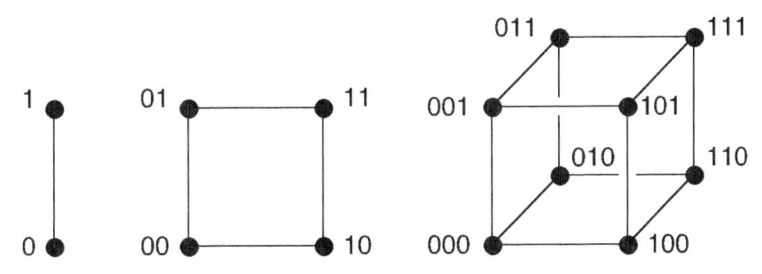

Figure 7.1.6 **The hypercube graphs** Q_1, Q_2, **and** Q_3.

Example 7.1.5: To see how the hypercube graph is related to a circulant graph, consider the operation of adding two binary strings of length n by adding each pair of corresponding coordinate values modulo 2. For instance,

$$
\begin{array}{r}
10110 \\
+\quad 01110 \\
\hline
11000
\end{array}
$$

Taking the set of n binary strings with one 1 as the connections has the effect of joining each binary string to each of the other strings from which it differs by one bit.

Example 7.1.5 suggests how to generalize the circulant-graph construction to an arbitrary *group* and any subset of elements of the group. (See Appendix A2.) This was developed by Max Dehn (1878-1952), building on a representation of complete graphs that appeared in a paper of Arthur Cayley (1821-1895), whose name has been given to the construction.

DEFINITION: To an arbitrary group $\mathcal{A}$ and to an arbitrary subset S of its domain, we associate the (undirected) **Cayley graph** $Cay\,(A:S)$, whose vertex set is the domain of the group $\mathcal{A}$, such that there is an edge between vertices u and v if there is an element $s \in S$ such that $u + s = v$ (for an additive group, or $u \star s = v$ if $\star$ is the group operation).

Example 7.1.5, continued: The hypercube graph Q_n may be described as the Cayley graph

$$Cay\,(\mathbb{Z}_2^n : E)$$

where E is the set of k-tuples

$$(1,0,0,\ldots,0), \quad (0,1,0,\ldots,0), \quad \ldots, \quad (0,\ldots,0,1)$$

Circular and Möbius Ladders

In general, a class of graphs is better conceptualized with the aid of both a picture and an algebraic specification, than by either one alone, if the graphs in that class have sufficient symmetry to be represented by a concise algebraic description.

DEFINITION: The **circular ladder graph** CL_n is representable geometrically in the plane as two concentric cycle graphs, each with n vertices spaced equally apart, such that there is a line joining the jth vertex on the outer cycle to the jth vertex on the inner cycle, for $j = 1, \ldots, n$.

Example 7.1.6:   Each circular ladder CL_n is 3-regular, as illustrated in Figure 7.1.7.

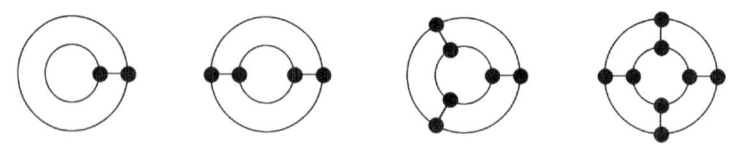

Figure 7.1.7   The circular ladders CL_1, CL_2, CL_3, and CL_4.

Alternatively, the circular ladder CL_n could be described as the Cayley graph

$$Cay\,(\mathbb{Z}_n \oplus \mathbb{Z}_2 : \{(1,0),(0,1)\})$$

The first component of each pair in $\mathbb{Z}_n \oplus \mathbb{Z}_2$ is added modulo n and the second component is added modulo 2. This alternative representation is illustrated for CL_4 in Figure 7.1.8.

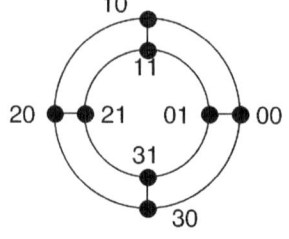

Figure 7.1.8   The Cayley graph $Cay\,(\mathbb{Z}_4 \oplus \mathbb{Z}_2 : \{(1,0),(0,1)\})$.

DEFINITION: The **Möbius ladder graph** ML_n is obtainable from the circular ladder CL_n by removing the edge joining vertex n and vertex 1 on both the inner cycle and the outer cycle, and then joining vertex n on the outer cycle to vertex 1 on the inner cycle and also joining vertex n on the inner cycle to vertex 1 on the outer cycle.

Example 7.1.7:   Each Möbius ladder ML_n is 3-regular, as illustrated in Figure 7.1.9. The name *Möbius ladder* reflects the resemblance of these drawings of the graph to the surface known as a Möbius band.

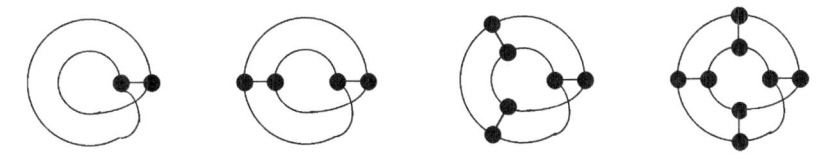

Figure 7.1.9   The Möbius ladders ML_1, ML_2, ML_3, and ML_4.

Alternatively, the Möbius ladder ML_n can be described as $circ\,(2n : 1,\, n)$ or, equivalently, as the Cayley graph

$$Cay\,(\mathbb{Z}_{2n} : \{1, n\})$$

as illustrated for ML_4 in Figure 7.1.10.

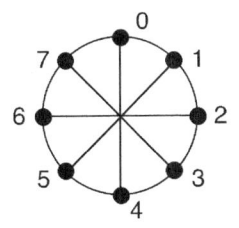

Figure 7.1.10   The Cayley graph $Cay\,(\mathbb{Z}_8 : \{1, 4\})$.

PREVIEW OF §7.4:   The formal sense in which we may say that two different drawings, or more generally, two specifications of any type, represent *equivalent* graphs (informally, the *same* graph) is called *graph isomorphism*.

Petersen Graph

In addition to the infinite families of regular graphs, there are many interesting special examples of regular graphs. The most frequently cited among these is almost certainly the *Petersen graph*.

DEFINITION: The **Petersen graph** is the 3-regular graph depicted in Figure 7.1.11.

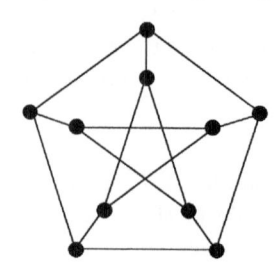

Figure 7.1.11   The Petersen graph.

EXERCISES for Section 7.1

In each of the Exercises 7.1.1 through 7.1.4, calculate the number of vertices and edges of the given graph.

7.1.1^S   Hypercube graph Q_4.   **7.1.2**   Hypercube graph Q_5.

7.1.3 $circ(10:1,2,5)$.   **7.1.4** $circ(2n:2,n)$.

DEFINITION: Two edges of a graph G are considered to be **adjacent edges** if they have at least one endpoint in common. The graph $L(G)$, called the **line graph** of G, has E_G as its vertex set. Each pair of edges that is adjacent in G is joined by an edge of $L(G)$.

7.1.5   Prove that the line graph of a regular simple graph is a regular simple graph.

7.1.6   Suppose that G is a d-regular graph with n vertices. How many vertices does the line graph $L(G)$ have? What is the degree of regularity of G?

7.1.7^S   What connected regular graph is a line graph of itself?

7.1.8   Of what graph is K_n a line graph?

In each of the Exercises 7.1.9 through 7.1.12, describe the given graph as a circulant graph.

7.1.9   Möbius ladder ML_n.   **7.1.10**   Circular ladder CL_n, n odd.

7.1.11   Complete graph K_n.   **7.1.12^S**   Complete bipartite graph $K_{n,n}$.

7.1.13   Is the Petersen graph describable as a circulant graph?

7.2 WALKS AND DISTANCE

A walk in a graph is a discrete analogue of a continuous curve. It is conceptualized as a sequence of consecutive edge-steps. It represents some kind of traversal, and it has an order of traversal, even if its edges are undirected. Applications are commonly concerned with shortest walks. The first few definitions about walks, paths, cycles, and distance are presented rather rapidly, in consideration of the likelihood that they have been encountered previously in a discrete mathematics course or, perhaps, in any one of various engineering or applied-mathematics courses.

Figure 7.2.1   A continuous curve and a discrete walk.

DEFINITION: A **walk** from vertex v_0 to vertex v_n in a graph G is an alternating sequence

$$W = \langle v_0, e_1, v_1, e_2, \ldots, e_n, v_n \rangle$$

of vertices and edges, such that edge e_j joins vertices v_{j-1} and v_j, for $j = 1, \ldots, n$. It is a **trivial walk** if $n = 0$. It is a **directed walk** if each edge e_j is directed from v_{j-1} to v_j. A **closed walk** is a walk that begins and ends at the same vertex. An **open walk** ends at a different vertex from the one at which it begins.

In a simple graph, since there is only one edge between two adjacent vertices, one can safely abbreviate that representation with a sequence of vertices,

$$W = \langle v_0, v_1, \ldots, v_n \rangle$$

without explicitly mentioning the edges.

DEFINITION: A **path** is a walk that has no repeated vertices (or edges), except that the last vertex may possibly be the same as the first. If so, it is a **closed path**, and if not, it is an **open path**.

DEFINITION: A closed path is also called a **circuit** or a **cycle**. A graph with no non-trivial circuits is said to be an **acyclic graph**.

Example 7.2.1:   Figure 7.2.2 below illustrates an open path. The last vertex, v_n, is not the same as the initial vertex, v_0.

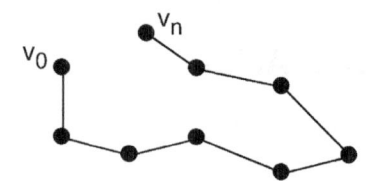

Figure 7.2.2   An open path.

DEFINITION: A graph in which there is an open path that traverses every edge is called a **path graph**. A path graph with n vertices is denoted P_n.

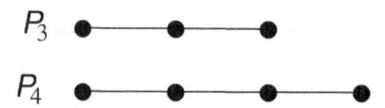

Figure 7.2.3   The path graphs P_3 and P_4.

Observe that we distinguish between a *path*, which is a sequence of vertices and edges, and a *path graph*, which is a graph. If in a graph G there is a *closed path* that traverses every edge, then that graph is a *cycle graph* and not a path graph.

DEFINITION: A graph is **connected** if for every pair of vertices u and v, there is a path from u to v.

Example 7.2.2:   The graph in Figure 7.2.4 is non-connected, because there is no path, for instance, between vertices u and v.

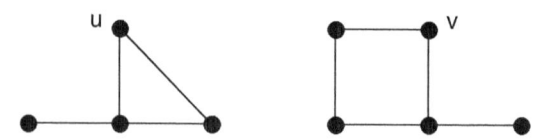

Figure 7.2.4   A non-connected graph.

PREVIEW OF §8.1:   A closed walk that traverses every edge of a graph exacly once is called an **eulerian tour**, after Leonhard Euler (1707-1783). An open walk that traverses every edge exactly once is called an **eulerian trail**. A circuit that traverses every vertex is called a **hamiltonian circuit**, after Sir William Rowan Hamilton (1805-1865).

Measuring Distance

DEFINITION: The **length of a walk**

$$W \; = \; \langle \, v_0, \, e_1, \, v_1, \, e_2, \, \ldots, \, e_n, \, v_n \, \rangle$$

is the number n of edge-steps in the walk. (This may be more than the number of different edges in the sequence, because one or more edges may be retraversed during the walk.)

DEFINITION: The **distance** $d(u,v)$ from vertex u to vertex v is the length of a shortest walk from u to v, if such a walk exists. Otherwise, the distance $d(u,v)$ is defined to be infinite.

Remark: Clearly, a shortest walk between two vertices must be a path. If a vertex w recurs on a walk from a vertex u to a vertex v, then one could excise the subwalk from w to itself and thereby obtain a shorter walk.

Remark: In an undirected graph, the distance from v to u must be the same as the distance from u to v, so we may refer to the *distance between two vertices*.

PREVIEW OF §7.8:   A standard algorithm for calculating shortest paths from a given vertex to all other vertices, due to E. Dijkstra [Dijk1959], is presented in §7.8. This algorithm applies to a more general sense of distance in a *weighted network*, in which arbitrarily large distances may be assigned to single edges.

COMPUTATIONAL NOTE: An elementary algorithm of R. W. Floyd [Floy1962] calculates all the pair distances in a weighted n-vertex graph, with total time proportional to n^3, based on work of S. Warshall [Wars1962]. An algorithm that has expected time of $n^2 \log^2 n$ is due to P. M. Spira [Spir1973].

Walks and Cycles in Bipartite Graphs

Proposition 7.2.1. *In a bipartite graph, there are no closed walks of odd length.*

Proof: Since there are no edges within either part of the bipartition, a walk from a vertex v is always in the other part after an odd number of edge-steps. $\Diamond$

Corollary 7.2.2. *In a bipartite graph, there are no cycles of odd length.* $\Diamond$

Theorem 7.2.3 [König, 1936]. *A graph with no odd cycles is bipartite.*

Proof: Proof for a connected graph G is sufficient. Let u be any vertex of G, and define

$$X = \{\, x \mid d(u,x) \text{ is even}\,\}$$
$$Y = \{\, y \mid d(u,y) \text{ is odd}\,\}$$

Suppose that both vertices z_1 and z_2 lie either in X or in Y, and that they are adjacent. Let P_1 be a shortest u-z_1 path and P_2 be a shortest u-z_2 path. By definition of X and Y, the lengths of these paths are either both even or both odd. Thus, they must meet at one or more vertices, as illustrated in Figure 7.2.5, for otherwise the concatenation of paths P_1, P_2 and edge $z_1 z_2$ would form an odd cycle.

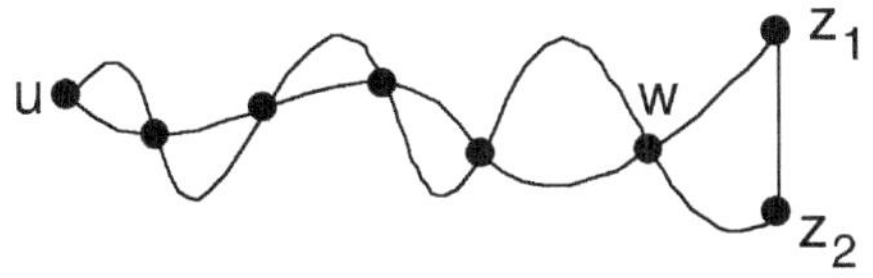

Figure 7.2.5   Shortest paths to z_1 and z_2.

Since P_1 and P_2 are both shortest paths, the distances along them to every inter-section vertex must be the same, and there must be some last intersection vertex w. Thus, the distance along the subpath of P_1 from w to z_1 must be of the same parity as the distance along the subpath of P_2 from w to z_2. Since these subpaths do not meet, their concatenation with edge z_1z_2 forms an odd cycle, a contradiction.   ◇

Diameter, Radius, and Girth

Various numerical invariants of a graph are based on the distances between vertices. They are of great importance to certain models for communications.

DEFINITION: The **diameter** of a graph G, denoted $diam(G)$, is the maximum of the distances between any two vertices. (The diameter may represent the maximum time needed for two nodes of a network to communicate.)

Example 7.2.3:   The diameter of the graph in Figure 7.2.6 is 3. In particular, vertex s is at distance 3 from vertices w, x, and z.

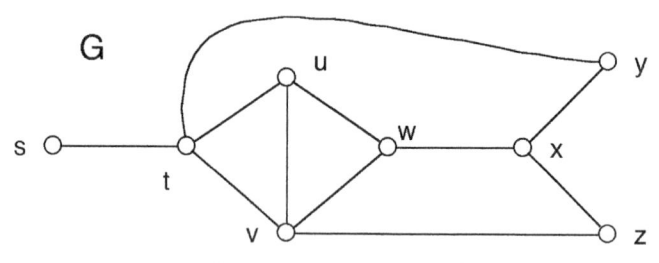

Figure 7.2.6   A graph.

DEFINITION: The **eccentricity** of a vertex v, denoted $ecc(v)$, is the maximum of the distances from v to other vertices. (The eccentricity may represent the worst-case time needed to reach any other node from a given node.)

Example 7.2.3, continued: In Figure 7.2.6, the vertices s, w, x, and z each have eccentricity 3, and the other four vertices have eccentricity 2.

DEFINITION: The **radius** of a graph G, denoted $rad(G)$, is the minimum of the eccentricities of the vertices. (A vertex whose eccentricity equals the radius of the network is regarded as **central**. It is a best node for originating or receiving a message within the network, in the sense that its worst-case time is least.)

Example 7.2.3, continued: The radius of the graph in Figure 7.2.6 is 2, the minimum of the eccentricities.

DEFINITION: The **girth** of a graph G is the length of a shortest non-trivial circuit, denoted $girth(G)$, if any such circuit exists.

Example 7.2.3, continued: The girth of the graph of Figure 7.2.6 is 3.

COMPUTATIONAL NOTE: The shortest path that includes a given edge uv can be calculated by a modification of Dijkstra's algorithm. Iterating this over all the edges leads to an elementary algorithm for calculating the girth.

The following table gives the diameter, radius, and girth of some of the graphs described earlier in this chapter. All of the graphs in the table are *vertex-transitive* (which is to be defined later in this chapter), which is a sufficient condition for the diameter and the radius to be equal.

graph	diameter	radius	girth
cycle graph C_n	$\left\lfloor \dfrac{n}{2} \right\rfloor$	$\left\lfloor \dfrac{n}{2} \right\rfloor$	n
hypercube Q_n	n	n	4, for $n \geq 2$
circular ladder CL_n	$\left\lceil \dfrac{n+1}{2} \right\rceil$	$\left\lceil \dfrac{n+1}{2} \right\rceil$	4, for $n \geq 4$
Möbius ladder ML_n	$\left\lceil \dfrac{n}{2} \right\rceil$	$\left\lceil \dfrac{n}{2} \right\rceil$	4, for $n \geq 2$
Petersen graph	2	2	5

Exercises

In each of the Exercises 7.2.1 through 7.2.4, calculate the radius and diameter of the given graph.

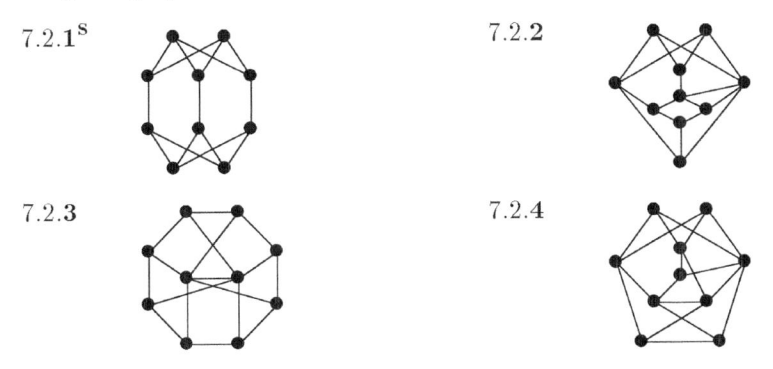

7.2.1[S]

7.2.2

7.2.3

7.2.4

In each of the Exercises 7.2.5 through 7.2.8, decide whether the specified graph is bipartite. Give a proof.

7.2.5[S]   The graph of Exercise 7.2.1.

7.2.6   The graph of Exercise 7.2.2.

7.2.7   The graph of Exercise 7.2.3.

7.2.8   The graph of Exercise 7.2.4.

In each of the Exercises 7.2.9 through 7.2.12, decide whether the girth of the specified graph is 3 or 4. Give a proof.

7.2.9[S]   The graph of Exercise 7.2.1.

7.2.10   The graph of Exercise 7.2.2.

7.2.11   The graph of Exercise 7.2.3.

7.2.12   The graph of Exercise 7.2.4.

In each of the Exercises 7.2.13 through 7.2.16, find all the central vertices of the specified graph.

7.2.13   The graph of Exercise 7.2.1.

7.2.14   The graph of Exercise 7.2.2.

7.2.15[S]   The graph of Exercise 7.2.3.

7.2.16   The graph of Exercise 7.2.4.

In each of the Exercises 7.2.17 through 7.2.20, calculate the radius and diameter of the indicated line graph.

7.2.17[S] $L(K_n)$.   **7.2.18** $L(Q_n)$.

7.2.19 $L(CL_n)$.   **7.2.20** $L(\text{Petersen graph})$.

7.2.21   Prove that the diameter of a connected graph is at most twice the radius, and give an example to illustrate that this upper bound can be realized.

7.3   TREES AND ACYCLIC DIGRAPHS

DEFINITION: A **tree** is a connected graph in which there are no non-trivial circuits. A tree with only one vertex is called **trivial**. A possibly non-connected graph with no non-trivial circuits is called a **forest**.

Example 7.3.1:   Figure 7.3.1 shows a tree and a graph that is not a tree, because it has a circuit.

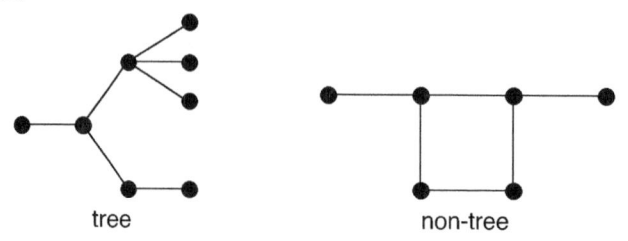

tree   non-tree

Figure 7.3.1   A tree and a non-tree.

DEFINITION: A *leaf* of a tree is a 1-valent vertex.

Example 7.3.1, continued: The graph on the left of Figure 7.3.1 is a tree with five leaves.

Trees are among the most commonly occurring graphs, both in pure mathematics and in applications. Their use in computer algorithms could be a book in itself. Our concern in this section is to lay the groundwork for such study. We establish their most significant characteristics, and we briefly introduce the two kinds of trees that arise most frequently as data structures for computer algorithms.

Properties of Trees

The following lemma is the key to establishing some of the most important characterizations of trees, because it permits inductive arguments on tree size, as in Proposition 7.3.2.

Lemma 7.3.1. *Every non-trivial tree T has at least two leaves.*

Proof: Let P be a path of maximum length in the tree T, and suppose that P runs from vertex s to vertex t. Since tree T is non-trivial, it follows that $s \neq t$ and, in turn, that there is a vertex $v \neq s$ immediately after s along path P. Suppose that vertex s is not a leaf. Then it must be adjacent to some other vertex w, in addition to v. The vertex w cannot lie on path P, because if it did, the walk from s to w along P, followed by the edge sw, would yield a circuit. Yet if w does not lie on path P, then preceding path P by edge sw would extend P to a longer path. By a similar argument, the vertex t must be a leaf.   ◇

Proposition 7.3.2. *The number of edges of a tree is one less than the number of vertices.*

Proof: A tree cannot have self-adjacencies, so if there is only one vertex, there must be no edges.

By way of induction, assume that a tree with $n - 1$ vertices must have $n - 2$ edges, for some $n \geq 2$, and let T be a tree with n vertices. By Lemma 7.3.1, there is a leaf in tree T. Deleting that leaf and the edge incident on it cannot introduce a circuit to T. Nor does it disconnect T. Thus, the graph obtained by that deletion is a tree. Since it has $n - 1$ vertices, it follows from the induction hypothesis that it has $n - 2$ edges. Accordingly, the tree T, with one more edge, has $n - 1$ edges.   ◇

Proposition 7.3.3. *Adding an edge to a tree T creates a cycle.*

Proof: If the new edge has only one endpoint, then that edge itself is a cycle. Therefore, suppose that it joins two vertices. Since the tree T is connected, there is already a path joining those two vertices, and adding the edge to that path transforms it into a cycle.   ◇

Proposition 7.3.4. *In a tree T, there is exactly one path joining any two vertices.*

Proof: Let P and Q be a pair of paths joining the same two vertices, say s and t, such that among all such instances of pairs of paths in T, the sum of their lengths is minimum. Minimality of that sum precludes the possibility that path Q intersects path P anywhere except vertices s and t. Accordingly, following a traversal of path P by a reverse traversal of Q is a circuit in T, a contradiction.   ◇

Rooted Trees and Binary Trees

DEFINITION: A **rooted tree** is a tree with one vertex designated as the root.

Computer programs use rooted trees as data structures. Access to the data is giving by passing the memory address of the root of the tree. There are, in fact, many other uses of rooted trees, including their capacity to represent a sequential decision process or a hierarchy. Designating some particular vertex of a tree as the root induces a host of additional properties. Some of the terminology for rooted trees is based on the metaphor of a family tree.

Remark: Although it is helpful to borrow terminology from the notion of family trees, it is provable that family trees have cycles. If one assumes that a generation is 25 years, then in a thousand years, there are 40 generations. The number 2^{40} of ancestors one would have had a thousand years ago, under the assumption of no cycles, vastly exceeds the total population of the earth.

DEFINITION: In a rooted tree T, the **parent** of a vertex v other than the root is the vertex immediately after v on the unique path in T from v to the root. The root itself may serve as a parent.

DEFINITION: Every vertex of which the vertex u is a parent is called a **child** of u. Two vertices with the same parent are called **siblings**.

DEFINITION: A **descendant of a vertex of a rooted tree** is defined recursively as that vertex itself, any child of that vertex, or any descendant of a child of that vertex. An **ancestor of a vertex of a rooted tree** is defined as any vertex of which that vertex is a descendant.

TERMINOLOGY NOTE: Sometimes it is convenient to distinguish a vertex itself from its other descendants and ancestors, in which case, its other descendants and ancestors are called *proper descendants* and *proper ancestors*, respectively.

DEFINITION: The **height of a rooted tree** is the maximum distance of any vertex from the root.

Example 7.3.2:   In the rooted tree of Figure 7.3.2, vertex u is the parent of vertices w, x, and y. The ancestors of vertex y are the vertices u and s. The height of the tree is 2.

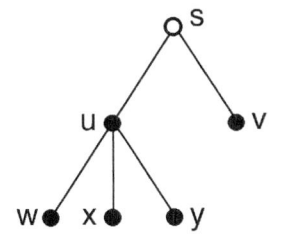

Figure 7.3.2   A rooted tree of height 2.

DEFINITION: A **binary tree** is a rooted tree in which each vertex has at most two children, such that each child (even an only child) is designated as a **left-child** or a **right-child**.

Example 7.3.3:   In the binary tree of Figure 7.3.3, vertex w is the left-child and vertex x the right-child of vertex u. Vertex x has no right-child. Its only child, the vertex a, is designated by the drawing as the left-child.

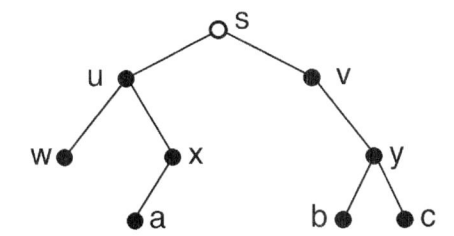

Figure 7.3.3   A binary tree.

Binary Search Trees

To illustrate one of the many ways that graphs are used as data structures in computer information systems, we now introduce vertex labeling as an augmentation of the graph model. Whereas the name of a vertex is a permanent part of its specification, a label is potentially a large record of information, that may change.

DEFINITION: A **labeling of the vertices** of a graph G in a set S is an assignment to each vertex of an element of S.

DEFINITION: The descendants of the left-child of a vertex v, plus the edges joining those vertices, is called the **left subtree** of v. Similarly, the descendants of the right-child of a vertex v, plus the edges joining those vertices, is called the **right subtree** of v.

DEFINITION: A **binary search tree** (abbr. **BST**) is a binary tree with a labeling of its vertices by distinct elements of a linearly ordered set, such that the label at each vertex follows all the labels in its left subtree and precedes all the labels in its right subtree. The label of a vertex v is called its **key**.

Example 7.3.4:   Two of the many ways in which the numbers

$$51 \quad 34 \quad 84 \quad 22 \quad 08 \quad 56 \quad 52 \quad 69 \quad 28 \quad 94 \quad 04 \quad 61$$

could be the keys of a binary search tree are illustrated in Figure 7.3.4. A data struc-tures and algorithms course presents algorithms for inserting and deleting keyed nodes from a BST, so that the resulting graph is also a BST.

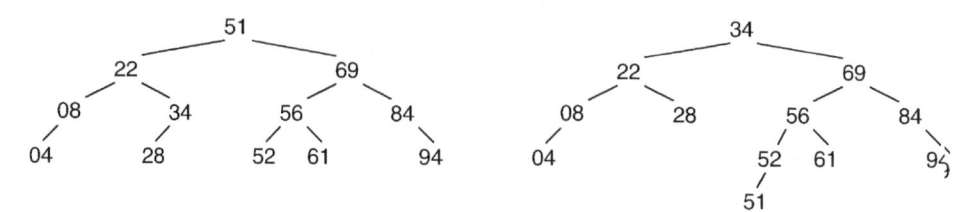

Figure 7.3.4   Two binary search trees with the same keys.

We observe that at node 22, for instance, the two nodes to its left have keys 04 and 08 that are less than 22, and the two nodes to its right have keys 28 and 34 that are greater than 22. This same relationship holds at every node.

To illustrate how one searches a BST, consider how we might seek a node with the target key 52 in the BST on the right, starting from the root. Target node 52, if in the BST, would have to be in the right subtree of node 34, that is, in the subtree rooted at node 69. Proceeding down a level, node 52 (if in the BST) would have to be in the left subtree of the subtree rooted at node 69, that is, in the subtree rooted at node 56. Node 52 would have to be in the left subtree of node 56, and indeed, it is the left-child of node 56. If we had been seeking a target node 54, we would have tried to go to the right subtree of node 52. Finding no right-child at 52, we would conclude that there is no node 54 in the BST. Algorithm 7.3.1 indicates how this would be done on a computer.

Algorithm 7.3.1:   Binary-Search-Tree Search

Input: a binary-search tree T and a target key t.
Output: a vertex v of T such that $key(v) = t$ if t is found,
         or a NULL vertex if no vertex has t as its key.

$\quad v := root(T)$
$\quad$ While $(v \neq \text{NULL})$ and $(t \neq key(v))$
$\quad\quad$ If $t > key(v)$
$\quad\quad\quad v := rightchild(v)$
$\quad\quad$ Else $v := leftchild(v)$
$\quad$ Return v.

In a computer implementation of a binary search tree, a node with a NULL key and no children is installed as an artificial left- or right-child, wherever the actual node has no left- or right-child, respectively. The following searching algorithm generalizes our description in Example 7.3.4. What is actually returned to the calling program is a pointer to a node. If the key at that node is NULL, this is interpreted as "not found". Otherwise, the search has located the node whose key matches the target key value, where there is the record with the data actually being sought.

Edge Directions

An edge between two vertices creates a connection in two opposite senses at once. Assigning a direction makes one of these senses *forward* and the other *backward*. In a line drawing, the choice of forward direction is indicated by placing an arrow on an edge. The option of assigning directions greatly enhances the modeling capability of graph models. The definitions for graphs with directed edges are in some ways similar, and in others, a little different.

DEFINITION: A **directed edge** (or **arc**) is an edge, one of whose endpoints is designated as the **tail**, and whose other endpoint is designated as the **head**. An arc is said to be **directed from** its tail to its head.

DEFINITION: A **multi-arc** is a collection of two or more arcs with the same tail and head.

DEFINITION: A **directed graph** (or **digraph**) is a graph each of whose edges is directed. If some edges are directed and some are not, it is called a **mixed graph**.

DEFINITION: A **simple digraph** is a digraph that has no self-loops and no multi-arcs.

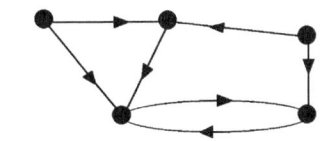

Figure 7.3.5   **A simple digraph.**

DEFINITION: The **underlying graph** of a directed or mixed graph G is the graph that results from removing all the designations of *head* and *tail* from the directed edges of G (i.e., deleting all the edge-directions).

Example 7.3.5:   Although the digraph in Figure 7.3.6 is simple, its underlying graph is not simple.

Figure 7.3.6   **A digraph D and its underlying graph G.**

DEFINITION: A **directed walk** in a graph is an alternating sequence

$$W \;=\; \langle\, v_0, e_1, v_1, e_2, \ldots, e_n, v_n \,\rangle$$

of vertices and arcs in which the head of the arc e_j is the vertex v_j, for $j = 1, \ldots, n$; it is a **directed path** if it is a path in the underlying graph.

DEFINITION: A **connected digraph** is a digraph whose underlying graph is connected. A **strongly connected digraph** is a digraph such that for every pair of vertices u and v, there is a directed walk from u to v.

Acyclic Digraphs

DEFINITION: An **acyclic digraph** (with acronym **dag**, for *directed acyclic graph*) is a digraph with no directed cycles.

Example 7.3.6:   The underlying graph of an acyclic digraph may contain cycles, as in Figure 7.3.7.

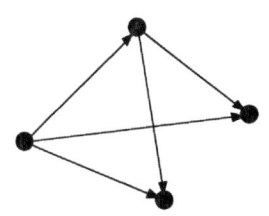

Figure 7.3.7   **A dag whose underlying graph is non-acyclic.**

Example 7.3.7:   Any binary relation R on a set U can be represented by a digraph in which there is an arc from u to v if and only if uRv. In this sense, every acyclic digraph may be regarded as a partial ordering on its vertices, by using the relation $u \preceq v$ to mean that there is a directed path from u to v. Trivial paths yield reflexivity, acyclicity implies antisymmetry, and concatenation of paths yields transitivity. Conversely, every partially ordered set can be represented as an acyclic digraph.

Transitive Digraphs

DEFINITION: A **transitive digraph** is a digraph such that whenever there is an arc from u to v and an arc from v to w, there is also an arc directly from u to w.

Proposition 7.3.5. *In a transitive digraph, if there is a directed path from vertex u to vertex v, then there is an arc $u \to v$.*

Proof: This follows readily from an induction argument. $\Diamond$

Example 7.3.6, continued: The dag of Figure 7.3.7 is transitive.

Corollary 7.3.6. *Let D be a transitive digraph whose underlying graph is simple. Then D is acyclic.*

Proof: Let $v_0, v_1, \ldots, v_{k-1}$ be a directed path in D, and consider an edge e joining v_0 and v_{k-1}, as illustrated in Figure 7.3.8.

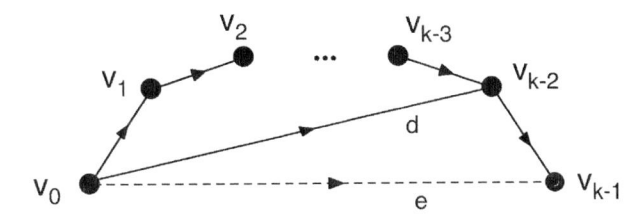

Figure 7.3.8 **A directed path in a transitive digraph D.**

By Proposition 7.3.5, there is an arc

$$d : v_0 \to v_{k-2}$$

Since there is an arc

$$v_{k-2} \to v_{k-1}$$

transitivity implies that the edge e joining v_0 and v_{k-1} must be directed from v_0 to v_{k-1}. Thus, the directed path cannot be completed to a directed cycle in D.   ◇

DEFINITION: The **transitive closure of a digraph** is the digraph obtained by joining each vertex u to every other vertex v such that there is a non-trivial directed path from u to v, but no pre-existing directed edge from u to v already.

Exercises

7.3.1   Prove that every vertex of maximum eccentricity in a tree is a leaf.

7.3.2   Let G be a connected graph in which the number of edges is one less than the number of vertices. Prove that G is a tree. This is a converse of Proposition 7.3.2.

7.3.3   Let G be a connected graph in which there is exactly one path joining any two vertices. Prove that G is a tree. This is a converse of Proposition 7.3.4.

7.3.4   Draw all possible binary search trees whose keys are 1, 2, and 3.

7.3.5   Draw all possible binary search trees whose keys are 1, 2, 3, and 4.

7.3.6[S]   How many different binary search trees are possible if there are n keys?

7.3.7   What is the minimum height of a binary search tree with n keys?

7.3.8   Prove that it is possible to assign directions to all the edges of a complete graph so that the resulting digraph is acyclic.

7.3.9[S]   Prove that it is possible to assign directions to all the edges of any simple graph so that the resulting digraph is acyclic.

7.3.10   Prove that the comparability digraph of a poset is transitive.

7.3.11   Prove that every transitive digraph is the comparability digraph of a poset.

DEFINITION: A digraph is a **tournament** if its underlying graph is a complete graph.

7.3.12   Prove that an acyclic tournament is transitive.

7.3.13   Prove that a transitive tournament is acyclic.

7.3.14[S]   Prove that a transitive tournament linearly orders its vertices.

7.3.15   Is the transitive closure of a connected acyclic digraph necessarily a tournament? Give a counterexample or a proof.

7.4   GRAPH ISOMORPHISM

We are often faced with the problem of deciding from drawings, incidence tables, or other descriptions of two graphs whether they specify mathematically equivalent graphs. Some features that seem to distinguish one graph from another are *invariant properties* of the graphs, and others are artifacts of the representation.

Example 7.4.1:   Each of the three graphs in Figure 7.4.1 has five vertices and seven edges. However, graphs G and H both have a 1-valent vertex, but graph J does not. We will see that the degree sequence is an invariant property.

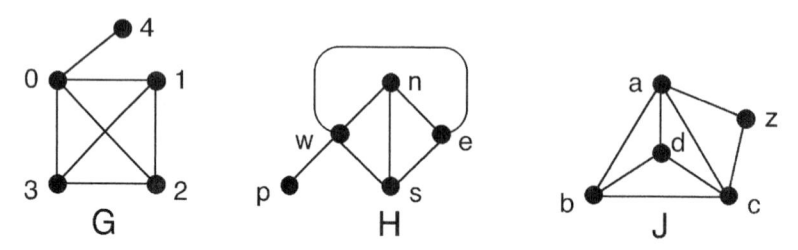

Figure 7.4.1   Three 5-vertex, 7-edge graphs.

However, the edge-crossing in graph G is merely a feature of the representation, since the edge 02 could be redrawn so that it does not cross edge 13. The names of the vertices are also features of the representation, not of the underlying structure.

Structural Equivalence for Simple Graphs

What turns an abstract set called *vertices* and another abstract set called *edges* into a graph is the *incidence structure*, that is, the specification of which vertices are the endpoints of which edge. For two graphs to be equivalent, it is not enough that they have the same number of vertices and the same number of edges. There must also be a correspondence of their incidence structures. In a simple graph, the incidence structure conveniently reduces to the adjacency relationship on the vertices.

DEFINITION: A bijection $f : V_G \to V_H$ on the vertex sets of two simple graphs is said to **preserve adjacency** if for every pair $\{u, v\}$ of adjacent vertices in G, the image pair $\{f(u), f(v)\}$ is adjacent in H. It is said to **preserve non-adjacency** if for every pair $\{u, v\}$ of non-adjacent vertices in G, the image pair $\{f(u), f(v)\}$ is non-adjacent in H.

The name for the confluence of bijection with structure preservation is *isomorphism*.

DEFINITION: Two simple graphs G and H are **isomorphic**, denoted $G \cong H$, if there is a bijection $f : V_G \to V_H$ that preserves both adjacency and non-adjacency. Such a bijection f is said to be an **isomorphism**. (The definitions of isomorphic and isomorphism are extended to general graphs by requiring that the number of edges joining each pair of vertices be preserved.)

Example 7.4.1, continued: The vertex bijection

$$0 \mapsto w \quad 1 \mapsto n \quad 2 \mapsto e \quad 3 \mapsto w \quad 4 \mapsto p$$

is an isomorphism from graph G to graph H of Figure 7.4.1.

Isomorphism is an equivalence relation on graphs. Clearly, a graph is isomorphic to itself, under the identity mapping. The following proposition establishes symmetry. Proving that the composition of two isomorphisms is an isomorphism is left as an exercise.

Proposition 7.4.1. *The inverse of a graph isomorphism is an isomorphism.*

Proof: The inverse of a bijection is a bijection, and the requirement that an isomorphism preserves non-adjacency as well as adjacency ensures that the inverse preserves adjacency and non-adjacency. $\diamond$

We can now establish two criteria for graphs to be isomorphic.

Theorem 7.4.2. *Let G and H be isomorphic graphs. Then G and H have the same numbers of vertices and edges.*

Proof: Since an isomorphism $f : G \to H$ is a bijection on the vertex sets, the numbers of vertices of G and H must be the same. Since an isomorphism preserves adjacency, the number of adjacent pairs in the codomain H must be at least as large as the number of adjacent pairs in the domain G. (Moreover, in case of multiple adjacencies, the sum of the multiplicities of adjacency in H must be at least as large as in G.) By Proposition 7.4.1, the number of adjacent pairs in H (or the sum of the multiplicities of adjacency) cannot be greater than in G. $\diamond$

Remark: A graph isomorphism f is best regarded as a bijection not only on the vertex set, but also as a bijection on the edge set of its domain graph, under the rule

$$uv \mapsto f(u)f(v)$$

The conceptualization of an isomorphism as a pair of mappings extends to general graphs by requiring that each edge e joining vertices u and v is mapped to a specific edge $f(e)$ joining vertices $f(u)$ and $f(v)$.

TERMINOLOGY: The number of vertices of a graph, and also the number of edges, is said to be a **graph invariant**, because two isomorphic graphs must have the same number of vertices (in view of Theorem 7.4.2), and also the same number of edges. This terminology is applied to any property of graphs that must have the same value for any two isomorphic graphs.

Graph Isomorphism Testing

Some of the necessary conditions for two graphs to be isomorphic are easily calculated. In particular, it is easy to determine their numbers of vertices, their numbers of edges, and their degree sequences. In the case of small graphs, it is not much more difficult to calculate their eccentricities, their diameters, and their radii, all of which are graph invariants.

Proposition 7.4.3. *A graph isomorphism* $f : G \to H$ *preserves the degree of every vertex in its domain.*

Proof: Let $v \in V_G$. Then the isomorphism f maps each neighbor u of v to a neighbor $f(u)$ of $f(v)$ (of the same multiplicity, in case these are general graphs). It follows that the degree of $f(v)$ equals the degree of d.   ◇

Theorem 7.4.4. *Two isomorphic graphs have the same degree sequence.*

Proof: This is an immediate corollary of Proposition 7.4.3.   ◇

Example 7.4.1, continued: As previously observed, in graphs G and H, the vertices labeled 4 and p have degree 1, but graph J has no vertex of degree 1. It follows from Theorem 7.4.4 that graphs G and H are not isomorphic to graph J.

Proposition 7.4.5. *A graph isomorphism* $f : G \to H$ *maps a walk of length* n

$$W = \langle v_0, e_1, v_1, e_2, \ldots, e_n, v_n \rangle$$

in its domain G *to a walk of length* n

$$W = \langle f(v_0), f(e_1), f(v_1), f(e_2), \ldots, f(e_n), f(v_n) \rangle$$

in its codomain H.

Proof: This is an immediate consequence of the requirement that a graph isomorphism preserves adjacency.   ◇

Theorem 7.4.6. *A graph isomorphism* $f : G \to H$ *maps a vertex* u *of* G *to a vertex* $f(u)$ *of the same eccentricity as* u.

Proof: This follows from Proposition 7.4.5.

Example 7.4.2:   The two trees T and U in Figure 7.4.2 have the same degree sequence. According to Theorem 7.4.4, an isomorphism from T to U would have to map the vertex x, the only 3-valent vertex of T, to the vertex y, the only 3-valent vertex of U. However, $ecc(x) = 3$ and $ecc(y) = 2$. Thus, by Theorem 7.4.6, the two trees are not isomorphic.

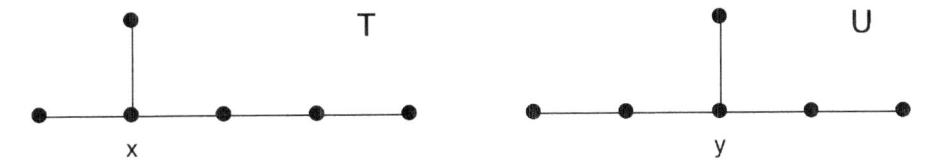

Figure 7.4.2   **These two trees with the same degree sequence are not isomorphic.**

Corollary 7.4.7.   *Two isomorphic graphs have the same diameter and radius.*

Proof: This follows from Theorem 7.4.6.   ◇

Example 7.4.3:   Since the circular ladder CL_4 and the Möbius ladder ML_4, shown in Figure 7.4.3, are both 3-regular, they cannot be distinguished by Theorem 7.4.4. However, their diameters are 3 and 2, respectively, so they are not isomorphic. This test works for all pairs CL_{2n} and ML_{2n}.

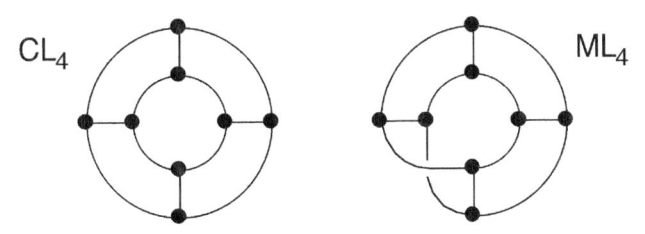

Figure 7.4.3   **The circular ladder CL_4 and the Möbius ladder ML_4 are not isomorphic.**

Proposition 7.4.8.   *A graph isomorphism maps a cycle in its domain to a cycle of the same length.*

Proof: This follows from Proposition 7.4.5.   ◇

Corollary 7.4.9.   *Two isomorphic graphs have the same girth.*

Proof: This follows from Proposition 7.4.8.   ◇

Example 7.4.4:   The circular ladder CL_3 and the Möbius ladder ML_3 are both 3-regular, and they both have diameter and radius 2. However, CL_3 has an odd cycle, but ML_3 does not. See Figure 7.4.4 below.

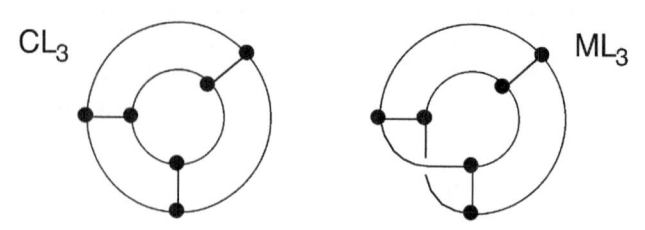

Figure 7.4.4 **The circular ladder CL_3 and the Möbius ladder ML_3 are not isomorphic.**

The pairs CL_{2n+1} and ML_{2n+1} can be distinguished, because CL_{2n+1} has some odd cycles and ML_{2n+1} does not, since it is bipartite.

By considering particular examples of isomorphism testing, one comes to realize the difficulties inherent in that pursuit. There is no known fixed list of short calculations that works on every pair of graphs. In general, one usually establishes isomorphism of two given graphs by constructing an explicit isomorphism, and one disproves isomorphism by finding incompatible properties in the two graphs.

Example 7.4.5:   The two graph drawings in Figure 7.4.5 both have a hexagon, but they look dissimilar because graph G has two overlapping triangles inside the hexagon, while graph H has a skewed 6-cycle inside.

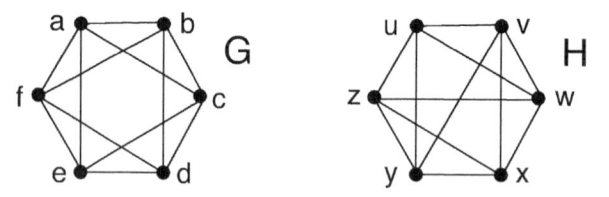

Figure 7.4.5   **Two 4-regular, 6-vertex graphs.**

It is a mistake to assume that the outer cycle in one drawing must be mapped to the outer cycle in the other. The first step of a more appropriate approach is to observe the total symmetry of graph G. Accordingly, if there is an isomorphism $G \to H$, then there is an isomorphism h in which vertex a of G is mapped to vertex u of H.

We observe that vertex d is the only vertex of G that is not adjacent to vertex a. It follows that $h(d)$ would have to be the vertex x, the only vertex of H that is not adjacent to u, because an isomorphism preserves non-adjacency.

Suppose that we now make the arbitrary choice of mapping vertex $b \in V_G$ to vertex $v \in V_H$. Since e is the only vertex of G not adjacent to b and since z is the only vertex of H not adjacent to v, it follows that $h(e)$ would have to be z. The progress so far in the attempted isomorphism is shown in Figure 7.4.6 below.

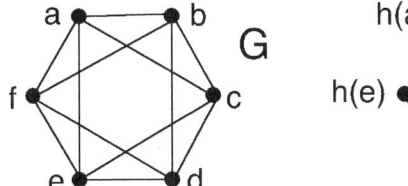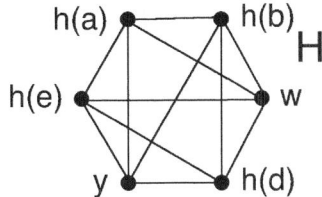

Figure 7.4.6   Attempted progress toward an isomorphism.

Vertex c is adjacent to vertices a, b, d, and e, and since the vertices w and y are both adjacent to each of the vertices $h(a)$, $h(b)$, $h(d)$, and $h(e)$, we can make the arbitrary choice of assigning $h(c) = w$ and $h(f) = y$. It can be directly verified that the resulting vertex bijection

$$a \mapsto u \quad b \mapsto v \quad c \mapsto w \quad d \mapsto x \quad e \mapsto z \quad f \mapsto y$$

preserves adjacency and non-adjacency everywhere. Thus, the graphs G and H are isomorphic.

Example 7.4.6:   The graphs $circ(8:1,2)$ and $circ(8:1,3)$ are not isomorphic. As seen in Figure 7.4.7, $circ(8:1,2)$ has some 3-cycles, but $circ(8:1,3)$ is bipartite, so it has no odd cycles.

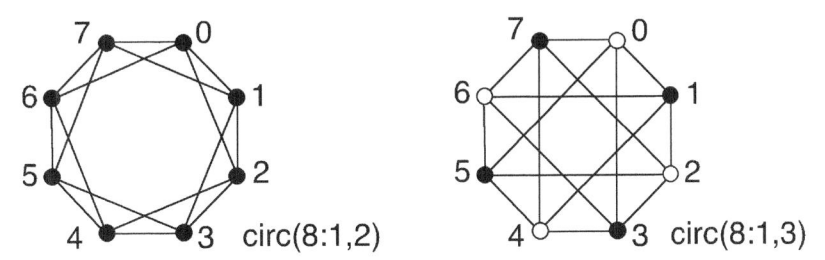

Figure 7.4.7 $circ(8:1,2)$ **and** $circ(8:1,3)$ **are not isomorphic.**

DEFINITION: The **graph isomorphism problem** is to invent a practical general algorithm to decide whether two given graphs are isomorphic or, alternatively, to prove that no such algorithm exists.

Since the number of possible bijections of the vertex sets of two n-vertex graphs is $n!$, the brute-force method of testing every one of them to see if it preserves adjacency and non-adjacency is not practical.

Isomorphism Type

Isomorphism is an equivalence relation on graphs. Each equivalence class under isomorphism is called an **isomorphism type**. A listing of all the isomorphism types of graphs meeting some requirement is commonly given by drawing a representative of each type.

Example 7.4.7:   Figure 7.4.8 depicts all the isomorphism types of simple graphs with four vertices and three edges.

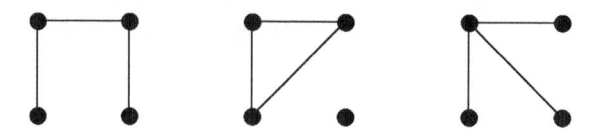

Figure 7.4.8   The three 4-vertex, 3-edge simple graphs.

When seeking to construct examples of all the graph isomorphism types that meet a given set of requirements, a systematic approach is helpful. Considering all plausible degree sequences is one kind of systematic approach. For instance, one may observe in this case that since there are three edges, the degree sum is 6. Since we are restricted to simple graphs, the maximum degree is 3. The ways to partition the number 6 into four non-negative integers, none greater than 3, are the following:

$$3300 \quad 3210 \quad 3111 \quad 2220 \quad 2211$$

The first two degree sequences cannot be achieved without multi-edges. Each of the other three sequences is represented by a single graph in Figure 7.4.8.

Remark: More generally, a single plausible degree sequence can be represented by two or more isomorphism types, as in Example 7.4.2.

TERMINOLOGY NOTE: Informally, one would usually say things like "these are the only two graphs such that ... ", instead of "these are representations of examples of the only two isomorphism types of graph such that ... ".

EXERCISES for Section 7.4

Exercises 7.4.1 through 7.4.6 are concerned with distinguishing the graphs A, B, C, and D. Hint: It is sufficient to consider eccentricities and cycle lengths.

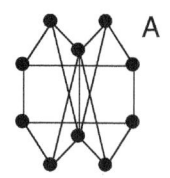 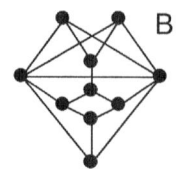 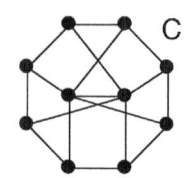 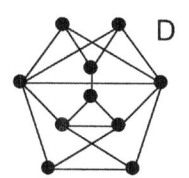

7.4.1[S] Prove that $A \not\cong B$.		**7.4.2** Prove that $A \not\cong C$.
7.4.3 Prove that $A \not\cong D$.		**7.4.4** Prove that $B \not\cong C$.
7.4.5 Prove that $B \not\cong D$.		**7.4.6** Prove that $C \not\cong D$.

Exercises 7.4.7 through 7.4.16 are concerned with distinguishing the graphs A, B, C, D, and E. Hint: It is sufficient to consider eccentricities and cycle lengths.

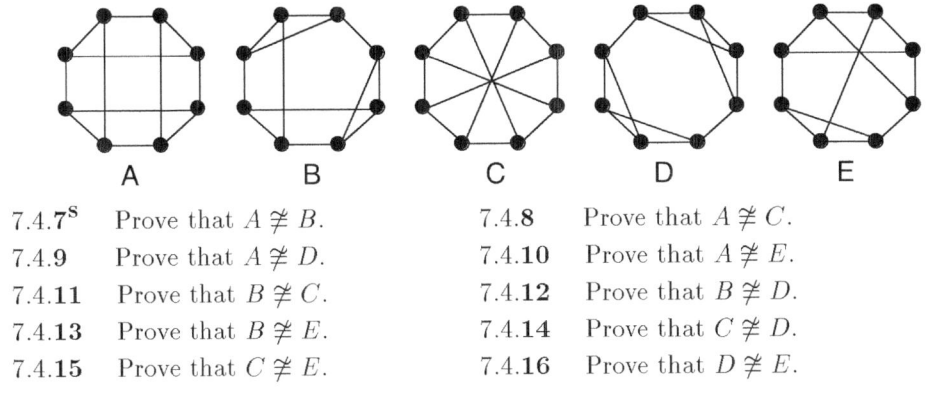

| A | B | C | D | E |

7.4.7[S]	Prove that $A \not\cong B$.	7.4.8	Prove that $A \not\cong C$.
7.4.9	Prove that $A \not\cong D$.	7.4.10	Prove that $A \not\cong E$.
7.4.11	Prove that $B \not\cong C$.	7.4.12	Prove that $B \not\cong D$.
7.4.13	Prove that $B \not\cong E$.	7.4.14	Prove that $C \not\cong D$.
7.4.15	Prove that $C \not\cong E$.	7.4.16	Prove that $D \not\cong E$.

In Exercises 7.4.17 through 7.4.24, construct one example (no duplicates) of each of the isomorphism types of graph meeting the given requirement.

7.4.17[S] Trees with 5 vertices.

7.4.18 Trees with 6 vertices.

7.4.19 Simple graphs with 4 vertices and 4 edges.

7.4.20 Simple graphs with 5 vertices and 4 edges.

7.4.21 Simple graphs with 5 vertices and 5 edges.

7.4.22 Simple graphs with 5 vertices and 6 edges.

7.4.23 General graphs with 3 vertices and 3 edges.

7.4.24 General graphs with 4 vertices and 3 edges.

7.4.25 Prove that the composition of two isomorphisms is an isomorphism.

7.5 GRAPH AUTOMORPHISM

What we perceive visually to be a symmetry in a drawing of a graph is representable combinatorially as a self-isomorphism, i.e., an isomorphism of the graph onto itself. In this sense of self-isomorphisms as symmetries, a graph may have symmetries beyond those that are visible in a particular drawing.

DEFINITION: An isomorphism from a graph to itself is called an **automorphism**.

An automorphism on a simple graph G can be specified by a permutation π on the vertex set. If the graph is complete, then every permutation of the vertex set specifies an automorphism on the graph. Otherwise, however, some of the permutations of the vertex set will map an adjacent pair to a non-adjacent pair.

As explained in §1.6, every permutation can be represented as a composition of disjoint cycles of the objects in the permuted set, and we employ the disjoint-cycles representation in what follows.

Sometimes, all the automorphisms of a graph correspond to geometric symmetries of a well-chosen drawing of the graph. In such a case, it may be reasonably straightforward to list all the automorphisms.

Example 7.5.1:   Figure 7.5.1 depicts the 3-cycle graph C_3, which is a complete graph. All six permutations of the vertex set $\{0, 1, 2\}$ are adjacency preserving, since all vertices are adjacent, and non-adjacency preserving as well, since there are no non-adjacent pairs.

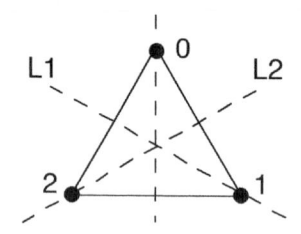

Figure 7.5.1   **The 3-cycle graph.**

3 rotations	3 reflections	
$0° = (0)(1)(2)$	thru vertical axis	$(0)(1\ 2)$
$120° = (0\ 1\ 2)$	thru axis L1	$(1)(0\ 2)$
$240° = (0\ 2\ 1)$	thru axis L2	$(2)(0\ 1)$

Proposition 7.5.1. *The inverse of a graph automorphism is a graph automorphism.*

Proof: This follows from Proposition 7.4.1.   ◇

Example 7.5.1, continued: We observe that rotation by $240°$ and rotation by $120°$ are inverses of each other. The identity is its own inverse. Moreover, each reflection is its own inverse.

TERMINOLOGY: The automorphisms on a graph form a group (see Appendix A2), under the operation of composition, called the **automorphism group of the graph**, which is denoted $\mathcal{A}ut(G)$. The identity mapping is the group identity. There are a number of routine details to verify.

Example 7.5.2:   The graph G of Figure 7.5.2 below has four vertices, so there are 24 vertex permutations. However, an automorphism preserves degree. Since vertices 0 and 2 are the only two 3-valent vertices, an automorphism must either fix both or swap them. Similarly, since 1 and 3 are the only 2-valent vertices, an automorphism must either fix both or swap them, as well. Only four vertex permutations satisfy both these restrictions. It is readily confirmed that all four such vertex permutations are adjacency preserving and non-adjacency preserving, so all four are automorphisms. We observe that each is its own inverse.

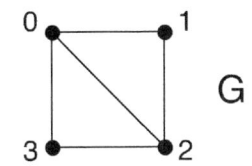

Figure 7.5.2   A graph with four automorphisms.

All four automorphisms are representable as rotations and reflections.

2 rotations	2 reflections	
$0° = (0)(1)(2)(3)$	thru SE diagonal	$(0)(2)(1\ 3)$
$180° = (0\ 2)(1\ 3)$	thru NW diagonal	$(1)(3)(0\ 2)$

Size of the Automorphism Group

It is thought to be quite difficult to produce a list of all the automorphisms of a given graph, even though it has not been proved that there is no reasonably quick way. Moreover, it is thought difficult even to determine the size of the automorphism group, because the graph isomorphism problem reduces to the latter problem. Here is the reduction, applied to two given connected graphs, G and H, assumed for simplicity to be disjoint.

Step 1: Calculate $|\mathcal{A}ut(G)|$ and $|\mathcal{A}ut(H)|$. If they are not equal, then decide that

$$G \not\cong H$$

and stop.

Step 2: Calculate $|\mathcal{A}ut(G \cup H)|$. Decide as follows:

$$G \not\cong H \quad \text{if } |\mathcal{A}ut(G \cup H)| = |\mathcal{A}ut(G)|^2$$
$$G \cong H \quad \text{if } |\mathcal{A}ut(G \cup H)| = 2\,|\mathcal{A}ut(G)|^2$$

Explanation: If $G \not\cong H$, then every automorphism on $G \cup H$ is representable as a union $\alpha \cup \beta$, where α is an automorphism on G and β an automorphism on H. We define

$$(\alpha \cup \beta)(v) = \begin{cases} \alpha(v) & \text{if } v \in V_G \\ \beta(v) & \text{if } v \in V_H \end{cases}$$

Since there are $|\mathcal{A}ut(G)|$ choices for α and equally many for β, there are $|\mathcal{A}ut(G)|^2$ automorphisms on $G \cup H$. However, if there exists an isomorphism $f : G \to H$, then there are additional $|\mathcal{A}ut(G)|^2$ automorphisms of the form

$$((\alpha \cup \beta) \cdot f)(v) = \begin{cases} \alpha(f^{-1}(v)) & \text{if } v \in V_H \\ \beta(f(v)) & \text{if } v \in V_G \end{cases}$$

thereby yielding a total of $2\,|\mathcal{A}ut(G)|^2$ automorphisms in all.

Orbits

DEFINITION: The **orbit of a vertex** v in a graph G is the set

$$\{ \alpha(v) \mid \alpha \in \mathcal{A}ut(G) \}$$

of vertices to which v is mapped by some automorphism of G.

Being co-orbital is clearly an equivalence relation that partitions the vertex set of a graph.

Example 7.5.2, continued: The orbits of the graph G in Figure 7.5.2 are

$$\{0, 2\} \quad \text{and} \quad \{1, 3\}$$

DEFINITION: A graph is **vertex-transitive** if every vertex is in the same orbit.

Example 7.5.2, continued: The graph G in Figure 7.5.2 is not vertex-transitive, because it has two orbits, not one.

Example 7.5.3:   The complete graph K_n is vertex-transitive.

Example 7.5.4:   The hypercube Q_n is vertex-transitive.

Example 7.5.5:   A complete bipartite graph $K_{m,n}$ has two orbits if $m \neq n$, one with m vertices, the other with n. If $m = n$, then there is only one orbit, because the partite sets could be swapped.

Example 7.5.6:   Every circulant graph $circ(n : S)$ is vertex-transitive. The automorphism that maps vertex i to vertex j is the operation of adding $j - i$ to every vertex.

Rigidity

Rigidity is the opposite of vertex-transitivity.

DEFINITION: A graph is **rigid** if every vertex has an orbit to itself (or, equivalently, the only automorphism is the identity mapping).

Example 7.5.7:   The graph G in Figure 7.5.3 is rigid.

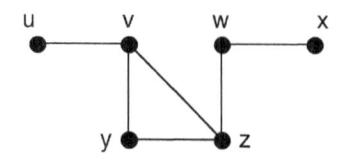

Figure 7.5.3   A rigid graph.

The vertices v and z are the only two of degree 3. However, whereas v has the 1-valent neighbor u, the vertex z has no 1-valent neighbors. Thus, both v and z

must have orbits to themselves. Similarly, vertices w and y are both 2-valent, but w has a 1-valent neighbor, the vertex x, and y does not. Finally, the vertices u and x are both 1-valent, but their lone neighbors, the vertices v and w, respectively, are 3-valent and 2-valent.

COMPUTATIONAL NOTE: Since we don't have an easy way even to calculate the number of automorphisms, much less to list the automorphisms, it may be unsurprising that calculating orbits is another of the problems commonly thought to be difficult.

EXERCISES for Section 7.5

In Exercises 7.5.1 through 7.5.12, write a list of the automorphisms of the indicated graph from Figure 7.5.4, giving each as a permutation of the vertex set.

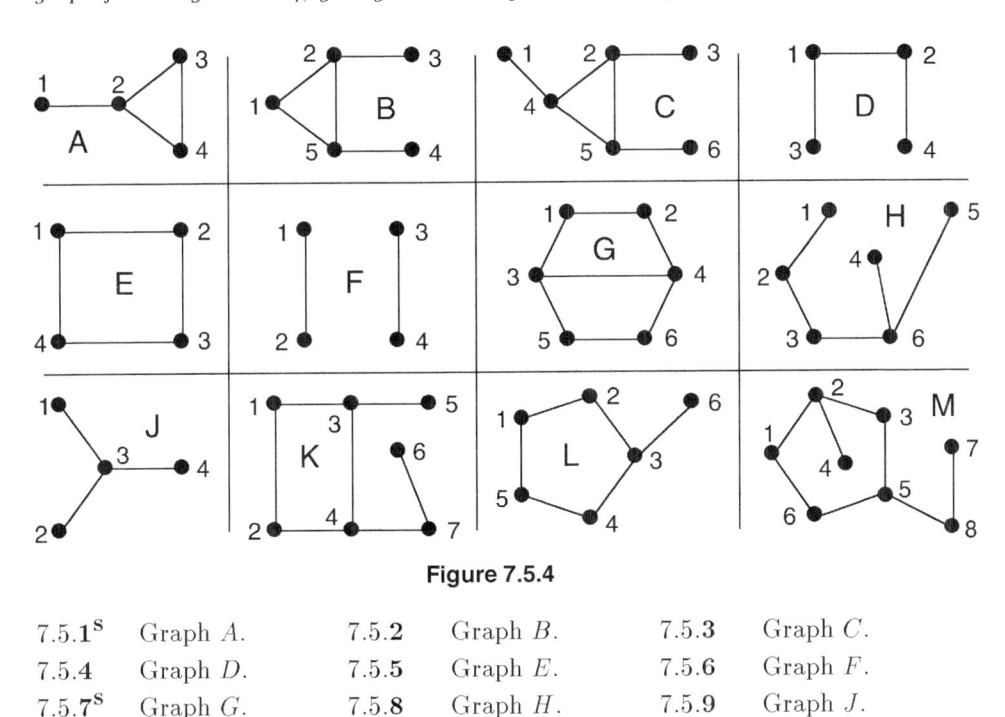

Figure 7.5.4

7.5.1[S]	Graph A.	7.5.2	Graph B.	7.5.3	Graph C.
7.5.4	Graph D.	7.5.5	Graph E.	7.5.6	Graph F.
7.5.7[S]	Graph G.	7.5.8	Graph H.	7.5.9	Graph J.
7.5.10	Graph K.	7.5.11	Graph L.	7.5.12	Graph M.

In Exercises 7.5.13 through 7.5.24, write a list of the vertex orbits of the indicated graph from Figure 7.5.4.

7.5.13[S]	Graph A.	7.5.14	Graph B.	7.5.15	Graph C.
7.5.16	Graph D.	7.5.17	Graph E.	7.5.18	Graph F.
7.5.19[S]	Graph G.	7.5.20	Graph H.	7.5.21	Graph J.
7.5.22	Graph K.	7.5.23	Graph L.	7.5.24	Graph M.

7.5.25   Which of the graphs in Figure 7.5.4 are vertex-transitive?

7.5.26   Which of the graphs in Figure 7.5.4 are rigid?

7.5.27   Draw a connected 3-regular graph that is not vertex-transitive, and prove that it is not vertex-transitive.

7.5.28   Draw a rigid tree.

7.6   SUBGRAPHS

This section is concerned with the *subgraphs* of a graph, that is, with the graphs that are contained within that graph. Questions of special interest include the existence of particular kinds of subgraphs. For instance, a path subgraph or a cycle subgraph might be used in a traversal. A tree subgraph might be used in searching a graph. Examination of the subgraphs is often useful in isomorphism testing.

DEFINITION: A **subgraph** of a graph G is a graph H whose vertex set and edge set are subsets of the vertex set and edge set, respectively, of G.

TERMINOLOGY NOTE: More generally, any graph that is isomorphic to a subgraph of G is called a "subgraph" of G, and one infers from context whether the cited subgraph is set-theoretically a part of G.

Example 7.6.1:   Figure 7.6.1 illustrates that C_6 and C_9 are subgraphs of the Petersen graph.

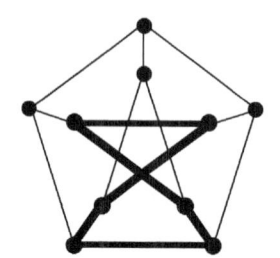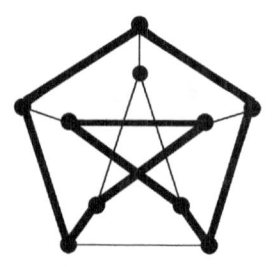

Figure 7.6.1   Two cycle subgraphs of the Petersen graph.

Spanning Paths and Cycles

DEFINITION: A subgraph of a graph G is a **spanning subgraph** of G if it contains every vertex of G.

Example 7.6.2:   Figure 7.6.2 illustrates that the path graph P_{10} spans the Petersen graph and that the cycle graph C_6 spans the octahedron graph. Clearly, any spanning cycle contains spanning paths.

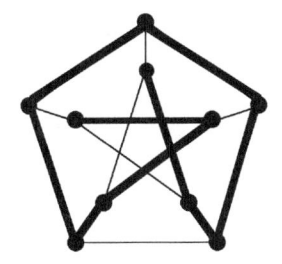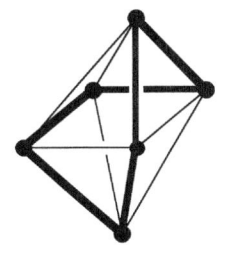

Figure 7.6.2   Path P_{10} and cycle C_6 as spanning graphs.

DEFINITION: A path subgraph that spans the graph in which it lies is called a **hamiltonian path**.

DEFINITION: A cycle subgraph that spans the graph in which it lies is called a **hamiltonian circuit**.

Example 7.6.2, continued: Thus, P_{10} is a hamiltonian path in the Petersen graph, and C_6 is a hamiltonian circuit in the octahedral graph.

Section 8.1 discusses hamiltonian circuits at greater length.

Components

One possible use of tree-growing is to find every vertex that is reachable from a given vertex. The relation of reachability is an equivalence relation. A *component* of a graph contains a maximum collection of mutually reachable vertices and all the edges that join them to each other.

DEFINITION: A **component** of a graph is a maximal connected subgraph. That is, it is a connected subgraph such that there is no path from any vertex within it to any vertex not within it.

In a drawing, it is possible to draw each component so that it is separate from the others.

Example 7.6.3:   In Figure 7.6.3, vertices u and v are in separate components, because they are not mutually reachable by paths from each other.

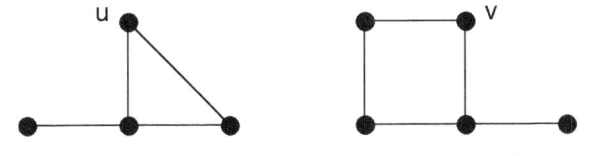

Figure 7.6.3   A non-connected graph.

Example 7.6.4:   Every component of a forest is a tree, as in Figure 7.6.4 below.

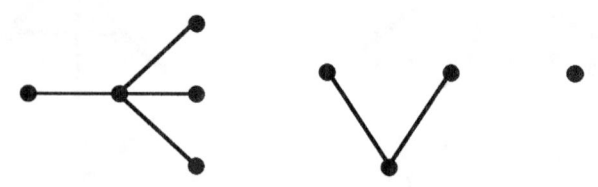

Figure 7.6.4   A forest.

Isomorphism Preserves Subgraphs

We have already used the presence and absence of cycles of a particular size to distinguish two graphs. This method of isomorphism testing can be generalized to the presence and absence of any particular subgraph.

Proposition 7.6.1.   Let J be a subgraph of a graph G and let $f : G \to H$ be an isomorphism. Then the subgraph $f(J)$ is isomorphic to J.

Proof: Since f maps V_G bijectively to V_H, it maps V_J bijectively to $V_{f(J)}$. Since f is adjacency preserving and non-adjacency preserving on all of the graph G, it is adjacency preserving and non-adjacency preserving on its subgraph J. $\diamond$

Example 7.6.5:   The two graphs in Figure 7.6.5 have the same degree sequence. However, graph G contains a subgraph that is isomorphic to K_4 and graph H does not. Thus, they are not isomorphic.

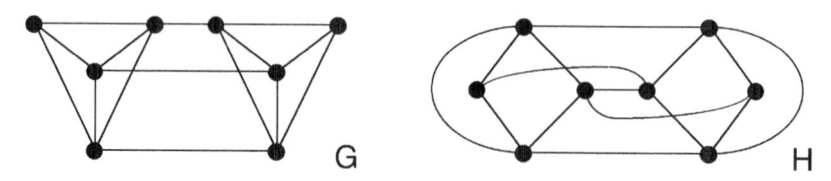

Figure 7.6.5   Two non-isomorphic graphs.

Induced Subgraphs

DEFINITION: Let G be a graph and U a subset of the vertex-set of G. The **induced subgraph** $G(U)$ is the subgraph whose vertex-set is U, and whose edge-set contains every edge of G whose endpoints are in U.

Example 7.6.6:   A subset U of the vertex set of the graph G in Figure 7.6.6 is indicated by solid black vertices. The edges of the induced subgraph $G(U)$ are indicated by thickening.

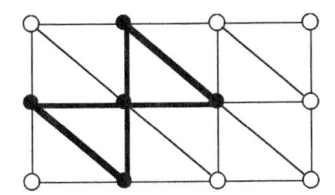

Figure 7.6.6 An induced subgraph in a graph.

EXERCISES for Section 7.6

In Exercises 7.6.1 through 7.6.4, draw the indicated graph, and draw a hamiltonian circuit within it.

7.6.1 Hypercube graph Q_3.

7.6.2[S] Wheel graph W_7.

7.6.3 Bipartite graph $K_{4,4}$.

7.6.4 Circulant graph $circ(8 : 2, 3)$.

In Exercises 7.6.5 through 7.6.8, show that there is no hamiltonian circuit in the indicated graph from Figure 7.6.7.

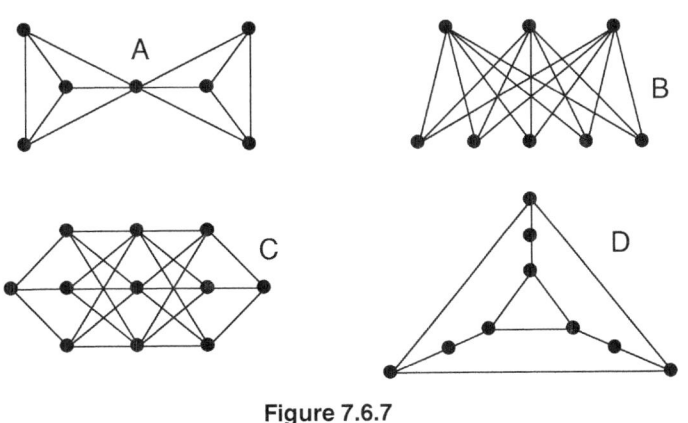

Figure 7.6.7

7.6.5 Graph A.

7.6.6[S] Graph B.

7.6.7 Graph C.

7.6.8 Graph D.

DEFINITION: The **center** of a graph is the subgraph induced on all the central vertices.

7.6.9 Find the center of graph A of Figure 7.6.7.

7.6.10 Find the center of graph C of Figure 7.6.7.

In each of the Exercises 7.6.11 through 7.6.14, find the center of the given graph.

7.6.11^S 7.6.12

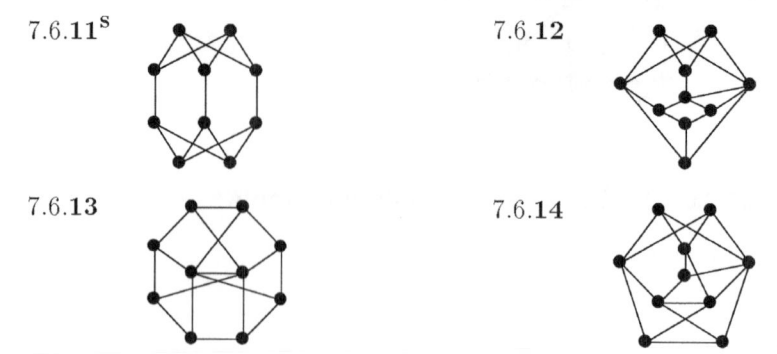

7.6.13   7.6.14

7.7 SPANNING TREES

Trees are among the most prevalent kinds of subgraphs that arise in graph-theoretic applications.

TERMINOLOGY: Let T be a tree subgraph of a graph G. A vertex of G that lies in the tree T is called a **tree vertex**. Every other vertex is called a **non-tree vertex**. Every edge of G that lies in the tree T is called a **tree edge**. Every other edge is called a **non-tree edge**. A **frontier edge** is an edge of the graph G with one endpoint in the tree T and the other not in T.

Example 7.7.1:   Figure 7.7.1 shows a tree subgraph in a graph. The tree edges are bold and the frontier edges ru, qu, tu, tv, and tw are dashed. Other non-tree edges are plain lines.

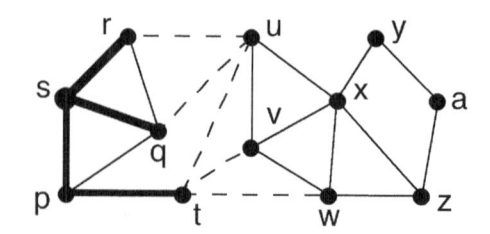

Figure 7.7.1   Frontier edges for a tree subgraph.

Remark: If a tree subgraph for a graph G spans G, then there are no frontier edges.

DEFINITION: A tree subgraph that spans a graph is called a **spanning tree**.

DEFINITION: The **cycle rank** of a connected graph G, denoted $\beta(G)$, is the number of edges that are non-tree edges for a spanning tree of G. Equivalently,

$$\beta(G) \;=\; |E_G| - |V_G| + 1$$

In view of Proposition 7.2.2, every spanning tree for a graph G has the same number of edges.

Example 7.7.2:   The two spanning trees for K_4 shown in Figure 7.7.2 both have three non-tree edges. We also observe that

$$|E_{K_4}| - |V_{K_4}| + 1 \;=\; 6 - 4 + 1 \;=\; 3$$

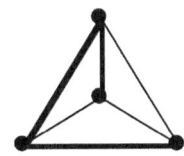 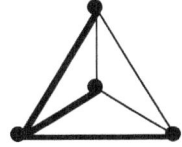

Figure 7.7.2   The cycle rank of K_4 is **3**.

Proposition 7.7.1. *Every connected graph G contains a spanning tree.*

Proof: If $\beta(G) = 0$, then G is already a tree, and it spans itself. Otherwise, removing an edge from any circuit reduces the cycle rank, without disconnecting the graph. An induction argument leads to the conclusion.   ◇

Tree-Growing

In computer science applications, spanning trees are commonly *grown* inside a connected graph. To initialize the growth of a spanning tree in a connected graph, a vertex is chosen as the root. Then at each iteration of the growth process, an edge is selected from among the set of frontier edges and added to the tree, along with its non-tree endpoint.

Example 7.7.1, continued: Figure 7.7.3 shows the result of adding the frontier edge tu to the growing tree. Then the edges ru and qu are no longer in the frontier, but edges uv and ux have been added to the frontier.

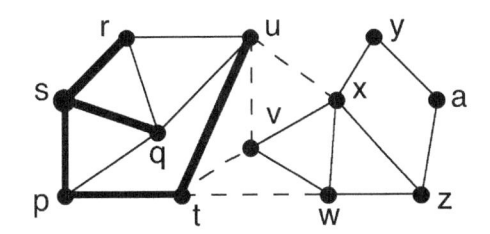

Figure 7.7.3   Frontier edges for a growing tree.

DEFINITION: The **discovery order** of the vertices of a graph under a tree-growing scheme is the order in which they are added to the tree. Each vertex acquires its **discovery number** in sequence, as it is added to the growing tree, starting with the number 0 for the root.

Depending on the application, there are different rules governing the selection of the next frontier edge to be added to the growing tree. We now consider two of the most common schemes for tree-growing, especially when visiting the nodes of a network.

Breadth-First and Depth-First Trees

The *breadth-first* and *depth-first* spanning trees represent two different objectives in reaching all vertices of a graph. First imagine a message being relayed along from the root vertex to the entire graph. A breadth-first spanning tree would tend to keep the paths from the root short, thereby having no vertex waiting too long to get the message. A depth-first tree would tend to reduce the number of vertices to which a non-leaf in the tree would have to relay the message.

DEFINITION: In a **breadth-first search** of a graph G, the frontier edges are prioritized primarily according to *smallest* discovery number of their tree endpoints. That is, an edge is selected whose tree endpoint has as small a discovery number as possible.

Among several frontier edges whose common tree endpoint (they all share the same tree endpoint) has lowest discovery number at some stage of a breadth-first search, the **secondary priority rule** is either random choice or based on some form of ordering of the edges incident on that vertex.

Example 7.7.3: Figure 7.7.4 below shows the outcome of a breadth-first search on the graph of Example 7.7.1 above, including the discovery numbers, starting from the vertex s. The secondary priority for competing frontier edges was alphabetic order of their non-tree endpoints.

DEFINITION: In a **depth-first search** of a graph G, the frontier edges are prioritized primarily according to *largest* discovery number of their tree endpoints.

Example 7.7.4: Figure 7.7.5 shows the outcome of a depth-first search on the graph of Example 7.7.1, including the discovery numbers, starting from the vertex s. The secondary priority rule used on competing frontier edges was once again alphabetic order of their non-tree endpoints.

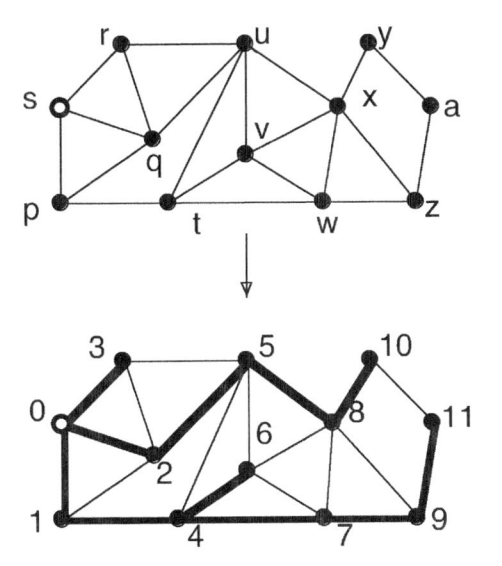

Figure 7.7.4   Outcome of a breadth-first search.

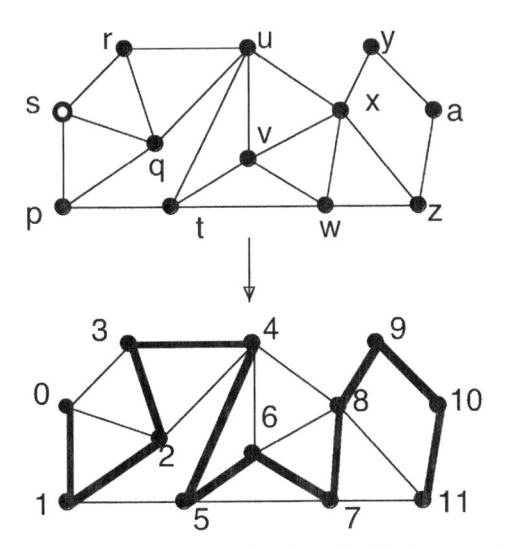

Figure 7.7.5   Outcome of a depth-first search.

We observe that in the breadth-first spanning tree, the maximum number of steps from the root to a leaf is 5 steps, and in the depth-first spanning tree, it is 11 steps. However, the root of the breadth-first tree must give the message to three neighbors, yet in the depth-first tree, each vertex passes the message along only to a single neighbor.

EXERCISES for Section 7.7

In Exercises 7.7.1 through 7.7.4, list the frontier edges for the indicated tree subgraph from Figure 7.7.6 below.

7.7.1[S]   Tree T_1.   **7.7.2**   Tree T_2.

7.7.3   Tree T_3.   **7.7.4**   Tree T_4.

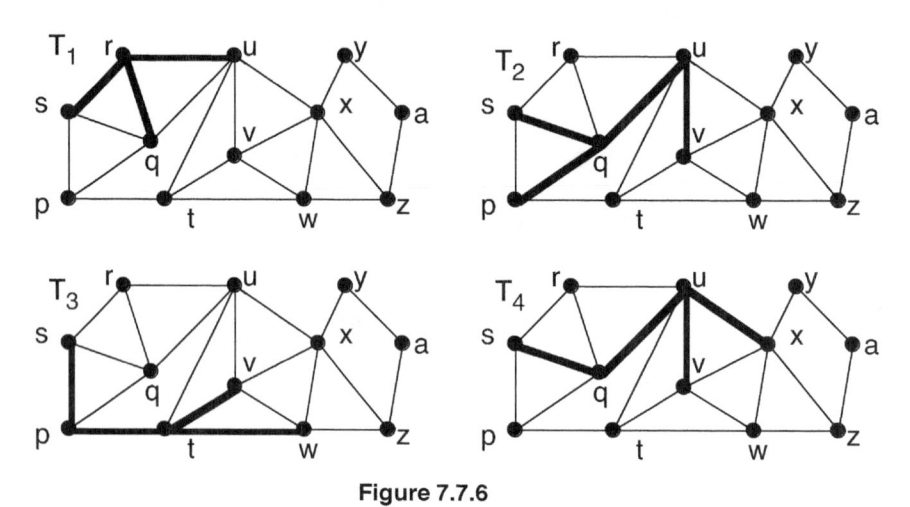

Figure 7.7.6

In each of the Exercises 7.7.5 through 7.7.8, mark the discovery numbers and draw the breadth-first spanning tree for the graph in Figure 7.7.7 and for indicated root vertex. The secondary priority for competing frontier edges is alphabetic order of the non-tree endpoints, as in Example 7.7.3.

7.7.5[S]   Root r.   **7.7.6**   Root t.

7.7.7   Root y.   **7.7.8**   Root z.

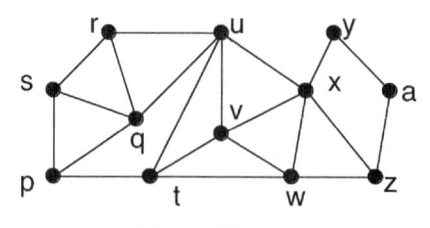

Figure 7.7.7

In each of the Exercises 7.7.9 through 7.7.12, mark the discovery numbers and draw the depth-first spanning tree for the graph in Figure 7.7.7 and for indicated root vertex. The secondary priority for competing frontier edges is alphabetic order of the non-tree endpoints, as in Example 7.7.4.

7.7.9[S]   Root r.   **7.7.10**   Root t.

7.7.11   Root y.   **7.7.12**   Root z.

7.8   EDGE WEIGHTS

Adding additional features to a graph often enhances the usefulness of graph models in applications. What is often in mind when *edge weights* are assigned is some form of cost associated with the edges, such as the cost of traversal. Accordingly, edge weights are assumed to be non-negative unless it is explicitly declared otherwise.

DEFINITION: An **edge-weight function** on a graph assigns a number to every edge.

Minimum-Weight Spanning Tree Problem

DEFINITION: The **minimum-weight spanning-tree problem** (abbr. **MST**) is to find a spanning tree in a graph with edge-weights such that the sum of the weights on its edges is the minimum possible for that graph.

Application 7.8.1   Suppose that the weight on an edge between two locations represents the cost of hard-wiring a cable between those two locations. Solving MST for such a configuration yields the minimum cost of having all locations in communication by cable (with relays).

Example 7.8.1:   Figure 7.8.1 below shows a minimum-weight spanning tree. If the five edges in that tree had the five minimum edge-weights in this 6-vertex graph, it would be immediately clear that that tree is a minimum-weight spanning tree. Such a combination of five edges could be obtained by replacing edge yz of the spanning tree with non-tree edge vx.

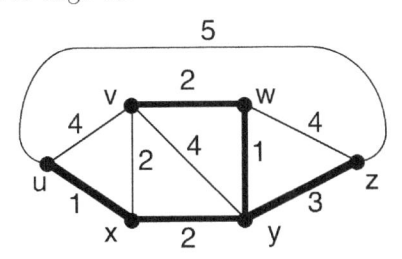

Figure 7.8.1   A minimum-weight spanning tree.

However, that resulting set of five edges is not a spanning tree, and the tree shown has the next smallest possible combination of five edge-weights. Thus, it has minimum-weight.

A graph may have many spanning trees. For instance, according to a result of Cayley, the complete graph K_n has n^{n-2} spanning trees. Thus, it is infeasible to consider all of them and their weights as a data set from which to select the minimum. Fortunately, there is a far more efficient way to find a minimum-weight spanning tree, attributed to R. C. Prim [Prim1957].

Prim's method begins by choosing any vertex as a root, and by then using smallest edge-weight iteratively as the primary prioritizing criterion for selection of a frontier edge. Random selection is a suitable tie-breaking rule.

Algorithm 7.8.1:   Prim's MST Algorithm

Input: a weighted connected graph G and starting vertex v.
Output: a minimum-weight spanning tree T.

  Initialize tree T as vertex v.
  Initialize S as the set of proper edges incident on v.
  While $S \neq \emptyset$
    Let e be a minimum-weight edge in S.
    Let w be the non-tree endpoint of edge e.
    Add edge e and vertex w to tree T.
    Update the frontier set S.
  Return tree T.

TERMINOLOGY: The tree in Prim's algorithm that grows to be the minimum spanning tree is called the *Prim tree*.

Proposition 7.8.1. *Let T_k be the Prim tree after k iterations of Prim's algorithm, for $0 \leq k \leq |V_G| - 1$. Then T_k is a subtree of a minimum spanning tree of G.*

Proof: The assertion is trivially true for $k = 0$, since T_0 is simply the starting vertex. This provides an inductive basis.

Assume for some number $k \leq |V_G| - 2$, that T_k is a subtree of a minimum spanning tree T of G. According to Prim's algorithm, the tree T_{k+1} is obtained by adding to tree T_k a frontier edge e of smallest weight. Let u and v be the endpoints of edge e, such that u is in tree T_k and v is not.

If spanning tree T contains edge e, then T_{k+1} is a subtree of T. If e is not an edge in tree T, then e is part of the unique cycle contained in $T + e$. Consider the path in T from u to v representing the "long way around the cycle". Let d be the first edge along this path that joins a vertex in T_k to a vertex not in T_k. In Figure 7.8.2, the black vertices and bold edges make up Prim tree T_k, the spanning tree T consists of everything except edge e, and the Prim tree T_{k+1} is $(T_k \cup v) + e$.

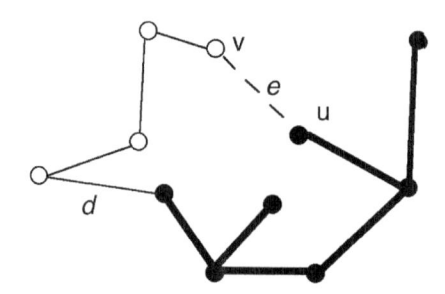

Figure 7.8.2   Alternative extensions of a Prim tree.

Since d was a frontier edge at the beginning of the $(k+1)^{st}$ iteration, it follows that $w(e) \leq w(d)$ (since the algorithm chose e). The tree

$$\hat{T} = T + e - d$$

is clearly a spanning tree of G, and T_{k+1} is a subtree of $\hat{T}$. Since

$$w(\hat{T}) \; = \; w(T) + w(e) - w(d) \; \leq \; w(T)$$

it follows that $\hat{T}$ is a minimum spanning tree of G. $\diamond$

Corollary 7.8.2. *When Prim's algorithm is applied to a connected graph, the result is a minimum spanning tree.* $\diamond$

Example 7.8.2: Figure 7.8.3 shows the minimum-weight spanning tree for an edge-weighted graph. The small number at each vertex is its discovery number from the tree-growing process of Prim's algorithm.

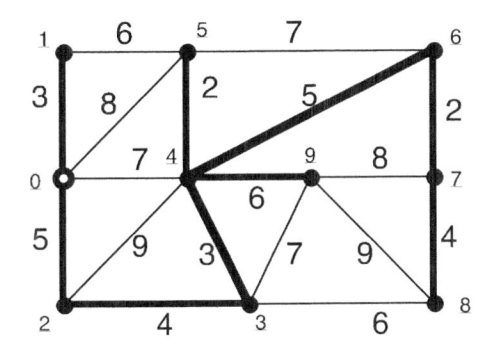

Figure 7.8.3   MST by Prim's algorithm, with discovery numbers.

Shortest Path Problem

DEFINITION: The **shortest path problem** is to find the path between two vertices s and t in a connected graph with edge-weights whose total edge-weight is minimum, i.e., a shortest s-t path.

Application 7.8.2 Beyond the obvious applications involving distance, suppose again that the weight on an edge represents the cost of hard-wiring a cable between its endpoints. Then solving the shortest-path problem yields the minimum cost of establishing cable communication between two specific locations.

Example 7.8.3: The graph in Figure 7.8.4 has edge weights and two designated vertices s and t. The s-t path in bold edges has total weight 14. There are many paths from s to t, but it can be proved that this path has the least total weight, and is, thus, the shortest.

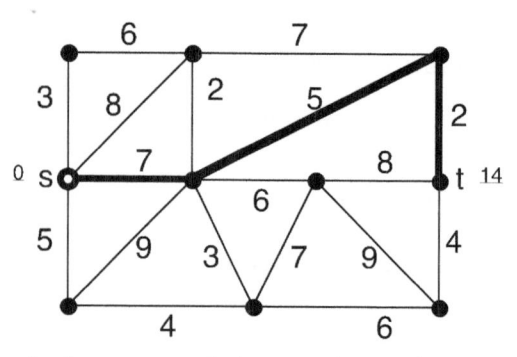

Figure 7.8.4   A shortest path in an edge-weighted graph.

An algorithm attributed to E. Dijkstra [Dijk1959] provides yet another instance of generic tree-growing. Instead of finding only the shortest path from a given vertex s to a single vertex, Dijkstra's algorithm finds a spanning tree, rooted at s, that contains the minimum path from s to every other vertex.

TERMINOLOGY: The dynamic tree that grows under Dijkstra's algorithm to a spanning tree with all the shortest paths is called the *Dijkstra tree*.

The *Dijkstra score* of a frontier edge is the sum of its own weight plus the sum of the weights along the unique path in the growing Dijkstra tree. An edge with least Dijkstra score is chosen. Random selection is an appropriate tie-breaking rule.

Algorithm 7.8.2:   Dijkstra's Shortest-Paths Algorithm

Input: a weighted connected graph G and starting vertex s.
Output: a shortest-path tree T with root s.

   Initialize tree T as vertex s.
   Initialize S as the set of proper edges incident on s.
   While $S \neq \emptyset$
      Select frontier edge e with lowest Dijkstra score.
      Let w be the non-tree endpoint of edge e.
      Add edge e and vertex w to tree T.
      Update the frontier set S.
   Return tree T.

Remark: Proof of the correctness of Dijkstra's algorithm is similar to the proof for Prim's algorithm. It is omitted.

Example 7.8.4:   Figure 7.8.5 shows the same edge-weighted graph as in Example 7.8.2 and the Dijkstra spanning tree with all the shortest paths. The small number at each vertex is its distance from the root. It is a different spanning tree from the Prim spanning tree, because a different prioritizing criterion is applied to the frontier edges.

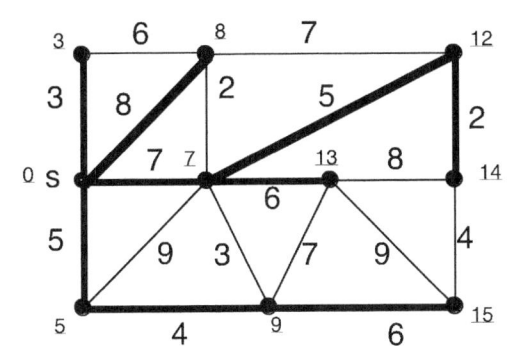

Figure 7.8.5 Dijsktra tree of shortest paths to all vertices.

EXERCISES for Section 7.8

Exercises 7.8.1 through 7.8.9 pertain to the edge-weighted graph in Figure 7.8.6. Use ascending lexicographic order of the non-tree vertex as a secondary priority in tree-growth, and lexicographic order of the tree vertex as a third priority (in case two different frontier edges lead to the same non-tree vertex).

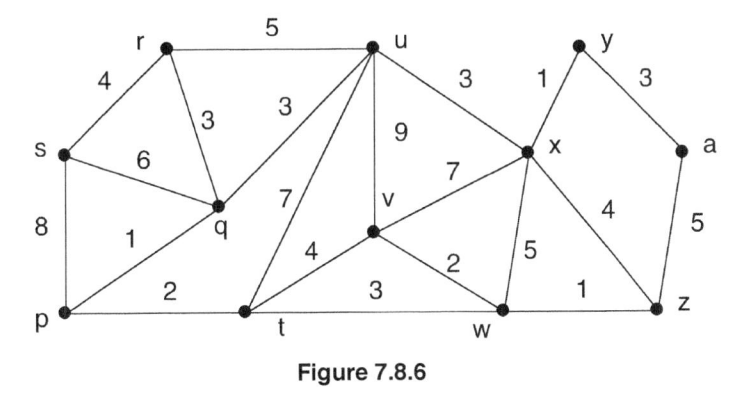

Figure 7.8.6

7.8.1[S] Draw the Prim tree and mark the discovery numbers of the vertices on the graph of Figure 7.8.6, using vertex s as the root.

7.8.2 Draw the Prim tree and mark the discovery numbers of the vertices on the graph of Figure 7.8.6, using vertex v as the root.

7.8.3 Draw the Prim tree and mark the discovery numbers of the vertices on the graph of Figure 7.8.6, using vertex z as the root.

7.8.4[S] Draw the Dijkstra tree and mark the discovery numbers of the vertices on the graph of Figure 7.8.6, using vertex s as the root.

7.8.5 Draw the Dijkstra tree and mark the discovery numbers of the vertices on the graph of Figure 7.8.6, using vertex v as the root.

7.8.6 Draw the Dijkstra tree and mark the discovery numbers of the vertices on the graph of Figure 7.8.6, using vertex z as the root.

7.8.7   Use Dijkstra's algorithm to find the minimum-weight cycle in the graph of Figure 7.8.6 that includes the edge tu.

7.8.8   Use Dijkstra's algorithm to find the minimum-weight cycle in the graph of Figure 7.8.6 that includes the edge vx.

7.8.9[S]   Use Dijkstra's algorithm to find the minimum-weight cycle in the graph of Figure 7.8.6 that includes the edge ux.

7.9   GRAPH OPERATIONS

Computationally, it is convenient to regard a graph as a variable, to which vertices and edges may be added, and from which vertices and edges may also be deleted.

Primary Operations

The maintenance operations of adding or deleting a vertex or an edge to or from a graph are called *primary operations* because they are the fundamental operations on a graph variable from which other operations are constructed.

Example 7.9.1:   Figure 7.9.1 illustrates the four primary operations.

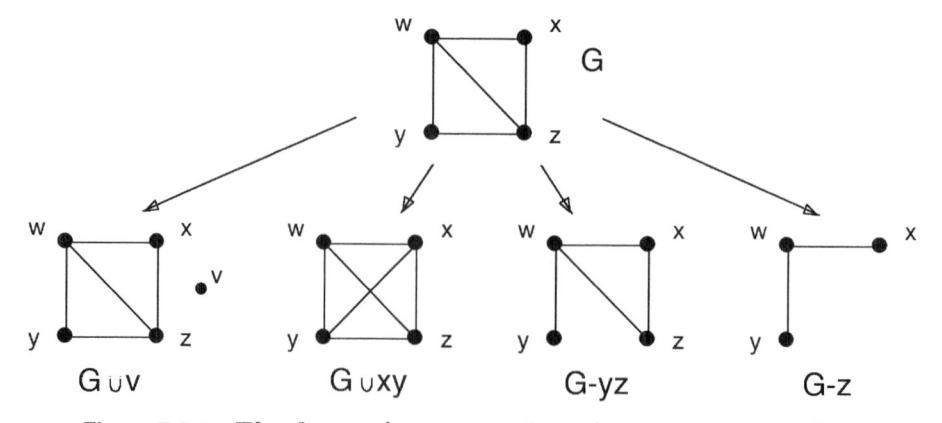

Figure 7.9.1   **The four primary graph maintenance operations.**

DEFINITION: Let G be a graph and v a new vertex, not in V_G. The **vertex-addition operation** yields the graph $G \cup v$ with vertex-set $V_G \cup \{v\}$ and edge-set E_G.

DEFINITION: Let G be a graph and e a new edge, not in E_G, whose endpoints are both in V_G. The **edge-addition operation** yields the graph $G \cup e$ with vertex-set V_G and edge-set $E_G \cup \{e\}$.

DEFINITION: Let e be an edge of a graph G. The **edge-deletion subgraph** $G - e$ is the graph with the vertex-set V_G and the edge-set $E_G - \{e\}$.

DEFINITION: Let v be a vertex of a graph G. The **vertex-deletion subgraph** $G - v$ is the subgraph of G whose vertex-set is $V_G - \{v\}$ and whose edge-set is $E_G - \{e \mid v \text{ is an endpoint of } e\}$.

Example 7.9.2:   A spanning tree of a connected graph can be grown by iteratively adding the non-tree vertex of a frontier edge and then adding the frontier edge itself. This is the idea behind the construction of the breadth-first tree, the depth-first tree, Prim's tree, and Dijkstra's tree.

Graph Reconstruction

The deletion operations lead immediately to the fascinating theoretical topic of *graph reconstruction*.

DEFINITION: Let G be a graph with vertices $v_1, v_2, \ldots, v_n$. The **vertex-deletion subgraph list** of G is the list of subgraphs

$$G - v_1, \quad G - v_2, \quad \ldots, \quad G - v_n$$

DEFINITION: The **reconstruction deck** of a graph G is its vertex-deletion subgraph list, with no labels on the vertices. (Each vertex-deletion subgraph is regarded as a card in the deck. It is possible that various cards in the deck are identical, since the deck is a list, not a set.)

The basic idea of graph reconstruction is to infer an unknown simple graph from its reconstruction deck.

Example 7.9.3:   Consider trying to infer a specification for a graph H with the reconstruction deck shown in Figure 7.9.2.

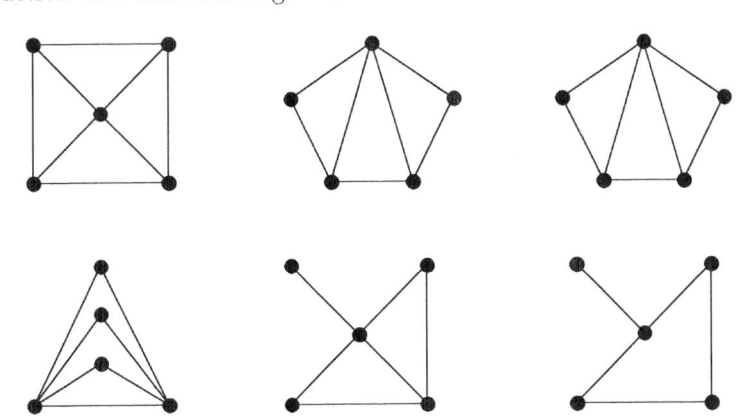

Figure 7.9.2   Reconstruction deck for a graph H.

Reconstructing the number of vertices is easy.

Proposition 7.9.1. *The number of vertices of a graph equals the number of cards in its reconstruction deck.* ◇

Proposition 7.9.2. *The number of edges of a simple graph G can be inferred from its reconstruction deck via the following formula.*

$$|E_G| \;=\; \frac{1}{n-2} \sum_v |E_G - v| \tag{7.9.1}$$

Proof: Each edge of the graph appears on $n - 2$ cards, i.e., on every card except the cards for its endpoints. Formula 7.9.1 is implemented by adding the numbers of edges on the cards and then dividing by $n - 2$. ◇

Example 7.9.3, continued: By Propositions 7.9.1 and 7.9.2, the graph H must have 6 vertices and 10 edges.

Corollary 7.9.3. *The degree sequence of edges of a graph G can be inferred from its reconstruction deck.*

Proof: For each card in the deck, the degree of the vertex that was deleted is the difference of the total number of edges in the graph and the number of edges on that card. ◇

Example 7.9.3, continued: By Proposition 7.9.3, the degree sequence of the graph H is

$$5 \quad 4 \quad 3 \quad 3 \quad 3 \quad 2$$

The two graphs with that degree sequence are shown in Figure 7.9.3.

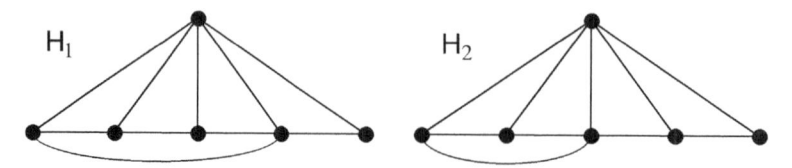

Figure 7.9.3   Two candidates for the reconstructed graph H.

The vertex-deletion subgraph list for the graph H_1 coincides with the given reconstruction deck, but that for the graph H_2 does not. Thus, graph H_1 is the correctly reconstructed graph.

The Graph-Reconstruction Problem

In seeking to reconstruct a graph as in Example 7.9.3, one obstacle that arises is that the number of graphs with a given degree sequence may be exponentially large in the number of edges. Another is that it is unknown whether a reconstruction deck uniquely characterizes a graph.

DEFINITION: The **graph-reconstruction problem** is to determine whether two non-isomorphic graphs with three or more vertices can have the same reconstruction deck.

There is a special case in which the graph is not only known to be unique, but also readily reconstructable.

Proposition 7.9.4. *Any regular graph can be reconstructed from its reconstruction deck.*

Proof: By Corollary 7.9.3, it can be determined from its reconstruction deck whether the graph is regular, and the degree d of regularity. Accordingly, one can reconstruct the graph by joining the vertices of degree $d - 1$ on any single card to a new vertex, which restores the missing vertex for that card.   ◇

Connectivity

The deletion operations also lead to the highly practical topic of *connectivity*.

DEFINITION: If deleting vertex v from graph G increases the number of components, then v is a **cut-vertex** of G.

DEFINITION: If deleting edge e from graph G increases the number of components, then v is a **cut-edge** of G.

Example 7.9.4:   In the graph of Figure 7.9.4, the vertices u, v, and w are cut-vertices. The edges uv and vw are cut-edges.

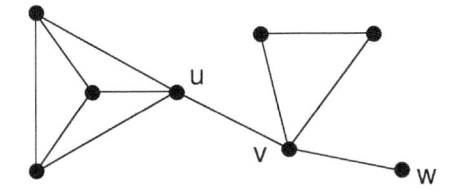

Figure 7.9.4   Graph with cut-vertices and cut-edges.

DEFINITION: Let G be a connected graph. The **connectivity** $\kappa(G)$ is the smallest number of vertices of G whose iterative deletion results either in a non-connected graph or in a 1-vertex graph.

DEFINITION: Let G be a connected graph. The **edge-connectivity** $\kappa'(G)$ is the smallest number of edges of G whose iterative deletion results in a non-connected graph, or 0 edges for a 1-vertex graph.

Example 7.9.5:   The graph G of Figure 7.9.5 has no cut-vertices, and it is disconnected by the removal of vertices w and x. Thus,

$$\kappa(G) \;=\; 2$$

Removal of edges a, b, and c disconnects graph G, and no smaller set of edges disconnects. Thus,

$$\kappa'(G) \;=\; 3$$

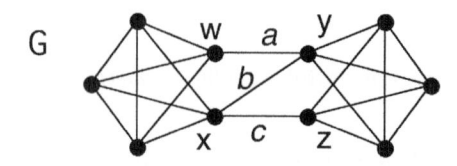

Figure 7.9.5   Graph G with $\kappa(G) = 2$ and $\kappa'(G) = 3$.

Graph Union

Beyond addition of a single vertex to an existing graph, there are several interesting ways to combine two existing graphs into a new graph. The simplest is *graph union*.

DEFINITION: The *(graph) union* $G \cup G'$ of two graphs

$$G = (V, E) \quad \text{and} \quad G' = (V', E')$$

is the graph whose vertex set and edge-set are the disjoint unions, respectively, of the vertex-sets and edge-sets of G and G', respectively.

NOTATION: The notation nG is used for the *n-fold self-union* of a graph G, that is, for the result of an iterated union of n disjoint copies of G.

Join

DEFINITION: The *join* $G + H$ of two graphs G and H is obtained from their disjoint union by joining each vertex of G to each vertex of H with a new edge.

DEFINITION: The *n-wheel* W_n is the join $C_n + K_1$. The n-wheel is said to be an *even wheel* or an *odd wheel*, depending on whether n is even or odd.

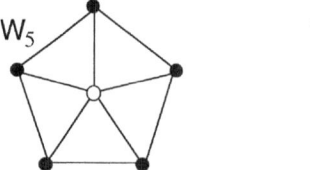 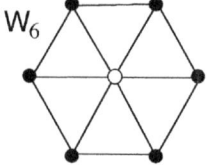

Figure 7.9.6   A 5-wheel and a 6-wheel.

NOTATION: In the special case where one of the two graphs in a join is a trivial graph, one may use the form $G + v$.

Cartesian Product

DEFINITION: The *(cartesian) product* $G \times H$ of the graphs G and H has as its vertex-set the cartesian product

$$V_{G \times H} \;=\; V_G \times V_H$$

and as its edge-set a union of two products

$$E_{G \times H} \;=\; (V_G \times E_H) \cup (E_G \times V_H)$$

If edge $e \in E_H$ has endpoints y and z and if $u \in V_G$ then edge $(u, e) \in V_G \times E_H$ has endpoints (u, y) and (u, z). If edge $d \in E_G$ has endpoints v and w and if $x \in V_H$ then edge $(d, x) \in E_G \times V_H$ has endpoints (v, x) and (w, x).

Example 7.9.6:   Figure 7.9.7 represents the cartesian product $C_3 \times K_{2,3}$ as a 3×5 grid in which the induced graph on each row is a copy of H and the induced graph on each column is a copy of G.

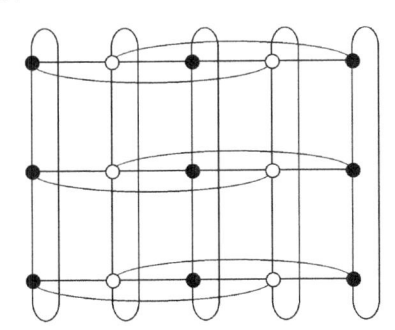

Figure 7.9.7   The cartesian product $C_3 \times K_{2,3}$.

Edge-Complementation

DEFINITION: A simple graph G has as its **edge-complement** (or **complement**) the graph $\overline{G}$ on the same vertex-set, such that two vertices of $\overline{G}$ are adjacent if and only if they are non-adjacent in G.

NOTATION: The notation $K_n - G$ is also commonly used for the edge-complement of an n-vertex graph G.

Example 7.9.7:   The edge-complement of the wheel W_5 is isomorphic to the graph union $K_1 \cup C_5$, as illustrated in Figure 7.9.8.

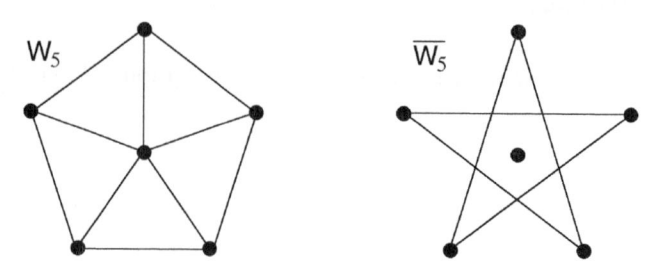

Figure 7.9.8   The wheel W_5 and its edge-complement.

Theorem 7.9.5. *Two simple graphs G and H are isomorphic if and only if their edge-complements are isomorphic.*

Proof: If $f : V_G \to V_H$ is a bijection that preserves adjacency and non-adjacency, then $f : V_{\overline{G}} \to V_{\overline{H}}$ preserves non-adjacency and adjacency, respectively, and vice versa.   ◇

Theorem 7.9.5 is useful in testing the isomorphism of two graphs whose average degree is at least half the number of vertices.

Example 7.9.8:   Figure 7.9.9 illustrates two 4-regular, 7-vertex graphs, called G and H. They both have girth 3 and eccentricity 2 at every vertex. However, the edge-complements differ, in that $\overline{G} \cong C_7$ and $\overline{H} \cong C_3 \cup C_4$.

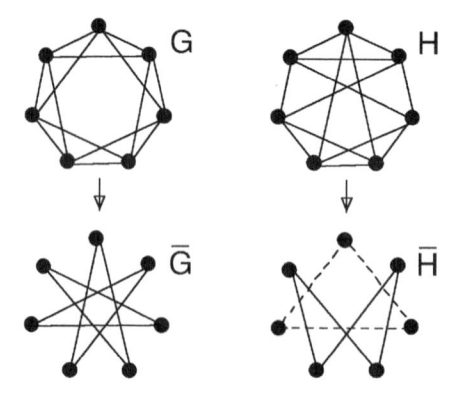

Figure 7.9.9   Two graphs and their edge-complements.

Exercises

In Exercises 7.9.1 through 7.9.6, reconstruct the graph with the given deck.

7.9.1^S

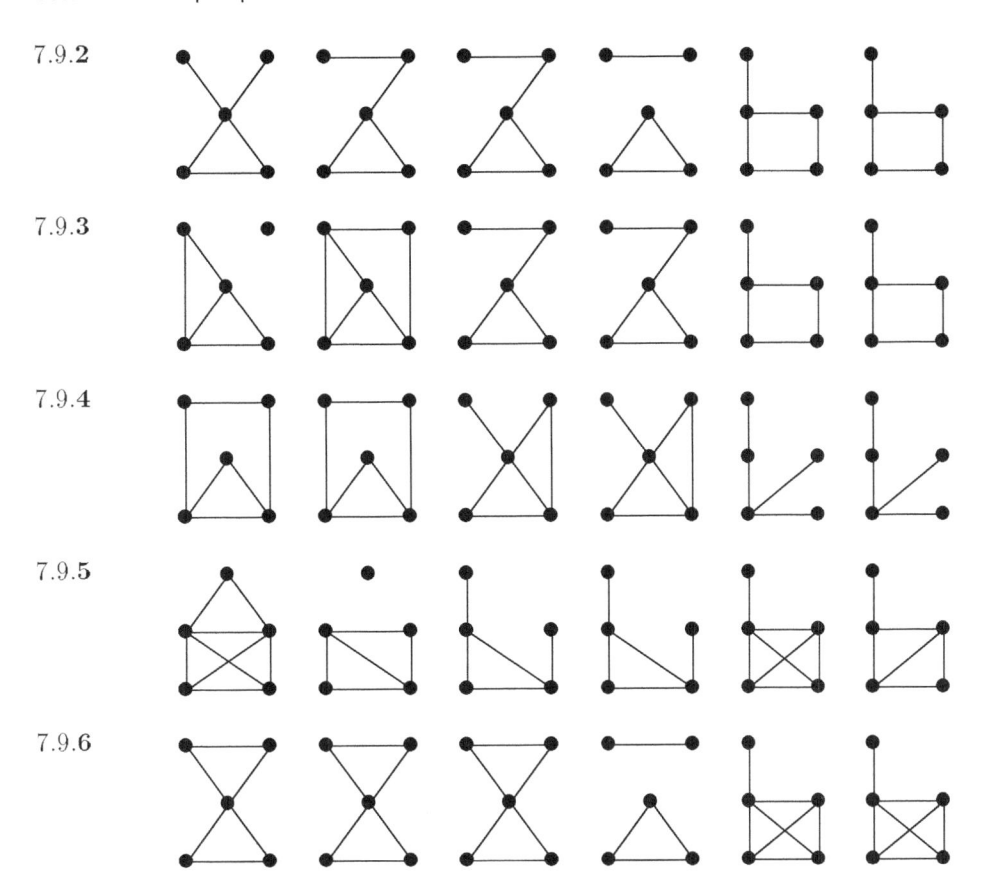

In Exercises 7.9.7 through 7.9.18, write a list of the cutpoints of the indicated graph from Figure 7.9.10, giving each as a permutation of the vertex set.

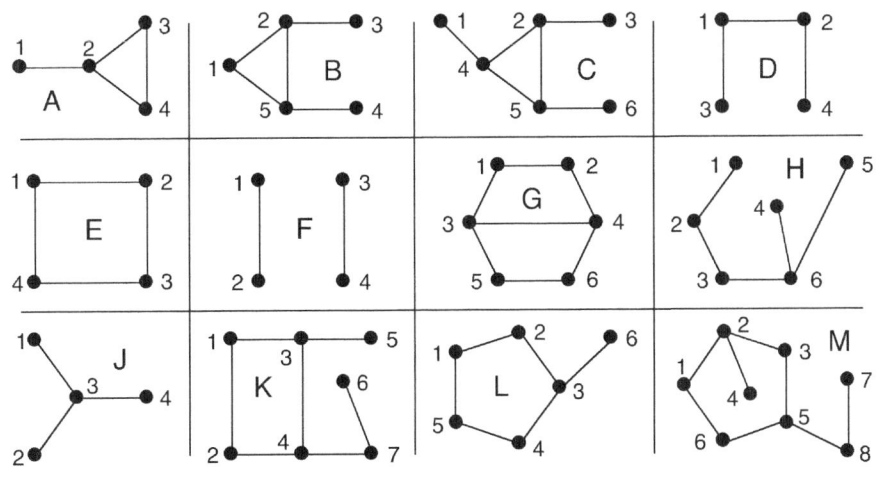

Figure 7.9.10

7.9.**7**	Graph A.	7.9.**8**	Graph B.	7.9.**9**	Graph C.
7.9.**10**	Graph D.	7.9.**11**	Graph E.	7.9.**12**	Graph F.
7.9.**13**	Graph G.	7.9.**14**	Graph H.	7.9.**15**	Graph J.
7.9.**16**$^\mathbf{S}$	Graph K.	7.9.**17**	Graph L.	7.9.**18**	Graph M.

In Exercises 7.9.19 through 7.9.30, write a list of the cutedges of the indicated graph from Figure 7.9.10.

7.9.**19**	Graph A.	7.9.**20**	Graph B.	7.9.**21**	Graph C.
7.9.**22**	Graph D.	7.9.**23**	Graph E.	7.9.**24**	Graph F.
7.9.**25**	Graph G.	7.9.**26**	Graph H.	7.9.**27**	Graph J.
7.9.**28**$^\mathbf{S}$	Graph K.	7.9.**29**	Graph L.	7.9.**30**	Graph M.

In Exercises 7.9.31 through 7.9.36, draw the edge-complement of the indicated graph from Figure 7.9.10.

7.9.**31**	Graph A.	7.9.**32**$^\mathbf{S}$	Graph B.	7.9.**33**	Graph D.
7.9.**34**	Graph E.	7.9.**35**	Graph F.	7.9.**36**	Graph G.

GLOSSARY

acyclic digraph: a digraph with no directed circuits.

acyclic graph: a graph with no circuits.

adjacent edges in a graph: edges that have an endpoint in common.

arc: a directed edge.

automorphism of a graph: an isomorphism from the graph to itself.

automorphism group $\mathcal{A}ut(G)$ of a graph G: the set of all automorphisms of the graph G, with the operation of composition.

binary search tree (abbr. BST):   a binary tree with a vertex labeling, such that the label at each vertex is greater than every label in its left subtree and less than every label in its right subtree.

binary tree: a rooted tree such that each vertex has a possible left-child, a possible right-child, and no other children.

breadth-first search: a method of constructing a spanning tree in a graph.

cartesian product $G \times H$:   the graph whose vertex-set is the cartesian product

$$V_{G \times H} \;=\; V_G \times V_H$$

and whose edge-set is the union

$$E_{G \times H} \;=\; (V_G \times E_H) \cup (E_G \times V_H)$$

If edge $e \in E_H$ has endpoints y and z and if $u \in V_G$ then edge $(u, e) \in V_G \times E_H$ has endpoints (u, y) and (u, z). If edge $d \in E_G$ has endpoints v and w and if $x \in V_H$ then edge $(d, x) \in E_G \times V_H$ has endpoints (v, x) and (w, x).

Cayley graph $Cay(A : S)$ for a group A and a subset S of its domain: a graph whose vertex set is the domain of the group A, such that there is an edge between vertices u and v if there is an element $s \in S$ such that $u + s = v$ (for an additive group, or $u \star s = v$ if $\star$ is the group operation).

center of a graph: the induced graph on all the *central vertices*.

central vertex in a graph: a vertex of minimum eccentricity.

child of a vertex v of a rooted tree: an adjacent vertex w such that v lies on the unique path from the root to w.

circuit: synonymous with closed path.

circulant graph $circ(n : S)$, where $S \subset \mathbb{Z}_n$: the graph whose vertex set is

$$\{ 0, \quad 1, \quad \ldots, \quad n - 1 \}$$

and in which two vertices i and j are adjacent if and only if there is a number $s \in S$ such that $i + s = j \bmod n$ or $j + s = i \bmod n$. The elements of the set S are called *connections*.

circular ladder graph CL_n: a graph isomorphic to the cartesian product $C_n \times K_2$.

complement of a simple graph G: the short form of *edge-complement*.

complete graph: a simple graph in which every pair of vertices is joined by an edge.

component of a graph: a maximal connected subgraph.

connected digraph: a digraph whose underlying graph is connected.

connected graph: a graph is which each pair of vertices is joined by a path.

connections: see *circulant graph*.

connectivity of a graph: the minimum number of vertices whose removal would either disconnect it or reduce it to a 1-vertex graph.

cut-edge: an edge whose removal would increase the number of components.

cut-vertex: a vertex whose removal would increase the number of components.

cycle: a closed path.

cycle graph C_n: a graph whose vertices and edges all lie on a single circuit of length n.

cycle rank of a connected graph G: the number of edges remaining after all the edges of a spanning tree are removed.

dag: acronym for *directed acyclic digraph*.

degree of a vertex: the total number of edge-ends incident on it; thus, the sum of the number of proper edges plus twice the number of self-loops of which it is an endpoint.

depth-first search: a method of constructing a spanning tree in a graph.

diameter of a graph: the maximum distance between any two vertices in the graph.

digraph: a directed graph.

___, **simple**:   a digraph with no multi-arcs and no self-loops.

Dijkstra tree: a spanning tree of all the shortest paths from its root.

directed graph: a graph in which every edge is directed.

directed path: a path in which every edge is directed.

directed walk: a walk in which every edge is directed.

discovery number of a vertex: its position in the discovery order.

discovery order of a vertex: the order in which it is added to a growing spanning tree.

distance between two vertices: the number of edges in a shortest walk between the vertices.

eccentricity of a vertex: the maximum distance to any other vertex.

edge-complement: the graph $\overline{G}$ on same vertex set as V, in which two vertices are adjacent if and only if they are not adjacent in G.

edge-connectivity of a graph: the minimum number of vertices whose removal would either disconnect it; zero for K_1.

edge-weight function: a function that assigns a number to every edge.

eulerian tour in a connected graph: a closed walk that contains every edge.

eulerian trail in a connected graph: an open walk that contains every edge.

forest: a graph whose every component is a tree.

frontier edge for a tree subgraph of a graph: an edge joining a tree vertex to a non-tree vertex.

girth of a non-acyclic graph: the size of the smallest circuit.

graph invariant: a measurement or property of a graph that is the same for every graph isomorphic to it.

graph isomorphism problem: the problem of designing a practical algorithm to test graph isomorphism or to prove that no such algorithm exists.

graph-reconstruction problem: the problem of deciding whether two non-isomorphic graphs could have the same *reconstruction deck*.

hamiltonian circuit in a graph: a spanning circuit.

hamiltonian cycle: same as *hamiltonian circuit*.

hamiltonian path: a spanning path.

height of a rooted tree: the maximum distance from the root to any leaf.

hypercube graph: the skeleton of any n-dimensional cube.

induced subgraph on a set U of vertices in a graph G: the graph $G(U)$ whose vertex set is U, in which two vertices are adjacent, if and only if they are adjacent in G.

isomorphism of graphs G and H: a bijection of their vertex sets that preserves all adjacencies and non-adjacencies.

isomorphism type of a graph: the class of all graphs that are isomorphic to it.

join $G + H$:   the graph obtained from the union $G \cup H$ by joining each vertex of G to each vertex of H.

labeling of the vertices of a graph in a set S: an assignment of label (or *key*) from S to each vertex of G.

leaf of a tree: any 1-valent vertex of an unrooted tree, or a childless vertex of a rooted tree.

left-child: see *binary tree*.

length of a walk: the number of edge-steps.

line graph of a graph G: the graph $L(G)$ whose vertices are the edges of G, using edge-adjacency in G as the adjacency rule in $L(G)$.

mixed graph: a graph in which some edges are directed and some are not.

Möbius ladder ML_n: a 3-regular graph something like the circular ladder CL_n except that it looks like it belongs on a Möbius band, instead of on a cylindrical band.

orbit of a vertex v of a graph G: the set of all vertices $u \in G$ such that there is an automorphism α with $\alpha(v) + u$.

parent of a vertex v in a rooted tree: an adjacent vertex on the unique path to v from the root.

path graph P_n: an n-vertex graph in which all the vertices lie on a single path.

path: a walk with no repeated vertices.

___, **closed**:   a path whose final vertex is the same as its initial vertex.

___, **open**:   a path whose final vertex is not the same as its initial vertex.

platonic graph: the skeleton of a platonic solid.

platonic solids: the five regular 3-dimensional polyhedra: tetrahedron, octahedron, cube, dodecahedron, isosahedron.

polyhedron: higher dimensional analog of a polygon.

Prim tree: a minimum-weight spanning tree grown by Prim's algorithm.

primary operations on a graph: adding and deleting a vertex or edge.

radius of a graph: the minimum eccentricity over all vertices.

reconstruction deck of a graph G: the list of all graphs $G - v$ such that $v \in V_G$.

regular graph: a graph whose vertices all have the same degree.

regular polygon: a (convex) polyhedron whose sides are of identical length, and whose interior angles are of identical measure.

regular polyhedron: a polyhedron whose sides are identical regular polygons.

right-child: see *binary tree*.

rigid graph: a graph whose only automorphism is the identity automorphism.

rooted tree: a tree with one vertex designated as the root.

self-union nG **of a graph**: the graph union of n disjoint copies of G.

shortest-path problems: problems concerned with finding the shortest paths between pairs of points.

siblings in a rooted tree: two vertices with the same *parent*.

simplex: generalization of a triangle to all dimensions.

skeleton of a polyhedron: the graph comprising its vertices and edges.

spanning subgraph of a graph: a subgraph that contains every vertex of the graph.

spanning tree: a subgraph that is a tree

strongly connected digraph: a digraph in which any two vertices are reachable from each other by directed walks.

subgraph of a graph G: a graph whose vertex set and edge set are subsets of V_G and E_G, or any graph isomorphic to such a subgraph.

tournament: a digraph whose underlying graph is a complete graph.

transitive digraph: a digraph in which whenever there is a directed walk from a vertex u to a vertex v, there is an arc from u to v.

tree: a connected acyclic graph.

trivial graph: a graph with one vertex and no edges.

trivial walk: a walk with no edges.

underlying graph of a digraph: the graph obtained by eliminating all the edge directions.

union of graphs: the graph whose vertex set and edge set are the disjoint union of their vertex sets and of their edge sets, respectively.

vertex-deletion subgraph of a graph G: in the graph reconstruction problem, a graph $G - v$.

vertex-transitive graph: a graph whose vertices are all in one orbit.

walk in a graph: an alternating sequence $W = \langle v_0, e_1, v_1, e_2, \ldots, e_n, v_n \rangle$ of vertices and edges, such that edge e_j joins vertices v_{j-1} and v_j, for $j = 1, \ldots, n$.

    ___, **closed**:   a walk whose final vertex is the same as its initial vertex.

    ___, **open**:   a walk whose final vertex is not the same as its initial vertex.

wheel graph W_n: the result of joining a new vertex with an n-cycle.

Chapter **8**

Graph Theory Topics

8.1 Traversability

8.2 Planarity

8.3 Coloring

8.4 Analytic Graph Theory

8.5 Digraph Models

8.6 Network Flows

8.7 Topological Graph Theory

The attraction of graph theory to most practitioners includes the accessibility of many of its most important branches and the interesting histories of their origins. Entire books have been written on each of these branches, so, of course, a single section of one chapter has to be quite selective. In this attempt at a broad view of graph theory, the emphasis is presenting the key results and methods that everyone who works in any aspect of graph theory or graph applications should know. A few of the classical results are stated without proof. The order of the sections is correlated with the order in which these large branches acquired their foundations and came to prominence.

8.1   TRAVERSABILITY

A *graph traversal* is a walk that uses either all the edges or all the vertices. In some historically celebrated graph traversal problems explored here, the edges are not weighted, and it is required that every edge or every vertex is to be visited only once. Algorithmists have subsequently focused on optimal traversals of various kinds in edge-weighted graphs.

The Bridges of Königsberg

Two branches of the River Pregel merge within the town of Königsberg, which is located in a part of present-day Russia previously known as East Prussia. There is an island just below the junction of the two branches, near the mouth of the river into the Baltic Sea. These four land areas were once connected by seven bridges, as illustrated in Figure 8.1.1. A problem arose of traversing all seven bridges of Königsberg without crossing any bridge more than once.

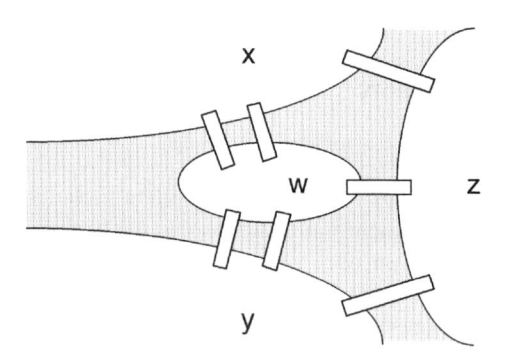

Figure 8.1.1   The seven bridges of Königsberg.

Proof in 1736 by the celebrated Swiss mathematician Leonhard Euler that no such traversal is possible is acclaimed as the origin of graph theory. The configuration of land areas and bridges is represented as a graph in Figure 8.1.2.

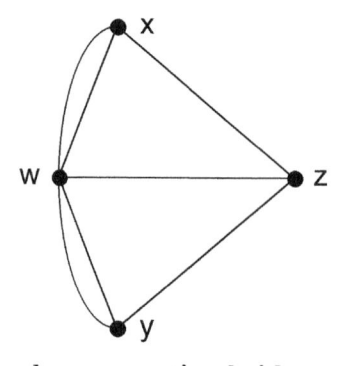

Figure 8.1.2   Graph representing bridges of Königsberg.

Eulerian Tours

Some terminology, some of it first mentioned some in §7.2, facilitates easy discussion of the kind of traversal required by the seven bridges problem.

DEFINITION: A **trail** is a walk with no repeated edges.

DEFINITION: An **eulerian graph** is a graph that has a closed trail that traverses every edge exactly once.

DEFINITION: An **eulerian trail** is a trail containing every edge of the graph. An **eulerian tour** is a closed eulerian trail.

Theorem 8.1.1. *A connected graph G has an eulerian tour if and only if the degree of every vertex is even.*

Proof: ($\Rightarrow$) First assume that G has an eulerian tour. When the tour starts, the number of available edge-ends at the terminus drops by one. Every time the tour enters and leaves a vertex along the way, the number of available edge-ends drops by two, preserving the parity. As the tour ends, the number of available edge-ends at the terminus drops by one. Thus, at each vertex, the number of available edge-ends when the tour began has the same parity as its final value, i.e., as zero. Thus, its initial value, i.e., its degree, must be even.

($\Leftarrow$) Next assume that the degree is everywhere even. This implies that the cycle rank $\beta(G)$ is at least 1, since a non-trivial tree would have two or more vertices of degree 1. If $\beta(G) = 1$, then G is a cycle and clearly has an eulerian tour. For $\beta(G) > 1$, we proceed inductively. Let C be any cycle subgraph of G. We observe that the degree at every vertex of $G - C$ is even, so it is even at every vertex of every component of $G - C$. By the induction hypothesis, each component has an eulerian tour. To construct an eulerian tour of G, start traversing anywhere on cycle C. As soon as a vertex of any component of $G - C$ is encountered, detour from cycle C with a eulerian tour of that component, and then continue the tour of C until the next component of $G - C$ is encountered. After detouring through the last such component, complete the traversal of cycle C, which also completes an eulerian tour of G. $\Diamond$

Remark: The proof of necessity of even degrees in Theorem 8.1.1 is due to Euler in 1736. However, the proof of sufficiency is by K. Hierholzer in 1873. See [BLW1986] for further historical details.

Corollary 8.1.2. *A connected graph G has an open eulerian trail if and only if there are exactly two vertices of odd degree.*

Proof: Let u and v be the two vertices of odd degree. Form a new graph G' by joining a new vertex s to vertices u and v. Since every vertex of G' has even degree, Theorem 8.1.1 implies that G' is eulerian. One eulerian tour starts and finishes at vertex s. The trail obtained by eliminating the edges su and sv is an open eulerian trail in G. $\Diamond$

Example 8.1.1: The graph in Figure 8.1.3 has an eulerian trail but no eulerian tour.

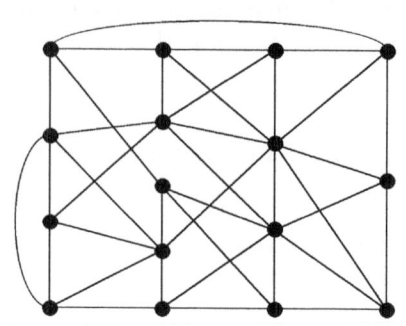

Figure 8.1.3   A graph with an eulerian trail, but no eulerian tour.

Hamiltonian Circuits

Sir William Rowan Hamilton (1805–1865), an Irish mathematician, observed in 1856 that the 1-skeleton of a dodecahedron (see §7.1), commonly called the *dodecahedral graph*, has a circuit that traverses every vertex, and he used this as the basis for a puzzle whose idea he sold for commercial distribution. The graph is reproduced in Figure 8.1.4.

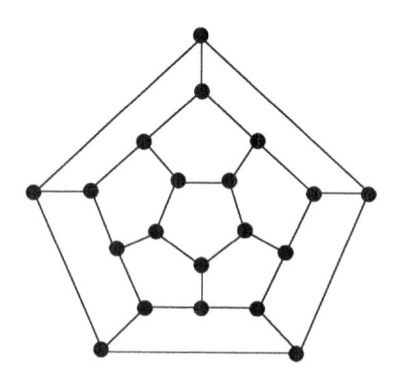

Figure 8.1.4   The dodecahedral graph.

DEFINITION: A **hamiltonian circuit** is a closed path that traverses every vertex.

DEFINITION: A **hamiltonian graph** is a graph with a hamiltonian circuit.

When is a Graph Hamiltonian?

There are certain circumstances under which it is clear that a graph cannot be hamiltonian. Some of the easiest to apply are identified by the following two propositions.

Proposition 8.1.3. *A graph with a cut-vertex is non-hamiltonian.*

Proof: Let v be a cut-vertex of a graph G, and let x and y be vertices that are in different components of $G - v$. A hamiltonian circuit in G would contain the vertices x and y, and there would be two paths joining them. Yet v lies on every path joining x and y, and it would thus occur twice on the same cycle, a contradiction ◇

Proposition 8.1.4. *A bipartite graph with partite sets of different sizes is non-hamiltonian.*

Proof: Let U and V be the two partite of a bipartite graph. On a hamiltonian circuit, the vertices of U and V would have to alternate, implying that there are equally many vertices in each. ◇

There is no known polynomial-time algorithm to decide whether a graph is hamiltonian. Nonetheless, for a reasonably small graph G, a useful approach may be to start identifying a set S of edges that would have to be a subset of a spanning cycle if any such cycle exists. Here is an outline of the steps.

- •1 If any vertex v has degree 2, then both incident edges should be added to the set S. If it is clear how to extend S to a spanning cycle, then the problem is solved.

- •2 If some vertex is an endpoint of two edges in S, then all other edges incident on that vertex are marked unusable, and excluded from the degree determination of step •1.

- •3 If the set S cannot be extended to a spanning cycle, for instance, if a subset of edges of S forms a cycle that does not span G, then the graph is not hamiltonian.

Example 8.1.2: In the graph $K_{2,3}$ of Figure 8.1.5, the edges ru and su are selected because vertex u is 2-valent, and then edges rv and sv because v is 2-valent. Since these four edges form a cycle that does not span, we conclude that the graph is not hamiltonian.

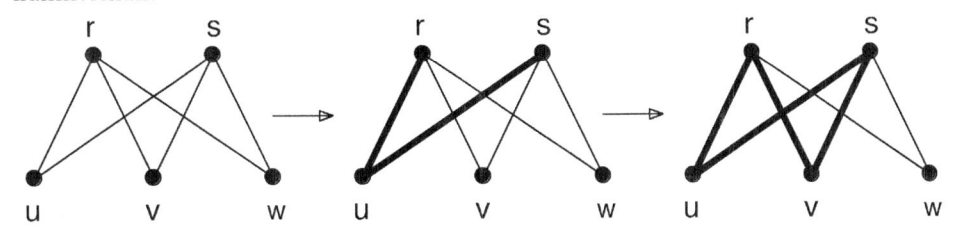

Figure 8.1.5 **Showing that a graph is not hamiltonian.**

Sufficiency Conditions

Intuitively, the more edges that a simple graph has, the more likely that it is hamiltonian. The classical results of Ore and Dirac support this intuition.

Theorem 8.1.5 [Ore1960]. *Let G be a simple n-vertex graph, where $n \geq 3$, such that*

$$deg(x) + deg(y) \geq n$$

for each pair of non-adjacent vertices x and y. Then the graph G is hamiltonian.

Proof: Suppose, to the contrary, that some graph satisfies the conditions of the theorem, yet is non-hamiltonian. Consider a maximal counterexample G — so that joining any two non-adjacent vertices of G would result in a hamiltonian graph. For $n \geq 3$, a complete graph K_n is hamiltonian, so there must exist two non-adjacent vertices s and t. It would be a contradiction to show that

$$deg(s) + deg(t) \leq n - 1$$

Since the graph $G + st$ contains a hamiltonian circuit, the graph G contains a hamiltonian path from s to t

$$\langle\, s = v_1, \quad v_2, \quad \ldots, \quad v_n = t \,\rangle$$

as in Figure 8.1.6.

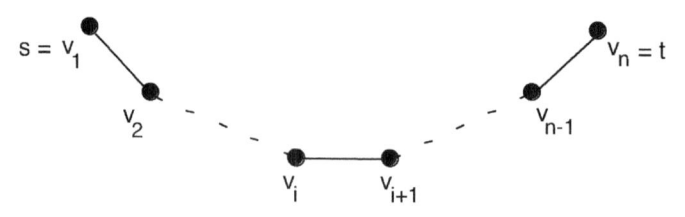

Figure 8.1.6   A hamiltonian path in G.

For each $i = 2, \ldots, n-1$, at least one of the pairs v_1, v_{i+1} and v_i, v_n is non-adjacent, since otherwise

$$\langle\, v_1, \quad v_2, \quad \ldots, \quad v_i, \quad v_n, \quad v_{n-1}, \quad \ldots, \quad v_{i+1}, \quad v_1 \,\rangle$$

would be a hamiltonian circuit in G (as illustrated in Figure 8.1.7). This means that if $A_G = [a_{i,j}]$ is the adjacency matrix for G, then $a_{1,i+1} + a_{i,n} \leq 1$, for $i = 2, \ldots, n-2$.

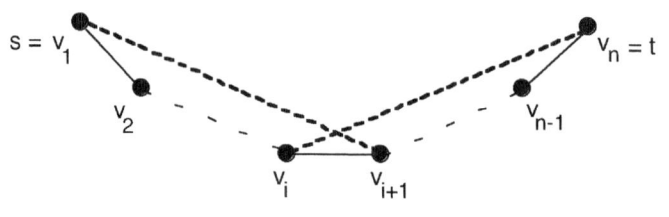

Figure 8.1.7

Thus,

$$deg(s) + deg(t) \;=\; \sum_{i=2}^{n-1} a_{1,i} + \sum_{i=2}^{n-1} a_{i,n}$$

$$=\; a_{1,2} + \sum_{i=3}^{n-1} a_{1,i} + \sum_{i=2}^{n-2} a_{i,n} + a_{n-1,n}$$

$$=\; 1 + \sum_{i=2}^{n-2} a_{1,i+1} + \sum_{i=2}^{n-2} a_{i,n} + 1$$

$$=\; 2 + \sum_{i=2}^{n-2} (a_{1,i+1} + a_{i,n})$$

$$\leq\; 2 + n - 3 \;=\; n - 1$$

which establishes the specified contradiction. $\diamond$

Corollary 8.1.6 [Dira1952]. *Let G be a simple n-vertex graph, where $n \geq 3$, such that*

$$deg(v) \;\geq\; \frac{n}{2}$$

for each vertex v. Then G is hamiltonian.

Proof: Dirac's theorem is easily derived as a consequence of Ore's theorem. Historically, it preceded Ore's theorem. $\diamond$

Postman and Traveling Salesman Problems

In an edge-weighted graph, the sum of the weights on the edge-steps traversed is taken as the cost of the traversal. Self-loops do not affect eulerian traversability. However, any connected graph can be made into an eulerian graph by selectively increasing the numbers of edges between two adjacent vertices.

DEFINITION: The **postman problem** in an edge-weighted graph is to find in a graph the least cost subset of edges whose doubling makes the graph eulerian.

The postman problem was introduced by **M-K Guan** [Guan1962]. He envisioned a postman who must traverse every street in a village and return to the starting point. This requires traversing some streets twice, if there are any junctions of an odd number of streets. Each weight is the cost of traversing a street. **J. Edmonds** and **E. Johnson** [EdJo1973] proved that the problem can be solved in time proportional to a polynomial function on the number of edges.

Example 8.1.3:   In Figure 8.1.8, the four white vertices, w, x, y, and z, have odd degree. If the edge wy at cost 9 is doubled and the edge xz at cost 8 is doubled, then the resulting graph is eulerian at a total cost of 17. There are several better choices. In particular, doubling the edge xy at cost 3 and the edges wu and uz at costs 2 and 4, respectively, thereby making the graph eulerian at a total cost of 9, is the optimal choice, which is provable by *ad hoc* methods.

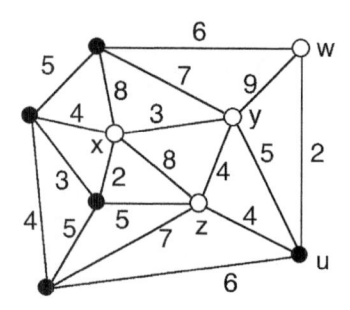

Figure 8.1.8   An edge-weighted graph with four odd-degree vertices.

Remark: The streets that are doubled in a postman tour can be partitioned into a set of paths that serve to match the odd-degree vertices. An exhaustive approach to the postman problem involves considering all possible pairings of the odd-degree vertices and finding, for each pairing, a set of least-distance paths that join the pairs. The Edmonds-Johnson algorithm is a great improvement.

DEFINITION: The **traveling salesman problem** in an edge-weighted graph is to find the hamiltonian tour with lowest cost.

A prototype version of this problem imagined a traveling salesman who sought to sell his product in the capital cities of each of the states of the United States. The distance between two capital cities was taken to be the cost.

Except for a graph with only one vertex, a self-loop is of no use to a hamiltonian tour. Moreover, if there is a multi-adjacency between two vertices, then a hamiltonian tour would either select from it the edge with lowest weight or skip that adjacency altogether.

Example 8.1.4:   The edge-weighted complete graph in Figure 8.1.9 has five vertices.

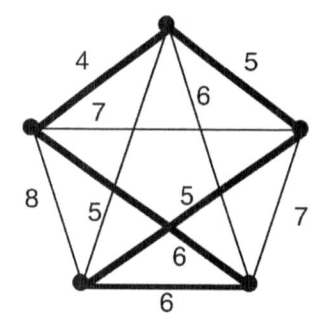

Figure 8.1.9   An edge-weighted complete graph.

There are 5! possible orderings in which all the vertices can be visited. However, two orderings correspond to the same hamiltonian tour if they differ only by cyclic permutation or by reversal of order. Thus, in K_5 there are

$$\frac{5!}{5 \cdot 2} = 12$$

distinct hamilton cycles. Of these, the hamilton cycle with bold edges is the one with lowest cost.

An *ad hoc* proof of this begins with the observation that the five edges with least weight have the weights

$$4 \quad 5 \quad 5 \quad 5 \quad 6$$

for a total cost of 25. However, the three edges with weight 5 form a 3-cycle, so that at most two of them could be used in a 5-cycle. Accordingly, the minimum cost of a 5-cycle is at least 26, which is the cost of the indicated 5-cycle.

EXERCISES for Section 8.1

In Exercises 8.1.1 through 8.1.4, write a list of the edges that might be traversed in sequence on an eulerian tour of the given graph.

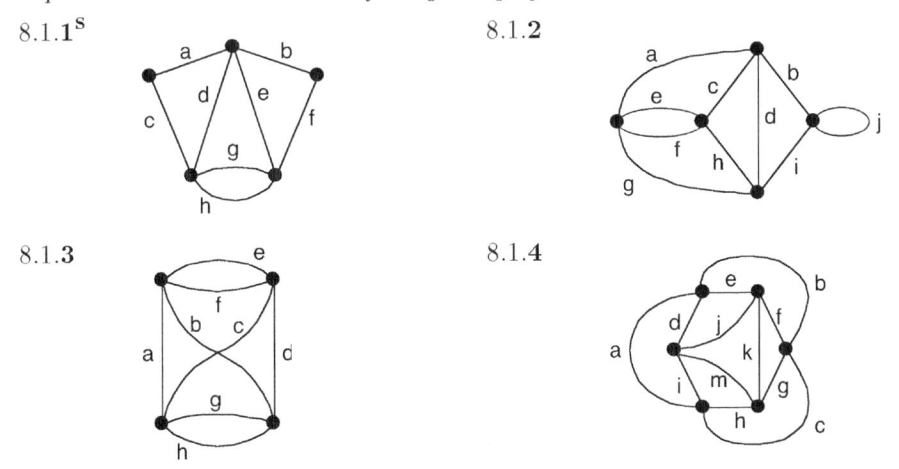

8.1.1[S]   **8.1.2**

8.1.3   **8.1.4**

In Exercises 8.1.5 through 8.1.8, thicken the edges that might be retraversed in an optimal postman tour of the given graph.

8.1.5[S]   **8.1.6**

8.1.**7**

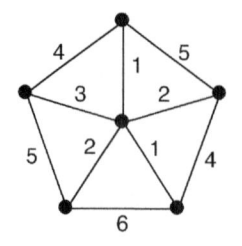

8.1.**8**

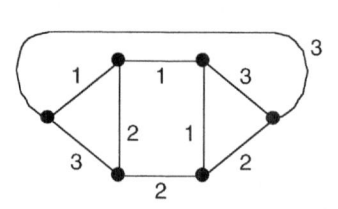

In Exercises 8.1.9 through 8.1.12, either thicken a hamiltonian circuit in the given graph or explain why it is not hamiltonian.

8.1.**9**[S]

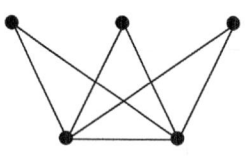

8.1.**10**

8.1.**11**

8.1.**12**

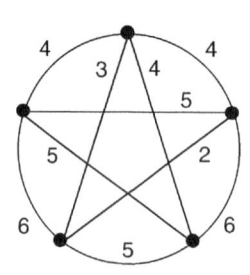

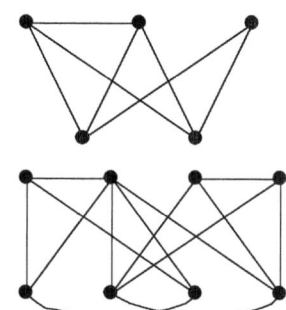

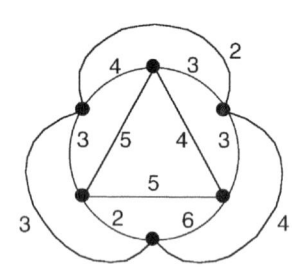

In Exercises 8.1.13 and 8.1.14, thicken an optimal tour for the traveling salesman.

8.1.**13**

8.1.**14**

8.2 PLANARITY

A basic method to draw a connected graph G on any surface, not just the plane, starts by selecting an edge e_1 and drawing it (and its endpoints, of course) on that surface. At each subsequent stage, assuming that edges $e_1, \ldots, e_{n-1}$ have been drawn, select and draw an additional edge e_n of G, such that at least one of its endpoints already appears in the drawing. More specific methods differ primarily in their criteria for selecting the next edge e_n to be drawn.

DEFINITION: A graph is a **planar graph** if it can be drawn in the plane so that the vertices are distinct points of the plane and so that the interior of each edge crosses through no edge or vertex.

This section presents two kinds of criteria for deciding whether a graph is planar.

Regions of a Graph Drawing

Intuitively, drawing a graph on a surface partitions the surface into *regions*, the pieces of surface that would result if one were to cut the surface along the entirety of every edge in the drawing.

DEFINITION: Let G be a graph drawn on a surface S. Two points $x, y \in S - G$ are *mutually reachable* if there is a continuous curve in the complement $S - G$ from x to y that does not cross the graph G. A **region of the drawing** is an equivalence class under the mutual reachability relation.

Example 8.2.1:   Figure 8.2.1 shows a bipartite graph and a crossing-free drawing in the plane. There are two *interior* regions, R and R', and also the *exterior region* R'', for a total of three regions.

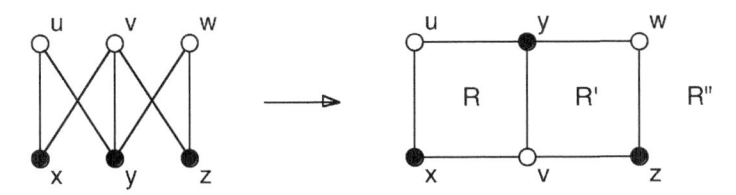

Figure 8.2.1   A graph with a crossing-free drawing.

DEFINITION: The **boundary walk** of a region of drawing of a connected graph on a surface is the closed walk that traverses the boundary. (There may be repeated vertices and edges.)

Example 8.2.1, continued: The boundary walk of region R in Figure 8.2.1 is

$$u, \quad y, \quad v, \quad x, \quad u$$

The boundary walk of region R' is

$$y, \quad w, \quad z, \quad v, \quad y$$

DEFINITION: A **face** of a drawing of a connected graph is the union of a region with the vertices and edges in its boundary walk.

DEFINITION: The **size of a face** is the number of edge-steps in its boundary walk.

Example 8.2.1, continued: In Figure 8.2.1,

$$size(R) \; = \; 4, \quad size(R') \; = \; 4, \quad \text{and} \quad size(R'') \; = \; 6$$

Remark: It is important to recognize that face-size counts edge-steps, not edges. Thus, a single edge may occur twice.

Example 8.2.2:   The face sizes in the drawing of Figure 8.2.2 are

$$1 \quad 2 \quad 4 \quad 6 \quad 7$$

The 6-sided face is the "exterior" face. In traversing the boundary of the 7-sided face R, one imagines walking slightly inside the region, just off the graph. Thus, the boundary walk includes a step from y to x, a step around the self-loop at vertex x, and a step from x back to y and also four steps around the other sides of the region.

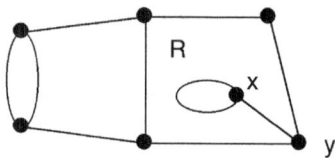

Figure 8.2.2   Another illustration of the Euler polyhedral formula.

Observe that the sum of the face sizes is 20, which is twice the number 10 of edges in the graph. This property holds for any graph drawing on any surface.

Theorem 8.2.1   [Edge-Face Equation].   *The sum of the sizes of the faces of a drawing of a graph G on a surface equals twice the number of edges.*

Proof: Each edge contributes 2 to this sum, since each of its two sides in a surface borders one and only one region. $\diamond$

DEFINITION: The **Jordan separation property** is that for any closed curve in the sphere or the plane, it is impossible to get from a point on one side of that curve to a point on the other side without crossing through the curve.

The effect of the Jordan separation property is that every cycle of a graph in a planar drawing bipartitions the set of regions.

Remark: There are infinitely many surfaces for which the Jordan separation property does not hold, such as the torus, which is the surface of a doughnut. In particular, a closed curve that goes once around the hole does not separate the surface into two parts. The Jordan separation property is precisely what makes it less difficult to decide which graphs have crossing-free drawings in the plane (or sphere) than to make that decision for other surfaces.

Euler Polyhedral Equation

DEFINITION: The **Euler polyhedral formula** for a drawing of a graph G in a surface S is the formula

$$|V| - |E| + |F|$$

where V, E, and F are the sets of vertices, edges, and faces. For a graph drawing in the plane, it is imperative to count the exterior region.

Example 8.2.2, continued: The formula $|V| - |E| + |F|$ applied to the drawing of Figure 8.2.2 has the evaluation

$$7 - 10 + 5 \; = \; 2$$

Our immediate goal is to establish the remarkable fact that for every crossing-free drawing of a connected graph in the plane or sphere — for any planar graph whatever, with or without self-loops and/or multiple edges, the value of the Euler polyhedral formula is always 2.

Remark: The value of the Euler polyhedral formula is not 2 for drawings of non-connected graphs. For instance, the disjoint union of two 3-cycles has six vertices, six edges, and three faces in a planar drawing, yielding $6 - 6 + 3 = 1$.

Theorem 8.2.2. *The equation*

$$|V| - |E| + |F| \; = \; 2 \tag{8.2.1}$$

holds for every crossing-free drawing of a connected graph G in the plane or sphere.

Proof: If a proper edge of the graph G is drawn first, the initial drawing has two vertices, one edge, and one region, in which case

$$|V| - |E| + |F| \; = \; 2 - 1 + 1 \; = \; 2$$

If a self-loop is drawn first, then (by the Jordan separation property) it separates the plane or sphere into two regions, and we have

$$|V| - |E| + |F| \; = \; 1 - 1 + 2 \; = \; 2$$

We continue inductively, following the basic method of drawing offered at the start of this section. We suppose that the Euler polyhedral equation holds, so far.

There are two cases for the next edge. In both cases, we let V', E', and F' denote the sets of vertices, edges, and faces, respectively, that result from adding that edge to the drawing.

Case 1. If one of its endpoints is new to the drawing, then that endpoint is drawn in the interior of some region and joined to the other endpoint by a line that does not separate the region. Thus,

$$|V'| - |E'| + |F'| \; = \; (|V| + 1) - (|E| + 1) + |F| \; = \; 2$$

Case 2. If the new edge adds no new endpoint to the drawing, then that edge (whether proper of not) is drawn through the interior of some region, thereby splitting the region into two regions (by the Jordan separation property). Hence,

$$|V'| - |E'| + |F'| \; = \; |V| - (|E| + 1) + (|F| + 1) \; = \; 2 \qquad\qquad \diamond$$

DEFINITION: Equation (8.2.1) is called the **Euler polyhedral equation** for the sphere.

The Kuratowski Graphs

Our next objective is to use this remarkable equation to prove that various graphs are not planar. We first illustrate for two graphs of special importance.

DEFINITION: The graphs K_5 and $K_{3,3}$ are called the **Kuratowski graphs**.

Figure 8.2.3   The Kuratowski graphs.

Proposition 8.2.3. *The complete graph K_5 is non-planar.*

Proof: If K_5 could be drawn in the plane, the Euler polyhedral equation

$$|V| - |E| + |F| = 5 - 10 + |F| = 2$$

implies that $|F| = 7$. Since K_5 is a simple graph, the minimum length of a non-trivial closed walk is 3, which is, accordingly, the minimum face-size. It follows that the sum of the face-sizes would be at least $3|F| = 21$. However, this contradicts the Edge-Face Equation (Theorem 8.2.1), which implies that the sum of the face-sizes must be $2|E| = 20$.   ◇

Prop 8.2.4. *The complete bipartite graph $K_{3,3}$ is non-planar.*

Proof: For a putative drawing of $K_{3,3}$ in the plane, the Euler polyhedral equation

$$|V| - |E| + |F| = 6 - 9 + |F| = 2$$

implies that $|F| = 5$. Since $K_{3,3}$ is simple and bipartite, the minimum length of a non-trivial closed walk is 4, which is, therefore, the minimum face-size. It follows that the sum of the face-sizes would be at least $4|F| = 20$. However, the Edge-Face Equation implies that the sum of the face-sizes must be only $2|E| = 18$, a contradiction.   ◇

Algebraic Planarity Tests

The method of proof used in the proofs of Theorems 8.2.3 and 8.2.4 yields a general inequality that is applicable to testing whether a graph can be drawn on a surface. This inequality, in turn, particularizes to the sphere and plane.

Theorem 8.2.5 [Edge-Face Inequality]. *Let G be a graph drawn in any surface. Then*

$$girth(G) \cdot |F| \leq 2 \cdot |E| \tag{8.2.2}$$

Proof: The girth of a graph is surely less than or equal to the minimum face-size for any drawing of that graph. It follows that $girth(G) \cdot |F|$ is less than or equal to the sum of the face-sizes, which (by the Edge-Face Equation) equals $2|E|$.   ◇

Corollary 8.2.6. *A simple graph G is non-planar if*

$$|E| > 3|V| - 6 \qquad\qquad (8.2.3)$$

Proof: Since 3 is less than or equal to the girth of any simple graph, the Edge-Face Inequality implies that

$$|F| \leq \frac{2 \cdot |E|}{girth(G)} = \frac{2}{3} \cdot |E| \qquad\qquad (8.2.4)$$

Rearranging the Euler polyhedral equation yields

$$|E| - |F| = |V| - 2 \qquad\qquad (8.2.5)$$

Combining (8.2.4) and (8.2.5) implies

$$|E| - \frac{2}{3} \cdot |E| \geq |V| - 2$$

which readily yields the result. $\diamond$

Example 8.2.3: Every 8-vertex simple graph with more than 18 edges is non-planar.

Corollary 8.2.7. *A simple bipartite graph G is non-planar if*

$$|E| > 2|V| - 4$$

Proof: Since 4 is the minimum possible girth of a simple bipartite graph, the Edge-Face Inequality implies that

$$|F| \leq \frac{2 \cdot |E|}{girth(G)} = \frac{2}{4} \cdot |E| \qquad\qquad (8.2.6)$$

Rearranging the Euler polyhedral equation yields

$$|E| - |F| = |V| - 2 \qquad\qquad (8.2.7)$$

Combining (8.2.6) and (8.2.7) implies

$$|E| - \frac{2}{4} \cdot |E| \geq |V| - 2$$

which readily yields the result. $\diamond$

Example 8.2.4: Every 12-vertex simple bipartite graph with more than 20 edges is non-planar.

Kuratowski's Theorem

DEFINITION: Let e be an edge with endpoints u and v in a graph G. The operation of **subdividing** edge e formally adds a new vertex x to the graph, joins it to vertices u and v, and deletes edge e. Intuitively, we place the new vertex x in the middle of edge e and call the resulting two new edges e' and e''.

Figure 8.2.4   Subdividing an edge.

DEFINITION: A **Kuratowski subgraph** of a graph is a subgraph that is isomorphic to the result of iterated subdivisions on K_5 or on $K_{3,3}$.

Figure 8.2.5   Results of some iterated subdivisions
on K_5 and $K_{3,3}$.

Theorem 8.2.8 [Kuratowski, 1930].   *A graph is planar if and only if it has no Kuratowski subgraph.* $\Diamond$ *(proof omitted)*

Remark: Since the Kuratowski graphs are non-planar, inserting a few extra dots (representing vertices) along their edges won't help make them planar. Moreover, appending extra edges and vertices to them won't help either. The more difficult direction is proving that a non-planar graph must have a Kuratowski subgraph. See [GrYe2006] for a detailed proof.

Example 8.2.5:   Figure 8.2.6 illustrates how a Kuratowski subgraph might be nestled into a graph. Three vertices of a subdivided $K_{3,3}$ are in white and three in black. Subdivision vertices are gray. Unused edges are dashed.

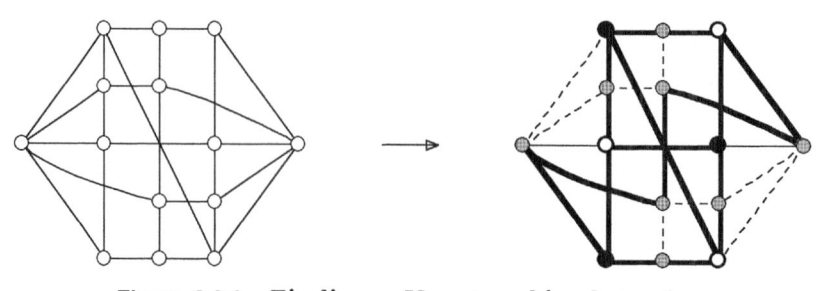

Figure 8.2.6   Finding a Kuratowski subgraph.

EXERCISES for Section 8.2

In each of the Exercises 8.2.1 through 8.2.4, list the numbers of vertices, edges, and faces shown in the designated planar graph drawing from Figure 8.2.7. Verify that the Euler polyhedral equation $|V| - |E| + |F| = 2$ is satisfied.

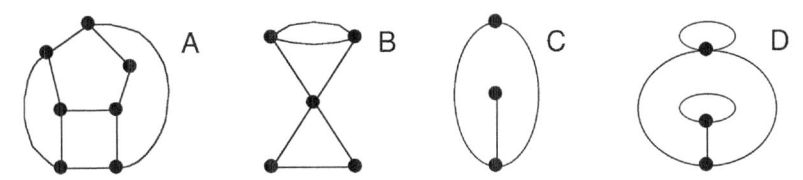

Figure 8.2.7   **Four planar graph drawings.**

8.2.1[S] Graph A.	8.2.2 Graph B.
8.2.3 Graph C.	8.2.4 Graph D.

In each of the Exercises 8.2.5 through 8.2.8, list the sizes of all the faces shown in the designated planar graph drawing from Figure 8.2.7. Verify that the Face-Size Equation is satisfied.

8.2.5[S] Graph A.	8.2.6 Graph B.
8.2.7 Graph C.	8.2.8 Graph D.

In each of the Exercises 8.2.9 through 8.2.12, verify that the designated graph from Figure 8.2.7 does not meet the non-planarity criterion of equation (8.2.3), but find a subgraph that is isomorphic to a subdivided $K_{3,3}$, thereby establishing non-planarity.

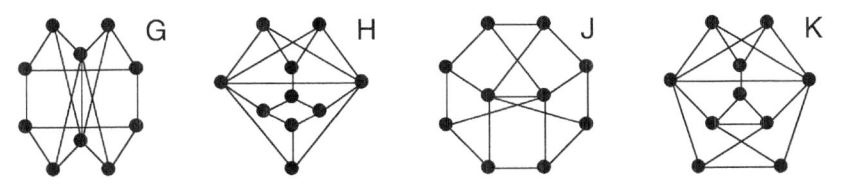

Figure 8.2.8   **Four 3-regular 10-vertex graphs.**

8.2.9[S] Graph G.	8.2.10 Graph H.
8.2.11 Graph J.	8.2.12 Graph K.

8.3   COLORING

DEFINITION: A **vertex k-coloring of a graph** (or simply **coloring**) is an assignment to the vertices either of the names of k different colors or, more usually, of integers $1, 2, \ldots, k$, that are spoken of as *colors* in this context. A graph coloring is a **proper coloring** if no two adjacent vertices have the same color.

Example 8.3.1:   As shown in Figure 8.3.1, vertex colorings may be indicated in a drawing either by placing numbers or other color designators as labels on the vertices, or by using graphic features directly on the vertices.

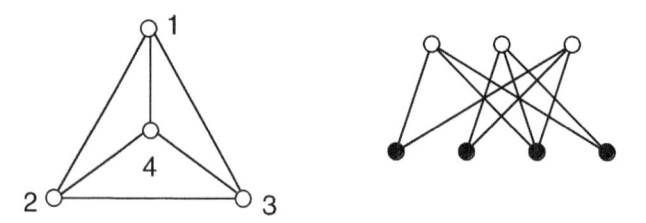

Figure 8.3.1   **Vertex colorings of two graphs.**

DEFINITION: A graph is k-**colorable** if it has a proper k-coloring.

DEFINITION: The **chromatic number** of a graph G, denoted $\chi(G)$, is the smallest number of colors required for a proper vertex coloring.

The chromatic number of a complete graph K_n is n, since assigning a distinct color to each vertex is clearly proper. The chromatic number of a connected bipartite graph is 2, since all the vertices in the same part can have the same color. Thus, both these vertex colorings are optimal.

There are two basic steps to the process of establishing a specific value k for the chromatic number of a graph G.

- *Upper Bound*: Show that $\chi(G) \le k$. Presenting a proper k-coloring for G is a standard way.

- *Lower Bound*: Show that $\chi(G) \ge k$. Proof by finding a subgraph of G that requires k colors is an elementary way to establish a lower bound.

Remark: In studying chromatic numbers, it is usually implicit that the graphs are simple. A multi-adjacency has no more effect on proper colorings than a simple adjacency. Since the endpoint of a self-loop is adjacent to itself, a graph with a self-loop has no proper colorings.

Quite a few practical problems can be modeled by finding the chromatic number of a graph.

Example 8.3.2:   Suppose there are n radio stations in various locations, and that broadcasts from pairs that are too close to each other would interfere with each other. The chromatic number of the network is the minimum number of frequencies that would have to be assigned so that no two stations would interfere.

Cliques and Independent Sets

If a graph has k mutually adjacent vertices, then the chromatic number must be at least k, since they must all have different colors. Thus, the number of vertices in the largest such set is a lower bound on the chromatic number.

DEFINITION: A **clique** in a graph G is maximal set of mutually adjacent vertices. The **clique number** $\omega(G)$ is the cardinality of the largest clique.

Proposition 8.3.1. *For any graph G*

$$\chi(G) \geq \omega(G) \qquad\qquad \Diamond$$

Example 8.3.3:   In the South America graph of Figure 8.3.2, the vertices Bo, Br, Pa, and Ar form a clique. A proper 4-coloring for the South America graph is shown, establishing an upper bound. That Bo, Br, Pa, and Ar are a clique yields a lower bound of 4. Thus, the chromatic number of the South America graph is 4.

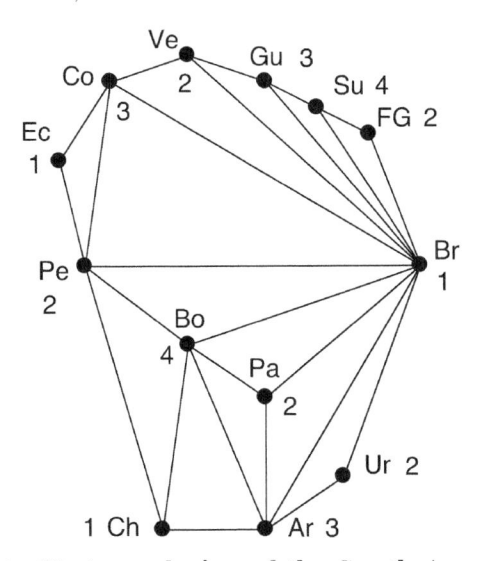

Figure 8.3.2   Vertex coloring of the South America graph.

DEFINITION: An **independent set** of vertices in a graph G is a set of mutually non-adjacent vertices. The **independence number** $\alpha(G)$ is the cardinality of the largest independent set.

Example 8.3.4:   Figure 8.3.3 shows a proper 4-coloring for the graph $circ(7 : 1, 2)$, thereby establishing an upper bound on its chromatic number.

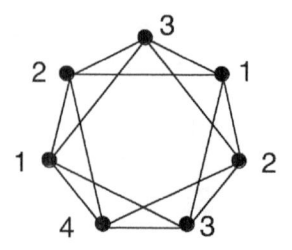

Figure 8.3.3   **Vertex 4-coloring of** $circ\,(7 : 1, 2)$.

Moreover, each vertex of the $circ(7 : 1, 2)$ is adjacent to all but two of the other vertices, and these two are adjacent to each other, which implies that

$$\alpha(circ\,(7 : 1, 2)) \;=\; 2$$

Since vertices with the same color must be non-adjacent, it follows that a proper coloring has at most two vertices of each color. Thus, there must be at least $\lceil 7/2 \rceil = 4$ colors. Accordingly,

$$\chi(circ\,(7 : 1, 2)) \;=\; 4$$

The method used in this example generalizes to the following proposition.

Proposition 8.3.2.   *For any graph* G

$$\chi(G) \;\geq\; \left\lceil \frac{|V_G|}{\alpha(G)} \right\rceil$$

Proof: Each color in a proper coloring is assigned to at most $\alpha(G)$ vertices, because two vertices with the same color must be non-adjacent.   ◇

Vertex- and Edge-Additions and Deletions

Joining a single new vertex to some or all of the vertices of a graph increases the chromatic number by at most 1, since the new vertex could simply be assigned a new color. Deleting a single vertex cannot decrease the chromatic number by more than 1, because otherwise, restoring that vertex would not recover what was lost in chromatic number.

Moreover, adding a single edge increases the chromatic number by at most 1, since a new color could be assigned to one of its endpoints. It follows that dropping one edge cannot decrease the chromatic number by more than 1.

DEFINITION: A graph is ***chromatically critical*** if no matter what edge is deleted, the chromatic number drops by 1.

Example 8.3.5:   The 5-wheel has chromatic number 4, since the three colors are needed for the vertices on the 5-cycle at the rim and a fourth color for the vertex at the hub. Figure 8.3.4 below illustrates that both of the isomorphism types of graph obtainable by deleting a single edge from W_5 have a proper 3-coloring. It follows that W_5 is chromatically critical. More generally, every odd wheel W_{2n+1} has chromatic number 4 and is chromatically critical.

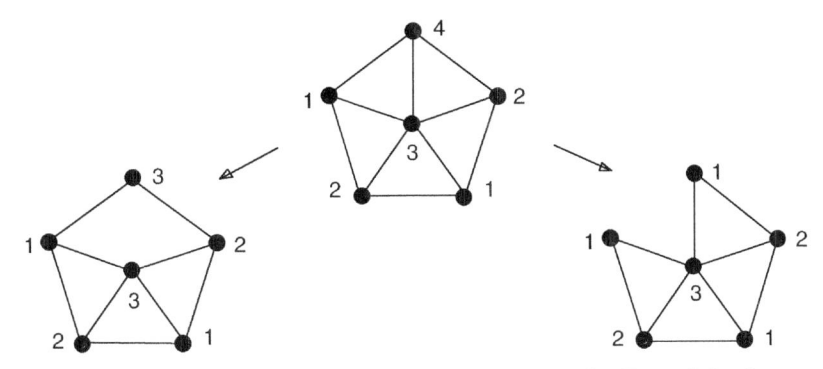

Figure 8.3.4 **The 5-wheel W_5 is chromatically critical.**

Wheel graphs illustrate a simple instance of two facts regarding the join operation and chromatic numbers.

Proposition 8.3.3. *For any two graphs G and H*

$$\chi(G + H) = \chi(G) + \chi(H)$$

Proof: In coloring the join $G + H$, if the vertices of G are properly colored with $\chi(G)$ colors and the vertices of H with $\chi(H)$ colors not used on vertices of G, the result is a proper $\chi(G) + \chi(H)$-coloring. Moreover, in any proper coloring of $G + H$, the colors used on vertices of G must all be distinct for the colors used on vertices of H, since every vertex of G is joined to every vertex of H. $\diamond$

Corollary 8.3.4. *The join of two chromatically critical graphs G and H is chromatically critical.*

Proof: This follows immediately from Proposition 8.3.3. $\diamond$

Complete Sets of Obstructions

DEFINITION: A graph G is called an **obstruction** to a possible property of graphs if no graph having that property contains the graph G.

DEFINITION: A set S of graphs is called a **complete set of obstructions** to a possible property of graphs if no graph having that property contains any graph in S as a subgraph, and if every graph lacking that property contains at least one of the graphs of S as a subgraph.

Example 8.3.6: By Theorem 8.2.8, the set of Kuratowski graphs is a complete set of obstructions to planarity.

Remark: When the context is colorability, the term *obstruction* is usually reserved for chromatically critical graphs.

Proposition 8.3.5. *The set $\{K_2\}$ is a complete set of obstructions to 1-colorability.*

Proof: This is equivalent to saying that a graph is 1-colorable if and only if it has no edges, which is clearly true. $\diamond$

Proposition 8.3.6. *The set of odd cycles is a complete set of obstructions to 2-colorability.*

Proof: A graph is 2-colorable if and only if it is bipartite. Thus, this follows from König's Theorem (Theorem 7.2.3) that a graph is bipartite if and only if it has no odd cycles. $\diamond$

Example 8.3.7: The odd wheels are obstructions to 3-colorability, but they do not form a complete set. The graph in Figure 8.3.5 is a chromatically critical graph that requires 4 colors.

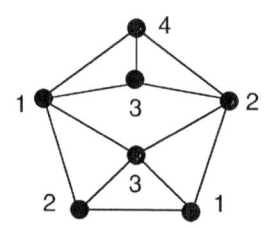

Figure 8.3.5   **A chromatically critical graph.**

Map-Colorings

Questions about map-colorings are readily reducible to questions about graph colorings, through the *Poincaré duality construction*.

DEFINITION: A **map** is a crossing-free drawing of a graph on a surface.

DEFINITION: A **map-coloring** is a function that assigns colors to the regions of a map. It is a **proper map-coloring** if two regions bordering each other on an edge must be assigned different colors.

DEFINITION: The **dual of a map** is the map drawn by inserting a (dual) vertex into each region (of that *primal map*) and then drawing, for every (primal) edge, a (dual) edge joining the (dual) vertices in the regions on the two sides of that (primal) edge.

Example 8.3.8: In Figure 8.3.6, a primal graph is drawn with black vertices and solid edges. The dual graph is shown with white vertices and dashed edges. Observe that although the primal graph is simple, the dual graph has a double adjacency and a self-loop. Observe also that the dual of the dual map is the primal map.

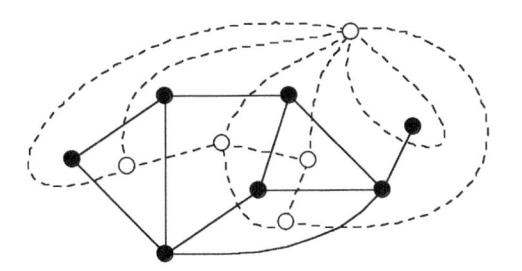

Figure 8.3.6   A map and the dual graph.

Example 8.3.9:   The South America graph of Example 8.3.3 is the dual graph of the map of South America.

Classical Map-Coloring Problems

The history of mathematical map-coloring problems dates back to a communication between Augustus DeMorgan and Sir William Rowan Hamilton in 1852, asking whether every planar map has a proper coloring with at most four colors. This is called the **Four-Color Map Problem**. It was solved in 1976 in the affirmative by Kenneth Appel and Wolfgang Haken, with the aid of a computer.

It is not difficult to prove, as now demonstrated, that six colors are sufficient to properly color any planar graph, hence, any planar map.

Proposition 8.3.7. *The average degree of a planar simple graph G is less than 6.*

Proof: By Corollary 8.2.6,

$$|E_G| \leq 3|V_G| - 6$$

By Euler's Degree-Sum Theorem,

$$average-degree(G) = \frac{2|E_G|}{|V_G|}$$

It follows that

$$average-degree(G) \leq \frac{6|V_G| - 12}{|V_G|} < 6 \qquad \diamondsuit$$

Theorem 8.3.8. *The chromatic number of a planar graph is at most 6.*

Proof: Let G be a chromatically critical planar graph, supposedly with chromatic number $k \geq 7$. (This could be obtained from any planar k-chromatic graph by deleting edges until further deletion would cause a decrease in the chromatic number.) By Proposition 8.3.7, it has a vertex v of degree 5 or less. Since G is chromatically critical, the graph $G - v$ has chromatic number at most $k - 1 \geq 6$. Consider a $(k\text{-}1)$-coloring of $G - v$. Since v has at most five neighbors, at least one of the $k - 1$ colors does not appear on a neighbor, so that $(k\text{-}1)$-coloring could be extended to G by assigning such a missing color to v. $\qquad \diamondsuit$

Remark: Percy Heawood proved in 1890 that five colors are sufficient. At the same time, he offered a formula for the maximum number of colors needed to properly color a map on any other surface, such as the torus. Verifying this formula became known as the *Heawood Map-Coloring Problem*. It was solved in 1968 by Gerhard Ringel and J. W. T. Youngs.

EXERCISES for Section 8.3

In each of the Exercises 8.3.1 through 8.3.6, give the clique number and the independence number of the indicated graph from Figure 8.3.7.

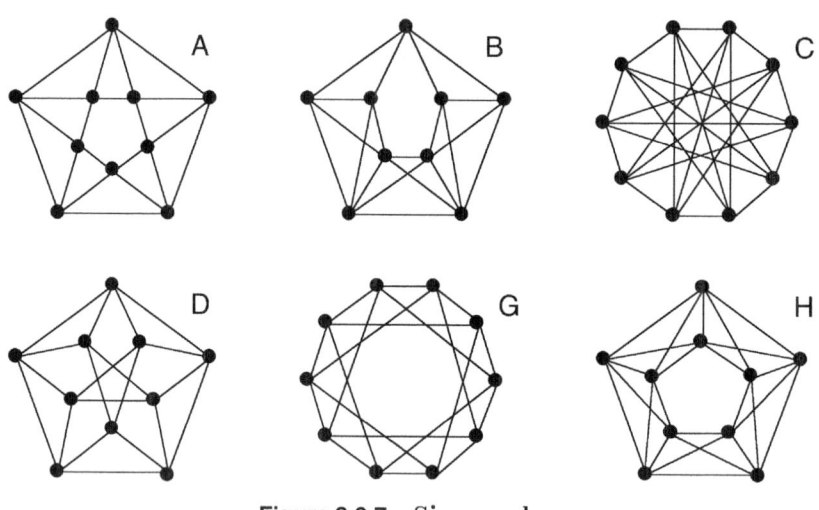

Figure 8.3.7   Six graphs.

$8.3.1^{S}$	Graph A.	**8.3.2**	Graph B.	**8.3.3**	Graph C.
8.3.4	Graph D.	**8.3.5**	Graph G.	**8.3.6**	Graph H.

In each of the Exercises 8.3.7 through 8.3.12, draw a minimum coloring for the indicated graph from Figure 8.3.7. Prove it is a minimum coloring.

$8.3.7^{S}$	Graph A.	**8.3.8**	Graph B.	**8.3.9**	Graph C.
8.3.10	Graph D.	**8.3.11**	Graph G.	**8.3.12**	Graph H.

In each of the Exercises 8.3.13 through 8.3.18, decide whether the indicated graph is chromatically critical, and give a proof. If not critical, also decide whether there is an edge whose removal reduces the chromatic number, and give a proof.

$8.3.13^{S}$	Graph A.	**8.3.14**	Graph B.	**8.3.15**	Graph C.
8.3.16	Graph D.	**8.3.17**	Graph G.	**8.3.18**	Graph H.

In each of the Exercises 8.3.19 through 8.3.22, give the chromatic number of the map of the indicated continent, where two adjacent countries must have different colors.

8.3.**19**[S]   Africa.   8.3.**20**   Asia.

8.3.**21**   Europe.   8.3.**22**   North America.

The Poincaré dual of each of the platonic graphs, as shown in Figure 8.3.8, is a platonic graph. In each of the Exercises 8.3.23 through 8.3.26, construct the dual of the indicated platonic graph, and identify the platonic solid of which it is the 1-skeleton.

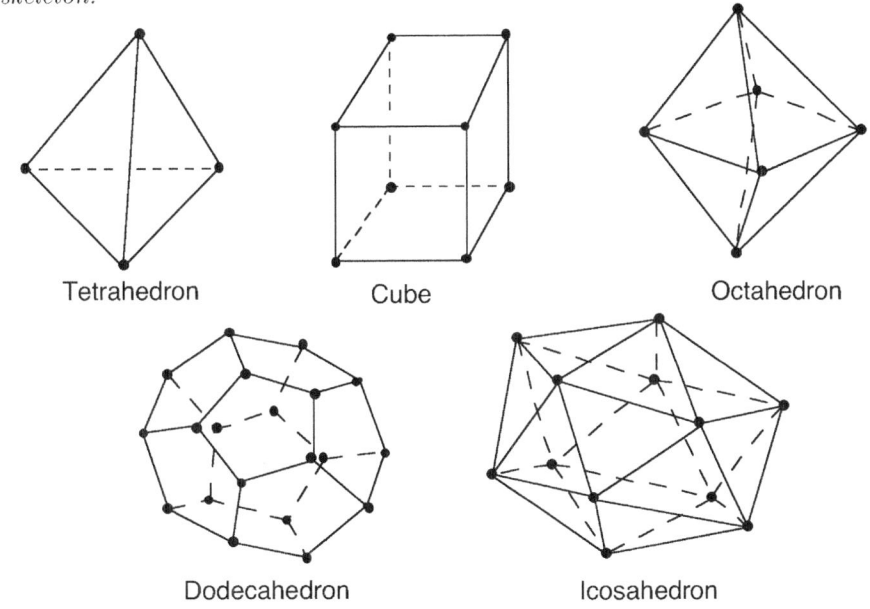

Figure 8.3.8   The five platonic graphs.

8.3.**23**[S]   Tetrahedron.   8.3.**24**   Cube.   8.3.**25**   Octahedron.

8.3.**26**   Dodecahedron.   8.3.**27**   Icosahedron.

In each of the Exercises 8.3.28 through 8.3.31, draw the dual of the indicated map.

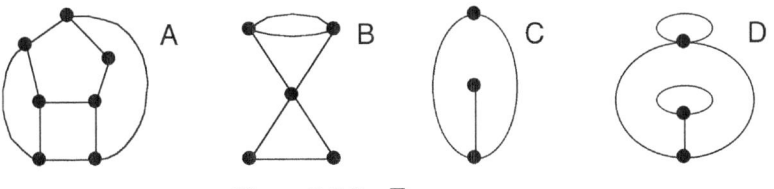

Figure 8.3.9   Four maps.

8.3.**28**   Map A.   8.3.**29**   Map B.

8.3.**30**[S]   Map C.   8.3.**31**   Map D.

8.4   ANALYTIC GRAPH THEORY

Analytic graph theory is a collective name for some related branches of graph theory that are largely concerned with contingent graph properties that occur as graphs get larger, in some sense. This section introduces two of these branches. In *extremal graph theory*, the contingent property of interest occurs after the number of edges is sufficiently large, relative to the number of vertices. *Ramsey graph theory* is concerned with a special type of property that occurs either in a simple graph G or in its edge-complement, regardless of the number of edges of G, once the number of vertices is sufficiently large.

In analytic graph theory, the implicit context is the category of simple graphs. The problems tend to be quite difficult. There are vey few easy results.

Extremal Graphs and Extremal Functions

DEFINITION: An n-vertex graph is said to be an **extremal graph** for a property $\mathcal{P}$ if it is a largest n-vertex graph without that property (usually, among simple graphs), in the sense of having the most edges.

DEFINITION: For a property $\mathcal{P}$, the value of the **extremal function**

$$ex(n, \mathcal{P})$$

is the number of edges in an extremal n-vertex graph for that property.

The main objectives of extremal graph theory are to identify extremal graphs and to derive formulas for extremal functions. A basic approach is to posit a plausible candidate for an extremal graph, and then to prove that the property holds for any larger graph.

Example 8.4.1:   Let $\mathcal{P}$ be the property of being non-acyclic, that is, of having at least one cycle. Since a tree is acyclic, and since an n-vertex tree has $n - 1$ edges, it is clear that

$$ex(n, \text{non-acyclic}) \geq n - 1$$

Since adding an edge to an n-vertex tree creates a cycle, it is reasonable to consider an n-vertex tree as a candidate for an extremal graph — a largest acyclic graph. In fact, some component of an n-edge, n-vertex graph must have at least as many edges as vertices, which implies that beyond the existence of a spanning tree in that component, there is at least one additional edge, which forms a cycle in the component. Accordingly,

$$ex(n, \text{non-acyclic}) \leq n - 1$$

Thus, the extremal n-vertex graphs for non-acyclicity are the acyclic graphs with $n - 1$ edges, that is, all the n-vertex trees.

Example 8.4.2:   Let $\mathcal{P}$ be the property of non-planarity. A planar n-vertex graph has at most $3n - 6$ edges, by Proposition 8.2.6. Thus,

$$ex(n, \text{non-planarity}) \ \leq \ 3n - 6$$

The graph $C_{n-2} + 2K_1$ has $3n - 6$ edges. Its planarity is illustrated in Figure 8.4.1. Thus,

$$ex(n, \text{planar}) \ \geq \ 3n - 6$$

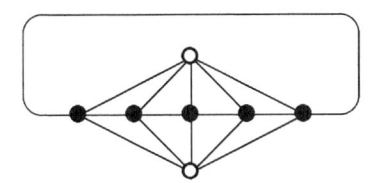

Figure 8.4.1   A planar drawing of $C_5 + 2K_1$.

The extremal graphs for non-planarity may be characterized as the graphs that triangulate the plane, that is, the simple graphs that can be drawn in the plane so that every region is 3-sided.

Turán's Theorem

Historically, the study of extremal graph theory began with the following theorem of Paul Turán. It concerns the property of having a 3-cycle. Turán proved this theorem while confined to a forced labor camp in Hungary.

Theorem 8.4.1 [Turán, 1941]. *Every simple n-vertex graph with more than $\lfloor n^2/4 \rfloor$ edges contains K_3 as a subgraph.*

Proof: We consider a graph G such that

$$|E_G| \ = \ \left\lfloor \frac{n^2}{4} \right\rfloor + 1$$

For $n = 3$ this implies that $|E_G| = 3$, in which case $G \cong K_3$, and for $n = 4$ it implies that $|E_G| = 5$, in which case G is obtainable by deleting a single edge from K_4. In either case, there is a K_3 in G. We continue inductively.

For arbitrary $n \geq 5$, let u and v be adjacent vertices in G. Suppose that there is no K_3 in $G - \{u, v\}$. Then, by the induction hypothesis, the number of edges in $G - \{u, v\}$ is at most

$$\left\lfloor \frac{(n-2)^2}{4} \right\rfloor$$

It follows that the number of edges joining the $n - 2$ vertices of $G - \{u, v\}$ to the vertices u and v is at least

$$\left(\left\lfloor \frac{n^2}{4} \right\rfloor + 1 \right) - \left\lfloor \frac{(n-2)^2}{4} \right\rfloor \ = \ n - 1$$

(Equality holds no matter whether n is even or odd.) By the pigeonhole principle, one of those $n - 2$ vertices must be joined both to u and to v, forming a K_3. $\diamond$

The Ramsey Puzzle

Frank Ramsey (1903-1930) was a British mathematician. His work has led to extensive mathematical development.

The classical Ramsey puzzle is to prove that among any six persons, there are either three mutual acquaintances or three mutual non-acquaintances. The next proposition is a graph-theoretic representation of the solution.

Proposition 8.4.2. *Let G be a simple graph with six vertices. Then either G or its edge-complement $\overline{G}$ contains K_3.*

Proof: Then let $v \in V_G$. Then either (Case 1) v has three neighbors in the graph G or (Case 2) v has three non-neighbors in G. These cases are illustrated in Figure 8.4.2, with dashed lines used here to represent non-adjacency.

Figure 8.4.2   Solving the Ramsey puzzle.

Case 1. If any two of the neighbors are adjacent to each other, then along with v they are spanned by a K_3 in G. Yet if no two are adjacent, then those three neighbors are spanned by a K_3 in $\overline{G}$.

Case 2. If any two of the non-neighbors are non-adjacent to each other, then along with v, they are spanned by a K_3 in $\overline{G}$. Yet if no two are non-adjacent, then the three non-neighbors are spanned by a K_3 in G.   ◇

Ramsey Numbers

DEFINITION: For any pair of positive integers s and t, the **Ramsey number** $r(s,t)$ is the minimum number n such that every simple graph G with at least n vertices either has K_s as a subgraph in itself or, alternatively, has K_t as a subgraph of its edge-complement $\overline{G}$.

Example 8.4.3:   Proposition 8.4.2 asserts that $r(3,3) = 6$.

Proposition 8.4.3. *For every positive integer s,*

$$r(s,1) = 1$$

Proof: The edge-complement of a non-empty contains K_1.   ◇

Proposition 8.4.4. *For every positive integer s,*

$$r(s,2) = s$$

Proof: If an s-vertex graph G is complete, then that graph itself is isomorphic to K_s. If it is not complete, then its complement $\overline{G}$ has at least one edge and contains K_2.   ◇

Proposition 8.4.5. *For all positive integers s and t,*

$$r(s,t) \; = \; r(t,s)$$

Proof: The definition of a Ramsey number $r(s,t)$ is symmetric in the two arguments s and t.   ◇

The Erdős-Szekeres Theorems

It is not obvious that the Ramsey numbers are well-defined. However, using Propositions 8.4.3, 8.4.4, and 8.4.5 to construct a basis, the following theorem establishes the existence of the other Ramsey numbers and an upper bound as well.

Theorem 8.4.6 [Erdős and Szekeres, 1935].
(a) For all integers $s, t \geq 2$,

$$r(s,t) \; \leq \; r(s-1,t) + r(s,t-1)$$

(b) Moreover, if $r(s-1,t)$ and $r(s,t-1)$ are both even, then

$$r(s,t) \; \leq \; r(s-1,t) + r(s,t-1) - 1$$

Proof: This builds by induction on Propositions 8.4.3 and 8.4.4.

(a) Let G be a simple graph on $r(s-1,t) + r(s,t-1)$ vertices, and let $v \in V_G$. Then either

    (Case 1) $deg_G(v) \geq r(s-1,t)$, or

    (Case 2) $deg_{\overline{G}}(v) \geq r(s,t-1)$.

This generalizes the proof of Proposition 8.4.2, as illustrated by Figure 8.4.3.

Figure 8.4.3 **Alternatives in an Erdős-Szekeres theorem.**

Case 1. If any $s-1$ of the neighbors are mutually adjacent, then along with vertex v, they are spanned by a K_s in G. Yet if at most $s-2$ are mutually adjacent, then by the definition of the Ramsey number, some t of these neighbors are mutually non-adjacent, yielding a K_t in $\overline{G}$.

Case 2. If any $t-1$ of the non-neighbors of v are mutually non-adjacent, then along with v they are spanned by a K_t in $\overline{G}$. Alternatively, if at most $t-2$ are mutually non-adjacent, then by the definition of the Ramsey number, some s of these non-neighbors are mutually adjacent, yielding a K_s in G.

(b) Now suppose that $r(s-1,t)$ and $r(s,t-1)$ are both even, and that

$$|V_G| \ = \ r(s-1,t) + r(s,t-1) - 1$$

Then

$$deg_G(v) \ + \ deg_{\overline{G}}(v) \ = \ r(s-1,t) + r(s,t-1) - 2$$

Since $|V_G|$ is an odd number, there is some vertex w of G with even degree.

Case 1. If $deg_G(w) \geq r(s-1,t)$, then we are done, as in part (a) Case 1 of above.

Case 2. If $deg_G(w) < r(s-1,t)$, then, since $deg_G(w)$ is even, it follows that $deg_G(w) \leq r(s-1,t) - 2$. Hence, $deg_{\overline{G}}(w) \geq r(t-1,s)$. The result now follows, as in Case 2 of part (a).   ◇

Corollary 8.4.7 [Erdős and Szekeres, 1935]. *For all positive integers s and t,*

$$r(s,t) \ \leq \ \binom{s+t-2}{s-1}$$

Proof: The assertion is trivially true for $s+t=2$, by Proposition 8.4.3. Continuing inductively, assume for some $k \geq 3$ that the inequality is true for all positive integers s and t such that $s+t < k$, and consider when $s+t = k$. Then

$$\begin{aligned}
r(s,t) \ &\leq \ r(s-1,t) + r(s,t-1) \quad \text{(by Theorem 8.4.6(a))} \\
&\leq \ \binom{(s-1)+t-2}{(s-1)-1} + \binom{s+(t-1)-2}{s-1} \quad \text{(ind hyp)} \\
&= \ \binom{s+t-3}{s-2} + \binom{s+t-3}{s-1} \\
&= \ \binom{s+t-2}{s-1} \quad \text{(Pascal's recursion)} \qquad\qquad ◇
\end{aligned}$$

Ramsey Number Calculations

Calculating Ramsey numbers $r(s,t)$ for $3 \leq s \leq t$ is notoriously difficult, except for the first few. The only nine known Ramsey numbers are shown in Table 8.4.1. Theorem 8.4.2 establishes that $r(3,3) = 6$. The next two results calculate $r(3,4)$ and $r(3,5)$, and a third result produces an upper bound for $r(4,4)$.

Table 8.4.1. The known Ramsey numbers $r(s,t)$ for $3 \leq s \leq t$.

	$t=3$	4	5	6	7	8	9
$s=3$	6	9	14	18	23	28	36
4		18	25				

Proposition 8.4.8. $r(3,4) = 9$.

Proof: Since the Ramsey numbers $r(2,4) = 4$ and $r(3,3) = 6$ are both even, Theorem 8.4.6(b) implies the upper bound

$$r(3,4) \leq r(2,4) + r(3,3) - 1 = 9$$

For the reverse inequality, observe that the 8-vertex circulant graph $circ\,(8;1,4)$, shown in Figure 8.4.4, has neither a K_3-subgraph nor an independent set of four vertices (i.e., its edge-complement has no K_4-subgraph). $\diamond$

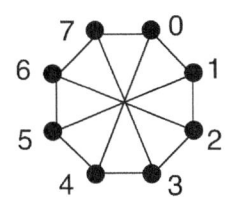

Figure 8.4.4   The circulant graph $circ\,(8;1,4)$.

Proposition 8.4.9. $r(3,5) = 14$.

Proof: The upper bound for $r(3,5)$ follows from Thm 8.4.6(a) and Propositions 8.4.4 and 8.4.8. In particular,

$$r(3,5) \leq r(2,5) + r(3,4) = 5 + 9 = 14$$

For the reverse inequality, consider the graph $circ\,(13;1,5)$, shown in Figure 8.4.5 below, which clearly has no K_3-subgraph. If it had a set S of five mutually non-adjacent vertices, then two of them would have to be within distance two of each other. By symmetry, we may take these two vertices to be 0 and 2. This excludes vertices 1, 3, 5, 7, 8, 10, and 12, leaving only 4, 6, 9, and 11 as possible members of S. But at most one of the vertices 4 and 9 can be in S, because they are adjacent. Likewise, at most one of 6 and 11 can be in S. $\diamond$

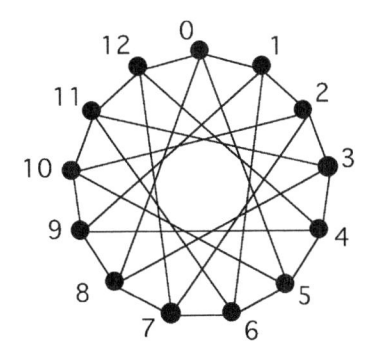

Figure 8.4.5   The circulant graph $circ\,(13;1,5)$.

Proposition 8.4.10. $r(4,4) \leq 18$.

Proof: By Theorem 8.4.6(a) and Propositions 8.4.5 and 8.4.8, we have

$$r(4,4) \leq r(3,4) + r(4,3) = 9 + 9 = 18 \qquad\qquad \diamondsuit$$

EXERCISES for Section 8.4

In each of the Exercises 8.4.1 through 8.4.5, derive a formula for the extremal number $ex(n, \mathcal{P})$ for the given property.

8.4.1[S] $\mathcal{P}$: has radius 1.

8.4.2 $\mathcal{P}$: contains P_4.

8.4.3 $\mathcal{P}$: is connected.

8.4.4 $\mathcal{P}$: contains $2K_2$, i.e., two disjoint copies of K_2.

8.4.5 $\mathcal{P}$: contains $3K_2$, i.e., three disjoint copies of K_2.

In each of the Exercises 8.4.6 through 8.4.11, use Theorem 8.4.6 of Erdős and Szekeres to calculate an upper bound for the indicated Ramsey number.

8.4.6[S] $r(3,6)$.   8.4.7 $r(3,7)$.   8.4.8 $r(3,8)$.

8.4.9 $r(4,4)$.   8.4.10 $r(4,5)$.   8.4.11 $r(5,5)$.

8.5   DIGRAPH MODELS

What imparts to the study of digraphs a special character is their incremental capacity for building mathematical models. A specialized theory has evolved from the need to understand such models. This section presents two examples of digraph modeling.

REVIEW FROM §7.3:   The **underlying graph** of a directed or mixed graph G is the graph that results from removing all the designations of *head* and *tail* from the directed edges of G (i.e., deleting all the edge-directions).

Robbins' Traffic Problem

Many streets and sections of streets of New York City have been designated for one-way traffic flow. In various parts of the city, some streets that appear on a map to be through streets have abrupt reversals of their one-way flow at various cross streets. Driving to a specified location in the city sometimes involves intricate weaving through the grid, thus inspiring the joke punchline, "You can't get there from here." In fact, a variation on Dijkstra's algorithm (see §7.8) will find a shortest directed path from any location in a digraph to any other, as well as determine whether there is any such path.

Herbert Robbins (1915-2001) posed and solved the problem of designing a traffic grid so that, even if every street is to be limited to one-way traffic, there is a way to assign directions that enables all pairs of locations to be mutually reachable. A few definitions from digraph theory facilitate the discussion.

DEFINITION: A digraph is a **connected digraph** if its underlying graph (§7.3) is connected (§7.2).

DEFINITION: A digraph is **strongly connected** if for any two vertices s and t, there is a directed path from s to t.

DEFINITION: An **orientation on a graph** is an assignment of directions to all of its edges. A graph is **strongly orientable** if it has an orientation that is strongly connected.

We observe that a digraph whose underlying graph has a cut-edge, as illustrated in Figure 8.5.1, cannot be strongly orientable, since whichever way it is oriented, there is no directed path from the head of a cutedge to its tail.

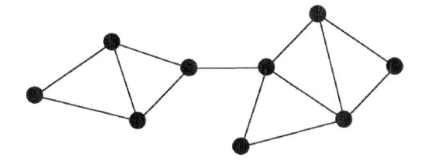

Figure 8.5.1   This graph with a cut-edge is not strongly orientable.

Robbins' solution inverts this observation: if a graph has no cut-edge, then it is possible to assign directions so that the resulting digraph is strongly connected. In digraph terminology, we are working toward Robbins' proof that every cut-edge-free graph is strongly orientable.

Synthesizing Cut-Edge-Free Graphs

DEFINITION: A **path addition** to a graph G is the addition to G of a path whose edges and internal vertices are not in G. If the path is closed, then it is a **cycle addition**. Otherwise it is an **open path addition**.

DEFINITION: A **Whitney-Robbins synthesis** of a graph G from a graph H is a sequence of graphs

$$H = G_0, \quad G_1, \quad \ldots, \quad G_k = G$$

such that for $i = 1, \ldots, k$, the graph G_i is derivable either as an open path addition or as a cycle addition to the graph G_{i-1}.

Theorem 8.5.1 [Whitney-Robbins Synthesis Theorem].   *A graph G is cut-edge-free if and only if G is either a cycle or derivable from a cycle by a Whitney-Robbins synthesis.*

Proof: ($\Leftarrow$) A cycle has no cut-edge. Moreover, the result of a path addition to a cut-edge-free graph is a cut-edge-free graph. Thus, by induction, every graph derivable by Whitney-Robbins synthesis from a cycle is cut-edge-free.

($\Rightarrow$) Suppose that G is a cut-edge-free graph. Since G is not a tree, there must exist a cycle C in G. Among all subgraphs of G that are Whitney-Robbins synthesizable from C, let R be one with the maximum number of edges. If $R = G$, then we are done.

Otherwise, consider an edge e with one endpoint s in subgraph R and the other endpoint t a vertex of $G - R$. Since the graph G has no cut-edges, the edge e is not a cut-edge, so there is a path P in $G - e$ from t to s. As one traverses path P from t toward s, let y (possibly $y = s$) be the first vertex in the subgraph R, and let W be the path from s to y that begins with edge e and continues with the subpath of P from t to y, as illustrated in Figure 8.5.2.

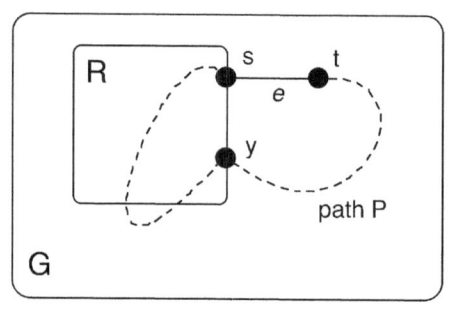

Figure 8.5.2   **A path addition in G to subgraph R.**

However, the graph obtained when subgraph R is extended by adding path W is a non-trivial Whitney-Robbins extension of R. This contradicts the choice of R as having the maximum possible number of edges.   ◇

Theorem 8.5.2 [Robbins, 1939]. *A connected graph is strongly orientable if and only if it has no cut-edges.*

Proof: ($\Rightarrow$) Let e be an edge of strongly orientable graph G, oriented so that G is strongly connected. Then there is a directed path from the head of e to the tail of e in the graph $G - e$. Accordingly, the edge e is not a cut-edge.

($\Leftarrow$) Let G be a connected graph with no cut-edges. By Theorem 8.5.1, it follows that G is either a cycle or derivable from a cycle by a Whitney-Robbins synthesis. Clearly, a cycle is strongly orientable. Strong orientability is preserved under path addition, simply by aligning directions of the edges of the path. It follows by induction that G is strongly orientable.   ◇

Tournaments

The graph-theoretic model called a *tournament* represents the outcome of what is familiarly known as a *round-robin tournament*, in which every player plays exactly one match with every other player. There is a vertex for each player, and an arc from x to y means that x beat y.

REVIEW FROM §7.3: A **simple digraph** is a digraph that has no self-loops and no multi-arcs.

DEFINITION: A **tournament** is a simple digraph whose underlying graph is a complete simple graph.

TERMINOLOGY: A **transitive tournament** is a tournament that is transitive as a digraph.

Proposition 8.5.3. *A tournament is transitive if and only if it is acyclic.*

Proof: Since the underlying graph of a tournament is simple, it follows from Corollary 7.3.6 that a transitive tournament is acyclic.

Conversely, if the tournament D is acyclic, that D has arcs $x \to y$ and $y \to z$, then the edge joining x and z must be directed from x to z. Thus, tournament D is transitive.   ◇

Of course, one would like to be able from the outcomes of the matches not only to say who is the winner, but to rank the players. The out-degree of a vertex represents the number of other players the corresponding player has beaten, so it is natural to rank the contestants according to out-degree.

Example 8.5.1:   Figure 8.5.3 depicts a non-transitive tournament. Observe that v beats x, x beats z, and z beats v. Vertices are labeled with their out-degrees.

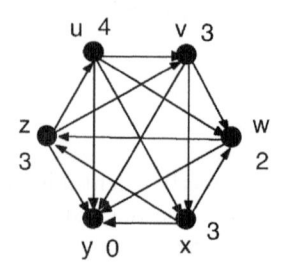

Figure 8.5.3   A non-transitive tournament.

DEFINITION: The **score sequence** of a tournament is the sequence of out-degrees.

Theorem 8.5.4. *An n-vertex tournament is transitive if and only if the score sequence is*

$$n-1, \quad n-2, \quad \ldots, \quad 1, \quad 0$$

Proof: ($\Rightarrow$) In a transitive tournament, if x beats y, then x beats everybody that y beats, so the out-degree of x is at least one more than the out-degree of y. Thus, no two vertices have the same out-degree. By the pigeonhole principle, the score sequence must be $n-1, n-2, \ldots, 0$.

($\Leftarrow$) Suppose that the score sequence is $n-1, n-2, \ldots, 0$. If $n = 1$, the tournament is transitive. Moreover, adding a player who beats everyone else to a transitive tournament yields a transitive tournament. $\Diamond$

Hamiltonian Paths in Tournaments

In a digraph, a *hamiltonian path* is usually understood to be a directed path.

Proposition 8.5.5. *A transitive tournament D has exactly one hamiltonian path.*

Proof: Suppose that D has n vertices, By Proposition 8.5.4, the score sequence is

$$n-1, \quad n-2, \quad \ldots, \quad 1, \quad 0$$

Clearly, the vertex with score 0 is beaten by every other vertex. Continuing inductive, it is clear that every vertex beats all the vertices with lower scores. Thus, there is a (directed) hamiltonian path from the vertex with score $n-1$ to the vertex with score 0. $\Diamond$

Theorem 8.5.6 [Rédei, 1934]. *Every tournament D has a hamiltonian path.*

Proof: Let $P = \langle v_0, v_1, \ldots, v_{k-1} \rangle$ be a directed path of maximum length in the tournament D. If D has no hamiltonian path, then there is a vertex z of D not on path P. For $j = 0, \ldots, k-1$, let e_j be the edge joining v_j and z, as shown in Figure 8.5.4.

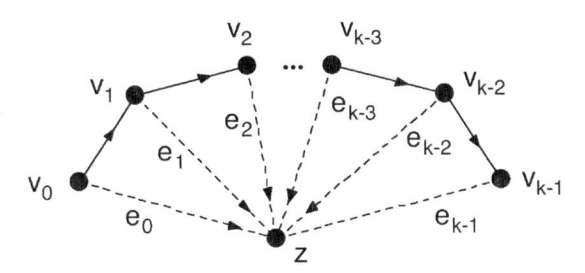

Figure 8.5.4   **Finding a hamilton path in a tournament.**

Arc e_0 must be directed from v_0 to z, as shown, lest preceding path P by that arc yield a longer directed path, contradicting the maximality of P. Moreover, for $j = 1, \ldots, k-1$, if arc e_{j-1} is directed toward z, then arc e_j must also be directed toward z, lest substituting the pair of arcs e_{j-1} and e_j into the path P for the arc $v_{j-1} \to v_j$ yield a longer path, thereby contradicting the maximality of P. However, having arc e_{k-1} directed toward z implies that the directed path P can be extended by arc e_{k-1}, contradicting the maximality of P. $\Diamond$

Kings

DEFINITION: A ***king in a tournament*** is a vertex such that every other vertex is reachable by a directed path of length 1 or 2.

Theorem 8.5.7. *Every tournament D has a king.*

Proof: If D has only one vertex, it is a king. Assume inductively that every $n-1$-vertex tournament has a king, and let u be any vertex in an n-vertex tournament. Let z be a king in the tournament $D - u$, and let S be the set containing the vertex z and all the vertices that z beats. If u beats every vertex in S, then u is king, as shown in Figure 8.5.5(a). Otherwise, some vertex in S beats u, in which case z is a king in tournament D, as in Figure 8.5.5(b).

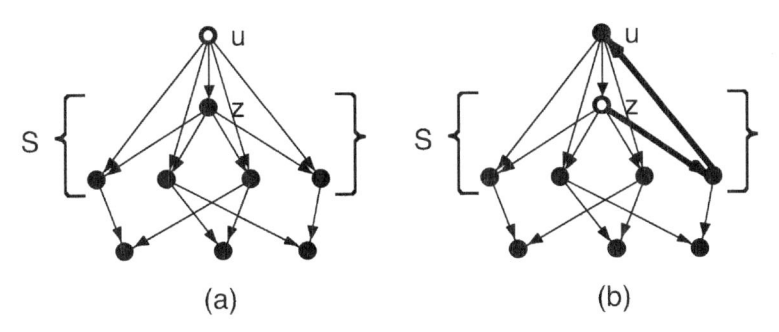

Figure 8.5.5   **Finding a king in every tournament.**

EXERCISES for Section 8.5

In Exercises 8.5.1 through 8.5.4, construct a Whitney-Robbins synthesis of the given graph.

8.5.1^S 8.5.2

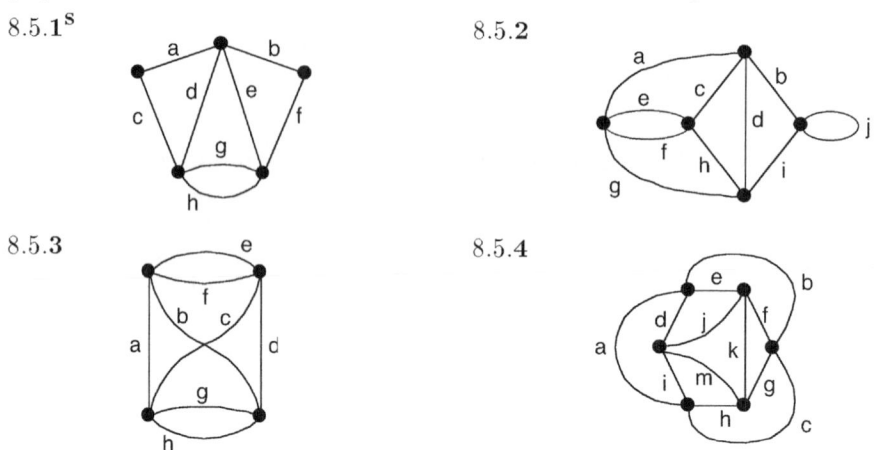

8.5.3   8.5.4

In each of the Exercises 8.5.5 through 8.5.10, construct the score sequence of the designated tournament of Figure 8.5.6.

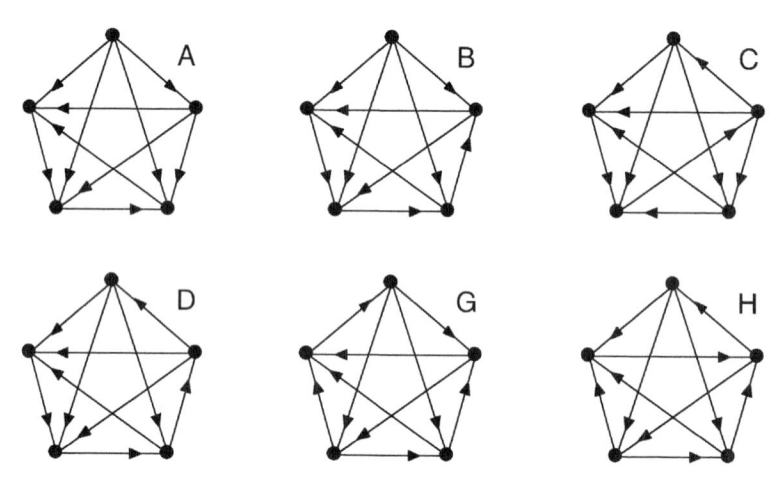

Figure 8.5.6   Six tournaments.

8.5.5^S Tournament A.   8.5.6   Tournament B.   8.5.7   Tournament C.

8.5.8   Tournament D.   8.5.9   Tournament G.   8.5.10   Tournament H.

In each of the Exercises 8.5.11 through 8.5.16, find a hamilton path in the designated tournament of Figure 8.5.7.

8.5.11^S Tournament A.   8.5.12   Tournament B.   8.5.13   Tournament C.

8.5.14   Tournament D.   8.5.15   Tournament G.   8.5.16   Tournament H.

8.6 NETWORK FLOWS

Network flows are used in a variety of mathematical optimization problems. The general idea is that there are one or more source nodes of some commodity, one or more target nodes where the commodity will be consumed, various relay nodes, and a set of connections among the nodes, each with a capacity for flow from one end to the other.

DEFINITION: A *source* in a connected digraph is a designated vertex with non-zero out-degree. A *sink* is a designated vertex with non-zero indegree. A network with a single source vertex s and a single sink vertex t (sometimes called the *target*) is called an *s-t network*.

DEFINITION: A *capacitated network* is a connected digraph in which each arc e is assigned a non-negative weight $cap(e)$, called its *capacity*.

Example 8.6.1:   A capacitated *s-t* network is shown in Figure 8.6.1.

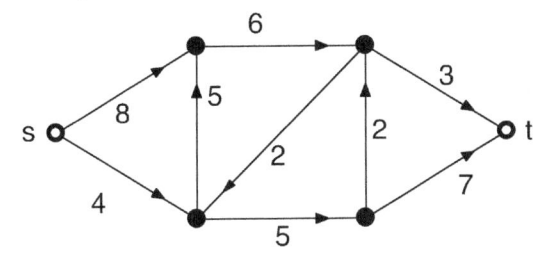

Figure 8.6.1   A capacitated network.

DEFINITION: Let v be a vertex of a digraph D. The *out-set* of v, denoted $Out(v)$, is the set of all arcs in D directed outward from v. The *in-set* of v, denoted $In(v)$, is the set of all arcs in D directed inward into v.

Feasible Flows

DEFINITION: A *feasible flow* on a capacitated *s-t*-network N is a function

$$f : E_N \to \mathbb{R}^+$$

that assigns to each arc e a non-negative real number $f(e)$ such that

1. (*capacity constraint*)   for every $e \in E_N$,

$$f(e) \ \leq \ cap(e)$$

2. (*conservation constraint*) for every $v \in V_N$, except source s and sink t,

$$\sum_{e \in In(v)} f(e) \ = \ \sum_{e \in Out(v)} f(e)$$

Example 8.6.1, continued: Figure 8.6.2 shows a flow for the network of Figure 8.6.1. The capacity of each arc is the first number, and the flow on that arc is the second number.

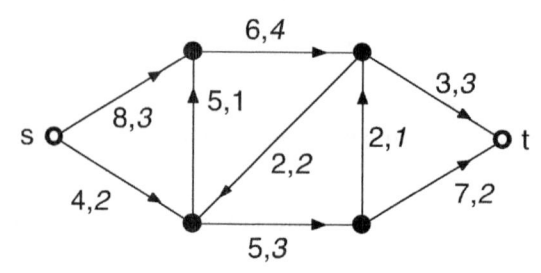

Figure 8.6.2   A feasible flow on a capacitated network.

DEFINITION: The **value of a flow** f on a capacitated s-t network, denoted $val(f)$, is the net flow out of the source

$$val(f) \; = \; \sum_{e \in Out(s)} f(e) \; - \; \sum_{e \in In(s)} f(e)$$

Example 8.6.1, continued: The value of the flow in Figure 8.6.2 is $2 + 3 = 5$.

DEFINITION: A **maximum flow** is a flow f such that $val\,(f)$ is greater than or equal to the value of any other flow on the same network.

Cuts

DEFINITION: In an s-t network N, let V_s and V_t be a partition of the vertex set V_N such that $s \in V_s$ and $t \in V_t$. Then the set of all arcs directed from a vertex of V_s to a vertex of V_t is called an **s-t cut** on network N. It is denoted $\langle V_s, V_t \rangle$.

DEFINITION: The **capacity of a cut** $\langle V_s, V_t \rangle$ is the sum of the capacities in its arcs.

Example 8.6.1, continued: In Figure 8.6.3, the capacity of the s-t cut

$$\langle \{s, u, w, x\}, \{v, t\} \rangle$$

(shown with dashed arcs) is $6 + 2 + 7 = 15$.

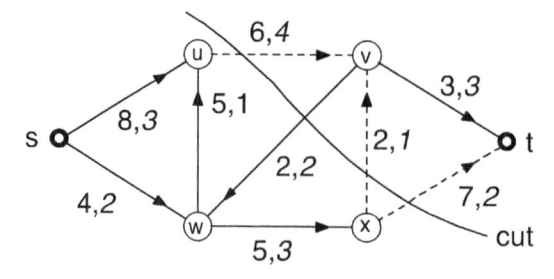

Figure 8.6.3   A feasible flow on a capacitated network.

DEFINITION: The **net flow through an s-t cut** is the sum of the flows on the arcs of the cut (the **forward flow**) minus the sum of the flows on arcs from vertices in V_t to vertices in V_s (the **backflow**).

Example 8.6.1, continued: The net flow through the cut in Figure 8.6.3 is $4 + 1 + 2 - 2 = 5$. We observe that this equals the value of the flow.

Proposition 8.6.1. *Let f be a flow on a network. The net flow through any s-t cut $\langle V_s, V_t \rangle$ equals val (f).*

Proof: The net flow through the cut $V_s = \{s\}$ is $val(f)$. For $V_s = \{s, v_1, \ldots, v_k\}$, the net flow through the cut is the sum of the net flows out of all the vertices in V_s. By the conservation constraint, the net flows out of each of these vertices except s is 0. Thus, the net flow through the cut $\langle V_s, V_t \rangle$ is $val(f)$. $\Diamond$

Proposition 8.6.2. *Let f be a flow on a network. The forward flow through any s-t cut $\langle V_s, V_t \rangle$ is less than or equal to the capacity of that cut.*

Proof: The flow through each arc in the cut is at most the capacity of that arc. Summing over the arcs yields the conclusion. $\Diamond$

Corollary 8.6.3. *Let f be a flow on a network. The net flow through any s-t cut $\langle V_s, V_t \rangle$ is less than or equal to the capacity of that cut.*

Proof: The net flow is less than or equal to the forward flow, since the backflow is non-negative. $\Diamond$

DEFINITION: A **minimum cut** is a cut whose capacity is less than or equal to the capacity of any other cut on the same network.

Theorem 8.6.4. *Let f be a flow on a network. Then the value of a maximum flow is less than or equal to the capacity of a minimum cut. Moreover, if the value of a flow f equals the capacity of some cut $\langle V_s, V_t \rangle$, then flow f is a maximum flow, and cut $\langle V_s, V_t \rangle$ is a minimum cut.*

Proof: This follows from Proposition 8.6.1 and Corollary 8.6.3. $\Diamond$

Increasing the Flow Along a Directed Path

The equilibrium in which the flow achieves its maximum, the capacity of the minimum cut, is always achievable, due to an algorithm of Ford and Fulkerson. We are now working toward a description of their optimization method.

We consider, as a preliminary, a flow f on an s-t network with a directed s-t path

$$P = \langle s, \quad e_1, \quad v_1, \quad e_2, \quad \ldots, \quad e_k, \quad t \rangle$$

such that for $j = 1, \ldots, k$

$$f(e_j) < cap(e_j)$$

Let

$$\Delta_P \;=\; \min\{\, cap(e_j) - f(e_j) \mid j = 1, \dots, k \,\}$$

and suppose that the flow on each of the arcs $e_1, \dots, e_k$ is increased by Δ_P. The resulting flow is feasible, since it preserves the conservation of flow at each internal vertex along path P, and by choice of the increment Δ_P, none of the resulting arc flows exceeds capacity.

Example 8.6.1, continued: When the flow in each of the arcs along the directed path $\langle\, s, w, x, t \,\rangle$ is increased by 2, the resulting network flow is as shown in Figure 8.6.4. The network flow is increased thereby from 5 to 7.

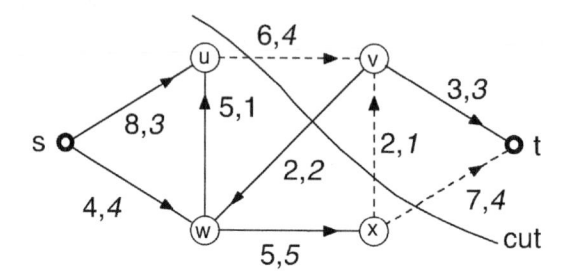

Figure 8.6.4 **Increasing the flow along a directed path.**

There is no other directed s-t path along which the flow could be increased. Since the arc sw is filled to capacity, the first arc in such a path would have to be su, after which the only possible next arc is uv. From vertex v, the arcs vw and vt are already at capacity.

Augmenting Along a Quasi-Path

Although increasing some forward flows is an obvious way to achieve an increased net flow, it is also possible to increase the net flow by decreasing some backward flows.

DEFINITION: An **s-t quasi-path** in an s-t network is a sequence

$$Q \;=\; \langle\, s, \;\; e_1, \;\; v_1, \;\; \dots, \;\; v_{k-1}, \;\; e_k, \;\; t \,\rangle$$

whose vertices and edges form an s-t path in the underlying graph. The arc e_j is a **forward arc** if it is directed from v_{i-1} to v_i and a **backward arc** if it is directed from v_i to v_{i-1}.

DEFINITION: Let f be a flow on an s-t network and Q a quasi-path as above. The **slack on an arc** e is given by

$$\Delta(e) \;=\; \begin{cases} cap(e) - f(e) & \text{if } e \text{ is a forward arc} \\ f(e) & \text{if } e \text{ is a backward arc} \end{cases}$$

The **slack on the quasi-path** Q is given by

$$\Delta(Q) \;=\; \min\{\, \Delta(e_j) \mid j = 1, \dots, k \,\}$$

DEFINITION: A quasi-path Q with positive slack for a flow f is called a **flow-augmenting quasi-path** or an **f-augmenting quasi-path**. **Augmenting the flow** on Q means increasing the flow by $\Delta(Q)$ on every forward arc and decreasing the flow by $\Delta(Q)$ on every backward arc.

Example 8.6.1, continued: Figure 8.6.5 shows a flow-augmenting quasi-path for our running example.

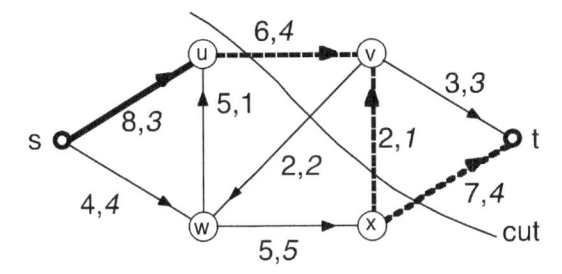

Figure 8.6.5   **A flow-augmenting quasi-path with $\Delta(Q) = 1$.**

Example 8.6.1, continued: Figure 8.6.6 shows the result of augmenting the flow on that quasi-path. Observe that the net flow through the revised cut equals the capacity of that cut. By Theorem 8.6.4, it follows that this is maximum flow and that the revised cut is a minimum cut.

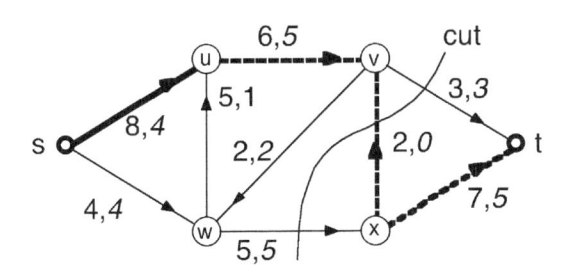

Figure 8.6.6   **Result of augmenting the flow.**

What remains to be proved here is that if no such quasi-path exists, then the exisiting flow is a maximum.

Achieving Maximum Flow

If there is a flow-augmenting s-t quasi-path in a network N, it can be found by a breadth-first search on a digraph D_N associated with network N. The vertex set of D_N is V_N. Each arc e of network N leads to one or two arcs in D_N.

If $f(e) = 0$, then there is an arc in D_N from the tail of e to the head, but none from the head to the tail.

If $0 < f(e) < cap(e)$, then there is an arc from the tail of e to the head, and another from the head to the tail.

If $f(e) = cap(e)$, then there is an arc from the head of e to the tail, but none from the tail to the head.

Making an iterative breadth-first search until no flow-augmenting quasi-path can be found is known as the **Ford-Fulkerson algorithm**.

Theorem 8.6.5. *A flow f in a network N is a maximum flow if and only if there does not exist an f-augmenting quasi-path.*

Proof: ($\Rightarrow$) Suppose that flow f is a maximum flow in network N. There cannot exist an f-augmenting quasi-path, since augmenting the flow on the that quasi-path would yield an increased flow.

($\Leftarrow$) Suppose there exists no f-augmenting flow, and let V_s be the set of vertices in a breadth-first tree and V_t the remaining vertices. Let e be a frontier arc. If e is directed from V_s to V_t, then $f(e) = cap(e)$; if e is directed from V_t to V_s, then $f(e) = 0$. It follows that the net flow through the cut equals the capacity. Therefore, by Theorem 8.6.4, the flow is maximum and the cut is minimum. $\diamond$

Bipartite Matching

Network flow theory is remarkably versatile. We now examine how it applies to a personnel assignment problem.

DEFINITION: A **matching** in a graph is a set of edges no two of which have an endpoint in common. A **maximum matching** is a matching with the greatest number of edges.

Suppose that in a bipartite graph, the vertices $p_1, \ldots, p_m$ represent available people and the vertices $w_1, \ldots, w_n$ represent jobs of different types. There is an edge from each person to each job of which that person is capable. A maximum matching covers as many jobs as possible and, simultaneously, assigns work to as many people as possible.

Example 8.6.2: Figure 8.6.7 shows an ad hoc matching in a bipartite graph with five matched pairs. It is maximal, in the sense that one of the endpoints of every remaining edge is already in this matching. It may be observed that if person p_1 switched to the unassigned job w_3, then the unassigned person p_2 could cover the job w_1 relinquished by p_1.

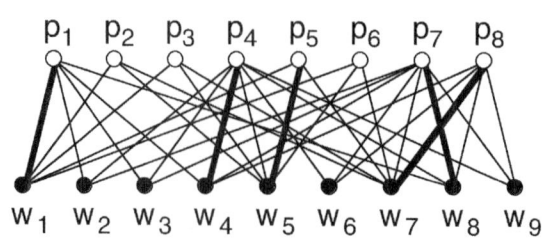

Figure 8.6.7 **A personnel assignment bipartite graph.**

A method to find swaps that increase the size of the matching is based on flow theory. All the edges are directed from the personnel toward the jobs. Then a new source vertex s is joined by arcs to each of the personnel, and all the personnel are joined by arcs to a new sink t, as illustrated in Figure 8.6.8. Dark edges represent a flow of 1. Lighter edges have a flow of 0.

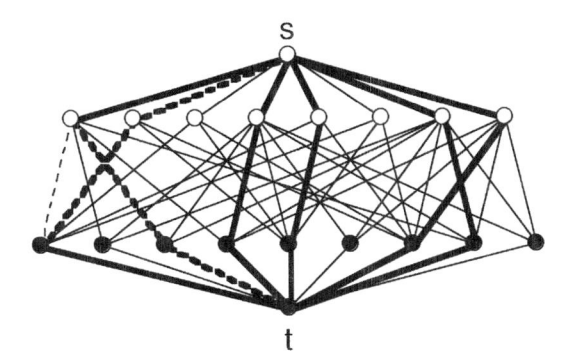

Figure 8.6.8   **A personnel assignment bipartite graph.**

Every arc is given a capacity of 1. The dashed arcs in Figure 8.6.8 indicate a flow augmenting path, relative to Figure 8.6.7. The net effect of a single flow augmentation is to increase the size of the matching by one edge. A maximum flow thereby yields a maximum matching.

EXERCISES for Section 8.6

In Exercises 8.6.1 through 8.6.6, construct a maximum flow for the given network.

8.6.**1**

8.6.**2**

8.6.**3**

8.6.**4**

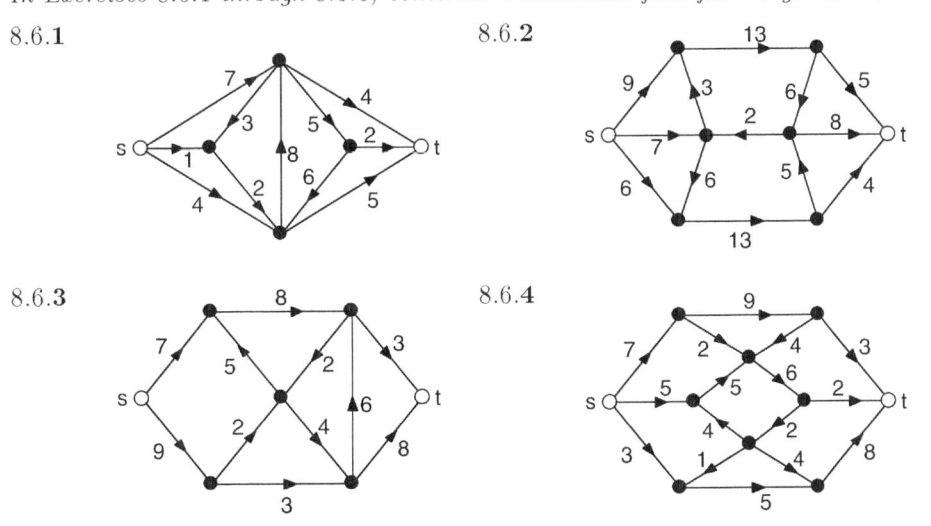

8.6.**5**   8.6.**6**

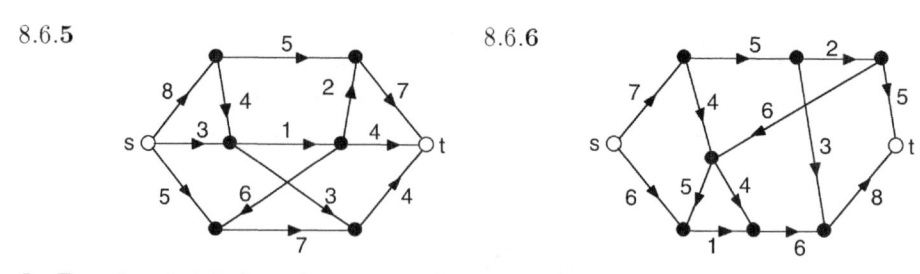

In Exercises 8.6.7 through 8.6.12, construct a minimum cut for the network of the given exercise.

8.6.**7**   The network of Exercise 8.6.1.

8.6.**8**   The network of Exercise 8.6.2.

8.6.**9**   The network of Exercise 8.6.3.

8.6.**10**   The network of Exercise 8.6.4.

8.6.**11**   The network of Exercise 8.6.5.

8.6.**12**   The network of Exercise 8.6.6.

8.7   TOPOLOGICAL GRAPH THEORY

Topological graph theory is concerned with topological methods on graphs, especially in the study of placements of graphs on surfaces. By the late 19th century, Percy Heawood (1861-1955) and Lothar Heffter (1862-1962) had taken initial steps to elevate the study of graphs on surfaces beyond the plane to the higher order surfaces. This section presents algebraic tests for imbeddability in higher surfaces that generalize some of the tests for planarity. It also presents some methods for representing and constructing imbeddings in higher order surfaces.

The exposition in this section presumes some basic acquaintance with surfaces. For further detail, see [GrYe2006].

Two Sequences of Surfaces

The *orientable surfaces* are represented in Figure 8.7.1 as an infinite sequence $S_0, S_1, S_2, \ldots$.

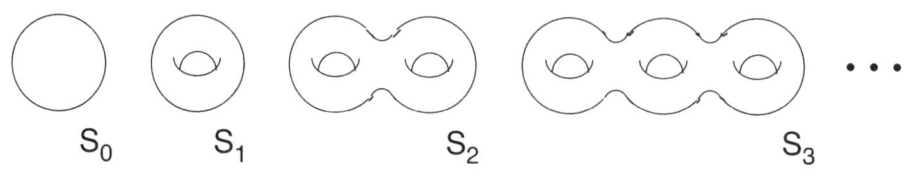

Figure 8.7.1   The orientable surfaces.

The initial surface in this sequence is the sphere. The next surface is the torus S_1. In general, the surface S_g might be positioned in 3-space so as to bound a g-hole doughnut, but it could also adapt to more exotic positioning, involving knotting and linkages of various holes wth each other.

The *non-orientable surfaces* form the sequence N_1, N_2, N_3, ... represented in Figure 8.7.2. The surface N_k is formed by excising the interiors of k disjoint closed disks from the sphere and capping each of them with a Möbius band.

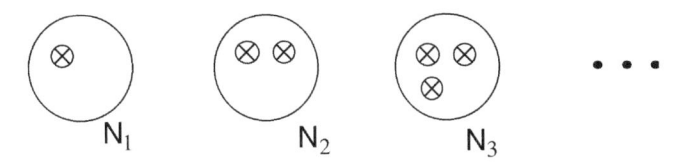

Figure 8.7.2   The non-orientable surfaces.

Flat-Polygon Representation of Surfaces

A closed surface in space may have contours that are hidden from the viewer behind other parts of the surface. Vertices or edges of a graph drawn on the "back" of the surface would also be hidden. A way to bring everything, in effect, to the front is to cut the surface open and flatten it out.

DEFINITION: A **flat-polygon representation** of a surface S is a drawing of a polygon and a matching of its sides, such that when the sides are pasted together according to the matching, the result is a surface that is topologically equivalent to the surface S. Figure 8.7.3 shows a flat polygon representation of the torus.

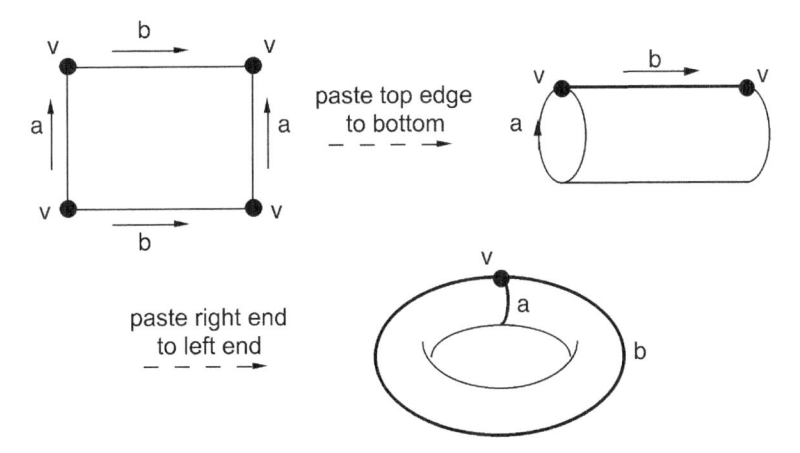

Figure 8.7.3   Flat-polygon representation of a torus.

Pasting the bottom of the rectangle to the top, as per the b-matching, yields a cylindrical tube. Then pasting the right end of the tube to the left end, as per the a-matching, yields a torus.

Example 8.7.1: Figure 8.7.4 is a drawing of the complete bipartite graph $K_{3,3}$ on the torus, represented as a flat rectangle. In this drawing, the sides of the rectangle are represented by dashed lines, to distinguish them from the solid edges of the graph. The edge 23 is construed to extend continuously between vertices 2 and 3 when the rectangle is pasted into a torus. Likewise, the edge 05 extends continuously between its endpoints.

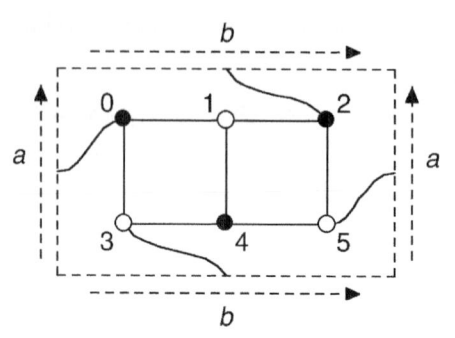

Figure 8.7.4 $K_{3,3}$ **on a torus.**

Remark: Although there are a few widely adopted practices in drawing graphs on flat-polygon representations of surfaces, various graphic devices are used *ad hoc* to enhance the clarity of individual drawings. The dashed lines for the sides of the flat polygon of Figure 8.7.4 are such a device.

Proposition 8.7.1. *Every orientable surface S_g has a flat-polygon representation.*

Proof: Draw a closed curve around each of the handles of the surface, with a single point on it designated to serve as corners of the polygon. Cutting the surface S_g open on all the closed curves changes it into a sphere with $2g$ holes, each of which has a designated point on it. Then draw a tree with $2g - 1$ edges on the surface-with-holes, whose vertices are the designated points. Cutting the surface open on the tree and then flattening the resulting surface yields a polygon with $6g - 2$ corners and sides.   ◇

The most usual form of flat-polygon representation of S_g is achieved by contracting the tree to a single point before cutting on the g closed curves. In that case, the flat polygon has $4g$ corners and sides. All of the corners are pasted into that single point when the surface is constructed from the polygon.

Remark: Non-orientable surfaces also have flat-polygon representations. The idea there is to draw a closed curve as the central circuit on each of the k Möbius bands of N_k, each with a designated point, and a tree joining the designated points.

Cellular Imbeddings

TERMINOLOGY: In the present context, the topological word *imbedding* is used to mean a crossing-free drawing of a graph.

DEFINITION: A region of an imbedding of a graph on a surface is **cellular** if it is topologically equivalent to an open disk. That means that every closed curve inside the region bounds a disk within the region. It is a **strongly cellular region** if its boundary is a cycle in the graph.

DEFINITION: A **(strongly) cellular imbedding** is an imbedding such that every region is (strongly) cellular.

Example 8.7.1, continued: The imbedding in Figure 8.7.4 is strongly cellular.

Example 8.7.2:   The flat-polygon drawing of $K_7 \to S_1$ in Figure 8.7.5 is strongly cellular. Notice that some edges and vertices of the graph are drawn along the boundary of the polygon.

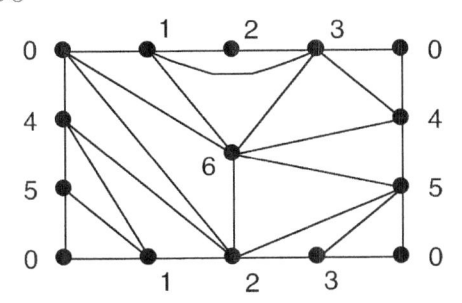

Figure 8.7.5 K_7 **on the torus.**

This tends to make the drawings seem simpler, once one gets used to the idea that the same edge (e.g., edges 12 or 45) may appear twice and that the same vertex (e.g., vertices 2 or 4) may appear twice if it occurs on a side of the polygon and even more than twice (e.g., vertex 0) if it is drawn at a corner of the polygon.

Example 8.7.3:   Figure 8.7.6 shows two non-cellular imbeddings on the torus. In either case, one could draw a closed curve in the "larger" region that does not separate that region.

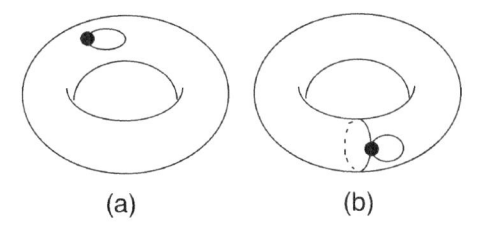

(a)   (b)

Figure 8.7.6   **Two non-cellular imbeddings on the torus.**

Generalized Euler Polyhedral Equation

In §8.2, the Euler polyhedral equation

$$|V| - |E| + |F| = 2 \tag{8.2.1}$$

is combined with the Edge-Face Inequality

$$girth(G) \cdot |F| \leq 2 \cdot |E| \tag{8.2.2}$$

to produce the result that a simple graph G is non-planar if

$$|E| \geq 3|V| - 6 \tag{8.2.3}$$

The following theorem generalizes the Euler polyhedral equation to all orientable surfaces, which will enable us to generalize inequality (8.2.3).

Theorem 8.7.2 [Generalized Euler Polyhedral Equation]. *The equation*

$$|V| - |E| + |F| = 2 - 2g \tag{8.7.1}$$

holds for every imbedding of a connected graph G in the surface S_g.

Proof: This generalizes Theorem 8.2.2. For a detailed proof, see [GrYe2006] or [GrTu2001].   ◇

Corollary 8.7.3. *Let $\rightarrow S_g$ be the property of imbeddabilty in the surface S_g. Then*

$$ex(n, \rightarrow S_g) = 3n - 6 + 6g \tag{8.7.2}$$

Proof: Since 3 is less than or equal to the girth of any simple graph, the Edge-Face Inequality (8.2.2) implies that

$$|F| \leq \frac{2 \cdot |E|}{girth(G)} = \frac{2}{3} \cdot |E| \tag{8.7.3}$$

Rearranging the Euler polyhedral equation (8.7.1) yields

$$|E| - |F| = |V| - 2 + 2g \tag{8.7.4}$$

Combining (8.7.3) and (8.7.4) implies

$$|E| - \frac{2}{3} \cdot |E| \geq |V| - 2 + 2g$$

which readily yields the conclusion.   ◇

Example 8.7.4:   Suppose that a simple graph G has 8 vertices and that we want to know if it has a crossing-free drawing on the torus S_1. Then

$$\begin{aligned}
3|V| - 6 + 6g &= 3 \cdot 8 - 6 + 6 \cdot 1 \\
&= 3 \cdot 8 - 6 + 6 \\
&= 24
\end{aligned}$$

Thus, every 8-vertex simple graph with more than 24 edges is non-toroidal.

Corollary 8.7.4. *A simple bipartite graph G cannot be imbedded on the orientable surface S_g if*

$$|E| > 2|V| - 4 + 4g \tag{8.7.5}$$

Proof: Since 4 is the minimum possible girth of a simple bipartite graph, the Edge-Face Inequality implies that

$$|F| \leq \frac{2 \cdot |E|}{girth(G)} = \frac{2}{4} \cdot |E| \tag{8.7.6}$$

Rearranging the Euler polyhedral equation yields

$$|E| - |F| = |V| - 2 + 2g \tag{8.7.7}$$

Combining (8.7.6) and (8.7.7) implies

$$|E| - \frac{2}{4} \cdot |E| \geq |V| - 2 + 2g$$

which readily yields the result. $\Diamond$

Example 8.7.5:   Every 10-vertex simple bipartite graph with more than 20 edges is non-planar. For instance, $K_{3,7}$, $K_{4,6}$, and $K_{5,5}$ are all non-toroidal.

Minimum Genus

DEFINITION: The **minimum genus of a graph** G is the smallest integer g such that G has an imbedding on the orientable surface S_g. It is denoted $\gamma_{\min}(G)$.

The landmark solution of the Heawood Map-Coloring Problem (see §8.3) by Ringel and Youngs [RiYo1968] is built on a reduction to calculating the minimum genus of all the complete graphs. Table 8.7.1 gives the genus of some complete graphs K_n for smaller values of n.

Table 8.7.1   The genus of some complete graphs.

n	4	5	6	7	8	9	10	11	12	13	14	15	$\cdots$
$\gamma_{\min}(K_n)$	0	1	1	1	2	3	4	5	6	8	10	11	$\cdots$

This achievement launched the emergence of topological graph theory as a major branch. Deriving formulas for the minimum genus of classes of graphs was initially the primary focus. The paradigm for proving such formulas was to establish a lower bound for the minimum genus, and then to describe how to construct a crossing-free drawing in a surface whose genus equals that lower bound. Lower bounds commonly relied on the following theorem.

Theorem 8.7.5. *Let G be a connected graph. Then*

$$\gamma_{\min}(G) \geq \left\lceil \frac{|E| \cdot (girth(G) - 2)}{2 \cdot girth(G)} - \frac{|V|}{2} + 1 \right\rceil \tag{8.7.8}$$

Proof: For a minimum imbedding $G \to S$, we have $\gamma(S) = \gamma_{\min}(G)$, and, thus, the Euler polyhedral equation is

$$|V| - |E| + |F| = 2 - 2\gamma_{\min}(G)$$

The Edge-Face Inequality implies that

$$|F| \leq \frac{2|E|}{girth(G)}$$

which implies that

$$|V| - \frac{|E| \cdot (girth(G) - 2)}{girth(G)} \geq 2 - 2\gamma_{\min}(G)$$

and, in turn, that

$$\gamma_{\min}(G) \geq \frac{|E| \cdot (girth(G) - 2)}{2 \cdot girth(G)} - \frac{|V|}{2} + 1$$

Since $\gamma_{\min}(G)$ is integer-valued, the conclusion follows.   ◇

Corollary 8.7.6. *Let G be a simple connected graph. Then*

$$\gamma_{\min}(G) \geq \left\lceil \frac{|E|}{6} - \frac{|V|}{2} + 1 \right\rceil \tag{8.7.9}$$

Proof: Since G is simple, it follows that $girth(G) \geq 3$. Substitution of this inequality into the inequality of Theorem 8.7.5 yields the conclusion.   ◇

Theorem 8.7.7. *The complete graph K_n has the lower bound*

$$\gamma_{\min}(K_n) \geq \left\lceil \frac{(n-3)(n-4)}{12} \right\rceil \tag{8.7.10}$$

for its minimum genus.

Proof: The substitutions into inequality (8.7.9) of

$$|V| = n \quad \text{and} \quad |E| = \frac{n^2 - n}{2}$$

quickly yield the inequality (8.7.10).   ◇

The challenge of the Heawood problem was to construct an imbedding for each complete graph in the surface whose genus realized the lower bound of inequality (8.7.10). The direct method of drawing graphs on flat polygons works well for the torus, but on higher genus surfaces, it is very difficult to check the connections of many edges that pass through a side of the polygon to the matched side.

L. Heffter (1891) introduced the idea of specifying an imbedding of a graph by listing the boundary-walks of its faces. For instance, consider the imbedding $K_7 \to S_1$ depicted in Figure 8.7.7. In Ringel's early work, he saw how algebra could be used to condense the specification of the boundary walks.

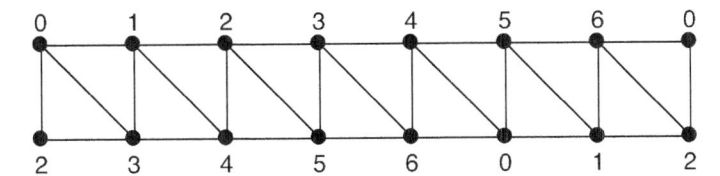

Figure 8.7.7   A symmetrical drawing of K_7 on the torus.

The boundary-walks of the faces (clockwise around each face) are as follows.

013	032
124	143
235	254
346	365
450	406
561	510
602	621

Remark 1:   In using a list of faces to specify a graph imbedding, one important requirement is that there are exactly two occurrences of each edge over all the face-boundary walks. For instance the edge 34 occurs on the edges 346 in the first column and 143 in the second column. Another is that the faces incident at each vertex form a single cycle around that vertex. For instance, Figure 8.7.8 shows how the faces incident at the vertex 0 form such a cycle.

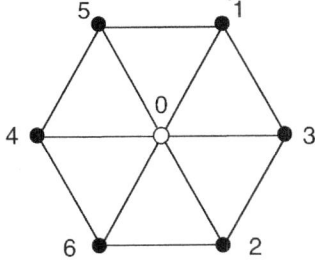

Figure 8.7.8   The cycle of faces incident on vertex 0.

Remark 2:   We observe that the other faces in each of the two columns are generated by developing the face in the top row modulo 7. Ringel used the development of a list of faces from a single principal face as the key to constructing several infinite classes of imbeddings of complete graphs.

An enormous breakthrough occurred in 1963 when W. Gustin introduced a combinatorial device called a *current graph* to generate the boundary of a principal face of an imbedding of a complete graph with all the faces 3-sided. Ringel and Youngs added many different possible augmentations to current graphs in their successful quest to construct minimum imbeddings of all the complete graphs.

A more general device for constructing graph imbeddings, called a *voltage graph*, was subsequently invented by J. L. Gross (see [Gros1974]) and extended by Gross and T. W. Tucker ([GrTu1977]). Voltage graphs are combinatorial specifications of coverings and branched coverings, which had been previously understood as topological abstractions of Riemann surfaces. They provide a capability to construct graph imbeddings for any kind of graph with adequately many fixed-point-free symmetries, i.e., far beyond complete graphs, and they have been used extensively toward that objective and various others. They unified (see [GrTu1974]) the many (about 15) forms of current graphs that had been used on the Heawood problem.

Remark: The monograph [GrTu1987] surveys topological graph theory up to about 1985. The edited volume [GrTu2008] describes the programmatic themes that have developed subsequently, including the distribution of imbeddings, algorithms and obstructions, graph minors, enumerating coverings, symmetrical maps, and connections between imbeddings and geometries and designs.

EXERCISES for Section 8.7

In Exercises 8.7.1 through 8.7.9, draw an imbedding of the given graph on a flat polygon representation of the torus.

8.7.1	$K_2 + C_6$.	8.7.2	$C_3 \times C_3$.	8.7.3	$circ(8 : 1, 2, 3)$.
8.7.4	$circ(8; 1, 2, 4)$.	8.7.5	$circ(9 : 1, 2, 3)$.	8.7.6	$C_3 + C_5$.
8.7.7	ML_5.	8.7.8	$K_{4,4}$.	8.7.9	Q_4.

In Exercises 8.7.10 through 8.7.15, prove that the given graph has no imbedding on the torus.

8.7.10	$K_2 + W_6$.	8.7.11	$K_{4,5}$.	8.7.12	Q_5.
8.7.13	$Q_4 + K_2$.	8.7.14	$C_3 + ML_5$.	8.7.15	$circ(10 : 1, 2, 3, 4)$.

In Exercises 8.7.16 through 8.7.21, use the girth to calculate a lower bound with inequality (8.7.8) for the minimum genus of the given graph.

8.7.16	Q_n.	8.7.17	$K_{m,n}$.	8.7.18	$K_1 + Q_n$.
8.7.19	$K_2 + Q_n$.	8.7.20	$C_3 + ML_n$.	8.7.21	$circ(n : 1, 2, 3, 4)$.

GLOSSARY

analytic graph theory: usually, the analysis of the class of all simple graphs for extremal or probabilistic phenomena, including Ramsey phenomena.

augmenting the flow: increasing the flow along a chain of arcs from source to target.

boundary walk of a face: a closed walk that traverses the entire perimeter of the face.

capacitated network: a digraph in which each arc is assigned a non-negative number called its capacity.

capacity constraint on a flow: the requirement that the flow in each arc not exceed the *capacity* of the arc.

capacity of a cut: the sum of the capacities of the arcs from the source side of the cut to the target side.

cellular imbedding of a graph in a surface: an imbedding such that every region is topologically equivalent to an open disk.

chromatic number of a graph: the minimum number of colors required for a proper coloring.

chromatically critical graph: a graph such that the removal of any edge would reduce the chromatic number.

clique in a graph: a maximal set of mutually adjacent vertices.

clique number: the cardinality of a largest clique.

coloring: an assignment to each vertex of a member of a set called *colors*.

complete set of obstructions to a graph property: a set of graphs lacking that property, of which at least one member is a subgraph of any graph that lacks the property.

connected digraph: a digraph whose underlying graph is connected.

conservation constraint: the requirement of a network flow that at every vertex except the source and the target, the sum of the flows on the in-arcs equals the sum of the flows on the out-arcs.

current graph: a device invented by W. Gustin, used in constructing the boundary walks of an imbedded complete graph in the solution of the Heawood problem.

dodecahedral graph: the 1-skeleton of the dodecahedron.

dual of a map: a map obtained by inserting a dual vertex in each region and drawing through each edge a dual edge joining the vertices whose regions are adjacent through that edge.

Euler polyhedral equation for an imbedding of a graph on a surface: for the sphere, the equation $|V| - |E| + |F| = 2$; for the surface S_g, the equation $|V| - |E| + |F| = 2 - 2g$.

eulerian graph: a graph with an eulerian tour.

eulerian tour in a graph: a closed walk that traverses every edge exactly once.

eulerian trail in a graph: a walk that traverses every edge exactly once.

extremal function $ex(n, \mathcal{P})$: the number of edges of an extremal graph for that property.

extremal graph for a property $\mathcal{P}$: a graph with the maximum number of edges, relative to its number of vertices, for any graph that lacks the property $\mathcal{P}$.

extremal graph theory: the study of extremal graphs and extremal functions.

face of an imbedding: the union of a region and the edges in its boundary walk.

feasible flow in a capacitated network: a flow that satisfies the capacity constraint and the conservation constraint.

flat-polygon representation of a surface S: representation by a polygon whose sides are paired in such a way that pasting the paired sides together yields a surface that is topologically equivalent to S.

flow in a network: an assignment of non-negative numbers to its arcs.

Ford-Fulkerson algorithm: an algorithm for maximizing the flow in a capacitated network.

Four-Color Map Problem: a famous problem from the 19th century, asking whether every map on the sphere requires at most four colors for a proper coloring; solved by K. Appel and W. Haken in 1976.

hamiltonian circuit: a cycle that is incident on every vertex.

hamiltonian graph: a graph that has a hamiltonian circuit.

Heawood Map-Coloring Problem: the problem of finding the chromatic number of all the surfaces except the sphere (and plane); solved in 1968 by G. Ringel and J. W. T. Youngs.

imbedding of a graph in a surface: a drawing without any edge-crossings.

independence number of a graph: the maximum number of mutually non-adjacent vertices.

independent set of vertices: a set on mutually non-adjacent vertices.

Jordan separation property: a property of the sphere and plane that every closed curve separates it.

Königsberg: a city in what was once East Prussia, where the *seven bridges problem* originated.

king in a tournament: a vertex from which every other vertex is reachable by a direced path of length at most 2.

Kuratowski graphs: the graphs K_5 and $K_{3,3}$, which K. Kuratowski proved in 1930 are a complete set of obstructions to planarity of a graph.

Kuratowski subgraph of a graph: a subgraph isomorphic to a subdivision either of K_5 or of $K_{3,3}$.

map: a crossing-free drawing of a graph on a surface.

map-coloring: an assignment to each region of a map a member of a set called *colors*.

matching in a graph: a set of edges in which no two edges share an endpoint.

maximum flow: a flow on a capacitated network such that the net outflow from the source is the maximum possible.

maximum matching: a matching that has the maximum possible number of edges.

minimum cut: a cut whose capacity is the minimum among all cuts that separate the source from the target.

minimum genus of a graph: the minimum number of handles needed on an orientable surface such that the graph can be drawn without edge-crossings.

mutually reachable vertices: two vertices in a digraph such that each is reachable from the other (by a directed path).

net flow through an $s - t$ **cut**:   the difference between the sum of the flows on arcs that cross from the s side to the t and the sum of the flows on the arcs that cross back.

non-orientable surface: a surface that contains a Möbius band.

open path addition to a graph: adding a path joining two different vertices.

orientable surface: a surface that does not contain a Möbius band; any surface in the infinite sequence $S_0, \ S_1, \ S_2, \cdots$.

orientation on a graph: an assignment of directions to its edges.

path addition: see *Whitney-Robbins synthesis*.

planar graph: a graph that can be drawn in the plane with no crossings.

Poincaré duality construction: see *dual of a map*.

postman problem: the problem of finding a closed walk in a weighted graph that traverses every edge at least once, such that the sum of the weights encountered in traversal is a minimum; invented by M-K Guan.

primal map: a name given to the existing map on a surface to distinguish it from the dual map that is constructed from it.

proper coloring of a graph: a coloring in which no two adjacent vertices are assigned the same color.

proper map-coloring: a map-coloring in which no two adjacent regions are assigned the same color.

Ramsey graph theory: the pursuit of Ramsey numbers and various generalizations.

Ramsey number $r(s,t)$:   the smallest number r such that if G is any simple graph with r vertices, then either G contains K_s or else $\overline{G}$ contains K_t.

region of a drawing: a topological component of the complement of the image of a graph on a surface.

round-robin tournament: a contest in which every player plays every other player once; modeled by a type of digraph called a *tournament*.

score sequence of a tournament: the sequence of out-degrees.

seven bridges problem: a problem whose solution by Euler is acclaimed as the origin of graph theory.

simple digraph: a digraph with at most one arc from u to v, for any pair of vertices u and v.

size of a face of a graph drawing: the number of edge-steps in a traversal of its boundary walk.

strongly cellular region: a region whose boundary walk is a cycle.

strongly connected digraph: a digraph in which in every pair of vertices, both vertices are reachable from the other.

subdividing a graph: inserting an additional vertex in the interior of an edge, or iterating this operation.

topological graph theory: the use of topological methods in studying graphs, especially of the placement of graphs on surfaces.

tournament: a digraph whose underlying graph is a complete graph.

trail: a walk with no repeated edges.

transitive tournament: a tournament that is transitive as a digraph.

traveling salesman problem in a weighted graph: finding the hamiltonian circuit with the least total weight.

underlying graph of a digraph: the result of eliminating all the edge directions.

value of a flow: the net flow out of the source vertex.

vertex k-coloring of a graph: a coloring with k colors.

voltage graph: a device invented by J. L. Gross and augmented in collaboration with T. W. Tucker, used to realize graph imbeddings by covering-space constructions, to count covering spaces of a graph, and to give an algebraic specification of a graph; circulant graphs and Cayley graphs are special cases; current graphs are duals to imbedded voltage graphs.

Whitney-Robbins synthesis of a 2-edge-connected graph G: a sequence of graphs, starting from a cycle graph and concluding with the graph G, in which each graph is obtained by adding an open or closed path to the preceding graph.

Chapter 9

Graph Enumeration

9.1 Burnside-Pólya Counting

9.2 Burnside's Lemma

9.3 Counting Small Simple Graphs

9.4 Partitions of Integers

9.5 Calculating a Cycle Index

9.6 General Graphs and Digraphs

When Pólya showed how the algebraic theorem called Burnside's Lemma could be augmented to count graph isomorphism types, he had in mind an application in physical chemistry, that of counting the number of distinct isomers with a given chemical formula. A couple of decades after Pólya, Harary launched an extensive program of developing additional augmentations, to permit wide-ranging application of the fundamental Burnside-Pólya method to the enumeration of graphs and other combinatorial objects, including, for instance, finite automata. The present chapter presents the basics of the method and should how it can be used to count the isomorphism types of various graph objects, including simple graphs, general graphs, and digraphs.

9.1   BURNSIDE-PÓLYA COUNTING

This section presents a way to algebraize a problem of counting equivalence classes under geometric or other symmetries, by representing the symmetries as a collection of permutations. Proof that the technique works is deferred to §9.2.

REVIEW FROM §1.6:

- A **permutation of a set** S is a one-to-one, onto function from S to itself.

- **Corollary 1.6.4.** Every permutation of a finite set S can be represented as the composition of disjoint cycles of elements of S.

FROM APPENDIX A2:

- An algebraic system $\langle U, \star \rangle$ is called a **group** if it has the following properties:

  the operation $\star$ is associative.

  there is an identity element.

  every element of U has an inverse.

Example 9.1.1: Suppose that each square of a 2×2-checkerboard is to be colored black or white. Then, since the number of squares is 4 and the number of colors is at most 2, the number of possible colorings is $2^4 = 16$. These 16 colorings are partitioned into 6 cells in Figure 9.1.1.

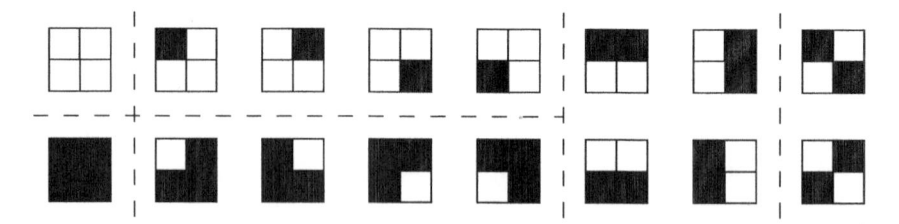

Figure 9.1.1   The sixteen 2×2-checkerboards.

Broken lines in the figure partition the set of 16 checkerboards into six cells. Each such cell represents an *equivalence class* under a *geometric symmetry*. That is, any checkerboard within each cell could be obtained from any other checkerboard in the cell by one of the clockwise rotations in the set

$$G = \{\, 0°, \quad 90°, \quad 180°, \quad 270° \,\}$$

Also, there are no two boards in different cells that are related by one of the clockwise rotations. Thus, up to rotational symmetry, the number of colorings is 6.

In this example, the composition of two rotations in G is representable as their addition modulo $360°$ and is also in G. The rotation of $0°$ serves as the identity. Moreover, the inverse of each rotation is in G. Thus, the set of four rotations under composition forms a *group*. Having a group of symmetries is a highly significant feature in counting problems.

Counting the equivalence classes of colorings of Example 9.1.1 is easy enough, for 2 colors on a 2×2 board, simply by drawing all cases and organizing them into equivalence classes, as in Figure 9.1.1. However, what if there were 5 possible colors and the board was 4×4, as in Figure 9.1.2?

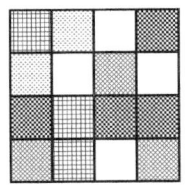

Figure 9.1.2   A 5-colored 4×4-checkerboard.

Since there are 16^5 such 5-colorings, and since the maximum size of an equivalence class is 4, the number of equivalence classes would be at least

$$\frac{5^{16}}{4}$$

There is a remarkable counting method, called *Burnside-Pólya counting*, that is used for such enumeration problems, to obtain the exact number of classes. Burnside-Pólya counting is based on a group-theoretic principle widely known as *Burnside's Lemma* and its enhancement by Pólya [Póly1937]. This method reduces such counting problems to evaluating a polynomial called the *cycle index*. Its capacity for widespread application was developed by Harary (see [HaPa1973]) in the 1950s and then further developed by his many students and others.

Permutations on Discrete Sets

The set-up for Burnside-Pólya counting under a group of symmetries is to represent the objects of the symmetries as a discrete set and to represent each symmetry as a permutation.

DEFINITION: A closed non-empty collection P of permutations on a set Y of objects that forms a group under the operation of composition is called a **permutation group**. The combined structure may be denoted $\mathcal{P} = [P : Y]$. It is often denoted P when the set Y of objects is understood from context.

Remark: Permutation groups can be non-commutative, and in practice, the phrase *permutation group* usually refers to a non-commutative group, even though some of our early examples are commutative.

Example 9.1.1, continued: The objects in the problem of counting checkerboards are the individual squares of the checkerboard. Each of the four squares is assigned a number, starting with the number 1 in the upper right, and proceeding clockwise in assigning 2, 3, and 4, as shown in Figure 9.1.3.

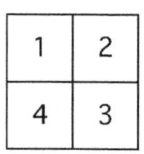

Figure 9.1.3   **Numbering the squares of a checkerboard.**

The four rotational symmetries of the group G that acts on the 2×2 checkerboard can be represented as permutations of the numbers assigned to its squares, as shown in the second column of Table 9.1.1.

Table 9.1.1   **Algebraizing the checkerboard symmetries.**

rotation	permutation
$0°$	$(1)(2)(3)(4)$
$90°$	$(1 \quad 2 \quad 3 \quad 4)$
$180° = 2 \times 90°$	$(1 \quad 2 \quad 3 \quad 4)^2 = (1 \quad 3)(2 \quad 4)$
$270° = 3 \times 90°$	$(1 \quad 2 \quad 3 \quad 4)^3 = (1 \quad 4 \quad 3 \quad 2)$

It is easily verified that this collection of permutations satisfies the group axioms. We may observe that it is commutative.

Cyclic Permutations

DEFINITION: A permutation on a set Y whose representation in disjoint cycle form has only one cycle containing more than one element of Y is called a **cyclic permutation**.

- The number of elements in that one cycle is called the **length of that cycle**. Also, a cycle of length k is called a k-cycle.

Proposition 9.1.1. *Let $n \in \mathbb{Z}^+$, and let*

$$\alpha = (1 \quad 2 \quad \cdots \quad n)$$

Then for $j = 1, \ldots, n - 1$ and for $r = 1, \ldots, n$, we have

$$\alpha^j(r) = \begin{cases} r + j & \text{if } r + j \leq n \\ r + j \bmod n & \text{otherwise} \end{cases}$$

Proof: This is provable by a straightforward induction on the power j. It is clearly true for $j = 1$. The inductive hypothesis is that

$$\alpha^{j-1}(r) = \begin{cases} r + j - 1 & \text{if } r + j - 1 \leq n \\ r + j - 1 \bmod n & \text{otherwise} \end{cases}$$

The inductive step is that

$$\alpha^j(r) = \alpha^1\alpha^{j-1}(r) = \begin{cases} \alpha^1(r+j-1) & \text{if } r+j-1 \le n \\ \alpha^1(r+j-1 \bmod n) & \text{otherwise} \end{cases}$$

$$= \begin{cases} r+j & \text{if } r+j \le n \\ r+j \bmod n & \text{otherwise} \end{cases} \qquad \diamond$$

Corollary 9.1.2. *Let p be a prime number, and let the permutation α be the p-cycle $(1 \quad 2 \quad \cdots \quad p)$. Then for $j = 1, \ldots, p-1$, the permutation*

$$\alpha^j = (1 \quad 2 \quad \cdots \quad p)^j$$

in the group $\mathbb{Z}_p$ is a p-cycle.

Proof: It is sufficient to show that each two elements of the sequence

$$1 \quad \alpha^j(1) \quad \alpha^{2j}(1) \quad \cdots \quad \alpha^{(p-1)j}(1) \tag{9.1.1}$$

are distinct. Suppose, to the contrary, that

$$\alpha^{uj}(1) = \alpha^{vj}(1)$$

for some pair u and v such that $1 \le u, v \le p-1$. Then, by Proposition 9.1.1, we have

$$uj + 1 \equiv vj + 1 \bmod p$$

It follows that

$$p \setminus uj - vj$$

and in turn, since p is prime, that

$$p \setminus j \quad \text{or} \quad p \setminus (u-v)$$

Since $1 \le j \le p-1$, it follows that

$$p \setminus (u-v)$$

Since $0 \le u, v \le p-1$, it now follows that

$$u = v$$

Therefore, by the pigeonhole principle, the sequence (9.1.1) contains all of the numbers $1, \ldots, p$. Accordingly, we have

$$\alpha^j = (1 \quad 1+j \quad 1+2j \quad \cdots \quad 1+(p-1)j) \tag{9.1.2}$$

That is, the permutation α^j is a p-cycle. $\qquad \diamond$

Example 9.1.2: Let α be the 5-cycle $(1 \quad 2 \quad 3 \quad 4 \quad 5)$. Then we have

$$\alpha^3 = (1 \quad 4 \quad 2 \quad 5 \quad 3) \quad \text{and} \quad \alpha^4 = (1 \quad 5 \quad 4 \quad 3 \quad 2)$$

Example 9.1.3: However, if α is the n-cycle $(1 \quad 2 \quad \cdots \quad n)$ and the number n is not prime, then some of the permutations α^j are not cyclic. For instance, for $\alpha = (1 \quad 2 \quad 3 \quad 4)$, we have

$$\alpha^2 = (1 \quad 3)(2 \quad 4)$$

as previously observed in Table 9.1.1.

Cyclic Permutation Groups

Table 9.1.2 generalizes the group of Table 9.1.1 to a permutation group on the set $\{1, 2, \ldots, n\}$. Geometrically, visualize the numbers as equally spaced in cyclic order $1, 2, \ldots, n$ around the unit circle in the xy-plane.

Table 9.1.2 The cyclic permutation group $\mathbb{Z}_n$.

rotation	permutation
0	$(\,1\,)(\,2\,)\ \cdots\ (\,n\,)$
$\frac{2\pi}{n}$	$(\,1\quad 2\quad \cdots\quad n\,)$
$2 \cdot \frac{2\pi}{n}$	$(\,1\quad 2\quad \cdots\quad n\,)^2$
$\cdots$	$\cdots$
$(n-1) \cdot \frac{2\pi}{n}$	$(\,1\quad 2\quad \cdots\quad n\,)^{n-1}$

DEFINITION: The group of permutations in Table 9.1.2 is called a **cyclic permutation group** on the set $\{1, 2, \ldots, n\}$. It can be denoted $\left[\mathbb{Z}_n : [1 : n] \right]$, but is more usually denoted, simply, $\mathbb{Z}_n$.

Proposition 9.1.3. *The permutation group $\mathbb{Z}_n$ is commutative, for $n \in \mathbb{Z}^+$.*

Proof: Let $\alpha = (\,1\quad 2\quad \cdots\quad n\,)$. Then, as indicated by Table 9.1.2, any two permutations in $\mathbb{Z}_n$ could be represented in the forms α^r and α^s. Then

$$\alpha^r \alpha^s \ =\ \alpha^{r+s} \ =\ \alpha^{s+r} \ =\ \alpha^s \alpha^r \qquad\qquad \diamondsuit$$

Corollary 9.1.2 establishes that if n is prime, then every permutation in the cyclic permutation group Z_n is cyclic. However, when $n = 4$, which is not prime, as first noted in Table 9.1.1, we have

$$(\,1\quad 2\quad 3\quad 4\,)^2 \ =\ (\,1\quad 3\,)(\,2\quad 4\,)$$

That is, the permutations in a cyclic permutation group need not all be cyclic permutations. The next proposition sharpens this observation.

Proposition 9.1.4. *Let $\alpha = (\,1\quad 2\quad \cdots\quad n\,)$ be an n-cycle in $\mathbb{Z}_n$. Then for $j = 1, \ldots, n-1$, the permutation α^j has $\gcd(j, n)$ cycles, each of length*

$$\frac{n}{\gcd(j, n)}$$

Proof: For an arbitrary object $k \in [1 : n]$, we observe that all the objects in the cycle containing k must lie in the sequence

$$k \quad \alpha^j(k) \quad \alpha^{2j}(k) \quad \cdots \quad \alpha^{(n-1)j/\gcd(j,n)}(k) \qquad\qquad (9.1.3)$$

This is because the next element in that sequence would be

$$\alpha^{nj/\gcd(j,n)}(k) \ = \ (\alpha^n)^{j/\gcd(j,n)}(k) \ = \ k$$

which holds because α^n is the identity permutation. Thus, the maximum length of a cycle of the permutation α^j is

$$\frac{n}{\gcd(j,n)}$$

We next assert that the elements of the sequence (9.1.3) are mutually distinct. To see this, suppose that

$$1 \le u \le v \le \frac{n-1}{\gcd(j,n)} \tag{9.1.4}$$

and that $\alpha^{uj}(k) = \alpha^{vj}(k)$. Then, by Proposition 9.1.1, we have

$$uj + k \ \equiv \ vj + k \bmod n$$

It follows that

$$n \setminus vj - uj$$

and, in turn, that $\dfrac{n}{\gcd(j,n)}$ divides $v - u$. By (9.1.4), it now follows that $u = v$. Thus, the minimum length of a cycle in α^j is

$$\frac{n}{\gcd(j,n)}$$

It follows that every cycle of the permutation α^j is of that length. $\diamond$

Corollary 9.1.5. *Let* $\alpha \ = \ (\,1 \quad 2 \quad \cdots \quad n\,)$ *be a permutation in* $\mathbb{Z}_n$. *Then for* $j = 1, \ldots, n-1$, *the permutation* α^j *is cyclic if and only if* $j \perp n$. $\diamond$

Example 9.1.4:   Let $\alpha = (\,1 \quad 2 \quad \cdots \quad 12\,)$ in the cyclic permutation group $\mathbb{Z}_{12}$. Since

$$\gcd(8,12) = 4 \quad \text{and} \quad \frac{12}{4} = 3$$

Proposition 9.1.4 implies that the permutation α^8 has four 3-cycles. In fact,

$$\alpha^8 \ = \ (\,1 \quad 9 \quad 5\,)(\,2 \quad 10 \quad 6\,)(\,3 \quad 11 \quad 7\,)(\,4 \quad 12 \quad 8\,)$$

Example 9.1.5:   Let $\alpha = (\,1 \quad 2 \quad \cdots \quad 10\,)$ in $\mathbb{Z}_{10}$. Since

$$\gcd(3,10) = 1 \quad \text{and} \quad \frac{10}{1} = 10$$

Proposition 9.1.4 implies that α^3 has one 10-cycle. In fact,

$$\alpha^3 \ = \ (\,1 \quad 4 \quad 7 \quad 10 \quad 3 \quad 6 \quad 9 \quad 2 \quad 5 \quad 8\,)$$

Example 9.1.6:   If the number of colors for the 2×2-board is increased to 3, then the number of boards is

$$\frac{1}{4}\left(3^4 + 3^2 + 2 \cdot 3\right) \;=\; \frac{1}{4}(81 + 9 + 6) \;=\; 24$$

If only one color is actually used, then there are 3 choices of a color. If exactly two colors are used, there are $\binom{3}{2} = 3$ choices of two colors and then 4 patterns (from Example 9.1.5) possible for each such choice, for a subtotal of $3 \cdot 4 = 12$. If all three colors are used, there are 3 choices of the color that is used on two squares and three possible patterns with those colors, up to rotation, as shown in Figure 9.1.4, for a subtotal of 9. The sum of the three subtotals 3, 12, and 9 is 24, thereby confirming the Burnside-Pólya calculation.

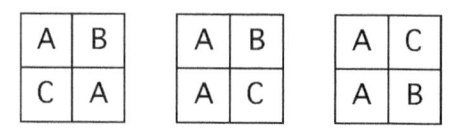

Figure 9.1.4   **The three 3-color patterns, up to rotation.**

Reflections

In addition to the four rotations on a checkerboard, there are also four reflections. They are not a closed collection of permutations, since the composition of any two of them yields a rotation. However, if they are included with the four rotations, there would be a permutation group of cardinality 8. For 2-coloring the 2×2-board of Figure 9.1.3,

1	2
4	3

there would be the following additional permutations and cycle structures in the cycle index.

Table 9.1.5   **Cycle structures of the four reflections.**

reflection	permutation	cycle structure
thru $x - $ axis	$(1 \quad 4)(2 \quad 3)$	t_2^2
thru $y - $ axis	$(1 \quad 2)(3 \quad 4)$	t_2^2
NE diagonal	$(2)(4)(1 \quad 3)$	$t_1^2 t_2$
SE diagonal	$(1)(3)(2 \quad 4)$	$t_1^2 t_2$

To generalize from Table 9.1.5 to reflections on a larger set of objects, we model the set of numbers $1, 2, \ldots, n$ as points evenly spaced around the unit circle. There are n possible reflections of the plane through a line through the origin that map this set of points bijectively onto itself. Each corresponds to a permutation of the set of numbers, as illustrated in Figure 9.1.5. Each cycle in each such permutation is either a 1-cycle or a 2-cycle.

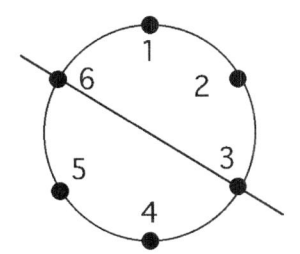

Figure 9.1.5 **The reflection** $(1\ 5)(2\ 4)(3)(6)$.

It is natural to regard the cyclic permutation

$$\alpha \;=\; \begin{pmatrix} 1 & 2 & \cdots & n \end{pmatrix}$$

as the *principal rotation* on the unit circle model of $[1:n]$. Suppose that we similarly regard the permutation

$$\beta(i) \;=\; n+1-i$$

as the *principal reflection* on $[1:n]$. We observe that β corresponds to a reflection of the unit circle through a line through the origin that bisects the arc between the points numbered 1 and n. If n is odd, this line also passes thru the point numbered $\frac{n+1}{2}$. If n is even, it bisects the arc between the points numbered $\frac{n}{2}$ and $\frac{n}{2}+1$. In the immediate context, we call this line the β-bisector.

Remark: We observe that the composition permutation

$$\alpha^j \beta(i) \;=\; \alpha^j(n+1-i) \;=\; \begin{cases} 1+j-i & \text{if } j \geq i \\ n+1+j-i & \text{otherwise} \end{cases}$$

represents reflection through the circle bisector obtained by rotating the β-bisector $\frac{j\pi}{n}$ radians clockwise.

Example 9.1.7: The reflection $(1\ 5)(2\ 4)(3)(6)$ depicted in Figure 9.1.5 is $\alpha^5\beta$.

Dihedral Permutation Groups

The composition of any two reflections on the unit circle model of

$$\{1, 2, \ldots, n\}$$

is a rotation, and the composition of a rotation and a reflection is a reflection. Accordingly, the union of the sets of rotations and reflections on $\{1, 2, \ldots, n\}$ is closed under composition. Thus, it forms a permutation group.

Example 9.1.8:   This composition of reflections

$$(1 \quad 5)(2 \quad 4)(3)(6) \circ (1 \quad 4)(2 \quad 3)(5 \quad 6)$$

is the rotation

$$(1 \quad 6 \quad 5 \quad 4 \quad 3 \quad 2)$$

DEFINITION: The permutation group

$$\{\alpha^j \mid j \in [1:n]\} \;\cup\; \{\alpha^j \beta \mid j \in [1:n]\}$$

is called the **dihedral group** on the cyclic set $\{1, 2, \ldots, n\}$. It is denoted $\mathbb{D}_n$.

Example 9.1.9:   Thus, the dihedral group $\mathbb{D}_4$ is given in Table 9.1.6, which is formed from the union of the rotations of Table 9.1.3 and the reflections of Table 9.1.5.

Table 9.1.6   The dihedral group $\mathbb{D}_4$.

symmetry	permutation	cycle structure
$0°$	$(1)(2)(3)(4)$	t_1^4
$90°$	$(1 \quad 2 \quad 3 \quad 4)$	t_4
$180°$	$(1 \quad 3)(2 \quad 4)$	t_2^2
$270°$	$(1 \quad 4 \quad 3 \quad 2)$	t_4
$x - \text{axis}$	$(1 \quad 4)(2 \quad 3)$	t_2^2
$y - \text{axis}$	$(1 \quad 2)(3 \quad 4)$	t_2^2
SE diag	$(1)(3)(2 \quad 4)$	$t_1^2 t_2$
NE diag	$(2)(4)(1 \quad 3)$	$t_1^2 t_2$

cycle index $\mathcal{Z}_{\mathbb{D}_4}$: $\frac{1}{8}\left(t_1^4 + 2t_1 t_2 + 3t_2^2 + 2t_4\right)$

Using the cycle index of $\mathbb{D}_4$ on its bottom line, one can calculate that the corresponding number of colorings would be

$$
\begin{aligned}
\mathcal{Z}_{\mathbb{D}_4}(2, \ldots, 2) &= \frac{1}{8}\left(2^4 + 2 \cdot 2^2 \cdot 2 + 3 \cdot 2^2 + 2 \cdot 2\right) \\
&= \frac{1}{8}\left(16 + 16 + 12 + 4\right) \\
&= \frac{48}{8} = 6
\end{aligned}
$$

This corresponds to the observation that the partition of 2-colorings in Figure 9.1.1 corresponds to dihedral symmetry as well as to pure rotational symmetry. By way of contrast, the middle and rightmost 3-coloring patterns of Figure 9.1.4 are related by reflection. Thus, one expects, as now calculated, that there would be fewer dihedral classes than the 24 classes of colorings (under rotation) that were counted in Example 9.1.6.

$$\mathcal{Z}_{\mathbb{D}_4}(3,\ldots,3) \;=\; \frac{1}{8}\left(3^4 + 2\cdot 3^2\cdot 3 + 3\cdot 3^2 + 2\cdot 3\right)$$

$$=\; \frac{1}{8}\left(81 + 54 + 27 + 6\right)$$

$$=\; \frac{168}{8} \;=\; 21$$

Larger Sets of Permuted Objects

When the same two groups of four or eight geometric symmetries act on larger checkerboards, the cycle structures of the permutations change, even though the terminology is preserved.

Example 9.1.10: The 3×3-checkerboard of Figure 9.1.6

1	2	3
8	9	4
7	6	5

Figure 9.1.6 The 3×3-checkerboard.

has the following table of permutations and cycle structures for its dihedral symmetry group.

Table 9.1.7 Symmetries on the 3×3-checkerboard.

symmetry	permutation	cycle structure
$0°$	$(1)(2)\cdots(9)$	t_1^9
$90°$	$(9)(1\;\;3\;\;5\;\;7)(2\;\;4\;\;6\;\;8)$	$t_1 t_4^2$
$180°$	$(5)(1\;\;5)(2\;\;6)(3\;\;7)(4\;\;8)$	$t_1 t_2^4$
$270°$	$(9)(1\;\;7\;\;5\;\;3)(2\;\;8\;\;6\;\;4)$	$t_1 t_4^2$
$x-\text{axis}$	$(4)(8)(9)(1\;\;7)(2\;\;6)(3\;\;5)$	$t_1^3 t_2^3$
$y-\text{axis}$	$(2)(6)(9)(1\;\;3)(4\;\;8)(5\;\;7)$	$t_1^3 t_2^3$
SE diag	$(1)(5)(9)(2\;\;8)(3\;\;7)(4\;\;6)$	$t_1^3 t_2^3$
NE diag	$(3)(7)(9)(1\;\;5)(2\;\;4)(6\;\;8)$	$t_1^3 t_2^3$
cycle index $=$	$\frac{1}{8}\left(t_1^9 + t_1 t_2^4 + 2t_1 t_4^2 + 4t_1^3 t_2^3\right)$	

Thus, as the number of ways to color the 3×3 checkerboard with at most 2 colors, Burnside-Pólya counting gives

$$\frac{1}{8}\left(2^9 + 2\cdot 2^4 + 2\cdot 2\cdot 2^2 + 4\cdot 2^3\cdot 2^3\right) \;=\; 102$$

EXERCISES for Section 9.1

In each of the Exercises 9.1.1 through 9.1.8, for the cyclic permutation

$$\alpha = \begin{pmatrix} 1 & 2 & 3 & 4 & 5 & 6 & 7 \end{pmatrix}$$

calculate the indicated permutation α^k.

9.1.1	α^2	**9.1.2**$^{\mathbf{S}}$	α^3	**9.1.3**	α^4	**9.1.4**	α^5
9.1.5	α^6	**9.1.6**	α^7	**9.1.7**	α^8	**9.1.8**	α^9

In each of the Exercises 9.1.9 through 9.1.16, for the cyclic permutation

$$\alpha = \begin{pmatrix} 1 & 2 & 3 & 4 & 5 & 6 & 7 & 8 & 9 & 10 & 11 & 12 \end{pmatrix}$$

calculate the indicated permutation α^k.

9.1.9	α^2	**9.1.10**$^{\mathbf{S}}$	α^3	**9.1.11**	α^4	**9.1.12**	α^5
9.1.13	α^6	**9.1.14**	α^7	**9.1.15**	α^8	**9.1.16**	α^{11}

In each of the Exercises 9.1.17 through 9.1.24, calculate the cyclic index polynomial for the indicated cyclic group.

9.1.17	$\mathbb{Z}_5$	**9.1.18**$^{\mathbf{S}}$	$\mathbb{Z}_6$	**9.1.19**	$\mathbb{Z}_7$	**9.1.20**	$\mathbb{Z}_8$
9.1.21	$\mathbb{Z}_9$	**9.1.22**	$\mathbb{Z}_{10}$	**9.1.23**	$\mathbb{Z}_{11}$	**9.1.24**	$\mathbb{Z}_{12}$

In each of the Exercises 9.1.25 through 9.1.32, substitute the number 2 into the cyclic index polynomial for the indicated cyclic group.

9.1.25	$\mathbb{Z}_5$	**9.1.26**$^{\mathbf{S}}$	$\mathbb{Z}_6$	**9.1.27**	$\mathbb{Z}_7$	**9.1.28**	$\mathbb{Z}_8$
9.1.29	$\mathbb{Z}_9$	**9.1.30**	$\mathbb{Z}_{10}$	**9.1.31**	$\mathbb{Z}_{11}$	**9.1.32**	$\mathbb{Z}_{12}$

In each of the Exercises 9.1.33 through 9.1.40, do a Pólya substitution of $b+w$ into the cyclic index polynomial for the indicated cyclic group.

9.1.33	$\mathbb{Z}_5$	**9.1.34**$^{\mathbf{S}}$	$\mathbb{Z}_6$	**9.1.35**	$\mathbb{Z}_7$	**9.1.36**	$\mathbb{Z}_8$
9.1.37	$\mathbb{Z}_9$	**9.1.38**	$\mathbb{Z}_{10}$	**9.1.39**	$\mathbb{Z}_{11}$	**9.1.40**	$\mathbb{Z}_{12}$

In each of the Exercises 9.1.41 through 9.1.44, calculate the cyclic index polynomial for the indicated dihedral group.

9.1.41	$\mathbb{D}_5$	**9.1.42**$^{\mathbf{S}}$	$\mathbb{D}_6$	**9.1.43**	$\mathbb{D}_7$	**9.1.44**	$\mathbb{D}_8$

In each of the Exercises 9.1.45 through 9.1.48, substitute the number 2 into the cyclic index polynomial for the indicated dihedral group.

9.1.45	$\mathbb{D}_5$	**9.1.46**$^{\mathbf{S}}$	$\mathbb{D}_6$	**9.1.47**	$\mathbb{D}_7$	**9.1.48**	$\mathbb{D}_8$

In each of the Exercises 9.1.49 through 9.1.52, do a Pólya substitution of $b+w$ into the cyclic index polynomial for the indicated dihedral group.

9.1.49	$\mathbb{D}_5$	**9.1.50**$^{\mathbf{S}}$	$\mathbb{D}_6$	**9.1.51**	$\mathbb{D}_7$	**9.1.52**	$\mathbb{D}_8$

9.2 BURNSIDE'S LEMMA

The mathematical principle that underlies the calculations of §9.1 has commonly been called *Burnside's Lemma*, after its appearance in Burnside's influential monograph [Burn1911]. A footnote in the first edition of that monograph attributes the result to Frobenius [Frob1887], whose derivation appears to have been preceded by Cauchy [Cauc1847].

DEFINITION: Let $\mathcal{P} = [P : Y]$ be a permutation group, and let $y \in Y$. The **orbit** of the object y under the action of P is the set $\{\pi(y) \mid \pi \in P\}$.

With the benefit of this definition, we may say that Burnside's Lemma counts orbits.

Proposition 9.2.1. *Let $\mathcal{P} = [P : Y]$ be a permutation group. Then being co-orbital is an equivalence relation.*

Proof: The identity permutation maps each object to itself, so each object is in its own orbit. If $\pi(y) = y'$, then $\pi^{-1}(y') = y$, so the relation is symmetric. If $\pi(y) = y'$ and $\pi'(y') = y''$, then $(\pi' \circ \pi)(y) = y''$, so the relation is transitive. $\diamond$

Orbits

Given a permutation group $\mathcal{P} = [P : Y]$, the orbit of an object y can be computed from a listing of the disjoint cycle form for every permutation $\pi \in P$. Quite simply, the orbit of y is the set of objects that appear in the same cycle as y in any of the permutations π. More efficiently to compute, the orbit of y is also the set of objects that appear immediately after y in some permutation of the group.

Example 9.2.1: When the group of 4 rotations acts on the set of squares of the 2×2-checkerboard, then all 4 squares are in the same orbit.

1	2
4	3

rotation	permutation
$0°$	$(1)(2)(3)(4)$
$90°$	$(1\ \ 2\ \ 3\ \ 4)$
$180°$	$(1\ \ 3)(2\ \ 4)$
$270°$	$(1\ \ 4\ \ 3\ \ 2)$

However, when the group of 4 rotations acts on the set of 16 colorings of the board with at most two colors, there are 6 orbits, as indicated in Figure 9.1.1. In this particular example, they happen to be identifiable with the cells of the $90°$ rotation, as illustrated in Figure 9.2.1.

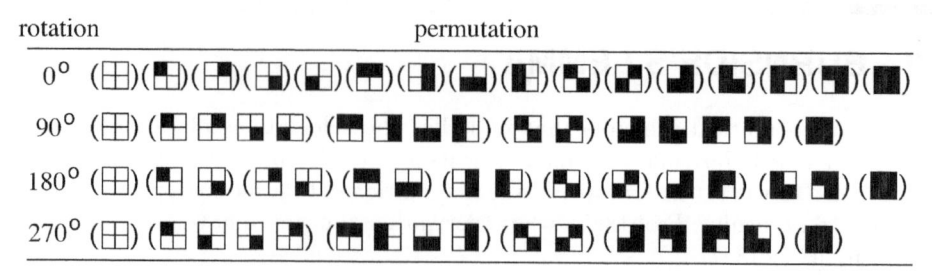

rotation   permutation

Figure 9.2.1   **Permutations of the colored 2×2 boards.**

Remark: Although the next lemma is expressed in terms of the orbits of permutation groups, it is really a fact about set partitions.

Lemma 9.2.2. *Let $\mathcal{P} = [P : Y]$ be a permutation group with n orbits. Then*

$$\sum_{y \in Y} \frac{1}{|orbit(y)|} = n \tag{9.2.1}$$

Proof: Suppose that $Y_1, \ldots, Y_n$ are the orbits. Since the orbits Y_j partition Y, it follows that

$$
\begin{aligned}
\sum_{y \in Y} \frac{1}{|orbit(y)|} &= \sum_{j=1}^{n} \sum_{y \in Y_j} \frac{1}{|orbit(y)|} \\
&= \sum_{j=1}^{n} \sum_{y \in Y_j} \frac{1}{|Y_j|} \\
&= \sum_{j=1}^{n} \frac{1}{|Y_j|} \sum_{y \in Y_j} 1 = \sum_{j=1}^{n} \frac{1}{|Y_j|} |Y_j| \\
&= \sum_{j=1}^{n} 1 = n \qquad \qquad \diamondsuit
\end{aligned}
$$

Example 9.2.2:   The left side of Equation (9.2.1) is the sum of the reciprocals of the sizes of the cells in a partition. The partition depicted at the left of Figure 9.2.2 has three cells. When the objects in the cells are converted to the reciprocals of the cell sizes, as on the right, then the sum of the fractions within each cell must equal 1. Thus, the sum of all the reciprocals must equal the number of cells, as on the right side of the equation in Lemma 9.2.2.

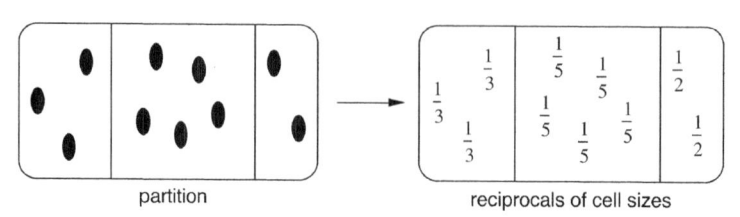

partition   reciprocals of cell sizes

Figure 9.2.2   **Reciprocals of cell sizes of a partition.**

Induced Actions on Colorings

Pólya's main contribution to the counting method was to discover a relationship between the cycle index of a permutation group action on a set Y and the number of orbits *induced on* the colorings of Y.

DEFINITION: A k-*coloring* of a set Y is a mapping f from Y *onto* the set

$$\{1, 2, \ldots, k\}$$

TERMINOLOGY: The value $f(y)$ is called the *color* of the object y. Often the names or initials of actual colors, such as black and white, are used in place of integer values.

DEFINITION: A $(\leq k)$-*coloring* of a set Y is a coloring that uses k or *fewer* colors, formally a mapping f from Y onto any set $\{1, 2, \ldots, t\}$ with $t \leq k$.

NOTATION: The set of all $(\leq k)$-colorings of the elements of a set Y is denoted $Col_k(Y)$.

Example 9.2.3:   Figure 9.2.3, a copy of Figure 9.1.1, represents all the (≤ 2)-colorings of $\{1, 2, 3, 4\}$ as 2×2-checkerboards.

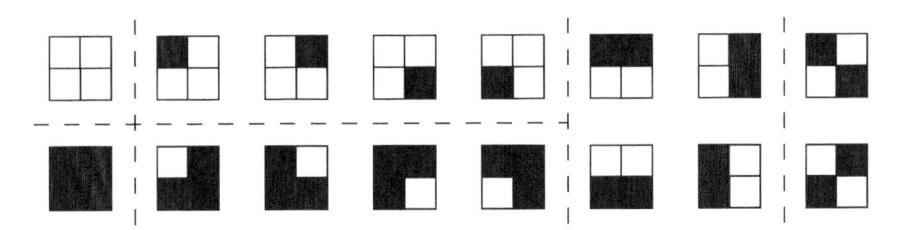

Figure 9.2.3   The 16 2×2-checkerboards.

Proposition 9.2.3. *Let Y be a set. Then $|Col_k(Y)| = k^{|Y|}$.*

Proof: This is a direct application of the Rule of Product (see §0.4).   ◇

Example 9.2.3, continued: There are 16 (≤ 2)-colorings of the set $\{1, 2, 3, 4\}$, as shown in Figure 9.2.3.

DEFINITION: Let $\mathcal{P} = [P : Y]$ be a permutation group acting on a set Y, and let f and g be $(\leq k)$-colorings of the objects in Y.

- Then the coloring f is $\mathcal{P}$-*equivalent* to the coloring g if there is a permutation $\pi \in P$ such that $g = f\pi$, that is, if for every object $y \in Y$, the color $g(y)$ is the same as the color $f(\pi(y))$.

- The mapping $f \mapsto f\pi$ is a permutation on the colorings, called the *induced action* of π on the set of colorings. The notation $[P : Col_k(Y)]$ distinguishes the permutation group of such induced actions from the group $[P : Y]$.

Example 9.2.3, continued: The 2-coloring

$$f: \quad 1 \mapsto b \quad 2 \mapsto w \quad 3 \mapsto w \quad 4 \mapsto w$$

is $\mathbb{Z}_4$-equivalent to the 2-coloring

$$g: \quad 1 \mapsto w \quad 2 \mapsto b \quad 3 \mapsto w \quad 4 \mapsto w$$

under the permutation

$$\pi: \quad (1 \quad 2 \quad 3 \quad 4)$$

This is the formal reason why the coloring with its one black square in the upper left corner of a 2×2-checkerboard, represented in this example as the mapping f, is equivalent to the coloring with its one black square at the upper right, represented by the mapping g, under the 90° rotation, represented by the permutation π.

Fixed Points

DEFINITION: Let $\mathcal{P} = [P : Y]$ be a permutation group, and let $\pi \in P$. A **fixed point** of π is an object $y \in Y$ such that $\pi(y) = y$.

NOTATION: $Fix(\pi)$ denotes the set of all fixed points of the permutation π.

Remark: The fixed point set $Fix(\pi)$ comprises the objects that lie in the 1-cycles of π.

Example 9.2.3, continued: The number of fixed points of each of the four rotations is different for different sets on which the rotation group acts.

rotation	# fixed pts in $\{1,2,3,4\}$	# fixed pts in $2 \times 2 -$ colorings
0°	4	16
90°	0	2
180°	0	4
270°	0	2
total # fixed pts	4	24

Stabilizers

DEFINITION: Let $\mathcal{P} = [P : Y]$ be a permutation group, and let $y \in Y$. The **stabilizer** of y is the subset

$$\mathcal{S}tab(y) \;=\; \{\pi \in P \mid \pi(y) = y\}$$

of the permutation group P.

Remark: Clearly, $\mathcal{S}tab(y)$ is a subgroup of P. It is non-empty, since it contains the identity permutation. It is closed under composition, because the composition of two permutations that both fix object y must also fix the object y.

Remark: $\mathcal{S}tab(y)$ can be constructed computationally as the set of all permutations in P in which the object y lies in a 1-cycle.

Example 9.2.4:   When the rotation group $\mathbb{Z}_4$ acts on the set $\{1, 2, 3, 4\}$, the stabilizer $\mathcal{S}tab(3)$ is the trivial subgroup

$$\{(1)(2)(3)(4)\}$$

Example 9.2.5:   When the dihedral group $\mathbb{D}_4$ acts on the set $\{1, 2, 3, 4\}$, the stabilizer $\mathcal{S}tab(3)$ is the subgroup

$$\{(1)(2)(3)(4),\ (1)(3)(2\ \ 4)\}$$

Example 9.2.6:   Consider the action of the rotation group $\mathbb{Z}_4$ on the set of 2×2-checkerboards. The stabilizer of the board with four white squares is the entire group.

Lemma 9.2.4. Let $\mathcal{P} = [P : Y]$ be a permutation group. Then

$$\sum_{y \in Y} |\mathcal{S}tab(y)| \ = \ \sum_{\pi \in P} |Fix(\pi)|$$

Proof: Consider a matrix M in which the rows are indexed by the objects of the set Y, and in which the columns are indexed by the permutations in P, such that

$$M[y, \pi] \ = \ \begin{cases} 1 & \text{if } \pi(y) = y \\ 0 & \text{otherwise} \end{cases}$$

Then the sum of row y is $|\mathcal{S}tab(y)|$ and the sum of column π is $|Fix(\pi)|$. This lemma simply asserts that the sum of the row sums of M equals the sum of its column sums, which is true of any matrix. $\diamond$

Lemma 9.2.5. Let $\mathcal{P} = [P : Y]$ be a permutation group and $y \in Y$. Then

$$|\mathcal{S}tab(y)| \ = \ \frac{|P|}{|orbit(y)|}$$

Proof: Suppose that

$$orbit(y) \ = \ \{y = y_1, y_2, \ldots, y_n\}$$

and that, for $j = 1, \ldots, n$, we let P_j be the subset of permutations of P that maps object y to object y_j. Then the subsets

$$P_1, \ P_2, \ \ldots, \ P_n$$

partition the permutation group P, and $P_1 = \mathcal{S}tab(y)$.

For $j = 1, \ldots, n$, let π_j be any permutation such that $\pi_j(y) = y_j$. Then the rule $\pi \mapsto \pi \circ \pi_j$ (composition with π_j) is a bijection from P_1 to P_j, which implies that

$$|P_j| = |P_1| = |Stab(y)|$$

for $j = 1, \ldots, n$.

Since each of the n partition cells $P_1, P_2, \ldots, P_n$ of group P has cardinality $|Stab(y)|$, it follows that

$$n \cdot |Stab(y)| = |P|$$

But $n = |orbit(y)|$, which completes the proof.   ◇

Proof of Burnside's Lemma

Theorem 9.2.6 [Burnside's Lemma].   **Let** $\mathcal{P} = [P : Y]$ **be a permutation group with** n **orbits.** *Then*

$$n = \frac{1}{|P|} \sum_{\pi \in P} |Fix(\pi)|$$

Proof: Lemmas 9.2.4, 9.2.5, and 9.2.2 establish the following chain of equalities, which proves Burnside's lemma.

$$\frac{1}{|P|} \sum_{\pi \in P} |Fix(\pi)| \;=\; \frac{1}{|P|} \sum_{y \in Y} |Stab(y)| \qquad \text{(Lemma 9.2.4)}$$

$$=\; \frac{1}{|P|} \sum_{y \in Y} \frac{|P|}{|orbit(y)|} \qquad \text{(Lemma 9.2.5)}$$

$$=\; \frac{1}{|P|} \, |P| \sum_{y \in Y} \frac{1}{|orbit(y)|}$$

$$=\; \sum_{y \in Y} \frac{1}{|orbit(y)|} \;=\; n \qquad \text{(Lemma 9.2.2)} \qquad ◇$$

Orbits of the Induced Action on Colorings

In several examples of the preceding section, the orbits of an induced action

$$[P : Col_k(Y)]$$

on ($\leq k$)-colorings were counted by substituting the number k of colors into the cycle index of the underlying action $[P : Y]$ on a set Y. The connection of this substitution into Burnside's Lemma — viz., that it counts the sum of the sizes of the fixed-point sets of the induced action — is now to be established.

Remark: The actions of $\mathbb{Z}_n$ and $\mathbb{D}_n$ on the set $\{1, \ldots, n\}$ have only one orbit, because they both have a permutation in which a single cycle contains every object. One does not need Burnside's Lemma to count this one orbit.

NOTATION: If $p(x_1, \ldots, x_n)$ is a multivariate polynomial, then $p(k, \ldots, k)$ denotes the result of substituting the value k for every variable x_j.

Lemma 9.2.7. *Let* $\mathcal{P} = [P : Y]$ *be a permutation group, and let* $\pi_Y \in P$, *with induced action* π_{CY} *on the coloring set* $Col_k(Y)$. *Then the number of* $(\leq k)$-*colorings of* Y *that are fixed by* π_{CY} *is given by*

$$|Fix(\pi_{CY})| = \zeta(\pi_Y)(k, \ldots, k)$$

Proof: A $(\leq k)$-coloring c is fixed by π_{CY} if and only if within each cycle of π_Y, all the objects are assigned the same color by c. Thus, there are k independent choices possible for each cycle of π_Y. Therefore,

$$|Fix(\pi_{CY})| = k^n$$

where n is the number of cycles in π_Y. But the number k^n is precisely the value of $\zeta(\pi_Y)(k, \ldots, k)$. $\diamond$

Theorem 9.2.8. *Let* $\mathcal{P} = [P : Y]$ *be a permutation group. Then the number of orbits of* $[P : Col_k(Y)]$ *is*

$$\mathcal{Z}_\mathcal{P}(k, \ldots, k)$$

Proof: Applying Burnside's lemma to the induced permutation group

$$[P : Col_k(Y)]$$

gives the number of orbits among the colorings as

$$\frac{1}{|P|} \sum_{\pi_{CY} \in P_C} |Fix(\pi_{CY})| = \frac{1}{|P|} \sum_{\pi_Y \in P_Y} \zeta(\pi_Y)(k, \ldots, k) \qquad \text{(Lemma 9.2.7)}$$

$$= \mathcal{Z}_\mathcal{P}(k, \ldots, k) \qquad\qquad\qquad \diamond$$

Pólya Inventory

As defined in §9.1, **Pólya substitution** means substituting the polynomial

$$c_1^j + \cdots + c_k^j$$

for the variable t_j $(j = 1, \ldots, n)$ into the cycle index of a permutation group $[P : Y]$. The resulting generating function in the indeterminates $c_1, \ldots, c_k$ is called a **Pólya inventory** for the color classes of the induced action $[P : Col_k(Y)]$.

Proposition 9.2.9. *For any permutation group* $\mathcal{P} = [P : Y]$, *every term in the Pólya inventory for* $[P : Col_k(Y)]$ *has* $|Y|$ *as its degree.*

Proof: Let $\pi \in P_Y$ and let $\zeta(\pi) = t_1^{r_1} \cdots t_n^{r_n}$ be its cycle structure. Then, clearly,

$$\sum_{j=1}^n j \cdot r_j = |Y|$$

because π permutes the set Y and the sum is the total number of objects of Y over all the cycles of π. Substituting the polynomial $c_1^j + \cdots + c_k^j$ for a factor $t_j^{r_j}$ $(j = 1, \ldots, n)$ in $\zeta(\pi)$ contributes $j \cdot r_j$ to the degree of each corresponding term of the generating function. $\diamond$

Corollary 9.2.10. *For any permutation group* $\mathcal{P} = [P : Y]$, *the Pólya inventory for* $[P : Col_k(Y)]$ *is a sum*

$$\sum_{s_1+\cdots+s_k=|Y|} p_{s_1,\cdots,s_k} c_1^{s_1} \cdots c_k^{s_k}$$

over the partitions of $|Y|$. $\diamond$

DEFINITION: Let $\mathcal{P} = [P : Y]$ be a permutation group. The **weight of a coloring** $f \in Col_k(Y)$ is a monomial $wt(f) = c_1^{s_1} \cdots c_k^{s_k}$ of degree $|Y|$ such that s_j is the number of objects of Y assigned color c_j, for $j = 1, \cdots, k$. The **weight of a coloring orbit** is the weight of any coloring in that orbit.

Remark: The definition of the induced permutation action $[P : Col_k(Y)]$ implies that two colorings in the same orbit must have the same weight. Indeed, the weights are quite simply a mathematical device that enables us to inventory orbits according to some significant characteristic, such as the number of squares of a given color.

Theorem 9.2.11 [Pólya Inventory Theorem]. *Let* $\mathcal{P} = [P : Y]$ *be a permutation group. Then every term*

$$p_{s_1,\cdots,s_k} c_1^{s_1} \cdots c_k^{s_k}$$

in a Pólya inventory for the induced action $[P : Col_k(Y)]$ *has as its coefficient* $p_{s_1,\cdots,s_k}$ *the number of coloring orbits whose weight is* $c_1^{s_1} \cdots c_k^{s_k}$.

Proof: A weighted form of Lemma 9.2.2 is that the coefficient of the weight $c_1^{s_1} \cdots c_k^{s_k}$ in the sum

$$\sum_{f \in Col_k(Y)} \frac{wt(f)}{|orbit(f)|}$$

is the number of orbits of weight $c_1^{s_1} \cdots c_k^{s_k}$. As before, it is really about partitions of sets. Similarly, a weighted version of Lemma 9.2.4 asserts that the sum over all permutations in $[P : Col_k(Y)]$ of the sums of the weights of their fixed-point sets equals the sum over all colorings $f \in Col_k(Y)$ of the products $wt(f) \cdot |Stab(f)|$. $\diamond$

Coloring Necklaces

We conclude this section by illustrating the application of the theory to a classic example, that of counting *necklaces*. Our model for an n-beaded necklace is a geometric conceptualization of the cycle graph C_n as a regular n-sided polygon, as illustrated in Figure 9.2.4. The numbering of the *beads* is used in specification of the permutations.

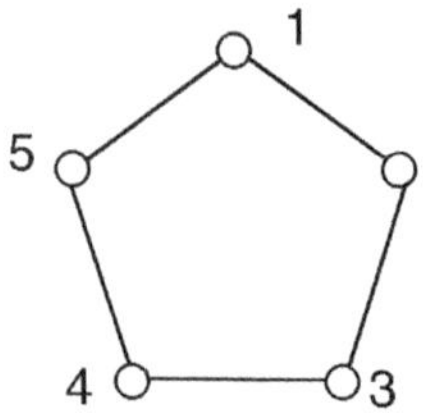

Figure 9.2.4   A 5-beaded necklace.

Example 9.2.7:   Suppose that each bead is to be colored either black or white, and that two necklaces are indistinguishable (i.e., equivalent) if one can be obtained from the other by a rotation of the polygon. We model these symmetries by the cyclic permutation group $\mathbb{Z}_5$, in which the clockwise rotation of $\frac{2\pi}{5}$ corresponds to the permutation

$$\alpha = (1 \quad 2 \quad 3 \quad 4 \quad 5)$$

The cycle structure of the identity permutation is t_1^5. By Corollary 9.1.2, the cycle structure of the permutations

$$\alpha, \quad \alpha^2, \quad \alpha^3, \quad \text{and} \quad \alpha^4$$

is t_5. Thus, the cycle index polynomial is

$$\mathcal{Z}_{\mathbb{Z}_5} = \frac{1}{5}\left(t_1^5 + 4t_5\right)$$

According to Theorem 9.2.8, the number of 2-colored, 5-beaded necklaces is

$$\frac{1}{5}\left(2^5 + 4\cdot 2\right) = \frac{40}{5} = 8$$

Table 9.2.1 provides the corresponding Pólya inventory.

Table 9.2.1   Inventory for 5-beaded necklaces under cyclic symmetry.

cycle structure	subst	b^5	b^4w	b^3w^2	b^2w^3	bw^4	w^5
t_1^5	$(b+w)^5$	1	5	10	10	5	1
$4t_5$	$4(b^5+w^5)$	4	0	0	0	0	4
sum		5	5	10	10	5	5
$\div 5$		1	1	2	2	1	1

Figure 9.2.5 illustrates the eight necklaces.

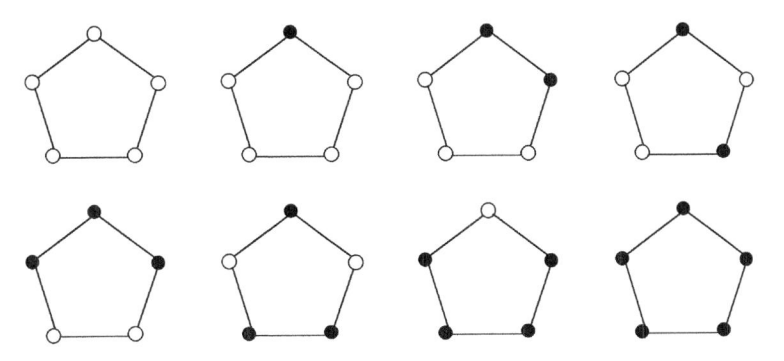

Figure 9.2.5   The eight 2-colored, 5-beaded necklaces.

Allowing reflections cannot possibly increase the number of equivalence classes, since any two necklaces that are related by a rotation remain related by that rotation when reflections are included.

Example 9.2.8:   In accordance with the description of reflections in §9.1, each of the five reflections has the cycle structure $t_1 t_2^2$. It follows that the dihedral group $\mathbb{D}_5$ that includes all the reflections as well as the rotations has the cycle index

$$\mathcal{Z}_{\mathbb{D}_5} \;=\; \frac{1}{10}\left(t_1^5 + 5t_1 t_2^2 + 4t_5\right)$$

Under dihedral symmetry the number of 2-colored, 5-beaded necklaces is

$$\frac{1}{10}\left(2^5 + 5 \cdot 2 \cdot 2^2 + 4 \cdot 2\right) \;=\; \frac{80}{10} \;=\; 8$$

In other words, for 2-colored, 4 beaded necklaces, dihedral symmetry yields no fewer equivalence classes of necklaces than cyclic symmetry.

However, allowing reflections as well as rotations might decrease the number of equivalence classes, since it is possible that two necklaces that are not equivalent under any rotation are equivalent under a reflection.

Example 9.2.9:   Consider the 3-colored, 5-beaded necklaces. Under cyclic symmetry, the number of equivalence classes is

$$\frac{1}{5}\left(3^5 + 4 \cdot 3\right) \;=\; \frac{255}{5} \;=\; 51$$

By way of contrast, under dihedral symmetry, the number of classes is only

$$\frac{1}{10}\left(3^5 + 5 \cdot 3 \cdot 3^2 + 4 \cdot 3\right) \;=\; \frac{390}{10} \;=\; 39$$

Figure 9.2.6 illustrates two necklaces that are equivalent under a vertical reflection, and thus under dihedral symmetry. However, none of the five rotations on one necklace produces the other necklace, so they are not equivalent under cyclic symmetry.

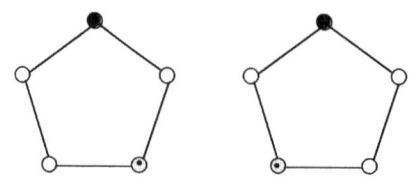

Figure 9.2.6   **Two 3-colored necklaces that are equivalent under dihedral symmetry, but not under cyclic symmetry.**

EXERCISES for Section 9.2

In each of the Exercises 9.2.1 through 9.2.4, use Theorem 9.2.8 to count the number of necklaces of the given type under cyclic symmetry.

9.2.1	4-colored, 5-beaded		**9.2.2**	2-colored, 6-beaded
9.2.3	3-colored, 6-beaded		**9.2.4**	2-colored, 7-beaded

In each of the Exercises 9.2.5 through 9.2.8, use Theorem 9.2.8 to count the number of necklaces of the given type under dihedral symmetry.

9.2.5	4-colored, 5-beaded		**9.2.6**	2-colored, 6-beaded
9.2.7	3-colored, 6-beaded		**9.2.8**	2-colored, 7-beaded

In each of the Exercises 9.2.9 through 9.2.12, construct a Pólya inventory of the necklaces of the given type under cyclic symmetry.

9.2.9	3-colored, 5-beaded		**9.2.10**	2-colored, 6-beaded
9.2.11	3-colored, 6-beaded		**9.2.12**	2-colored, 7-beaded

In each of the Exercises 9.2.13 through 9.2.16, construct a Pólya inventory of the necklaces of the given type under dihedral symmetry.

9.2.13	3-colored, 5-beaded		**9.2.14**	2-colored, 6-beaded
9.2.15	3-colored, 6-beaded		**9.2.16**	2-colored, 7-beaded

In each of the Exercises 9.2.17 through 9.2.20, draw two necklaces of the given type that are equivalent under dihedral symmetry but not under cyclic symmetry.

9.2.17	3-colored, 3-beaded.		**9.2.18**	2-colored, 6-beaded
9.2.19	3-colored, 6-beaded.		**9.2.20**	2-colored, 7-beaded

9.3   COUNTING SMALL SIMPLE GRAPHS

Some of the best illustrations of Burnside-Pólya counting are in graph theory.

REVIEW FROM §0.6:

- A **graph** $G = (V, E)$ is a mathematical structure consisting of two finite sets V and E. The elements of V are called **vertices** (or **nodes**), and the elements of E are called **edges**. Each edge has a set of one or two vertices associated to it, which are called its **endpoints**.

- The vertex set and edge set of a graph G are sometimes denoted V_G and E_G, respectively.

- An edge is said to **join** its endpoints. A vertex joined by an edge to a vertex v is said to be a **neighbor** of v. The endpoints of an edge are said to be **adjacent vertices**.

- A graph is a **simple graph** if every edge has two distinct endpoints and if no two edges have the same two endpoints. In a simple graph, an edge with endpoints u and v may be denoted uv.

DEFINITION: An **automorphism of a simple graph** $G = (V, E)$ is a permutation $\pi : V \to V$ that preserves adjacency and non-adjacency. That is,

- if u and v are adjacent, then so are $\pi(u)$ and $\pi(v)$.
- if u and v are non-adjacent, then so are $\pi(u)$ and $\pi(v)$.

Vertex Automorphisms and Colorings

Proposition 9.3.1. *The set of all automorphisms of a simple graph* $G = (V, E)$ *acts as a permutation group on its vertex set.*

Proof: The composition of two automorphisms preserves both adjacency and non-adjacency. Accordingly, the composition is an automorphism.   ◇

NOTATION: The automorphism group of the vertex set of a graph G is denoted $Aut_V(G)$.

Example 9.3.1:   The graph W_5 in Figure 9.3.1 is called a *wheel graph* because it looks something like a wheel with five spokes.

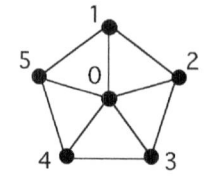

Figure 9.3.1   The wheel graph W_5.

The automorphism group $\mathcal{A}ut_V(W_5)$ has 5 rotations and 5 reflections, for a total of 10 permutations. Its cycle index is

$$\mathcal{Z}_{\mathcal{A}ut_V(W_5)}(t_1,\ldots,t_6) \;=\; \frac{1}{10}\left(t_1^6 + 5t_1^2t_2^2 + 4t_1t_5\right)$$

The number of (≤ 2)-colorings is

$$\mathcal{Z}_{\mathcal{A}ut_V(W_5)}(2,\ldots,2) \;=\; \frac{1}{10}\left(2^6 + 5\cdot 2^2\cdot 2^2 + 4\cdot 2\cdot 2\right) \;=\; 16$$

Table 9.3.1 calculates the Pólya inventory. Due to symmetry, the coefficients of

$$b^2w^4, \quad bw^5, \quad \text{and} \quad w^6$$

must be the same as the coefficients of

$$b^4w^2, \quad b^5w, \quad \text{and} \quad b^6$$

respectively. Thus, they may be omitted from the table.

Table 9.3.1   Pólya inventory for (≤ 2)-colorings of $V(W_5)$.

$\zeta(\pi)$	subst	b^6	b^5w	b^4w^2	b^3w^3
t_1^6	$(b+w)^6$	1	6	15	20
$5t_1^2t_2^2$	$5(b+w)^2(b^2+w^2)^2$	5	10	15	20
$4t_1t_5$	$4(b+w)(b^5+w^5)$	4	4	0	0
sum		10	20	30	40
$\div 10$		1	2	3	4

Figure 9.3.2 illustrates the four colorings for the case of three black vertices and three white.

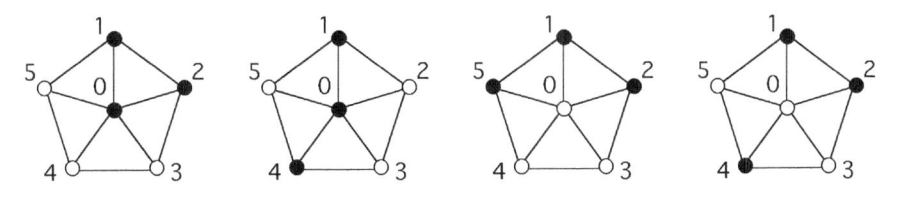

Figure 9.3.2   Four (≤ 2)-colorings of the wheel graph W_5.

Edge Automorphisms and Colorings

DEFINITION: For any automorphism $\pi : V \to V$ of the vertex set of a simple graph $G = (V, E)$, there is an **induced edge automorphism** π_E given by the rule

$$\pi_E(uv) \;=\; \pi(u)\pi(v)$$

Example 9.3.1, continued: The following translations of vertex automorphisms into edge automorphisms represent all three kinds of transformation of cycle structure.

$$(0)(1)(2)(3)(4)(5) \mapsto (01)(02)(03)(04)(05)(12)(23)(34)(45)(50)$$
$$(0)(1\ 2\ 3\ 4\ 5) \mapsto (01\ 02\ 03\ 04\ 05)(12\ 23\ 34\ 45\ 51)$$
$$(0)(1)(3\ 4)(2\ 5) \mapsto (01)(34)(02\ 05)(03\ 04)(23\ 54)(12\ 15)$$

Proposition 9.3.2.   *The set of all automorphisms of a simple graph $G = (V, E)$ acts as a permutation group on its edge set.*

Proof: The composition of two edge automorphisms is an edge automorphism. $\diamondsuit$

NOTATION: The automorphism group of the edge set of a graph G is denoted $Aut_E(G)$.

Example 9.3.1, continued: The cycle index of $Aut_E(W_5)$ is

$$\mathcal{Z}_{Aut_E(W_5)}(t_1, \ldots, t_{10}) \ = \ \frac{1}{10}\left(t_1^{10} + 5t_1^2 t_2^4 + 4t_5^2\right)$$

The number of (≤ 2)-colorings is

$$\mathcal{Z}_{Aut_E(W_5)}(2, \ldots, 2) \ = \ \frac{1}{10}\left(2^{10} + 5 \cdot 2^2 \cdot 2^4 + 4 \cdot 2^2\right) \ = \ 136$$

Table 9.3.2 gives a partial inventory.

Table 9.3.2 Partial inventory for (≤ 2)-colorings of $E(W_5)$.

$\zeta(\pi)$	subst	a^{10}	$a^9 b$	$a^8 b^2$	$a^7 b^3$
t_1^{10}	$(a+b)^{10}$	1	10	45	120
$5t_1^2 t_2^4$	$5(a+b)^2(a^2+b^2)^4$	5	10	25	40
$4t_5^2$	$4(a^5+b^5)^2$	4	0	0	0
sum		10	20	70	160
$\div 10$		1	2	7	16

Figure 9.3.3 shows the seven colorings for the case of eight light (a^8) edges and two dark (b^2) edges.

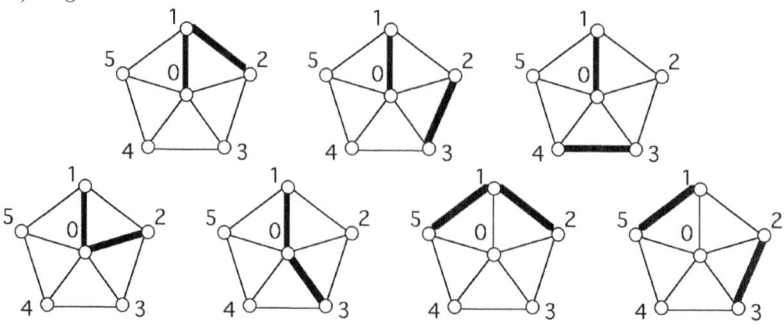

Figure 9.3.3   Seven edge-$a^8 b^2$-colorings of the graph W_5.

Orbits of Labeled Graphs

DEFINITION: A **standard labeled graph** on n vertices is a simple graph whose vertices are identified with the numbers $1, 2, \ldots, n$.

Proposition 9.3.3. *The number of standard labeled simple graphs on n vertices is*

$$2^{\binom{n}{2}}$$

Proof: Each possible edge ij is either absent or present. $\diamond$

DEFINITION: An **isomorphism of two simple graphs** G and H is a bijection $V_G \to V_H$ that preserves adjacency and non-adjacency of all pairs of vertices in V_G.

It is clear that any simple graph on n vertices is isomorphic to at least one of the standard labeled graphs. However, there are isomorphism relations among the standard labeled simple graphs. For instance, any two of the $\binom{n}{2}$ n-vertex simple graphs with only one edge are isomorphic. Proposition 9.3.4 indicates how Burnside-Pólya enumeration is used to count the *isomorphism types* of n-vertex simple graphs.

Proposition 9.3.4. *The number of isomorphism types of n-vertex simple graphs equals the number of orbits of the action*

$$[\mathcal{A}ut_E(K_n) : Col_2(E_{K_n})]$$

Proof: If the two colors used for coloring edges of K_n are regarded as *absent* and *present*, then the full set of 2^n edge-(≤ 2)-colorings is in one-to-one correspondence with the full set of n-vertex standard labeled simple graphs, and in this regard, the coloring classes correspond to the isomorphism types. $\diamond$

Proposition 9.3.4 suggests the following general strategy for counting the isomorphism types of the n-vertex simple graphs. It is illustrated by a series of propositions for counting the isomorphism types of simple graphs with 4 and 5 vertices.

Step 1: *Calculate the cycle-index polynomial of $\mathcal{A}ut_V(K_n)$.*

Since knowing the cycle-index polynomial is sufficient for algebraic counting, writing out all the permutations in a large permutation group can be avoided.

Step 2: *Calculate the cycle-index polynomial of $\mathcal{A}ut_E(K_n)$.*

The cycle-index polynomial of $\mathcal{A}ut_E(K_n)$ is obtained by considering each cycle size and each pair of cycle sizes in the cycle-index polynomial of $\mathcal{A}ut_V(K_n)$.

Step 3: *Apply Theorem 9.4.8.*

This final step in counting the isomorphism types of graphs with n vertices is to simply substitute the number 2 for every variable in the cycle-index polynomial $\mathcal{Z}_{\mathcal{A}ut_E(K_n)}$.

Simple Graphs on 4 Vertices

The following three propositions implement the general counting strategy for the 4-vertex simple graphs.

Proposition 9.3.5. *The permutation action* $[\mathcal{A}ut_V(K_4) : V_{K_4}]$ *has the following cycle-index polynomial.*

$$\mathcal{Z}_{\mathcal{A}ut_V(K_4)}(t_1, t_2, t_3, t_4) \;=\; \frac{1}{24}\left(t_1^4 + 6t_1^2 t_2 + 8t_1 t_3 + 3t_2^2 + 6t_4\right)$$

Proof: The 24 vertex-permutations in $\mathcal{A}ut_V(K_4)$ are naturally partitioned according to the five possible cycle structures:

$$t_1^4 \quad t_1^2 t_2 \quad t_1 t_3 \quad t_2^2 \quad t_4$$

Each cell in this partition is to be counted.

t_1^4: 1 automorphism.
Only the identity permutation has this cycle structure.

$t_1^2 t_2$: 6 automorphisms.
The number of ways to choose two vertices for the 2-cycle is

$$\binom{4}{2} = 6$$

$t_1 t_3$: 8 automorphisms.
The number of ways to choose three vertices for the 3-cycle is

$$\binom{4}{3} = 4$$

and the number of ways to arrange them in a cycle is $(3-1)! = 2$.

t_2^2: 3 automorphisms.
There are three ways to group four objects into two cycles, when it does not matter which cycle is written first.

t_4: 6 automorphisms.
They correspond to the $(4-1)! = 6$ ways that four objects can be arranged in a cycle. $\diamond$

Proposition 9.3.6. *The permutation action* $[\mathcal{A}ut_E(K_4) : E_{K_4}]$ *has the following cycle-index polynomial.*

$$\mathcal{Z}_{\mathcal{A}ut_E(K_4)}(t_1, t_2, t_3, t_4) \;=\; \frac{1}{24}\left(t_1^6 + 9t_1^2 t_2^2 + 8t_3^2 + 6t_2 t_4\right)$$

Proof: The size of the cycle to which an edge belongs is determined by the cycles to which its endpoints belong. Thus, for every automorphism $\pi \in \mathcal{A}ut_E(K_n)$, the cycle structure $\zeta(\pi_E)$ of the edge-permutation is determined by the the cycle structure $\zeta(\pi_V)$ of the vertex-permutation.

Case 1. If $\zeta(\pi_V) = t_1^4$, then $\zeta(\pi_E) = t_1^6$.

Justification: If both endpoints of a given edge e are in a 1-cycle of the vertex-permutation π_V, then they are both fixed points. In a simple graph, the corresponding edge-permutation must map that edge to itself.

Case 2. If $\zeta(\pi_V) = t_1^2 t_2$, then $\zeta(\pi_E) = t_1^2 t_2^2$.

Justification: An edge of K_4 is mapped to itself if both its endpoints are in a 2-cycle or if each endpoint is in a 1-cycle. Thus, two edges of K_4 are fixed by π_V. Each of the other four edges has one endpoint in a 1-cycle, which is fixed by π_V, and the other in a 2-cycle of π_V, which is mapped by π_V to the other vertex in that 2-cycle. It follows that such an edge lies in a 2-cycle of π_E.

Case 3. If $\zeta(\pi_V) = t_1 t_3$, then $\zeta(\pi_E) = t_3^2$.

Justification: The three edges of K_4 that have both their ends in the 3-cycle of π_V lie in a 3-cycle of π_E. The three edges of K_4 that have one endpoint in a 1-cycle of π_V and the other endpoint in a 3-cycle all lie in another 3-cycle of π_E.

Case 4. If $\zeta(\pi_V) = t_2^2$, then $\zeta(\pi_E) = t_1^2 t_2^2$.

Justification: The two edges that have both endpoints in the same 2-cycle of π_V are both fixed by π_E. If an edge has one endpoint in one 2-cycle of π_V and the other endpoint in another 2-cycle of π_V, then that edge lies in a 2-cycle of π_E with the edge whose respective endpoints are the other vertices of those 2-cycles of π_V.

Case 5. If $\zeta(\pi_V) = t_4$, then $\zeta(\pi_E) = t_2 t_4$.

Justification: The four edges whose endpoints are consecutive vertices in the 4-cycle of π_V form a cycle of π_E. The two edges whose endpoints are spaced 2 apart in the 4-cycle of π_V form a 2-cycle of π_E. $\diamond$

Corollary 9.3.7. *There are exactly 11 isomorphism types of simple graph with 4 vertices.*

Proof: Proposition 9.3.6 asserts that

$$\mathcal{Z}_{Aut_E(K_4)}(2,2,2,2) \;=\; \frac{1}{24}\big(2^6 + 9 \cdot 2^2 \cdot 2^2 + 8 \cdot 2^2 + 6 \cdot 2 \cdot 2\big) \;=\; 11$$

The 11 graphs promised by this calculation are shown in Figure 9.3.4. $\diamond$

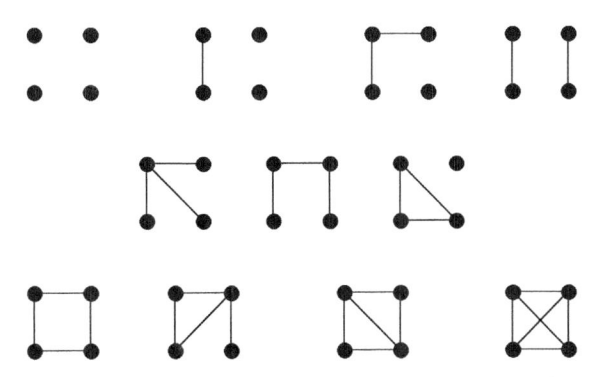

Figure 9.3.4   The 11 simple 4-vertex graphs.

Observe that the Pólya inventory in Table 9.3.3 is consistent with Figure 9.3.4. The indeterminates a and p stand for *absent* and *present*. They are Pólya-substituted into the cycle index to produce the table.

$$\mathcal{Z}_{Aut_E(K_4)}(t_1, t_2, t_3, t_4) \;=\; \frac{1}{24}\left(t_1^6 + 9t_1^2 t_2^2 + 8t_3^2 + 6t_2 t_4\right)$$

For instance, the coefficient of 2 for the monomial $a^4 p^2$ at the bottom of the column labeled $a^4 p^2$ signifies that there are exactly two 4-vertex graphs with exactly 2 edges present.

Table 9.3.3 Isomorphism types of 4-vertex graph.

$\zeta(\pi_E)$	subst	a^6	$a^5 p$	$a^4 p^2$	$a^3 p^3$
t_1^6	$(a+p)^6$	1	6	15	20
$9t_1^2 t_2^2$	$9(a+p)^2(a^2+p^2)^2$	9	18	27	36
$8t_3^2$	$8(a^3+p^3)^2$	8	0	0	16
$6t_2 t_4$	$6(a^2+p^2)(a^4+p^4)$	6	0	6	0
sum		24	24	48	72
$\div 24$		1	1	2	3

Simple Graphs with 5 Vertices

Proposition 9.3.8. *There are exactly 34 isomorphism types of simple graph with 5 vertices.*

Proof: By using the same approach as in the calculation of the number of 4-vertex simple graphs, it can be shown (an explicit general method for any number of vertices is given in the next section) that

$$\mathcal{Z}_{Aut_V(K_5)}(t_1, \ldots, t_5) \;=\;$$
$$\frac{1}{120}\left(t_1^5 + 10t_1^3 t_2 + 15t_1 t_2^2 + 20t_1^2 t_3 + 20t_2 t_3 + 30t_1 t_4 + 24t_5\right)$$

Therefore

$$\mathcal{Z}_{Aut_E(K_5)}(t_1, \ldots, t_{10}) \;=\;$$
$$\frac{1}{120}\left(t_1^{10} + 10t_1^4 t_2^3 + 15t_1^2 t_2^4 + 20t_1 t_3^3 + 20t_1 t_3 t_6 + 30t_2 t_4^2 + 24t_5^2\right)$$

and accordingly

$$\mathcal{Z}_{Aut_E(K_5)}(2, \ldots, 2) \;=\;$$
$$\frac{1}{120}\left(2^{10} + 10 \cdot 2^7 + 15 \cdot 2^6 + 20 \cdot 2^4 + 20 \cdot 2^3 + 30 \cdot 2^3 + 24 \cdot 2^2\right) \;=\; 34 \quad \Diamond$$

Partial Pólya Inventories

Doing a hand calculation of a complete Pólya inventory can be quite tedious. We complete this section by considering how to calculate the number of 5-vertex graphs with 4 edges, to illustrate how to do a selective partial inventory. Table 9.3.4 summarizes the calculation.

Table 9.3.4 Counting the 5-vertex graphs with 3 edges.

$\zeta(\pi_E)$	substitute	$a^7 p^3$ − contribution	evaluate
t_1^{10}	$(a+p)^{10}$	$1 \cdot \binom{10}{3}$	120
$10 t_1^4 t_2^3$	$10(a+p)^4(a^2+p^2)^3$	$10 \cdot \left[\binom{4}{1}\binom{3}{1} + \binom{4}{3}\binom{3}{0} \right]$	160
$15 t_1^2 t_2^4$	$15(a+p)^2(a^2+p^2)^4$	$15 \cdot \binom{2}{1}\binom{4}{1}$	120
$20 t_1 t_3^3$	$20(a+p)(a^3+p^3)^3$	$20 \cdot \binom{1}{0}\binom{3}{1}$	60
$20 t_1 t_3 t_6$	$20(a+p)(a^3+p^3)(a^6+p^6)$	$20\binom{1}{0}\binom{1}{1}\binom{1}{0}$	20
$30 t_2 t_4^2$	$30(a^2+p^2)(a^4+p^4)^2$	0	
$24 t_5^2$	$24(a^5+p^5)^2$	0	
sum			480
$\div 120$			4

Example 9.3.2: When the expression $10(a+p)^4(a^2+p^2)^3$ on the second line of the table is fully expanded, there are two contributions to the term with the monomial $a^7 p^3$. One results from the product $12 a^7 p^3$ of $4 a^3 p^1$ and $3 a^4 p^2$ in the expansions of $(a+p)^4$ and $(a^2+p^2)^3$, respectively, and the other results from the product $4 a^7 p^3$ of $4 a^1 p^3$ and $1 a^6 p^0$ in the expansions of $(a+p)^4$ and $(a^2+p^2)^3$, respectively. The sum 12+4 is multiplied by the coefficient 10 to yield 160.

EXERCISES for Section 9.3

In each of the Exercises 9.3.1 through 9.3.8, construct the cycle index for the vertex automorphism group of the given graph, and then use it to count the number of (≤ 2)-vertex-colorings.

9.3.1	Path graph P_{2n}.	9.3.2	Path graph P_{2n+1}.
9.3.3^{S}	Bipartite graph $K_{2,3}$.	9.3.4	Bipartite graph $K_{2,4}$.
9.3.5	Wheel graph W_4.	9.3.6	Wheel graph W_6.
9.3.7	Circular ladder CL_3.	9.3.8	Circular ladder CL_4.

In each of the Exercises 9.3.9 through 9.3.16, construct the cycle index for the edge automorphism group of the given graph, and then use it to count the number of (≤ 2)-edge-colorings.

9.3.9	Path graph P_{2n}.	9.3.10	Path graph P_{2n+1}.

9.3.11[S]   Bipartite graph $K_{2,3}$.   9.3.12   Bipartite graph $K_{2,4}$.

9.3.13   Wheel graph W_4.   9.3.14   Wheel graph W_6.

9.3.15   Circular ladder CL_3.   9.3.16   Circular ladder CL_4.

In each of the Exercises 9.3.17 through 9.3.19, use a partial inventory table to calculate the number of graphs with 5 vertices and the given number of edges.

9.3.17[S]   2 edges.   9.3.18   4 edges.   9.3.19   5 edges.

9.4   PARTITIONS OF INTEGERS

In the cycle index $\mathcal{Z}_{\mathcal{A}ut_V(K_n)}(t_1, \ldots, t_n)$, each term is the product of a cycle structure

$$t_1^{e_1} t_2^{e_2} \cdots t_n^{e_n}$$

and a coefficient giving the number of permutations in the group $\mathcal{A}ut_V(K_n)$ having that cycle structure. The sum of the coefficents is $n!$, since there are $n!$ permutations in all.

Example 9.4.1:   In the previous section, we calculated that

$$\mathcal{Z}_{\mathcal{A}ut_V(K_4)}(t_1, t_2, t_3, t_4) \;=\; \frac{1}{24}\left(t_1^4 + 6t_1^2 t_2 + 8t_1 t_3 + 3t_2^2 + 6t_4\right) \tag{9.4.1}$$

and that

$$\mathcal{Z}_{\mathcal{A}ut_V(K_5)}(t_1, \ldots, t_5) \;=\;$$
$$\frac{1}{120}\left(t_1^5 + 10t_1^3 t_2 + 15t_1 t_2^2 + 20t_1^2 t_3 + 20t_2 t_3 + 30t_1 t_4 + 24t_5\right) \tag{9.4.2}$$

DEFINITION: A **partition of an integer** n (with n positive) is a sum

$$s_1 + s_2 + \cdots + s_k$$

whose value is n and whose summands are positive integers. It is usually represented with the summands in nonincreasing order, without the addition signs.

The cycle structure $t_1^{e_1} t_2^{e_2} \cdots t_n^{e_n}$ of a permutation corresponds to the integer partition

$$\overbrace{n\,n\cdots n}^{e_n\ n's} \cdots \overbrace{2\,2\cdots 2}^{e_2\ 2's} \overbrace{1\,1\cdots 1}^{e_1\ 1's}$$

Example 9.4.1, continued: The five partitions of the integer 4 are

$$1111 \quad 211 \quad 22 \quad 31 \quad 4$$

Observe that they correspond to the terms of the cycle index polynomial

$$\mathcal{Z}_{\mathcal{A}ut_V(K_4)}(t_1, t_2, t_3, t_4) \;=\; \frac{1}{24}\big(t_1^4 + 6t_1^2 t_2 + 8t_1 t_3 + 3t_2^2 + 6t_4\big)$$

Similarly, the seven partitions of the integer 5

$$11111 \quad 2111 \quad 221 \quad 311 \quad 32 \quad 41 \quad 5$$

correspond to the terms of the cycle index polynomial

$$\mathcal{Z}_{\mathcal{A}ut_V(K_5)}(t_1, \ldots, t_5) \;=$$
$$\frac{1}{120}\big(t_1^5 + 10t_1^3 t_2 + 15t_1 t_2^2 + 20t_1^2 t_3 + 20t_2 t_3 + 30t_1 t_4 + 24t_5\big)$$

Listing all Partitions of an Integer

As an aid in calculating the cycle index polynomial $\mathcal{Z}_{\mathcal{A}ut_V(K_n)}(t_1, \ldots, t_n)$ for the automorphism group of a complete graph of general size, we now introduce a systematic way to list all the partitions of an arbitrary integer n.

In Example 9.4.1, the partitions of the integer 5 are given in ascending order. Algorithm 9.4.1 lists all the partitions of n in descending order. The key step of constructing the next partition after $s_1 s_2 \cdots s_k$ has a relatively easy intuitive description. It is assumed that the parts s_i of the partition given as input are written in order of descending size.

Algorithm 9.4.1:   Next Integer Partition (S)

Input: an integer partition $S = s_1 \cdots s_k$, not all 1's.
Output: The next integer partition, in descending order.

  Assign $b := \max\{j \mid s_j \neq 1\}$.
  Assign $M := s_b - 1$.
  Assign $s_b := M$.
  Replace suffix $s_{b+1} \cdots s_k$ by string of $\lfloor k - j + 1/M \rfloor$ M's.
  If $k - j + 1 \bmod M > 0$
    then append integer $k - j + 1 \bmod M$ to string S.
  Return (S)

Example 9.4.2:   We apply the algorithm to the partition

$$8666411111$$

of the integer 35. Then index b is 5, the location of the rightmost non-1, whose value s_5 is 4, so $M = 3$. The assignment $s_5 := M$ decreases the value of s_5 to 3.

The suffix of five 1's is replaced by two 3's, since $\lfloor 6/3 \rfloor = 2$. The division has zero remainder. Thus, the partition returned is

$$8666333$$

To obtain a list all the partitions of an integer n, that integer itself is supplied to the algorithm Next Integer Partition to initiate the list. Then the output is supplied to the algorithm iteratively, until a string of n 1's is obtained as the final partition.

Example 9.4.3: For $n = 6$, iterative application of Next Integer Partition produces the sequence

$$6 \rightarrow 51 \rightarrow 42 \rightarrow 411 \rightarrow 33 \rightarrow 321 \rightarrow 3111$$
$$\rightarrow 222 \rightarrow 2211 \rightarrow 21111 \rightarrow 111111$$

Ferrers Diagrams

DEFINITION: The **Ferrers diagram** of a partition $s_1 s_2 \cdots s_k$ is an array of k rows of dots, with s_j dots in row j.

Example 9.4.4: The partition 6441 has the Ferrers diagram

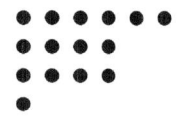

Figure 9.4.1 Ferrers diagram for the integer partition 6441.

DEFINITION: The **conjugate of a Ferrers diagram** for the partition $s_1 s_2 \cdots s_k$ is the Ferrers diagram with k columns of dots, with s_j dots in column j.

Example 9.4.4, continued: The conjugate of the Ferrers diagram in Figure 9.4.1 is the diagram

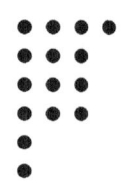

Figure 9.4.2 Conjugate of the previous Ferrers diagram.

DEFINITION: The **conjugate of a partition** S is the partition whose Ferrers diagram is the conjugate of the Ferrers diagram of S.

Ferrers diagrams are the easiest way to prove the following kind of theorem about partitions.

Theorem 9.4.1. *The number of partitions of an integer n into at most k parts equals the number of partitions of n into parts of size at most k.*

Proof: The operation of conjugation on Ferrers diagrams is a bijection from the set of partitions of the number n into at most k parts to the set of partitions of size at most k.   ◇

Partition Lattices

DEFINITION: The ***inclusion relation on integer partitions*** is given by

$$s_1 s_2 \cdots s_k \preceq u_1 u_2 \cdots u_\ell \quad \text{if } k \leq \ell$$
$$\text{and } s_j \leq u_j, \ \text{ for } j = 1, \ldots, k$$

Its name reflects the fact that the Ferrers diagram of the first is contained in the Ferrers diagram of the second.

DEFINITION: For an integer partition $S = s_1 s_2 \cdots s_k$, the **Young's lattice** $\mathcal{Y}_S$ has as its domain the set of all integer partitions included in S. The partial ordering is inclusion.

Example 9.4.5:   The Hasse diagram for the Young's lattice for the integer partition 3221 is shown in Figure 9.4.3.

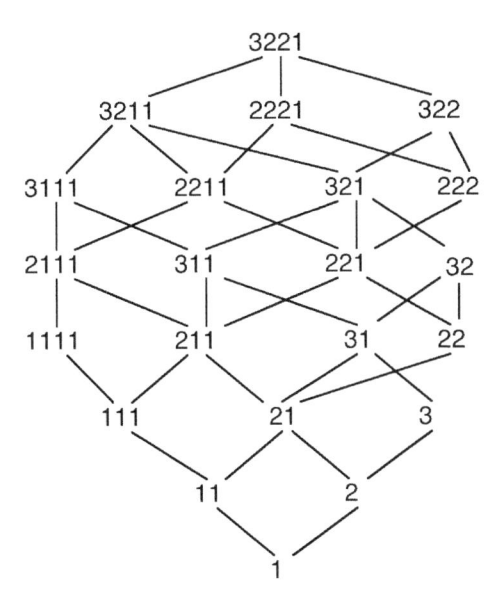

Figure 9.4.3   Young's lattice $\mathcal{Y}_{3221}$.

A Young's lattice is graded. The rank of each partition is the integer it partitions. There is a second kind of lattice for partitions of integers, in which all the elements partition the same integer.

DEFINITION: Let $B = b_1 b_2 \cdots b_k$ and $U = u_1 u_2 \cdots u_\ell$ be any two partitions of the integer n, such that

$$b_1 + \cdots + b_j \leq u_1 + \cdots + u_j \quad \text{for all } j \leq \min(k, \ell)$$

Then the partition U has **summation dominance** over the partition B. The resulting poset $\mathcal{SD}_n$ is called a **summation dominance lattice**.

Proposition 9.4.2. *The order of the partitions of an integer n produced by iterative application of the algorithm Next Integer Partition is a linear extension of the summation dominance partial ordering.*

Proof: Since the input to the algorithm has lexicographic dominance over the output, it cannot be summation dominated by the output. $\diamond$

Example 9.4.6: A Hasse diagram for the summation dominance lattice $\mathcal{D}_7$ is shown in Figure 9.4.4. Observe that it is unranked.

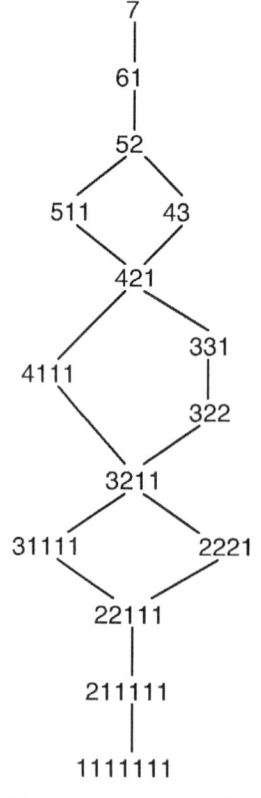

Figure 9.4.4 Summation dominance lattice $\mathcal{SD}_7$.

EXERCISES for Section 9.4

In each of the Exercises 9.4.1 through 9.4.4, list all the partitions of the given integer, with at most 3 parts.

9.4.1^S 8   9.4.**2**   9   9.4.**3**   10   9.4.**4**   11

In each of the Exercises 9.4.5 through 9.4.8, draw the Ferrers diagram for the given partition and draw its conjugate.

9.4.**5**^{S} 5322   9.4.**6**   53211   9.4.**7**   64322   9.4.**8**   64331

In each of the Exercises 9.4.9 through 9.4.12, draw the Youngs lattice for the given partition.

9.4.**9**^{S} 332   9.4.**10**   422   9.4.**11**   431   9.4.**12**   2222

9.4.**13**   Draw the summation dominance lattice $\mathcal{D}_8$.

9.5   CALCULATING A CYCLE INDEX

Beyond writing a cycle structure monomial $t_1^{e_1} \cdots t_n^{e_n}$ corresponding to each partition of an integer n, calculating the cycle index of $\mathcal{A}ut_V(K_n)$ requires writing the coefficient of each such monomial. When calculating $\mathcal{A}ut_E(K_n)$, another step is transforming a cycle structure for the automorphism action on the vertices of the complete graph K_n to the corresponding cycle structure for the automorphism action on the edges of K_n.

Multinomial Coefficients

DEFINITION: Let $r_1, r_2, \ldots, r_k$ be a list of non-negative integers whose sum is n. The value of the **multinomial coefficient**

$$\left(\begin{matrix} & & n & & \\ r_1 & r_2 & \cdots & r_k \end{matrix} \right)$$

is the number of ways that a set of n distinct objects can be distributed into k distinct boxes

$$B_1, B_2, \ldots, B_k$$

so that, for $j = 1, \ldots, k$, there are r_j objects in box B_j. The concept of a multinomial coefficient generalizes the binomial coefficient

$$\binom{n}{r}$$

which is the special case with two boxes, the first of size r, and the second of size $n - r$.

Example 9.5.1:   Suppose that an 8-member executive board of an organization is to be partitioned into a 2-member recruiting committee, a 3-member finance committee, and a 3-member events committee. The three committees are distinguished by their organizational missions.

Proposition 9.5.1.   *Let* $r_1, r_2, \ldots, r_k$ *be a list of non-negative integers whose sum is* n. *Then*

$$\begin{pmatrix} & & n & & \\ r_1 & r_2 & \cdots & r_k \end{pmatrix} = \frac{n!}{r_1!\, r_2! \cdots r_k!}$$

Proof: For $k = 2$, the right side is a familiar formula for the value of the binomial coefficient. We continue inductively.

Clearly, the number of ways to distribute n objects into the boxes

$$B_1, B_2, \ldots, B_k$$

of sizes $r_1, r_2, \ldots, r_k$ equals the product of the number of ways to select r_1 objects for box B_1 and the number of ways to distribute the remaining $n - r_1$ objects into the boxes $B_2, B_3, \ldots, B_k$. That is,

$$\begin{pmatrix} & & n & & \\ r_1 & r_2 & \cdots & r_k \end{pmatrix} = \begin{pmatrix} n \\ r_1 \end{pmatrix} \begin{pmatrix} & & n - r_1 & & \\ r_2 & r_3 & \cdots & r_k \end{pmatrix} \qquad (9.5.1)$$

By the induction hypothesis,

$$\begin{pmatrix} & n - r_1 & & \\ r_2 & r_3 & \cdots & r_k \end{pmatrix} = \frac{(n - r_1)!}{r_2!\, r_3! \cdots r_k!} \qquad (9.5.2)$$

Thus, combining Equations (9.5.1) and (9.5.2),

$$\begin{pmatrix} & & n & & \\ r_1 & r_2 & \cdots & r_k \end{pmatrix} = \begin{pmatrix} n \\ r_1 \end{pmatrix} \frac{(n - r_1)!}{r_2!\, r_3! \cdots r_k!}$$

$$= \frac{n!}{r_1!\, r_2! \cdots r_k!} \qquad\qquad \diamond$$

Example 9.5.1, continued: It follows from Proposition 9.5.1 that the number of ways to partition the executive board into the three committees of those prescribed sizes is

$$\frac{8!}{2!\,3!\,3!} = 560$$

Non-distinct Boxes

When a permutation π partitions n objects into disjoint cycles, the cycles are non-distinct. For instance, the permutations

$$(1\ 2)(3\ 7\ 5)(4\ 6\ 8) \ \text{ and } \ (1\ 2)(4\ 6\ 8)(3\ 7\ 5)$$

are identical, since the order in which the 3-cycles are written in a disjoint cycle representation has no bearing on the effect of the permutation.

Proposition 9.5.2. *Let* $r_1, r_2, \ldots, r_k$ *and* $e_1, e_2, \ldots, e_k$ *be list of non-negative integers such that*

$$n \ = \ e_1 r_1 + e_2 r_2 + \cdots + e_k r_k$$

Then the number of ways to partition n *distinct objects into non-distinct boxes, with* e_j *boxes of size* r_j, *for* $j = 1, \ldots, k$, *is*

$$\frac{n!}{(r_1!)^{e_1}(r_2!)^{e_2}\cdots(r_k!)^{e_k}\,e_1!e_2!\cdots e_k!}$$

Proof: If the boxes were distinct, the number of possible distributions would be

$$\frac{n!}{(r_1!)^{e_1}(r_2!)^{e_2}\cdots(r_k!)^{e_k}}$$

by Proposition 9.5.1. Application of the Rule of Quotient motivates division by

$$e_1!e_2!\cdots e_k!$$

for non-distinct boxes. $\diamondsuit$

Example 9.5.2:   The three ways to partition the integer six into three parts are 411, 321, and 222. The Stirling subset number

$$\left\{ \begin{matrix} 6 \\ 3 \end{matrix} \right\}$$

is the total number of ways to partition 6 objects into 3 cells. Thus, in view of Proposition 9.5.2, we anticipate the result of the following computation.

$$
\begin{aligned}
\left\{ \begin{matrix} 6 \\ 3 \end{matrix} \right\} &= \binom{6}{4\ 1\ 1}\frac{1}{2!} + \binom{6}{3\ 2\ 1} + \binom{6}{2\ 2\ 2}\frac{1}{3!} \\
&= \frac{6!}{4!}\cdot\frac{1}{2!} + \frac{6!}{3!\,2!} + \frac{6}{2!\,2!\,2!}\cdot\frac{1}{3!} \\
&= 15 + 60 + 15 \\
&= 90
\end{aligned}
$$

Corollary 9.5.3. *The number of permutations of cycle structure* $t_1^{e_1} \cdots t_n^{e_n}$ *in* $Aut_V(K_n)$ *is*

$$
\left(\underbrace{\overbrace{r_1 \cdots r_1}^{e_1 r_1' s}}_{} \cdots \underbrace{\overbrace{r_k \cdots r_k}^{e_k r_k' s}}_{} \right)^{\!n} \frac{((r_1 - 1)!)^{e_1} \cdots ((r_k - 1)!)^{e_k}}{e_1! \cdots e_k!}
$$

Proof: Proposition 9.5.2 accounts for the multinomial coefficient and for the denominator of the fraction. The numerator of the fraction accounts for the objects within each cell into a cycle of the permutation. The number of ways to organize s objects into a cycle is $(s - 1)!$.   ◇

Example 9.5.2, continued: The Stirling cycle number $\begin{bmatrix} 6 \\ 3 \end{bmatrix}$ is the number of permutations of 6 objects with three cycles. Thus, consistent with Corollary 9.5.3, we obtain

$$
\begin{aligned}
\begin{bmatrix} 6 \\ 3 \end{bmatrix} &= \begin{pmatrix} 6 \\ 4\ 1\ 1 \end{pmatrix} \frac{3!\,0!\,0!}{1!\,2!} + \begin{pmatrix} 6 \\ 3\ 2\ 1 \end{pmatrix} \frac{2!\,1!\,0!}{1!\,1!\,1!} + \begin{pmatrix} 6 \\ 2\ 2\ 2 \end{pmatrix} \frac{1!\,1!\,1!}{3!} \\
&= \frac{6!}{4!} \cdot 3 + \frac{6!}{3!\,2!} \cdot 2 + \frac{6!}{2!\,2!\,2!} \cdot \frac{1}{6} \\
&= 90 + 120 + 15 \\
&= 225
\end{aligned}
$$

Proposition 9.5.4. *The cycle index of the group* $Aut_V(K_6)$ *is*

$$
\begin{aligned}
\mathcal{Z}_{Aut_V(K_6)}(t_1, \ldots, t_6) &= \frac{1}{720} \big(t_1^6 + 15 t_1^4 t_2 + 45 t_1^2 t_2^2 + 15 t_2^3 + 40 t_1^3 t_3 \\
&\quad + 120 t_1 t_2 t_3 + 40 t_3^2 + 90 t_1^2 t_4 + 90 t_2 t_4 + 144 t_1 t_5 + 120 t_6 \big)
\end{aligned}
$$

Proof: This follows from Corollary 9.5.3.   ◇

Transforming the Cycle Index

When counting isomorphism types of graphs, one further step is to transform each vertex automorphism cycle structure into the corresponding edge automorphism cycle structure. This step may vary, according to the kind of graphs being counted. What follows here is for counting simple graphs.

We continue to use variables t_j in the cycle structure of a graph automorphism action on the set of vertices. For clarity, we will use variables y_j in the cycle structure of the corresponding action on the set of edges.

Theorem 9.5.5. *Let* π *be an automorphism of the complete graph* K_n, *represented by permutations* π_V *on* V_{K_n} *and* π_E *on* E_{K_n}. *Then*

   (i) $t_{2p+1} \to y_{2p+1}^p$:

   Each $(2p + 1)$-*cycle in* π_V *corresponds to* p $(2p + 1)$-*cycles in* π_E.

(ii) $t_{2p} \to y_{2p}^{p-1} y_p$:

Each $(2p)$-cycle in π_V corresponds to $p-1$ $(2p)$-cycles and one p-cycle in π_E.

(iii) $t_p t_q \to y_{\mathrm{lcm}\,(p,q)}^{\gcd\,(p,q)}$:

Each $(p$-cycle, q-cycle$)$-pair in the permutation π_V, with $p \neq q$, corresponds to $\gcd(p,q)$ $\mathrm{lcm}(p,q)$-cycles in π_E.

(iv) $t_p t_p \to y_p^{p}$:

Each pair of p-cycles in π_V corresponds to p p-cycles in π_E.

Proof: The key to all four parts of the proof is that if ij is any edge in K_n, then the endpoints of the edge $\pi_E(ij)$ are $\pi_V(i)$ and $\pi_V(j)$. We begin each part with an illustrative example.

(i) $t_{2p+1} \to y_{2p+1}^{p}$:

The case $p = 2$, where we consider a factor t_5 corresponding to some 5-cycle $(1\ 2\ 3\ 4\ 5)$ in π_V, is illustrated in Figure 9.5.1. The copy of K_5 in Figure 9.5.1 is the induced subgraph of K_n on the five vertices 1, 2, 3, 4, and 5. Since π_E maps the edge 12 to the edge 23, maps 23 to 34, and so on, it must contain the cycle $(12\ 23\ 34\ 45\ 15)$, whose edges are solid lines in the figure.

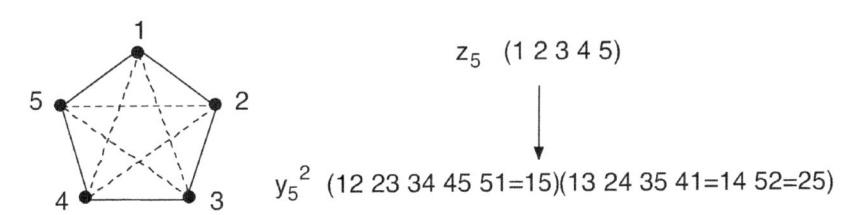

$$z_5 \quad (1\ 2\ 3\ 4\ 5)$$
$$\downarrow$$
$$y_5{}^2 \quad (12\ 23\ 34\ 45\ 51{=}15)(13\ 24\ 35\ 41{=}14\ 52{=}25)$$

Figure 9.5.1   Transformation of t_5 into $y_5{}^2$.

Since π_E maps the edge 13 to the edge 24, maps 24 to 35, and so on, it must also contain the cycle $(12\ 23\ 34\ 45\ 15)$, whose edges are dashed lines in the figure.

More generally, for a cycle in π_V of the form

$$\left(\,v_1 \quad v_2 \quad \cdots \quad v_{2p+1}\,\right)$$

the $p-1$ corresponding $(2p+1)$-cycles in π_E are

$$\begin{array}{cccc}
\left(\,v_1 v_2 & v_2 v_3 & \cdots & v_{2p+1} v_1\,\right) \\
\left(\,v_1 v_3 & v_2 v_4 & \cdots & v_{2p+1} v_2\,\right) \\
& \vdots & & \\
\left(\,v_1 v_p & v_2 v_{p+1} & \cdots & v_{2p+1} v_{p-1}\,\right) \\
\left(\,v_1 v_{p+1} & v_2 v_{p+2} & \cdots & v_{2p+1} v_p\,\right)
\end{array}$$

(ii) $t_{2p} \to y_{2p}^{p-1} y_p$:

For the case $p = 3$, we consider a factor t_6 corresponding to a 6-cycle $(1\ 2\ 3\ 4\ 5\ 6)$ in π_V, as illustrated in Figure 9.5.2. The dark solid edges form one 6-cycle in π_E.

Even though the dashed edges form two 3-cycles in the copy of K_6 in the figure, they all lie in the same 6-cycle of π_E. Even though the three light solid edges form no cycles in K_6, they form a 3-cycle in π_E.

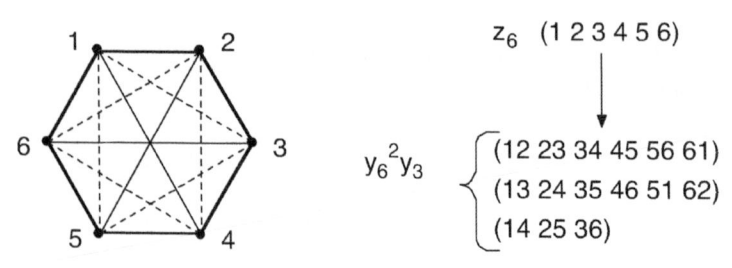

Figure 9.5.2   **Transformation of t_6 into $y_6{}^2 y_3$.**

In general, for a cycle in π_V of the form

$$(v_1 \quad v_2 \quad \cdots \quad v_{2p})$$

the corresponding cycles in π_E are

$$
\left.
\begin{array}{l}
(v_1 v_2 \quad v_2 v_3 \quad \cdots \quad v_{2p} v_1) \\
(v_1 v_3 \quad v_2 v_4 \quad \cdots \quad v_{2p} v_2) \\
\qquad\qquad \vdots \\
(v_1 v_p \quad v_2 v_{p+1} \quad \cdots \quad v_{2p} v_{p-1})
\end{array}
\right\} \quad p-1 \ \ (2p)\text{-cycles}
$$

$$(v_1 v_{p+1} \quad v_2 v_{p+2} \quad \cdots \quad v_p v_{2p})\} \quad \text{one } p\text{-cycle}$$

(iii) $t_p t_q \to y_{\mathrm{lcm}\,(p,q)}^{\gcd\,(p,q)}$:

We consider two cycles $(1 \ \cdots \ 4)$ and $(\underline{1} \ \cdots \ \underline{6})$ in π_V, which yields $\gcd\,(4,6) = 2$ cycles in π_E, each of length $\mathrm{lcm}\,(4,6) = 12$. For clarity in Figure 9.5.3, these two cycles in π_E are shown separately. Neither is a cycle in the graph, only in the permutation.

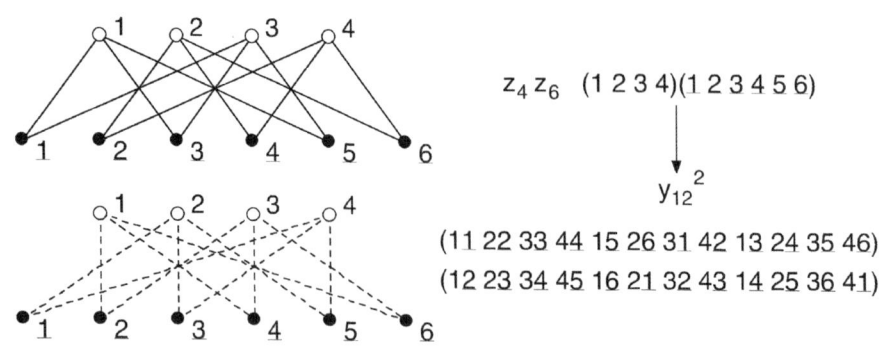

Figure 9.5.3   **Transformation of $t_4 t_6$ into y_{12}^2.**

In general, for a pair of cycles in π_V

$$(u_1 \quad u_2 \quad \cdots \quad v_p) \quad \text{and} \quad (v_1 \quad v_2 \quad \cdots \quad v_q)$$

the gcd (p, q) corresponding lcm (p, q)-cycles in π_E are

$$
\begin{pmatrix} u_1 v_1 & u_2 v_2 & \cdots & u_p v_q \end{pmatrix}
$$
$$
\begin{pmatrix} u_1 v_2 & u_2 v_3 & \cdots & u_p v_1 \end{pmatrix}
$$
$$
\vdots
$$
$$
\begin{pmatrix} u_1 v_{\gcd(p,q)} & u_2 v_{\gcd(p,q)+1} & \cdots & u_p v_{\gcd(p,q)-1} \end{pmatrix}
$$

In the special case of two p-cycles in π_V, there are p p-cycles in π_E.

This completes the proof of the theorem.   ◇

Proposition 9.5.6. *There are exactly 156 isomorphism types of simple graph with 6 vertices.*

Proof: By Proposition 9.5.4,

$$
\mathcal{Z}_{\mathcal{A}ut_V(K_6)}(t_1, \ldots, t_6) = \frac{1}{720} \big(t_1^6 + 15t_1^4 t_2 + 45t_1^2 t_2^2 + 15t_2^3 + 40t_1^3 t_3 + 120t_1 t_2 t_3
$$
$$
+ 40t_3^2 + 90t_1^2 t_4 + 90t_2 t_4 + 144t_1 t_5 + 120t_6 \big)
$$

By Theorem 9.5.5,

$$
\mathcal{Z}_{\mathcal{A}ut_E(K_6)}(y_1, \ldots, y_{15}) = \frac{1}{720} \big(y_1^{15} + 15y_1^7 y_2^4 + 45y_1^3 y_2^6 + 15y_1^3 y_2^6 + 40y_1^3 y_3^4
$$
$$
+ 120y_1 y_2 y_3^2 y_6 + 40y_3^5 + 90y_1 y_2 y_4^3 + 90y_1 y_2 y_4^3 + 144y_5^3 + 120y_3 y_6^2 \big)
$$

Substituting 2 for each of the variables $y_1, \ldots, y_{15}$ in

$$
\mathcal{Z}_{\mathcal{A}ut_E(K_6)}(y_1, \ldots, y_{15})
$$

yields 156.   ◇

EXERCISES for Section 9.5

In each of the Exercises 9.5.1 through 9.5.6, use a partial inventory table to calculate the number of isomorphism types of 6-vertex graphs with the given number of edges.

9.5.1   2 edges.   9.5.2   3 edges.   9.5.3^S 4 edges.
9.5.4   5 edges.   9.5.5   6 edges.   9.5.6   7 edges.

9.5.7   Calculate the cycle index polynomial of $\mathcal{A}ut_V(K_7)$.

9.5.8   Calculate the cycle index polynomial of $\mathcal{A}ut_E(K_7)$.

9.5.9^S Calculate the number of isomorphism types of simple graphs with 7 vertices.

In each of the Exercises 9.5.10 through 9.5.15, use a partial inventory table to calculate the number of isomorphism types of 7-vertex graphs with the given number of edges.

9.5.10   2 edges.   9.5.11   3 edges.   9.5.12^S 4 edges.
9.5.13   5 edges.   9.5.14   6 edges.   9.5.15   7 edges.

9.6   GENERAL GRAPHS AND DIGRAPHS

With some small variations, the same techniques used to count the isomorphism types of simple graphs can be applied to counting graphs with multiple edges, self-loops, and edge directions.

Counting Multigraphs

DEFINITION: A **multigraph** is a graph model in which multiple edges are permitted, but not self-loops. A *c*-**multigraph** has at most c edges joining any two vertices.

In counting isomorphism types of simple n-vertex graphs, the model was edge 2-colorings of K_n. The color 1 represented the presence of an edge and the color 0 the absence. If the number of edges between two vertices is permitted to rise from 1 to c, then the number of colors in the model is increased from 2 to $c + 1$.

Proposition 9.6.1. *The 3-vertex 2-multigraphs fall into exactly 10 isomorphism types.*

Proof: The first step is calculating the cycle index

$$\mathcal{Z}_{Aut_V(K_3)}(t_1, t_2, t_3) = \frac{1}{6}(t_1^3 + 3t_1t_2 + 2t_3)$$

The second is transforming it to

$$\mathcal{Z}_{Aut_E(K_3)}(y_1, y_2, y_3) = \frac{1}{6}(y_1^3 + 3y_1y_2 + 2y_3)$$

The final step of substituting 3 yields

$$\mathcal{Z}_{Aut_E(K_3)}(3, 3, 3) = \frac{1}{6}(3^3 + 3 \cdot 3 \cdot 3 + 2 \cdot 3)$$
$$= \frac{60}{6} = 10 \qquad \diamond$$

The result of Proposition 9.6.1 is confirmed by the list of multigraphs in Figure 9.6.1.

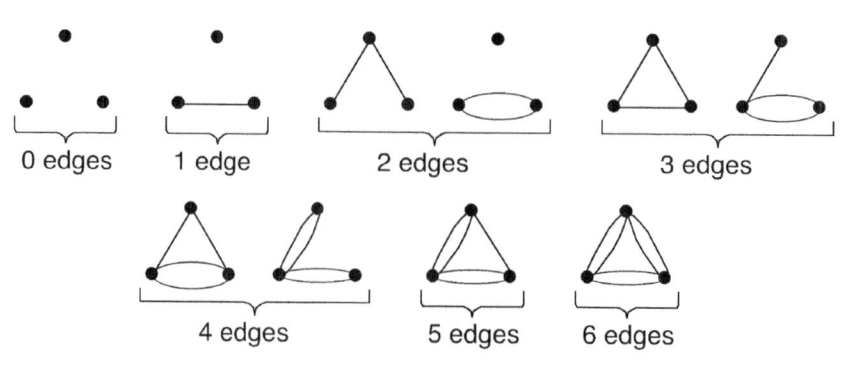

Figure 9.6.1   The ten 3-vertex 2-multigraphs.

Counting General Graphs

Augmenting the complete graph by adding a self-loop at every vertex enables us to solve a counting problem for general graphs.

DEFINITION: The **very complete graph** K_n° on n vertices has an edge joining each pair of vertices and a self-loop at each vertex.

The cycle index for $\mathcal{A}ut_V(K_n^\circ)$ is the same as for $\mathcal{A}ut_V(K_n)$. However there is a slightly different rule for transforming a cycle structure. Whatever cycle structure would have been obtained for $\mathcal{A}ut_E(K_n)$ is augmented by a factor of y_j for each factor t_j.

Proposition 9.6.2. *There are exactly 20 isomorphism types of 3-vertex general graph with edge multiplicity at most 1.*

Proof: Under the modified transformation rule, the cycle index

$$\mathcal{Z}_{\mathcal{A}ut_V(K_3^\circ)}(t_1, t_2, t_3) \;=\; \frac{1}{6}\big(t_1^3 + 3t_1 t_2 + 2t_3 \big)$$

is transformed into

$$\mathcal{Z}_{\mathcal{A}ut_E(K_3^\circ)}(y_1, y_2, y_3) \;=\; \frac{1}{6}\big(y_1^6 + 3y_1^2 y_2^2 + 2y_3^2 \big)$$

Substituting 2 yields

$$
\begin{aligned}
\mathcal{Z}_{\mathcal{A}ut_E(K_3^\circ)}(2,2,2) &\;=\; \frac{1}{6}\big(2^6 + 3\cdot 2^2 \cdot 2^2 + 2\cdot 2^2 \big) \\
&\;=\; \frac{120}{6} \;=\; 20 \hspace{4em} \diamond
\end{aligned}
$$

Figure 9.6.2 shows the isomorphism types for 0 to 3 edges. The graphs with 4, 5, and 6 edges are the complements in K_n° of the graphs with 2, 1, and 0 edges, respectively.

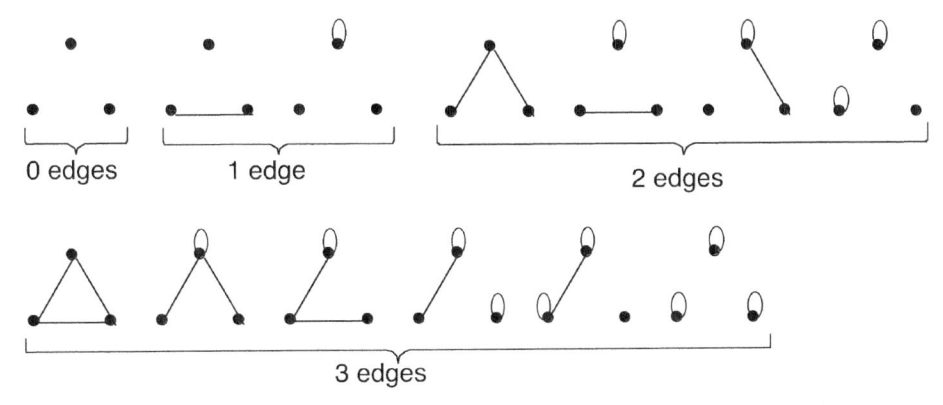

Figure 9.6.2   **The 3-vertex general graphs with edge multiplicity at most 1 and at most 3 edges.**

Counting Simple Digraphs

Another modification of the main model enables us to count isomorphism types of digraphs.

DEFINITION: The **complete digraph** $\vec{K}_n$ has n vertices and an arc from each vertex to each other vertex.

The cycle index for $\mathcal{A}ut_V(\vec{K}_n)$ is the same as for $\mathcal{A}ut_V(K_n)$. However there are different rules for transforming a cycle structure, as follows.

Theorem 9.6.3. *Let π be an automorphism of the complete digraph $\vec{K}_n$, represented by permutations π_V on $V_{\vec{K}_n}$ and π_E on $E_{\vec{K}_n}$. Then*

(i) $t_{2p+1} \to y_{2p+1}^{2p}$: Each $(2p+1)$-cycle in π_V corresponds to $2p$ $(2p+1)$-cycles in π_E.

(ii) $t_{2p} \to y_{2p}^{2p-1}$: Each $(2p)$-cycle in π_V corresponds to $2p-1$ $(2p)$-cycles in π_E.

(iii) $t_p t_q \to y_{\text{lcm}\,(p,q)}^{2\gcd\,(p,q)}$: Each $(p$-cycle, q-cycle$)$-pair in π_V, with $p \neq q$, corresponds to $2\gcd\,(p,q)$ $\text{lcm}\,(p,q)$-cycles in π_E.

(iv) $t_p t_p \to y_p^{2p}$: Each pair of p-cycles in π_V corresponds to $2p$ p-cycles in π_E.

Proof: Omitted. Analogous to the proof of Theorem 9.5.5. $\diamond$

Proposition 9.6.4. *There are exactly 16 isomorphism types of 3-vertex simple digraph.*

Proof: Under Theorem 9.6.3, the cycle index

$$\mathcal{Z}_{\mathcal{A}ut_V(\vec{K}_3)}(t_1, t_2, t_3) \;=\; \frac{1}{6}\big(t_1^3 + 3t_1 t_2 + 2t_3\big)$$

is transformed into

$$\mathcal{Z}_{\mathcal{A}ut_E(\vec{K}_3)}(y_1, y_2, y_3) \;=\; \frac{1}{6}\big(y_1^6 + 3y_2^3 + 2y_3^2\big)$$

Substituting 2 yields

$$\mathcal{Z}_{\mathcal{A}ut_E(\vec{K}_3)}(2, 2, 2) \;=\; \frac{1}{6}\big(2^6 + 3\cdot 2^3 + 2\cdot 2^2\big)$$
$$\;=\; \frac{96}{6} \;=\; 16 \qquad\qquad \diamond$$

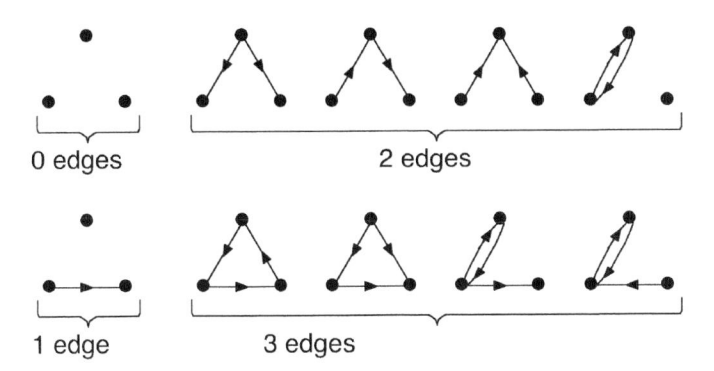

Figure 9.6.3   **The 3-vertex simple digraphs with at most 3 edges.**

EXERCISES for Section 9.6

9.6.1[S]   Count the 3-vertex 3-multigraphs.

9.6.2   Draw all the isomorphism types of 3-vertex 3-multigraphs, and compare to your answer to Exercise 9.6.1.

9.6.3   Count the 4-vertex 2-multigraphs.

9.6.4   Draw all the isomorphism types of 4-vertex 2-multigraphs, and compare to your answer to Exercise 9.6.3.

9.6.5   Count the 2-vertex general graphs with edge-multiplicity at most 2.

9.6.6   Draw all the isomorphism types of 2-vertex general graphs with edge-multiplicity at most 2, and compare to your answer to Exercise 9.6.5.

9.6.7   Count the 2-vertex general graphs with edge-multiplicity at most 3.

9.6.8   Draw all the isomorphism types of 2-vertex general graphs with edge-multiplicity at most 3, and compare to your answer to Exercise 9.6.7.

9.6.9   Count the 4-vertex simple digraphs.

9.6.10   Draw all the isomorphism types of 4-vertex simple digraphs, and compare to your answer to Exercise 9.6.9.

GLOSSARY

automorphism of a simple graph: a bijection on the vertex set that preserves adjacencies and non-adjacencies as well.

Burnside-Pólya counting: an algebraic method of counting induced equivalence classes; the most important method of graphical enumeration.

Burnside's Lemma: a theorem for counting the equivalence classes of objects under a permutation group.

($\leq k$)-coloring: a coloring that uses at most k colors, but possibly fewer.

complete digraph: a digraph such that from each vertex to each other vertex there is an arc.

conjugate of a Ferrers diagram: the result of reflecting the diagram through its southwest diagonal.

conjugate of a partition S: the partition whose Ferrers diagram is the conjugate of the diagram of S.

cycle index of a permutation group: a multivariate polynomial that gives the distribution of cycle structures of the permutations in that group.

cycle structure of a permutation π: a monomial $t_1^{r_1} t_2^{r_2} \cdots t_n^{r_n}$ such that r_j is the number of cycles of length j in the disjoint cycle form of π.

cyclic permutation: a permutation with only one cycle in its disjoint cycle form.

cyclic permutation group: a group in which there is a permutation π such that every other element is obtainable as an iterated composition of π with itself.

dihedral group $\mathbb{D}_n$:   the group of symmetries of a regular n-sided polygon, or any group isomorphic to that group.

Ferrers diagram of a partition $s_1 s_2 \cdots s_k$ of an integer: an array of k rows of dots, with s_j dots in row j, for $j = 1, \ldots, k$.

fixed point of a permutation π: an object y such that $\pi(y) = y$.

group: an algebraic structure whose domain is closed under a binary operation, called either *addition* or *multiplication*, such that there is an identity element and that every element has an inverse.

inclusion relation on integer partitions: we write $s_1 s_2 \cdots s_k \preceq u_1 u_2 \cdots u_\ell$ if $k \leq \ell$ and if $s_j \leq u_j$ for $j = 1, \ldots, k$.

induced action of a permutation $\pi : Y \to Y$ on the set $f : Y \to [1 : k]$ of colorings of Y: the permutation that maps any coloring f to the coloring $f\pi$.

induced edge automorphism of an automorphism $\pi : V \to V$ on the vertex set of a simple graph: the automorphism $uv \mapsto \pi(u)\pi(v)$.

isomorphism of two simple graphs G and H: a bijection $V_G \to V_H$ that preserves all adjacencies and all non-adjacencies as well.

length of a cycle: the number of edge-steps (or, equivalently, of vertices).

monomial: a polynomial with only one term.

multigraph: a graph in which there may exist at least one pair of edges with the same set of endpoints.

multinomial coefficient: generalization of a binomial coefficient.

orbit of an object y under a permutation group $\mathcal{P} = [P : Y]$: the set

$$\{\pi(y) \mid \pi \in P\}$$

partition of an integer n: a sum $s_1 + s_2 + \cdots + s_k$ of positive integers, usually written in non-increasing order, often without the plus signs, i.e., as $s_1 s_2 \cdots s_k$.

permutation group $\mathcal{P} = [P : Y]$: a set Y of objects and a set P of permutations that is a group under composition; for a finite set of objects, a closed non-empty set of permutations is a group.

permutation of a set: a bijection.

Pólya Inventory Theorem: proof of the correctness of a method for obtaining a breakdown of the total count given by Burnside counting into counts for subclasses or orbits, according to features of the objects within the orbits.

Pólya substitution: a rule for substituting monomials for indeterminates in the cycle index polynomial, so as to produce a Pólya inventory polynomial.

simple graph: a graph with no multi-edges or self-loops.

stabilizer of an object x in a permutation group: that subgroup of permutations that map x to itself.

summation dominance lattice: a lattice of all the partitions of an integer n.

wheel graph W_n: the graph obtained by joining a new vertex to an n-cycle.

Young's lattice a lattice of partitions by inclusion.

Chapter 10

Combinatorial Designs

10.1 Latin Squares

10.2 Block Designs

10.3 Classical Finite Geometries

10.4 Projective Planes

10.5 Affine Planes

A *combinatorial design* (or alternatively, an *incidence structure*) consists of a domain set X and another set B, commonly represented as subsets of that domain, analogous to the way in which the edges of a simple graph can be represented as pairs of vertices. This final chapter studies several kinds of combinatorial designs, each with additional axioms and/or mathematical structure on the domain and/or on the subsets.

Remark: Some constructions in this chapter that are described here only for a prime field $\mathbb{Z}_p$ are extendible to a prime power, since there is a finite field for every prime power.

10.1   LATIN SQUARES

A *Latin square* is a type of combinatorial design most easily described as an $n \times n$ array.

DEFINITION: A **Latin square** on a set X of n objects is an $n \times n$ array such that each object in X occurs once in each row and once in each column.

Example 10.1.1: A Latin square on a set of four graphic patterns is shown in Figure 10.1.1.

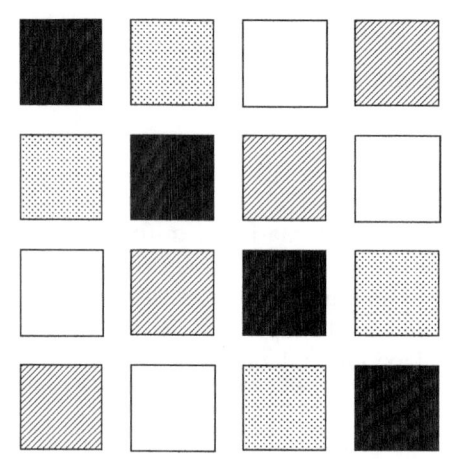

Figure 10.1.1   A 4×4 **Latin square.**

The standard symbols for an $n \times n$ Latin square are the integers modulo n. The rows and columns of a Latin square on $\mathbb{Z}_n$ are commonly indexed in $\mathbb{Z}_n$, so that there is a row 0 and a column 0. In particular, the following 4×4 Latin square on $\mathbb{Z}_4$ is obtainable from the Latin square of Figure 10.1.1 by a bijection of the symbol sets.

$$\begin{pmatrix} 0 & 1 & 2 & 3 \\ 1 & 0 & 3 & 2 \\ 2 & 3 & 0 & 1 \\ 3 & 2 & 1 & 0 \end{pmatrix} \tag{10.1.1}$$

TERMINOLOGY NOTE: Euler used Latin letters as the objects when he introduced the idea, whence the name *Latin squares*. In small examples, sometimes the objects are colors or patterns.

Remark: Thus, a sudoku is a form of 9×9 Latin square on the numbers 1 to 9, with an additional requirement that each number occur exactly once in certain 3×3 sub-arrays.

It is easy enough to construct a Latin square of any given size.

Proposition 10.1.1. *For every positive integer n, there exists an $n \times n$ Latin square with $\mathbb{Z}_n$ as the set of objects.*

Proof: Let $L[i,j] = i + j$ modulo n. Thus,

$$L = \begin{pmatrix} 0 & 1 & 2 & \cdots & n-2 & n-1 \\ 1 & 2 & 3 & \cdots & n-1 & 0 \\ 2 & 3 & 4 & \cdots & 0 & 1 \\ \vdots & \vdots & \vdots & \vdots & \vdots & \vdots \\ n-2 & n-1 & 0 & \cdots & n-4 & n-3 \\ n-1 & 0 & 1 & \cdots & n-3 & n-2 \end{pmatrix}$$

Clearly the array L is a Latin square. $\diamondsuit$

Example 10.1.2: For $n = 4$, the construction of Proposition 10.1.1 yields the following Latin square.

$$\begin{pmatrix} 0 & 1 & 2 & 3 \\ 1 & 2 & 3 & 0 \\ 2 & 3 & 0 & 1 \\ 3 & 0 & 1 & 2 \end{pmatrix} \tag{10.1.2}$$

A *Latin square* can be recognized as a type of combinatorial design $\langle X, B \rangle$ with additional structure. The set B is ordered, corresponding to the order of the rows in the array. Each member $B_j \in B$ contains every object of X, is construed to be ordered, corresponding to the order of the elements of a row. Moreover, the number of subsets in B equals the number of objects in X, and for each object x and each possible position within a row, there is a unique row in which x occupies that position.

Product of Latin Squares

The next definition indicates a method of construction of a new Latin square, starting from two given Latin squares.

DEFINITION: Let $A = (a_{ij})$ and $B = (b_{ij})$ be Latin squares on $\mathbb{Z}_r$ and $\mathbb{Z}_s$, respectively. Then the **product square** $A \otimes B$ is the Latin square on $\mathbb{Z}_r \times \mathbb{Z}_s$

$$A \otimes B = \begin{pmatrix} a_{00} \times B & a_{01} \times B & \cdots & a_{0(r-1)} \times B \\ a_{10} \times B & a_{11} \times B & \cdots & a_{1(r-1)} \times B \\ \vdots & \vdots & \cdots & \vdots \\ a_{(r-1)0} \times B & a_{01} \times B & \cdots & a_{(r-1)(r-1)} \times B \end{pmatrix}$$

where the $s \times s$ submatrix $a_{ij} \times B$ is given by

$$a_{ij} \times B = \begin{pmatrix} (a_{ij}, b_{00}) & (a_{ij}, b_{01}) & \cdots & (a_{ij}, b_{0(s-1)}) \\ (a_{ij}, b_{10}) & (a_{ij}, b_{11}) & \cdots & (a_{ij}, b_{1(s-1)}) \\ \vdots & \vdots & \cdots & \vdots \\ (a_{ij}, b_{(s-1)0}) & (a_{ij}, b_{(s-1)1}) & \cdots & (a_{ij}, b_{(s-1)(s-1)}) \end{pmatrix}$$

Proposition 10.1.2. Let $A = (a_{ij})$ and $B = (b_{ij})$ be Latin squares on $\mathbb{Z}_r$ and $\mathbb{Z}_s$, respectively. Their product $A \otimes B$ is a Latin square.

Proof: Since each row of A contains each number in $\mathbb{Z}_r$ and each row of B contains each number in $\mathbb{Z}_s$, it follows that each row of $A \otimes B$ contains each pair in $\mathbb{Z}_r \times \mathbb{Z}_s$. The same fact holds for the columns.   ◇

Example 10.1.3:   If

$$A = \begin{pmatrix} 0 & 1 \\ 1 & 0 \end{pmatrix} \quad \text{and} \quad B = \begin{pmatrix} 0 & 1 & 2 \\ 1 & 2 & 0 \\ 2 & 0 & 1 \end{pmatrix}$$

then

$$A \otimes B = \begin{pmatrix} (0,0) & (0,1) & (0,2) & (1,0) & (1,1) & (1,2) \\ (0,1) & (0,2) & (0,0) & (1,1) & (1,2) & (1,0) \\ (0,2) & (0,0) & (0,1) & (1,2) & (1,0) & (1,1) \\ (1,0) & (1,1) & (1,2) & (0,0) & (0,1) & (0,2) \\ (1,1) & (1,2) & (1,0) & (0,1) & (0,2) & (0,0) \\ (1,2) & (1,0) & (1,1) & (0,2) & (0,0) & (0,1) \end{pmatrix}$$

which we observe is equivalent to the Latin square

$$\begin{pmatrix} 0 & 1 & 2 & 3 & 4 & 5 \\ 1 & 2 & 0 & 4 & 5 & 3 \\ 2 & 0 & 1 & 5 & 3 & 4 \\ 3 & 4 & 5 & 0 & 1 & 2 \\ 4 & 5 & 3 & 1 & 2 & 0 \\ 5 & 3 & 4 & 2 & 0 & 1 \end{pmatrix}$$

under the bijection $\mathbb{Z}_2 \times \mathbb{Z}_3 \to \mathbb{Z}_6$ given by

$$(0,0) \mapsto 0 \quad (0,1) \mapsto 1 \quad (0,2) \mapsto 2$$
$$(1,0) \mapsto 3 \quad (1,1) \mapsto 4 \quad (1,2) \mapsto 5$$

Orthogonal Latin Squares

DEFINITION: Two $n \times n$ Latin squares $A = (a_{i,j})$ and $B = (b_{i,j})$ are **orthogonal Latin squares** if the n^2 ordered pairs $(a_{i,j}, b_{i,j})$ are mutually distinct.

Remark: By the pigeonhole principle, two $n \times n$ Latin squares are orthogonal if each possible ordered pair of domain elements occurs.

Example 10.1.4:   It is easy enough to construct the pair of orthogonal 4×4 Latin squares in Figure 10.1.2 by ad hoc methods. One Latin square is represented pictorially by the outer pattern in an array location, and the other Latin square by the inner pattern.

$$\begin{pmatrix} 0 & 1 & 2 & 3 \\ 1 & 0 & 3 & 2 \\ 2 & 3 & 0 & 1 \\ 3 & 2 & 1 & 0 \end{pmatrix}$$
outer

$$\begin{pmatrix} 3 & 0 & 1 & 2 \\ 2 & 1 & 0 & 3 \\ 0 & 3 & 2 & 1 \\ 1 & 2 & 3 & 0 \end{pmatrix}$$
inner

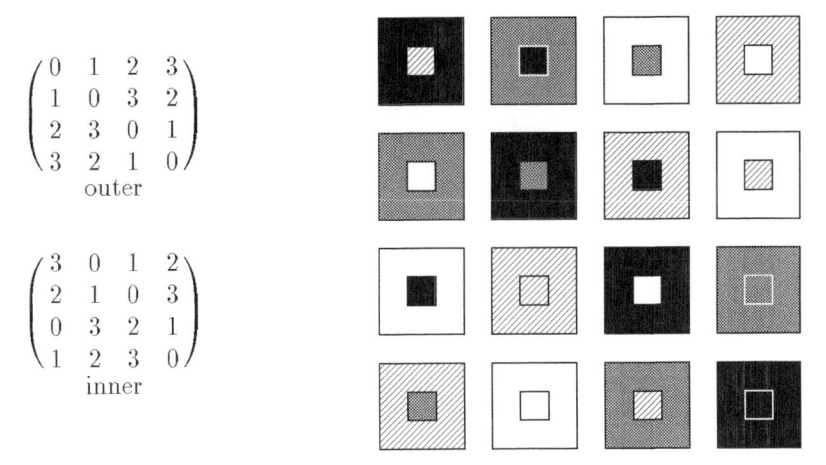

Figure 10.1.2 Two orthogonal Latin squares.

The next proposition indicates how to construct a family of mutually orthogonal Latin squares.

Proposition 10.1.3. *For $k = 1, \ldots, p-1$, where p is a prime number, let L_p^k be the $p \times p$ array such that*

$$L_p^k[i, j] = ki + j \bmod p \qquad 0 \le i, j \le p-1$$

Then the $p-1$ arrays

$$L_p^1, \quad L_p^2, \quad \ldots, \quad L_p^{p-1}$$

are mutually orthogonal Latin squares.

Proof: The entries in row i of the array L_p^k are

$$ki, \; ki+1, \; ki+2, \; \ldots, \; ki+(p-1)$$

which are clearly distinct. The entries in column j are

$$j, \; j+k, \; j+2k, \; \ldots, \; j+(p-1)k$$

Two of these entries differ by some number ck with $0 < c, k < p$. Since p is prime, $ck \not\equiv 0$ modulo p. Therefore, each of the arrays L_p^k is a Latin square.

Now suppose that the pairs of entries

$$\left(L_p^k[i, j], \; L_p^{k'}[i, j] \right) \quad \text{and} \quad \left(L_p^k[\hat{i}, \hat{j}], \; L_p^{k'}[\hat{i}, \hat{j}] \right)$$

are identical. Then

$$ki + j = k\hat{i} + \hat{j} \tag{10.1.3}$$

and

$$k'i + j = k'\hat{i} + \hat{j} \tag{10.1.4}$$

If $i \neq \hat{i}$, then $i - \hat{i}$ has a multiplicative inverse in $\mathbb{Z}_p$ (see Corollary 6.4.2). Hence,

$$k \; = \; \frac{\hat{j} - j}{i - \hat{i}} \qquad \text{from (10.1.3)}$$

and

$$k' \; = \; \frac{\hat{j} - j}{i - \hat{i}} \qquad \text{from (10.1.4)}$$

Therefore, $k = k'$.

Example 10.1.5:   The arrays L_5^2 and L_5^3 of Proposition 10.1.3 are orthogonal.

$$L_5^2 \; = \; \begin{pmatrix} 0 & 1 & 2 & 3 & 4 \\ 2 & 3 & 4 & 0 & 1 \\ 4 & 0 & 1 & 2 & 3 \\ 1 & 2 & 3 & 4 & 0 \\ 3 & 4 & 0 & 1 & 2 \end{pmatrix} \qquad L_5^3 \; = \; \begin{pmatrix} 0 & 1 & 2 & 3 & 4 \\ 3 & 4 & 0 & 1 & 2 \\ 1 & 2 & 3 & 4 & 0 \\ 4 & 0 & 1 & 2 & 3 \\ 2 & 3 & 4 & 0 & 1 \end{pmatrix}$$

Remark: If p is not a prime, then L_p^k might not be a Latin square. For instance, row 2 of the array L_6^3 is identical to row 0.

Theorem 10.1.4 [MacNeish, 1922]. *Let*

$$A^{(1)}, \; A^{(2)}, \; \ldots, \; A^{(r)}$$

be r mutually orthogonal $m \times m$ Latin squares, and let

$$B^{(1)}, \; B^{(2)}, \; \ldots, \; B^{(r)}$$

be r mutually orthogonal $n \times n$ Latin squares. Then the Latin squares

$$A^{(1)} \otimes B^{(1)}, \; A^{(2)} \otimes B^{(2)}, \; \cdots, \; A^{(r)} \otimes B^{(r)}$$

are mutually orthogonal.

Proof: Suppose that the pair of entries at location $ij \times k\ell$ of the Latin square $A^{(x)} \times B^{(x)}$ and of the Latin square $A^{(y)} \times B^{(y)}$, i.e.,

$$(a_{ij}^{(x)}, \, b_{k\ell}^{(x)}) \quad \text{and} \quad (a_{ij}^{(y)}, \, b_{k\ell}^{(y)})$$

is the same as the pair in location $pq \times uv$ of those two Latin squares, i.e., as the pair

$$(a_{pq}^{(x)}, \, b_{uv}^{(x)}) \quad \text{and} \quad (a_{pq}^{(y)}, \, b_{uv}^{(y)})$$

Then the pairs

$$(a_{ij}^{(x)}, \, a_{ij}^{(y)}) \quad \text{and} \quad (a_{pq}^{(x)}, \, a_{pq}^{(y)})$$

are identical, which implies, since $A^{(x)}$ and $A^{(y)}$ are orthogonal, that

$$i = p \quad \text{and} \quad j = q$$

Similarly,

$$k = u \quad \text{and} \quad \ell = v$$

Therefore, $A^{(x)} \times B^{(x)}$ and $A^{(y)} \times B^{(y)}$ are orthogonal. $\diamond$

Proposition 10.1.5. *For every odd number $n > 1$, there is a pair of orthogonal $n \times n$ Latin squares.*

Proof: This follows from Proposition 10.1.3 and Theorem 10.1.4, since every odd number factors into a product of odd primes. $\diamond$

Proposition 10.1.6. *Let $n = 2^k$ with $k \geq 2$. Then there is a pair of orthogonal $n \times n$ Latin squares.*

Proof: Example 10.1.4 gives a pair of orthogonal 4×4 Latin squares. The following is a pair of orthogonal 8×8 Latin squares.

$$
\begin{pmatrix}
0 & 1 & 2 & 3 & 4 & 5 & 6 & 7 \\
1 & 0 & 3 & 2 & 5 & 4 & 7 & 6 \\
2 & 3 & 0 & 1 & 6 & 7 & 4 & 5 \\
3 & 2 & 1 & 0 & 7 & 6 & 5 & 4 \\
4 & 5 & 6 & 7 & 0 & 1 & 2 & 3 \\
5 & 4 & 7 & 6 & 1 & 0 & 3 & 2 \\
6 & 7 & 4 & 5 & 2 & 3 & 0 & 1 \\
7 & 6 & 5 & 4 & 3 & 2 & 1 & 0
\end{pmatrix}
\qquad
\begin{pmatrix}
0 & 1 & 2 & 3 & 4 & 5 & 6 & 7 \\
7 & 6 & 5 & 4 & 3 & 2 & 1 & 0 \\
3 & 2 & 1 & 0 & 7 & 6 & 5 & 4 \\
4 & 5 & 6 & 7 & 0 & 1 & 2 & 3 \\
6 & 7 & 4 & 5 & 2 & 3 & 0 & 1 \\
1 & 0 & 3 & 2 & 5 & 4 & 7 & 6 \\
5 & 4 & 7 & 6 & 1 & 0 & 3 & 2 \\
2 & 3 & 0 & 1 & 6 & 7 & 4 & 5
\end{pmatrix}
$$

If k is even, then n is a power of 4, and if k is odd, then n is a product of 8 with a power of 4. It follows from the base cases 4×4 and 8×8 and Theorem 10.1.4 that there is a pair of orthogonal $n \times n$ Latin squares. $\diamond$

There are only two possible 2×2 Latin squares in $\mathbb{Z}_2$, and they are not orthogonal. Euler conjectured in 1782 that for n odd, there is no orthogonal pair of $2n \times 2n$ Latin squares. In 1901, Gaston Tarry [Tarr1901] proved by exhaustion that there is no 6×6 pair. However, Ernest Parker [Park1959] produced a 10×10 pair in 1960, and then Bose, Shrikhande, and Parker [BSP1960] proved that there is a $2n \times 2n$ orthogonal pair except for $n = 1$ or 3.

Summary. *For every positive integer n except 1, 2, and 6, there is a pair of orthogonal $n \times n$ Latin squares.*

Isotopic Latin Squares

DEFINITION: The Latin squares $L[i, j]$ and $L'[i, j]$ on $\mathbb{Z}_n$ are **isotopic Latin squares** if L' can be obtained from L by a sequence of transformations, each chosen from any of the following three types.

- A permutation of the rows.

- A permutation of the columns.

- Applying a permutation $\sigma : \mathbb{Z}_n \to \mathbb{Z}_n$ to the symbols of the array.

Example 10.1.6:   Swapping rows 0 and 1 of the Latin square

$$
\begin{pmatrix}
0 & 1 & 2 & 3 \\
1 & 2 & 3 & 0 \\
2 & 3 & 0 & 1 \\
3 & 0 & 1 & 2
\end{pmatrix}
\tag{10.1.2}
$$

yields the Latin square

$$
\begin{pmatrix}
1 & 2 & 3 & 0 \\
0 & 1 & 2 & 3 \\
2 & 3 & 0 & 1 \\
3 & 0 & 1 & 2
\end{pmatrix}
$$

Example 10.1.7:   Swapping the symbols 0 and 1 in the Latin square (10.1.2) yields this Latin square.

$$
\begin{pmatrix}
1 & 0 & 2 & 3 \\
0 & 2 & 3 & 1 \\
2 & 3 & 1 & 0 \\
3 & 1 & 0 & 2
\end{pmatrix}
$$

Remark: Clearly, isotopy on Latin squares is an equivalence relation.

DEFINITION: A Latin square on $\mathbb{Z}_n$ is said to be **normalized** if its initial row is

$$0 \quad 1 \quad \cdots \quad n-1$$

and its initial column is

$$
\begin{matrix}
0 \\
1 \\
\vdots \\
n-1
\end{matrix}
$$

Clearly, every Latin square is isotopic to a normalized Latin square.

Abstract Latin Squares

Isotopy allows three natural kinds of transformation on Latin squares that may be regarded as natural equivalences. The following alternative conceptualization of a Latin square allows some additional equivalences.

DEFINITION: An **abstract Latin square** on $\mathbb{Z}_n$ is a set L of triples

$$(r, \ c, \ s)$$

in $\mathbb{Z}_n \times \mathbb{Z}_n \times \mathbb{Z}_n$ such that

- For any $(i, j) \in \mathbb{Z}_n \times \mathbb{Z}_n$ there is a unique triple $(r,\ c,\ s)$ in L such that $i = r$ and $j = c$.

- For any $(i, k) \in \mathbb{Z}_n \times \mathbb{Z}_n$ there is a unique triple $(r,\ c,\ s)$ in L such that $i = r$ and $k = s$.

- For any $(j, k) \in \mathbb{Z}_n \times \mathbb{Z}_n$ there is a unique triple $(r,\ c,\ s)$ in L such that $j = c$ and $k = s$.

Proposition 10.1.7. *Every abstract Latin square corresponds to a unique concrete Latin square (i.e., the array form). Conversely, for every concrete Latin square, there is a unique abstract Latin square.*   ◇

We observe that the operation of transposition on the array form of a Latin square has as its abstract counterpart the operation of swapping the first and second entry in each triple. Yet from the abstract perspective, we could equally well swap the first and third entry of each triple. Indeed, we equally apply any of the six possible permutations uniformly to all the triples. This motivates the following definition.

DEFINITION: Let π be a permutation on the set $\{1,\ 2,\ 3\}$. The operation of transforming a Latin square by applying π to the coordinates of the triples is called a **conjugacy operation**. The array resulting from applying π to a Latin square L is called the π-**conjugate** of L. It may be denoted L^π.

Example 10.1.8:   Consider the following Latin square in array and abstract form.

$$
L = \begin{pmatrix} 0 & 3 & 1 & 2 \\ 1 & 2 & 0 & 3 \\ 3 & 0 & 2 & 1 \\ 2 & 1 & 3 & 0 \end{pmatrix}
\qquad
\begin{matrix}
(0,0,0) & (0,1,3) & (0,2,1) & (0,3,2) \\
(1,0,1) & (1,1,2) & (1,2,0) & (1,3,3) \\
(2,0,3) & (2,1,0) & (2,2,2) & (2,3,1) \\
(3,0,2) & (3,1,1) & (3,2,3) & (3,3,0)
\end{matrix}
$$

Applying the permutation $(1,2)(3)$ to the set of triples means swapping the first and second coordinates of each triple, thereby obtaining

$$
\begin{matrix}
(0,0,0) & (1,0,3) & (2,0,1) & (3,0,2) \\
(0,1,1) & (1,1,2) & (2,1,0) & (3,1,3) \\
(0,2,3) & (1,2,0) & (2,2,2) & (3,2,1) \\
(0,3,2) & (1,3,1) & (2,3,3) & (3,3,0)
\end{matrix}
$$

which is the abstract form of the Latin square

$$
L^{(1,2)(3)} = \begin{pmatrix} 0 & 1 & 3 & 2 \\ 3 & 2 & 0 & 1 \\ 1 & 0 & 2 & 3 \\ 2 & 3 & 1 & 0 \end{pmatrix}
$$

Observing that $L^{(1,2)(3)}$ is simply the transpose of L, we recognize that the transformation $L \mapsto L^{(1,2)(3)}$ simply swaps the roles of rows and columns.

Alternatively, applying the permutation $(1,3)(2)$ to the set of triples means swapping the first and third coordinates of each triple, thereby obtaining

$$
\begin{array}{cccc}
(0,0,0) & (3,1,0) & (1,2,0) & (2,3,0) \\
(1,0,1) & (2,1,1) & (0,2,1) & (3,3,1) \\
(3,0,2) & (0,1,2) & (2,2,2) & (1,3,2) \\
(2,0,3) & (1,1,3) & (3,2,3) & (0,3,3)
\end{array}
$$

which is the abstract form of the Latin square

$$
L^{(1,3)(2)} = \begin{pmatrix}
0 & 2 & 1 & 3 \\
1 & 3 & 0 & 2 \\
3 & 1 & 2 & 0 \\
2 & 0 & 3 & 1
\end{pmatrix}
$$

Remark 1: We observe that conjugacy is an equivalence relation on the Latin squares. The possible class sizes are 1, 2, 3, and 6.

Remark 2: For $n \leq 5$, the conjugacy operations on a Latin square produce only Latin squares that could be obtained by isotopy operations. However, for $n \geq 6$, they produce additional Latin squares.

DEFINITION: Two Latin squares L and L' are **main class isotopic** if L is isotopic to any conjugate of L'.

EXERCISES for Section 10.1

10.1.1^S Draw two orthogonal 3×3 Latin squares.

10.1.2   Prove that there are only two Latin squares on $\mathbb{Z}_3$ whose first row is

$$0 \quad 1 \quad 2$$

10.1.3   Prove that the number of non-identical Latin squares on $\mathbb{Z}_3$ is 12.

10.1.4   Prove that there is only one normalized Latin square on $\mathbb{Z}_3$.

10.1.5   Draw the four mutually non-identical normalized Latin squares on $\mathbb{Z}_4$.

In each of the Exercises 10.1.6 through 10.1.9, construct the designated Latin square, as specified in Proposition 10.1.3.

10.1.6^S L_5^4 10.1.7 L_7^2 10.1.8 L_7^3 10.1.9 L_7^5

In each of the Exercises 10.1.10 through 10.1.13, let

$$
A = \begin{pmatrix} 1 & 0 \\ 0 & 1 \end{pmatrix} \quad \text{and} \quad B = \begin{pmatrix} 2 & 1 & 0 \\ 1 & 0 & 2 \\ 0 & 2 & 1 \end{pmatrix}
$$

Construct the specified product square.

10.1.10^S $A \otimes A$ 10.1.11^S $A \otimes B$ 10.1.12 $B \otimes B$ 10.1.13 $B \otimes A$

10.1.**14**   Draw four mutually orthogonal 5×5 Latin squares.

10.1.**15**   Draw three mutually orthogonal 7×7 Latin squares.

In each of the Exercises 10.1.16 through 10.1.18, let

$$L \;=\; \begin{pmatrix} 0 & 3 & 1 & 2 \\ 1 & 2 & 0 & 3 \\ 3 & 0 & 2 & 1 \\ 2 & 1 & 3 & 0 \end{pmatrix}$$

Construct the specified conjugate.

10.1.**16**[S] $L^{(1)(2,3)}$ 10.1.**17** $L^{(1,2,3)}$ 10.1.**18** $L^{(1,3,2)}$

10.2   BLOCK DESIGNS

A generic *block design* can be regarded as a generalization of a graph, in which a *block* is a generalized edge.

DEFINITION: A **block design** $\mathcal{B}$ has a non-empty domain

$$X \;=\; \{\, x_1,\ x_2,\ \ldots,\ x_v \,\}$$

whose elements are sometimes called *varieties* and a non-empty collection

$$B \;=\; \{B_1,\ B_2,\ \ldots,\ B_b\}$$

of subsets of X called **blocks**. It is a **simple design** if no two blocks are identical.

DEFINITION: The number of blocks in which an element x appears is called the **valence of that element of the design**.

DEFINITION: The number of blocks in which a pair of elements x and y appears is called the **covalence of that pair**.

Thus, a graph could be regarded as a block design in which every block has size 2. The valence of an element within the block design would be its degree as a vertex of the graph. The covalence of a pair of elements of the design would be their multiplicity of adjacency as vertices of the graph. To allow self-loops in a graph, one would allow the blocks to be multisets of elements of the design and make suitable revisions in the definition of valence and covalence.

DEFINITION: A block design is **regular** if the following two conditions hold:

- every block is the same size $k \geq 2$, which is called the **blocksize**;
- each element x_j has the same valence; that is, each appears in the same number r of blocks, which is called the **replication number**.

Thus, a d-regular graph is a regular block design with blocksize 2 and replication number d.

Balanced Designs

The notion of *balancing* a design with *incomplete blocks* arose with Sir Ronald Fisher (1890-1962) in his theoretical study of the design of experiments in agriculture.

DEFINITION: A regular block design $\mathcal{B}$ with v varieties and b blocks is said to be **balanced** and is called either a *(v, b, r, k, λ)-design* or a *(v, k, λ)-design* if each pair of elements x_i and x_j has the same covalence, that is, if each pair appears in the same number λ of blocks, which is called the **index of the design**.

A balanced design is **complete** if $k = v$, so that each block contains all of X. If $k < v$, then it is **incomplete**.

TERMINOLOGY: A balanced incomplete block design is commonly called a **BIBD**.

Example 10.2.1:   For $X = \{0, 1, 2, 3\}$, the blocks

$$B_1 : 012 \quad B_2 : 013 \quad B_3 : 023 \quad B_4 : 123$$

form a $(4, 4, 3, 3, 2)$-design.

Example 10.2.2:   For $X = \{0, 1, \ldots, 8, 9, A\}$, the blocks

$$02348 \quad 13459 \quad 2456A \quad 35670 \quad 46781 \quad 57892$$
$$689A3 \quad 79A04 \quad 8A015 \quad 90126 \quad A1237$$

form a $(v = 11, b = 11, r = 5, k = 5, \lambda = 2)$-design. In this design, the initial block generates all of the others, if we regard the elements of X as integers modulo 11, with a standing for 10 modulo 11. Then each other block is obtained by adding 1 modulo 11 to each of the elements of the previous block.

Example 10.2.3:   For every integer $n \geq 2$, setting $X = [1 : n]$ and $B_1 = X$ yields a complete design with $v = n$, $b = 1$, $r = 1$, $k = n$, and $\lambda = 1$.

Example 10.2.4:   For every integer $n \geq 2$, setting $X = [1 : n]$ and having the pairs of elements from X as blocks yields a balanced design with

$$v = n, \quad b = \binom{n}{2}, \quad r = n - 1, \quad k = 2, \quad \lambda = 1$$

Thus, the complete graph K_n is representable as a BIBD.

Example 10.2.5:   When a simple graph is drawn on an arbitrary surface without crossings, each edge lies on exactly two faces. If the graph is K_n, and if all faces are k-sided, then this drawing may be regarded as a BIBD with $v = n$, blocksize k, and $\lambda = 2$, in which a block is the set of corners of a face.

Necessary Conditions

The examples above establish that BIBD's exist for certain combinations of the parameters v, b, r, k, and λ. However, there are no BIBD's for various other combinations. Our immediate concern is to derive some necessary conditions for the existence of a (v, b, r, k, λ)-design.

Proposition 10.2.1. *For every non-empty (v, b, r, k, λ)-BIBD*

$$(a)\ \lambda \geq 1 \quad \text{and} \quad (b)\ k < v$$

Proof: Since there is at least one block, and since it has at least two elements, some pair has at least once occurrence. Since all pairs occur equally often, it follows that $\lambda \geq 1$.

Since a block is a subset of the domain, its size cannot exceed the size of the domain. Thus, $k \leq v$. Since a BIBD is *incomplete*, it follows that $k < v$. $\diamond$

Proposition 10.2.2. *The parameters of a (v, b, r, k, λ)-design on*

$$X = \{\, x_1,\ x_2,\ \ldots,\ x_v \,\}$$

satisfy the following two conditions:

(a)
$$bk = vr$$

(b)
$$r(k-1) = \lambda(v-1)$$

Proof: First consider the $v \times b$ incidence matrix

$$I = \begin{array}{c|ccc} & B_1 & \cdots & B_b \\ \hline x_1 & \iota_{1,1} & \cdots & \iota_{1,b} \\ \vdots & \vdots & \vdots & \vdots \\ x_v & \iota_{v,1} & \cdots & \iota_{v,b} \end{array} \quad \text{with } \iota_{i,j} = \begin{cases} 1 & \text{if } x_i \in B_j \\ 0 & \text{otherwise} \end{cases}$$

There are v rows, each with row-sum r, and there are b columns, each with column-sum k. Therefore, $bk = vr$.

Next consider the $\binom{v}{2} \times b$ pair-incidence matrix

$$I' = \begin{array}{c|ccc} & B_1 & \cdots & B_b \\ \hline x_1 x_2 & \iota'_{12,1} & \cdots & \iota'_{12,b} \\ \vdots & \vdots & \vdots & \vdots \\ x_{v-1} x_v & \iota'_{(v-1)v,1} & \cdots & \iota'_{(v-1)v,b} \end{array}$$

with

$$\iota'_{ij,\ell} = \begin{cases} 1 & \text{if } x_i x_j \in B_\ell \\ 0 & \text{otherwise} \end{cases}$$

There are $\binom{v}{2}$ rows, each with row-sum λ, and there are b columns, each with column-sum $\binom{k}{2}$. Therefore,

$$\lambda \binom{v}{2} = b \binom{k}{2}$$

Accordingly,

$$\lambda v(v-1) = bk(k-1)$$
$$\Rightarrow \lambda v(v-1) = vr(k-1) \quad \text{since } bk = vr$$
$$\Rightarrow \lambda(v-1) = r(k-1) \qquad\qquad \diamond$$

TERMINOLOGY NOTE: The inferrability (from Proposition 10.2.2) of the parameters b and r from the parameters v, k, and λ justifies optionally calling a (v, b, r, k, λ)-design a (v, k, λ)-design.

Corollary 10.2.3. *For every non-empty BIBD,*

$$\lambda < r$$

Proof: Since $\lambda (v - 1) = r (k - 1)$ (from Theorem 10.2.2) and $k < v$ (from Proposition 10.2.1), it follows that $\lambda < r$.   ◇

REVIEW FROM LINEAR ALGEBRA:

- If AB is the product of the matrices A and B then

$$rank(AB) \; \leq \; \min\{rank(A),\, rank(B)\}$$

NOTATION: The *transpose* of a matrix M is denoted M^T.

Theorem 10.2.4 [Fisher's Inequality]. *In any BIBD, $b \geq v$.*

Proof: Let I be the incidence matrix of the BIBD. Then

$$I\,I^T \;=\; \begin{pmatrix} r & \lambda & \lambda & \lambda & \cdots & \lambda \\ \lambda & r & \lambda & \lambda & \cdots & \lambda \\ \lambda & \lambda & r & \lambda & \cdots & \lambda \\ \lambda & \lambda & \lambda & r & \cdots & \lambda \\ \vdots & \vdots & \vdots & \vdots & \ddots & \vdots \\ \lambda & \lambda & \lambda & \lambda & \cdots & r \end{pmatrix}$$

Subtracting the first column of a matrix from the other columns does not change the determinant. Hence,

$$det(I\,I^T) \;=\; \begin{vmatrix} r & \lambda - r & \lambda - r & \lambda - r & \cdots & \lambda - r \\ \lambda & r - \lambda & 0 & 0 & \cdots & 0 \\ \lambda & 0 & r - \lambda & 0 & \cdots & 0 \\ \lambda & 0 & 0 & r - \lambda & \cdots & 0 \\ \vdots & \vdots & \vdots & \vdots & \ddots & \vdots \\ \lambda & 0 & 0 & 0 & \cdots & r - \lambda \end{vmatrix}$$

Adding the other rows of a matrix to the first row does not change the determinant. Hence,

$$det(I\,I^T) \;=\; \begin{vmatrix} r + (v - 1)\lambda & 0 & 0 & 0 & \cdots & 0 \\ \lambda & r - \lambda & 0 & 0 & \cdots & 0 \\ \lambda & 0 & r - \lambda & 0 & \cdots & 0 \\ \lambda & 0 & 0 & r - \lambda & \cdots & 0 \\ \vdots & \vdots & \vdots & \vdots & \ddots & \vdots \\ \lambda & 0 & 0 & 0 & \cdots & r - \lambda \end{vmatrix}$$

Since the upper triangle of this matrix is all zeroes, the determinant is the product of the diagonal entries. Thus,

$$det(I\,I^T) \;=\; [r + (v - 1)\lambda](r - \lambda)^{v-1}$$

By Corollary 10.2.3, $r - \lambda > 0$. Moreover, $r + (v - 1)\lambda$ is positive. Thus, $det(I\,I^T)$ is non-zero. Accordingly, the rank of the $v \times v$-matrix $I\,I^T$ is v. Since the rank of the $v \times b$ incidence matrix I is at most b, and since the rank, v, of the product matrix $I\,I^T$ cannot exceed the rank of the matrix I, it follows that $v \leq b$.   ◇

Steiner Triple Systems

DEFINITION: A $(v, 3, 1)$-design is also called a **Steiner triple system**.

Example 10.2.6: The complete balanced block design

$$\mathcal{A} = \begin{cases} \text{domain} & X = \{0,\ 1,\ 2\} \\ 1 \text{ block} & B = \{012\} \end{cases} \qquad (10.2.1)$$

is a Steiner triple system. (A Steiner triple system on a domain with more than three elements is a BIBD.)

Example 10.2.7: The BIBD

$$\mathcal{B} = \begin{cases} \text{domain} & Y = \{0,\ 1,\ 2,\ 3,\ 4,\ 5,\ 6\} \\ 7 \text{ blocks} & C = \{013,\ 124,\ 235,\ 346,\ 450,\ 561,\ 602\} \end{cases} \qquad (10.2.2)$$

is a $(7, 3, 1)$-design. As in Example 10.2.2, the first block generates the others.

Proposition 10.2.5. In a $(v, 3, 1)$-design,

$$(a)\ r = \frac{v - 1}{2} \quad \text{and} \quad (b)\ b = \frac{v(v - 1)}{6}$$

Proof: Part (a) follows from Proposition 10.2.2(b):

$$r(k - 1) = \lambda(v - 1)$$

Simply substitute $k = 3$ and $\lambda = 1$.

For part (b), start with the equation

$$bk = rv$$

from Proposition 10.2.2(a). Then substitute 3 for k and $(v - 1)/2$ for r to obtain

$$3b = v \frac{v - 1}{2}$$

which leads immediately to the desired formula. $\diamond$

Corollary 10.2.6. In a $(v, 3, 1)$-design,

$$v \equiv 1 \text{ or } 3 \text{ modulo } 6$$

Proof: Proposition 10.2.5(a) implies that v is odd. Thus,

$$v \equiv 1,\ 3 \text{ or } 5 \text{ modulo } 6$$

However, if $v \equiv 5$ modulo 6, then $v(v-1) \equiv 2$ modulo 6, contradicting Proposition 10.2.5(b). $\diamond$

Constructing Designs

Jakob Steiner (1796-1893) asked in 1853 whether for every positive v such that $v \equiv 1$ or 3 modulo 6, there exists a $(v, 3, 1)$-design. He was unaware that in 1847, the Rev. Thomas P. Kirkman (1806-1895) had proved they always exist. Kirkman's methods are beyond the present scope. We presently offer some elementary methods that can also be used for constructing BIBD's with larger blocksize. The first such method generalizes Example 10.2.7.

DEFINITION: A set of numbers

$$S = \{a_1, \ a_2, \ \ldots, \ a_k\}$$

in $\mathbb{Z}_n$ is a **perfect difference set** of index λ for $\mathbb{Z}_n$ if each non-zero number in $\mathbb{Z}_n$ occurs exactly λ times in the list

$$\langle x_{ij} = a_i - a_j \mid a_i, a_j \in S; \ i \neq j \rangle$$

It is simply called a **perfect difference set** if $\lambda = 1$.

Proposition 10.2.7. *A perfect difference set B of cardinality k and index λ for $\mathbb{Z}_v$ generates a (v, k, λ)-design.*

Proof: For $j = 0, \ldots, v - 1$, let $B_j = \{j + b \mid b \in B\}$. By the definition of a perfect difference set, these blocks form a (v, k, λ)-design. $\Diamond$

Example 10.2.7, continued: The set $\{0, 1, 3\} \subset \mathbb{Z}_7$ is a perfect difference set of index 1, since

$$
\begin{array}{lll}
1 = 1 - 0 & 2 = 3 - 1 & 3 = 3 - 0 \\
4 = 0 - 3 & 5 = 1 - 3 & 6 = 0 - 1
\end{array}
$$

DEFINITION: A family $\mathcal{S}$ of sets $S_1, \ldots, S_f \subset \mathbb{Z}_n$ is a **perfect difference family** of index λ if each non-zero number in $\mathbb{Z}_n$ occurs exactly λ times in the list

$$\langle x_{ijk} = a_i - a_j \mid a_i, a_j \in S_k; \ i \neq j; \ 1 \leq k \leq f \rangle$$

It is called a **perfect difference family** if $\lambda = 1$.

Proposition 10.2.8. *If the sets of a perfect difference family of index λ for $\mathbb{Z}_v$ are all of the same size k, then they generate a (v, k, λ)-design.* $\Diamond$

Example 10.2.8:   We construct a perfect difference family for $\mathbb{Z}_{13}$

$$\{0, \ 1, \ 4\} \text{ with differences } \{1, \ 3, \ 4, \ 9, \ 10, \ 12\}$$
$$\{0, \ 2, \ 8\} \text{ with differences } \{2, \ 5, \ 6, \ 7, \ 8, \ 11\}$$

These two blocks together generate the following $(13, 3, 1)$-design.

$$X = \{0, \ 1, \ 2, \ 3, \ 4, \ 5, \ 6, \ 7, \ 8, \ 9, \ A, \ B, \ C\}$$

$$
B = \left\{
\begin{array}{lllllllll}
014 & 125 & 236 & 347 & 458 & 569 & 67A & 78B & 89C \\
    & 9A0 & AB1 & BC2 & C03 & & & & \\
028 & 139 & 24A & 35B & 46C & 570 & 681 & 792 & 8A3 \\
    & 9B4 & AC5 & B06 & C17 & & & &
\end{array}
\right\}
$$

Example 10.2.9:   The set $\{0, 1, 4, 6\}$ is a perfect difference set for $\mathbb{Z}_{13}$. Thus, with the domain

$$X = \{0,\ 1,\ 2,\ 3,\ 4,\ 5,\ 6,\ 7,\ 8,\ 9,\ A,\ B,\ C\}$$

the set of blocks

$$B = \left\{ \begin{array}{ccccccc} 0146 & 1257 & 2368 & 3479 & 458A & 569B & 67AC \\ 78B0 & 89C1 & 9A02 & AB13 & BC24 & C035 & \end{array} \right\}$$

forms a $(13, 4, 1)$-design.

The next example offers a way to construct a new Steiner triple system from two (possibly identical) smaller systems.

Example 10.2.10:   The cartesian product of the domain of the $(3, 3, 1)$-design $\mathcal{A}$ of Example 10.2.6 and the domain of the $(7, 3, 1)$-design $\mathcal{B}$ of Example 10.2.7 is representable as the following array.

$$\begin{array}{c|ccccccc} & 0 & 1 & 2 & 3 & 4 & 5 & 6 \\ \hline 0 & 00 & 01 & 02 & 03 & 04 & 05 & 06 \\ 1 & 10 & 11 & 12 & 13 & 14 & 15 & 16 \\ 2 & 20 & 21 & 22 & 23 & 24 & 25 & 26 \end{array}$$

To obtain a $(21, 3, 1)$-design $\mathcal{A} \times \mathcal{B}$ on the set of elements of that array, we choose as blocks

  (i) every column;

  (ii) from each row, each triple $\{ri,\ rj,\ rk\}$ such that $\{i,\ j,\ k\}$ is a block of $\mathcal{B}$;

  (iii) each triple $\{0i,\ 1j,\ 2k\}$ such that $\{i,\ j,\ k\}$ is a block of $\mathcal{B}$.

Observe that the number of blocks we have chosen is

$$7 + 21 + 42 = 70$$

Two elements xy and $x'y'$ of $\mathcal{A} \times \mathcal{B}$ appear in one and only one block. There are three cases.

  (i) $x \neq x'$ and $y = y'$: only in the block arising from column y.

  (ii) $x = x'$ and $y \neq y'$: only in the block arising from row x and the unique block of $\mathcal{B}$ in which y and y' are paired.

  (iii) $x \neq x'$ and $y \neq y'$: let x'' be the remaining row, and let y'' be the third entry in the unique block of $\mathcal{B}$ that contains both y and y'. Then $\{xy,\ x'y',\ x''y''\}$ is the unique block containing xy and $x'y'$.

DEFINITION: The **product of two Steiner triple systems** $\mathcal{A}$ and $\mathcal{B}$ is the triple system whose domain is the set of entries of the array representing the product of the domains of $\mathcal{A}$ and $\mathcal{B}$, with blocks as follows:

  (i) from each column of the array, each triple $\{rc,\ sc,\ tc\}$ such that $\{r,\ s,\ t\}$ is a block of $\mathcal{A}$;

(ii) from each row, each triple $\{ri,\ rj,\ rk\}$ such that $\{i,\ j,\ k\}$ is a block of $\mathcal{B}$;

(iii) each triple $\{ri,\ sj,\ tk\}$ such that $\{r,\ s,\ t\}$ is a block of $\mathcal{A}$ and $\{i,\ j,\ k\}$ is a block of $\mathcal{B}$.

Theorem 10.2.9. *Let $\mathcal{A}$ and $\mathcal{B}$ be Steiner triple systems with u and v varieties, respectively. Then their product is a Steiner triple system with uv varieties.*

Proof: The proof for the general case is essentially the same as for Example 10.2.10. $\diamond$

Remark: The definition and theorem just above are generalizable to a product of BIBD's and a theorem that the result is a new BIBD.

Isomorphism of Designs

DEFINITION: A bijection $f : X \to Y$ of the domains of two block designs

$$\mathcal{B} = \langle X, \{B_i\} \rangle \quad \text{and} \quad \mathcal{C} = \langle Y, \{C_j\} \rangle$$

is called an **isomorphism of block designs** if for every block C_j of design $\mathcal{C}$, there is a block B_i of design $\mathcal{B}$, such that the restriction $f : B_i \to C_j$ is onto.

Proposition 10.2.10. *Let $\mathcal{B} = \langle X, \{B_i\} \rangle$ be a $(7, 3, 1)$ Steiner system. Then $\mathcal{B}$ is isomorphic to the $(7, 3, 1)$ Steiner system with elements $0, 1, 2, 3, 4, 5, 6$ and blocks*

$$013 \quad 124 \quad 235 \quad 346 \quad 450 \quad 561 \quad 602$$

Proof: Choose an arbitrary element of X and call it x_0. Since each of the six other elements of X must appear with x_0 exactly once, there must be exactly three blocks of $\mathcal{B}$ that contain x_0. Call the other two elements in one of these blocks x_1 and x_3, and call the other two in a second of these blocks x_2 and x_6. Partially specify the bijection f by

$$x_0 \mapsto 0 \quad x_1 \mapsto 1 \quad x_2 \mapsto 2 \quad x_3 \mapsto 3 \quad x_6 \mapsto 6$$

which ensures some block preservation, namely,

$$x_0 x_1 x_3 \mapsto 013 \quad x_0 x_2 x_6 \mapsto 026 \quad x_0 x_4 x_5 \mapsto 045$$

The elements x_1 and x_2 appear together in a unique block of $\mathcal{B}$. Since the third element of that block cannot be x_0, x_3, or x_6, each of which appears in another block with x_1 or x_2, it can be called x_4, with the remaining element of X to be x_5. Completing the bijection specification with

$$x_4 \mapsto 4 \quad x_5 \mapsto 5$$

immediately ensures further block preservation

$$x_1 x_2 x_4 \mapsto 124$$

Moreover, given that $x_0 x_1 x_3$ and $x_1 x_2 x_4$ are blocks, it follows that the third block containing x_1 must be $x_1 x_5 x_6$. Similarly, the third block containing x_2 must be $x_2 x_3 x_5$. Since the elements x_3, x_4, and x_6 have so far appeared in only two blocks each, the seventh block must be $x_3 x_4 x_6$. Thus all blocks are preserved by the bijection f. $\diamond$

Remark: There is essentially only one $(7, 3, 1)$-design, as established by Proposition 10.2.10, and also only one $(9, 3, 1)$-design. There are two non-isomorphic $(13, 3, 1)$-designs and 80 mutually non-isomorphic $(15, 3, 1)$-designs. See the table on p764 of [CoDi2000a].

EXERCISES for Section 10.2

In each of the Exercises 10.2.1 through 10.2.6, construct a (v, k, λ)-design corresponding to the given values of v, k, and λ.

10.2.1[S]	$(4, 3, 2)$	10.2.2	$(5, 3, 3)$	10.2.3	$(6, 3, 4)$
10.2.4	$(5, 4, 3)$	10.2.5	$(6, 3, 2)$	10.2.6	$(7, 3, 2)$

In each of the Exercises 10.2.7 through 10.2.12, prove that there is no (v, k, λ)-design corresponding to the given values of v, k, and λ.

10.2.7[S]	$(5, 4, 2)$	10.2.8	$(8, 3, 1)$	10.2.9	$(8, 4, 1)$
10.2.10	$(8, 4, 2)$	10.2.11	$(8, 5, 2)$	10.2.12	$(9, 4, 1)$

For each of the Exercises 10.2.13 through 10.2.21, where

$$A = \{012\}, \quad B = \{012, 013, 023, 123\}, \quad \text{and} \quad C = \{0123\}$$

are the sets of blocks of the BIBD's A, B, and C, respectively, construct the parameter values v, b, r, k, and λ for the product of the two specfied BIBD's.

10.2.13[S]	$A \times A$	10.2.14	$A \times B$	10.2.15	$A \times C$
10.2.16	$B \times A$	10.2.17	$B \times B$	10.2.18	$B \times C$
10.2.19	$C \times A$	10.2.20	$C \times B$	10.2.21	$C \times C$

For each of the Exercises 10.2.22 through 10.2.30, where

$$A = \{012\}, \quad B = \{012, 013, 023, 123\}, \quad \text{and} \quad C = \{0123\}$$

are the sets of blocks of the BIBD's A, B, and C, respectively, construct the product of the two specfied BIBD's.

10.2.22[S]	$A \times A$	10.2.23	$A \times B$	10.2.24	$A \times C$
10.2.25	$B \times A$	10.2.26	$B \times B$	10.2.27	$B \times C$
10.2.28	$C \times A$	10.2.29	$C \times B$	10.2.30	$C \times C$

10.3 CLASSICAL FINITE GEOMETRIES

Many properties of the Euclidean spaces $\mathbb{R}^n$ can be derived purely from a short list of axioms about points and lines, without consideration of distance or angles, and without consideration that a line in $\mathbb{R}^n$ contains infinitely many points. In this spirit, various kinds of combinatorial designs on a finite set of elements have been called ***finite geometries***. The elements of their domains are traditionally called the ***points of the geometry***, and their distinguished subsets are called the ***lines of the geometry***. The following two general axioms are standard for geometries.

G1. Two distinct points are contained in at most one line.

G2. Two distinct lines intersect in at most one point.

NOTATION: In view of Axiom G1, we may denote the line containing two distinct points u and v by uv.

TERMINOLOGY: Two disjoint lines of a geometry are often said to be ***parallel lines***.

DEFINITION: The ***incidence matrix of a geometry*** $\langle X, L\rangle$ with p points

$$X = \{x_1, \ldots, x_p\}$$

and ℓ lines

$$L = \{L_1, \ldots, L_\ell\}$$

is the $p \times \ell$ matrix

$$M_{\langle X,L\rangle}[i,j] = \begin{cases} 1 & \text{if } x_i \in L_j \\ 0 & \text{otherwise} \end{cases}$$

A geometry is commonly specified by its incidence matrix.

Example 10.3.1: Figure 10.3.1 illustrates a geometry with a drawing of its four points and its six lines.

$$\begin{array}{c} \\ x_0 \\ x_1 \\ x_2 \\ x_3 \end{array} \begin{array}{c} \begin{array}{cccccc} L_0 & L_1 & L_2 & L_3 & L_4 & L_5 \end{array} \\ \begin{pmatrix} 1 & 1 & 0 & 0 & 1 & 0 \\ 0 & 1 & 0 & 1 & 0 & 1 \\ 1 & 0 & 1 & 0 & 0 & 1 \\ 0 & 0 & 1 & 1 & 1 & 0 \end{pmatrix} \end{array}$$

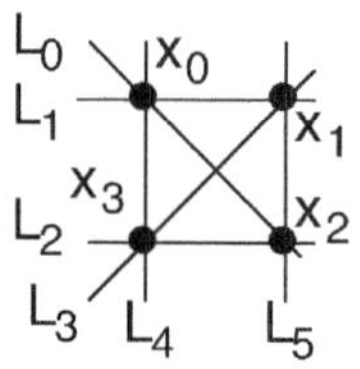

Figure 10.3.1 **A geometry with 4 points and 6 lines.**

DEFINITION: The ***dual of a geometry*** $\langle X, L\rangle$ is the geometry $\langle X^*, L^*\rangle$ with

$$X^* = L \quad \text{and} \quad L^* = X$$

whose incidence matrix is the transpose of the incidence matrix of $\langle X, L\rangle$. (In view of the reciprocity of Axioms G1 and G2, the dual design satisfies both of them.)

Example 10.3.1, continued: Figure 10.3.2 illustrates the dual of the geometry specified by Figure 10.3.1.

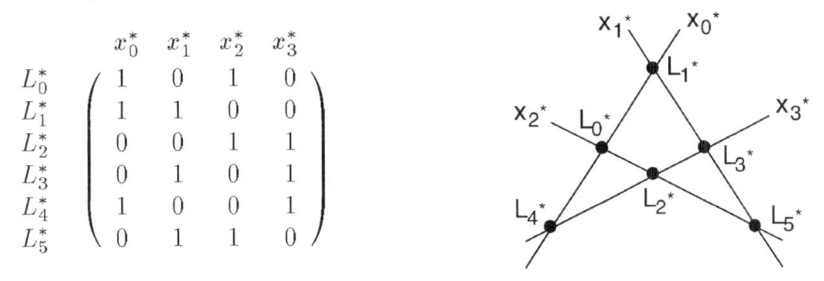

$$
\begin{array}{c@{\quad}c}
\begin{array}{c}
\\
L_0^*\\
L_1^*\\
L_2^*\\
L_3^*\\
L_4^*\\
L_5^*
\end{array}
&
\begin{array}{cccc}
x_0^* & x_1^* & x_2^* & x_3^* \\
1 & 0 & 1 & 0 \\
1 & 1 & 0 & 0 \\
0 & 0 & 1 & 1 \\
0 & 1 & 0 & 1 \\
1 & 0 & 0 & 1 \\
0 & 1 & 1 & 0
\end{array}
\end{array}
$$

Figure 10.3.2   **The dual geometry has 6 points and 4 lines.**

The Fano Plane

A design named for the Italian geometer Gino Fano (1871-1952) is the first of three widely cited classical geometries that we now consider.

DEFINITION: The **Fano plane** is defined by the incidence matrix

$$
\begin{array}{c@{\quad}c}
\begin{array}{c}
\\
0\\
1\\
2\\
3\\
4\\
5\\
6
\end{array}
&
\begin{array}{ccccccc}
L_0 & L_1 & L_2 & L_3 & L_4 & L_5 & L_6 \\
0 & 0 & 0 & 1 & 0 & 1 & 1 \\
1 & 1 & 0 & 1 & 0 & 0 & 0 \\
1 & 0 & 1 & 0 & 0 & 0 & 1 \\
0 & 0 & 1 & 1 & 1 & 0 & 0 \\
1 & 0 & 0 & 0 & 1 & 1 & 0 \\
0 & 1 & 1 & 0 & 0 & 1 & 0 \\
0 & 1 & 0 & 0 & 1 & 0 & 1
\end{array}
\end{array}
$$

It is depicted in the diagram in Figure 10.3.3, in which the line L_0 is represented by a circle.

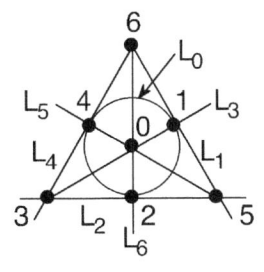

Figure 10.3.3   The Fano plane.

We observe that as a design, the Fano plane is precisely the Steiner triple system of Example 10.2.7.

The Pappus Geometry

A second classical geometry is named for Pappus of Alexandria (c. 300-350 C.E.), who proved the following theorem of Euclidean geometry.

Theorem of Pappus. *Let 0, 1, and 2 be three distinct points on a line L_1 and 3, 4, and 5 three distinct points on line $L_2 \neq L_1$, such that there are points of intersection*

$$6 = 04 \cap 13 \quad 7 = 05 \cap 23 \quad \text{and} \quad 8 = 15 \cap 24$$

Then the points 6, 7, and 8 are colinear.   ◇

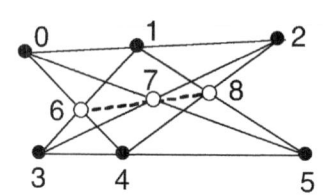

Figure 10.3.4   The geometry of Pappus.

DEFINITION: The **Pappus geometry** is the following finite geometry

$$X = \{0, 1, 2, 3, 4, 5, 6, 7, 8\}$$
$$L = \{012, 345, 064, 075, 163, 185, 273, 284, 678\}$$

or any other geometry of the same isomorphism type.

The Pappus geometry has uniform blocksize 3 and uniform replication number 3. As in Euclidean plane geometry, no pair of points occurs more than once in a line. However, in the Pappus geometry, and unlike Euclidean geometry, some pairs of points do not lie on any line. This implies that the Pappus geometry is not a Steiner triple system or a BIBD. The Pappus geometry shares the following property with Euclidean plane geometry.

Proposition 10.3.1. *Let L_i be any line of the Pappus geometry, and let p be a point that is not on that line. Then there is a unique line L_j containing the point p and parallel to the line L_i.*

Proof: The lines of the Pappus geometry are resolvable into three classes of parallel lines.

$$C_1 = \{012 \quad 345 \quad 678\}$$
$$C_2 = \{064 \quad 185 \quad 273\}$$
$$C_3 = \{075 \quad 163 \quad 284\}$$

If the given line L_i lies in the class C_k, then choose line L_j to be the unique line in class C_k that contains point p.   ◇

The Desargues Geometry

Another theorem of plane Euclidean geometry is due to Girard Desargues (1591-1661).

Theorem of Desargues. *Let 123 and 456 be triangles such that the lines 14, 25, and 36 meet at point 0. Let*

$$7 \ = \ 13 \cap 46 \quad 8 \ = \ 23 \cap 56 \quad \text{and} \quad 9 \ = \ 12 \cap 45$$

Then 7, 8, and 9 are colinear.   ◇

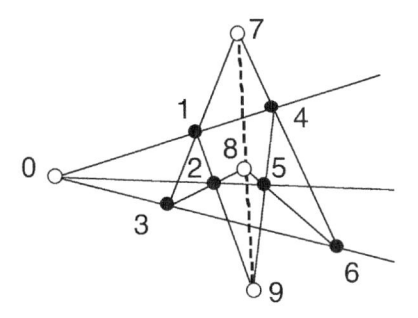

Figure 10.3.5   The geometry of Desargues.

DEFINITION: The **Desargues geometry** is the following finite geometry

$$X \ = \ \{\, 0,\ 1,\ 2,\ 3,\ 4,\ 5,\ 6,\ 7,\ 8,\ 9 \,\}$$
$$L \ = \ \{014,\ 025,\ 036,\ 137,\ 129,\ 238,\ 467,\ 459,\ 568,\ 789\,\}$$

or any other geometry of the same isomorphism type.

In the Desargues geometry, as in the Pappus geometry, there is a uniform blocksize of 3 and a uniform replication number of 3. As in Euclidean geometry and the Pappus geometry, no pair of points occurs more than once in a block. As in the Pappus geometry, and unlike Euclidean geometry, some pairs do not occur on any line. Accordingly, it is not a Steiner triple system or a BIBD.

Remark 1:   Observe that Proposition 10.3.1 does not apply to the Desargues geometry. In fact, for every line L_i in the Desargues geometry, there is a point p such that no line containing p intersects the line L_i. Such a point p is called a *pole* of the line L_i.

Example 10.3.2:   In the Desargues geometry, the point 8 is a pole of the line 014, and the point 1 is a pole of the line 568.

Remark 2:   Another interesting property in which Desargues geometry differs from Euclidean geometry is that in the Desargues geometry, two lines that are parallel to the same line are *not* parallel to each other.

Example 10.3.3:   The lines that are parallel to the line 789 of the Desargues geometry are 014, 036, and 025. Observe that any pair of them intersects in the point 0.

Partially Balanced Designs

DEFINITION: A $(v, b, r, k; \lambda_1, \lambda_2)$-**PBIBD** (stands for **partially balanced incomplete block design**) is a design with v elements and b blocks, in which

(i) each element lies in exactly r blocks;

(ii) each block contains exactly k elements;

(iii) each pair of distinct elements occurs either in λ_1 or λ_2 blocks.

Example 10.3.4:   The Pappus geometry is a $(9, 9, 3, 3; 1, 0)$-PBIBD.

Example 10.3.5:   The Desargues geometry is a $(10, 10, 3, 3; 1, 0)$-PBIBD.

EXERCISES for Section 10.3

10.3.1   Which pairs of points do not occur on any line of the Pappus geometry?

10.3.2   Which pairs of points do not occur on a line of the Desargues geometry?

DEFINITION: The **Levi graph** of a design $\langle X, B \rangle$ is the bipartite graph whose vertex set is $X \cup B$, such that there is an edge joining $x_i \in X$ and $B_j \in B$ whenever $x_i \in B_j$.

In each of the Exercises 10.3.3 through 10.3.6, construct the Levi graph for the specified design.

10.3.3   The BIBD with $B = \{012, 013, 023, 123\}$.

10.3.4[S]   The Fano plane.

10.3.5   The Pappus geometry.

10.3.6   The Desargues geometry.

DEFINITION: The **Menger graph** of a design $\langle X, B \rangle$ is the graph with vertex set X and an edge joining each pair of points of X that lie on a line of B.

In each of the Exercises 10.3.7 through 10.3.10, construct the Menger graph for the specified design.

10.3.7   The BIBD with $B = \{012, 013, 023, 123\}$.

10.3.8[S]   The Fano plane.

10.3.9   The Pappus geometry.

10.3.10   The Desargues geometry.

In each of the Exercises 10.3.11 through 10.3.14, prove that the Menger graph for the specified design is vertex-transitive.

10.3.11   The BIBD with $B = \{012, 013, 023, 123\}$.

10.3.12[S] The Fano plane.

10.3.13   The Pappus geometry.

10.3.14   The Desargues geometry.

10.4   PROJECTIVE PLANES

A *projective plane* is a type of finite geometry, and thus, a type of combinatorial design. Toward the end of this section, there is given a method for constructing projective planes from 3-dimensional vector spaces. This construction is what motivates calling these designs *projective planes*.

DEFINITION: A ***projective plane*** $\mathcal{P}$ has a domain X, whose elements are called *points*, and a collection of subsets of X that are called *lines*, such that the following axioms hold:

   PP1. For each pair of distinct points, there is exactly one line containing them.

   PP2. Each pair of distinct lines intersects in exactly one point.

   PP3. There exist four points, no three of which lie on the same line.

These three simple axioms have many implications.

Some Basic Examples

Example 10.4.1:   The Fano plane is a projective plane. This can be verified by checking its definition as a design.

Proposition 10.4.1. *For* $k \geq 3$, *any* $(v, k, 1)$-*design* $\mathcal{B}$ *generated by a perfect difference set* $S = \{a_1, \ldots, a_k\} \subseteq \mathbb{Z}_v$ *is a projective plane. Moreover,* $v = k^2 - k + 1$.

Proof: Let $i, j \in \mathbb{Z}_v$ with $i \leq j$. To find a block in $\mathcal{B}$ that contains both i and j, let a_r and a_s be the unique pair in the difference set S such that $a_s - a_r = j - i$. Then the block $S + (i - a_r)$ contains

$$a_r + (i - a_r) \; = \; i \quad \text{and}$$
$$a_s + (i - a_r) \; = \; i + (a_s - a_r) \; = \; i + (j - i) \; = \; j$$

No other pair from S has difference $j - i$, so no other pair can translate to i and j in the same block. Moreover, a_r and a_s translate to i and j only in the block $S + (i - a_r)$. This establishes Axiom PP1.

Next, consider two arbitrary blocks of $\mathcal{B}$, say

$$S + i \; = \; \{a_t + i \mid a_t \in S\} \quad \text{and} \quad S + j \; = \; \{a_t + j \mid a_t \in S\}$$

There is a unique pair a_r, a_s in the difference set S such that

$$j - i \; = \; a_r - a_s$$

It follows that the number $j + a_s = i + a_r$ is the unique point in the intersection

$$(S + i) \cap (S + j)$$

This establishes Axiom PP2.

To prove the third axiom, let B_1 and B_2 be any two blocks. By PP2, they intersect in a single point. Since $k \geq 3$, there are at least two points $x_1, x_2 \in B_1 - B_2$ and at least two points $x_3, x_4 \in B_2 - B_1$. The four points x_1, x_2, x_3, x_4 satisfy the condition of PP3.

The method of block generation yields v blocks. Thus, when $\mathcal{B}$ is represented as a (v, b, r, k, λ)-BIBD, we have $b = v$. Hence, the equation

$$bk \;=\; rv \qquad\qquad \text{Prop. 10.2.2(a)}$$

implies that $r = k$. Using that fact and the specification $\lambda = 1$, the equation

$$r(k - 1) \;=\; \lambda(v - 1) \qquad \text{Prop. 10.2.2(b)}$$

further implies that $v = k^2 - k + 1$. $\hspace{2cm}$ $\diamond$

Example 10.4.2: The 9-point Pappus geometry is not a projective plane, since, for instance, the lines 012 and 345 do not meet. Some projective planes do satisfy the Theorem of Pappus, but some do not.

Example 10.4.3: The 10-point Desargues geometry is not a projective plane, since (as observed previously) there are pairs of points with no lines through them. Some non-Desarguesian projective planes exist, but most of the familiarly encountered projective planes do satisfy the Theorem of Desargues.

The Duality Principle for Projective Planes

We observe that Axioms PP1 and PP2 are absolute duals of each other. The following proposition establishes a dual to Axiom PP3.

Proposition 10.4.2. *In a projective plane $\mathcal{P}$, there exist four lines, no three of which contain the same point.*

Proof: By Axiom PP3, there exist four points 0, 1, 2, and 3, no three on the same line. By Axiom PP1, there exist lines 01, 12, 23, and 03, as shown in Figure 10.4.1.

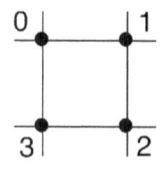

Figure 10.4.1   Proving the dual to Axiom PP3.

By Axiom PP2 none of these lines contains a third point from the set $\{0, 1, 2, 3\}$. Moreover, since among any three of these four lines, there are two with a common point in $\{0, 1, 2, 3\}$, it follows from Axiom PP2 that there cannot be some other point common to all three. $\hspace{1cm}$ $\diamond$

Duality Principle. *The dual of any valid statement about projective planes is also a valid statement about projective planes.*

Proof: Axioms PP1 and PP2 are dual to each other, and Proposition 10.4.2 is dual to Axiom PP3.   ◇

Projective Planes as BIBD's

Lemma 10.4.3. *For any two distinct lines L and L' of a projective plane $\mathcal{P}$, there is a point x such that $x \notin L \cup L'$.*

Proof: Let y be the intersection of the lines L and L', let $0, 1 \in L - y$ and $2, 3 \in L' - y$, as shown in Figure 10.4.2.

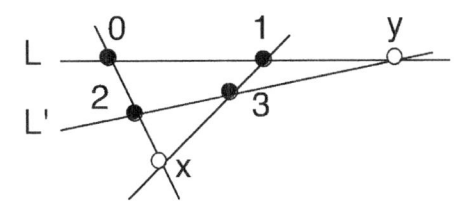

Figure 10.4.2 A point x not in the union of two lines.

Let x be the intersection of the lines 02 and 13. Since 0 is the unique interesction point of 02 and L, it follows that $x \notin L$. Since 3 is the unique interesection point of 13 and L', it follows that $x \notin L'$.   ◇

Proposition 10.4.4. *Any two lines of a projective plane $\mathcal{P}$ have the same number of points.*

Proof: Let L and L' be two distinct lines. By Lemma 10.4.3, there is a point $x \notin L \cup L'$. Now suppose that

$$L = \{y_0, \ldots, y_{k-1}\} \quad \text{with} \quad y_0 = L \cap L'$$

and that, for $j = 1, \ldots, k - 1$, the intersection of the line xy_j with line L' is the point z_j, as in Figure 10.4.3.

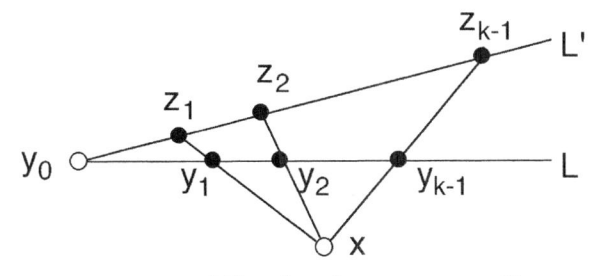

Figure 10.4.3 A bijection between two lines.

Then none of the points z_j, with $j = 1, \ldots, k - 1$, coincides with y_0, because the lines L and xy_j meet only at y_j. If $z_j = y_0$, then line xy_j would also meet line L

at y_0, which would be a second point in their intersection, since $y_j \neq y_0$. Moreover, if $i \neq j$, then the lines xy_i and xy_j are distinct, and then meet only at x. If $z_i = z_j$, then they would also meet there, contradicting Axiom PP2. Thus, the correspondence $y_i \mapsto z_i$ is a bijection of $L - y_0$ to $L' - y_0$. $\Diamond$

DEFINITION: The **order of a projective plane** is defined to be one less than its blocksize as a design. (Significantly, the order is *not* defined to be the number of elements.)

Corollary 10.4.5. *In a projective plane of order n, every point lies on exactly $n + 1$ lines.*

Proof: Using the definition of *order* just given, this assertion is simply the dual of Proposition 10.4.4. $\Diamond$

TERMINOLOGY: The set of all lines that meet at a point x of a projective plane is called the **pencil of lines** at x. (See Figure 10.4.4).

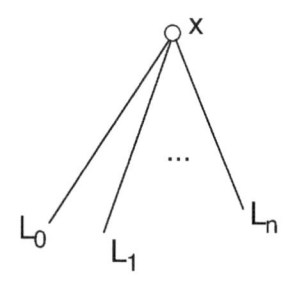

Figure 10.4.4   **The pencil of lines at point x.**

Proposition 10.4.6. *In a projective plane $\mathcal{P}$ of order n, the number of points is*

$$n^2 + n + 1$$

Proof: Let x be any point, and let

$$L_0, L_1, \ldots, L_n$$

be the pencil of lines that meets at x. Since every point of $\mathcal{P}$ lies on some line containing x, by Axiom PP1, the union of these lines is the entire domain of $\mathcal{P}$. Since no two of these lines intersect anywhere except x, by Axiom PP2, it follows that the number of points in $\mathcal{P}$ equals 1 for x plus n points on each of the $n + 1$ lines unique to that line, that is,

$$1 + n(n + 1) = n^2 + n + 1$$

points in all. $\Diamond$

Corollary 10.4.7. *In a projective plane $\mathcal{P}$ of order n, the number of lines is*

$$n^2 + n + 1$$

Proof: This is the dual of Proposition 10.4.6. $\Diamond$

Theorem 10.4.8. *A projective plane of order n is a BIBD with parameters*

$$(v = n^2 + n + 1, \ b = n^2 + n + 1, \ r = n, \ k = n, \ \lambda = 1)$$

Proof: This summarizes the results above.   ◇

Constructing Projective Planes

Much of the elementary theory of finite vector spaces is the same as for real vector spaces. The row-reduction algorithm is the key to establishing some additional facts to be used in the construction of some projective planes. After presenting some of the basics, we will use various such results from elementary linear algebra withour proof.

FROM APPENDIX A3:

- The *vector space* $\mathbb{Z}_p^3$, with p prime, is the set of triples (x_1, x_2, x_3) (called *points*) in $\mathbb{Z}_p$ under *vector addition*

$$(x_1, x_2, x_3) + (y_1, y_2, y_3) \ = \ (x_1 + y_1, x_2 + y_2, x_3 + y_3)$$

and *scalar multiplication*

$$c(x_1, x_2, x_3) \ = \ (cx_1, cx_2, cx_3)$$

- A *line in the finite vector space* $\mathbb{Z}_p^3$ is the set of all scalar multiples of a non-zero point (x_1, x_2, x_3), i.e., the set

$$\{(0, 0, 0), (x_1, x_2, x_3), (2x_1, 2x_2, 2x_3), \ \ldots$$
$$\ldots, ((p-1)x_1, (p-1)x_2, (p-1)x_3)\}$$

Proposition 10.4.9. *Every non-zero point (x_1, x_2, x_3) of the vector space $\mathbb{Z}_p^3$ lies in a unique line of $\mathbb{Z}_p^3$.*

Proof: Certainly, (x_1, x_2, x_3) lies in the line comprising all of its own scalar multiples. Since the modulus p is prime, every non-zero scalar in $\mathbb{Z}_p$ is a multiple modulo p of any other scalar. It follows that any line containing (x_1, x_2, x_3) must be that same line.   ◇

Corollary 10.4.10. *The number of lines in the vector space $\mathbb{Z}_p^3$ is $p^2 + p + 1$.*

Proof: Clearly, the number of non-zero points in $\mathbb{Z}_p^3$ is $p^3 - 1$. Since each line contains $p - 1$ non-zero points, and since two distinct lines meet only at $(0, 0, 0)$, it follows from the Rule of Quotient (§0.3) that the number of lines is

$$\frac{p^3 - 1}{p - 1} \ = \ p^2 + p + 1$$ ◇

A *plane in the finite vector space* $\mathbb{Z}_p^3$ is the set of sums of the scalar multiples of two points not on the same line.

DEFINITION: The **projective geometry** $PG(2,p)$ has as its points the set of all lines of the vector space $\mathbb{Z}_p^3$ and as its lines the set of all planes of $\mathbb{Z}_p^3$.

Example 10.4.4: In Figure 10.4.5, the seven points of the projective geometry $PG(2,2)$ are shown as lines through the origin 000 in Z_2^3.

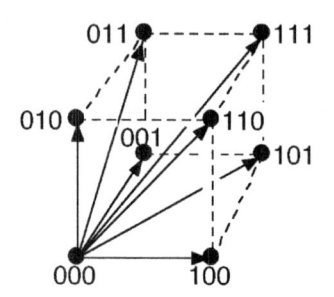

Figure 10.4.5   The projective geometry $PG(2,2)$.

Proposition 10.4.11. *The projective geometry $PG(2,p)$ is a projective plane of order p.*

Proof: Axiom PP1 holds because two distinct lines in the vector space $\mathbb{Z}_p^3$ determine a unique plane. Axiom PP2 holds because two distinct planes in $\mathbb{Z}_p^3$ meet in a line. Axiom PP3 holds because each combination of three of the following four vectors

$$(1,0,0) \quad (0,1,0) \quad (0,0,1) \quad (1,1,1)$$

lies on a line of $\mathbb{Z}_p^3$, is a linearly independent set, from which it follows that they and the lines they generate cannot all lie in the same plane of $\mathbb{Z}_p^3$. Hence, $PG(2,p)$ is a projective plane. Since a plane in $\mathbb{Z}_p^3$ has $p^2 - 1$ non-zero points and a line has $p - 1$ non-zero points, it follows that the number of points in a line of $PG(2,p)$ is

$$\frac{p^2 - 1}{p - 1} \;=\; p + 1$$

Thus, its order as a projective plane is p.   ◇

EXERCISES for Section 10.4

10.4.1^{S} Construct a list of all the points of the projective geometry $PG(2,2)$.

10.4.2^{S} Construct a list of all the lines of the projective geometry $PG(2,2)$.

10.4.3   Construct an isomorphism between $PG(2,2)$ and the Fano plane.

10.4.4   Construct a list of all the points of the projective geometry $PG(2,3)$.

10.4.5   Construct a list of all the lines of the projective geometry $PG(2,3)$.

10.4.6 Construct an isomorphism between $PG(2,3)$ and the BIBD generated by the difference set $\{0, 1, 3, 9\}$.

10.4.7 Prove that a BIBD generated by a difference set is a projective plane.

10.5 AFFINE PLANES

An *affine plane* is another kind of finite geometry. There is a close correspondence between affine planes and projective planes.

DEFINITION: An **affine plane** $\mathcal{A}$ has a domain X, whose elements are called *points*, and a collection of subsets of X that are called *lines*, such that the following axioms hold:

AP1. For each pair of distinct points, there is exactly one line containing them.

AP2. For any given line L_i and any point x not on L_i there is a line through x that is parallel to L_i.

AP3. There exist four points, no three of which lie on the same line.

Example 10.5.1: The following geometry, seen previously in §10.3, is an affine plane called $AG(2,2)$. The name is explained later in this section.

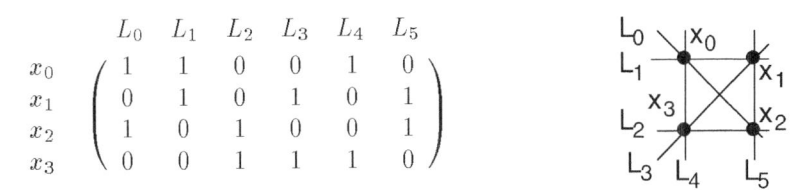

$$\begin{array}{c}
\quad\; L_0 \;\; L_1 \;\; L_2 \;\; L_3 \;\; L_4 \;\; L_5 \\
\begin{array}{c} x_0 \\ x_1 \\ x_2 \\ x_3 \end{array}
\left(\begin{array}{cccccc}
1 & 1 & 0 & 0 & 1 & 0 \\
0 & 1 & 0 & 1 & 0 & 1 \\
1 & 0 & 1 & 0 & 0 & 1 \\
0 & 0 & 1 & 1 & 1 & 0
\end{array}\right)
\end{array}$$

Figure 10.5.1 **The affine plane AG(2,2).**

Axioms AP1 and AP3 are easily verified for $AG(2,2)$ either from the incidence matrix or from the diagram. To verify Axiom AP2 from the drawing, one recognizes that lines L_0 and L_3 are parallel, in the sense of finite geometry, even though they cross each other in the drawing.

Example 10.5.2: The Fano plane is not an affine plane. In general, a projective plane has no pair of parallel lines. Thus, it cannot satisfy Axiom AP2.

Example 10.5.3: We observe that lines of the Pappus geometry can be partitioned into three cells of three lines each, as represented by the three columns to the left of Figure 10.5.2, such that within each cell, each point of the geometry occurs exactly once.

$$L = \left\{ \begin{matrix} 012, & 046, & 057 \\ 345, & 158, & 136 \\ 678, & 237, & 248 \end{matrix} \right\}$$

Figure 10.5.2   Resolving the geometry of Pappus.

If a point of the Pappus geometry does not lie on a given line, then it lies on another line in the same column as the given line, which is parallel to the given line. Thus, Axiom AP2 holds. However, the Pappus geometry does not satisfy AP1, so it is not an affine plane.

DEFINITION: A **resolvable geometry** is a geometry whose lines can be partitioned into cells such that the lines within each cell partition the domain.

Proposition 10.5.1. *A finite geometry satisfies Axiom AP2 if and only if it is resolvable.* ◇

The Affine Plane AG(2,p)

DEFINITION: An **affine line** in the finite vector space $\mathbb{Z}_p^2$ is the set of p points produced by adding a fixed pair (c_1, c_2) to every point on a line of $\mathbb{Z}_p^2$,

$$K = \{(0,0), (x_1, x_2), (2x_1, 2x_2), \ldots, ((p-1)x_1, (p-1)x_2)\}$$

thereby obtaining a set of the form

$$L = \{(c_1, c_2), (x_1 + c_1, x_2 + c_2), (2x_1 + c_1, 2x_2 + c_2), \ldots$$
$$\ldots, ((p-1)x_1 + c_1, (p-1)x_2 + c_2)\}$$

This is conceptualized like adding a fixed vector to every point on a line through the origin in the real plane $\mathbb{R}^2$, thereby translating the line to a parallel line, as illustrated in Figure 10.5.3.

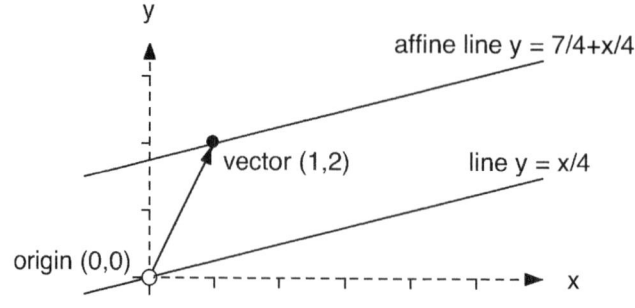

Figure 10.5.3   An affine line in the plane $\mathbb{R}^2$.

An alternative perspective is to choose numbers $a, b \in Z_p$ such that $ax_1 + bx_2 = 0$, so that the vector (a, b) is normal to the line K. Then the affine line L is the line normal to (a, b) that contains the point (c_1, c_2), i.e.,

$$L = \{(y_1, y_2) \mid ay_1 + by_2 = ac_1 + bc_2\}$$

Example 10.5.4:   Adding the fixed pair $(1, 2)$ in $\mathbb{Z}_5^2$ to the line

$$K = \{(0, 0),\ (1, 3),\ (2, 1),\ (3, 4),\ (4, 2)\}$$

yields the affine line

$$L = \{(1, 2),\ (2, 0),\ (3, 3),\ (4, 1),\ (0, 4)\}$$

We observe that $3 \cdot 1 + 4 \cdot 3 = 0$ and that $3 \cdot 1 + 4 \cdot 2 = 1$, so the affine line L is also specifiable as the set of pairs $(y_1,\ y_2)$ such that $3y_1 + 4y_2 = 1$.

DEFINITION: The **affine geometry** $AG(2, p)$ is the geometry whose points are the points of the vector space $\mathbb{Z}_p^2$ and whose lines are the affine lines of $\mathbb{Z}_p^2$.

Example 10.5.5:   In Figure 10.5.4, each of the four classes of parallel lines is represented by a different graphic – thin solid curve, thin dashed line, bold solid curve, and bold dashed line.

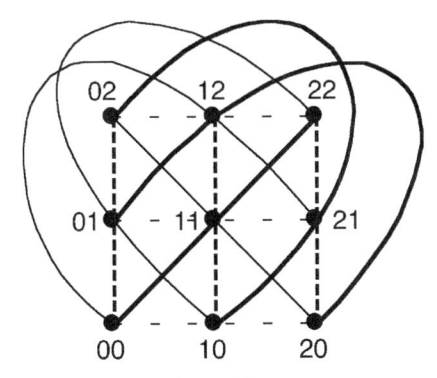

Figure 10.5.4   The affine geometry $AG(2, 3)$.

It may be observed in Figure 10.5.4 that every affine line in $AG(2, 3)$ has the same number of points – representing a common blocksize of $k = 3$ as a design, and that each pair of points occurs in exactly one affine line – representing a common index of $\lambda = 1$ as a design. These properties are verified shortly for every affine geometry $AG(2, p)$.

Proposition 10.5.2.   *The affine geometry $AG(2, p)$ is an affine plane.*

Proof: (Axiom AP1): Given any two points (x_1, x_2) and (y_1, y_2) in $AG(2, p)$, the affine line

$$\{(x_1, x_2) + j\,(y_1 - x_1, y_2 - x_2) \mid j = 0,\ \ldots,\ p - 1\}$$

contains both (x_1, x_2) (when $j = 0$) and (y_1, y_2) (when $j = 1$), and it is the only such affine line.

(Axiom AP2): Suppose that the affine line

$$L = \{(c_1, c_2) + j(x_1, x_2) \mid j \in Z_p\}$$

does not contain the point (y_1, y_2). Then the affine line

$$\{(y_1, y_2) + j(x_1, x_2) \mid j \in Z_p\}$$

is parallel to L and contains (y_1, y_2).

(Axiom AP3): No three of the points

$$(0,0), \quad (0,1), \quad (1,0), \quad \text{and} \quad (1,1)$$

lie on the same affine line in Z_p^2. $\diamond$

Proposition 10.5.3. *The affine plane $AG(2,p)$ is a $(p^2, p, 1)$-design, and, thus, a $(p^2, p^2 + p, p + 1, p, 1)$-BIBD.*

Proof: The number of points in $\mathbb{Z}_p^2$ is p^2, and thus the number of points in $AG(2, p)$ is p^2. Moreover, the number of points in every affine line in $\mathbb{Z}_p^2$, and, thus, in every line of $AG(2, p)$ is p. Since $AG(2, p)$ satisfies Axiom AP1, every pair of points of $AG(2, p)$ lies in exactly one line, so $\lambda = 1$. $\diamond$

Affine Planes from Projective Planes

Suppose that a particular block B is deleted from a combinatorial design $\mathcal{B}$, and that each point of B is deleted from the domain. This is called a *restriction of the design $\mathcal{B}$* to the complement of block B. Then the incidence matrix of the resulting design can be obtained from the incidence matrix for design $\mathcal{B}$ by deleting column B and also deleting each row corresponding to an element of B.

Example 10.5.6: If we delete column L_0 and rows 1, 2, and 4 from the incidence matrix for the Fano plane

	L_0	L_1	L_2	L_3	L_4	L_5	L_6
0	0	0	0	1	0	1	1
1	1	1	0	1	0	0	0
2	1	0	1	0	0	0	1
3	0	0	1	1	1	0	0
4	1	0	0	0	1	1	0
5	0	1	1	0	0	1	0
6	0	1	0	0	1	0	1

then the resulting incidence matrix is

$$
\begin{array}{c}
\quad L_1 \;\; L_2 \;\; L_3 \;\; L_4 \;\; L_5 \;\; L_6 \\
\begin{array}{c} 0 \\ 3 \\ 5 \\ 6 \end{array}
\left(
\begin{array}{cccccc}
0 & 0 & 1 & 0 & 1 & 1 \\
0 & 1 & 1 & 1 & 0 & 0 \\
1 & 1 & 0 & 0 & 1 & 0 \\
1 & 0 & 0 & 1 & 0 & 1
\end{array}
\right)
\end{array}
$$

Referring back to Example 10.5.1, we recognize that the corresponding design is isomorphic to the affine plane $AG(2,2)$.

Theorem 10.5.4. *The geometry $\mathcal{G}$ resulting from restricting a projective plane $PG(2,p)$ to the complement of any given line L is an affine plane.*

Proof: By Axiom PP2, the line L intersects every other line of $PG(2,p)$ exactly once. Since every line of $PG(2,p)$ has $p+1$ points, it follows that every line of $\mathcal{G}$ has p points.

(Axiom AP1):   Each pair of points of $PG(2,p)$ not on line L lies on exactly one line of $PG(2,p)$, by Axiom PP1. This implies that each pair of points of $\mathcal{G}$ lies on exactly one line.

(Axiom AP2):   For each of the points $x_0, \ldots, x_p \in L$, the pencil of lines of $PG(2,p)$ meeting at x_j partitions $PG(2,p) - x_j$, by Axiom PP1. By Axiom PP2, none of the points on line L is on any line of this pencil other than line L. Thus, after x is deleted from the remaining lines of that pencil, the resulting subsets partition the set $PG(2,p) - L$, which is precisely the domain of the geometry $\mathcal{G}$. Thus, the lines of $\mathcal{G}$ can be partitioned into $p+1$ sets of p points each. In other words, $\mathcal{G}$ is a resolvable geometry, from which it follows (by Proposition 10.5.1) that it satisfies Axiom AP2.

(Axiom AP3):   Choose the first two points w and x of the needed four from any line of $\mathcal{G}$. Then choose two other points y and z from any parallel line. Any subset of three points from this foursome must include either the pair w and x or the pair y and z. Since there is only one line through either pair, it follows from Axiom AP1 that no line can go through three of these points. $\diamond$

Projective Planes from Affine Planes

Now suppose that the $p+1$ classes of parallel lines in $AG(2,p)$ are

$$C_0, \; C_1, \; \ldots, \; C_p$$

Suppose further that $p+1$ distinct new points

$$\infty_0, \; \infty_1, \; \ldots, \; \infty_p$$

are added to the domain of $AG(2,p)$, that the point ∞_j is added to each of the lines in class C_j, and that a new line

$$L_\infty \;=\; \{\infty_0, \; \infty_1, \; \ldots, \; \infty_p\}$$

is added.

TERMINOLOGY: We adopt the name *projective extension* for each artifact of the construction just described.

Theorem 10.5.5. *The geometry $\mathcal{G}$ resulting from projective extension of the affine plane $AG(2,p)$ is a projective plane.*

Proof: Since $AG(2,p)$ has p^2 points and p^2+p lines of p points each, the geometry $\mathcal{G}$ has p^2+p+1 points and p^2+p+1 lines of $p+1$ points each.

(Axiom PP1): If two points of the geometry $\mathcal{G}$ are already in $AG(2,p)$, then they are on some line of $AG(2,p)$, and accordingly, they lie on the extension of that line in $\mathcal{G}$. If the two points are ∞_i and ∞_j, then they lie on the line L_∞. If one point x is from $AG(2,p)$ and the other is ∞_j, then since each of class C_j partitions $AG(2,p)$, the point x lies on some line of class C_j, and thus on the extension of that line in geometry $\mathcal{G}$.

(Axiom PP2): The line L_∞ evidently meets every other line of $\mathcal{G}$. Moreover, the intersection of two lines not in the same parallel class of $AG(2,p)$ is a single point, by Axiom $AP1$, from which it follows that their extensions meet only at that same point. If two lines of $AG(2,p)$ are in the same parallel class C_j, then their extensions meet only at ∞_j.

(Axiom PP3): If four points satisfy Axiom AP3 in $AP(2,p)$, then those same four points satisfy Axiom PP3 in $\mathcal{G}$.   ◇

Example 10.5.7: Figure 10.5.5 shows how the projective plane $PG(2,3)$ with 13 points can be constructed by extending the 9-point affine plane $AG(2,3)$.

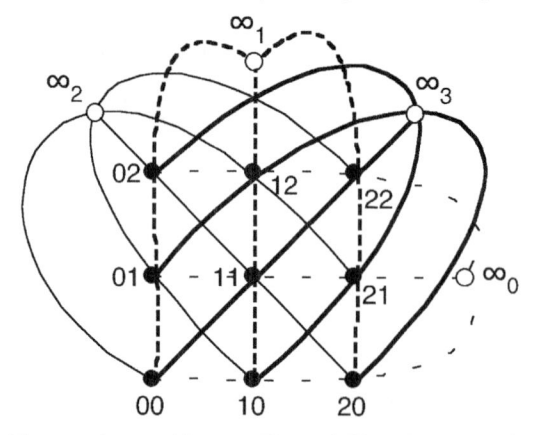

Figure 10.5.5   **Extending** $AG(2,3)$ **to** $PG(2,3)$.

EXERCISES for Section 10.5

10.5.1$^\text{S}$ Construct a list of all the points of the affine geometry $AG(2,2)$.

10.5.2$^\text{S}$ Construct a list of all the lines of the affine geometry $AG(2,2)$.

10.5.3   Calculate the replication number r of the affine geometry $AG(2,2)$.

10.5.4   Construct a list of all the points of the affine geometry $AG(2,3)$.

10.5.5   Construct a list of all the lines of the affine geometry $AG(2,3)$.

10.5.6   Calculate the replication number r of the affine geometry $AG(2,3)$.

GLOSSARY

abstract Latin square: representation of a Latin square as a set of triples.

affine geometry $AG(2,p)$: the finite geometry whose points are the points of the vector space $\mathbb{Z}_p^2$ and whose lines are the affine lines of $\mathbb{Z}_p^2$.

affine line in $\mathbb{Z}_p^2$: the set of points obtained by adding a fixed vector (c_1, c_2) to a one-dimensional subspace of $\mathbb{Z}_p^2$.

affine plane: a finite geometry such that the following three axioms hold:

AP1. For each pair of distinct points, there is exactly one line containing them.

AP2. For any given line L_i and any point x not on L_i there is a line through x that is parallel to L_i.

AP3. There exist four points, no three of which lie on the same line.

balanced block design: a regular design such that every pair of objects appears in the same number of blocks.

BIBD: balanced incomplete block design.

block design: a combinatorial design with no ordering of the blocks or within the blocks.

blocksize: the cardinality of a block of a regular block design.

combinatorial design: any mathematical structure involving a domain and a set of subsets of that domain.

complete block: a block that contains every object of the domain.

conjugacy operation on a Latin square: on an abstract Latin square, a permutation on the coordinates within each of its triples; otherwise, the Latin square corresponding to the effect of such an operation on its abstract counterpart.

covalence of a pair of elements of a block design: the number of blocks in which the pair lies; for a BIBD, this is the *index*.

Desargues geometry: a classical finite geometry with ten points and ten lines.

dual of a geometry $\langle X, \mathcal{B} \rangle$: the geometry whose incidence matrix is the transpose of the matrix for $\langle X, \mathcal{B} \rangle$.

Fano plane: a classical finite geometry with seven points and seven lines; equivalent to the projective plane $PG(2,2)$.

finite geometry: a combinatorial design $\langle X, \mathcal{L} \rangle$, in which X is said to be a set of *points* and $\mathcal{L}$ a set of *lines*, such that

G1. Two distinct points are contained in at most one line.

G2. Two distinct lines intersect in at most one point.

incidence matrix of a geometry $\langle X, \mathcal{L} \rangle$: a matrix with a row for each point x_i and a column for each line L_j, such that

$$M_{\langle X,L \rangle}[i,j] = \begin{cases} 1 & \text{if } x_i \in L_j \\ 0 & \text{otherwise} \end{cases}$$

incidence structure: a discrete structure involving two sets X and Y and a rule for the incidence of a member of x and a member of y; if the value of the incidence is restricted to be either 0 or 1, then the members of the set Y can be represented as subsets of X.

incomplete block design: a design with incomplete blocks.

index of a BIBD: the number λ of blocks in which each pair of objects appears.

isomorphism of block designs $\mathcal{B} = \langle X, B \rangle$ and $\mathcal{C} = \langle Y, C \rangle$:   a bijection $f : X \to Y$ such that each block of $\mathcal{B}$ is mapped onto a block of $\mathcal{C}$.

isotopy of Latin squares: a permutation of the rows, a permutation of the columns, or a permutation of the symbol set, or any iterated composition of three such operations.

Latin square: an $n \times n$ array of n symbols, arranged so that each symbol appears exactly once in each row and exactly once in each column.

Levi graph of a design $\langle X, B \rangle$:   a bipartite graph whose partite sets are X and B, with an edge joining $x_j \in X$ to $B_j \in B$ whenever $x_i \in B_j$.

line in a finite vector space: a 1-dimensional subspace.

lines of a geometry $\langle X, L \rangle$:   the members of L.

main class isotopic Latin squares: two Latin squares such that one is *isotopic* to a *conjugate* of the other.

Menger graph of a design $\langle X, L \rangle$:   the graph whose vertex set is X, such that two vertices are joined if they are contained in a block of B.

normalized Latin square: an $n \times n$ Latin square whose initial row and initial column both give the elements of $\mathbb{Z}_n$ in the order 0, 1, …, $n-1$.

order of a projective plane: one less than the cardinality of a line.

orthogonal Latin squares: Latin squares $A = (a_{i,j})$ and $B = (b_{i,j})$ such that the n^2 ordered pairs $(a_{i,j}, b_{i,j})$ are mutually distinct.

Pappus geometry: a classical finite geometry with nine points and nine lines.

parallel lines: two lines with no points in common.

partially balanced incomplete block design: a combinatorial design that is in most ways like a BIBD, except that the number of blocks in which a pair of objects occurs may have either of two values, λ_1 or λ_2, rather than only one possible value.

PBIBD: partially balanced incomplete block design.

pencil of lines at a point x: the set of all lines that contain x.

perfect difference family: generalization of a perfect difference set.

perfect difference set of index λ: a set of numbers $S = \{a_1, a_2, \ldots, a_k\}$ in $\mathbb{Z}_n$ such that each non-zero number in $\mathbb{Z}_n$ occurs exactly λ times in the list $\langle x_{ij} = a_i - a_j \mid a_i, a_j \in S; \ i \neq j \rangle$.

plane in a finite vector space: a 2-dimensional subspace.

points of a geometry $\langle X, L \rangle$:   the members of L.

pole of a line L in a geometry: a point p such that no line through p intersects L.

projective extension: the construction by which an affine plane is extended to a projective plane.

projective geometry $PG(2,p)$:   has as its points the set of all lines of the vector space $\mathbb{Z}_p^3$ and as its lines the set of all planes of $\mathbb{Z}_p^3$.

projective plane: a finite geometry with the following properties:

PP1. For each pair of distinct points, there is exactly one line containing them.

PP2. Each pair of distinct lines intersects in exactly one point.

PP3. There exist four points, no three of which lie on the same line.

regular block design: a block design with common blocksize, and in which each element appears in the same number of blocks.

replication number in a BIBD: the number of blocks in which each element occurs.

resolvable geometry: a finite geometry $\langle X, L \rangle$ whose lines can be partitioned so that within each cell, the lines partition the domain X.

restriction of the design: the design that results from deleting a set of points from the domain and from all the blocks.

scalar multiplication: multiplying every component of a vector by a scalar from the field.

Steiner triple system: a $(v, 3, 1)$-design.

transpose of a matrix: the result of reflecting the matrix through its main diagonal, so that its rows become columns and vice versa.

(v, b, r, k, λ)-design: a BIBD with v points, b blocks, replication number r, blocksize k, and index λ.

(v, k, λ)-design: a BIBD with v points, blocksize k, and index λ.

valence of an element of a block design: the number of blocks in which it lies.

vector addition: adding two vectors over the same field, coordinate by coordinate.

Appendix

A1 Relations and Functions

A2 Algebraic Systems

A3 Finite Fields and Vector Spaces

A1   RELATIONS AND FUNCTIONS

DEFINITION: A **binary relation** R from a set U to a set V may be regarded intuitively as a predicate on arguments $u \in U$ and $v \in V$ that is *true* or *false*, with the notation $u\,R\,v$ indicating *true*, i.e., that u is related to v. Set-theoretically the relation R may be represented as a subset of the cartesian product $U \times V$, in which $(u, v) \in R$ means u is related to v.

Some of the most important relations are from some set to itself.

Example A1.1:   A binary relation on the positive integers is defined by the rule

$$m \setminus n \ \text{ if and only if } \ \frac{n}{m} \ \text{ is an integer}$$

For instance, $7 \setminus 42$. One says, "7 *divides* 42".

Example A1.2:   For any positive integer m, the relation of *arithmetic congruence* modulo m on all the integers is defined by the rule

$$b \equiv a \text{ modulo } m \ \text{ if and only if } \ m \setminus b - a$$

For instance, $25 \equiv 3$ modulo 11. One says, "25 is *congruent modulo* 11 to 3" or "25 is *congruent* to 3 *modulo* 11".

Equivalence Relations

DEFINITION: A binary relation R from a set U to itself is an **equivalence relation** if it has the following three properties:

- *reflexivity*: uRu, for all $u \in U$.
- *symmetry*: if u_1Ru_2 then u_2Ru_1, for all u_1, $u_2 \in U$.
- *transitivity*: if u_1Ru_2 and u_2Ru_3 then u_1Ru_3, for all u_1, u_2, $u_3 \in U$.

Example A1.3:   *Congruence modulo m* on any subset of the integers is reflexive, since the quotient $\frac{b-b}{m}$ is the integer 0. It is symmetric, since, if the quotient $\frac{b-a}{m}$ is the integer k, then the quotient $\frac{a-b}{m}$ is the integer $-k$. To show it is transitive, suppose that $a \equiv b$ modulo m and $b \equiv c$ modulo m. Then there are integers r and s such that

$$\frac{a-b}{m} = r \quad \text{and} \quad \frac{b-c}{m} = s$$

It follows that

$$\frac{a-c}{m} = \frac{a-b}{m} + \frac{b-c}{m} = r + s$$

Thus, $a \equiv c$ modulo m.

Any binary relation R on a set U can be represented by a directed graph whose vertex set is U. If aRb, then there is an arc from a to b.

Example A1.3, continued: Figure A1.1 shows the directed graph for congruence modulo 3 the subset $\{0, 1, \ldots, 8\}$.

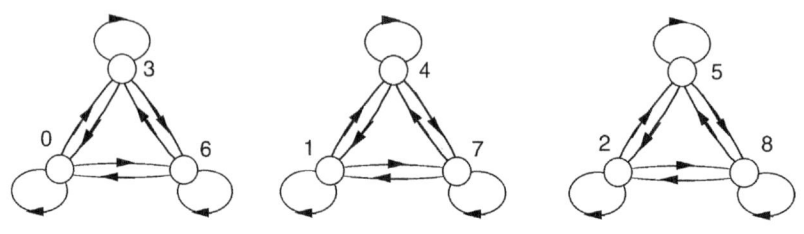

Figure A1.1   **Digraphic representation of congruence modulo 3.**

Proposition A1.1. *Let G be the digraph that represents a relation R on a set U.*

- *R is reflexive iff there is a self-loop at every vertex.*

- *R is symmetric iff for every arc, there is an arc with opposite direction joining the same two vertices.*

- *R is transitive iff whenever there is an arc from a vertex u to a vertex v and also an arc from vertex v to a vertex w, there is an arc from u to w.*   ◇

Example A1.3, continued: Observe that the digraph of Figure A1.1 has the three graph properties cited in Proposition A1.1.

Example A1.4:   The relation *divides* is reflexive, since the quotient $\frac{n}{n}$ is the integer 1. To show it is transitive, suppose that $a \setminus b$ and $b \setminus c$. Then there are integers r and s such that

$$\frac{b}{a} = r \quad \text{and} \quad \frac{c}{b} = s$$

It follows that

$$\frac{c}{a} = \frac{c}{b} \cdot \frac{b}{a} = sr$$

However, it is not an equivalence relation, because, although it is reflexive and transitive, it is not symmetric. For instance, 3 divides 6, but 6 does not divide 3.

Partial Orderings

DEFINITION: A binary relation R from a set U to itself is a **partial ordering** if it has the following three properties:

- *reflexivity*: uRu, for all $u \in U$.
- *anti-symmetry*: if $u_1 R u_2$ and $u_2 R u_1$ then $u_1 = u_2$, for all $u_1, u_2 \in U$.
- *transitivity*: if $u_1 R u_2$ and $u_2 R u_3$ then $u_1 R u_3$, for all $u_1, u_2, u_3 \in U$.

Example A1.4, continued: *Divides* is anti-symmetric on the positive integers, since, if the quotient $\frac{b}{a}$ is an integer, then $a \leq b$, and if the quotient $\frac{a}{b}$ is also an integer, then $b \leq a$. Thus, if both quotients are integers, then $a = b$. Since it is reflexive, anti-symmetric, and transitive, *divides* is a partial ordering.

Example A1.5: The relation *congruence modulo m* is not a partial ordering, because it is not anti-symmetric.

Properties of Some Functions

DEFINITION: A **function** from a set U to a set V is a relation from U to V such that for each $u \in U$, there is a unique $v \in V$, such that $(u, v) \in U \times V$.

- The set U to whose elements a function $f : U \to V$ assigns values is called the *domain*.
- The set V in which the values are assigned is called the *codomain*.

DEFINITION: A function $f : U \to V$ is **one-to-one** (or *injective*) if for all $u_1, u_2 \in U$,

$$u_1 \neq u_2 \;\Rightarrow\; f(u_1) \neq f(u_2)$$

DEFINITION: A function $f : U \to V$ is **onto** (or *surjective*) if for all $v \in V$,

$$(\exists u \in U)\,[\,f(u) = v\,]$$

DEFINITION: A function $f : U \to V$ is **bijective** if it is both one-to-one and onto.

Any function $f : U \to V$ can be represented by a digraph whose vertex set is $U \cup V$. There is an arc from u to v if and only if $f(u) = v$.

Proposition A1.2. *Let D be the digraph that represents a function $f : U \to V$. Then*

- *f is one-to-one if and only if no vertex $v \in V$ is the head of more than one arc of the digraph D.*
- *f is onto if and only if every vertex $v \in V$ is the head of at least one arc of the digraph D.*
- *f is bijective if every vertex $v \in V$ is the head of exactly one arc of D.* $\Diamond$

A2 ALGEBRAIC SYSTEMS

The algebraic system of most frequent concern in the application of combinatorial methods is the set of integers, along with the operations of addition and multiplication.

NOTATION: The set of *integers* is denoted $\mathbb{Z}$. The corresponding algebraic system is formally denoted $\langle \mathbb{Z}, +, \cdot \rangle$. However, in practice, both the algebraic system and its domain are familiarly denoted by $\mathbb{Z}$.

NOTATION: The *greatest common divisor* of two numbers is denoted $\gcd(m, n)$. Sometimes the gcd operator is applied to more than two numbers at a time.

Binary Operations

DEFINITION: A **binary operation** on a set U is a function from the cartesian product $U \times U$ to the set U itself.

DEFINITION: In the present context, an **algebraic system** means a non-empty set together with one or more binary operations.

- The set U is called the **domain** of that algebraic system.

Example A2.1: Addition and multiplication are binary operations on the integers, since the result of adding or multiplying two integers is an integer.

NOTATION: The operations of addition and multiplication are usually represented in *infix form*, i.e.,

$$x + y \quad \text{or} \quad x \cdot y$$

respectively, rather than the *prefix form* $+ (x, y)$ or $\cdot (x, y)$.

Integers Modulo n

DEFINITION: The **mod operator** is the arithmetic rule that assigns to a non-negative integer x (called the **moduland**) and a positive integer n (called the **modulus**) the number

$$x \bmod n$$

that is the remainder of operation of dividing x by n.

EXTENSION OF DEFINITION: For a negative moduland x, the function

$$x \mapsto x \bmod n$$

is extended from the domain $\mathbb{N}$ to the domain $\mathbb{Z}$ inductively, by the rule

$$x \bmod n = (x + n) \bmod n \qquad (A.2.1)$$

Example A2.2: $-5 \bmod 3 \; = \; -2 \bmod 3 \; = \; 1 \bmod 3 \; = \; 1.$ More generally, the following table illustrated the periodicity of the mod operator.

x		$\cdots$	-7	-6	-5	-4	-3	-2	-1	0	1	2	3	4	5	6	$\cdots$
$x \bmod 3$		$\cdots$	2	0	1	2	0	1	2	0	1	2	0	1	2	0	$\cdots$

Proposition A2.1. *Let x be a negative integer. Then*

$$x \bmod n \; = \; n - (|x| \bmod n) \qquad\qquad (A.2.2)$$

Proof: The proof is by induction. Details are omitted. $\diamond$

Example A2.3:   For instance,

$$\begin{aligned}
-5 \bmod 3 &= 3 - (\,|-5| \bmod 3\,) \\
&= 3 - (5 \bmod 3) \\
&= 3 - 2 \\
&= 1
\end{aligned}$$

Remark: The *mod operator* is now used to define an algebraic system called the *integers modulo n*. In both concepts, which are clearly distinct, taking the remainder of a division has a fundamental role.

DEFINITION: The domain of the algebraic system (with $n \geq 2$) known as the **integers modulo n**, denoted $\mathbb{Z}_n$, is the set of numbers

$$\{\, 0, \quad 1, \quad \ldots, \quad n-1 \,\}$$

There are two binary operations $(+)$ and $(\cdot)$ on $\mathbb{Z}_n$, given by the rules

$$b + c \; = \; b + c \bmod n$$
$$b \cdot c \; = \; b \cdot c \bmod n$$

In other words, in the algebraic system $\mathbb{Z}_n$, if adding or multiplying two numbers as usual for integers happens to exceed $n-1$, then divide by n and use the remainder as the result of the operation.

Associativity and Commutativity

In the rest of this chapter, we adopt the symbol $\star$ as a generic binary operation, to be written in infix form.

DEFINITION: A binary operation $\star$ on a set U is an **associative operation** if

$$(u \star v) \star w \; = \; u \star (v \star w) \qquad (\forall u, v, w \in U)$$

Example A2.3, continued: Addition and multiplication are both associative operations on the integers and also on the integers modulo n.

DEFINITION: A binary operation $\star$ on a set U is a **commutative operation** if

$$u \star v \; = \; v \star u \qquad (\forall u, v \in U)$$

Example A2.3, continued: Addition and multiplication are commutative operations on the integers and also on the integers modulo n.

Identity Element

DEFINITION: An element $\epsilon \in U$ is the **identity element** with respect to a binary operation $\star$ if

$$\epsilon \star u \ = \ u \star \epsilon \ = \ u \qquad (\forall u \in U)$$

There is at most one identity element with respect to a given binary operation.

Example A2.3, continued:

- The number 0 is the additive identity of the integers and also the additive identity of the integers modulo n.

- The number 1 is the multiplicative identity of $\mathbb{Z}$ and also the multiplicative identity of the integers modulo n.

Inverse Elements

DEFINITION: Suppose that the set U has an identity ϵ with respect to a binary operation $\star$, and let $u \in U$. An element u^{-1} is the *inverse* of u with respect to $\star$ if

$$u^{-1} \star u \ = \ u \star u^{-1} \ = \ \epsilon$$

There is at most one inverse for an element with respect to a given binary operation.

Example A2.3, continued:

- Every integer n has $-n$ as its additive inverse.

- The number 1 is the only integer whose multiplicative inverse is an integer.

- In the set of positive rational numbers, every element $\frac{p}{q}$ has the multiplicative inverse $\frac{q}{p}$.

- Some numbers have multiplicative inverses in $\mathbb{Z}_n$. For instance, 11 is the inverse of 5 in $\mathbb{Z}_{18}$, since

$$11 \cdot 5 \ = \ 55 \ \equiv \ 1 \bmod 18$$

It is proved in §6.4 that a number b has a multiplicative inverse modulo n if and only if $\gcd(b, n) = 1$.

Groups

DEFINITION: An algebraic system $\langle U, \ \star \rangle$ is called a **group** if it has the following properties:

> the operation $\star$ is associative.

> there is an identity element.

> every element of U has an inverse.

If $\star$ is commutative, then $\langle U, \ \star \rangle$ is called a *commutative group* or an *abelian group*. In this case, it is also commonly called an **additive group**, and its operation is written as addition, using the operational symbol $+$. Moreover, the additive identity is typically called **zero** and denoted 0.

Example A2.3, continued:

- The integers are a group under the operation of addition. They are not a group under multiplication, because most integers do not have multiplicative inverses.

- The positive rationals are a group under multiplication.

Rings

DEFINITION: An algebraic system $\langle U, +, \cdot \rangle$ is called a **ring** if it has the following properties:

the system $\langle U, + \rangle$ (ignoring $\cdot$) is an additive group.

the operation $\times$ is associative.

$z \cdot (x + y) = z \cdot x + z \cdot x$, for any $x, y, z \in U$. We say that the operation $(\cdot)$ *distributes* over the operation $(+)$.

The operation $(\cdot)$ is typically called **multiplication**.

Example A2.3, continued:

- The integers are a ring under the operations of addition and multiplication, as are the integers modulo n. In both cases, we commonly represent the multiplication operation by *juxtaposition*, that is,

$$ab \quad \text{means} \quad a \cdot b$$

- In some rings, including $\mathbb{Z}$ and $\mathbb{Z}_n$, multiplication is commutative and the number 1 serves as the multiplicative identity. No number except 1 or -1 has a multiplicative inverse in $\mathbb{Z}$.

Example A2.4:   The 2×2 real matrices are a ring, whose multiplication (i.e., matrix multiplication) is non-commutative. A square matrix has a multiplicative inverse if its determinant is non-zero.

Fields

DEFINITION: A **field** is a ring with commutative multiplication and a multiplicative identity, such that every non-zero element has a multiplicative inverse.

Example A2.5:

- The rational numbers $\mathbb{Q}$ are a field.
- The real numbers $\mathbb{R}$ are a field.
- The complex numbers $\mathbb{C}$ are a field.

Example A2.6:

- The ring $\mathbb{Z}_2$ is a field, since the number 1 is its own multiplicative inverse.
- The ring $\mathbb{Z}_3$ is a field, since both 1 and 2 are their own inverses.
- The ring $\mathbb{Z}_5$ is a field since

$$1 \cdot 1 = 1 \quad 2 \cdot 3 = 1 \quad 3 \cdot 2 = 1 \quad 4 \cdot 4 = 1$$

Since, according to §6.4, every number r such that $\gcd(r, n) = 1$ has a multiplicative inverse in $\mathbb{Z}_n$, it follows that for prime p, the ring $\mathbb{Z}_p$ is a field.

k-Tuples of Numbers

DEFINITION: The **direct sum** $\mathbb{Z}_{n_1} \oplus \mathbb{Z}_{n_2}$ of the additive groups $\mathbb{Z}_{n_1}$ and $\mathbb{Z}_{n_2}$ is a group whose domain is the set of 2-tuples

$$\left\{ (r, s) \;\middle|\; r \in \mathbb{Z}_{n_1}, \, s \in \mathbb{Z}_{n_2} \right\}$$

Its operation is coordinate-wise addition. That is,

$$(r, s) + (r', s') \;=\; (r + r' \bmod n_1, \, s + s' \bmod n_2)$$

Example A2.7:   In $\mathbb{Z}_3 \oplus \mathbb{Z}_4$, the elements are

$$
\begin{array}{cccc}
(0,0) & (0,1) & (0,2) & (0,3) \\
(1,0) & (1,1) & (1,2) & (1,3) \\
(2,0) & (2,1) & (2,2) & (2,3)
\end{array}
$$

Some examples of addition are

$$
\begin{aligned}
(1,3) + (2,2) &= (0,1) \\
(1,1) + (2,3) &= (0,0)
\end{aligned}
$$

Remark: The direct sum construction can be iterated

$$\mathbb{Z}_{n_1} \oplus \mathbb{Z}_{n_2} \cdots \oplus \mathbb{Z}_{n_k}$$

to k-tuples of numbers, with each coordinate having its respective modulus. All these direct sums are groups.

NOTATION: If every modulus over all k coordinates is the same number r, then the group may be denoted $\mathbb{Z}_r{}^k$.

Generators for an Algebraic System

DEFINITION: An element x is an **additive generator** for $\mathbb{Z}_n$ if every number r in $\mathbb{Z}_n$ can be written as an iterated sum

$$r = x + x + \cdots + x$$

Example A2.8:   The number 3 is an additive generator for $\mathbb{Z}_{10}$, but the number 2 is not. Whereas 3 generates the complete sequence

$$3 \quad 6 \quad 9 \quad 2 \quad 5 \quad 8 \quad 1 \quad 4 \quad 7 \quad 0$$

(where $6 = 3 + 3$, $9 = 3 + 3 + 3$, $2 = 3 + 3 + 3 + 3 \bmod 10$, etc.) the number 2 generates only the five numbers

$$2 \quad 4 \quad 6 \quad 8 \quad 0$$

before it repeats.

DEFINITION: A number x is a **multiplicative generator** for $\mathbb{Z}_n$ if every non-zero number r in $\mathbb{Z}_n$ can be written as an iterated product

$$r = x \cdot x \cdot \cdots \cdot x$$

Example A2.9:   The number 3 is a multiplicative generator for $\mathbb{Z}_7$, but the number 2 is not. Whereas 3 generates the complete sequence

$$\left\langle\, 3^j \bmod n \mid j = 1, \ldots, 6 \,\right\rangle \;=\; 3 \quad 2 \quad 6 \quad 4 \quad 5 \quad 1$$

the number 2 generates only the sequence

$$2 = 2^1 \quad 4 = 2^2 \quad 1 = 2^3$$

before it repeats.

DEFINITION: A set of elements $S = \{x_1, \ x_2, \ \ldots, \ x_s\}$ of an algebraic system $\mathcal{A}$ is a **generating set** for that system if every other element of that system can be obtained by applying the operations of the system $\mathcal{A}$ to some combination of the elements in S.

Example A2.10:   The numbers 3 and 5 form an additive generating set for $\mathbb{Z}_{15}$. (So does any number j such that $\gcd(j, 15) = 1$, by itself.)

Example A2.11:   The 2-tuples $(1, 0)$ and $(0, 1)$ generate the group $\mathbb{Z}_{n_1} \oplus \mathbb{Z}_{n_2}$. No single 2-tuple generates such a group.

Prime Factorization

Although the entire set of integers is generated additively by the number 1, it takes the infinite set of all the *prime numbers* to generate the integers multiplicatively.

DEFINITION: A **divisor** of an integer n is an integer d such that the equation $xd = n$ has an integer solution $x = q$.

DEFINITION: A **prime number** is a positive integer $p > 1$ that has no divisors except 1 and itself.

The following theorem from elementary number theory is commonly called the **Fundamental Theorem of Arithmetic**.

Theorem A2.2 [Prime-Power Factorization Theorem]. *Every positive integer n can be written uniquely as a product of powers of distinct, non-decreasing primes*

$$n = p_1^{e_1} p_2^{e_2} \cdots p_r^{e_r}$$

A3   FINITE FIELDS AND VECTOR SPACES

Example A2.6 asserts that for prime p, all non-zero numbers have multiplicative inverses in the ring $\mathbb{Z}_p$ (as proved in §6.4), which implies that $\mathbb{Z}_p$ is a field. In order to present various algebraic methods to persons who have not previously had at least the equivalent of a semester course on groups, rings, and fields, Chapter 10 develops various design constructions explicitly for prime fields $\mathbb{Z}_p$, even though they are readily generalizable to all finite fields. This section is intended for readers wanting to understand the methods of Chapter 10 in greater generality. For other readers, it is optional. Our main concern in this present section is to describe the construction of a field $GF(p^r)$ for every prime power p^r.

DEFINITION: An additive group $\langle V, + \rangle$ is called a **vector space** over the field $\mathcal{F}$ if there is a scalar product $\mathcal{F} \times V \to V$ such that

(i) $\qquad\qquad c(\mathbf{v} + \mathbf{v}') = c\mathbf{v} + c\mathbf{v}'$ for all $c \in \mathcal{F}$ and all $\mathbf{v}, \mathbf{v}' \in V$

(ii) $\qquad\qquad (a + b)\mathbf{v} = a\mathbf{v} + b\mathbf{v}$ for all $a, b \in \mathcal{F}$ and all $\mathbf{v} \in V$

(iii) $\qquad\qquad (ab)\mathbf{v} = a(b\mathbf{v})$ for all $a, b \in \mathcal{F}$ and all $\mathbf{v} \in V$

(iv) $\qquad\qquad 1\mathbf{v} = \mathbf{v}$ for mult. identity $1 \in \mathcal{F}$ and all $\mathbf{v} \in V$

An element of the domain V is called a **vector**. An element of the field $\mathcal{F}$ is called a **scalar**.

Example A3.1:   The vector space $\mathbb{R}^d$ is the set of d-tuples with entries in $\mathbb{R}$ (i.e., real numbers), where coordinate-wise addition in $\mathbb{R}$ is the vector addition. The scalar product is

$$c(x_1, x_2, \ldots, x_d) = (cx_1, cx_2, \ldots, cx_d)$$

Bases and Dimension

DEFINITION: Let $\mathbf{v}_1, \mathbf{v}_2, \ldots, \mathbf{v}_k$ be vectors in a vector space $\langle V, + \rangle$ over a field $\mathcal{F}$ and $c_1, c_2, \ldots, c_k$ scalars in $\mathcal{F}$. The iterated sum

$$c_1\mathbf{v}_1 + c_2\mathbf{v}_2 + \cdots + c_k\mathbf{v}_k$$

is called a **linear combination** of those vectors.

Example A3.1, continued: In the vector space $\mathbb{R}^d$, every vector

$$(x_1, \, x_2, \, x_3, \, \ldots, \, x_d)$$

can be expressed as a linear combination of the vectors

$$(1, 0, 0, 0, \, \ldots, 0, 0)$$
$$(0, 1, 0, 0, \, \ldots, 0, 0)$$
$$(0, 0, 1, 0, \, \ldots, 0, 0)$$
$$\vdots$$
$$(0, 0, 0, 0, \, \ldots, 0, 1)$$

That is,

$$
\begin{aligned}
(x_1, \, x_2, \, x_3, \, \ldots, \, x_d) \;=\; & \; x_1 \, (1, 0, 0, 0, \, \ldots, 0, 0) \\
& +\; x_2 \, (0, 1, 0, 0, \, \ldots, 0, 0) \\
& +\; x_3 \, (0, 0, 1, 0, \, \ldots, 0, 0) \\
& \qquad\qquad \vdots \\
& +\; x_d \, (0, 0, 0, 0, \, \ldots, 0, 1)
\end{aligned}
$$

DEFINITION: If every vector in a vector space $\langle V, + \rangle$ is expressible as a linear combination of the vectors $\mathbf{v}_1, \mathbf{v}_2, \ldots, \mathbf{v}_k$, then those vectors are said to be a **spanning set of vectors** for $\mathcal{F}$ or to **span** $\mathcal{F}$.

DEFINITION: A set of vectors $\mathbf{v}_1, \mathbf{v}_2, \ldots, \mathbf{v}_k$ that span the vector space $\langle V, + \rangle$ such that the linear combination expressing each vector $\mathbf{v}$ is unique is called a **basis for the vector space** $\langle V, + \rangle$.

Example A3.1, continued: In the vector space $\mathbb{R}^d$, the vectors

$$(1, 0, 0, 0, \, \ldots, 0, 0)$$
$$(0, 1, 0, 0, \, \ldots, 0, 0)$$
$$(0, 0, 1, 0, \, \ldots, 0, 0)$$
$$\vdots$$
$$(0, 0, 0, 0, \, \ldots, 0, 1)$$

form a basis.

Theorem A3.1 [Invariance of Dimension]. *Every basis for a vector space has the same cardinality.*

Proof: We omit details of the proof. It follows from the *row reduction algorithm*, which amounts to a systematization of the standard method of solving simultaneous linear equations by iterative substitution. Moreover, the row reduction algorithm is used in proving that every vector space has a basis.   ◇

Example A3.2: The polynomials of finite degree with coefficients in the field $\mathbb{Z}_p$ form an infinite-dimensional vector space, denoted $\mathbb{Z}_p[x]$. We observe that every such polynomial can be expressed uniquely as a linear combination of the polynomials

$$1 \quad x \quad x^2 \quad x^3 \quad \cdots$$

Moreover, the polynomials of degree at most $d-1$ with coefficients in $\mathbb{Z}_p$ form a d-dimensional vector space over the field $\mathbb{Z}_p$, with the vectors

$$1 \quad x \quad x^2 \quad \cdots \quad x^{d-1}$$

serving as a basis.

Irreducible Polynomials

DEFINITION: A polynomial in $\mathbb{Z}_p[x]$ is said to be **irreducible over the prime field** $\mathbb{Z}_p$ if it cannot be factored into two polynomials (in $\mathbb{Z}_p[x]$) of smaller degree.

Example A3.3:   The polynomial $x^2 + 1$ factors over $\mathbb{Z}_2$ into

$$\begin{aligned}(x+1)(x+1) \;&=\; x^2 + 2x + 1 \\ &\equiv\; x^2 + 1 \text{ modulo } 2\end{aligned}$$

and the polynomial $x^2 + x + 1$ is irreducible (which could be proved by trying all possible factorizations).

Example A3.4:   The polynomial $x^2 + x + 1$ factors over $\mathbb{Z}_3$ into

$$\begin{aligned}(x+2)(x+2) \;&=\; x^2 + 4x + 4 \\ &\equiv\; x^2 + x + 1 \text{ modulo } 3\end{aligned}$$

and the polynomial $x^2 + x + 2$ is irreducible.

Remark: Tables of irreducible polynomials can be downloaded from various websites.

Polynomials Modulo a Polynomial

TERMINOLOGY: When a polynomial $f(x)$ is divided by a polynomial $g(x)$ of degree d, there is a **quotient polynomial** $q(x)$ and a **remainder polynomial** $r(x)$ of degree less than d such that

$$f(x) \;=\; q(x)g(x) + r(x)$$

DEFINITION: The algebraic structure $\mathcal{F}[x]\,/\,g(x)$ of **polynomials modulo a polynomial** over the field $\mathcal{F}$ has polynomial addition and multiplication modulo $g(x)$ as its operations, under which it is a ring.

Example A3.5: Here are the addition and multiplication tables for the ring $\mathbb{Z}_2[x] / x^2 + x + 1$. It can be verified that $\mathbb{Z}_2[x] / x^2 + x + 1$ satisfies all of the axioms for a field.

Table A3.1   Arithmetic tables for $\mathbb{Z}_2[x] / x^2 + x + 1$.

+	0	1	x	$x+1$
0	0	1	x	$x+1$
1	1	0	$x+1$	x
x	x	$x+1$	0	1
$x+1$	$x+1$	x	1	0

$\cdot$	0	1	x	$x+1$
0	0	0	0	0
1	0	1	x	$x+1$
x	0	x	$x+1$	1
$x+1$	0	$x+1$	1	x

Basic Facts about Finite Fields

Proofs of the following basic facts about finite fields can be found in many textbooks on abstract algebra. We presently assert them without proof.

Proposition A3.2. *For any irreducible polynomial $g(x)$ of degree d over $\mathbb{Z}_p[x]$, where p is prime, the ring $\mathbb{Z}_p[x]/g(x)$ is a field of order p^d.*

Proposition A3.3. *There exists a field of order n if and only if $n = p^d$ for some prime number p.*

Proposition A3.4. *All fields of order p^d are isomorphic.*

Proposition A3.5. *The additive group of the field of order p^d is isomorphic to the vector space $\mathbb{Z}_p^d$.*

Proposition A3.6. *The multiplicative group of the field of order p^d is generated by some single element of the field.*

Bibliography

B1 General Reading

B2 References

B1   GENERAL READING

In addition to the references listed below, the reader may wish to consult the *Handbook of Discrete and Combinatorial Mathematics*, edited by K. H. Rosen, which contains many contributed sections with summaries of the most important research results and extensive bibliographies.

ALGEBRA

[Gall2006] J. A. Gallian, *Contemporary Abstract Algebra*, Sixth Edition, Houghton-Mifflin, 2006.

[Hers2001] I. N. Herstein, *Abstract Algebra*, Third Edition, J. Wiley & Sons, 2001.

[MaBi1967] S. Maclane and G. Birkhoff, *Algebra*, Macmillan, 1967.

ALGORITHMS and COMPUTATION

[AHU1983] A. V. Aho, J. E. Hopcroft, and J. D. Ullman, *Data Structures and Algorithms*, Addison-Wesley, 1983.

[CLRS2001] T. H. Cormen, C. E. Leiserson, R. L. Rivest, and C. Stein, *Introduction to Algorithms*, Second Edition, MIT Press, 2001.

COMBINATORIAL and DISCRETE MATHEMATICS

[BeQu2003] A. T. Benjamin and J. J. Quinn, *Proofs that Really Count*, Mathematical Association of America, 2003.

[BeWi1991] E. A. Bender and S. G. Williamson, *Foundations of Applied Combinatorics*, Addison-Wesley, 1991. (See especially Section 12.4 of this text for further discussion of asymptotic estimation.)

[Bona2007] M. Bóna, *Introduction to Enumerative Combinatorics*, McGraw-Hill, 2007.

[Brua2004] R. Brualdi, *Introductory Combinatorics*, Fourth Edition, Prentice-Hall, 2004.

[Devo2004], J. L. Devore, *Probability and Statistics*, Sixth Edition, Brooks/Cole, 2004.

[GKP1994] R. L. Graham, D. E. Knuth, and O. Patashnik, *Concrete Mathematics: A Foundation for Computer Science*, Second Edition, Addison-Wesley, 1994.

[Grim2004] R. P. Grimaldi, *Discrete and Combinatorial Mathematics*, Fifth Edition, Addison-Wesley, 2004.

[Liu1968] C. L. Liu, *Introduction to Combinatorial Mathematics*, McGraw-Hill, 1968.

[MiRo1991] J. G. Michaels and K. H. Rosen, editors, *Applications of Discrete Mathematics*, McGraw-Hill, 1991.

[PeWiZe1996] M. Petkovšek, H. S. Wilf, and D. Zeilberger, *A=B*, A. K. Peters, 1996.

[PoSt1990] A. Polimeni and H. J. Straight, *Foundations of Discrete Mathematics*, Brooks/Cole, 1990.

[RoTe2005] F. S. Roberts and B. Tesman, *Applied Combinatorics*, Second Edition, Prentice-Hall, 2005.

[Rose2007] K. H. Rosen, *Discrete Mathematics and Its Applications*, Sixth Edition, McGraw-Hill, 2007.

[SlPl1995] N. J. A. Sloane and S. Plouffe, *The Encyclopedia of Integer Sequences*, Second Edition, Academic Press, 1995. (A comprehensive reference volume on integer sequences.)

[StWh1986] D. Stanton and D. White, *Constructive Combinatorics*, Springer-Verlag, 1986.

[Stra1993] H. J. Straight, *Combinatorics: An Invitation*, Brooks/Cole 1993.

[Tuck2001] A. Tucker, *Applied Combinatorics*, Fourth Edition, John Wiley & Sons, 2001.

[Wilf1993] H. S. Wilf, *generatingfunctionology*, Second Edition, Academic Press, 1993.

COMBINATORIAL DESIGNS

[CoDi2000a] C. J. Colbourn and J. H. Dinitz, Block designs, §12.1 in *Handbook of Combinatorial Designs*, edited by K. H. Rosen, CRC Press, 2000.

[CoDi2000b] C. J. Colbourn and J. H. Dinitz, editors, Symmetric designs and finite geometries, §12.2 in *Handbook of Combinatorial Designs*, edited by K. H. Rosen, CRC Press, 2000.

[CoDi2000c] C. J. Colbourn and J. H. Dinitz, editors, Latin squares and orthogonal arrays, §12.3 in *Handbook of Combinatorial Designs*, edited by K. H. Rosen, CRC Press, 2000.

[CoDi2007] C. J. Colbourn and J. H. Dinitz, editors, *Handbook of Combinatorial Designs*, Second Edition, CRC Press, 2007.

[Wall2008] W. D. Wallis, *Introduction to Combinatorial Designs*, CRC Press, 2008, to appear.

GRAPH THEORY

[BLW1986] N. L. Biggs, E. K. Lloyd, and R. J. Wilson, *Graph Theory 1736-1936*, Oxford, 1986.

[Boll1998] B. Bollobás, *Modern Graph Theory*, Springer, 1998.

[ChLe2004] G. Chartrand and L. Lesniak, *Graphs & Digraphs*, Fourth Edition, CRC Press, 2004.

[ChZh2005] G. Chartrand and P. Zhang, *Introduction to Graph Theory*, McGraw Hill, 2005.

[Dist2000] R. Diestel, *Graph Theory*, Springer-Verlag, Second Edition, 2000.

[Gibb1985] A. Gibbons, *Algorithmic Graph Theory*, Cambridge University Press, 1985.

[Goul1988] R. Gould, *Graph Theory*, Benjamin/Cummings, 1988.

[GRS1990] R. L. Graham, B. L. Rothschild, and J. H. Spencer, *Ramsey Theory*, John Wiley & Sons, 1990.

[GrYe2006] J. L. Gross and J. Yellen, *Graph Theory and Its Applications*, 2nd Edition, CRC Press, 2006.

[GrYe2004] J. L. Gross and J. Yellen, editors, *Handbook of Graph Theory*, CRC Press, 2004.

[Hara1969] F. Harary, *Graph Theory*, Addison-Wesley, 1969.

[HaPa1973] F. Harary and E. Palmer, *Graphical Enumeration*, Academic Press, 1973.

[West2001] D. B. West, *Introduction to Graph Theory*, Second Edition, Prentice-Hall, 2001.

[Wils1996] R. J. Wilson, *Introduction to Graph Theory*, Addison Wesley Longman, 1996.

[Wils2004] R. J. Wilson, *History of Graph Theory*, §1.3 in *Handbook of Graph Theory*, edited by J. L. Gross and J. Yellen, CRC Press, 2004.

NUMBER THEORY

[Andr1994] G. E. Andrews, *Number Theory*, Dover. (First Edition, Saunders, 1971.)

[NiZu1991] I. Niven, H. S. Zuckerman, and H. L. Montgomery, *An Introduction to the Theory of Numbers*, Fifth Edition, Wiley.

[Rose2005] K. H. Rosen, *Elementary Number Theory, Fifth Edition*, Addison-Wesley, 2005.

SURFACES and TOPOLOGICAL GRAPH THEORY

[GrTu1987] J. L. Gross and T. W. Tucker, *Topological Graph Theory*, Dover, 2001. (First Edition, Wiley-Interscience, 1987.)

[Mass1967] W. S. Massey, *Algebraic Topology: An Introduction*, Harbrace, 1967.

[Ring1974] G. Ringel, *Map Color Theorem*, Springer, 1974.

[Whit2001] A. T. White, *Graphs of Groups on Surfaces*, North-Holland, 2001.

B2 REFERENCES

The following references are cited in the text.

[AhHoUl1974] A. V. Aho, J. E. Hopcroft, and J. D. Ullman, *Design and Analysis of Computer Algorithms*, Addison-Wesley, 1974.

[BaCo1987], W. W. Rouse Ball and H. S. M. Coxeter, *Mathematical Recreations and Essays*, Dover, 1987.

[BoNi2004] B. Bollobás and V. Nikiforov, *Extremal Graph Theory*, §8.1 in *Handbook of Graph Theory*, edited by J. L. Gross and J. Yellen, CRC Press, 2004.

[BSP1960] R. C. Bose, S. K. Shrikhande, and E. T. Parker, Further results on the construction of mutually orthogonal Latin squares and the falsity of Euler's conjecture, *Canad. J. Math.* **12** (1960), 189–203.

[Burn1911] W. Burnside, *Theory of Groups of Finite Order*, Second Edition, Dover, 1955. Original edition: Cambridge Univ. Press, 1911.

[Cauc1845] A. Cauchy, Mémoire sur diverses propriétés remarquables des substitutions réguliéres ou irrégulières, et des systémes de substitutiones conjugées, *Comptes Rendus Acad. Sci. Paris*, **21** (1845), 835.

[CoDi2000a] C. J. Colbourn and J. H. Dinitz, *Block Designs*, §12.1 in *Handbook of Combinatorial Designs*, edited by K. H. Rosen, CRC Press, 2000.

[Dijk1959] E. W. Dijkstra, A note on two problems in connexion with graphs, *Numer. Math.* **1** (1959), 269–271.

[Dilw1950] R. P. Dilworth, A decomposition theorem for partially ordered sets, *Ann. Math.* 51 (1950), 161–166.

[Faud2004] R. Faudree, *Ramsey Graph Theory*, §8.3 in *Handbook of Graph Theory*, edited by J. L. Gross and J. Yellen, CRC Press, 2004.

[Floy1962] R. W. Floyd, Algorithm 97: Shortest path, *Communications of Asso. Comp. Mach.* **5** (1962), 345.

[Frob1887] F. G. Frobenius, Über die Congruenz nach einem aus zwei endlichen Gruppen gebildeten Doppelmodul, *J. Reine Angew. Math.* **101** (1887), 273–299.

[Gros1974] J. L. Gross, Voltage graphs, *Discrete Math.* **9** (1974), 239–246.

[GrHa1980] J. L. Gross and F. Harary, Some problems in topological graph theory, *J. Graph Theory* **4** (1980), 253–263.

[GrTu1977] J. L. Gross and T. W. Tucker, Generating all graph coverings by permutation voltage assignments, *Discrete Math.* **18** (1977), 273–283.

[GrTu1987] J. L. Gross and T. W. Tucker, *Topological Graph Theory*, Dover, 2001. (First Edition, Wiley-Interscience, 1987.)

[GrTu2008] J. L. Gross and T. W. Tucker, eds., *Topics in Topological Graph Theory*, series editors: L. W. Beineke and R. J. Wilson, Cambridge Univ Press., projected for publication in 2008.

[GrYe2006] J. L. Gross and J. Yellen, *Graph Theory and Its Applications*, 2nd Edition, CRC Press, 2006.

[Gust1963] W. Gustin, Orientable embedding of Cayley graphs, *Bull. Amer. Math. Soc.* **69** (1963), 272–275.

[Hall1956] M. Hall, *Proc. Symp. Applied Math.* **6**, 203, Amer. Math. Soc., 1956.

[Heff1891] L. Heffter, Über das Problem der Nachbargebiete, *Math. Annalen* **38** (1891), 477–580.

[Knut1973] D. E. Knuth, *Sorting and Searching*, Volume 3 of *The Art of Computer Programming*, Addison-Wesley, 1973.

[Liu1968] C. L. Liu, *Introduction to Combinatorial Mathematics*, McGraw-Hill, 1968.

[Maur04] S. B. Maurer, *Directed Acyclic Graphs*, §3.2 in *Handbook of Graph Theory*, edited by J. L. Gross and J. Yellen, CRC Press, 2004.

[Park1960] E. T. Parker, Orthogonal Latin squares, *Proc. Nat. Acad. Sci.* **45** (1959), 859–862.

[Póly1973] G. Pólya, Kombinatorische Anzahlbestimmungen für Gruppen, Graphen und chemische Verbinungen, *Acta Math.* **68** (1937), 145–254.

[Prim1957] R. C. Prim, Shortest connection networks and some generalizations, *Bell System Tech. J.* **36** (1957), 1389–1401.

[RiYo1968] G. Ringel and J. W. T. Youngs, Solution of the Heawood map-coloring problem, *Proc. Acad. Nat. Sci. USA* **60** (1968), 438–445.

[Spir1973] P. M. Spira, A new algorithm for finding all shortest paths in a graph of positive arcs in average time $O(n^2 \log^2 n)$, *SIAM J. Computing* **2** (1973), 28–32.

[Tarr1900], G. Tarry, Le problème des 36 officiers, *Comptes Rend. Assoc. Fr.* **1** (1900), 122–123.

[Tarr1901], G. Tarry, Le problème des 36 officiers, *Comptes Rend. Assoc. Fr.* **2** (1901), 170–203.

[Wars1962] S. Warshall, A theorem on Boolean matrices, *J. of Assoc. Comp. Mach.* **9** (1962), 11–12.

[Wils2004] R. J. Wilson, *History of Graph Theory*, §1.3 in *Handbook of Graph Theory*, edited by J. L. Gross and J. Yellen, CRC Press, 2004.

SOLUTIONS AND HINTS

Chapter 0   Introduction to Combinatorics

0.3   Some Rules for Counting

0.3.6: $\lfloor x + y \rfloor \geq \lfloor \lfloor x \rfloor + \lfloor y \rfloor \rfloor = \lfloor x \rfloor + \lfloor y \rfloor$

0.3.10:   Use the Rule of Quotient iteratively: $\dfrac{7!}{3!\,2!\,1!\,1!}$

0.3.16:   Use the Pigeonhole Principle. Let the students be the pigeons and the seven days of the week be the pigeonholes. It would take 8 students to be certain of a match.

0.4   Counting Selections

0.4.3:   This amounts to selection of 8 objects from 3 distinct types, with repetitions permitted, since an initial choice of one object of each type is required:

$$\binom{10}{2}$$

0.4.6: $\dbinom{15}{3}$

0.4.8: $\dbinom{7}{3\ 2\ 2} = \dfrac{7!}{3!\,2!\,2!} = 210$

0.4.12:   105

0.4.16:   A partition of 5 objects into three parts is of type 311 or type 221.

$$\binom{5}{3\ 1\ 1} \cdot \frac{1}{2!} + \binom{5}{2\ 2\ 1} \cdot \frac{1}{2!} = 10 + 15 = 25$$

0.5   Permutations

0.5.1: $\begin{pmatrix} 1 & 2 & 3 & 4 & 5 \\ 3 & 5 & 2 & 4 & 1 \end{pmatrix} = (1 \ \ 3 \ \ 2 \ \ 5)(4)$

0.5.7: $\begin{pmatrix} 1 & 2 & 3 & 4 & 5 \\ 5 & 3 & 1 & 4 & 2 \end{pmatrix} = (1 \ \ 5 \ \ 2 \ \ 3)(4)$

0.5.13: $(1 \ \ 2 \ \ 3) \circ (2 \ \ 4 \ \ 5) = (1 \ \ 4 \ \ 5 \ \ 2 \ \ 3)$

0.5.21: $\begin{pmatrix} 7 \\ 4 \ 2 \ 1 \end{pmatrix} \cdot 3!$

0.6   Graphs

0.6.1:   There are 7 edges. The degree sum is 14.

0.6.9: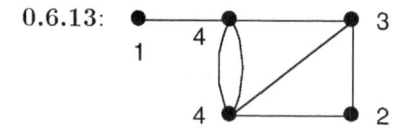

0.6.13: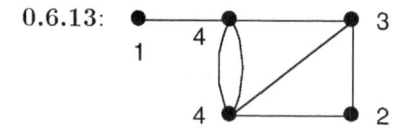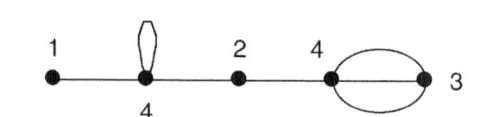

Chapter 1   Sequences

1.1   Sequences as Lists

1.1.1: $\langle x_n = 2^n \bmod 7 \rangle \Rightarrow$ 1 2 4 1 2 4 1 2 4 1 2 4 ...
If $n = 3k + j$, then $2^n = (2^3)^k \cdot 2^j = (1 + 7)^k \cdot 2^j \equiv 1^k \cdot 2^j \equiv 2^j$ modulo 7.

1.1.7:   Use induction.

1.1.13:   Period $P = 3$. Sequence $a_n = n^2 \bmod 3$. One way to derive this is to assume there is a polynomial $ax^2 + bx + c$ that satisfies it. These three linear equations

$$a(3k)^2 + b(3k) + c = 0$$
$$a(3k + 1)^2 + b(3k + 1) + c = 1$$
$$a(3k + 2)^2 + b(3k + 2) + c = 1$$

are equivalent modulo 3 to the equations

$$c = 0$$
$$a + b + c = 1$$
$$a + 2b + c = 1$$

whose solution is $a = 1, b = 0, c = 0$

1.1.20:   Polynomial $f(x) = x^2 + x(x-1)(x-2)(x-3)$

1.1.21: $\displaystyle\sum_{k=1}^{n} \frac{1}{n^2} < 1 + \int_{x=1}^{n} \frac{dx}{x^2} = 2 - \frac{1}{n}$

1.1.25:   Compare $\lg x_n$ to $\lg y_n$.

1.2   Recurrences

1.2.6:   Compare $2 \lg n$ to $\lg(n+1) + \lg(n-1)$. Exponentiating both sides yields n^2 on the left, which is larger than $(n+1)(n-1)$ on the right.

1.2.7: $(n+1)^2 + (n-1)^2 = (n^2 + 2n + 1) + (n^2 - 2n + 1) = 2n^2 + 2 > n^2$.

1.2.13:   Use induction. The induction step is

$$
\begin{aligned}
c_{n+2} + c_n &= (c_0 c_{n+1} + \cdots + c_{n+1} c_0) + (c_0 c_{n-1} + \cdots + c_{n-1} c_0) \\
&= c_0(c_{n+1} + c_{n-1}) + \cdots + c_{n-1}(c_2 + c_0) + c_n c_1 + c_{n+1} c_0 \\
&\geq 2 c_0 c_n + \cdots + 2 c_{n-1} c_1 + c_n c_1 + c_{n+1} c_0 \quad \text{by ind hyp} \\
&\geq 2 c_0 c_n + \cdots + 2 c_{n-1} c_1 + 2 c_n c_0 \quad \text{since } c_{n+1} \geq c_n \text{ and } c_1 = c_0 = 1 \\
&= c_{n+1}
\end{aligned}
$$

1.3   Pascal's Recurrence

1.3.2: $\displaystyle \binom{n-1}{r-1} + \binom{n-1}{r} = \left(\binom{n-2}{r-2} + \binom{n-2}{r-1} \right) + \left(\binom{n-2}{r-1} + \binom{n-2}{r} \right)$

1.3.5:   Evaluate $(1+x)^n |_{x=1}$

1.4   Differences and Partial Sums

1.4.1:   Difference table for n^4.

n^4	0	1	16	81	256	625	1296	2401	$\cdots$
		1	15	65	175	369	671	1105	$\cdots$
			14	50	110	194	302	434	$\cdots$
				36	60	84	108	132	$\cdots$
					24	24	24	24	$\cdots$

1.4.7:   Difference sequence for c^n.

$$
\begin{array}{c|cccccc}
c^n & 1 & c & c^2 & c^3 & c^4 & \cdots \\
& c-1 & (c-1)c & (c-1)c^2 & (c-1)c^3 & \cdots \\
& (c-1)^2 & (c-1)^2 c & (c-1)^2 c^2 & \cdots \\
& (c-1)^3 & (c-1)^3 c & \cdots
\end{array}
$$

1.4.13: $6^2 - 3^2 = 27$

$$\sum_{n=3}^{5} \triangle(n^2) = \sum_{n=3}^{5}(2n+1) = 7 + 9 + 11 = 27$$

1.5   Falling Powers

1.5.1:   For $r > xy$, both expressions have the value 0. For $\min(x,y) < r \le xy$, we have $x^{\underline{r}}y^{\underline{r}} = 0$ and $(xy)^{\underline{r}} > 0$. In the main case $2 \le r \le \min(x,y)$, since $xy - j > (x-j)(y-j)$, for $1 \le j \le r$, it follows that

$$(xy)^{\underline{r}} = \prod_{j=0}^{r-1}(xy-j) > \prod_{j=0}^{r-1}(x-j)(y-j) = x^{\underline{r}}y^{\underline{r}}$$

1.5.5: $\displaystyle\binom{-\frac{1}{2}}{4} = \frac{(-\frac{1}{2})^{\underline{4}}}{4!} = \frac{(-\frac{1}{2})(-\frac{3}{2})(-\frac{5}{2})(-\frac{7}{2})}{24} = \frac{35}{128}$

1.5.14:   To compare $\log(n^2)$ to

$$\frac{\log((n-1)^2) + \log((n+1)^2)}{2}$$

directly from the definition, we multiply both sides by 2 and exponentiate. These operations are both monotonically increasing. Since

$$n^2 > (n-1)(n+1)$$

we conclude that n^2 is log-concave. Equivalently, with less effort, we could use inequalities (1.5.3) and (1.5.4).

1.6   Stirling Numbers: A Preview

1.6.1: $x^{\underline{6}} = x^6 - 15x^5 + 85x^4 - 225x^3 + 274x^2 - 120x$

1.6.5: $x^6 = x^{\underline{6}} + 15x^{\underline{5}} + 65x^{\underline{4}} + 90x^{\underline{3}} + 31x^{\underline{2}} + x^{\underline{1}}$

1.6.11: $\displaystyle\binom{7}{1\,2\,4} = 105$

1.7 Ordinary Generating Functions

1.7.1 :
$$(1+b)(1+d)(1+n+n^2)(1+a+a^2+a^3)$$
$$\longrightarrow (1+z)^2(1+z+z^2)(1+z+z^2+z^3)$$
$$= 1 + 4z + 8z^2 + 11z^3 + 11z^4 + 8z^5 + 4z^6 + z^7$$

1.7.7 :
$$(1+b)(1+d)\left(1+n+\frac{n^2}{2!}\right)\left(1+a+\frac{a^2}{2!}+\frac{a^3}{3!}\right)$$
$$\longrightarrow (1+z)^2\left(1+z+\frac{z^2}{2!}\right)\left(1+z+\frac{z^2}{2!}+\frac{z^3}{3!}\right)$$
$$= 1 + 4\frac{z}{1!} + 14\frac{z^2}{2!} + 43\frac{z^3}{3!} + 114\frac{z^4}{4!} + 250\frac{z^5}{5!} + 420\frac{z^6}{6!} + 420\frac{z^7}{7!}$$

1.7.13 : $\quad 1 + 5z + 19z^2 + 65z^3 + 211z^4 + \cdots$

1.7.21 : $\quad$ (a) $1 + z + z^2 + \cdots$ $\quad$ (b) $\left.\dfrac{n!}{(1-z)^{n+1}}\right|_{z=0} \dfrac{z^n}{n!} = z^n$

1.8 Synthesizing Generating Functions

1.8.1 : $\quad \dfrac{1}{1+z}$

1.8.9 : $\quad \dfrac{1}{(1-z)^2}$

1.8.19 : The third difference row begins with 1, 2, 4, ... Therefore, try subtracting the sequence $\langle 2^n \rangle$. The resulting sequence is 0, 1, 4, 9, 16, 125, ..., apparently n^2. The sequence $n^2 + 2^n$ perfectly matches the given sequence.

1.9 Asymptotic Estimates

1.9.1 : Given the polynomial $f(n) = a_{d-1}n^{d-1} + \cdots + a_1 n + a_0$, let M be a number larger than absolute value of any of the coefficients a_j. Then, for $j = 0, \ldots, d-1$ and $n \geq 1$, we have $a_j n^j \leq M n^d$. Therefore, for $n \geq 1$,

$$f(n) = \sum_{j=0}^{d-1} a_j n^j \leq \sum_{j=0}^{d-1} M n^d = M d n^d$$

2.8 Divide-and-Conquer Relations

2.8.1: $(lo, \ hi) \ = \ (1, \ 7) \to (4, \ 7) \to (4, \ 5) \to (5, \ 5)$

2.8.5 :
$$\begin{array}{l}
[\,(\,92\,) \ \ (\,56\,) \ \ (\,83\,) \ \ (\,97\,) \ \ (\,72\,) \ \ (\,78\,) \ \ (\,15\,)\,] \\
[\,(\,56 \ \ 92\,) \ \ (\,83 \ \ 97\,) \ \ (\,72 \ \ 78\,) \ \ (\,15\,)\,] \\
[\,(\,56 \ \ 83 \ \ 92 \ \ 97\,) \ \ (\,15 \ \ 72 \ \ 78\,)\,] \\
[\,(\,15 \ \ 56 \ \ 72 \ \ 78 \ \ 83 \ \ 92 \ \ 97\,)\,]
\end{array}$$

2.8.9: $t_n \ = \ 3cn^{\lg 3} - 2cn$

2.8.15:   Select $n!$. Actually, it is sufficient to select the least common multiple of the numbers 1 to n.

Chapter 3   Evaluating Sums

3.1 Normalizing Summation

3.1.1:
$$\sum_{7 \le k^2 \le 45} \frac{1}{k} \ = \ \sum_{k=\lceil \sqrt{7} \rceil}^{\lfloor \sqrt{45} \rfloor} \frac{1}{k} \ = \ \sum_{k=3}^{6} \frac{1}{k} \ = \ \sum_{k=0}^{3} \frac{1}{k+3}$$

3.1.7:   Proposition 3.1.4(a) asserts that $\ln(n+1) < H_n < \ln(n) + 1$. Therefore,
$$\ln(n) \ < \ H_n \ < \ \ln(n) + 1$$

Divide through by $\ln n$ to obtain the desired result.

3.2 Perturbation

3.2.1 :
$$S \ = \ \sum_{k=0}^{n} 3^k$$

$$\Rightarrow \quad S + 3^{n+1} \ = \ 3^0 + \sum_{k=1}^{n+1} 3^k \ = \ 3^0 + 3 \sum_{k=0}^{n} 3^k \ = \ 3^0 + 3S$$

$$\Rightarrow \quad S \ = \ \frac{3^{n+1} - 3^0}{2}$$

3.2.15:   Perturbing $\displaystyle\sum_{k=0}^{n} k^4$ yields $\displaystyle\sum_{k=0}^{n} k^3 \ = \ \frac{1}{4}\left[n^4 + 2n^3 + n^2\right]$.

3.3 Summing with Generating Functions

3.3.1 : (a) $$\frac{1}{1-3z} = \sum_{n=0}^{\infty} 3^n z^n$$

(b) $\Rightarrow$ $$\frac{1}{1-z} \cdot \frac{1}{1-3z} = \sum_{n=0}^{\infty} \left(\sum_{k=0}^{n} 3^k \right) z^n$$

(c) $$\frac{1}{1-z} \cdot \frac{1}{1-3z} = \frac{3/2}{1-3z} + \frac{-1/2}{1-z}$$

(d) $\Rightarrow$ $$\sum_{k=0}^{n} 3^k = \frac{3}{2} \cdot 3^n - \frac{1}{2} \cdot 1^n$$

3.3.7 : (a) $$\frac{1}{1-z/3} = \sum_{n=0}^{\infty} \frac{1}{3^n} z^n$$

(b) $\Rightarrow$ $$\frac{1}{1-z} \cdot \frac{1}{1-z/3} = \sum_{n=0}^{\infty} \left(\sum_{k=0}^{n} \frac{1}{3^k} \right) z^n$$

(c) $$\frac{1}{1-z} \cdot \frac{1}{1-z/3} = \frac{-1/2}{1-z/3} + \frac{3/2}{1-z}$$

(d) $\Rightarrow$ $$\sum_{k=0}^{n} \frac{1}{3^k} = \frac{-1}{2} \cdot \frac{1}{3^n} + \frac{3}{2} \cdot 1^n$$

3.3.13 : (a) $$\frac{1}{1-z\sqrt{2}} = \sum_{n=0}^{\infty} \sqrt{2}^n z^n$$

(b) $\Rightarrow$ $$\frac{1}{1-z} \cdot \frac{1}{1-z\sqrt{2}} = \sum_{n=0}^{\infty} \left(\sum_{k=0}^{n} \sqrt{2}^k \right) z^n$$

(c) $$\frac{1}{1-z} \cdot \frac{1}{1-z\sqrt{2}} = \frac{2+\sqrt{2}}{1-z\sqrt{2}} - \frac{1+\sqrt{2}}{1-z}$$

(d) $\Rightarrow$ $$\sum_{k=0}^{n} \sqrt{2}^k = (2+\sqrt{2}) \cdot \sqrt{2}^n - (1+\sqrt{2}) \cdot 1^n$$

3.3.17 : (a) $$\frac{1}{(1-3z)^2} = \sum_{n=0}^{\infty} \binom{n+1}{1} 3^n z^n$$

(b) $\Rightarrow$ $$\frac{1}{1-z} \cdot \frac{1}{(1-3z)^2} = \sum_{n=0}^{\infty} \left(\sum_{k=0}^{n} \binom{k+1}{1} 3^k \right) z^n$$

(c) $$\frac{1}{1-z} \cdot \frac{1}{(1-3z)^2} = \frac{1/4}{1-z} + \frac{-3/4}{1-3z} + \frac{3/2}{(1-3z)^2}$$

(d) $\Rightarrow$ $$\sum_{k=0}^{n} \binom{k+1}{1} 3^k = \frac{1}{4} - \frac{3}{4} \cdot 3^n + \frac{3}{2} \cdot \binom{n+1}{1} \cdot 3^n$$

3.4   Finite Calculus

3.4.1 :
$$3^k = \triangle\left(\frac{3^k}{2}\right)$$

$$\Rightarrow \sum_{k=0}^{n} 3^k = \frac{3^k}{2}\bigg|_{k=0}^{n+1} = \frac{3^{n+1}}{2} - \frac{1}{2} = \frac{3^{n+1}-1}{2}$$

3.4.4:   To sum $k4^k$ by parts, use $g(k) = k$ and $\triangle f(k) = 4^k$. Then

$$\sum_{k=0}^{n} k4^k = k\frac{4^k}{3}\bigg|_{k=0}^{n+1} - \sum_{k=0}^{n}\frac{4^{k+1}}{3} = k\frac{4^k}{3}\bigg|_{k=0}^{n+1} - \frac{4^{k+1}}{9}\bigg|_{k=0}^{n+1}$$

$$= \left((n+1)\frac{4^{n+1}}{3} - 0\right) - \left(\frac{4^{n+2}}{9} - \frac{4}{9}\right)$$

3.4.19:   To sum kH_k by parts, use $g(k) = H_k$ and $\triangle f(k) = k$. Then

$$\sum_{k=0}^{n} kH_k = \frac{k^{\underline{2}}}{2}H_k\bigg|_{k=0}^{n+1} - \sum_{k=0}^{n}\frac{(k+1)^{\underline{2}}}{2}\frac{1}{k+1}$$

$$= \frac{k^{\underline{2}}}{2}H_k\bigg|_{k=0}^{n+1} - \sum_{k=0}^{n}\frac{k}{2} = \frac{k^{\underline{2}}}{2}H_k\bigg|_{k=0}^{n+1} - \frac{k^{\underline{2}}}{4}\bigg|_{k=0}^{n+1}$$

$$= \left(\frac{n^{\underline{n}}+n}{2}H_{n+1} - 0\right) - \left(\frac{n^2+n}{4} - 0\right) = \frac{n^{\underline{n}}+n}{2}H_{n+1} - \frac{n^2+n}{4}$$

3.4.24: $\triangle k^{\underline{-2}} = -2k^{\underline{-3}}$

3.4.30: $-k^{\underline{-1}}$

3.4.33:   (a) 103, 140.   (b) $3n^2 - 2n + 7$

3.5   Iteration and Partitioning of Sums

3.5.5: $\dfrac{(n+1)^{\underline{3}}}{2} + \dfrac{3}{2}(n+1)^{\underline{2}}$

3.5.14: $n + (n \equiv 2 \bmod 3)$

3.6   Inclusion-Exclusion

3.6.1: $\phi(48) = 48 - 24 - 16 + 8 = 16.$

3.6.7 : $2!\begin{Bmatrix}5\\2\end{Bmatrix} = 2^5 - \begin{pmatrix}2\\1\end{pmatrix}\cdot 1^5 + \begin{pmatrix}2\\2\end{pmatrix}\cdot 0^5$

$$= 2^5 - 2 + 0 = 30$$

$$\Rightarrow \begin{Bmatrix}5\\2\end{Bmatrix} = 15$$

3.6.11: $\dfrac{6!\left\{^8_6\right\}}{8^6}$

3.6.16: $\displaystyle\sum_{k=0}^{n}\binom{n}{k}(2n-k)!(-1)^k$

3.6.17: $\binom{n}{k}D_{n-k}$

3.6.18 : $|A_1|=0 \quad |A_2|=|A_3|=1\cdot 2!=2 \quad \Rightarrow \quad S_1=4$

$\qquad |A_2A_3|=1 \quad |A_1A_2|=|A_1A_3|=0 \quad \Rightarrow \quad S_2=1$

$\qquad\qquad\qquad |A_1A_2A_3|=0 \quad \Rightarrow \quad S_3=0$

$\qquad\qquad\qquad |U|=3!=6$

$\Rightarrow \quad |\overline{A_1}\,\overline{A_2}\,\overline{A_3}| \;=\; |U|-S_1+S_2-S_3 \;=\; 6-4+1-0 \;=\; 3$

3.6.22: $\quad |U|=t^4 \quad S_1=4t^3 \quad S_2=6t^2 \quad S_3=t^2+3t \quad S_4=t\,.$

$\Rightarrow \quad P(G,t) \;=\; t^4-4t^3+6t^2-(t^2+3t)+t$

$\qquad\qquad\quad = \; t^4-4t^3+5t^2-2t$

Chapter 4 Binomial Coefficients

4.1 Binomial Coefficient Identities

4.1.4 : $\quad \dbinom{-\frac{1}{2}}{3} \;=\; \dfrac{\left(-\frac{1}{2}\right)^{\underline{3}}}{3!} \;=\; \dfrac{\left(-\frac{1}{2}\right)\left(-\frac{3}{2}\right)\left(-\frac{5}{2}\right)}{3!} \;=\; -\dfrac{5}{16}$

4.1.7 : $\quad \dbinom{n+2}{n-2} \;=\; \dbinom{n+2}{4} \;=\; \dfrac{(n+2)(n+1)\,n\,(n-1)}{4!}$

$\qquad\qquad\qquad\qquad\qquad = \; \dfrac{n^4+2n^3-n^2-2n}{4!}$

4.1.9 : $(n-r)\dbinom{n}{r} \;=\; (n-r)\dbinom{n}{n-r} \qquad \text{(symmetry)}$

$\qquad\qquad\qquad = \; n\dbinom{n-1}{n-r-1} \qquad \text{(absorption)}$

$\qquad\qquad\qquad = \; n\dbinom{n-1}{r} \qquad\qquad \text{(symmetry)}$

4.1.17: For $n = 0$ and $n = 1$, the equation is clearly correct. For $n \geq 2$,

$$\sum_{k=0}^{n} (-1)^k \, k \binom{n}{k} = \sum_{k=1}^{n} (-1)^k \, n \binom{n-1}{k-1} \qquad \text{(absorption)}$$

$$= -n \sum_{k=1}^{n} (-1)^{k-1} \binom{n-1}{k-1}$$

$$= -n \, (x-1)^{n-1} \Big|_{x=1} = 0$$

4.1.22 : $\displaystyle \binom{x}{r} \, r = \frac{x^{\underline{r}}}{r!} \, r = \frac{x \cdot (x-1)^{\underline{r-1}}}{(r-1)!} = x \binom{x-1}{r-1}$

4.1.24: First method: iteratively reduce upper and lower index.

$$\binom{80}{48} \equiv \binom{40}{24} \equiv \binom{20}{12} \equiv \binom{10}{6} \equiv \binom{5}{3} \equiv \binom{2}{1} \equiv 0 \bmod 2$$

Second method: align binary numerals flush right; scan for 0 over 1.

$$\begin{aligned} 80_{10} &= 1\,0\,1\,0\,0\,0\,0_2 \quad \Rightarrow \quad \text{even} \\ 48_{10} &= 0\,1\,1\,0\,0\,0\,0_2 \end{aligned}$$

4.2   Binomial Inversion Operation

4.2.5 : $\displaystyle g_n = \binom{n}{2} b - \binom{n}{1} a$

4.2.12: $\displaystyle g_n = (-2)^n$

4.2.13: $g_2 = 2$, and $g_j = 0$ for $j \neq 2$.

4.3   Application to Statistics

4.3.4 : $\displaystyle \Pr(X = j) = \frac{\binom{M}{j}\binom{N-M}{n-j}}{\binom{N}{n}}$

4.3.5 : $\displaystyle E(X) = n \cdot \frac{M}{N}$

4.3.7 : $\displaystyle V(X) = \left(\frac{N-n}{N-1}\right) \cdot n \cdot \frac{M}{N} \cdot \left(1 - \frac{M}{N}\right)$

4.3.11 : $\displaystyle \Pr(X = j) = \binom{j+r-1}{r-1} p^r \, (1-p)^j$

4.3.13 : $E(X) \;=\; \dfrac{r\,(1-p)}{p}$

4.3.14 : $V(X) \;=\; \dfrac{r\,(1-p)}{p^2}$

4.3.18 : $E(X) \;=\; \lambda$

4.3.20 : $V(X) \;=\; \lambda$

4.4   The Catalan Recurrence

4.4.1:   This is Proposition 4.4.3.

4.4.4: $q(n, n-2) \;=\; c_n - c_{n-1}.$

4.4.9:   There are $\binom{2n}{n-2}$ paths in $[0 : n+2] \times [0 : n-2]$. By reflection of suffixes, this provides the correct count.

4.4.15:   There are

$$\binom{2n}{n-k-1}$$

NE-paths that enter the NW k-triangle, and equally many that enter the SE k-triangle. No two paths enter both extreme triangles, since $k \geq n/2$. Thus, the total number of paths that enter either of them is

$$2\binom{2n}{n-k-1}$$

Chapter 5   Partitions and Permutations

5.1   Stirling Subset Numbers

5.1.2: $\left\{ {n \atop k} \right\} + \left\{ {n \atop k-1} \right\} + \cdots + \left\{ {n \atop k-r} \right\}$

5.1.5: $k! \left\{ {n \atop k} \right\} + (k-1)! \left\{ {n \atop k-1} \right\} + \cdots + (k-r)! \left\{ {n \atop k-r} \right\}$

5.1.9: $\left\{ {6 \atop 4} \right\} + 5 \left\{ {6 \atop 5} \right\} \;=\; 65 + 5 \cdot 15 \;=\; 140 \;=\; \left\{ {7 \atop 5} \right\}$

5.1.17 : $\binom{6}{4} \left\{ {4 \atop 4} \right\} + \binom{6}{5} \left\{ {5 \atop 4} \right\} + \binom{6}{6} \left\{ {6 \atop 4} \right\} \;=\; 15 \cdot 1 + 6 \cdot 10 + 1 \cdot 65$

$$= 140 \;=\; \left\{ {7 \atop 5} \right\}$$

5.1.25: $5^2 \begin{Bmatrix} 4 \\ 4 \end{Bmatrix} + 5^1 \begin{Bmatrix} 5 \\ 4 \end{Bmatrix} + 5^0 \begin{Bmatrix} 6 \\ 4 \end{Bmatrix} = 25 \cdot 1 + 5 \cdot 10 + 1 \cdot 65 = 140 = \begin{Bmatrix} 7 \\ 5 \end{Bmatrix}$

5.1.33 : $0 \begin{Bmatrix} 1 \\ 0 \end{Bmatrix} + 1 \begin{Bmatrix} 2 \\ 1 \end{Bmatrix} + 2 \begin{Bmatrix} 3 \\ 2 \end{Bmatrix} + 3 \begin{Bmatrix} 4 \\ 3 \end{Bmatrix} + 4 \begin{Bmatrix} 5 \\ 4 \end{Bmatrix} + 5 \begin{Bmatrix} 6 \\ 5 \end{Bmatrix}$

$$= 0 \cdot 0 + 1 \cdot 1 + 2 \cdot 3 + 3 \cdot 6 + 4 \cdot 10 + 5 \cdot 15 = 140 = \begin{Bmatrix} 7 \\ 5 \end{Bmatrix}$$

5.2   Stirling Cycle Numbers

5.2.3 : $\begin{bmatrix} 6 \\ 4 \end{bmatrix} + 6 \begin{bmatrix} 6 \\ 5 \end{bmatrix} = 85 + 6 \cdot 15 = 175 = \begin{bmatrix} 7 \\ 5 \end{bmatrix}$

5.2.16 : $\begin{pmatrix} 4 \\ 4 \end{pmatrix} \begin{bmatrix} 6 \\ 4 \end{bmatrix} + \begin{pmatrix} 5 \\ 4 \end{pmatrix} \begin{bmatrix} 6 \\ 5 \end{bmatrix} + \begin{pmatrix} 6 \\ 4 \end{pmatrix} \begin{bmatrix} 6 \\ 6 \end{bmatrix} = 1 \cdot 85 + 5 \cdot 15 + 15 \cdot 1 = 175 = \begin{bmatrix} 7 \\ 5 \end{bmatrix}$

5.2.19 : $6^{\underline{2}} \begin{bmatrix} 4 \\ 4 \end{bmatrix} + 6^{\underline{1}} \begin{bmatrix} 5 \\ 4 \end{bmatrix} + 6^{\underline{0}} \begin{bmatrix} 6 \\ 4 \end{bmatrix} = 30 \cdot 1 + 6 \cdot 10 + 1 \cdot 85 = 175 = \begin{bmatrix} 7 \\ 5 \end{bmatrix}$

5.2.27 : $0 \begin{bmatrix} 1 \\ 0 \end{bmatrix} + 1 \begin{bmatrix} 2 \\ 1 \end{bmatrix} + 2 \begin{bmatrix} 3 \\ 2 \end{bmatrix} + 3 \begin{bmatrix} 4 \\ 3 \end{bmatrix} + 4 \begin{bmatrix} 5 \\ 4 \end{bmatrix} + 5 \begin{bmatrix} 6 \\ 5 \end{bmatrix}$

$$= 1 \cdot 0 + 2 \cdot 1 + 3 \cdot 3 + 4 \cdot 6 + 5 \cdot 10 + 6 \cdot 15 = 175 = \begin{bmatrix} 7 \\ 5 \end{bmatrix}$$

5.3   Inversions and Ascents

5.3.1:   3124   1423   2143   2314   1342

5.3.13: $I_7(8) = 455$

5.3.17:   532164

5.3.22 : $3 \left\langle \begin{matrix} 6 \\ 2 \end{matrix} \right\rangle + 5 \left\langle \begin{matrix} 6 \\ 1 \end{matrix} \right\rangle = 3 \cdot 302 + 5 \cdot 57$

$$= 1191 = \left\langle \begin{matrix} 7 \\ 2 \end{matrix} \right\rangle$$

5.5   Exponential Generating Functions

5.5.1: ze^z

5.5.5: $(e^z - 1 - z)^2 = e^{2z} - 2ze^z - 2e^z + z^2 + 1 + 2z$

5.5.9: $u_n = \begin{cases} 0 & \text{if } 0 \le n \le 3 \\ 2^n - 2n - 2 & \text{if } n \ge 4 \end{cases}$

5.5.13: $a_n = n \cdot n!$

5.5.15: $a_n = \dfrac{1}{n!} \displaystyle\sum_{j=0}^{n} n^{\underline{j}}$

5.6   Posets and Lattices

5.6.4:   Maximum chain: $\emptyset \subset 1 \subset 12 \subset 123$

5.6.10:   Maximum anti-chain: 1, 2, 3

5.6.16:   Partition into minimum number of chains:

   (1) $\emptyset \subset 1 \subset 12 \subset 123$;   (2) $2 \subset 23$;   (3) $3 \subset 13$

5.6.22:   Partition into minimum number of anti-chains:

   (1) $\emptyset$;   (2) 1, 2, 3;   (3) 12, 13, 23;   (4) 123

5.6.28:   There are 48 linear extensions. $\emptyset$ must be first and 123 last. If the three singletons 1,2,3 precede the three doubletons 12, 13, 23, then there are 3! possible ordering on the singletons and 3! of the doubletons, for a total of 36. There are also 12 linear extensions in which some doubleton precedes the other two doubletons.

Chapter 6   Integer Operators

6.1   Euclidean Algorithm

6.1.1 : 89 mod 71 $= 18$

   71 mod 18 $= 17$

   18 mod 17 $= 1$

   17 mod 1 $= 0$ $gcd\,(89,\,71) = 1$

6.1.9: $4 \cdot 89 + (-5) \cdot 71 = 1$ derived as follows:

j	n_j	m_j	q_j	
0	89	71	1	$4 \cdot 89 - 5 \cdot 71$
1	71	18	3	$(-1) \cdot 71 + 1 \cdot 18 = (-1) \cdot 71 + 4 \cdot (89 - 1 \cdot 71)$
2	18	17	1	$1 \cdot 18 - 1 \cdot 17 = 1 \cdot 18 - 1 \cdot (71 - 3 \cdot 18)$
3	17	1	17	$1 \cdot 1 + 0 \cdot 0 = 1 \cdot (18 - 1 \cdot 17)$
4	1	0	$STOP$	

6.2   Chinese Remainder Theorem

6.2.1: $(6 \bmod 8,\ 3 \bmod 9) + (7 \bmod 8,\ 5 \bmod 9) = (5 \bmod 8,\ 8 \bmod 9)$

6.2.5: $29 \bmod (5,\ 7) = (4 \bmod 5,\ 1 \bmod 7)$

6.2.11: $(6 \bmod 8,\ 3 \bmod 9) = 30$, derived as follows:
$$(-1) \cdot 8 + 1 \cdot 9 = 1$$
$$3 \cdot (-1) \cdot 8 + 6 \cdot 1 \cdot 9 = 30$$

6.2.20:   38

6.3   Polynomial Divisibility

6.3.1: $x^2 - 4x + 2$ is prime, since the roots are irrational.

6.3.9: $x^2 + 3x + 7 \bmod x - 2 = 17$

6.3.19: $gcd\left(x^3 - 6x^2 + 11x - 6,\ x^2 - 3x + 2\right) = x^2 - 3x + 2$

6.4   Prime and Composite Moduli

6.4.3:   The inverse of 21 mod 25 is 6.

6.4.11: $221^{64} \bmod 25 = 6$

6.4.17: $4! \bmod 5 = -1$, since 5 is prime.

6.4.25:   The quadratic residues of 14 are
$$1 = 1^2,\ 9 = 3^2,\ 11 = 5^2$$

6.4.31:   The solutions to $x^2 \equiv 1(\text{ modulo } 24)$ are 1, 5, 7, 11, 13, 17, 19, and 23.

6.5   Euler Phi-Function

6.5.5: $\phi(30) = 1 \cdot 2 \cdot 4 = 8$

6.5.11:   Since $\phi(25) = 20$, we have $221^{64} \bmod 25 = 21^{64} = 21^4 = 6$.

6.5.19 : $\phi(1) + \phi(2) + \phi(3) + \phi(5) + \phi(6) + \phi(10) + \phi(15) + \phi(30) =$
$$1 + 1 + 2 + 4 + 2 + 4 + 8 + 8 = 30$$

6.5.25: The set-difference $F_n - F_{n-1}$ is the set of reduced fractions $\frac{r}{n}$. The number of such fractions is $\phi(n)$.

6.6 The Möbius Function

6.6.3: $\mu(18) = 0$

6.6.11: $\sigma(n) = \displaystyle\sum_{d \backslash n} \iota(d)$

Chapter 7 Graph Fundamentals

7.1 Regular Graphs

7.1.1: 16 vertices, 32 edges.

7.1.7: The complete graph K_3.

7.1.12: $circ\,(2n : 1,\ 3,\ 5,\ \cdots,\ 2n-1)$

7.2 Walks and Distance

7.2.1: radius $= 3$ and diameter $= 3$.

7.2.5: Color the top two vertices white, the next three black, the next three white, and the bottom two black. The color classes are a bipartition.

7.2.9: It is easy enough to find a cycle of length 4. Since the graph is bipartite, it has no odd cycles, and, thus, no cycle of length 3. Thus, the girth is 4.

7.2.15: The two vertices inside the outer octagon have eccentricity 2 and they are the central vertices.

7.2.17: radius $= 2$ and diameter $= 2$.

7.3 Trees and Acyclic Digraphs

7.3.6: The Catalan number c_n.

7.3.9: Label the vertices with consecutive integers. Then direct each edge from its lower-numbered endpoint to the higher-numbered endpoint.

7.3.14:   We may interpret it as reflexive. It is obviously anti-symmetric. It is transitive by hypothesis. Thus, the vertices are partially ordered. Since the underlying graph is complete, it follows that every pair of elements is comparable, which implies that the ordering is complete.

7.4   Graph Isomorphism

7.4.1: A has two vertices of eccentricity 2, and B has five.

7.4.7: A is bipartite (thus, no odd cycles), and B has two 3-cycles.

7.4.17:   Due to Euler's theorem on degree-sum and Proposition 7.2.5, the plausible degree sequences are 22211, 32111, and 41111. Each corresponds to exactly one tree, as shown:

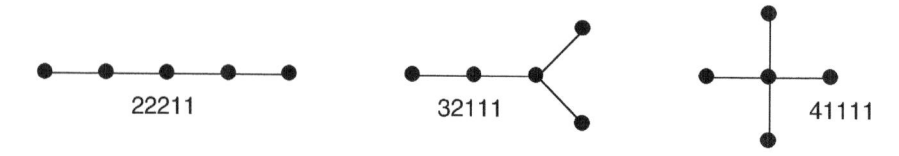

7.5   Graph Automorphism

7.5.1:   There are two automorphisms: $(1)(2)(3)(4)$ and $(1)(2)(3\ 4)$.

7.5.7:   There are four automorphisms:

$$(1)(2)(3)(4)(5)(6),\quad (1\ 5)(3)(4)(2\ 6)$$
$$(1,\ 2)(3,\ 4)(5,\ 6),\quad (1,\ 6)(3,\ 4)(2,\ 5)$$

7.5.13:   The vertex orbits are $\{1\}$, $\{2\}$, and $\{3,\ 4\}$.

7.5.19:   The vertex orbits are $\{1,\ 2,\ 5,\ 6\}$ and $\{3,\ 4\}$.

7.6   Subgraphs

7.6.2:   Hamiltonian cycle in the wheel graph W_7.

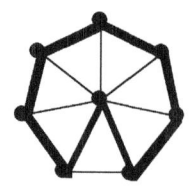

7.6.6:   In any cycle in a bipartite graph, the vertices must alternate between one partite set and the other. Thus, the number of vertices in a cycle in $K_{3,5}$ is at most 6.

7.6.13: The two vertices inside the outer octagon have eccentricity 2 and they are the central vertices. The edge joining them is the only edge in the center.

7.7 Spanning Trees

7.7.1: The frontier edges are sp, qp, ut, uv, and ux.

7.7.5:

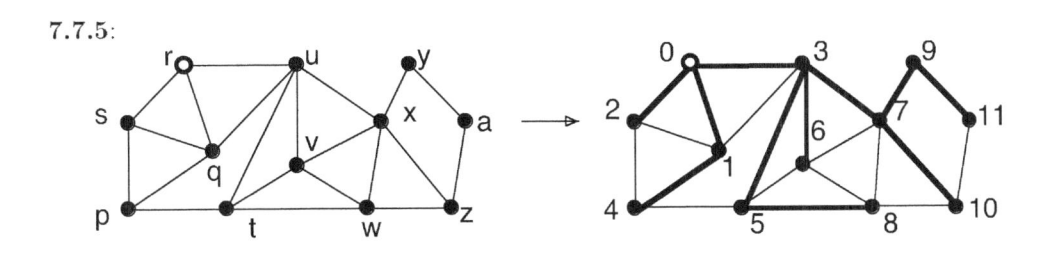

7.7.9:

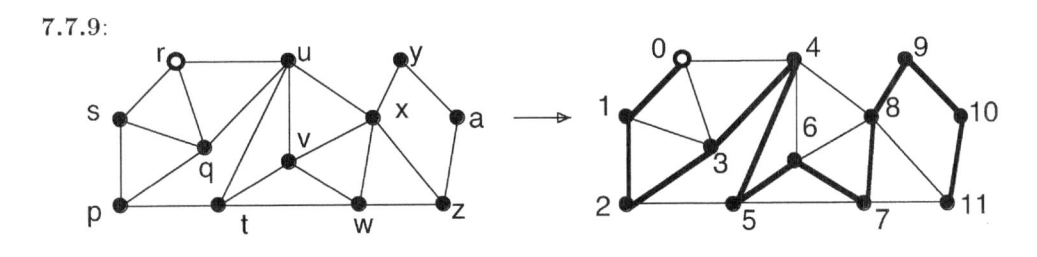

7.8 Edge Weights

7.8.1:

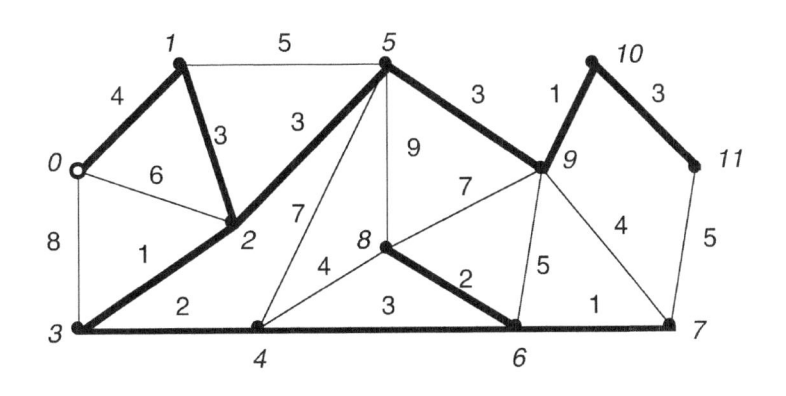

7.8.4:

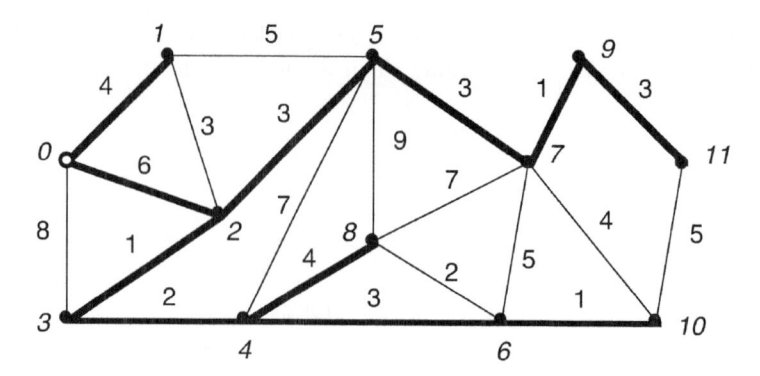

7.8.9:   Delete edge ux and run Dijkstra's algorithm on the resulting graph to get a shortest path in it between vertices u and x.

7.9   Graph Operations

7.9.1:   The degree sequence is 522221. When the vertex of degree 5 is deleted, the result is the sixth card in the given deck.

7.9.16:   The cut-vertices are vertices 3, 4, and 7.

7.9.28:   The cut-edges are 35, 47, and 67.

7.9.32:   We observe that this graph is isomorphic to its edge-complement.

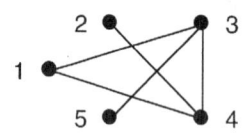

Chapter 8   Graph Theory Topics

8.1   Traversability

8.1.1: $a\ b\ f\ h\ d\ e\ g\ c.$

8.1.5:   Select a matching of minimum weight.

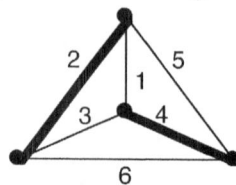

8.1.9:   The three vertices of degree 2 are mutually non-adjacent. Thus, they cannot lie on a 5-cycle.

8.2 Planarity

8.2.1: 7 vertices, 10 edges, 5 faces: $7 - 10 + 5 = 2$.

8.2.5: The face-sizes are 34445: $3 + 4 + 4 + 4 + 5 = 2 \cdot 10$.

8.2.9: $|E| = 17 \not> 24 = 3|V| - 6$.

8.3 Coloring

8.3.1: $\omega(A) = 3 \quad \alpha(A) = 3$.

8.3.7: Four colors is a lower bound, by Proposition 8.3.2. The shape of a vertex is used in this drawing as another way of indicating color classes.

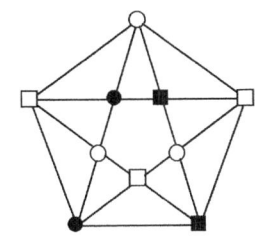

8.3.13: It is not chromatically critical. Indeed, no matter what edge is removed, the independence number remains 3. Thus, by Proposition 8.3.2, the chromatic number remains at least 4.

8.3.19: The chromatic number of Africa is 4, since (for instance) Malawi is the hub of a 3-wheel.

8.3.23: The dual of a tetrahedron is a tetrahedron.

8.3.30:

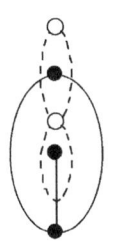

8.4 Analytic Graph Theory

8.4.1: $ex(n, \mathcal{P}) = \left\lfloor \dfrac{n(n-2)}{2} \right\rfloor.$

8.4.6: $r(3, 6) \le r(2, 6) + r(3, 5) - 1 = 6 + 14 - 1 = 19$.

8.5   Digraph Models

8.5.1:   Start (for instance) with acd, then attach the closed path gh, then the open paths e and bf.

8.5.5:   43111.

8.5.11:

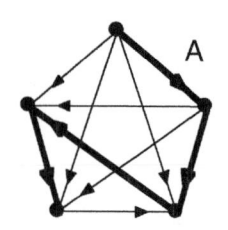

8.6   Network Flows

8.6.3:

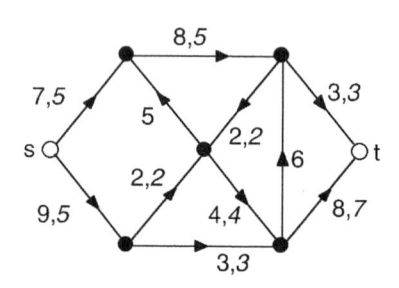

8.6.9:

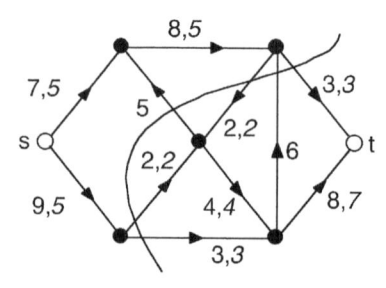

8.7   Topological Graph Theory

8.7.1:

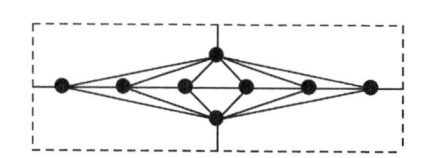

8.7.10: $|V| = 10$ and $|E| = 36$. Use Corollary 8.7.6.

8.7.16: $girth = 4$ and $\gamma_{\min}(Q_4) = n2^{n-3} - 2^{n-1} + 1$.

Chapter 9 Graph Enumeration

9.1 Burnside-Pólya Counting

9.1.2: $\alpha^3 = \begin{pmatrix} 1 & 4 & 7 & 3 & 6 & 2 & 5 \end{pmatrix}$

9.1.10: $\alpha^3 = \begin{pmatrix} 1 & 4 & 7 & 10 \end{pmatrix}\begin{pmatrix} 2 & 5 & 8 & 11 \end{pmatrix}\begin{pmatrix} 3 & 6 & 9 & 12 \end{pmatrix}$

9.1.18: $\mathscr{Z}_{\mathbb{Z}_6} : \dfrac{1}{6}\left(t_1^6 + t_2^3 + 2t_3^2 + 2t_6\right)$

9.1.26: 14

9.1.34: $b^6 + b^5 w + 3b^4 w^2 + 4b^3 w^3 + 3b^2 w^4 + b w^5 + b^6$

9.1.42: $\mathscr{Z}_{\mathbb{Z}_6} : \dfrac{1}{12}\left(t_1^6 + 3t_1^2 t_2^2 + 4t_2^3 + 2t_3^2 + 2t_6\right)$

9.1.46: 13

9.1.50: $b^6 + b^5 w + 3b^4 w^2 + 3b^3 w^3 + 3b^2 w^4 + b w^5 + b^6$

9.2 Burnside's Lemma

9.2.1: 208

9.2.5: 136

9.2.10: $b^6 + b^5 w + 3b^4 w^2 + 4b^3 w^3 + 3b^2 w^4 + b w^5 + b^6$

9.2.14: $b^6 + b^5 w + 3b^4 w^2 + 3b^3 w^3 + 3b^2 w^4 + b w^5 + b^6$

9.2.17:

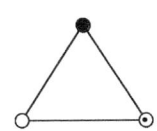 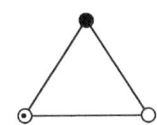

9.3 Counting Small Simple Graphs

9.3.3: $\mathscr{Z}_{\mathcal{A}ut_V(K_{2,3})} : \dfrac{1}{12}\left(t_1^5 + 4t_1^3 t_2 + 2t_1^2 t_3 + 3t_1 t_2^2 + 2t_2 t_3\right)$
12 (≤ 2)-vertex-colorings.

9.3.11: $\mathscr{Z}_{\mathcal{A}ut_E(K_{2,3})} : \dfrac{1}{12}\left(t_1^6 + 3t_1^2 t_2^2 + 4t_2^3 + 2t_3^2 + 2t_6\right)$
13 (≤ 2)-edge-colorings.

9.3.17: 2 graphs with 5 vertices and 2 edges.

9.4 Partitions of Integers

9.4.1: 8, 71, 62, 53, 44, 611, 521, 431, 422, 332

9.4.5:

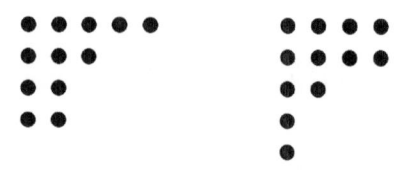

9.4.9:

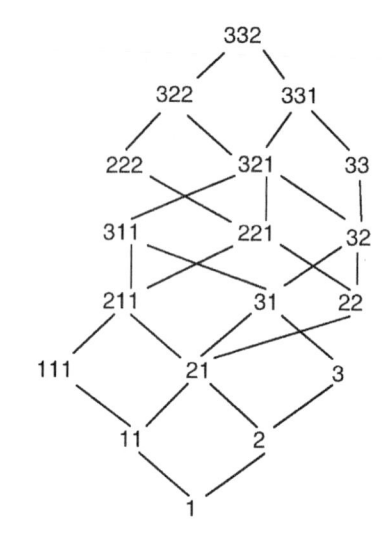

9.5 Calculating a Cycle Index

9.5.3: 9 graphs.

9.5.9: 1044 graphs.

9.5.12: 10 graphs.

9.6 General Graphs and Digraphs

9.6.1: Substituting 4 into the cycle index

$$\mathcal{Z}_{\mathcal{A}ut_E(K_3)}(y_1, y_2, y_3) = \frac{1}{6}(y_1^3 + 3y_1 y_2 + 2y_3)$$

yields

$$\mathcal{Z}_{\mathcal{A}ut_E(K_3)}(4, 4, 4) = \frac{1}{6}(4^3 + 3 \cdot 4 \cdot 4 + 2 \cdot 4) = \frac{120}{6} = 20$$

Chapter 10   Designs

10.1   Latin Squares

10.1.1:
$$\begin{pmatrix} 0 & 1 & 2 \\ 1 & 2 & 0 \\ 2 & 0 & 1 \end{pmatrix} \qquad \begin{pmatrix} 0 & 1 & 2 \\ 2 & 0 & 1 \\ 1 & 2 & 0 \end{pmatrix}$$

10.1.6: $L_5^4 = \begin{pmatrix} 0 & 1 & 2 & 3 & 4 \\ 4 & 0 & 1 & 2 & 3 \\ 3 & 4 & 0 & 1 & 2 \\ 2 & 3 & 4 & 0 & 1 \\ 1 & 2 & 3 & 4 & 0 \end{pmatrix}$

10.1.10: $A \otimes A = \begin{pmatrix} (1,1) & (1,0) & (0,1) & (0,0) \\ (1,0) & (1,1) & (0,0) & (0,1) \\ (0,1) & (0,0) & (1,1) & (1,0) \\ (0,0) & (0,1) & (1,0) & (1,1) \end{pmatrix}$

10.1.15: $L^{(1)(2,3)} = \begin{pmatrix} 0 & 2 & 3 & 1 \\ 2 & 0 & 1 & 3 \\ 1 & 3 & 2 & 0 \\ 3 & 1 & 0 & 2 \end{pmatrix}$

10.2   Block Designs

10.2.1:   varieties: 0, 1, 2, 3; blocks: 012, 013, 023, 123.

10.2.7:   The replication number r calculated by applying Proposition 10.2.2(b) would not be an integer.

10.2.13: $v = 9$, $b = 12$, $r = 4$, $k = 3$, and $\lambda = 1$.

10.2.22:   varieties: 00, 01, 02, 10, 11, 12, 20, 21, 22.

blocks :   {00, 10, 20}   {00, 11, 21}   {02, 12, 22}
            {00, 01, 02}   {10, 11, 12}   {20, 21, 22}
            {00, 11, 22}   {00, 12, 21}   {01, 10, 22}
            {01, 12, 20}   {02, 10, 21}   {02, 12, 20}

10.3   Classical Finite Geometries

10.3.4:   These are two drawings of the Levi graph for the Fano plane. The one on the right illustrates that it is vertex-transitive.

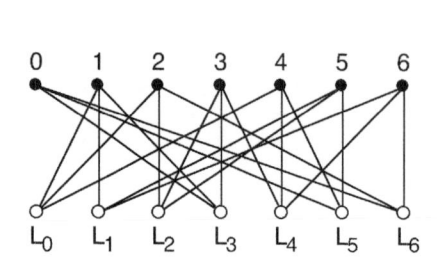
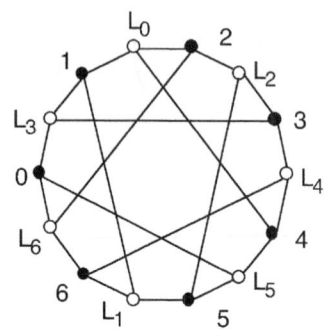

10.3.8:   Since $\lambda = 1$, the Menger graph for the Fano plane is K_7.

10.3.12:   See the solution to Exercise 10.3.8 above.

10.4   Projective Planes

10.4.1:   points of $PG(2,2)$: $\{000, 001\}$, $\{000, 010\}$, $\{000, 100\}$, $\{000, 011\}$, $\{000, 101\}$, $\{000, 110\}$, $\{000, 111\}$.

10.4.2:   lines of $PG(2,2)$:

$\{000,\ 001,\ 010,\ 011\}$
$\{000,\ 001,\ 100,\ 101\}$
$\{000,\ 100,\ 010,\ 110\}$
$\{000,\ 001,\ 110,\ 111\}$
$\{000,\ 010,\ 101,\ 111\}$
$\{000,\ 100,\ 011,\ 111\}$
$\{000,\ 101,\ 110,\ 011\}$

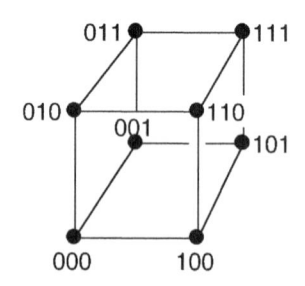

10.5   Affine Planes

10.5.1:   points of $AG(2,2)$: 00, 01, 10, 11.

10.5.2:   lines of $AG(2,2)$:

$\{00,\ 01\}$
$\{00,\ 10\}$
$\{00,\ 11\}$
$\{01,\ 10\}$
$\{01,\ 11\}$
$\{10,\ 11\}$

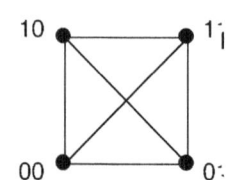

INDEXES

I1 Index of Notations
I2 General Index

I1 INDEX OF NOTATIONS

The notations listed here are generally those that occur at multiple locations in the text. They are grouped according to the context in which they occur. Some totally standard notations are omitted.

Numbers and Arithmetic

algebraic and arithmetic domains

$\mathbb{C}$	the complex numbers, 50
$\mathbb{D}_n$	the dihedral group, 500
$\mathbb{N}$	the non-negative integers (natural numbers), 50
$\mathbb{Q}$	the rational numbers, 50
$\mathbb{R}$	the real numbers, 50
$\mathbb{Z}$	the integers, 50
$\mathbb{Z}^+$	the positive integers, 50
$\mathbb{Z}_n$	the integers mod n, 585
$\mathbb{Z}[x]$	polynomial ring with integer coefficients, 342
$[k:m]$	integer interval $\{k, k+1, \ldots, m\}$, 54

arithmetic operators

$\lfloor x \rfloor$	floor of a real number, 12
$\lceil x \rceil$	ceiling of a real number, 12
$x^{\underline{r}}$	falling power of a real number, 14
$x^{\overline{r}}$	rising power of a real number, 79
$\ln x$	natural logarithm of a real number, 51
$\lg x$	base-2 logarithm of a real number, 51
$\gcd(m, n)$	greatest common divisor, 43
$\mathrm{lcm}(m, n)$	least common multiple, 43
$n \bmod d$	mod operator, 53
$\phi(n)$	Euler phi-function, 356
(predicate)	Iverson truth function, 165
$\partial g(x)$	degree of a polynomial, 342

629

Probability and Statistics

Ω	sample space for probability, 239
$Pr(A)$	probability of the event A, 239
$E(X)$ or μ_X	mean of the random variable X, 240
$SD(X)$ or σ_X	standard deviation of the random variable X, 240
$V(X)$ or σ_X^2	variance of the random variable X, 240
$\overline{X}$	sample mean of a random variable X, 244
$\hat{\theta}$	estimator of a statistic θ, 243

Partially Orders Sets and Lattices

$x \preceq y$	generic partial ordering relation, 306
$\mathcal{B}_n$	boolean poset, 307
$\mathcal{D}_n$	divisibility poset, 308
$\mathcal{D}$	infinite divisibility poset, 308
$\mathcal{I}_n$	inversion poset, 310
$\mathcal{P}_n$	partition poset, 309
$\mathcal{SD}_n$	summation dominance lattice, 526
$\mathcal{Y}_S$	Young's lattice for the integer partition S, 525
$N_k(P)$	the kth Whitney number of a ranked poset, 316
$I_n(k)$	the inversion coefficient, 289
$lub(x, y)$	least upper bound under a partial ordering, 311
$glb(x, y)$	greatest lower bound under a partial ordering, 312

Graph Theory

basic notations

$G = (V, E)$	graph G with vertex-set V and edge-set E, 35
E_G or $E(G)$	edge-set of graph G, 35
V_G or $V(G)$	vertex-set of graph G, 35
uv	in a simple graph, an edge joining vertices u and v, 36
$deg(v)$	degree of vertex v, 36
I_G	incidence matrix of graph G, 40
A_G	adjacency matrix of graph G, 40
W	walk $\langle v_0, e_1, v_1, e_2, \ldots, e_n, v_n \rangle$ in a graph, 41
$\overline{G}$	edge-complement of a simple graph, 423
$G \cong H$	isomorphism relation between graphs G and H, 393

special families

C_n	cycle graph with n vertices, 372
$Cay(A : S)$	Cayley graph, 376
$circ(n : S)$	circulant graph, 374
CL_n	circular ladder with n rungs, 376
K_n	complete graph on n vertices, 38
$K_{m,n}$	complete bipartite graph with bipartition subsets of sizes m and n, 38
K_n	complete graph on n vertices, 373

ML_n	Möbius ladder with n rungs, 377	
P_n	path graph on n vertices, 380	
Q_n	hypercube graph, 375	
W_n	n-wheel, 422	

graph invariants

$\mathcal{A}ut(G)$	automorphism group of a graph, 400
$\mathcal{A}ut_E(G)$	automorphism group on the edge set of a graph, 516
$\mathcal{A}ut_V(G)$	automorphism group on the vertex set of a graph, 514
$diam(G)$	diameter of a graph G, 382
$ecc(G)$	eccentricity of a vertex v, 382
$girth(G)$	girth of a graph G, 383
$rad(G)$	radius of a graph G, 382
$P(G, t)$	chromatic polynomial of the graph G, 210

graph operations

$G \cup H$	graph union, 422
nG	n-fold self-union of graph G, 422
$G + H$	join of graphs G and H, 422
$G \times H$	product of graphs, 423

graph analytic functions

$ex(n, \mathcal{P})$	extremal function, 456
$r(s, t)$	Ramsey number, 458

flows in networks

$cap(e)$	capacity on an arc of a network, 469
$val(f)$	value of a flow f, 470
$\langle V_s, V_t \rangle$	an s-t cut, 470
$\Delta(e)$	the slack on arc e, 472
$\Delta(Q)$	the slack on quasi-path Q, 472

graphs on surfaces

N_k	the non-orientable surface of crosscap number k, 477
S_g	the orientable surface of genus g, 476
$\gamma_{\min}(G)$	minimum genus of a graph, 481

permutations and Burnside-Pólya counting

$\begin{pmatrix} 1 & 2 & \cdots & n \\ \pi(1) & \pi(2) & \cdots & \pi(n) \end{pmatrix}$	2-line representation of a permutation, 27
$(\,\pi(1)\ \pi(2)\ \cdots\ \pi(n)\,)$	1-line representation of permutation, 288
$Col_k(Y)$	set of all $(\leq k)$-colorings of a set Y, 505
$Fix(\pi)$	the set of all objects fixed by permutation π, 506
$\mathcal{P} = [P : Y]$	a group of permutations on a set Y of objects, 491
$Stab(y)$	the set of all permutations that fix object y, 506

$$\mathcal{Z}_{\mathcal{P}}(t_1, \ldots, t_n) \qquad \text{cycle index of a group, 496}$$
$$\zeta(\pi) = t_1^{r_1} t_2^{r_2} \cdots t_n^{r_n} \qquad \text{cycle structure of permutation } \pi, 496$$

Combinatorial Designs

$A \otimes B$ product of two Latin squares, 543

L_p^k a particular Latin Square, 544

L^π a conjugate of the Latin square L, 549

$(v,\, b,\, r,\, k,\, \lambda)$ parameters of a BIBD, 552

I2 GENERAL INDEX

Appendix

A1 Relations and Functions

A2 Algebraic Systems

A3 Finite Fields and Vector Spaces

A1 RELATIONS AND FUNCTIONS

DEFINITION: A **binary relation** R from a set U to a set V may be regarded intuitively as a predicate on arguments $u \in U$ and $v \in V$ that is *true* or *false*, with the notation $u\,R\,v$ indicating *true*, i.e., that u is related to v. Set-theoretically the relation R may be represented as a subset of the cartesian product $U \times V$, in which $(u, v) \in R$ means u is related to v.

Some of the most important relations are from some set to itself.

Example A1.1: A binary relation on the positive integers is defined by the rule

$$m \setminus n \ \text{ if and only if } \ \frac{n}{m} \ \text{ is an integer}$$

For instance, $7 \setminus 42$. One says, "7 *divides* 42".

Example A1.2: For any positive integer m, the relation of *arithmetic congruence* modulo m on all the integers is defined by the rule

$$b \equiv a \ \text{modulo} \ m \ \text{ if and only if } \ m \setminus b - a$$

For instance, $25 \equiv 3$ modulo 11. One says, "25 is *congruent modulo* 11 to 3" or "25 is *congruent* to 3 *modulo* 11".

Equivalence Relations

DEFINITION: A binary relation R from a set U to itself is an **equivalence relation** if it has the following three properties:

- *reflexivity*: uRu, for all $u \in U$.
- *symmetry*: if $u_1 R u_2$ then $u_2 R u_1$, for all $u_1,\ u_2 \in U$.
- *transitivity*: if $u_1 R u_2$ and $u_2 R u_3$ then $u_1 R u_3$, for all $u_1,\ u_2,\ u_3 \in U$.

Example A1.3: *Congruence modulo m* on any subset of the integers is reflexive, since the quotient $\frac{b-b}{m}$ is the integer 0. It is symmetric, since, if the quotient $\frac{b-a}{m}$ is the integer k, then the quotient $\frac{a-b}{m}$ is the integer $-k$. To show it is transitive, suppose that $a \equiv b$ modulo m and $b \equiv c$ modulo m. Then there are integers r and s such that

$$\frac{a-b}{m} = r \quad \text{and} \quad \frac{b-c}{m} = s$$

It follows that

$$\frac{a-c}{m} = \frac{a-b}{m} + \frac{b-c}{m} = r+s$$

Thus, $a \equiv c$ modulo m.

Any binary relation R on a set U can be represented by a directed graph whose vertex set is U. If aRb, then there is an arc from a to b.

Example A1.3, continued: Figure A1.1 shows the directed graph for congruence modulo 3 the subset $\{0, 1, \ldots, 8\}$.

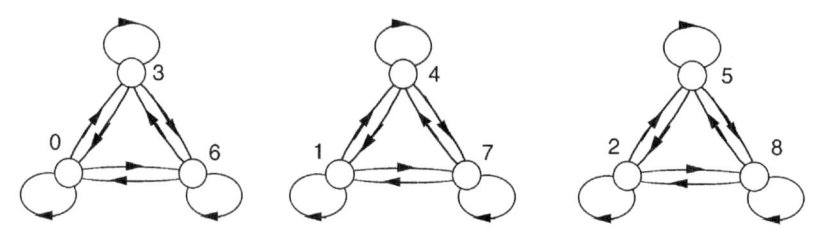

Figure A1.1   Digraphic representation of congruence modulo 3.

Proposition A1.1. *Let G be the digraph that represents a relation R on a set U.*

- *R is reflexive iff there is a self-loop at every vertex.*
- *R is symmetric iff for every arc, there is an arc with opposite direction joining the same two vertices.*
- *R is transitive iff whenever there is an arc from a vertex u to a vertex v and also an arc from vertex v to a vertex w, there is an arc from u to w.*   ◇

Example A1.3, continued: Observe that the digraph of Figure A1.1 has the three graph properties cited in Proposition A1.1.

Example A1.4: The relation *divides* is reflexive, since the quotient $\frac{n}{n}$ is the integer 1. To show it is transitive, suppose that $a \backslash b$ and $b \backslash c$. Then there are integers r and s such that

$$\frac{b}{a} = r \quad \text{and} \quad \frac{c}{b} = s$$

It follows that

$$\frac{c}{a} = \frac{c}{b} \cdot \frac{b}{a} = sr$$

However, it is not an equivalence relation, because, although it is reflexive and transitive, it is not symmetric. For instance, 3 divides 6, but 6 does not divide 3.

Partial Orderings

DEFINITION: A binary relation R from a set U to itself is a **partial ordering** if it has the following three properties:

- *reflexivity*: uRu, for all $u \in U$.
- *anti-symmetry*: if $u_1 R u_2$ and $u_2 R u_1$ then $u_1 = u_2$, for all $u_1, u_2 \in U$.
- *transitivity*: if $u_1 R u_2$ and $u_2 R u_3$ then $u_1 R u_3$, for all $u_1, u_2, u_3 \in U$.

Example A1.4, continued: *Divides* is anti-symmetric on the positive integers, since, if the quotient $\frac{b}{a}$ is an integer, then $a \leq b$, and if the quotient $\frac{a}{b}$ is also an integer, then $b \leq a$. Thus, if both quotients are integers, then $a = b$. Since it is reflexive, anti-symmetric, and transitive, *divides* is a partial ordering.

Example A1.5: The relation *congruence modulo m* is not a partial ordering, because it is not anti-symmetric.

Properties of Some Functions

DEFINITION: A **function** from a set U to a set V is a relation from U to V such that for each $u \in U$, there is a unique $v \in V$, such that $(u, v) \in U \times V$.

- The set U to whose elements a function $f : U \to V$ assigns values is called the *domain*.
- The set V in which the values are assigned is called the *codomain*.

DEFINITION: A function $f : U \to V$ is **one-to-one** (or *injective*) if for all $u_1, u_2 \in U$,

$$u_1 \neq u_2 \implies f(u_1) \neq f(u_2)$$

DEFINITION: A function $f : U \to V$ is **onto** (or *surjective*) if for all $v \in V$,

$$(\exists u \in U) \, [\, f(u) = v \,]$$

DEFINITION: A function $f : U \to V$ is **bijective** if it is both one-to-one and onto.

Any function $f : U \to V$ can be represented by a digraph whose vertex set is $U \cup V$. There is an arc from u to v if and only if $f(u) = v$.

Proposition A1.2. *Let D be the digraph that represents a function $f : U \to V$. Then*

- *f is one-to-one if and only if no vertex $v \in V$ is the head of more than one arc of the digraph D.*
- *f is onto if and only if every vertex $v \in V$ is the head of at least one arc of the digraph D.*
- *f is bijective if every vertex $v \in V$ is the head of exactly one arc of D.* $\diamond$

A2 ALGEBRAIC SYSTEMS

The algebraic system of most frequent concern in the application of combinatorial methods is the set of integers, along with the operations of addition and multiplication.

NOTATION: The set of *integers* is denoted $\mathbb{Z}$. The corresponding algebraic system is formally denoted $\langle\,\mathbb{Z},\,+,\,\cdot\,\rangle$. However, in practice, both the algebraic system and its domain are familiarly denoted by $\mathbb{Z}$.

NOTATION: The *greatest common divisor* of two numbers is denoted $\gcd(m,n)$. Sometimes the gcd operator is applied to more than two numbers at a time.

Binary Operations

DEFINITION: A **binary operation** on a set U is a function from the cartesian product $U \times U$ to the set U itself.

DEFINITION: In the present context, an **algebraic system** means a non-empty set together with one or more binary operations.

- The set U is called the **domain** of that algebraic system.

Example A2.1:   Addition and multiplication are binary operations on the integers, since the result of adding or multiplying two integers is an integer.

NOTATION: The operations of addition and multiplication are usually represented in *infix form*, i.e.,

$$x + y \quad \text{or} \quad x \cdot y$$

respectively, rather than the *prefix form* $+(x,y)$ or $\cdot\,(x,y)$.

Integers Modulo n

DEFINITION: The **mod operator** is the arithmetic rule that assigns to a non-negative integer x (called the **moduland**) and a positive integer n (called the **modulus**) the number

$$x \bmod n$$

that is the remainder of operation of dividing x by n.

EXTENSION OF DEFINITION: For a negative moduland x, the function

$$x \mapsto x \bmod n$$

is extended from the domain $\mathbb{N}$ to the domain $\mathbb{Z}$ inductively, by the rule

$$x \bmod n \;=\; (x + n) \bmod n \qquad\qquad (A.2.1)$$

Example A2.2: $-5 \bmod 3 = -2 \bmod 3 = 1 \bmod 3 = 1$. More generally, the following table illustrated the periodicity of the mod operator.

x	$\cdots$	-7	-6	-5	-4	-3	-2	-1	0	1	2	3	4	5	6	$\cdots$
$x \bmod 3$	$\cdots$	2	0	1	2	0	1	2	0	1	2	0	1	2	0	$\cdots$

Proposition A2.1. *Let x be a negative integer. Then*

$$x \bmod n = n - (|x| \bmod n) \qquad (A.2.2)$$

Proof: The proof is by induction. Details are omitted. $\qquad\qquad\qquad \Diamond$

Example A2.3:   For instance,

$$\begin{aligned}
-5 \bmod 3 &= 3 - (\,|{-5}| \bmod 3\,) \\
&= 3 - (5 \bmod 3) \\
&= 3 - 2 \\
&= 1
\end{aligned}$$

Remark: The *mod operator* is now used to define an algebraic system called the *integers modulo n*. In both concepts, which are clearly distinct, taking the remainder of a division has a fundamental role.

DEFINITION: The domain of the algebraic system (with $n \geq 2$) known as the ***integers modulo n***, denoted $\mathbb{Z}_n$, is the set of numbers

$$\{\,0, \quad 1, \quad \ldots, \quad n-1\,\}$$

There are two binary operations $(+)$ and $(\cdot)$ on $\mathbb{Z}_n$, given by the rules

$$b + c = b + c \bmod n$$
$$b \cdot c = b \cdot c \bmod n$$

In other words, in the algebraic system $\mathbb{Z}_n$, if adding or multiplying two numbers as usual for integers happens to exceed $n-1$, then divide by n and use the remainder as the result of the operation.

Associativity and Commutativity

In the rest of this chapter, we adopt the symbol $\star$ as a generic binary operation, to be written in infix form.

DEFINITION: A binary operation $\star$ on a set U is an ***associative operation*** if

$$(u \star v) \star w = u \star (v \star w) \qquad (\forall u, v, w \in U)$$

Example A2.3, continued: Addition and multiplication are both associative operations on the integers and also on the integers modulo n.

DEFINITION: A binary operation $\star$ on a set U is a ***commutative operation*** if

$$u \star v = v \star u \qquad (\forall u, v \in U)$$

Example A2.3, continued: Addition and multiplication are commutative operations on the integers and also on the integers modulo n.

Identity Element

DEFINITION: An element $\epsilon \in U$ is the **identity element** with respect to a binary operation $\star$ if

$$\epsilon \star u = u \star \epsilon = u \qquad (\forall u \in U)$$

There is at most one identity element with respect to a given binary operation.

Example A2.3, continued:

- The number 0 is the additive identity of the integers and also the additive identity of the integers modulo n.
- The number 1 is the multiplicative identity of $\mathbb{Z}$ and also the multiplicative identity of the integers modulo n.

Inverse Elements

DEFINITION: Suppose that the set U has an identity ϵ with respect to a binary operation $\star$, and let $u \in U$. An element u^{-1} is the *inverse* of u with respect to $\star$ if

$$u^{-1} \star u = u \star u^{-1} = \epsilon$$

There is at most one inverse for an element with respect to a given binary operation.

Example A2.3, continued:

- Every integer n has $-n$ as its additive inverse.
- The number 1 is the only integer whose multiplicative inverse is an integer.
- In the set of positive rational numbers, every element $\frac{p}{q}$ has the multiplicative inverse $\frac{q}{p}$.
- Some numbers have multiplicative inverses in $\mathbb{Z}_n$. For instance, 11 is the inverse of 5 in $\mathbb{Z}_{18}$, since

$$11 \cdot 5 = 55 \equiv 1 \bmod 18$$

It is proved in §6.4 that a number b has a multiplicative inverse modulo n if and only if $\gcd(b, n) = 1$.

Groups

DEFINITION: An algebraic system $\langle U, \star \rangle$ is called a **group** if it has the following properties:

> the operation $\star$ is associative.
>
> there is an identity element.
>
> every element of U has an inverse.

If $\star$ is commutative, then $\langle U, \star \rangle$ is called a *commutative group* or an *abelian group*. In this case, it is also commonly called an **additive group**, and its operation is written as addition, using the operational symbol $+$. Moreover, the additive identity is typically called **zero** and denoted 0.

Example A2.3, continued:

- The integers are a group under the operation of addition. They are not a group under multiplication, because most integers do not have multiplicative inverses.
- The positive rationals are a group under multiplication.

Rings

DEFINITION: An algebraic system $\langle U, +, \cdot \rangle$ is called a **ring** if it has the following properties:

> the system $\langle U, + \rangle$ (ignoring $\cdot$) is an additive group.
>
> the operation $\times$ is associative.
>
> $z \cdot (x + y) = z \cdot x + z \cdot x$, for any $x, y, z \in U$. We say that the operation $(\cdot)$ *distributes* over the operation $(+)$.

The operation $(\cdot)$ is typically called **multiplication**.

Example A2.3, continued:

- The integers are a ring under the operations of addition and multiplication, as are the integers modulo n. In both cases, we commonly represent the multiplication operation by *juxtaposition*, that is,

$$ab \quad \text{means} \quad a \cdot b$$

- In some rings, including $\mathbb{Z}$ and $\mathbb{Z}_n$, multiplication is commutative and the number 1 serves as the multiplicative identity. No number except 1 or -1 has a multiplicative inverse in $\mathbb{Z}$.

Example A2.4:   The 2×2 real matrices are a ring, whose multiplication (i.e., matrix multiplication) is non-commutative. A square matrix has a multiplicative inverse if its determinant is non-zero.

Fields

DEFINITION: A **field** is a ring with commutative multiplication and a multiplicative identity, such that every non-zero element has a multiplicative inverse.

Example A2.5:

- The rational numbers $\mathbb{Q}$ are a field.
- The real numbers $\mathbb{R}$ are a field.
- The complex numbers $\mathbb{C}$ are a field.

Example A2.6:

- The ring $\mathbb{Z}_2$ is a field, since the number 1 is its own multiplicative inverse.
- The ring $\mathbb{Z}_3$ is a field, since both 1 and 2 are their own inverses.
- The ring $\mathbb{Z}_5$ is a field since

$$1 \cdot 1 = 1 \quad 2 \cdot 3 = 1 \quad 3 \cdot 2 = 1 \quad 4 \cdot 4 = 1$$

Since, according to §6.4, every number r such that $\gcd(r, n) = 1$ has a multiplicative inverse in $\mathbb{Z}_n$, it follows that for prime p, the ring $\mathbb{Z}_p$ is a field.

k-Tuples of Numbers

DEFINITION: The **direct sum** $\mathbb{Z}_{n_1} \oplus \mathbb{Z}_{n_2}$ of the additive groups $\mathbb{Z}_{n_1}$ and $\mathbb{Z}_{n_2}$ is a group whose domain is the set of 2-tuples

$$\left\{ (r, s) \mid r \in \mathbb{Z}_{n_1}, \, s \in \mathbb{Z}_{n_2} \right\}$$

Its operation is coordinate-wise addition. That is,

$$(r, s) + (r', s') = (r + r' \bmod n_1, \, s + s' \bmod n_2)$$

Example A2.7:   In $\mathbb{Z}_3 \oplus \mathbb{Z}_4$, the elements are

$$
\begin{array}{cccc}
(0,0) & (0,1) & (0,2) & (0,3) \\
(1,0) & (1,1) & (1,2) & (1,3) \\
(2,0) & (2,1) & (2,2) & (2,3)
\end{array}
$$

Some examples of addition are

$$
\begin{aligned}
(1,3) + (2,2) &= (0,1) \\
(1,1) + (2,3) &= (0,0)
\end{aligned}
$$

Remark: The direct sum construction can be iterated

$$\mathbb{Z}_{n_1} \oplus \mathbb{Z}_{n_2} \cdots \oplus \mathbb{Z}_{n_k}$$

to k-tuples of numbers, with each coordinate having its respective modulus. All these direct sums are groups.

NOTATION: If every modulus over all k coordinates is the same number r, then the group may be denoted $\mathbb{Z}_r{}^k$.

Generators for an Algebraic System

DEFINITION: An element x is an **additive generator** for $\mathbb{Z}_n$ if every number r in $\mathbb{Z}_n$ can be written as an iterated sum

$$r = x + x + \cdots + x$$

Example A2.8:   The number 3 is an additive generator for $\mathbb{Z}_{10}$, but the number 2 is not. Whereas 3 generates the complete sequence

$$3 \quad 6 \quad 9 \quad 2 \quad 5 \quad 8 \quad 1 \quad 4 \quad 7 \quad 0$$

(where $6 = 3 + 3$, $9 = 3 + 3 + 3$, $2 = 3 + 3 + 3 + 3 \bmod 10$, etc.) the number 2 generates only the five numbers

$$2 \quad 4 \quad 6 \quad 8 \quad 0$$

before it repeats.

DEFINITION: A number x is a **multiplicative generator** for $\mathbb{Z}_n$ if every non-zero number r in $\mathbb{Z}_n$ can be written as an iterated product

$$r = x \cdot x \cdots \cdot x$$

Example A2.9:   The number 3 is a multiplicative generator for $\mathbb{Z}_7$, but the number 2 is not. Whereas 3 generates the complete sequence

$$\left\langle 3^j \bmod n \mid j = 1, \ldots, 6 \right\rangle = 3 \quad 2 \quad 6 \quad 4 \quad 5 \quad 1$$

the number 2 generates only the sequence

$$2 = 2^1 \quad 4 = 2^2 \quad 1 = 2^3$$

before it repeats.

DEFINITION: A set of elements $S = \{x_1, x_2, \ldots, x_s\}$ of an algebraic system $\mathcal{A}$ is a **generating set** for that system if every other element of that system can be obtained by applying the operations of the system $\mathcal{A}$ to some combination of the elements in S.

Example A2.10:   The numbers 3 and 5 form an additive generating set for $\mathbb{Z}_{15}$. (So does any number j such that $\gcd(j, 15) = 1$, by itself.)

Example A2.11:   The 2-tuples $(1, 0)$ and $(0, 1)$ generate the group $\mathbb{Z}_{n_1} \oplus \mathbb{Z}_{n_2}$. No single 2-tuple generates such a group.

Prime Factorization

Although the entire set of integers is generated additively by the number 1, it takes the infinite set of all the *prime numbers* to generate the integers multiplicatively.

DEFINITION: A **divisor** of an integer n is an integer d such that the equation $xd = n$ has an integer solution $x = q$.

DEFINITION: A **prime number** is a positive integer $p > 1$ that has no divisors except 1 and itself.

The following theorem from elementary number theory is commonly called the **Fundamental Theorem of Arithmetic**.

Theorem A2.2 [Prime-Power Factorization Theorem]. *Every positive integer n can be written uniquely as a product of powers of distinct, non-decreasing primes*

$$n = p_1^{e_1} p_2^{e_2} \cdots p_r^{e_r}$$

A3   FINITE FIELDS AND VECTOR SPACES

Example A2.6 asserts that for prime p, all non-zero numbers have multiplicative inverses in the ring $\mathbb{Z}_p$ (as proved in §6.4), which implies that $\mathbb{Z}_p$ is a field. In order to present various algebraic methods to persons who have not previously had at least the equivalent of a semester course on groups, rings, and fields, Chapter 10 develops various design constructions explicitly for prime fields $\mathbb{Z}_p$, even though they are readily generalizable to all finite fields. This section is intended for readers wanting to understand the methods of Chapter 10 in greater generality. For other readers, it is optional. Our main concern in this present section is to describe the construction of a field $GF(p^r)$ for every prime power p^r.

DEFINITION: An additive group $\langle V, + \rangle$ is called a **vector space** over the field $\mathcal{F}$ if there is a scalar product $\mathcal{F} \times V \to V$ such that

(i) $c\,(\mathbf{v} + \mathbf{v}') = c\mathbf{v} + c\mathbf{v}'$ for all $c \in \mathcal{F}$ and all $\mathbf{v}, \mathbf{v}' \in V$

(ii) $(a + b)\,\mathbf{v} = a\mathbf{v} + b\mathbf{v}$ for all $a, b \in \mathcal{F}$ and all $\mathbf{v} \in V$

(iii) $(ab)\,\mathbf{v} = a(b\mathbf{v})$ for all $a, b \in \mathcal{F}$ and all $\mathbf{v} \in V$

(iv) $1\mathbf{v} = \mathbf{v}$ for mult. identity $1 \in \mathcal{F}$ and all $\mathbf{v} \in V$

An element of the domain V is called a **vector**. An element of the field $\mathcal{F}$ is called a **scalar**.

Example A3.1:   The vector space $\mathbb{R}^d$ is the set of d-tuples with entries in $\mathbb{R}$ (i.e., real numbers), where coordinate-wise addition in $\mathbb{R}$ is the vector addition. The scalar product is

$$c\,(x_1, x_2, \ldots, x_d) = (cx_1, cx_2, \ldots, cx_d)$$

Bases and Dimension

DEFINITION: Let $\mathbf{v}_1, \mathbf{v}_2, \ldots, \mathbf{v}_k$ be vectors in a vector space $\langle V, + \rangle$ over a field $\mathcal{F}$ and $c_1, c_2, \ldots, c_k$ scalars in $\mathcal{F}$. The iterated sum

$$c_1\mathbf{v}_1 + c_2\mathbf{v}_2 + \cdots + c_k\mathbf{v}_k$$

is called a **linear combination** of those vectors.

Example A3.1, continued: In the vector space $\mathbb{R}^d$, every vector

$$(x_1,\, x_2,\, x_3,\, \ldots,\, x_d)$$

can be expressed as a linear combination of the vectors

$$(1,\, 0,\, 0,\, 0,\, \ldots,\, 0,\, 0)$$
$$(0,\, 1,\, 0,\, 0,\, \ldots,\, 0,\, 0)$$
$$(0,\, 0,\, 1,\, 0,\, \ldots,\, 0,\, 0)$$
$$\vdots$$
$$(0,\, 0,\, 0,\, 0,\, \ldots,\, 0,\, 1)$$

That is,

$$
\begin{aligned}
(x_1,\, x_2,\, x_3,\, \ldots,\, x_d) \;=\; & \; x_1\,(1,\, 0,\, 0,\, 0,\, \ldots,\, 0,\, 0) \\
& + \; x_2\,(0,\, 1,\, 0,\, 0,\, \ldots,\, 0,\, 0) \\
& + \; x_3\,(0,\, 0,\, 1,\, 0,\, \ldots,\, 0,\, 0) \\
& \qquad\quad \vdots \\
& + \; x_d\,(0,\, 0,\, 0,\, 0,\, \ldots,\, 0,\, 1)
\end{aligned}
$$

DEFINITION: If every vector in a vector space $\langle V,\, +\rangle$ is expressible as a linear combination of the vectors $\mathbf{v}_1, \mathbf{v}_2, \ldots, \mathbf{v}_k$, then those vectors are said to be a **spanning set of vectors** for $\mathcal{F}$ or to **span** $\mathcal{F}$.

DEFINITION: A set of vectors $\mathbf{v}_1, \mathbf{v}_2, \ldots, \mathbf{v}_k$ that span the vector space $\langle V,\, +\rangle$ such that the linear combination expressing each vector $\mathbf{v}$ is unique is called a **basis for the vector space** $\langle V,\, +\rangle$.

Example A3.1, continued: In the vector space $\mathbb{R}^d$, the vectors

$$(1,\, 0,\, 0,\, 0,\, \ldots,\, 0,\, 0)$$
$$(0,\, 1,\, 0,\, 0,\, \ldots,\, 0,\, 0)$$
$$(0,\, 0,\, 1,\, 0,\, \ldots,\, 0,\, 0)$$
$$\vdots$$
$$(0,\, 0,\, 0,\, 0,\, \ldots,\, 0,\, 1)$$

form a basis.

Theorem A3.1 [Invariance of Dimension]. *Every basis for a vector space has the same cardinality.*

Proof: We omit details of the proof. It follows from the *row reduction algorithm*, which amounts to a systematization of the standard method of solving simultaneous linear equations by iterative substitution. Moreover, the row reduction algorithm is used in proving that every vector space has a basis. $\diamond$

Example A3.2: The polynomials of finite degree with coefficients in the field $\mathbb{Z}_p$ form an infinite-dimensional vector space, denoted $\mathbb{Z}_p[x]$. We observe that every such polynomial can be expressed uniquely as a linear combination of the polynomials

$$1 \quad x \quad x^2 \quad x^3 \quad \cdots$$

Moreover, the polynomials of degree at most $d - 1$ with coefficients in $\mathbb{Z}_p$ form a d-dimensional vector space over the field $\mathbb{Z}_p$, with the vectors

$$1 \quad x \quad x^2 \quad \cdots \quad x^{d-1}$$

serving as a basis.

Irreducible Polynomials

DEFINITION: A polynomial in $\mathbb{Z}_p[x]$ is said to be **irreducible over the prime field** $\mathbb{Z}_p$ if it cannot be factored into two polynomials (in $\mathbb{Z}_p[x]$) of smaller degree.

Example A3.3: The polynomial $x^2 + 1$ factors over $\mathbb{Z}_2$ into

$$\begin{aligned}
(x + 1)(x + 1) &= x^2 + 2x + 1 \\
&\equiv x^2 + 1 \text{ modulo } 2
\end{aligned}$$

and the polynomial $x^2 + x + 1$ is irreducible (which could be proved by trying all possible factorizations).

Example A3.4: The polynomial $x^2 + x + 1$ factors over $\mathbb{Z}_3$ into

$$\begin{aligned}
(x + 2)(x + 2) &= x^2 + 4x + 4 \\
&\equiv x^2 + x + 1 \text{ modulo } 3
\end{aligned}$$

and the polynomial $x^2 + x + 2$ is irreducible.

Remark: Tables of irreducible polynomials can be downloaded from various websites.

Polynomials Modulo a Polynomial

TERMINOLOGY: When a polynomial $f(x)$ is divided by a polynomial $g(x)$ of degree d, there is a **quotient polynomial** $q(x)$ and a **remainder polynomial** $r(x)$ of degree less than d such that

$$f(x) \;=\; q(x)g(x) + r(x)$$

DEFINITION: The algebraic structure $\mathcal{F}[x] \,/\, g(x)$ of **polynomials modulo a polynomial** over the field $\mathcal{F}$ has polynomial addition and multiplication modulo $g(x)$ as its operations, under which it is a ring.

Example A3.5: Here are the addition and multiplication tables for the ring $\mathbb{Z}_2[x]\,/\,x^2 + x + 1$. It can be verified that $\mathbb{Z}_2[x]\,/\,x^2 + x + 1$ satisfies all of the axioms for a field.

Table A3.1   Arithmetic tables for $\mathbb{Z}_2[x]\,/\,x^2 + x + 1$.

$+$	0	1	x	$x + 1$
0	0	1	x	$x + 1$
1	1	0	$x + 1$	x
x	x	$x + 1$	0	1
$x + 1$	$x + 1$	x	1	0

$\cdot$	0	1	x	$x + 1$
0	0	0	0	0
1	0	1	x	$x + 1$
x	0	x	$x + 1$	1
$x + 1$	0	$x + 1$	1	x

Basic Facts about Finite Fields

Proofs of the following basic facts about finite fields can be found in many textbooks on abstract algebra. We presently assert them without proof.

Proposition A3.2. *For any irreducible polynomial $g(x)$ of degree d over $\mathbb{Z}_p[x]$, where p is prime, the ring $\mathbb{Z}_p[x]/g(x)$ is a field of order p^d.*

Proposition A3.3. *There exists a field of order n if and only if $n = p^d$ for some prime number p.*

Proposition A3.4. *All fields of order p^d are isomorphic.*

Proposition A3.5. *The additive group of the field of order p^d is isomorphic to the vector space $\mathbb{Z}_p^d$.*

Proposition A3.6. *The multiplicative group of the field of order p^d is generated by some single element of the field.*

Bibliography

B1 General Reading

B2 References

B1 GENERAL READING

In addition to the references listed below, the reader may wish to consult the *Handbook of Discrete and Combinatorial Mathematics*, edited by K. H. Rosen, which contains many contributed sections with summaries of the most important research results and extensive bibliographies.

ALGEBRA

[Gall2006] J. A. Gallian, *Contemporary Abstract Algebra*, Sixth Edition, Houghton-Mifflin, 2006.

[Hers2001] I. N. Herstein, *Abstract Algebra*, Third Edition, J. Wiley & Sons, 2001.

[MaBi1967] S. Maclane and G. Birkhoff, *Algebra*, Macmillan, 1967.

ALGORITHMS and COMPUTATION

[AHU1983] A. V. Aho, J. E. Hopcroft, and J. D. Ullman, *Data Structures and Algorithms*, Addison-Wesley, 1983.

[CLRS2001] T. H. Cormen, C. E. Leiserson, R. L. Rivest, and C. Stein, *Introduction to Algorithms*, Second Edition, MIT Press, 2001.

COMBINATORIAL and DISCRETE MATHEMATICS

[BeQu2003] A. T. Benjamin and J. J. Quinn, *Proofs that Really Count*, Mathematical Association of America, 2003.

[BeWi1991] E. A. Bender and S. G. Williamson, *Foundations of Applied Combinatorics*, Addison-Wesley, 1991. (See especially Section 12.4 of this text for further discussion of asymptotic estimation.)

[Bona2007] M. Bóna, *Introduction to Enumerative Combinatorics*, McGraw-Hill, 2007.

[Brua2004] R. Brualdi, *Introductory Combinatorics*, Fourth Edition, Prentice-Hall, 2004.

[Devo2004], J. L. Devore, *Probability and Statistics*, Sixth Edition, Brooks/Cole, 2004.

[GKP1994] R. L. Graham, D. E. Knuth, and O. Patashnik, *Concrete Mathematics: A Foundation for Computer Science*, Second Edition, Addison-Wesley, 1994.

[Grim2004] R. P. Grimaldi, *Discrete and Combinatorial Mathematics*, Fifth Edition, Addison-Wesley, 2004.

[Liu1968] C. L. Liu, *Introduction to Combinatorial Mathematics*, McGraw-Hill, 1968.

[MiRo1991] J. G. Michaels and K. H. Rosen, editors, *Applications of Discrete Mathematics*, McGraw-Hill, 1991.

[PeWiZe1996] M. Petkovšek, H. S. Wilf, and D. Zeilberger, *A=B*, A. K. Peters, 1996.

[PoSt1990] A. Polimeni and H. J. Straight, *Foundations of Discrete Mathematics*, Brooks/Cole, 1990.

[RoTe2005] F. S. Roberts and B. Tesman, *Applied Combinatorics*, Second Edition, Prentice-Hall, 2005.

[Rose2007] K. H. Rosen, *Discrete Mathematics and Its Applications*, Sixth Edition, McGraw-Hill, 2007.

[SlPl1995] N. J. A. Sloane and S. Plouffe, *The Encyclopedia of Integer Sequences*, Second Edition, Academic Press, 1995. (A comprehensive reference volume on integer sequences.)

[StWh1986] D. Stanton and D. White, *Constructive Combinatorics*, Springer-Verlag, 1986.

[Stra1993] H. J. Straight, *Combinatorics: An Invitation*, Brooks/Cole 1993.

[Tuck2001] A. Tucker, *Applied Combinatorics*, Fourth Edition, John Wiley & Sons, 2001.

[Wilf1993] H. S. Wilf, *generatingfunctionology*, Second Edition, Academic Press, 1993.

COMBINATORIAL DESIGNS

[CoDi2000a] C. J. Colbourn and J. H. Dinitz, Block designs, §12.1 in *Handbook of Combinatorial Designs*, edited by K. H. Rosen, CRC Press, 2000.

[CoDi2000b] C. J. Colbourn and J. H. Dinitz, editors, Symmetric designs and finite geometries, §12.2 in *Handbook of Combinatorial Designs*, edited by K. H. Rosen, CRC Press, 2000.

[CoDi2000c] C. J. Colbourn and J. H. Dinitz, editors, Latin squares and orthogonal arrays, §12.3 in *Handbook of Combinatorial Designs*, edited by K. H. Rosen, CRC Press, 2000.

[CoDi2007] C. J. Colbourn and J. H. Dinitz, editors, *Handbook of Combinatorial Designs*, Second Edition, CRC Press, 2007.

[Wall2008] W. D. Wallis, *Introduction to Combinatorial Designs*, CRC Press, 2008, to appear.

GRAPH THEORY

[BLW1986] N. L. Biggs, E. K. Lloyd, and R. J. Wilson, *Graph Theory 1736-1936*, Oxford, 1986.

[Boll1998] B. Bollobás, *Modern Graph Theory*, Springer, 1998.

[ChLe2004] G. Chartrand and L. Lesniak, *Graphs & Digraphs*, Fourth Edition, CRC Press, 2004.

[ChZh2005] G. Chartrand and P. Zhang, *Introduction to Graph Theory*, McGraw Hill, 2005.

[Dist2000] R. Diestel, *Graph Theory*, Springer-Verlag, Second Edition, 2000.

[Gibb1985] A. Gibbons, *Algorithmic Graph Theory*, Cambridge University Press, 1985.

[Goul1988] R. Gould, *Graph Theory*, Benjamin/Cummings, 1988.

[GRS1990] R. L. Graham, B. L. Rothschild, and J. H. Spencer, *Ramsey Theory*, John Wiley & Sons, 1990.

[GrYe2006] J. L. Gross and J. Yellen, *Graph Theory and Its Applications*, 2nd Edition, CRC Press, 2006.

[GrYe2004] J. L. Gross and J. Yellen, editors, *Handbook of Graph Theory*, CRC Press, 2004.

[Hara1969] F. Harary, *Graph Theory*, Addison-Wesley, 1969.

[HaPa1973] F. Harary and E. Palmer, *Graphical Enumeration*, Academic Press, 1973.

[West2001] D. B. West, *Introduction to Graph Theory*, Second Edition, Prentice-Hall, 2001.

[Wils1996] R. J. Wilson, *Introduction to Graph Theory*, Addison Wesley Longman, 1996.

[Wils2004] R. J. Wilson, *History of Graph Theory*, §1.3 in *Handbook of Graph Theory*, edited by J. L. Gross and J. Yellen, CRC Press, 2004.

NUMBER THEORY

[Andr1994] G. E. Andrews, *Number Theory*, Dover. (First Edition, Saunders, 1971.)

[NiZu1991] I. Niven, H. S. Zuckerman, and H. L. Montgomery, *An Introduction to the Theory of Numbers*, Fifth Edition, Wiley.

[Rose2005] K. H. Rosen, *Elementary Number Theory, Fifth Edition*, Addison-Wesley, 2005.

SURFACES and TOPOLOGICAL GRAPH THEORY

[GrTu1987] J. L. Gross and T. W. Tucker, *Topological Graph Theory*, Dover, 2001. (First Edition, Wiley-Interscience, 1987.)

[Mass1967] W. S. Massey, *Algebraic Topology: An Introduction*, Harbrace, 1967.

[Ring1974] G. Ringel, *Map Color Theorem*, Springer, 1974.

[Whit2001] A. T. White, *Graphs of Groups on Surfaces*, North-Holland, 2001.

B2 REFERENCES

The following references are cited in the text.

[AhHoUl1974] A. V. Aho, J. E. Hopcroft, and J. D. Ullman, *Design and Analysis of Computer Algorithms*, Addison-Wesley, 1974.

[BaCo1987], W. W. Rouse Ball and H. S. M. Coxeter, *Mathematical Recreations and Essays*, Dover, 1987.

[BoNi2004] B. Bollobás and V. Nikiforov, *Extremal Graph Theory*, §8.1 in *Handbook of Graph Theory*, edited by J. L. Gross and J. Yellen, CRC Press, 2004.

[BSP1960] R. C. Bose, S. K. Shrikhande, and E. T. Parker, Further results on the construction of mutually orthogonal Latin squares and the falsity of Euler's conjecture, *Canad. J. Math.* **12** (1960), 189–203.

[Burn1911] W. Burnside, *Theory of Groups of Finite Order*, Second Edition, Dover, 1955. Original edition: Cambridge Univ. Press, 1911.

[Cauc1845] A. Cauchy, Mémoire sur diverses propriétés remarquables des substitutions régulières ou irrégulières, et des systémes de substitutiones conjugées, *Comptes Rendus Acad. Sci. Paris*, **21** (1845), 835.

[CoDi2000a] C. J. Colbourn and J. H. Dinitz, *Block Designs*, §12.1 in *Handbook of Combinatorial Designs*, edited by K. H. Rosen, CRC Press, 2000.

[Dijk1959] E. W. Dijkstra, A note on two problems in connexion with graphs, *Numer. Math.* **1** (1959), 269–271.

[Dilw1950] R. P. Dilworth, A decomposition theorem for partially ordered sets, *Ann. Math.* 51 (1950), 161–166.

[Faud2004] R. Faudree, *Ramsey Graph Theory*, §8.3 in *Handbook of Graph Theory*, edited by J. L. Gross and J. Yellen, CRC Press, 2004.

[Floy1962] R. W. Floyd, Algorithm 97: Shortest path, *Communications of Asso. Comp. Mach.* **5** (1962), 345.

[Frob1887] F. G. Frobenius, Über die Congruenz nach einem aus zwei endlichen Gruppen gebildeten Doppelmodul, *J. Reine Angew. Math.* **101** (1887), 273–299.

[Gros1974] J. L. Gross, Voltage graphs, *Discrete Math.* **9** (1974), 239–246.

[GrHa1980] J. L. Gross and F. Harary, Some problems in topological graph theory, *J. Graph Theory* **4** (1980), 253–263.

[GrTu1977] J. L. Gross and T. W. Tucker, Generating all graph coverings by permutation voltage assignments, *Discrete Math.* **18** (1977), 273–283.

[GrTu1987] J. L. Gross and T. W. Tucker, *Topological Graph Theory*, Dover, 2001. (First Edition, Wiley-Interscience, 1987.)

[GrTu2008] J. L. Gross and T. W. Tucker, eds., *Topics in Topological Graph Theory*, series editors: L. W. Beineke and R. J. Wilson, Cambridge Univ Press., projected for publication in 2008.

[GrYe2006] J. L. Gross and J. Yellen, *Graph Theory and Its Applications*, 2nd Edition, CRC Press, 2006.

[Gust1963] W. Gustin, Orientable embedding of Cayley graphs, *Bull. Amer. Math. Soc.* **69** (1963), 272–275.

[Hall1956] M. Hall, *Proc. Symp. Applied Math.* **6**, 203, Amer. Math. Soc., 1956.

[Heff1891] L. Heffter, Über das Problem der Nachbargebiete, *Math. Annalen* **38** (1891), 477–580.

[Knut1973] D. E. Knuth, *Sorting and Searching*, Volume 3 of *The Art of Computer Programming*, Addison-Wesley, 1973.

[Liu1968] C. L. Liu, *Introduction to Combinatorial Mathematics*, McGraw-Hill, 1968.

[Maur04] S. B. Maurer, *Directed Acyclic Graphs*, §3.2 in *Handbook of Graph Theory*, edited by J. L. Gross and J. Yellen, CRC Press, 2004.

[Park1960] E. T. Parker, Orthogonal Latin squares, *Proc. Nat. Acad. Sci.* **45** (1959), 859–862.

[Póly1973] G. Pólya, Kombinatorische Anzahlbestimmungen für Gruppen, Graphen und chemische Verbinungen, *Acta Math.* **68** (1937), 145–254.

[Prim1957] R. C. Prim, Shortest connection networks and some generalizations, *Bell System Tech. J.* **36** (1957), 1389–1401.

[RiYo1968] G. Ringel and J. W. T. Youngs, Solution of the Heawood map-coloring problem, *Proc. Acad. Nat. Sci. USA* **60** (1968), 438–445.

[Spir1973] P. M. Spira, A new algorithm for finding all shortest paths in a graph of positive arcs in average time $O(n^2 \log^2 n)$, *SIAM J. Computing* **2** (1973), 28–32.

[Tarr1900], G. Tarry, Le problème des 36 officiers, *Comptes Rend. Assoc. Fr.* **1** (1900), 122–123.

[Tarr1901], G. Tarry, Le problème des 36 officiers, *Comptes Rend. Assoc. Fr.* **2** (1901), 170–203.

[Wars1962] S. Warshall, A theorem on Boolean matrices, *J. of Assoc. Comp. Mach.* **9** (1962), 11–12.

[Wils2004] R. J. Wilson, *History of Graph Theory*, §1.3 in *Handbook of Graph Theory*, edited by J. L. Gross and J. Yellen, CRC Press, 2004.

SOLUTIONS AND HINTS

Chapter 0   Introduction to Combinatorics

0.3   Some Rules for Counting

0.3.6: $\lfloor x + y \rfloor \geq \lfloor \lfloor x \rfloor + \lfloor y \rfloor \rfloor = \lfloor x \rfloor + \lfloor y \rfloor$

0.3.10:   Use the Rule of Quotient iteratively: $\dfrac{7!}{3!\,2!\,1!\,1!}$

0.3.16:   Use the Pigeonhole Principle. Let the students be the pigeons and the seven days of the week be the pigeonholes. It would take 8 students to be certain of a match.

0.4   Counting Selections

0.4.3:   This amounts to selection of 8 objects from 3 distinct types, with repetitions permitted, since an initial choice of one object of each type is required:

$$\binom{10}{2}$$

0.4.6: $\dbinom{15}{3}$

0.4.8: $\dbinom{7}{3\ 2\ 2} = \dfrac{7!}{3!\,2!\,2!} = 210$

0.4.12:   105

0.4.16:   A partition of 5 objects into three parts is of type 311 or type 221.

$$\binom{5}{3\ 1\ 1} \cdot \frac{1}{2!} + \binom{5}{2\ 2\ 1} \cdot \frac{1}{2!} = 10 + 15 = 25$$

0.5   Permutations

0.5.1: $\begin{pmatrix} 1 & 2 & 3 & 4 & 5 \\ 3 & 5 & 2 & 4 & 1 \end{pmatrix} = (1\ 3\ 2\ 5)(4)$

0.5.7: $\begin{pmatrix} 1 & 2 & 3 & 4 & 5 \\ 5 & 3 & 1 & 4 & 2 \end{pmatrix} = (1\ 5\ 2\ 3)(4)$

0.5.13: $(1\ 2\ 3) \circ (2\ 4\ 5) = (1\ 4\ 5\ 2\ 3)$

0.5.21: $\begin{pmatrix} 7 \\ 4\ 2\ 1 \end{pmatrix} \cdot 3!$

0.6   Graphs

0.6.1:   There are 7 edges. The degree sum is 14.

0.6.9: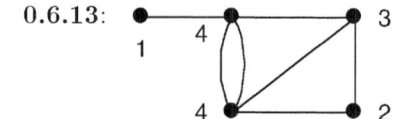

0.6.13: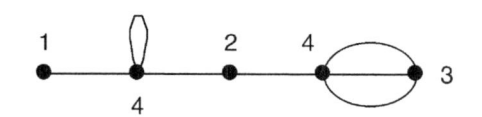

Chapter 1   Sequences

1.1   Sequences as Lists

1.1.1: $\langle x_n = 2^n \bmod 7 \rangle \Rightarrow 1\ \ 2\ \ 4\ \ 1\ \ 2\ \ 4\ \ 1\ \ 2\ \ 4\ \ 1\ \ 2\ \ 4\ \ \ldots$
If $n = 3k + j$, then $2^n = (2^3)^k \cdot 2^j = (1+7)^k \cdot 2^j \equiv 1^k \cdot 2^j \equiv 2^j$ modulo 7.

1.1.7:   Use induction.

1.1.13:   Period $P = 3$. Sequence $a_n = n^2 \bmod 3$. One way to derive this is to assume there is a polynomial $ax^2 + bx + c$ that satisfies it. These three linear equations

$$a(3k)^2 + b(3k) + c = 0$$
$$a(3k+1)^2 + b(3k+1) + c = 1$$
$$a(3k+2)^2 + b(3k+2) + c = 1$$

are equivalent modulo 3 to the equations

$$c = 0$$
$$a + b + c = 1$$
$$a + 2b + c = 1$$

whose solution is $a = 1, b = 0, c = 0$

1.1.20: Polynomial $f(x) = x^2 + x(x-1)(x-2)(x-3)$

1.1.21: $\displaystyle\sum_{k=1}^{n} \frac{1}{n^2} < 1 + \int_{x=1}^{n} \frac{dx}{x^2} = 2 - \frac{1}{n}$

1.1.25: Compare $\lg x_n$ to $\lg y_n$.

1.2 Recurrences

1.2.6: Compare $2 \lg n$ to $\lg(n+1) + \lg(n-1)$. Exponentiating both sides yields n^2 on the left, which is larger than $(n+1)(n-1)$ on the right.

1.2.7: $(n+1)^2 + (n-1)^2 = (n^2 + 2n + 1) + (n^2 - 2n + 1) = 2n^2 + 2 > n^2$.

1.2.13: Use induction. The induction step is
$$\begin{aligned}
c_{n+2} + c_n &= (c_0 c_{n+1} + \cdots + c_{n+1} c_0) + (c_0 c_{n-1} + \cdots + c_{n-1} c_0) \\
&= c_0(c_{n+1} + c_{n-1}) + \cdots + c_{n-1}(c_2 + c_0) + c_n c_1 + c_{n+1} c_0 \\
&\geq 2c_0 c_n + \cdots + 2c_{n-1} c_1 + c_n c_1 + c_{n+1} c_0 \quad \text{by ind hyp} \\
&\geq 2c_0 c_n + \cdots + 2c_{n-1} c_1 + 2c_n c_0 \quad \text{since } c_{n+1} \geq c_n \text{ and } c_1 = c_0 = 1 \\
&= c_{n+1}
\end{aligned}$$

1.3 Pascal's Recurrence

1.3.2: $\dbinom{n-1}{r-1} + \dbinom{n-1}{r} = \left(\dbinom{n-2}{r-2} + \dbinom{n-2}{r-1}\right) + \left(\dbinom{n-2}{r-1} + \dbinom{n-2}{r}\right)$

1.3.5: Evaluate $\left.(1+x)^n\right|_{x=1}$

1.4 Differences and Partial Sums

1.4.1: Difference table for n^4.

n^4	0		1		16		81		256		625		1296		2401		$\cdots$
		1		15		65		175		369		671		1105			$\cdots$
			14		50		110		194		302		434			$\cdots$	
				36		60		84		108		132			$\cdots$		
					24		24		24		24			$\cdots$			

1.4.7:   Difference sequence for c^n.

$$
\begin{array}{c|cccccc}
c^n & 1 & c & c^2 & c^3 & c^4 & \cdots \\
    & c-1 & (c-1)c & (c-1)c^2 & (c-1)c^3 & \cdots \\
    & (c-1)^2 & (c-1)^2c & (c-1)^2c^2 & \cdots \\
    & (c-1)^3 & (c-1)^3c & \cdots
\end{array}
$$

1.4.13: $6^2 - 3^2 = 27$

$$\sum_{n=3}^{5} \triangle(n^2) = \sum_{n=3}^{5} (2n+1) = 7+9+11 = 27$$

1.5   Falling Powers

1.5.1:   For $r > xy$, both expressions have the value 0. For $\min(x,y) < r \le xy$, we have $x^{\underline{r}}y^{\underline{r}} = 0$ and $(xy)^{\underline{r}} > 0$. In the main case $2 \le r \le \min(x,y)$, since $xy - j > (x-j)(y-j)$, for $1 \le j \le r$, it follows that

$$(xy)^{\underline{r}} = \prod_{j=0}^{r-1}(xy-j) > \prod_{j=0}^{r-1}(x-j)(y-j) = x^{\underline{r}}y^{\underline{r}}$$

1.5.5: $\displaystyle \binom{-\frac{1}{2}}{4} = \frac{(-\frac{1}{2})^{\underline{4}}}{4!} = \frac{(-\frac{1}{2})(-\frac{3}{2})(-\frac{5}{2})(-\frac{7}{2})}{24} = \frac{35}{128}$

1.5.14:   To compare $\log(n^2)$ to

$$\frac{\log((n-1)^2) + \log((n+1)^2)}{2}$$

directly from the definition, we multiply both sides by 2 and exponentiate. These operations are both monotonically increasing. Since

$$n^2 > (n-1)(n+1)$$

we conclude that n^2 is log-concave. Equivalently, with less effort, we could use inequalities (1.5.3) and (1.5.4).

1.6   Stirling Numbers: A Preview

1.6.1: $x^{\underline{6}} = x^6 - 15x^5 + 85x^4 - 225x^3 + 274x^2 - 120x$

1.6.5: $x^6 = x^{\underline{6}} + 15x^{\underline{5}} + 65x^{\underline{4}} + 90x^{\underline{3}} + 31x^{\underline{2}} + x^{\underline{1}}$

1.6.11: $\displaystyle \binom{7}{1\ 2\ 4} = 105$

1.7   Ordinary Generating Functions

1.7.1 : $(1 + b)(1 + d)(1 + n + n^2)(1 + a + a^2 + a^3)$

$$\longrightarrow (1 + z)^2(1 + z + z^2)(1 + z + z^2 + z^3)$$
$$= 1 + 4z + 8z^2 + 11z^3 + 11z^4 + 8z^5 + 4z^6 + z^7$$

1.7.7 : $(1 + b)(1 + d)\left(1 + n + \dfrac{n^2}{2!}\right)\left(1 + a + \dfrac{a^2}{2!} + \dfrac{a^3}{3!}\right)$

$$\longrightarrow (1 + z)^2\left(1 + z + \dfrac{z^2}{2!}\right)\left(1 + z + \dfrac{z^2}{2!} + \dfrac{z^3}{3!}\right)$$
$$= 1 + 4\dfrac{z}{1!} + 14\dfrac{z^2}{2!} + 43\dfrac{z^3}{3!} + 114\dfrac{z^4}{4!} + 250\dfrac{z^5}{5!} + 420\dfrac{z^6}{6!} + 420\dfrac{z^7}{7!}$$

1.7.13: $1 + 5z + 19z^2 + 65z^3 + 211z^4 + \cdots$

1.7.21:   (a) $1 + z + z^2 + \cdots$ (b) $\left.\dfrac{n!}{(1 - z)^{n+1}}\right|_{z=0}\dfrac{z^n}{n!} = z^n$

1.8   Synthesizing Generating Functions

1.8.1: $\dfrac{1}{1 + z}$

1.8.9: $\dfrac{1}{(1 - z)^2}$

1.8.19:   The third difference row begins with 1, 2, 4, ... Therefore, try subtracting the sequence $\langle 2^n \rangle$. The resulting sequence is 0, 1, 4, 9, 16, 125, ..., apparently n^2. The sequence $n^2 + 2^n$ perfectly matches the given sequence.

1.9   Asymptotic Estimates

1.9.1:   Given the polynomial $f(n) = a_{d-1}n^{d-1} + \cdots + a_1 n + a_0$, let M be a number larger than absolute value of any of the coefficients a_j. Then, for $j = 0, \ldots, d - 1$ and $n \geq 1$, we have $a_j n^j \leq Mn^d$. Therefore, for $n \geq 1$,

$$f(n) = \sum_{j=0}^{d-1} a_j n^j \leq \sum_{j=0}^{d-1} Mn^d = Mdn^d$$

Chapter 2 Solving Recurrences

2.1 Types of Recurrences

2.1.1: $a_n = 2a_{n-1} - 1;\ a_0 = 3 \Rightarrow a_1 = 5 \quad a_2 = 9 \quad a_3 = 17 \quad a_4 = 33 \quad a_5 = 65$
Guess: $a_n = 2^{n+1} + 1$. Now prove using induction.
Basis: true for $n = 0$. Ind hyp: assume true for $n - 1$, with $n \geq 1$.
Ind step: $a_n = 2a_{n-1} - 1$ (given recursion)
$\quad = 2 \cdot (2^n + 1) - 1$ (by ind hyp) $= 2^{n+1} + 1.$

2.1.5: $a_n = c_1 a_{n-1} + c_2 a_{n-2} \Rightarrow 3^n - 2^n = c_1(3^{n-1} - 2^{n-1}) + c_2(3^{n-2} - 2^{n-2})$
Therefore, $3^n = c_1 3^{n-1} + c_2 3^{n-2}$ and $2^n = c_1 2^{n-1} + c_2 2^{n-2}$. Simplify to the equations

$$3^2 = 3c_1 + c_2 \quad \text{and} \quad 2^2 = 2c_1 + c_2$$

and solve them:

$$c_1 = 5 \text{ and } c_2 = -6$$

Thus, the recurrence is $a_n = 5a_{n-1} - 6a_{n-2}.$

2.1.15: $t_0 = 1;\ t_1 = 3;\ t_n = 2t_{n-1}$ for $n \geq 2$

2.1.18: Ans. $s_0 = 1;\ s_1 = 1;\ s_n = s_{n-1} - s_{n-2} + f_{n-1}$ for $n \geq 2$, where f_n is the nth Fibonacci number.

2.2 Finding Generating Functions

2.2.1: $A(z) = \dfrac{3}{1 - 2z}$

2.2.9: $A(z) = \dfrac{z + 2z^2}{(1 + z)^2(1 - 8z)}$

2.2.16: Clearly $t_0 = 0$. Suppose it takes t_{n-1} moves to transfer $n - 1$ disks from source to target. To get n disks from source to target, first $n - 1$ disks must be transferred to the vertex of degree 1 that is neither the source nor the target, which requires t_{n-1} steps. Then it takes 2 steps to transfer the largest disk to the target, followed by another t_{n-1} steps to transfer the stack of $n - 1$ disks to the target. Thus, the recursion is $t_n = 2t_{n-1} + 2$. The solution is $t_n = 2^{n+1} - 2$.

2.3 Partial Fractions

2.3.1 : $$A(z) = \frac{3}{1 - 2z} \Rightarrow a_n = 3 \cdot 2^n$$

2.3.9 : $$A(z) = \frac{-(10 + 19z)/81}{(1 + z)^2} + \frac{10/81}{1 - 8z}$$

$$\Rightarrow a_n = \frac{1}{81}\left(10 \cdot 8^n + (9n - 10)(-1)^n\right)$$

2.4 Characteristic Roots

2.4.1: Root is $\tau = 2$.
Solution: $a_n = 3 \cdot 2^n$

2.4.9: Roots are $\tau_1 = -1$ and $\tau_2 = 8$.
Solution: $a_n = \dfrac{1}{81}\left(10 \cdot 8^n + (9n - 10)(-1)^n\right)$

2.4.15: Roots are $\tau_1 = 1 + i$ and $\tau_2 = 1 - i$.
Solution: $a_n = \dfrac{1 - i}{2}(1 + i)^n + \dfrac{1 + i}{2}(1 - i)^n$

2.5 Simultaneous Recursions

2.5.2: $x_n = \dfrac{1}{4}\left[-(-1)^n + 3^n\right] \quad y_n = \dfrac{1}{2}\left[(-1)^n + 3^n\right]$

2.5.7: $L_n: \quad 2 \quad 1 \quad 3 \quad 4 \quad 7 \quad 11 \quad 18 \quad 29 \quad 47 \quad 76 \quad \ldots$

2.5.9: $r_n = f_{n+1}$, for $n \geq 0$

2.5.14: The queen numbers are $q_n = f_n$. The drone numbers are $d_n = f_{n-1}$, for $n \geq 1$; this even makes sense at $n = 0$, since running the Fibonacci recurrence backwards yields $f_{-1} = f_1 - f_0 = 1 - 0 = 1$.

2.6 Fibonacci Number Identities

2.6.5: $202 = 144 + 55 + 3$

2.6.9: $f_n^2 = (f_{n-1} + f_{n-2})^2 = f_{n-1}^2 + 2f_{n-1}f_{n-2} + f_{n-2}^2$, etc.

2.6.16: One method of proof is first to observe that $f_{2n+1} = f_{(n+1)+n}$ and then to apply the forward-shift identity. Another is to observe that $f_{2n+1} = t_{2n}$, and that a $1 \times 2n$ chessboard covered with 1-tiles and 2-tiles may either have a break between tiles at position n or not. If so, then that break separates two tile sequences of length n, each of which can be achieved in $t_n = f_{n+1}$ ways, for a total of $t_n^2 = f_{n+1}^2$ ways. If not, then a 2-tile at the middle is sandwiched between two tile sequences of length t_{n-1}, which can occur in $t_{n-1}^2 = f_n^2$ ways.

2.6.23: $f_{2n+1} - 1 = f_2 + f_4 + \cdots + f_{2n}$

2.7 Non-Constant Coefficients

2.7.1: Multiplying by n yields the recursion $nx_n = 3(n-1)x_{n-1} - 2(n-2)x_{n-2}$. Substituting $y_n = nx_n$ now yields $y_n = 3y_{n-1} - 2y_{n-2}$, with $y_0 = 0$ and $y_1 = 3$, with the solution $y_n = 3 \cdot 2^n - 3$. Thus, $x_n = (3 \cdot 2^n - 3)/n$, for $n > 0$.

2.8   Divide-and-Conquer Relations

2.8.1: $(lo, hi) = (1, 7) \to (4, 7) \to (4, 5) \to (5, 5)$

2.8.5 :
$$\begin{array}{l}
[\,(92)\ \ (56)\ \ (83)\ \ (97)\ \ (72)\ \ (78)\ \ (15)\,] \\
[\,(56\ \ 92)\ \ (83\ \ 97)\ \ (72\ \ 78)\ \ (15)\,] \\
[\,(56\ \ 83\ \ 92\ \ 97)\ \ (15\ \ 72\ \ 78)\,] \\
[\,(15\ \ 56\ \ 72\ \ 78\ \ 83\ \ 92\ \ 97)\,]
\end{array}$$

2.8.9: $t_n = 3cn^{\lg 3} - 2cn$

2.8.15:   Select $n!$. Actually, it is sufficient to select the least common multiple of the numbers 1 to n.

Chapter 3   Evaluating Sums

3.1   Normalizing Summation

3.1.1:
$$\sum_{7 \le k^2 \le 45} \frac{1}{k} = \sum_{k=\lceil \sqrt{7} \rceil}^{\lfloor \sqrt{45} \rfloor} \frac{1}{k} = \sum_{k=3}^{6} \frac{1}{k} = \sum_{k=0}^{3} \frac{1}{k+3}$$

3.1.7:   Proposition 3.1.4(a) asserts that $\ln(n+1) < H_n < \ln(n) + 1$. Therefore,

$$\ln(n) < H_n < \ln(n) + 1$$

Divide through by $\ln n$ to obtain the desired result.

3.2   Perturbation

3.2.1 :
$$S = \sum_{k=0}^{n} 3^k$$

$$\Rightarrow \quad S + 3^{n+1} = 3^0 + \sum_{k=1}^{n+1} 3^k = 3^0 + 3\sum_{k=0}^{n} 3^k = 3^0 + 3S$$

$$\Rightarrow \quad S = \frac{3^{n+1} - 3^0}{2}$$

3.2.15:   Perturbing $\displaystyle\sum_{k=0}^{n} k^4$ yields $\displaystyle\sum_{k=0}^{n} k^3 = \frac{1}{4}\left[n^4 + 2n^3 + n^2\right]$.

3.3 Summing with Generating Functions

3.3.**1** : (a) $\dfrac{1}{1-3z} = \displaystyle\sum_{n=0}^{\infty} 3^n z^n$

(b) $\Rightarrow$ $\dfrac{1}{1-z}\cdot\dfrac{1}{1-3z} = \displaystyle\sum_{n=0}^{\infty}\left(\sum_{k=0}^{n} 3^k\right) z^n$

(c) $\dfrac{1}{1-z}\cdot\dfrac{1}{1-3z} = \dfrac{3/2}{1-3z} + \dfrac{-1/2}{1-z}$

(d) $\Rightarrow$ $\displaystyle\sum_{k=0}^{n} 3^k = \dfrac{3}{2}\cdot 3^n - \dfrac{1}{2}\cdot 1^n$

3.3.**7** : (a) $\dfrac{1}{1-z/3} = \displaystyle\sum_{n=0}^{\infty}\dfrac{1}{3^n} z^n$

(b) $\Rightarrow$ $\dfrac{1}{1-z}\cdot\dfrac{1}{1-z/3} = \displaystyle\sum_{n=0}^{\infty}\left(\sum_{k=0}^{n}\dfrac{1}{3^k}\right) z^n$

(c) $\dfrac{1}{1-z}\cdot\dfrac{1}{1-z/3} = \dfrac{-1/2}{1-z/3} + \dfrac{3/2}{1-z}$

(d) $\Rightarrow$ $\displaystyle\sum_{k=0}^{n}\dfrac{1}{3^k} = \dfrac{-1}{2}\cdot\dfrac{1}{3^n} + \dfrac{3}{2}\cdot 1^n$

3.3.**13** : (a) $\dfrac{1}{1-z\sqrt{2}} = \displaystyle\sum_{n=0}^{\infty}\sqrt{2}^{\,n} z^n$

(b) $\Rightarrow$ $\dfrac{1}{1-z}\cdot\dfrac{1}{1-z\sqrt{2}} = \displaystyle\sum_{n=0}^{\infty}\left(\sum_{k=0}^{n}\sqrt{2}^{\,k}\right) z^n$

(c) $\dfrac{1}{1-z}\cdot\dfrac{1}{1-z\sqrt{2}} = \dfrac{2+\sqrt{2}}{1-z\sqrt{2}} - \dfrac{1+\sqrt{2}}{1-z}$

(d) $\Rightarrow$ $\displaystyle\sum_{k=0}^{n}\sqrt{2}^{\,k} = (2+\sqrt{2})\cdot\sqrt{2}^{\,n} - (1+\sqrt{2})\cdot 1^n$

3.3.**17** : (a) $\dfrac{1}{(1-3z)^2} = \displaystyle\sum_{n=0}^{\infty}\binom{n+1}{1} 3^n z^n$

(b) $\Rightarrow$ $\dfrac{1}{1-z}\cdot\dfrac{1}{(1-3z)^2} = \displaystyle\sum_{n=0}^{\infty}\left(\sum_{k=0}^{n}\binom{k+1}{1} 3^k\right) z^n$

(c) $\dfrac{1}{1-z}\cdot\dfrac{1}{(1-3z)^2} = \dfrac{1/4}{1-z} + \dfrac{-3/4}{1-3z} + \dfrac{3/2}{(1-3z)^2}$

(d) $\Rightarrow$ $\displaystyle\sum_{k=0}^{n}\binom{k+1}{1} 3^k = \dfrac{1}{4} - \dfrac{3}{4}\cdot 3^n + \dfrac{3}{2}\cdot\binom{n+1}{1}\cdot 3^n$

3.4   Finite Calculus

3.4.1 :
$$3^k = \triangle\left(\frac{3^k}{2}\right)$$

$$\Rightarrow \sum_{k=0}^{n} 3^k = \frac{3^k}{2}\Big|_{k=0}^{n+1} = \frac{3^{n+1}}{2} - \frac{1}{2} = \frac{3^{n+1} - 1}{2}$$

3.4.4:   To sum $k4^k$ by parts, use $g(k) = k$ and $\triangle f(k) = 4^k$. Then

$$\sum_{k=0}^{n} k4^k = k\frac{4^k}{3}\Big|_{k=0}^{n+1} - \sum_{k=0}^{n}\frac{4^{k+1}}{3} = k\frac{4^k}{3}\Big|_{k=0}^{n+1} - \frac{4^{k+1}}{9}\Big|_{k=0}^{n+1}$$

$$= \left((n+1)\frac{4^{n+1}}{3} - 0\right) - \left(\frac{4^{n+2}}{9} - \frac{4}{9}\right)$$

3.4.19:   To sum kH_k by parts, use $g(k) = H_k$ and $\triangle f(k) = k$. Then

$$\sum_{k=0}^{n} kH_k = \frac{k^{\underline{2}}}{2}H_k\Big|_{k=0}^{n+1} - \sum_{k=0}^{n}\frac{(k+1)^{\underline{2}}}{2}\frac{1}{k+1}$$

$$= \frac{k^{\underline{2}}}{2}H_k\Big|_{k=0}^{n+1} - \sum_{k=0}^{n}\frac{k}{2} = \frac{k^{\underline{2}}}{2}H_k\Big|_{k=0}^{n+1} - \frac{k^{\underline{2}}}{4}\Big|_{k=0}^{n+1}$$

$$= \left(\frac{n^{\underline{n}}+n}{2}H_{n+1} - 0\right) - \left(\frac{n^2+n}{4} - 0\right) = \frac{n^{\underline{n}}+n}{2}H_{n+1} - \frac{n^2+n}{4}$$

3.4.24: $\triangle k^{\underline{-2}} = -2k^{\underline{-3}}$

3.4.30: $-k^{\underline{-1}}$

3.4.33:   (a) 103, 140.   (b) $3n^2 - 2n + 7$

3.5   Iteration and Partitioning of Sums

3.5.5: $\dfrac{(n+1)^{\underline{3}}}{2} + \dfrac{3}{2}(n+1)^{\underline{2}}$

3.5.14: $n + (n \equiv 2 \bmod 3)$

3.6   Inclusion-Exclusion

3.6.1: $\phi(48) = 48 - 24 - 16 + 8 = 16.$

3.6.7 : $2!\begin{Bmatrix}5\\2\end{Bmatrix} = 2^5 - \binom{2}{1}\cdot 1^5 + \binom{2}{2}\cdot 0^5$

$$= 2^5 - 2 + 0 = 30$$

$$\Rightarrow \begin{Bmatrix}5\\2\end{Bmatrix} = 15$$

3.6.11: $\quad \dfrac{6! \, \left\{ {8 \atop 6} \right\}}{8^6}$

3.6.16: $\quad \displaystyle\sum_{k=0}^{n} \binom{n}{k} (2n - k)!(-1)^k$

3.6.17: $\quad \dbinom{n}{k} D_{n-k}$

3.6.18 : $\quad |A_1| = 0 \quad |A_2| = |A_3| = 1 \cdot 2! = 2 \quad \Rightarrow \quad S_1 = 4$

$\qquad\qquad |A_2 A_3| = 1 \quad |A_1 A_2| = |A_1 A_3| = 0 \quad \Rightarrow \quad S_2 = 1$

$\qquad\qquad\qquad\qquad\qquad |A_1 A_2 A_3| = 0 \quad \Rightarrow \quad S_3 = 0$

$\qquad\qquad\qquad\qquad\qquad\qquad |U| = 3! = 6$

$\quad \Rightarrow \quad |\overline{A_1}\,\overline{A_2}\,\overline{A_3}| \;=\; |U| - S_1 + S_2 - S_3 \;=\; 6 - 4 + 1 - 0 \;=\; 3$

3.6.22: $\quad |U| = t^4 \quad S_1 = 4t^3 \quad S_2 = 6t^2 \quad S_3 = t^2 + 3t \quad S_4 = t \,.$

$\quad \Rightarrow \quad P(G,t) \;=\; t^4 - 4t^3 + 6t^2 - (t^2 + 3t) + t$

$\qquad\qquad\qquad\quad = \; t^4 - 4t^3 + 5t^2 - 2t$

Chapter 4 Binomial Coefficients

4.1 Binomial Coefficient Identities

4.1.4 : $\quad \dbinom{-\frac{1}{2}}{3} \;=\; \dfrac{\left(-\frac{1}{2}\right)^{\underline{3}}}{3!} \;=\; \dfrac{\left(-\frac{1}{2}\right)\left(-\frac{3}{2}\right)\left(-\frac{5}{2}\right)}{3!} \;=\; -\dfrac{5}{16}$

4.1.7 : $\quad \dbinom{n+2}{n-2} \;=\; \dbinom{n+2}{4} \;=\; \dfrac{(n+2)(n+1)\, n\, (n-1)}{4!}$

$\qquad\qquad\qquad\qquad\qquad\qquad = \; \dfrac{n^4 + 2n^3 - n^2 - 2n}{4!}$

4.1.9 : $\quad (n-r)\dbinom{n}{r} \;=\; (n-r)\dbinom{n}{n-r} \qquad \text{(symmetry)}$

$\qquad\qquad\qquad\qquad = \; n \dbinom{n-1}{n-r-1} \qquad \text{(absorption)}$

$\qquad\qquad\qquad\qquad = \; n \dbinom{n-1}{r} \qquad\qquad \text{(symmetry)}$

4.1.17:   For $n = 0$ and $n = 1$, the equation is clearly correct. For $n \geq 2$,

$$\sum_{k=0}^{n} (-1)^k \, k \, \binom{n}{k} \;=\; \sum_{k=1}^{n} (-1)^k \, n \, \binom{n-1}{k-1} \qquad \text{(absorption)}$$

$$= \; -n \sum_{k=1}^{n} (-1)^{k-1} \binom{n-1}{k-1}$$

$$= \; -n \, (x-1)^{n-1} \big|_{x=1} \; = \; 0$$

4.1.22 : $\displaystyle \binom{x}{r} \, r \;=\; \frac{x^{\underline{r}}}{r!} \, r \;=\; \frac{x \cdot (x-1)^{\underline{r-1}}}{(r-1)!} \;=\; x \binom{x-1}{r-1}$

4.1.24:   First method: iteratively reduce upper and lower index.

$$\binom{80}{48} \equiv \binom{40}{24} \equiv \binom{20}{12} \equiv \binom{10}{6} \equiv \binom{5}{3} \equiv \binom{2}{1} \equiv 0 \bmod 2$$

Second method: align binary numerals flush right; scan for 0 over 1.

$$\overset{*}{}$$
$$80_{10} \; = \; 1\,0\,1\,0\,0\,0\,0_2 \quad \Rightarrow \quad \text{even}$$
$$48_{10} \; = \; 0\,1\,1\,0\,0\,0\,0_2$$

4.2   Binomial Inversion Operation

4.2.5 : $\displaystyle g_n \;=\; \binom{n}{2} b \;-\; \binom{n}{1} a$

4.2.12: $g_n \;=\; (-2)^n$

4.2.13: $g_2 \;=\; 2$, and $g_j \;=\; 0$ for $j \neq 2$.

4.3   Application to Statistics

4.3.4 : $\displaystyle \Pr(X = j) \;=\; \frac{\binom{M}{j} \binom{N-M}{n-j}}{\binom{N}{n}}$

4.3.5 : $\displaystyle E(X) \;=\; n \cdot \frac{M}{N}$

4.3.7 : $\displaystyle V(X) \;=\; \left(\frac{N-n}{N-1} \right) \cdot n \cdot \frac{M}{N} \cdot \left(1 - \frac{M}{N} \right)$

4.3.11 : $\displaystyle \Pr(X = j) \;=\; \binom{j+r-1}{r-1} p^r \, (1-p)^j$

4.3.13 : $\qquad E(X) = \dfrac{r(1-p)}{p}$

4.3.14 : $\qquad V(X) = \dfrac{r(1-p)}{p^2}$

4.3.18 : $\qquad E(X) = \lambda$

4.3.20 : $\qquad V(X) = \lambda$

4.4 The Catalan Recurrence

4.4.1: This is Proposition 4.4.3.

4.4.4: $q(n, n-2) = c_n - c_{n-1}$.

4.4.9: There are $\binom{2n}{n-2}$ paths in $[0 : n+2] \times [0 : n-2]$. By reflection of suffixes, this provides the correct count.

4.4.15: There are

$$\binom{2n}{n-k-1}$$

NE-paths that enter the NW k-triangle, and equally many that enter the SE k-triangle. No two paths enter both extreme triangles, since $k \geq n/2$. Thus, the total number of paths that enter either of them is

$$2\binom{2n}{n-k-1}$$

Chapter 5   Partitions and Permutations

5.1   Stirling Subset Numbers

5.1.2: $\displaystyle \left\{ {n \atop k} \right\} + \left\{ {n \atop k-1} \right\} + \cdots + \left\{ {n \atop k-r} \right\}$

5.1.5: $\displaystyle k! \left\{ {n \atop k} \right\} + (k-1)! \left\{ {n \atop k-1} \right\} + \cdots + (k-r)! \left\{ {n \atop k-r} \right\}$

5.1.9: $\displaystyle \left\{ {6 \atop 4} \right\} + 5 \left\{ {6 \atop 5} \right\} = 65 + 5 \cdot 15 = 140 = \left\{ {7 \atop 5} \right\}$

5.1.17 : $\displaystyle \binom{6}{4}\left\{ {4 \atop 4} \right\} + \binom{6}{5}\left\{ {5 \atop 4} \right\} + \binom{6}{6}\left\{ {6 \atop 4} \right\} = 15 \cdot 1 + 6 \cdot 10 + 1 \cdot 65$

$$= 140 = \left\{ {7 \atop 5} \right\}$$

5.1.25: $\quad 5^2 \begin{Bmatrix} 4 \\ 4 \end{Bmatrix} + 5^1 \begin{Bmatrix} 5 \\ 4 \end{Bmatrix} + 5^0 \begin{Bmatrix} 6 \\ 4 \end{Bmatrix} = 25 \cdot 1 + 5 \cdot 10 + 1 \cdot 65 = 140 = \begin{Bmatrix} 7 \\ 5 \end{Bmatrix}$

5.1.33: $\quad 0 \begin{Bmatrix} 1 \\ 0 \end{Bmatrix} + 1 \begin{Bmatrix} 2 \\ 1 \end{Bmatrix} + 2 \begin{Bmatrix} 3 \\ 2 \end{Bmatrix} + 3 \begin{Bmatrix} 4 \\ 3 \end{Bmatrix} + 4 \begin{Bmatrix} 5 \\ 4 \end{Bmatrix} + 5 \begin{Bmatrix} 6 \\ 5 \end{Bmatrix}$

$$= 0 \cdot 0 + 1 \cdot 1 + 2 \cdot 3 + 3 \cdot 6 + 4 \cdot 10 + 5 \cdot 15 = 140 = \begin{Bmatrix} 7 \\ 5 \end{Bmatrix}$$

5.2 Stirling Cycle Numbers

5.2.3: $\quad \begin{bmatrix} 6 \\ 4 \end{bmatrix} + 6 \begin{bmatrix} 6 \\ 5 \end{bmatrix} = 85 + 6 \cdot 15 = 175 = \begin{bmatrix} 7 \\ 5 \end{bmatrix}$

5.2.16: $\quad \binom{4}{4} \begin{bmatrix} 6 \\ 4 \end{bmatrix} + \binom{5}{4} \begin{bmatrix} 6 \\ 5 \end{bmatrix} + \binom{6}{4} \begin{bmatrix} 6 \\ 6 \end{bmatrix} = 1 \cdot 85 + 5 \cdot 15 + 15 \cdot 1 = 175 = \begin{bmatrix} 7 \\ 5 \end{bmatrix}$

5.2.19: $\quad 6^2 \begin{bmatrix} 4 \\ 4 \end{bmatrix} + 6^1 \begin{bmatrix} 5 \\ 4 \end{bmatrix} + 6^0 \begin{bmatrix} 6 \\ 4 \end{bmatrix} = 30 \cdot 1 + 6 \cdot 10 + 1 \cdot 85 = 175 = \begin{bmatrix} 7 \\ 5 \end{bmatrix}$

5.2.27: $\quad 0 \begin{bmatrix} 1 \\ 0 \end{bmatrix} + 1 \begin{bmatrix} 2 \\ 1 \end{bmatrix} + 2 \begin{bmatrix} 3 \\ 2 \end{bmatrix} + 3 \begin{bmatrix} 4 \\ 3 \end{bmatrix} + 4 \begin{bmatrix} 5 \\ 4 \end{bmatrix} + 5 \begin{bmatrix} 6 \\ 5 \end{bmatrix}$

$$= 1 \cdot 0 + 2 \cdot 1 + 3 \cdot 3 + 4 \cdot 6 + 5 \cdot 10 + 6 \cdot 15 = 175 = \begin{bmatrix} 7 \\ 5 \end{bmatrix}$$

5.3 Inversions and Ascents

5.3.1: $\quad 3124 \quad 1423 \quad 2143 \quad 2314 \quad 1342$

5.3.13: $\quad I_7(8) = 455$

5.3.17: $\quad 532164$

5.3.22: $\quad 3 \left\langle \begin{matrix} 6 \\ 2 \end{matrix} \right\rangle + 5 \left\langle \begin{matrix} 6 \\ 1 \end{matrix} \right\rangle = 3 \cdot 302 + 5 \cdot 57$

$$= 1191 = \left\langle \begin{matrix} 7 \\ 2 \end{matrix} \right\rangle$$

5.5 Exponential Generating Functions

5.5.1: ze^z

5.5.5: $(e^z - 1 - z)^2 \; = \; e^{2z} - 2ze^z - 2e^z + z^2 + 1 + 2z$

5.5.9: $u_n \; = \; \begin{cases} 0 & \text{if } 0 \le n \le 3 \\ 2^n - 2n - 2 & \text{if } n \ge 4 \end{cases}$

5.5.13: $a_n \; = \; n \cdot n!$

5.5.15: $a_n \; = \; \dfrac{1}{n!} \sum_{j=0}^{n} n^{\underline{j}}$

5.6 Posets and Lattices

5.6.4: Maximum chain: $\emptyset \subset 1 \subset 12 \subset 123$

5.6.10: Maximum anti-chain: 1, 2, 3

5.6.16: Partition into minimum number of chains:

(1) $\emptyset \subset 1 \subset 12 \subset 123$;   (2) $2 \subset 23$;   (3) $3 \subset 13$

5.6.22: Partition into minimum number of anti-chains:

(1) $\emptyset$;   (2) 1, 2, 3;   (3) 12, 13, 23;   (4) 123

5.6.28: There are 48 linear extensions. $\emptyset$ must be first and 123 last. If the three singletons 1,2,3 precede the three doubletons 12, 13, 23, then there are 3! possible ordering on the singletons and 3! of the doubletons, for a total of 36. There are also 12 linear extensions in which some doubleton precedes the other two doubletons.

Chapter 6 Integer Operators

6.1 Euclidean Algorithm

6.1.1 : $89 \bmod 71 \; = \; 18$

$71 \bmod 18 \; = \; 17$

$18 \bmod 17 \; = \; 1$

$17 \bmod 1 \; = \; 0 \qquad gcd\,(89, 71) \; = \; 1$

6.1.9: $4 \cdot 89 + (-5) \cdot 71 = 1$ derived as follows:

j	n_j	m_j	q_j	
0	89	71	1	$4 \cdot 89 - 5 \cdot 71$
1	71	18	3	$(-1) \cdot 71 + 1 \cdot 18 = (-1) \cdot 71 + 4 \cdot (89 - 1 \cdot 71)$
2	18	17	1	$1 \cdot 18 - 1 \cdot 17 = 1 \cdot 18 - 1 \cdot (71 - 3 \cdot 18)$
3	17	1	17	$1 \cdot 1 + 0 \cdot 0 = 1 \cdot (18 - 1 \cdot 17)$
4	1	0	$STOP$	

6.2   Chinese Remainder Theorem

6.2.1: $(6 \bmod 8, 3 \bmod 9) + (7 \bmod 8, 5 \bmod 9) = (5 \bmod 8, \ 8 \bmod 9)$

6.2.5: $29 \bmod (5, 7) = (4 \bmod 5, 1 \bmod 7)$

6.2.11: $(6 \bmod 8, 3 \bmod 9) = 30$, derived as follows:
$$(-1) \cdot 8 + 1 \cdot 9 = 1$$
$$3 \cdot (-1) \cdot 8 + 6 \cdot 1 \cdot 9 = 30$$

6.2.20:   38

6.3   Polynomial Divisibility

6.3.1: $x^2 - 4x + 2$ is prime, since the roots are irrational.

6.3.9: $x^2 + 3x + 7 \bmod x - 2 = 17$

6.3.19: $gcd\left(x^3 - 6x^2 + 11x - 6, \ x^2 - 3x + 2\right) = x^2 - 3x + 2$

6.4   Prime and Composite Moduli

6.4.3:   The inverse of 21 mod 25 is 6.

6.4.11: $221^{64} \bmod 25 = 6$

6.4.17: $4! \bmod 5 = -1$, since 5 is prime.

6.4.25:   The quadratic residues of 14 are
$$1 = 1^2, \ 9 = 3^2, \ 11 = 5^2$$

6.4.31:   The solutions to $x^2 \equiv 1(\text{ modulo } 24)$ are 1, 5, 7, 11, 13, 17, 19, and 23.

6.5   Euler Phi-Function

6.5.5: $\phi(30) = 1 \cdot 2 \cdot 4 = 8$

6.5.11:   Since $\phi(25) = 20$, we have $221^{64} \bmod 25 = 21^{64} = 21^4 = 6$.

6.5.19 : $\phi(1) + \phi(2) + \phi(3) + \phi(5) + \phi(6) + \phi(10) + \phi(15) + \phi(30) =$
$$1 + 1 + 2 + 4 + 2 + 4 + 8 + 8 = 30$$

6.5.25: The set-difference $F_n - F_{n-1}$ is the set of reduced fractions $\frac{r}{n}$. The number of such fractions is $\phi(n)$.

6.6   The Möbius Function

6.6.3: $\mu(18) = 0$

6.6.11: $\sigma(n) = \displaystyle\sum_{d \setminus n} \iota(d)$

Chapter 7   Graph Fundamentals

7.1   Regular Graphs

7.1.1: 16 vertices, 32 edges.

7.1.7: The complete graph K_3.

7.1.12: $circ\,(2n : 1,\ 3,\ 5,\ \cdots,\ 2n - 1)$

7.2   Walks and Distance

7.2.1: radius $= 3$ and diameter $= 3$.

7.2.5: Color the top two vertices white, the next three black, the next three white, and the bottom two black. The color classes are a bipartition.

7.2.9: It is easy enough to find a cycle of length 4. Since the graph is bipartite, it has no odd cycles, and, thus, no cycle of length 3. Thus, the girth is 4.

7.2.15: The two vertices inside the outer octagon have eccentricity 2 and they are the central vertices.

7.2.17: radius $= 2$ and diameter $= 2$.

7.3   Trees and Acyclic Digraphs

7.3.6: The Catalan number c_n.

7.3.9: Label the vertices with consecutive integers. Then direct each edge from its lower-numbered endpoint to the higher-numbered endpoint.

7.3.14: We may interpret it as reflexive. It is obviously anti-symmetric. It is transitive by hypothesis. Thus, the vertices are partially ordered. Since the underlying graph is complete, it follows that every pair of elements is comparable, which implies that the ordering is complete.

7.4 Graph Isomorphism

7.4.1: A has two vertices of eccentricity 2, and B has five.

7.4.7: A is bipartite (thus, no odd cycles), and B has two 3-cycles.

7.4.17: Due to Euler's theorem on degree-sum and Proposition 7.2.5, the plausible degree sequences are 22211, 32111, and 41111. Each corresponds to exactly one tree, as shown:

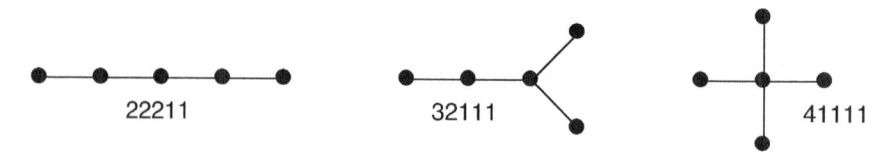

7.5 Graph Automorphism

7.5.1: There are two automorphisms: $(1)(2)(3)(4)$ and $(1)(2)(3\,4)$.

7.5.7: There are four automorphisms:
$$(1)(2)(3)(4)(5)(6), \quad (1\,5)(3)(4)(2\,6)$$
$$(1,\,2)(3,\,4)(5,\,6), \quad (1,\,6)(3,\,4)(2,\,5)$$

7.5.13: The vertex orbits are $\{1\}$, $\{2\}$, and $\{3,\,4\}$.

7.5.19: The vertex orbits are $\{1,\,2,\,5,\,6\}$ and $\{3,\,4\}$.

7.6 Subgraphs

7.6.2: Hamiltonian cycle in the wheel graph W_7.

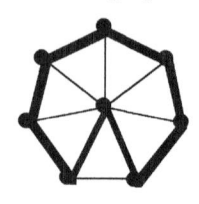

7.6.6: In any cycle in a bipartite graph, the vertices must alternate between one partite set and the other. Thus, the number of vertices in a cycle in $K_{3,5}$ is at most 6.

7.6.13: The two vertices inside the outer octagon have eccentricity 2 and they are the central vertices. The edge joining them is the only edge in the center.

7.7 Spanning Trees

7.7.1: The frontier edges are sp, qp, ut, uv, and ux.

7.7.5:

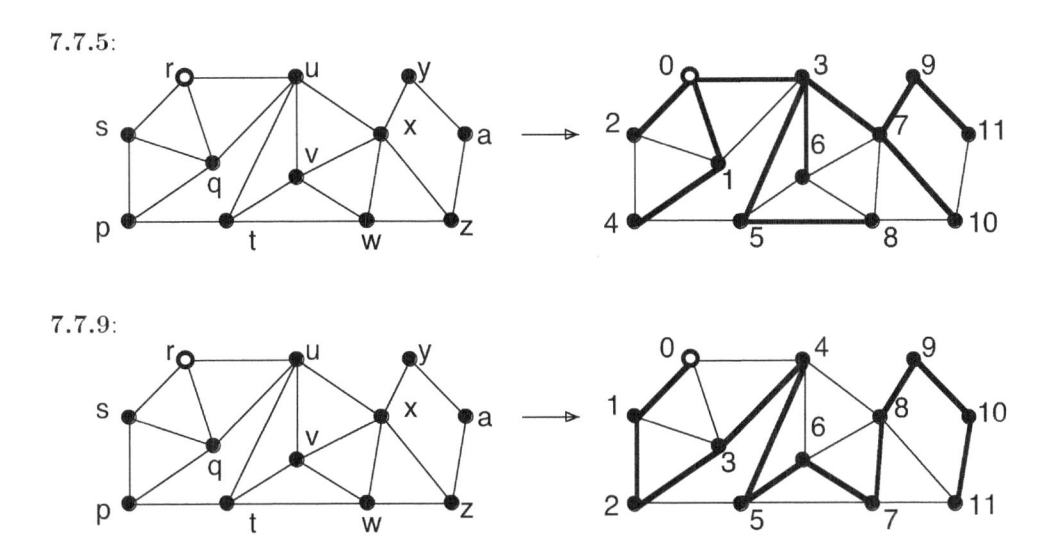

7.7.9:

7.8 Edge Weights

7.8.1:

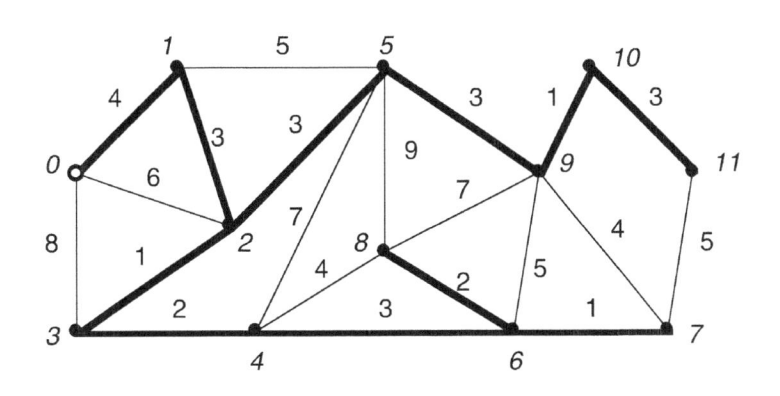

7.8.4:

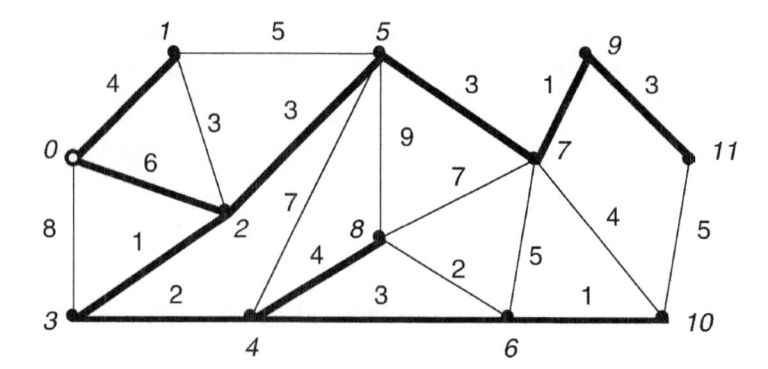

7.8.9:   Delete edge ux and run Dijkstra's algorithm on the resulting graph to get a shortest path in it between vertices u and x.

7.9   Graph Operations

7.9.1:   The degree sequence is 522221. When the vertex of degree 5 is deleted, the result is the sixth card in the given deck.

7.9.16:   The cut-vertices are vertices 3, 4, and 7.

7.9.28:   The cut-edges are 35, 47, and 67.

7.9.32:   We observe that this graph is isomorphic to its edge-complement.

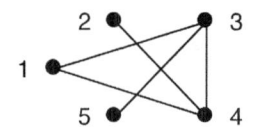

Chapter 8   Graph Theory Topics

8.1   Traversability

8.1.1: $a\ b\ f\ h\ d\ e\ g\ c$.

8.1.5:   Select a matching of minimum weight.

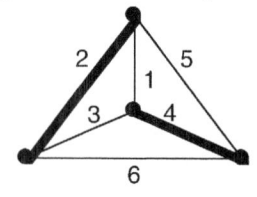

8.1.9:   The three vertices of degree 2 are mutually non-adjacent. Thus, they cannot lie on a 5-cycle.

8.2   Planarity

8.2.1:   7 vertices, 10 edges, 5 faces: $7 - 10 + 5 = 2$.

8.2.5:   The face-sizes are 34445: $3 + 4 + 4 + 4 + 5 = 2 \cdot 10$.

8.2.9: $|E| = 17 \not> 24 = 3|V| - 6$.

8.3   Coloring

8.3.1: $\omega(A) = 3$ $\alpha(A) = 3$.

8.3.7:   Four colors is a lower bound, by Proposition 8.3.2. The shape of a vertex is used in this drawing as another way of indicating color classes.

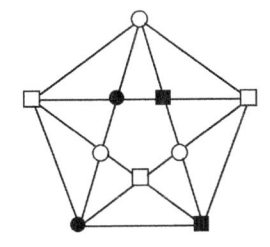

8.3.13:   It is not chromatically critical. Indeed, no matter what edge is removed, the independence number remains 3. Thus, by Proposition 8.3.2, the chromatic number remains at least 4.

8.3.19:   The chromatic number of Africa is 4, since (for instance) Malawi is the hub of a 3-wheel.

8.3.23:   The dual of a tetrahedron is a tetrahedron.

8.3.30:

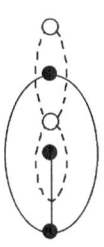

8.4   Analytic Graph Theory

8.4.1: $ex(n, \mathcal{P}) = \left\lfloor \dfrac{n(n-2)}{2} \right\rfloor$.

8.4.6: $r(3,6) \leq r(2,6) + r(3,5) - 1 = 6 + 14 - 1 = 19$.

8.5 Digraph Models

8.5.1: Start (for instance) with acd, then attach the closed path gh, then the open paths e and bf.

8.5.5: 43111.

8.5.11:

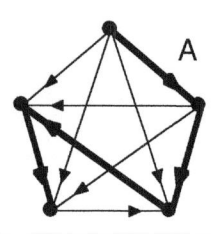

8.6 Network Flows

8.6.3:

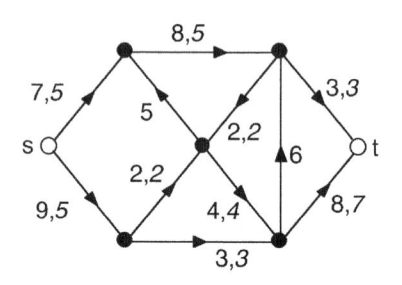

8.6.9:

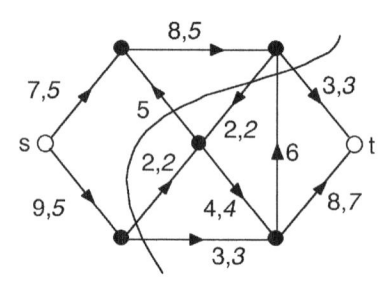

8.7 Topological Graph Theory

8.7.1:

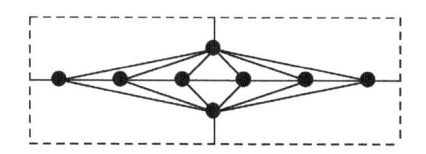

8.7.10: $|V| = 10$ and $|E| = 36$. Use Corollary 8.7.6.

8.7.16: $girth = 4$ and $\gamma_{\min}(Q_4) = n2^{n-3} - 2^{n-1} + 1$.

Chapter 9   Graph Enumeration

9.1   Burnside-Pólya Counting

9.1.2: $\alpha^3 = \begin{pmatrix} 1 & 4 & 7 & 3 & 6 & 2 & 5 \end{pmatrix}$

9.1.10: $\alpha^3 = \begin{pmatrix} 1 & 4 & 7 & 10 \end{pmatrix}\begin{pmatrix} 2 & 5 & 8 & 11 \end{pmatrix}\begin{pmatrix} 3 & 6 & 9 & 12 \end{pmatrix}$

9.1.18: $\mathscr{Z}_{\mathbb{Z}_6} : \dfrac{1}{6}\left(t_1^6 + t_2^3 + 2t_3^2 + 2t_6\right)$

9.1.26: 14

9.1.34: $b^6 + b^5w + 3b^4w^2 + 4b^3w^3 + 3b^2w^4 + bw^5 + b^6$

9.1.42: $\mathscr{Z}_{\mathbb{Z}_6} : \dfrac{1}{12}\left(t_1^6 + 3t_1^2t_2^2 + 4t_2^3 + 2t_3^2 + 2t_6\right)$

9.1.46: 13

9.1.50: $b^6 + b^5w + 3b^4w^2 + 3b^3w^3 + 3b^2w^4 + bw^5 + b^6$

9.2   Burnside's Lemma

9.2.1: 208

9.2.5: 136

9.2.10: $b^6 + b^5w + 3b^4w^2 + 4b^3w^3 + 3b^2w^4 + bw^5 + b^6$

9.2.14: $b^6 + b^5w + 3b^4w^2 + 3b^3w^3 + 3b^2w^4 + bw^5 + b^6$

9.2.17:

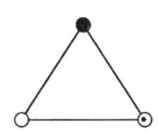 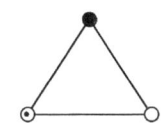

9.3   Counting Small Simple Graphs

9.3.3: $\mathscr{Z}_{\mathcal{A}ut_V(K_{2,3})} : \dfrac{1}{12}\left(t_1^5 + 4t_1^3t_2 + 2t_1^2t_3 + 3t_1t_2^2 + 2t_2t_3\right)$
$12\ (\leq 2)$-vertex-colorings.

9.3.11: $\mathscr{Z}_{\mathcal{A}ut_E(K_{2,3})} : \dfrac{1}{12}\left(t_1^6 + 3t_1^2t_2^2 + 4t_2^3 + 2t_3^2 + 2t_6\right)$
$13\ (\leq 2)$-edge-colorings.

9.3.17:   2 graphs with 5 vertices and 2 edges.

9.4 Partitions of Integers

9.4.1: 8, 71, 62, 53, 44, 611, 521, 431, 422, 332

9.4.5:

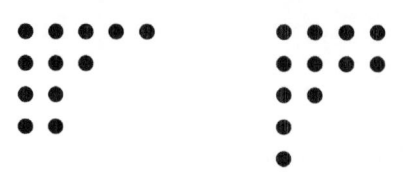

9.4.9:

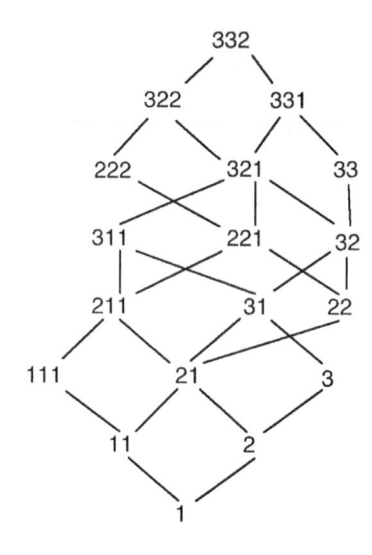

9.5 Calculating a Cycle Index

9.5.3: 9 graphs.

9.5.9: 1044 graphs.

9.5.12: 10 graphs.

9.6 General Graphs and Digraphs

9.6.1: Substituting 4 into the cycle index

$$\mathcal{Z}_{\mathcal{A}ut_E(K_3)}(y_1, y_2, y_3) = \frac{1}{6}(y_1^3 + 3y_1y_2 + 2y_3)$$

yields

$$\mathcal{Z}_{\mathcal{A}ut_E(K_3)}(4, 4, 4) = \frac{1}{6}(4^3 + 3 \cdot 4 \cdot 4 + 2 \cdot 4) = \frac{120}{6} = 20$$

Chapter 10   Designs

10.1   Latin Squares

10.1.1: $\begin{pmatrix} 0 & 1 & 2 \\ 1 & 2 & 0 \\ 2 & 0 & 1 \end{pmatrix}$ $\begin{pmatrix} 0 & 1 & 2 \\ 2 & 0 & 1 \\ 1 & 2 & 0 \end{pmatrix}$

10.1.6: $L_5^4 = \begin{pmatrix} 0 & 1 & 2 & 3 & 4 \\ 4 & 0 & 1 & 2 & 3 \\ 3 & 4 & 0 & 1 & 2 \\ 2 & 3 & 4 & 0 & 1 \\ 1 & 2 & 3 & 4 & 0 \end{pmatrix}$

10.1.10: $A \otimes A = \begin{pmatrix} (1,1) & (1,0) & (0,1) & (0,0) \\ (1,0) & (1,1) & (0,0) & (0,1) \\ (0,1) & (0,0) & (1,1) & (1,0) \\ (0,0) & (0,1) & (1,0) & (1,1) \end{pmatrix}$

10.1.15: $L^{(1)(2,3)} = \begin{pmatrix} 0 & 2 & 3 & 1 \\ 2 & 0 & 1 & 3 \\ 1 & 3 & 2 & 0 \\ 3 & 1 & 0 & 2 \end{pmatrix}$

10.2   Block Designs

10.2.1:   varieties: 0, 1, 2, 3; blocks: 012, 013, 023, 123.

10.2.7:   The replication number r calculated by applying Proposition 10.2.2(b) would not be an integer.

10.2.13: $v = 9$, $b = 12$, $r = 4$, $k = 3$, and $\lambda = 1$.

10.2.22:   varieties: 00, 01, 02, 10, 11, 12, 20, 21, 22.

blocks :   {00, 10, 20}   {00, 11, 21}   {02, 12, 22}
          {00, 01, 02}   {10, 11, 12}   {20, 21, 22}
          {00, 11, 22}   {00, 12, 21}   {01, 10, 22}
          {01, 12, 20}   {02, 10, 21}   {02, 12, 20}

10.3 Classical Finite Geometries

10.3.4: These are two drawings of the Levi graph for the Fano plane. The one on the right illustrates that it is vertex-transitive.

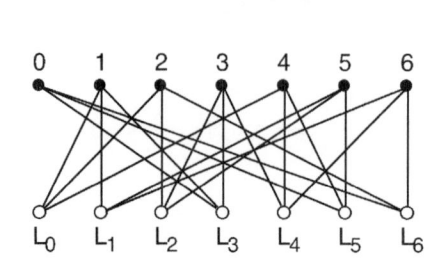 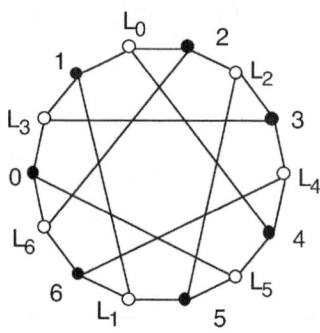

10.3.8: Since $\lambda = 1$, the Menger graph for the Fano plane is K_7.

10.3.12: See the solution to Exercise 10.3.8 above.

10.4 Projective Planes

10.4.1: points of $PG(2,2)$: $\{000, 001\}$, $\{000, 010\}$, $\{000, 100\}$, $\{000, 011\}$, $\{000, 101\}$, $\{000, 110\}$, $\{000, 111\}$.

10.4.2: lines of $PG(2,2)$:

$\{000, 001, 010, 011\}$
$\{000, 001, 100, 101\}$
$\{000, 100, 010, 110\}$
$\{000, 001, 110, 111\}$
$\{000, 010, 101, 111\}$
$\{000, 100, 011, 111\}$
$\{000, 101, 110, 011\}$

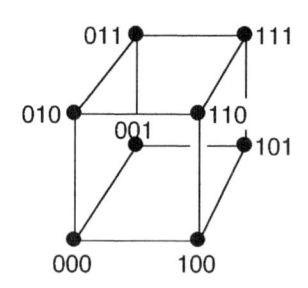

10.5 Affine Planes

10.5.1: points of $AG(2,2)$: 00, 01, 10, 11.

10.5.2: lines of $AG(2,2)$:

$\{00, 01\}$
$\{00, 10\}$
$\{00, 11\}$
$\{01, 10\}$
$\{01, 11\}$
$\{10, 11\}$

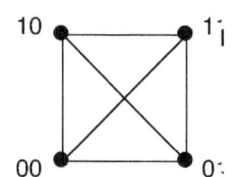

INDEXES

I1 Index of Notations
I2 General Index

I1   INDEX OF NOTATIONS

The notations listed here are generally those that occur at multiple locations in the text. They are grouped according to the context in which they occur. Some totally standard notations are omitted.

Numbers and Arithmetic

algebraic and arithmetic domains

$\mathbb{C}$	the complex numbers, 50
$\mathbb{D}_n$	the dihedral group, 500
$\mathbb{N}$	the non-negative integers (natural numbers), 50
$\mathbb{Q}$	the rational numbers, 50
$\mathbb{R}$	the real numbers, 50
$\mathbb{Z}$	the integers, 50
$\mathbb{Z}^+$	the positive integers, 50
$\mathbb{Z}_n$	the integers mod n, 585
$\mathbb{Z}[x]$	polynomial ring with integer coefficients, 342
$[k:m]$	integer interval $\{k, k+1, \ldots, m\}$, 54

arithmetic operators

$\lfloor x \rfloor$	floor of a real number, 12
$\lceil x \rceil$	ceiling of a real number, 12
$x^{\underline{r}}$	falling power of a real number, 14
$x^{\overline{r}}$	rising power of a real number, 79
$\ln x$	natural logarithm of a real number, 51
$\lg x$	base-2 logarithm of a real number, 51
$\gcd(m, n)$	greatest common divisor, 43
$\operatorname{lcm}(m, n)$	least common multiple, 43
$n \bmod d$	mod operator, 53
$\phi(n)$	Euler phi-function, 356
(predicate)	Iverson truth function, 165
$\partial g(x)$	degree of a polynomial, 342

arithmetic relations

$a \equiv b \,(\text{modulo } m)$	congruence relation, 335
$d \setminus n$	d divides n, 168
$m \perp n$	m and n are relatively prime, 168

Sets and Sequences

sets and multisets

$\emptyset$	empty set, 15
$|S|$	cardinality of a set S, 9
$\overline{S}$	complement of a set S, 202
(S, ι)	multiset, 15
$|(S, \iota)|$	cardinality of multiset, 16

sequences

$\langle a_j \rangle$	the sequence $a_0, a_1, a_2, \ldots, 50$
$\triangle a_n$	for a sequence $\langle a_n \rangle$, the difference $a_{n+1} - a_n$, 67
a_n'	for a sequence $\langle a_n \rangle$, the difference $a_{n+1} - a_n$, 67
B_n	the nth Bell number, 266
c_n	the nth Catalan number, 60
D_n	derangement number, 114
f_n	the nth Fibonacci number, 59
H_n	the nth harmonic number, 52

generalized sequences

$\binom{n}{k}$	binomial coefficient (also called a combination coefficient), 8
$P(n, k)$	the number of ordered selections of k objects from n objects, 18
$\binom{n}{r_1 \; r_2 \; \cdots \; r_k}$	multicombination coefficient, 22
$s_{n,k}$	Stirling number of the first kind, 80
$S_{n,k}$	Stirling number of the second kind, 82
$\left\{ \begin{matrix} n \\ k \end{matrix} \right\}$	Stirling subset number, 25
$\left[\begin{matrix} n \\ k \end{matrix} \right]$	Stirling cycle number, 84
$\left\langle \begin{matrix} n \\ k \end{matrix} \right\rangle$	Eulerian number, 292

recurrences and generating functions

$\hat{g}_n$	homogeneous part of a solution to a recurrence, 127
$\dot{g}_n$	particular part of a solution to a recurrence, 127
$G(z)$	ordinary generating function, 85
$\hat{G}(z)$	exponential generating function, 85

Probability and Statistics

Ω	sample space for probability, 239
$Pr(A)$	probability of the event A, 239
$E(X)$ or μ_X	mean of the random variable X, 240
$SD(X)$ or σ_X	standard deviation of the random variable X, 240
$V(X)$ or σ_X^2	variance of the random variable X, 240
$\overline{X}$	sample mean of a random variable X, 244
$\hat{\theta}$	estimator of a statistic θ, 243

Partially Orders Sets and Lattices

$x \preceq y$	generic partial ordering relation, 306
$\mathcal{B}_n$	boolean poset, 307
$\mathcal{D}_n$	divisibility poset, 308
$\mathcal{D}$	infinite divisibility poset, 308
$\mathcal{I}_n$	inversion poset, 310
$\mathcal{P}_n$	partition poset, 309
$\mathcal{SD}_n$	summation dominance lattice, 526
$\mathcal{Y}_S$	Young's lattice for the integer partition S, 525
$N_k(P)$	the kth Whitney number of a ranked poset, 316
$I_n(k)$	the inversion coefficient, 289
$lub(x, y)$	least upper bound under a partial ordering, 311
$glb(x, y)$	greatest lower bound under a partial ordering, 312

Graph Theory

basic notations

$G = (V, E)$	graph G with vertex-set V and edge-set E, 35
E_G or $E(G)$	edge-set of graph G, 35
V_G or $V(G)$	vertex-set of graph G, 35
uv	in a simple graph, an edge joining vertices u and v, 36
$deg(v)$	degree of vertex v, 36
I_G	incidence matrix of graph G, 40
A_G	adjacency matrix of graph G, 40
W	walk $\langle v_0, e_1, v_1, e_2, \ldots, e_n, v_n \rangle$ in a graph, 41
$\overline{G}$	edge-complement of a simple graph, 423
$G \cong H$	isomorphism relation between graphs G and H, 393

special families

C_n	cycle graph with n vertices, 372
$Cay(A : S)$	Cayley graph, 376
$circ(n : S)$	circulant graph, 374
CL_n	circular ladder with n rungs, 376
K_n	complete graph on n vertices, 38
$K_{m,n}$	complete bipartite graph with bipartition subsets of sizes m and n, 38
K_n	complete graph on n vertices, 373

$$\mathcal{Z}_\mathcal{P}(t_1, \ldots, t_n) \qquad \text{cycle index of a group, 496}$$
$$\zeta(\pi) = t_1^{r_1} t_2^{r_2} \cdots t_n^{r_n} \qquad \text{cycle structure of permutation } \pi, 496$$

Combinatorial Designs

$$A \otimes B \qquad \text{product of two Latin squares, 543}$$
$$L_p^k \qquad \text{a particular Latin Square, 544}$$
$$L^\pi \qquad \text{a conjugate of the Latin square } L, 549$$
$$(v, b, r, k, \lambda) \qquad \text{parameters of a BIBD, 552}$$

I2 GENERAL INDEX

12A04AF - #0021 - 120617 - C0 - 254/178/38 [40] - CB - 9781584887430